Evolution of Global Electricity Markets

New Paradigms, New Challenges, New Approaches

T0329112

This book is dedicated to Paul L. Joskow in recognition of his invaluable contributions to the study of electricity markets.

 Paul L. Joskow is an economist whose research has focused on topics in industrial organization, government regulation, the organization of firms, law and economics, energy and environmental economics. He was on the faculty of the Department of Economics at MIT from 1972 to 2010 and is now the Elizabeth and James Killian Professor of Economics, Emeritus there. He has been President of the Alfred P. Sloan Foundation since January 1, 2008. Professor Joskow received a B.A. from Cornell University in 1968 and a PhD in Economics from Yale University in 1972. He served as Head of the MIT Department of Economics from 1994 to 1998 and was the Director of the MIT Center for Energy and Environmental Policy Research from 1999 to 2007. Professor Joskow has published 6 books and over 125 articles and papers. His papers have appeared in the *American Economic Review, Bell Journal of Economics, Rand Journal of Economics, Journal of Political Economy, Journal of Law and Economics, Journal of Law, Economics and Organization, International Economic Review, Review of Economics and Statistics, Journal of Econometrics, Journal of Applied Econometrics, Yale Law Journal, New England Journal of Medicine, Foreign Affairs, Energy Journal, Electricity Journal, Oxford Review of Economic Policy*, and other journals and books.

Professor Joskow is a Director of Exelon Corporation, a Director of TransCanada Corporation, and a Trustee of the Putnam Mutual Funds. He is a Trustee of Yale University and a member of the Board of Overseers of the Boston Symphony Orchestra. He previously served as a Director of New England Electric System, State Farm Indemnity Company, National Grid plc, and the Whitehead Institute of Biomedical Research. Professor Joskow has served on the US EPA's Acid Rain Advisory Committee and on the Environmental Economics Committee of the EPA's Science Advisory Board. He is a member of the Scientific Advisory Board of the Fondation Jean-Jacques Laffont (Toulouse, France). He served as President of the Yale University Council from 1993 to 2006. Professor Joskow is a past-President of the International Society for New Institutional Economics, a Distinguished Fellow of the Industrial Organization Society, a Fellow of the Econometric Society and of the American Academy of Arts and Sciences, a Distinguished Fellow of the American Economic Association, and a member of the Council on Foreign Relations.

Evolution of Global Electricity Markets

New Paradigms, New Challenges, New Approaches

Edited by

Fereidoon P. Sioshansi
Menlo Energy Economics

AMSTERDAM • BOSTON • HEIDELBERG • LONDON
NEW YORK • OXFORD • PARIS • SAN DIEGO
SAN FRANCISCO • SINGAPORE • SYDNEY • TOKYO
Academic Press is an imprint of Elsevier

Academic Press is an imprint of Elsevier
225 Wyman Street, Waltham, MA 02451, USA
525 B Street, Suite 1800, San Diego, CA 92101-4495, USA
32 Jamestown Road, London NW1 7BY, UK

First edition 2013

Notice

Knowledge and best practice in this field are constantly changing. As new research and
experience broaden our understanding, changes in research methods, professional practices,
or medical treatment may become necessary.

Practitioners and researchers must always rely on their own experience and knowledge
in evaluating and using any information, methods, compounds, or experiments described herein.
In using such information or methods they should be mindful of their own safety and the
safety of others, including parties for whom they have a professional the fullest extent of the
law, neither the Publisher nor the authors, contributors, or editors, assume any liability for any
injury and/or damage to persons or property as a matter of products liability, negligence or
otherwise, or from any use or operation of any methods, products, instructions, or ideas
contained in the material herein.

Library of Congress Cataloging-in-Publication Data
A catalog record for this book is available from the Library of Congress

British Library Cataloguing-in-Publication Data
A catalogue record for this book is available from the British Library

ISBN: 978-0-12-397891-2

For information on all Academic Press publications
visit our website at http://store.elsevier.com

Printed and bound in United States of America

13 14 15   10 9 8 7 6 5 4 3 2 1

Working together
to grow libraries in
developing countries

www.elsevier.com • www.bookaid.org

Contents

About the Authors   ix

Foreword   xxvii
Stephen Littlechild

Preface   xxxi
Andrew Reeves

Introduction   xxxv
Fereidoon P. Sioshansi

Part I
The Evolution of European Electricity Markets

1. **Evolution of the British Electricity Market and the Role of Policy for the Low-Carbon Future**   3
   David Newbery

2. **Electricity Market Reform in Britain: Central Planning Versus Free Markets**   31
   Malcolm Keay, John Rhys and David Robinson

3. **The French Paradox: Competition, Nuclear Rent, and Price Regulation**   59
   Jacques Percebois

4. **Turnaround in Rough Sea—Electricity Market in Germany**   93
   Wolfgang Pfaffenberger and Esther Chrischilles

5. **The Growing Impact of Renewable Energy in European Electricity Markets**   125
   Reinhard Haas, Hans Auer, Gustav Resch and Georg Lettner

6   Renewable Energy, Efficient Electricity Networks, and
    Sector-Specific Market Power Regulation   147
    Günter Knieps

7   From Niche to Mainstream: The Evolution of Renewable
    Energy in the German Electricity Market   169
    Dierk Bauknecht, Gert Brunekreeft and Roland Meyer

8   The Challenges of Electricity Market Regulation in the
    European Union   199
    Walter Boltz

Part II
The Evolution of Electricity Markets in the Americas

9   The Evolution of the PJM Capacity Market: Does
    It Address the Revenue Sufficiency Problem?   227
    Joseph E. Bowring

10 Texas Electricity Market: Getting Better   265
    Parviz Adib, Jay Zarnikau and Ross Baldick

11 From the Brink of Abyss to a Green, Clean, and Smart
    Future: The Evolution of California's Electricity Market   297
    Lorenzo Kristov and Stephen Keehn

12 Unfinished Business: The Evolution of US Competitive
    Retail Electricity Markets   331
    Young Kim

13 Fragmented Markets: Canadian Electricity Sectors'
    Underperformance   363
    Pierre-Olivier Pineau

14 Latin American Energy Integration: An Outstanding
    Dilemma   393
    Ricardo Raineri, Isaac Dyner, José Goñi, Nivalde Castro,
    Yris Olaya and Carlos Franco

Part III
The Evolution of BRICs Electricity Markets

15 The Evolution of Brazilian Electricity Market   435
    Luiz Pinguelli Rosa, Neilton Fidelis da Silva, Marcio Giannini
    Pereira and Luciano Dias Losekann

16 The Russian Electricity Market Reform: Toward the
    Reregulation of the Liberalized Market?   461
    Anatole Boute

17 Not Seeing the Wood for the Trees? Electricity Market
    Reform in India   497
    Anupama Sen and Tooraj Jamasb

18 Reform Postponed: The Evolution of China's Electricity
    Markets   531
    Philip Andrews-Speed

Part IV
The Evolution of Electricity Markets in Australasia

19 Evolution of Australia's National Electricity Market   571
    Alan Moran and Rajat Sood

20 Is Electricity Industry Reform the Right Answer to the
    Wrong Question? Lessons from Australian Restructuring
    and Climate Policy   615
    Iain MacGill and Stephen Healy

21 Weak Regulation, Rising Margins, and Asset
   Revaluations: New Zealand's Failing Experiment in
   Electricity Reform   645
   Geoff Bertram

22 The Korean Electricity Market: Stuck in Transition   679
   Suduk Kim, Yungsan Kim and Jeong Shik Shin

23 After Fukushima: The Evolution of Japanese Electricity
   Market   715
   Hiroshi Asano and Mika Goto

24 The Singapore Electricity Market: From Partial to Full
   Competition   739
   Youngho Chang and Yanfei Li

25 Market Design for the Integration of Variable
   Generation   757
   Jenny Riesz, Joel Gilmore and Magnus Hindsberger

Epilogue   791
Jean-Michel Glachant

Index   799

Parviz M. Adib is the founder of Pionergy offering regulatory, economic, and strategic advices on energy-related topics including the development and pricing of value added services under smart meters. He has more than 30 years of experience with energy markets and previously worked with APX Inc., where he identified strategies for clients' successful operation in competitive markets. Earlier, Dr. Adib worked at the Public Utility Commission of Texas where he served in various capacities, including the director of the Market Oversight Division where he played a key role in design of electricity market in ERCOT, advised the Commission on emerging market and regulatory issues, and performed the duties of the first ERCOT Market Monitor.

Dr. Adib's main research interests include the development of competitive markets and customer choice, quantitative assessing of customers behavior, and energy and public policy.

Dr. Adib has a PhD in economics from the University of Texas at Austin, where he taught graduate and undergraduate courses.

Philip Andrews-Speed is a principal fellow at the Energy Studies Institute of the National University of Singapore. Until 2010, he was a professor of energy policy and director of the Centre of Energy, Petroleum and Mineral Law and Policy at the University of Dundee.

The focus of his research is on energy policy, regulation, and reform in China, on the interface between China's energy policy and international relations, and on the global and regional governance of energy and natural resources. Recent books include *China, Oil and Global Politics* with Roland Dannreuther (Routledge, 2011) and *The Governance of Energy in China: Transition of a Low-Carbon Economy* (Palgrave MacMillan, 2012).

He has BA and PhD degrees in geology from the University of Cambridge, and an LLM in Natural Resources Law and Policy from the University of Dundee.

Hiroshi Asano is a deputy associate vice president, the Central Research Institute of Electric Power Industry (CRIEPI) in Tokyo, Japan, a visiting professor, the University of Tokyo, and a visiting researcher, Research Institute of Advanced Network Technology, Waseda University. In his multiple capacities, he is engaged in research, teaching, and variety of managerial and advisory duties.

Dr. Asano has been engaged in research projects on demand-side management, systems approaches, and energy policy analysis since joining the

CRIEPI in 1984. His interests include systems analysis of distributed energy resources, demand response, and power markets.

He has BEng, MEng, and Doctor of Engineering degrees in Electrical Engineering from the University of Tokyo.

Hans Auer is associate professor at Energy Economics Group, Vienna University of Technology, Austria. He is mainly responsible for supervision of master's degree and PhD students as well as acquisition and coordination of European projects in the field of market integration of renewable technologies into future, innovative grid infrastructures.

Hans is interested in energy systems analyses, energy modeling and energy policies with special focus on future energy market and grid infrastrucutre design. In this context, he is interested in the role of and interdependences between hybrid energy grids (electricity, gas, heating/cooling) for providing different energy services in particular.

Hans holds a master's degree in electrical engineering and a PhD in energy economics from Vienna University of Technology.

Ross Baldick is a professor and Leland Barclay fellow in the department of electrical and computer engineering at the University of Texas at Austin. Earlier, he was a postdoctoral fellow at the Lawrence Berkeley Laboratory and an assistant professor at Worcester Polytechnic Institute.

His current research interests include optimization and economic theory applied to electric power system operations, public policy, transmission planning, electricity market restructuring, electric system security, electric transportation, and the economic implications of integration of renewables. Dr. Baldick has over 50 refereed publications and his book, *Applied Optimization*, is used as a textbook in his Optimization of Engineering Systems course. He is a former editor of *IEEE Transactions on Power Systems* and a fellow of the IEEE and director of the NSF I/UCRC on electric vehicles.

He received his BSc and BE degrees from the University of Sydney, Australia and his MS and PhD from the University of California, Berkeley.

Dierk Bauknecht is a senior researcher with the Öko-Institut's Energy and Climate Division, which he joined in 2001. He coordinates various projects on the system integration of renewables.

His main research interest is the integration of renewables into electricity networks and markets. This includes regulatory issues, system modeling, and scenario analysis as well as questions on the governance of system transformation. He has been involved in the development and application of electricity dispatch and investment models. He coordinates the institute's contribution to the German Smart Grid model region eTelligence and is engaged in the evaluation and comparison of different flexibility options in the power system.

Dr. Bauknecht graduated in political science at the Freie Universität Berlin and holds an MSc and PhD from the Science and Technology Policy Research at the University of Sussex, UK.

Geoffrey Bertram is a senior associate at the Institute of Policy Studies, Victoria University of Wellington, and a director of the energy and environmental consultancy Simon Terry Associates Research Ltd. Until 2009, he taught in the School of Economics and Finance at Victoria University.

Dr. Bertram has over 30 years of experience researching and consulting on New Zealand public utility industries, including electricity, gas, water and telecommunications, and has published extensively on the deregulation and restructuring policies pursued by successive New Zealand governments since the 1980s. He has served on the editorial boards of *World Development*, *Environment and Development Economics*, and *Asia Pacific Viewpoint*.

Dr. Bertram holds BA and BA honors degrees from Victoria University of Wellington, and MPhil. and DPhil. (in economics) degrees from Oxford University.

Walter Boltz is executive director of E-Control, the Austrian Energy Regulator. Before joining E-Control he was active in banking and held management positions in consulting companies.

His recent activities entail several chairmanships within the Council of Energy Regulators. Since 2006, Mr. Boltz has been the chairman of the Gas Working Group and since 2010 vice president of ACER, the Agency for the Cooperation of Energy Regulators.

Walter Boltz holds a degree in technical physics from Vienna University of Technology.

Anatole Boute is a lecturer in law at the University of Aberdeen's Centre for Energy Law and legal adviser to the IFC Russia Renewable Energy Program affiliated with the World Bank. Earlier, he was part of the Energy Law practice of the Brussels law firm Janson Baugniet.

His research focuses on the regulatory aspects of the energy transition toward sustainability, with a special interest for the Russian electricity and heating system. Anatole Boute has advised the United Nations Conference on Trade and Development for the World Investment Report 2010 on investing in a low-carbon economy. As member of the Brussels bar, he advised on and was involved in litigations concerning energy market liberalization, renewable energy, energy efficiency, emissions trading and nuclear energy.

Anatole Boute has MA degrees in law and political sciences and an LLM in energy law from the University of Leuven. He also holds a PhD from the University of Groningen.

Joseph E. Bowring is the president of Monitoring Analytics, LLC. He has been the Independent Market Monitor for PJM Interconnection since 1999, responsible for all aspects of market monitoring. His prior experience includes serving as senior staff economist for the New Jersey Board of Public Utilities, as chief economist for the New Jersey Department of the Public Advocate's Division of Rate Counsel, as an economist at the US Department of Energy, Energy Information Administration, and as an independent consulting economist. He has taught economics as a member of the faculty at Bucknell University and Villanova University.

He received a PhD in economics from the University of Massachusetts.

Gert Brunekreeft is a professor of energy economics at Jacobs University Bremen in Germany. Before joining Jacobs University, he was senior economist for the energy company EnBW AG and held research positions in applied economics at Tilburg University, the University of Cambridge and at Freiburg University. He is an associate researcher to a number of research centers and is an associate editor for the *Competition and Regulation in Network Industries*.

Professor Brunekreeft's main research interests are in industrial economics, regulation theory, and competition policy of network industries, especially electricity and gas markets. Current issues concern the economics of vertical unbundling and the relation between regulation and investment. He is the author of several books and published widely in academic journals, including *Journal of Regulatory Economics*, *Utilities Policy*, *Oxford Review of Economic Policy*, and *Energy Journal*.

He holds a degree from the University of Groningen and a PhD from Freiburg University, both in economics.

Nivalde Castro is a professor at the Institute of Economics of the Federal University of Rio de Janeiro (FURJ) and the coordinator of Study Group of the Electric Power Sector ("Grupo de Estudos do Setor Elétrico"—GESEL).

Professor Castro's research interests include the Brazilian electric power sector and the Latin American countries on issues related to renewable power—wind, biomass, solar, and hydroelectricity—funding models for new projects, sustainable matrix, green economy, among others.

He has bachelor's and master's degrees in economics and a doctorate in education from UFRJ—Universidade Federal do Rio de Janeiro.

Youngho Chang is an assistant professor of economics at the division of economics and an adjunct senior fellow at the S. Rajaratnam School of International Studies (RSIS), Nanyang Technological University, Singapore. He is a member of technical committee for Clean Development Mechanism (CDM), Designated National Authority (DNA), National Environment Agency, Singapore.

He specializes in the economics of climate change, energy and security, oil and macroeconomy, and the economics of electricity market deregulation. His current research interests include the impact of oil price fluctuations in macroeconomic performance, energy security, energy use, and climate change. He has numerous publications and coedited two books—*Energy Conservation in East Asia*: *Towards Greater Energy Security* (World Scientific, 2011) and *Energy Security*: *Asia Pacific Perspectives* (Manas Publications, 2010).

Dr. Chang received a BASc in landscape architecture from the Seoul National University, an MA in economics from the Yonsei University and a PhD in economics from the University of Hawaii at Manoa.

Esther Chrischilles is an economist at the Cologne Institute for Economic Research within the department of environment, energy and resources where she studies recent developments in German energy and climate politics. Currently she is engaged in a transdisciplinary project dealing with the adaptation to inevitable impacts of climate change.

Her main research interests concern economic instruments of climate protection as well as energy use patterns within the industrial sector and the impact of increasing energy cost on industrial value added. She has published several articles and studies concerning different aspects of climate change and the German government's recent *turnaround* policy.

Ms. Chrischilles holds a diploma from the University of Cologne, where she studied economics specializing in energy economics and political science.

Isaac Dyner is a professor of operational research and energy policy at the Universidad Nacional de Colombia. He has been a visiting professor of the British Academy and academic visitor at Imperial College London, City University, London Business School, Warwick University, University of Lugano, and several Latin American universities.

He has over 175 publications in energy, operational research, and system dynamics in scientific journals as well as five books and four book chapters. He has been the director of the Centre of Excellence CeiBA, the Institutes for Decision Sciences and Energy, a member of both Colombian Councils for Energy Research and for Natural Sciences and Mathematics. He has consulted for businesses and government agencies in the areas of strategy, deregulation, modeling, and energy.

He obtained his PhD from London Business School and Master of Sciences from both University of Warwick and Southampton University, UK.

Carlos J. Franco is a professor in the Department of Computer and Decision Sciences at the National University of Colombia in the Medellin campus where he teaches in subjects such as complex systems, system modeling, and energy markets. Currently, he is the chair of the undergraduate and graduate programs in the systems and computing areas.

His research area is on energy systems analysis, including policies evaluation and strategies formulation. His recent work includes low carbon economies, electricity markets integration and biofuels, among others. He worked for the power exchange of Colombia (XM) and the main grid operator of Colombia (ISA).

He has a degree in civil engineering, a master's degree in water resources management, and a PhD in engineering, all from the National University of Colombia.

Joel Gilmore is the principal of renewable energy and climate policy at ROAM Consulting, a leading provider of expert services in energy market modeling. With more than 9 years of experience in energy market modeling and analysis, Joel leads a team to model and advise on topics related to the impacts of climate policy and renewable energy on electricity systems, such as renewable energy target schemes, carbon pricing, and the integration of wind and solar energy.

Dr. Gilmore's accomplishments includes work for market participants in the National Electricity Market (NEM) and South-West Interconnected System (SWIS) in Australia. He is interested in solar technologies and their integration into electricity grids.

Dr. Gilmore has a bachelor of science and a PhD in physics from the University of Queensland, Australia.

Jean-Michel Glachant is director of the Florence School of Regulation and holder of the Loyola de Palacio Chair. He is a research partner of CEEPR (MIT, USA), EPRG (Cambridge University, UK); chief editor of *Economics of Energy & Environmental Policy*, a journal of the International Association for Energy Economics, and a member of the board of the IAEE.

Earlier, he was a professor of economics at La Sorbonne University and head of the department of economics at the University of Paris-Sud 11. He has been adviser to DG TREN, DG COMP, DG RESEARCH, and DG ENERGY of the European Commission.

Jean-Michel obtained his PhD in economics from La Sorbonne University, France.

José Goñi is a professor at the Universidad de Concepción. His extensive diplomatic carrier including being ambassador in Sweden, Italy, Mexico, and the United States, and has represented Chile at FAO and other international institutions in the UN system. He has also served as Minister of Defense.

His academic, teaching, writing, and speaking interests are diverse including a number of teaching and speaking engagements at universities and research institutions around the world. Mr. Goñi has edited several books and has published articles on economic, social, and political issues notably "Olof Palme y América Latina," edit. Nueva Sociedad, Caracas (1986); "Desarrollo con equidad", editorial Azul, Montevideo (1989); "200 years" on USA–Chile bilateral relationships (2010); and articles in newspapers and other publications. He also writes short stories and is working on a novel.

He has bachelor and business engineering degrees from Universidad de Concepción.

Mika Goto is a senior research economist at Central Research Institute of Electric Power Industry (CRIEPI), where she is engaged in a wide range of projects including performance assessment and empirical analysis of industrial organization of electric utilities using applied econometrics and mathematical programming.

Dr. Goto's research interests include regulatory reform. She has published papers in regulatory reform of energy industries, R&D policy, and management.

She has published in *Energy Economics, Energy Policy, European Journal of Operational Research, Decision Support Systems* and *Omega*.

She has PhD in economics from Nagoya University.

Reinhard Haas is a professor at Energy Economics Group, Vienna University of Technology in Vienna, Austria. He is responsible for teaching, supervising master's degree and PhD students, research and managerial duties. He advises governmental institutions, municipalities and utilities in energy policy issues.

He is engaged in energy economics and energy policy issues focusing on heading toward sustainable energy systems. His special interests are the promotion of renewable energy and energy efficiency measures. Regarding these topics he is on the editorial board of various international journals.

He holds a master's degree in mechanical engineering and doctor of energy economics at Vienna University of Technology.

Stephen Healy is a senior lecturer in the School of History and Philosophy and co-coordinator of the Environmental Studies Program of UNSW's Faculty of Arts and Social Science. He is also a research coordinator of the Centre for Energy and Environmental Markets.

Dr. Healy has worked for Greenpeace International, the NSW EPA and Middlesex University, UK, where he led the Science, Technology and Society Programme. His research interests include climate change, energy, risk and uncertainty and public participation. He is interested in historical development of contemporary systems of energy provision, energy institutions, governance and politics and reenvisioning energy consumption. He has published across a number of fields and coauthored a Guide to Environmental Risk Management, published in 2006.

He has a BSc (Hons) in physics and an electrical engineering PhD, in Photovoltaics, from UNSW.

Magnus Hindsberger is senior manager market modeling at the Australian Energy Market Operator (AEMO), responsible for market simulations of future electricity and gas market outcomes supporting the long-term electricity planning process and cost-benefit assessments of new investments. Previously, he worked at Transpower New Zealand, the Nordic consulting company ECON, as well as Elkraft System, now merged into Energinet.dk, which is a Danish transmission system operator.

He has a strong background in electricity market modeling and its application for long-term scenario planning and cost–benefit assessments. He is interested in renewable generation and has been working on high-penetration renewables scenarios and in particular wind integration issues in Denmark, New Zealand and Australia.

Dr. Hindsberger holds Ms and PhD degrees in engineering (operations research) from the Technical University of Denmark.

Tooraj Jamasb is Professor of Energy Economics at Durham University Business School. Prior to his current post, he was SIRE chair in energy economics in Heriot-Watt University, Edinburgh. He is a research associate at Electricity Policy Research Group, Cambridge University, Centre for Energy and Environmental Policy Research, MIT and a member of the academic advisory panel of the Northern Ireland Utility Regulator. He has participated on projects for the Council of European Energy Regulators, European energy regulators, energy companies, and the World Bank.

His research includes energy sector reform, energy networks, incentive regulation, energy policy, energy demand and public acceptance and energy technology. He is coeditor of *The Future of Electricity Demand: Customers, Citizens and Loads*, 2011; *Delivering a Low-Carbon Electricity System*, 2008 and *Future Electricity Technologies and Systems*, 2006.

He holds a PhD in energy economics from University of Cambridge, and masters' degrees in Energy Management and Policy from University of Pennsylvania, French Institute of Petroleum, and Norwegian School of Management, and a BBA from Tehran Business School.

Malcolm Keay is a senior research fellow at the Oxford Institute for Energy Studies where he works on issues connected with electricity and the transition to a low carbon energy system.

He has had a wide ranging career in the energy sector, including energy policy development (he was Director of Energy Policy at the UK Department of Trade Industry in 1996−99), international energy affairs, energy regulation, and energy consultancy. He has acted as an adviser on many energy studies, and was special adviser to the House of Lords Committee Inquiry into Energy Security in Europe and Director of the Energy and Climate Change Study for the World Energy Council.

Malcolm has an MA degree from Cambridge University.

Stephen Keehn is a senior advisor in the market and infrastructure policy department of the California ISO. He has worked on market design changes to integrate renewables, resource adequacy changes to ensure flexible resources and how the smart grid and distributed generation will impact the grid and market structures.

Stephen previously worked for San Diego Gas & Electric, Sempra Energy and Southern California Edison. While at Sempra he helped champion the idea of a central capacity market as the best solution for California's resource adequacy problem. He was also involved in some of the initial incentive ratemaking attempts in California, and worked on various other policy issues, including California's greenhouse gas cap and trade program. He has experience in both the electricity and natural gas industries.

Stephen has a bachelor's degree in mathematics and economics from the University of Wisconsin, and a master's degree in economics from Stanford University.

Suduk Kim is a former professor at the department of energy studies, Ajou University, Korea. He undertakes interdisciplinary research in the quantitative analysis and modeling of energy systems for the associated implication and policy recommendation to private and public sectors.

His career in energy sector includes that he is currently one of the lead authors of energy system for fifth assessment report of IPCC, a member of scenario study group of World Energy Council, and a board member of Korea Institute of Energy Technology Evaluation and Planning.

He has BA and PhD degrees in economics from Seoul National University and Rutgers, respectively.

Young Kim is former associate director of DNV KEMA's retail energy practice. He has 10 years of experience in the energy industry, working primarily on business and policy issues related to competitive energy markets. His primary areas of expertise include retail business planning and market analysis, performance management, financial analysis, and policy analysis. As head of research of DNV KEMA's Retail Energy Markets advisory service, he is responsible for developing and executing a wide range of analysis and intelligence gathering on competitive energy businesses.

Young has advised the senior management of numerous retailers, utilities, and investors on market conditions, investment opportunities and entry strategies in the competitive energy business. He has conducted over a dozen valuations of unregulated energy firms and has a deep understanding of the ways in which value is created in the competitive retail energy sector.

Young has a bachelor's degree in economics from the Massachusetts Institute of Technology and was a PhD candidate (ABD) at the University of Chicago.

Yungsan Kim is a professor at department of economics and finance of Hanyang Univesity in Seoul, Korea, and the director of Hanyang Economic Research Institute. He is the president of the Korean Academic Society of Industrial Organization (2013).

His main research interests are industrial organization, corporate finance, and energy economics with a focus on power industry restructuring. He has conducted numerous consulting works on the issues of power industry restructuring for the Korean government, Korea Power Exchange, and KEPCO. He has published in journals, such as the *Journal of Law, Economics and Organization* and the *Journal of Finance*.

He has a PhD degree in economics from UCLA.

Günter Knieps is a professor of economics at the University of Freiburg, Germany, where he is engaged in research in the field of network economics. Prior to joining the Faculty of Economics in Freiburg he held a chair of microeconomics at the University of Groningen, the Netherlands. He is a member of the Scientific Council of the Federal Ministry of Economics and Technology as well as of the Federal Ministry of Transport, Building and Urban Development.

Professor Knieps' main research interests include study of network economics, deregulation, competition policy, industrial economics, and sector studies on energy, telecommunications, and transportation. He has published widely in academic and professional journals.

Dr. Knieps has diplomas in economics and mathematics from the University of Bonn, Germany, and a PhD in mathematical economics from the University of Bonn.

Lorenzo Kristov is a principal in the market and infrastructure development division of the California ISO engaged in market design, transmission planning and new generator interconnection, focusing on implementation of environmental mandates such as California's renewable portfolio standard and other public policy goals that affect the wholesale power markets.

In 2012, he led an initiative to integrate generator interconnection into the transmission planning process; in 2010, he led the redesign of the transmission planning process. He was also a primary designer of the ISO's LMP-based market structure in 2009. Prior to CAISO he worked at the California Energy Commission involved in development of the rules for retail direct access. Before that, he was a Fulbright scholar in Indonesia working on creating a commercial and regulatory framework to advance private power development.

He received a bachelor's in mathematics from Manhattan College, a master's degree in statistics from North Carolina State University and a PhD in economics from the University of California at Davis.

Georg Lettner is a junior researcher and PhD candidate at Energy Economics Group, Vienna University of Technology, Austria.

Georg joined Energy Economics Group in 2009 and is a key expert in grid and market integration policies of DG/RES-E technologies and smart grids concepts. Georg's research interests are focused on simulation software development in the field of smart grid and market integration of renewable electricity generation technologies into electricity markets in general, and the consideration of economic interactions between the different stakeholders in this context ("business models") in this context in particular.

Georg holds a master's degree in electrical engineering from Vienna University of Technology.

Yanfei Li is a research fellow at the Energy Research Institute of Nanyang Technological University, Singapore, where he conducts research in energy economics and economics of technological change, serving both academic and consulting constituents. He has collaborated with government agencies of Singapore such as Energy Market Authority, Land Transport Authority, National Climate Change Secretariat, and Economic Research Institute for ASEAN and East Asia.

Dr. Yanfei Li's current research covers oil prices, regional gas trade modeling, regional power generation planning and trade, economic impacts of information and communication technology, and roadmapping for electrical vehicle technology adoption in Singapore. He has numerous publications in these areas.

Dr. Yanfei Li received BA in economics from Peking University and PhD in economics from Nanyang Technological University.

Stephen Littlechild is fellow, Judge Business School, University of Cambridge, emeritus professor, University of Birmingham, and an international consultant on regulation, competition, and privatization.

Professor Littlechild advised the UK government on the regulation of the privatized industries during the 1980s, including designing RPI-X regulation for British Telecoms in 1983. He was the first Director General of Electricity Supply and head of the Office of Electricity Regulation (OFFER) from 1989–98. Previously he was professor of commerce at the University of Birmingham (1975–89), member of the Monopolies and Mergers Commission (1983–89), expert member of the Airport Price Control Advisory Group of the Civil Aviation Authority (2005–9), and commissioner at Postcomm (2008–11).

Professor Littlechild has a bachelor of commerce degree from University of Birmingham, a PhD from University of Texas at Austin, and honorary degrees DSc from University of Birmingham, and D Civ Law from University of East Anglia.

Luciano Losekann is associate professor in the department of economics at Fluminense Federal University, where he teaches microeconomics and energy economics. Currently he is the head of the economics department and a member of the Energy Economics Group, located at the Federal University of Rio de Janeiro.

His research focuses on the Brazilian electricity and natural gas market, particularly on their institutional aspects.

Luciano holds a PhD in economics from the Federal University of Rio de Janeiro.

Iain MacGill is an associate professor in the School of Electrical Engineering and Telecommunications at the University of NSW, and joint director (Engineering) for the University's Centre for Energy and Environmental Markets (CEEM). Iain's teaching and research interests include electricity industry restructuring and the Australian National Electricity Market, sustainable energy technologies, and energy and climate policy.

CEEM undertakes interdisciplinary research in the monitoring, analysis and design of energy and environmental markets and their associated policy frameworks. It brings together UNSW researchers from the Faculties of Engineering, Business, Science, Law and Arts and Social Sciences. Dr. MacGill leads work in two of CEEM's three research areas, sustainable energy transformation, including energy technology assessment and renewable energy integration, distributed energy systems including "smart grids" and "smart" homes, distributed generation and demand-side participation. He has published widely in these and related areas.

Dr. MacGill has a bachelor of engineering and a master's of engineering science from the University of Melbourne and a PhD on electricity market modeling from UNSW.

Roland Meyer is a postdoctoral research fellow at Jacobs University Bremen and is associated to the Bremer Energie Institut.

His main research interest is regulation and market design. In his PhD, he analyzed vertical relations and integration economies in the electricity supply industry. Since then he participated in several research projects on regulation, smart grids and investment in networks and generation capacity. As a research fellow at Jacobs University, Roland's main research field is now market design issues raised by the ongoing transition of the electricity industry to low-carbon generation.

Dr. Meyer holds a diploma in economics from the University of Kiel and a PhD in energy economics at Jacobs University Bremen.

Alan Moran is director of the Deregulation Unit at the Melbourne-based Institute of Public Affairs. He has written and presented extensively on energy policy issues and particularly on climate change policies.

Alan joined the IPA in 1996 after a career in the Federal Departments of Trade and Industry & Commerce, the Productivity Commission (where he headed the Business Regulation Review Unit and was first assistant commissioner) and in the Victorian Kennett government where he was deputy secretary of energy from 1994–1996. Alan was also a member of the National Electricity Code Administrator (NECA) Code Change Panel from the commencement of the National Electricity Market in 1998 until it was dissolved in 2005.

Alan has a PhD degree from Liverpool University, UK.

David Newbery, FBA, is a research fellow in the Control and Power Research Group at Imperial College London, a director of the Cambridge Electricity Policy Research Group, and emeritus professor of applied economics, University of Cambridge.

His interests include the design, regulation, and performance of liberalized electricity markets and transmission networks. He was the president of the European Economic Association in 1996 and president of the IAEE 2013. Occasional economic advisor to Ofgem, Ofwat, and ORR, former member of the Competition Commission and chairman of the Dutch Electricity Market Surveillance Committee, currently a panel member of Ofgem's Low Carbon Network Fund. He has advised DECC and the House of Commons on Electricity Market Reform, Ofgem on transmission access pricing and DG ENER on Transmission Rights.

He was educated at the University of Cambridge with undergraduate degrees in mathematics and economics, and a PhD and ScD in economics.

Yris Olaya is an associate professor at the computer science department, Universidad Nacional de Colombia, Medellín, Colombia, where she teaches graduate and undergraduate courses on operations research and energy policy, and conducts research in energy economics.

Her research focuses on developing optimization and simulation models that help assessing energy policies and regulations, particularly for the natural gas and power sectors.

She holds a BS degree in petroleum engineering from Universidad Nacional de Colombia and a PhD degree in mineral economics from Colorado School of Mines.

Jacques Percebois is a professor and honorary dean of the faculty of economics at the University of Montpellier, where he teaches public and energy economics and serves as the head of Research Centre in Energy Economics, CREDEN. He is also in charge of lectures at Ecole des Mines de Paris and Institut Français du Pétrole (IFPEN).

He is the author of numerous publications including a recent book with J. P. Hansen, *Energie: économie et politiques*, which won two prizes in France. He is the recipient of IAEE's Award for outstanding contributions to the profession of energy economics in 2006 and most recently in charge of a major report on future of nuclear energy in France, titled Energies 2050, published in February 2012, by the French Minister of Energy.

He was awarded master's and PhD degrees in economics and political science from University of Grenoble and PhD in public economics from University of Paris, Nanterre.

Marcio Giannini Pereira is a research economist at Electric Power Research Center (ELETROBRAS CEPEL) with more than 12 years of experience in energy market, where he is engaged in a wide range of projects including renewable energy on electricity systems, energy poverty, rural electrification, sustainable development, monitoring and evaluations of energy projects, including social, environmental, and economics issues. Dr. Pereira's research interests include energy planning, public policy, economics of regulation, market restructuring and renewable energy market.

He has PhD in energy planning from Federal University of Rio de Janeiro.

Wolfgang Pfaffenberger is an emeritus professor of economics at University of Oldenburg in Germany and adjunct professor at Jacobs University in Bremen, where he is engaged in the school's economics and European Utility Management programs. He is honorary doctor of the faculty of economies of the State University of Novosibirsk in Russia.

His main research interests include economics of competition in network industries and the electricity and natural gas industries in particular. He has published a textbook on electricity economies and coauthored a textbook on energy economies, both in German, and has written numerous studies and articles on macroeconomic implications of energy and environmental policy, on various aspects of market opening in the electricity supply industry and on the problems in transformation countries. He was the coeditor of a prior volume to the present book, *Electricity Market Reform: An International Perspective*, published in 2006.

Professor Pfaffenberger holds a diploma in economics from Freie Universität Berlin where he also got his PhD.

Pierre-Olivier Pineau is an associate professor at HEC Montréal, Canada and president of the Canadian Association for Energy Economics. He was awarded the HEC Montreal professorship in energy management in 2010 and oversees energy management courses at HEC Montréal.

His main research focus is on electricity market integration, especially economic and environmental benefits resulting from market rules harmonization across jurisdictions. His research topics include electricity reforms in developing countries, especially in West and Central Africa. Professor Pineau also has interests in energy consumption reduction, energy poverty, and sustainable development.

He obtained an MA in philosophy from the University of Montreal and a PhD in administration from HEC Montreal.

Ricardo Raineri Bernain is a professor at Pontificia Universidad Católica de Chile and former Chilean Energy Minister. He has consulted to Latin American governments and the private sector, has served as chairman of the Chilean State Oil Company ENAP, and of the Council of the State holding in charge of the strategic direction of more than 20 Chilean public enterprises. Currently, he is an Alternate Executive Director of the World Bank Group.

His main research and interest areas include the performance of the energy sector, international relations, regional energy integration, strategy and competition policy, economics of regulation, market restructuring, pricing, and corporate governance. His experience also extends to Corporate Governance issues and to other service sectors such as telecommunications, healthcare and ports, among others. Professor Raineri has published in a variety of academic journals.

Ricardo obtained MA and PhD in economics from the University of Minnesota under the direction of Edward C. Prescott, 2004 Nobel laureate in economics.

Andrew Reeves is the Chairman of Australia's national independent regulator, the Australian Energy Regulator (AER).

Andrew brings over 30 years of energy experience to his role with previous government appointments as an AER Board member, an Associate Commissioner in the Energy Division of the Australian Competition and Consumer Commission (ACCC), the Commissioner of the Tasmanian Government Prices Oversight Commission and Regulator of the Tasmanian electricity supply industry. Andrew has written extensively on energy policy issues in Australia and has been involved in the economic regulation of the Australian energy sector since the commencement of the AER's responsibility for national regulation of electricity and gas markets. The AER works with the government, the regulated businesses and consumers to implement energy market reforms that promote investment and outcomes that are in the long term interests of consumers.

Mr. Reeves has a Bachelor of Engineering from the University of Queensland, with post graduate qualifications in Economics from Macquarie University.

Gustav Resch is working as senior researcher at Energy Economics Group, Vienna University of Technology, Austria. He joined the research group in 2000 and is currently leading a team dealing with policy-related research in the area of energy policy and energy economics with a focus on renewable energy technologies—with proven expertise on international level.

His fields of activity include techno-economic assessments of (renewable) energy technologies, design and evaluation of energy policy strategies and energy modeling (focusing on policy interactions and technology dynamics). He has led or contributed to several European and international studies and policy assessments related to renewable energies, assessing for example the feasibility and impacts of EU targets for renewables by 2020 and beyond.

He holds a master's degree in electrical engineering (energy technology) and a PhD in energy economics at Vienna University of Technology.

John Rhys is a senior research fellow at the Oxford Institute for Energy Studies, where he works on issues related to energy policy, with particular reference to the electricity sector, climate change, and market reforms. He currently acts as secretary to the British Institute of Energy Economics (BIEE) Climate Change Policy Group.

Dr. Rhys is a former managing director of NERA Economic Consulting, and a former chief economist at the UK Electricity Council. At NERA he was closely involved in UK electricity restructuring and privatization, led the firm's international energy consulting work, and was personally involved in a number of energy policy and power sector reform studies in Eastern Europe and Asia, carried out for national governments under the auspices of the World Bank, ADB and other agencies.

He has a mathematics degree from the University of Oxford and an MSc and PhD from the London School of Economics.

Jenny Riesz is a senior consultant in Strategic Energy Advisory Services at AECOM, based in Sydney, Australia. With more than 8 years of experience in energy market modeling, analysis, and analytical research, her work focuses on the integration of renewable technologies into electricity markets. With clients such as the Australian Energy Market Commission, and the Independent Market Operator of Western Australia her research has explored the impact of intermittent renewables on electricity systems.

Dr. Riesz is interested in multidisciplinary research at the nexus between technical and economic factors. She is an active member of the CIGRE C5-11 International Working Group, contributing to research on market design for large-scale integration of intermittent renewable energy sources and demand-side management.

Dr. Riesz has a PhD in physics from the University of Queensland, Australia, and is a chartered engineer with the Institution of Engineering and Technology (IET).

David Robinson is a senior research fellow at the Oxford Institute for Energy Studies.

He specializes in the field of public policy and corporate strategy in the energy sector, with a particular interest in the role of markets and regulation in meeting the challenges of climate change. He has worked for 25 years as a consultant for governments, regulators, energy companies and international organizations. He was co-chair of the European operations of NERA Economic Consulting and led the European practice of *The Brattle Group*. He is based in Madrid, where he runs his own consulting firm.

He earned his DPhil in economics at the University of Oxford.

Luiz Pinguelli Rosa is director of COPPE, the Institute of Post-Graduate Studies and Research in Engineering at Federal University of Rio de Janeiro, professor of the Energy Planning Program and secretary general of the Brazilian Forum on Climate Change. Formerly he was president of Centrais Elétricas Brasileiras SA (Eletrobras). He served on the UN's Intergovernmental Panel on Climate Change (IPCC) from 1998 to 2001 and is a member of the Brazilian Academy of Sciences.

His current research interests include energy planning, energy technology, and climate change. Through his long professional career, he has been engaged in theoretical physics, nuclear engineering, and studies of the contribution of the energy sector to the greenhouse effect, including measurements of greenhouse gas emissions from hydro reservoirs.

Professor Rosa has PhD in physics from Pontifical Catholic University (PUC-RJ).

Anupama Sen is a senior research fellow at the Oxford Institute for Energy Studies and a visiting fellow at Wolfson College, Cambridge University.

Her main research interests lie in the economic and applied econometric analysis of the outcomes of electricity sector reform in developing countries, and particularly on India. Her past research has investigated the economic impacts of autonomous electricity sector reform in Indian states, while accounting for their inherent economic and political diversity, published in *The Energy Journal*. Her current work focuses on India's energy sector, including oil, gas, climate change and energy pricing. She has also consulted for international organizations working on the natural resources sector, such as Revenue Watch.

She holds a PhD degree from Cambridge University, an MSc degree from the London School of Economics, and political science and a BA (Hons) from the University of Mumbai.

Jeong Shik Shin is a Chair Professor of School of Economics at Chung-Ang University. He is also Director of Research Center for Future Energy Environment Industries.

Dr. Shin's professional experience includes economist positions at the Ohio Energy Division, at Southern California Edison and at the Korea Energy Economics Institute. Dr. Shin worked for several energy projects in the Asian region for the World Bank and for the Asian Development Bank. Dr. Shin also served for several government positions in Korea including the following: President of the Korea Energy Economics Institute, a commissioner at the Korea Electricity Commission and a member of the Presidential Committee for Green Growth Korea.

He received a BA and a MA in economics from Seoul National University and a PhD in economics from the Ohio State University.

Neilton Fidelis Silva is a professor at the Federal Institute of Education, Science and Technology of Rio Grande do, Brazil, and a researcher at COPPE, the Post-graduate Engineering School of the Federal University of Rio de Janeiro (UFRJ).

He is currently working for the presidential office of the Brazilian Government as assistant to the executive secretary of the Brazilian Forum on Climate Change. His research focuses on energy planning, renewable energy, and climate change.

Professor Silva's qualifications include a DSc from the Federal University of Rio de Janeiro and a BSc and MSc in electrical engineering from the Federal University of Rio Grande do Norte.

Fereidoon P. Sioshansi is president of Menlo Energy Economics, a consulting firm and the editor and publisher of *EEnergy Informer*, a newsletter with international circulation. His professional experience includes working at Southern California Edison Company, the Electric Power Research Institute, National Economic Research Associates, and Global Energy Decisions (GED), acquired by ABB.

His interests include energy efficiency, renewable energy technologies, smart grid, dynamic pricing, regulatory policy, and integrated resource planning. He has edited six books prior to this: *Electricity Market Reform: An International Perspective*, with W. Pfaffenberger (2006), *Competitive Electricity Markets: Design, Implementation, Performance* (2008), *Generating Electricity in a Carbon Constrained World* (2009), *Energy, Sustainability and the Environment: Technology, Incentives, Behavior* (2011), *Smart Grid: Integrating Renewable, Distributed & Efficient Energy* (2011) and *Energy Efficiency: Towards the End of Electricity Demand Growth* (2013).

He has degrees in engineering and economics, including an MS and PhD in economics from Purdue University.

Rajat Sood is a member of the Energy Practice within Frontier Economics, a microeconomics consulting firm based in Melbourne, Australia.

Rajat has advised policy-makers, regulators, businesses, consumer groups and law firms operating in the Australasian energy sector for over a dozen years. His clients have included the Australian Energy Market Commission, the Australian Energy Regulator, the New Zealand Electricity Authority, the Singapore Energy Market Authority and most Australian State Governments and energy businesses. Rajat's key areas of expertise are the restructuring and reform of state-owned utilities, market design, network pricing and regulation and trade practices (antitrust) economics. Prior to consulting, Rajat was a solicitor with Freehills.

Rajat has honors degrees in economics and law from the University of Melbourne.

Jay Zarnikau is president of Frontier Associates, a consulting firm, where he provides consulting services to utilities, retail electric providers, and energy consumers in the design and evaluation of energy efficiency programs, retail market strategies, electricity pricing, demand forecasting, and energy policy. Jay is also an adjunct professor of public policy and statistics at the University of Texas, where he teaches graduate courses in regression analysis and quantitative methods.

Jay formerly served as a program manager at the University of Texas at Austin Center for Energy Studies, and was the director of Electric Utility Regulation at the Public Utility Commission of Texas. His publications include numerous articles in academic and trade journals.

Jay has a PhD degree in economics from the University of Texas at Austin.

Foreword

Electricity market reform was intended to increase the role of market forces throughout the electricity sector and correspondingly to reduce the role of political forces. The main components were private ownership, restructuring, competition, and incentive regulation. These were to replace state ownership and control in a monopoly sector characterized by political rather than economic decisions, inefficiency, a cost-plus mentality, and unresponsiveness to the wishes of consumers.

Properly implemented, this reform could be expected to bring greater productive efficiency—better planned and implemented capital expenditure and lower operating costs. There would be greater responsiveness to customer preferences—improved reliability and better customer service. The ability to choose supplier would stimulate a greater variety of tariffs, contractual terms, and methods of payment. Across the board there would be more innovation.

What about prices? They could be expected to be more cost-reflective. Where prices already reflected costs, and competition and/or regulation was effective in driving costs down, prices would fall in real terms. But where prices were initially below costs, or where competition and/or regulation was ineffective, prices could be expected to increase. And there is the rub: what would be the outcome in each particular case?

Sioshansi and Pfaffenberger (2006)[1] painted the picture in more than a dozen countries after some 15 years of electricity market reform. Some countries took the plunge, some paddled at the edges, and some stayed out of the water. What were the outcomes? Broadly speaking, the predictions proved correct. Politics were by no means absent, but economic factors played a greater role than before. New investment took place, sometimes on an unprecedented scale. Efficiency, responsiveness, and innovation improved. Properly specified, the predictions about prices also proved correct. Not all prices fell, but neither could they have been expected to do so. Some prices needed to rise if they were to reflect costs.

Of course, it was all more complex than this. There was often a great deal of uncertainty and controversy. Does a particular increase in wholesale price reflect the removal of a previous policy that held prices below cost, or an increase in underlying costs, or a shift in demand, or the recognition of a scarcity that requires a higher price to bring forth the necessary capacity? Or does it reflect the exercise of market power by generators, or inappropriate

---

1. Sioshansi, F.P., Pfaffenberger, W., 2006. Electricity Market Reform: An International Perspective. Elsevier.

rules used to calculate market price, or lax regulation? In some cases, these questions remained unanswered.

Nonetheless, Paul Joskow's Introduction to the 2006 volume noted the emergence of a "textbook model" for electricity reform. We could begin to see which countries had followed the model and to what extent. We could also assess their performance in the light of it, bearing in mind the initial conditions just noted. Broadly speaking, the textbook model seemed the way to go.

Two years later, Sioshansi (2008)[2] provided more information and evidence, looking in more detail at "design, implementation and performance." It became even more apparent that the road to reform was a rocky one. Some countries were still making progress toward the textbook model. And there was some attempt to extend reform from local or national markets to regional or international markets. But some markets were falling by the wayside, others were bedevilled by problems of insufficient restructuring or inadequate design. It also became apparent that design and implementation could be more complex than the textbook model implied. Was a capacity mechanism needed? How to incorporate demand participation and distributed generation? Maybe there were multiple roads to reform, multiple textbook models.

Nevertheless, despite the additional complexity of the issues related to restructuring and regulation, there seemed no serious challenge to the fundamental drivers of reform, namely private ownership and competition. Evidence was adduced from Texas and Australia, in particular, to suggest that concerns about resource adequacy were overplayed. Nuclear energy was not yet an issue. Yet, in retrospect, subsidized and/or mandated renewables were being deposited like cuckoos in many market nests. And on the horizon appeared global climate change, that small cloud as yet no bigger than a man's hand, presaging the storm to come.

Here we are, another 5 years later. What does the present volume tell us about the state of affairs and of reform in the world's electricity markets today?

In the United States, most of the regional power systems have accepted reform and created competitive wholesale markets, though the design and implementation of capacity markets remain a work in progress. The retail sector—that last bastion of political resistance to reform—is now somewhat more open to competition. Perhaps about half the US population can now choose their supplier, but whether retail markets will extend further is yet uncertain. Texas, the leader in US retail markets, also has a vigorous energy-only market that calls into question the arguments for capacity markets. In addition, Texas seems to have taken in its stride the challenge of significant

2. Sioshansi, F.P., 2008. Competitive Electricity Markets: Design, Implementation, Performance. Elsevier.

renewables and transmission expansion. In contrast, California has abandoned the textbook model and reverted to central planning.

Canada's electricity market remains fragmented. Alberta is still swimming well and Ontario has made some effort, but most of the other provinces have not yet ventured into the water. Vertically integrated incumbents, particularly government-owned hydropower companies, have been preferred to competition. Political factors have prevented regional integration in Canada, though the case for it is argued herein.

Latin America, like Canada, has very diverse electricity sectors. Some countries are untouched by reform, but markets continue to develop in Chile, Colombia, and Peru. Brazil began to reform but has increasingly reverted to central planning. Argentina, once at the very forefront of reform, is systematically undermining the market, and not only in the electricity sector. Across Latin America, again as in Canada, political differences have precluded any significant regional integration.

Australia's reform continues, with Victoria still the leader. Retail markets are very active. There is increasing evidence of the advantages of private over public ownership. Support for privatization may be increasing, even in New South Wales. But as elsewhere, political involvement has hindered a more effective national market. New Zealand is described as drifting in a regulatory vacuum.

Asia is positively dispiriting. Japan: reform still under consideration and with nuclear obviously an issue. Korea: reform frustrated. India: reform too difficult, though little pockets of market are emerging. China: reform postponed. Russia: reform repealed. Where prices are held below costs, reform offers few attractions to customers. But how long can the costs of such subsidies be sustained?

In Europe, a common energy framework and policy have been actively developed. New transmission links are gradually impacting on wholesale markets. Retail markets are slowly evolving. Norway and Sweden (not discussed in this volume) appear to be doing well. Yet France, still dominated by an unrestructured and predominantly state-owned incumbent, has no more than a toe in the water of market reform, seemingly only for form's sake. Politically determined prices of nuclear energy are argued to preclude further reform. Germany struggles with political decisions on nuclear and renewables. The United Kingdom is characterized by even more challenging targets for renewables and carbon; concerns about security of supply that seem to be created or exacerbated by government policy; active government involvement in detailed market design, not least of a capacity mechanism and related contracts; increasing subsidy for renewables and the prospect of subsidy on an unprecedented scale for nuclear; and the prospect of retail market regulation that will virtually extinguish retail competition.

In sum, the beacons of electricity market reform shine brightly in Texas and Victoria. They continue to burn in several other states in the United

States and Australia, and in Chile, Colombia, and Peru. The lights are flickering, nearly extinguished, or as yet unlit, in most of Asia. Africa remains the Dark Continent. And Europe? In 1914, Sir Edward Grey remarked that, "The lamps are going out all over Europe. We shall not see them lit again in our time." The European picture is mixed. But what was once the first and brightest lamp of all is indeed being dismantled and not because it failed to provide good light.

Where it was implemented, electricity market reform, based on private ownership and competitive markets, has generally achieved what was predicted for it. And it could continue to do so. Many economists would argue that, with an appropriate framework of carbon taxes or property rights, competitive markets could meet the challenge of climate change along with the challenge of technological innovation. Indeed, competitive markets could adapt to, and help to shape, the future needs of customers in a more efficient way, providing investment to meet demand with less dependence on expensive and problematic renewables, and a less erratic role for nuclear. Andrew Reeves' excellent Preface to this volume outlines in more detail many of the challenges and needed responses.

At the same time, Jean-Michel Glachant's stimulating Epilogue forcefully reminds us that the energy world is inevitably characterized by unexpected developments. In the event, hopes and fears—not all obviously realistic—about security of supply, renewables, nuclear, and climate change have led governments to superimpose their own ideas on electricity market reform. They now intend to provide direction and supervision of the electricity sector. We must assume that this will be consistent with government direction and supervision of that sector in the past—there is no reason to suppose that leopards can change their spots. The consequence will be electricity bills and/or taxes higher than they otherwise would be. Thanks to electricity market reform, and the increased transparency that it has brought, the causal links are now clearer than they were before. Businesses and voters are already expressing concerns about rising electricity bills. These concerns will grow. In response to political pressures, governments will increasingly reconsider their policies. We may hope that the lamps of European electricity market reform will be lit again in our time.

Stephen Littlechild

Fellow, Judge Business School, University of Cambridge, Emeritus Professor, University of Birmingham, Member of Competition Commission, and Former Director General of the Office of Electricity Regulation (OFFER)

Preface

Over the past 15 years international electricity markets have been transformed from centralized to more open and competitive arrangements. Typically, the old vertically integrated utilities were broken up into discrete elements of the supply chain, with competing generators and retailers and a regulated network sector. Most of the changes aimed to widen markets, facilitate energy trade across national and state borders, and optimize infrastructure use. As set out in this timely compendium, the reforms in the main have been successful in delivering the generation investment, when and where required. There have also been significant retail reforms, with more consumers enjoying choice in their supply arrangements.

However, energy markets are changing rapidly and their design and regulation face new challenges. Faltering economic growth, wavering commitment to nuclear generation and changes in the relative cost of fuel—particularly from the development of unconventional gas—is creating an uncertain future for generation investment. In markets with mature infrastructure, the combination of aging networks, rising peak demand, and more stringent reliability standards are driving higher costs. These drivers, on top of the higher costs of low-carbon investment, may crowd out the social acceptance of further reform, slowing the rate of transformation.

At the same time, lower communication costs and technological advances in photovoltaics and demand management mean the traditional one-way generation-to-consumer supply chain is evolving. While energy market reform has to date focussed on competition in generation and retail, the industry will soon see the consumer at the center of competition between local or self-generation, demand management, storage, and grid services. New institutions will develop in response to these changing needs. The emergence of competitively priced electric vehicles would make such change inevitable. Just as the telecommunications sector has rapidly transformed from the monopoly supply of an integrated service, so the electricity supply industry will transform to a new interaction with consumers and a diversity of service and supply.

While there are clear opportunities to develop new services such as integrating the smart home with dynamic load management, local generation and storage, it remains unclear who will provide such services. Extension of monopoly network regulation into these areas risks forgoing the opportunities for innovation that would naturally flow from a competitive market for services. A key challenge for market design and regulation is finding the right mix of incentives for efficient investment to deliver the energy services required by consumers. The challenge is to allow markets to evolve but at the same time ensure there is a level playing field with no one party—whether distributor or retailer—enjoying the privilege of incumbency.

In this complex environment, customer empowerment is an essential ingredient. Energy customers are increasingly bewildered by technological change and escalating prices, especially when price increases are not matched by appreciable improvements in service quality. For example, the rollout of smart meters has garnered strong opposition in several markets. That opposition in some cases reflected a lack of groundwork in communicating how smart meters may empower the consumer to better manage their energy costs. More direct engagement with consumers in the decision and rollout process may have led to more positive outcomes. Similarly, informed conversations with the consumers about time of use pricing and demand-related pricing is essential.

As noted, we are starting to see distributors and retailers use smart meter and communications capability to provide customers with sensible and timely information about use and cost. While only a small proportion of customers may use this information to actively manage load, it does have application in advising customers of progress with their bill, of alerts of high rates of use, and of greater frequency of billing to put customers in greater control of their usage and cost.

Some markets are doing better than others in regard to customer engagement. Consumer representatives in the North American system are actively engaged throughout the regulatory process; for example, by challenging network businesses' submissions and in some respects directly negotiating the outcomes, such as the terms of rollout for smart meters and compulsory demand management programs. Under these conditions, customer representatives are well informed and well resourced and sufficient time is allowed for this engagement.

In addition to these changes, a different set of challenges faces wholesale markets and producer-end network development. A low-carbon economy and changes in comparative costs of fuel would change the pattern of investment in both generation and the networks; incentives across the two sectors must be better aligned to ensure an efficient investment mix. As noted in several chapters of this volume, this means generators, particularly the renewable sort, should have clear pricing signals to locate in the most efficient parts of the network.

In the retail sector, regulatory reform remains incomplete in most markets, with a mix of approaches to pricing that range from full deregulation to government administered prices. To achieve competitive and transparent markets, governments should minimize policy intervention, such as the prescribing of retail prices that constrain retailers from developing innovative tariffs to meet consumer needs.

To sum up, there is the need and prospect over the next 10 years or so for enhanced sophistication in the delivery of electricity services. It is not yet clear how regulation of the sector may facilitate or impede that transition, but it is important that regulators adopt structures that encourage and reward efficient innovation, adopt open systems to allow participation by multiple agents in activities that are not strictly monopoly activities, and, most importantly, be open and communicate with customers and the community about the issues, needs, and benefits of reform. This collection, reflecting the diversity of international experience in reform and perspectives on the future, contains important lessons for the future development and evolution of electricity markets.

Andrew Reeves
Chairman, Australian Energy Regulator

Introduction

BACKGROUND

A 2006 volume edited by Professor Wolfgang Pfaffenberger and me, titled *Electricity Market Reform: An International Perspective,*[1] covered the experiences of a number of countries and jurisdictions around the world who had introduced market reform initiatives in their electricity markets. Following the successful experiences of Chile in the 1980s and England and Wales in 1990, the topic was in vogue as countries and many states in the United States were contemplating similar moves.

During the 1990s, European Union, Australia, New Zealand and a number of other countries, states and provinces in Americas and elsewhere introduced market reform initiatives. At the time, there was speculation in the United States for a federal policy to introduce retail competition nationwide while the Federal Energy Regulatory Commission (FERC) was pushing to create organized wholesale markets that would span the entire country. A number of developing and centrally planned countries also became intrigued by the purported efficiency gains promised by market reform and the introduction of competition in generation.

Following the much publicized 2000–01 California electricity crisis and the not-so-successful experience in Ontario, Canada, interest in market reform cooled considerably, to put it mildly. With the exception of Texas, which opened its market to competition in 2002, no other state has introduced competition in the retail sector in the United States after the collapse of the California market while a number of states with the intention to do so abandoned these plans. And while FERC's efforts have succeeded in placing roughly half of the country in the hands of organized wholesale market operators, market reform in the United States remains an unfinished business. As described in Chapter 12 by Kim, retail competition is not universally available to consumers in the United States.

Likewise, while continental Europe has succeeded in introducing market reform initiatives on paper, its electricity markets are *not* integrated as originally envisioned, nor is retail competition as successful in a number of key markets despite the official European Commission's Directives. Several chapters in this volume cover the developments in European markets, some of which have been sidetracked by other priorities, such as Germany, now dealing with a challenging nuclear phase out.

---

1. *Electricity Market Reform: An International Perspective*, Elsevier, 2006.

Internationally, efforts to introduce market reform have been stalled, postponed and/or stopped by those who prefer to maintain the status quo in countries such as South Korea, India, China, South Africa, Thailand and Malaysia, just to name a few. In a number of other countries, notably Japan, the reform initiatives have been successfully frustrated by the incumbent monopoly utilities, with the result that less than 3% of Japanese generation is competitively provided, its power exchange handles little trade, and retail competition is virtually nonexistent as described in Chapter 23 in this volume.

In some cases—in a number of states in the United States, for example—market reform has not been as successful as initially assumed or promised. These types of experiences have collectively given politicians and policy makers ample reasons to pause or postpone any efforts to introduce market reform, as in the case of California, where things have essentially reverted to where they stood prior to the 2000–01 crisis with the exception of a functional independent system operator, as described in Chapter 11 by Kristov and Keehn. The expected gains no longer seem overwhelming when measured against the potential risks, the uncertainties and the costs.

Setting aside the ongoing policy debate on the pros and cons of electricity market reform (EMR), there are continuing discussions on what is the best form and structure of competitive markets, including issues such as energy-only markets versus markets with capacity payments. A companion volume published in 2008 dealt with market design and performance issues.[2]

The passage of time alone warranted the updating of the 2006 volume. But that is not all. A number of *new* issues and priorities have emerged in the recent past, which were not critical or important when the 2006 volume was being assembled. A few of these issues now dominate any discussion of market reform—both in markets that have introduced reform and those that have not including the following:

- *Low-carbon energy mix*—An increasing number of governments have already adopted or are considering adopting mandatory low-carbon targets for their energy sectors. Since electricity can be generated from a variety of fuels/technologies, it is typically the sector most likely to be affected by such initiatives. Among the issues to consider is how best to achieve certain targets—say 20%, 50%, or 80% less carbon emissions by a certain date, say 2020, 2030, or 2050[3]? In the absence of relatively high and stable carbon prices, competitive markets do not appear capable of meeting such ambitious targets on their own.[4]

---

2. *Competitive Electricity Markets: Design, Implementation, Performance*, Elsevier, 2008.
3. EU, for example, has a 20-20-20 target requiring 20% of primary energy to come from renewables, 20% reduction in greenhouse gas emissions relative to 1990 levels, and 20% improvement in energy efficiency by 2020 further described in this volume.
4. MacGill and Healy go even further, suggesting that market reform initiatives might inadvertently have made it harder for the stakeholders to reduce greenhouse gas emissions due to the increased commercial pressures implied by market competition.

- *Renewable energy*—Likewise, a growing number of governments favor a significant role for renewable energy resources in their generation mix, not only to meet low-carbon targets but also for other reasons, including job creation, sustainable growth, energy security, and so on. Ambitious mandatory renewables targets are becoming the norm as in the case of California, 33% by 2020 or 80% by 2050 for Germany. There is ongoing debate on how best to meet such ambitious targets, through generous feed-in-tariffs, renewable portfolio standards, production tax credits, government loan guarantees, or a combination thereof. Needless to say, the growing penetration of renewables in the generation mix leads to new challenges for market operators, a subject covered in several chapters in the book.
- *Energy efficiency*—Another top priority is to increase the efficiency of energy utilization in general, and in the electric sector, in particular. The prior two objectives would be so much easier to meet if future demand growth could be lowered. Lowering demand, however, typically increases per unit costs since the industry's significant fixed costs would have to be recovered from a declining sale volume.
- *Distributed generation*—Rapid technological advances and rising electricity prices have combined to make distributed generation cost-effective in a growing number of places. As more consumers gradually become *prosumers*—potentially generating more power than they consume at least during certain periods—the industry's traditional paradigm will be further perturbed.
- *Demand participation*—Another important issue receiving increased attention globally is to engage customers to become active participants in how supply-and-demand are balanced in real time, as opposed to the traditional model where they were passive consumers, disengaged, and unresponsive to the market price signals.

There are many others: smart meters, dynamic pricing, smart grid, and so on. These are among a long list of emerging issues gaining momentum and expected to have pronounced effect on electricity markets in the future, whether they are restructured or not. These are the *new paradigms*, and *new challenges* that require *new approaches*, this volume's subtitle.

OBJECTIVES AND ORGANIZATION OF THE BOOK

This book, which includes contributions from world-class experts, scholars and academics from different parts of the world, is focused on providing a comprehensive assessment of the relevant issues in *today's* rapidly changing electricity markets, facing a host of new challenges and issues that were not prevalent, or in some cases barely present, when the 2006 volume was compiled.

The book is organized into four parts focused on regional markets and issues as outlined below:

Part I covers developments in the European markets.

In Chapter 1, David Newbery points out that the United Kingdom was in the forefront of liberalizing the electricity market in 1990, but abandoned the Pool model in 2001. Its energy policy has fluctuated markedly over this period, with varying degrees of support or resistance to gas, coal, and nuclear power.

The author explains that the United Kingdom's current energy policy objectives are to deliver secure, sustainable, and affordable electricity, while meeting challenging targets for decarbonization and renewable energy. The UK government has consulted and concluded that the present electricity market arrangements will not deliver all three goals, and has proposed a major electricity market reform (EMR).

The chapter's main contribution is to describe the background to, reasons for, and the nature of, the EMR, pointing to the need for further market and institutional reforms.

In Chapter 2, Malcolm Keay, John Rhys, and David Robinson describe how climate change policy is influencing EMRs, in particular in the United Kingdom and Europe.

The authors highlight the difficulty of providing clear signals for investment in low-carbon generation in liberalized markets. Many countries are resorting to more centralized approaches, in order to accelerate investment.

The authors consider alternative policies and conclude that unless more innovative approaches are introduced, the trend away from liberalization is likely to continue and strengthen.

In Chapter 3, Jacques Percebois points out that the liberalization of the French electricity market has been hampered by two peculiarities not uncommon in a number of European countries. Firstly, due to the large market share held by EDF, the incumbent utility, it is difficult for competitors to gain market shares. Secondly, due to the duality of prevailing tariffs—regulated and market rates—competition is not fair.

The author explains that the former are set according to the production costs of French generation, which is mostly from low-cost nuclear power stations. The market prices paid by customers who have exercised their eligibility, however, are determined by prices observed on the European wholesale electricity market, which largely depend on the higher cost of thermal power stations.

The chapter's main contribution is to describe efforts underway to address these market issues, notably the implementation of Nouvelle Organisation du Marché de l'Electricité (NOME), a new law allowing competitors access to a portion of the low-cost nuclear generation during off-peak hours at a regulated price, and the future of nuclear power in France.

In Chapter 4, Wolfgang Pfaffenberger and Esther Chrischilles provide a short account of the structure of the German electricity market after the reform of 2005. The focus of the chapter is on the consequences of the political decisions taken in March 2011 to abandon nuclear power, mostly replaced with renewable energy as the main source of electricity in the future.

This turnaround implies fundamental changes in all segments of the power system: Generation will need to cope with the higher share of fluctuating renewables; market rules will need to create incentives for investors to provide backup power; the transmission system will have to cope with the growing regional disparities—deficit in the south, surplus in the north—and changes in power flows within European network caused by fluctuating renewables; and the distribution system will have to adapt to an increasing share of decentralized, intermittent generation feeding the network.

The chapter highlights three challenges: The necessity to enlarge the transmission network; the difficulties to achieve ambitious greenhouse gas reduction targets and the increased cost of electricity and its impact on the German economy.

In Chapter 5, Reinhard Haas, Hans Auer, Gustav Resch, and Georg Lettner examine the historical growth, the current status and the future prospects for renewable electricity generation in Europe in view of the mandatory goals and targets.

Specifically, the authors highlight how the volatility and the intermittency of renewable resources like wind and photovoltaic generation impacts the merit order and the dispatching of resources and how a growing renewable portion will impact how wholesale prices are determined and their impact on thermal generation.

The chapter's conclusions are that spot market prices will become more volatile in the future leading to increased role of intraday trade, increased flexibility offered by a smarter grid and the need for storage capabilities.

In Chapter 6, Günter Knieps analyzes the challenges for electricity transmission networks caused by increasing generation of renewable energy.

Given the institutional background of liberalized European energy markets, the author develops a disaggregated nodal pricing framework consisting of electricity injection and extraction prices under variable infrastructure. The author examines transmission pricing schemes to maximize social welfare and compares them with those resulting under a monopoly pricing regime—both based on system externalities of network usage. The resulting generalized merit order provides the proper incentives for decentralized generators.

The chapter's main conclusion is that entrepreneurial incentives for electricity network providers arise taking into account efficient transmission charges and efficient signals for infrastructure investments creating proper incentives for decentralized generators. Therefore, regulatory instruments should not disturb entrepreneurial search for pricing structures and investment for electricity transmission networks.

In Chapter 7, Dierk Bauknecht, Gert Brunekreeft, and Roland Meyer examine the effect of growing share of electricity from renewable resources in the context of the original competitive electricity market on two dimensions. First, in operational terms, renewables affect the supply-demand balance and wholesale prices. Second, the large-scale integration of intermittent renewables exacerbates the "missing money problem" in generation investment and the efficacy of an *energy-only* market.

The authors discuss the integration of renewables both in terms of support schemes and market design including the debate on introducing a capacity market in Germany.

The chapter suggests moving beyond the current debate about feed-in-tariffs—with no market integration—versus a quota system—with full market integration. The chapter also examines the wisdom of introducing a capacity mechanism to address generation adequacy in a world where the role of conventional generators shifts toward providing reserves instead of significant amounts of energy.

In Chapter 8, Walter Boltz covers the history of European electricity market regulation from the first directive in the 1990s until the third directive in 2009.

Specifically the author demonstrates the inability of member states to solve national competition issues in electricity markets leading to the process of more and more detailed structural provisions in EU law as well as the implementation of a light version of an EU regulatory body. The most recent legal act on market integrity in energy markets (REMIT) goes even further and uses the Agency for the Cooperation of Energy Regulators (ACER) as the core institution of the surveillance of wholesale markets.

The chapter's main contribution is to relate the developments in electricity markets such as more interventionist tendencies in view of increasing subsidy elements, intermittency, high investment needs in transport, etc. to the present regulatory system.

Part II is focused on markets in North and South America.

In Chapter 9, Joseph E. Bowring examines the development of the PJM capacity market and the extent to which it has succeeded in addressing the revenue sufficiency problem that affects all wholesale power markets.

The author's main contribution is a critical review of the role of capacity markets in competitive wholesale power markets, the development of the PJM capacity market including its positive features and a number of remaining problems associated with capacity market design. The issues are illustrated with data from PJM markets.

The chapter's conclusions are that while capacity markets are a critical part of the design of the PJM power market, the design and implementation of capacity markets remains a work in progress.

In Chapter 10, Parviz Adib, Jay Zarnikau, and Ross Baldick describe how the Texas electricity market, considered the best competitive retail market in North America, has successfully addressed challenges and introduced refinements in recent years to improve operational efficiency and transparency.

The authors examine wholesale and retail markets, the transition to nodal pricing, steps taken to eliminate default retail prices, transmission expansion rules to accommodate renewable resources, policies to address market power, the introduction of advanced metering infrastructure and efforts to integrate price responsive loads in dispatching.

The chapter's main conclusions are that, despite the successful implementation of many refinements, a number of challenges remain, including concerns about resource adequacy and reliability. These require addressing ineffective scarcity pricing, and creating a level playing field for demand side resources and price responsive loads in the market.

In Chapter 11, Lorenzo Kristov and Stephen Keehn describe how California's electricity market, in less than a decade, went from being the industry's poster child for a failed experiment to a stable, efficient market and an innovative platform for integrating renewable resources.

The authors describe California's original market structure and the reasons that led to the 2000–01 crisis. More important, they explain the subsequent transition to a robust and efficient market, completed in April 2009 with the introduction of locational marginal pricing, and how the California Independent System Operator (CAISO) has addressed a host of new challenges arising from California's ambitious environmental and energy policies and the adoption of new technologies.

Among the chapter's main contributions is to offer a broad perspective on market evolution that encompasses grid operations, grid infrastructure development, distribution-connected resources and initiatives to achieve the 33% new renewable energy target by 2020.

In Chapter 12, Young Kim examines the development of retail choice in states where it is allowed, how these markets have evolved and, in some cases, thrive despite many remaining barriers and challenges.

The author describes what makes certain markets successful and vibrant, such as in Texas, and compares important design parameters such as the default service and identifies some of the most important barriers to further development of retail markets in the United States.

The chapter's main contribution is to demonstrate that retail competition is viable and can thrive given appropriate market design. It also provides a perspective on how to expand retail markets in both current restructured US markets and those that remain closed.

In Chapter 13, Pierre-Olivier Pineau provides an overview of the current status of electricity markets in Canada. The author explains that while provinces endowed with large hydropower resources—British Columbia, Manitoba, Quebec, and Newfoundland and Labrador—have kept their electricity sector largely government-owned and regulated, other provinces—

Alberta, Ontario and New Brunswick—have liberalized to some extent, while the Maritime Provinces struggle with high-cost, privately owned, thermal power plants.

This patchwork of electricity markets creates a landscape with hugely divergent price and emission levels and a large number of parallel and introverted regulatory agencies. Contrary to many regions across the world, such as the EU, no integration or harmonization reform is presently being considered.

The chapter's contribution is two-fold: First, to provide a perspective on the status of various electricity markets in Canada and their evolution; Second, to highlight the important economic and environmental costs resulting from the lack of integration of provincial electricity markets with their Canadian and US neighbors.

In Chapter 14, Ricardo Raineri, Isaac Dyner, Nivalde Castro, José Goñi, Yris Olaya, and Carlos Franco point out that the Latin America region is blessed with ample energy resources including large deposits of coal, oil, natural gas, and hydroelectric power. The problem is that these resources are unequally distributed among neighboring countries and the region remains poorly integrated, the main obstacle remains historic geopolitical reasons.

The chapter describes the region's growing need for clean and competitive energy and examines the reasons for lack of integration, which is paradoxical when compared to other parts of the world where energy networks— electricity, oil, and natural gas pipelines—are better integrated. The chapter examines the technical, regulatory, commercial and geopolitical reasons for this, which leads to persistence of the "haves" and the "have-nots" within the region.

The chapter's main contribution is an assessment of the scope of opportunities for increased integration that have not been captured to date and the benefits that can be gained including a discussion of possible remedies.

Part III covers the rise of four growing economies of Brazil, Russia, India, and China, the so-called BRICs.

In Chapter 15, Luiz Pinguelli Rosa, Neilton Fildelis Silva, Marcio Giannini Pereira, and Luciano Losekann provide an overview of the evolution of the Brazilian EMR starting in the 1990s. The authors point out that the first round of liberalization directives were not effective resulting in insufficient investment in the power sector and partially to blame for the power shortages and rationing of 2001/2002.

During a subsequent round of reforms introduced in 2004, the state assumed a central role in resource planning while obliging distribution companies to secure their needs through long-term power purchase contracts through competitive auctions, which remains the main instrument in implementing government's energy policy.

The chapter describes the major challenges facing the Brazilian power sector as finding the means to provide adequate supplies of electricity to a growing economy at affordable prices while moving toward a cleaner energy mix.

In Chapter 16, Anatole Boute describes the evolution of Russian electricity industry since the collapse of the Soviet Union and examines how the liberalization and privatization reforms adopted in 2003 have been implemented to attract private investment to modernize the infrastructure.

The author highlights that although the electricity market is supposed to function since January 2011 on a fully liberalized basis, the government has reintroduced different mechanisms to protect consumers from price increases. Despite an existing capacity remuneration mechanism, the risk of political interference with prices, together with constant regulatory changes, affect much-needed private investment in the sector and discourages the development of renewable energy.

The chapter concludes that, given the instability of investment conditions and the political sensitivity of electricity prices, long-term regulated capacity tariffs will continue to be needed in the future to attract the necessary investments. This, however, contradicts with the stated goals of liberalization.

In Chapter 17, Anupama Sen and Tooraj Jamasb discuss India's attempts to reform its electricity sector over the past two decades with mixed results.

The authors identify three main phases of market reform since the early 1990s and discuss the outcome of these to date. Despite ambitious goals, electricity reforms have consistently failed to deliver even on basic objectives such as reductions in technical and commercial losses. Instead, they have resulted in a highly divergent and fragmented system and persistent chronic power shortages. The chapter discusses the main causes of these failures and the emergence of a parallel market for renewable energy, which appears to be functioning.

The chapter identifies the main challenges to continued EMR in India as they relate to regulation, security of supply, and the political and institutional rigidities that act as countervailing forces to reform.

In Chapter 18, Philip Andrews-Speed describes the evolution of China's electricity industry and market since the mid-1990s. Many structural reforms were undertaken between 1997 and 2003, but since that time reform has stalled. The chapter examines the interactions between the partially reformed power market and other parts of the energy sector.

The author's main contribution is to identify the factors which determine the direction and pace of electricity sector reform in China. He also shows how stalled reforms in the power market creates serious problems for the effective governance of the sector.

The chapter's conclusion is that China's government is unlikely to undertake further radical reform of the power sector in the near future, though it will seek to make minor adjustments in order to promote the transition to a low-carbon economy.

Part IV covers a number of important markets in Australia, New Zealand, Japan, South Korea, and Singapore.

In Chapter 19, Alan Moran and Rajat Sood track the progress of reforms to the Australian electricity supply industry over the past two decades.

The authors explain that many durable reforms have been achieved, including substantial vertical and horizontal disaggregation and the introduction of independent economic regulation. However, the reforms remain incomplete in key areas like retail price regulation, elements of which remain in most states, and ongoing public ownership in some states. Moreover, the Australian electricity sector faces intensified and fresh challenges as a result of policies aimed at reducing greenhouse gas emissions.

The chapter's conclusions are that Australia's current energy policy settings risk jeopardizing the fruits of its earlier reforms and forfeiting the benefits of its rich fossil-fuel endowments.

In Chapter 20, Iain MacGill and Stephen Healy question the widely held view that the Australian National Electricity Market (NEM) is a successful example of electricity industry restructuring. While it has delivered reliable electricity supply with apparent economic efficiency gains over previous arrangements despite challenges such as peak demand growth, it has largely failed in reducing greenhouse emissions although this was an explicit target of public policy when restructuring initially commenced.

The authors explore the reasons for this divergent performance and possible options to address it given growing alarm regarding climate change. While commonly ascribed to a lack of political will on environmental aims, restructuring itself has also created difficulties for effective climate action through a narrow economic focus that resulted in: the creation of large politically powerful market incumbents opposed in many cases to improved environmental outcomes; retail market arrangements encouraging competition in energy sales rather than in efficient provision of desired energy services, and poor governance of external climate policies.

The chapter provides recommendations on how the environmental performance of the NEM might be improved and concludes that there are also more fundamental challenges with the current market framework that will need to be addressed for low-carbon industry transformation.

In Chapter 21, Geoffrey Bertram describes how New Zealand's restructuring agenda has lost momentum as the five dominant generator-retailers consolidated their grip on the market after 2000, excluding new entrants and blocking developments that threatened their profitability such as distributed generation, feedback tariffs, and smart metering.

The author suggests that the regulatory oversight has been weak, further undermined by repeated changes in the agencies and goalposts. The result has been an ineffective authority facing a powerful industry alliance that has left small and residential consumers powerless in the face of steeply rising electricity charges.

The chapter's main conclusion is that despite the lack of coherent energy or regulatory policy from government, New Zealand's natural resource endowment, combined with the incumbent generators' desire to lock up investment opportunities, has finally begun to produce progress toward a low-carbon electricity future with large-scale development of wind and geothermal resources.

In Chapter 22, Suduk Kim, Yungsan Kim, and Jeong Shik Shin provide an overview of the evolution of the Korean electricity industry over the past three decades focusing on the more recent restructuring of the market and the changes in the energy mix.

The authors explain how the restructuring process started and stalled amid fierce political controversy, leaving things in virtual limbo. The chapter discusses the problems associated with the unfinished reforms and measures introduced to address them as well as policies focused on influencing Korea's long-term energy mix, energy efficiency, energy security, and sustainable economic growth.

The chapter's conclusion is that the current market structure and design are not comprehensive and self-contained, and hence need be upgraded to meet current and future challenges.

In Chapter 23, Hiroshi Asano and Mika Goto provide an overview of developments in Japan's power sector since the 1990s when market liberalization process was initially introduced.

The authors describe the effects of the liberalization to date and what has been achieved, not only from an economic and competitive perspective but also in terms of its environmental impact. The authors also examine the current debate on the future direction of energy policy in Japan in the aftermath of the devastating March 2011 earthquake and nuclear accident at Fukushima Daiichi plant. The country faces multiple challenges on multiple fronts including the need for institutional reforms to further encourage competition in both wholesale and retail markets, address limited interutility transfer capacity as well as policies to promote renewable energy integration and demand side participation.

The chapter concludes that the time has arrived for Japan to set a new energy policy direction that combines energy security, economic prosperity, and environmental sustainability enabled by the most advanced combination of supply-side options and efficient end-use technologies.

In Chapter 24, Youngho Chang and Yanfei Li present how Singapore has deregulated its electricity market in which the price of almost three quarters of total demand is being determined by real-time pricing while generation companies, which are subject to vesting contracts, are required to provide about 60% of demand.

The authors describe how the introduction of competition with the contractual obligation has performed in the partially deregulated electricity market in Singapore and suggest it is time to extend competition to the full market.

This chapter's main conclusion is that the Singapore electricity market is in a transition mode, poised to advance beyond the legacy of past regulation with high concentration of assets among three large generating companies and burdened by the stranded cost of old and inefficient plant—into a fully competitive and transparent mode.

In Chapter 25, Jenny Riesz, Joel Gilmore, and Magnus Hindsberger point out that future electricity markets will need to effectively integrate variable renewable technologies. This can be facilitated by features such as short dispatch intervals and short delay times from gate closure to dispatch, significantly reducing forecast errors and the amount of balancing reserves required. At the same time, balancing areas should be maximized, to increase geographical diversity, allow access to the maximum possible number of generation assets and facilitate sharing of balancing errors between regions.

The authors describe these and other features of markets that effectively integrate variable generation. For example, they argue that secondary reserves should be procured efficiently and set dynamically. Exposure of all generators to negative prices when they occur is also important, incentivizing curtailment during periods when there is too much generation on the network.

The chapter's main contribution is to illustrate these features using case studies and document how they contribute to the effective integration of variable technologies into future electricity markets.

In the book's Epilogue, Jean-Michel Glachant provides a perspective on the evolution of the global electricity markets over the past two decades including a list of big surprises that were not originally recognized as being significant.

As often happens in an edited book with multiple contributions from multiple experts from different countries, different disciplines with different perspectives, this volume provides a *mosaic* of ideas, insights, and approaches to how the challenges of complex electricity markets are being and/or should be addressed. As the editor, I have personally gained tremendously from these contributions and am indebted to them for their efforts in making this possible.

Fereidoon P. Sioshansi
Menlo Energy Economics

Part I

The Evolution of European Electricity Markets

Chapter 1

Evolution of the British Electricity Market and the Role of Policy for the Low-Carbon Future☆

David Newbery

Imperial College, London and Electricity Policy Research Group, University of Cambridge, Cambridge, UK

1 INTRODUCTION

The United Kingdom was the first European country to liberalize its electricity market in 1990 and set the model for subsequent European Union (EU) Electricity Directives that have driven EU liberalization and compelled unbundling of generation from transmission and distribution. In many ways, the United Kingdom has remained at the forefront of further reforms, and although subsequent policies often seem poorly designed compared with the initial restructuring, that reflects the difficulties of making further changes to an already privatized industry, with its explicit and implicit property rights. Most reforms create winners and losers, who can be more easily accommodated within public ownership, but are bitterly contested when the winners and losers are different legal entities.

The UK experience is instructive for at least five reasons:

1. As the United Kingdom has devolved administrations, it adopted three different restructuring models (one each in Scotland, Northern Ireland, and England and Wales) that can be compared.
2. Britain abandoned the original compulsory gross power pool model with capacity payments for a bilateral self-dispatched energy-only market in 2001, with questionable results.
3. Ofgem, the Great Britian (GB) regulator, repeatedly wrestled, largely unsuccessfully, with reforming the transmission pricing arrangements but

☆This paper draws heavily on Newbery (2012a).

© 2013 Elsevier Inc. All rights reserved.

failed to recognize the importance of compensating losers to avoid legal challenges, and so failed to make the desired changes.

4. It demonstrates the difficulty of insulating a privately owned industry from continued political interference and lobbying. This can be seen in the remarkable number of *Energy White Papers* (reports addressing problems and/or proposing legislative changes) since 1990, four since 2003 alone.

5. Finally, UK energy policy is constrained by EU Directives and legislation, and their interaction is often problematic. This can be seen over targets for renewable energy supply (RES), the form of the EU emissions trading system (ETS), the financing of merchant interconnectors, and most recently, in the design of the EU target electricity model. Indeed, the challenge of meeting legally binding carbon and EU renewable energy targets has led the Government to propose far-reaching reforms to support for low-carbon generation in its *Electricity Market Reform* (EMR) (although not, misleadingly, in the electricity market itself).

This chapter sets out briefly the evolution of the British[1] electricity supply industry and the reasons the Government embarked on a fundamental rethink in 2010 with its consultation on *EMR* (DECC, 2010, 2011) and the publication of the subsequent *Draft Energy Bill* (HMG, 2012). It then discusses various criticisms made of the proposals and suggests better solutions.

2 THE FIRST TWO DECADES[2]

Figure 1.1 gives a map of GB showing the high-tension transmission system or grid. The map also shows the three transmission zones, two in Scotland and one in England and Wales owned by National Grid Electricity Transmission plc (NGET). As the boundary between Scotland and England is already heavily congested, the map shows the proposed reinforcements that will be required by the United Kingdom's ambitious renewables targets which require a large number of wind farms in Scotland.

The United Kingdom had a population of 62.3 million in 2011 (GB had a population of 60.4 million), and in that year its peak electricity demand was just under 60 GW. Generation capacity was 79.8 GW generating 348 TWh, which with imports gave a total supply of 354 TWh,[3] down from its previous peak, as shown in Figure 1.2. Of the available supply, transmission and distribution losses amounted to 9%, and of the 318 TWh consumed, domestic customers took 112 TWh or 35%, ahead of industrial consumption at 32%. Total consumption per head of population was therefore 5104 kWh/person/

---

1. Thus largely ignoring the experience of Northern Ireland that eventually integrated with the Republic of Ireland to create the radically different All-Island Single Electricity Market.
2. This draws on the account of restructuring and privatization in Newbery (2006).
3. Sources for transmission http://www.nationalgrid.com and electricity http://www.decc.gov.uk/en/content/cms/statistics/energy_stats/source/electricity/electricity.aspx.

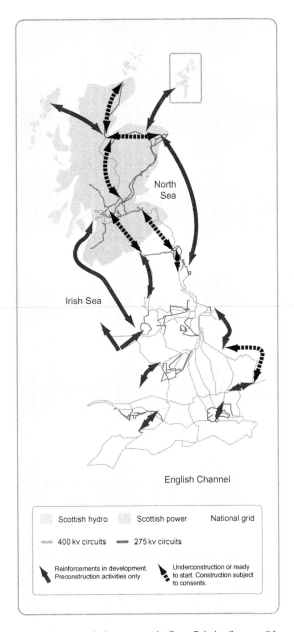

North
Sea

Irish Sea

English Channel

Scottish hydro   Scottish power   National grid

400 kv circuits   275 kv circuits

Reinforcements in development.   Underconstruction or ready
Preconstruction activities only.   to start. Construction subject
                                        to consents.

FIGURE 1.1   High-tension transmission systems in Great Britain. *Source: Ofgem.*

year, with peak demand slightly less than 1 kW/head (both 88% of their
previous peak in 2005). Figure 1.1 also shows the evolution of the fuel mix
in generation. At the end of the 1990s, roughly one-third of generation came
from coal, gas, and nuclear, but the coal share of generation has fallen from

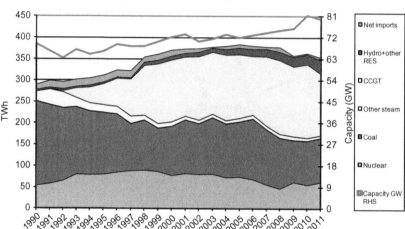

Electricity supplied by, and capacity of, UK generators, 1990–2011

FIGURE 1.2   Evolution of fuel mix in United Kingdom's electricity supply and installed capacity.

its privatization peak of 70% to less than 30%, nuclear has declined from its peak share of 29% to 18%, while gas in modern combined cycle gas turbines (CCGT) rose from 0% to 48% in 2010, but fell in 2011, partly as renewable electricity rose to 10% in 2011.

The United Kingdom adopted three liberalization models—full ownership unbundling in England and Wales, retained vertical integration for the two companies in Scotland (whose locations are shown in Figure 1.1), and a form of single buyer in Northern Ireland, but the focus here will be on the model for England and Wales that now extends to Scotland, and which forms GB.

In 1990, the Government restructured and privatized the British electricity supply industry, with the exception of the nuclear power stations, where the newer stations were sold in 1995 (Table 1.1 lists a time line). In England and Wales, the Central Electricity Generating Board (CEGB) was broken up into four companies: National Power (NP), with 60% of the fossil capacity, PowerGen (PG), with 40% of the fossil capacity, Nuclear Power, state-owned with the nuclear stations, and National Grid Co. (NGC) with the transmission assets and pumped storage (which was subsequently sold). The 12 area boards (the distribution and supply companies) were sold to become the regional electricity companies (RECs) with a time-limited franchise to supply all customers below a certain size (initially 1 MW, then reduced to 100 kW in 1994, and finally abolished in 1998−1999). Initially, the RECs collectively owned NGC, which they sold in 1995 at the same time that limitations on takeovers were relaxed. Almost immediately aggressive takeovers demonstrated the leniency of the first reset price controls.

TABLE 1.1 Major Events in the British Electricity Supply Industry

Event	Date	Comments
Sale of RECs	1990	British ESI restructured, CEGB split up, electricity pool created, NGC transferred to RECs, RECs sold to public. 5000 1 MW customers free to buy in pool
Sale of generation	1991	60% National Power (NP) and PowerGen (PG) sold to public
Second-tier market	1994	100 kW market contestable (additional 45,000 eligible)
End of golden share	1995	Remaining 40% of NP and PG sold. Government golden share in RECs expired, RECs subject to acquisitions
First price control	1995	RECs subject to new price control, reopened after takeover wave
British energy	1996	The nuclear power company privatized
Third-tier market	1998–9	All 22 million customers contestable starting May 1998
EU Electricity Directive	1999	EU Electricity Directive effective February, Offer report on pool prices (one of several). Merger of Offer and Ofgas into Ofgem
NETA	2001	New electricity trading arrangements introduced 27th March
Energy Bill, *White Paper*	2003	BETTA proposed, *White Paper* on low-carbon future
ETS	2005	EU emissions trading starts Jan 1 without banking
BETTA	2005	BETTA go live
Energy *White Paper*	2007	*Meeting the energy challenge* consults on future of nuclear power
ETS Phase 2	2008	Second period of emissions trading allows banking
DECC established, Climate Change Act	2008	Department of energy and Climate Change and CCC established by Labour government
EU Renewables Directive	2008	Raises target EU renewable energy from 12.5% to 20% by 2020
EU Third Energy package	2009	Proposed 2007, adopted 2009, effected 2011, creates ACER, mandates ownership unbundling

(Continued)

TABLE 1.1 (Continued)

Event	Date	Comments
Project Discovery	2009	Ofgem concerned over security of supply
Energy Act 2010	2010	Labour government passed on April 8, 2010, requires reporting on decarbonization of generation and development and use of CCS
Electricity Market Reform consultation	2010	Change of government in May. Conservative and Liberal Democrat coalition launches consultation
CPF	2011	HM Treasury sets trajectory of carbon prices for electricity
DECC *White Paper*	2011	*Planning our electric future: a White Paper for secure, affordable, and low-carbon electricity*
EU Target Electricity Model	2012	Development and implementation of the target model in electricity with market coupling by 2014
Draft Energy Bill	2012	Published in May with *Policy Overview*

The centerpiece of the restructuring was the creation of a wholesale electricity market, the electricity pool. All but the smallest generators were required to offer day-ahead supply schedules specifying start-up costs and prices they would be willing to accept for operating at various levels, as well as technical parameters such as ramp rate and minimum output. These offers were fed into the old CEGB scheduling program, GOAL, to determine the unconstrained dispatch (i.e., ignoring transmission constraints, as all generators had paid for and had rights to firm access to the grid). This (typically infeasible) dispatch determined the system marginal cost (SMC), for each half hour, which all unconstrained plant dispatched received. The security-constrained dispatch schedule determined which plants that were "in merit" (i.e., in the unconstrained dispatch program) must be "constrained off" the system and paid their lost profit, taken as the difference between SMC and their offer price. Out-of-merit plant that required to balance the system were "constrained on" and paid their offer price. As all companies had copies of the GOAL program and the grid capacities they could determine which plant would likely be constrained on or off and manipulate these bids to maximize profits, as they could in determining their supply functions. All plants declared available also received a capacity payment, equal to the value of lost load (VOLL) *less* the SMC and times the loss of load probability (LOLP), added to the SMC to give the pool purchase price.

All suppliers (retailers) were assured supply at the wholesale price (the pool selling price, which included the cost of constraints and other ancillary services), which they could hedge with financial contracts. Although the initial market structure was very concentrated, with the two incumbent fossil generators setting the price in England and Wales almost all the time, it was open to entry, as any merchant generator could offer its power at avoidable cost and receive the same price as the incumbents. Favorable financial conditions for gas-fired plant (attractive long-term contracts by the RECs who could pass on the costs to captive domestic customers until 1998) encouraged massive entry.

The timing of the privatization was fortunate, in that efficient and relatively cheap CCGTs became available in modest sizes, while rapid expansion of North Sea gas offered attractive fuel prices, particularly when compared with expensive British coal. From 1990−2000, new gas-fired entry (shown in Figure 1.2) and the incumbents' sale of coal-fired plant created a workably competitive industry that was considered an ideal model by many observers and was influential in stimulating European electricity liberalization through a sequence of EU Directives. Newbery and Pollitt (1997) argued that unbundling and liberalization had delivered substantial efficiency gains and net social benefits in England and Wales, as a result of making the wholesale market contestable and facilitating competitive entry, despite the initially flawed duopolistic structure. However, UK regulatory concerns over market power and political worries over the displacement of coal by what some thought were artificially favorable entry conditions for gas-fired plant led to pressures for market reform.

In response, the Government and Ofgem (the energy regulator) introduced the new electricity trading arrangements (NETA) in 2001, replacing the mandatory gross pool and central dispatch with a model of voluntary bilateral contracting, self-dispatch, and an opaque two-price balancing mechanism that was designed to encourage forward contracting before dispatch by penalizing both long and short positions. Generators had to seek out buyers for the physical generation contracts and retailers had to secure supply before dispatch, each managing demand and supply separately rather than delegating it to the system operator (SO) for central dispatch.

The theory was that contracting privately and fully ahead of time would encourage more competitive behavior and address concerns over market power. The theory had already been criticized (Newbery, 1998), and the post-2001 evidence suggested that the fall in wholesale prices claimed as a major NETA success was a result of reduced concentration that occurred *before* NETA (Bower, 2002; Evans and Green, 2005; Newbery, 2005). The lack of a liquid wholesale market and the penal imbalance charges adversely affected wind generators and combined heat and power (CHP) plant and strengthened the case for vertical integration between generators and supply (retail) companies, so that in short order the British electricity market had consolidated into six vertically integrated utilities, who between them supplied 99% of the domestic retail market.

NETA was subsequently replaced in 2005 by the British Electricity Trading and Transmission Arrangements (BETTA) that put Scottish transmission under NGET, the GB SO (acting as an independent SO in Scotland). It created a single-price area for the whole of GB, despite serious congestion on the border between England and Scotland that significantly raised redispatch costs.

Nevertheless, by comparison with the continent, the British electricity market seemed to signal the success of the liberalized, unbundled, and competitive model for wholesale electricity markets, notably in comparison with some countries whose energy markets were investigated by the European Commission in its Energy Sector Inquiry (DG Comp, 2007).

3 THE CLIMATE CHALLENGE

Meanwhile, concerns over climate change were rapidly moving up the political agenda both in the United Kingdom and in Europe. The EU response was to set up the ETS, a cap-and-trade system designed to limit emissions of CO_2 from the covered sectors (about half the total, but notably including the power sector). Trade in the EU allowances (EUAs) put a price on CO_2, but uncertainty about demand for the EUAs led to high price volatility, as shown in Figure 1.3. The first trading period did not allow banking of EUAs beyond 2007, and the growing realization that their allocation had been overgenerous

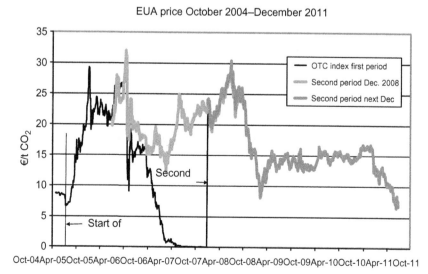

FIGURE 1.3 Price of EU emissions allowances. *Source: EEX.*

led to a collapse in the carbon price from late 2006, which fell to zero by mid-2007. The second period did allow banking (and their EUAs could be traded ahead of the start of the second period) and initially their price rose to over €25/t CO_2.

In October 2008, the then UK Labour Government established the Department of Energy and Climate Change (DECC), shortly before the *Climate Change Act* received Royal Assent in November 2008. The Act provides a legal framework for ensuring that Government meets its commitments to tackle climate change. It also set up the Committee on Climate Change (CCC) as an independent body to advise and monitor the Government's carbon commitment. The Act requires that emissions are reduced by at least 80% by 2050 compared to 1990 levels and that the Government commit to a series of 5-year carbon budgets.[4] Figure 1.4 shows the evolution of UK greenhouse gas (GHG) emissions, distinguishing between CO_2 emissions from the named sectors, including power, and the remaining GHG emissions. Until 2010, these are the non-CO_2 emissions, but in 2020, they are the CCC budget total GHG less the CCC projection for the power sector. The average allowed GHG emissions under the various carbon budgets are also shown from 2012 to 2020. Finally, Figure 1.4 shows the EU ETS allowances in averages per year in the second and projected third period, which only apply to CO_2 and then only to the covered sector, roughly half the total.

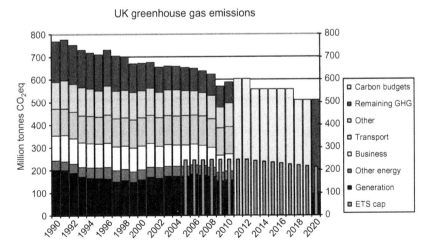

FIGURE 1.4 Greenhouse gas emissions to date and projected under the carbon budgets. *Source: DECC climate change statistics, CCC.*

4. http://www.theccc.org.uk/about-the-ccc/climate-change-act.

In the same year, the EU proposed the *20-20-20 Renewables Directive* (introduced in 2009 as 2009/28/EC). This increased the share of EU *energy* (not electricity) that must be generated from renewables by 2020 from 12.5% to 20% but without reducing the cap on CO_2 emissions. The expected increase in the supply of zero-carbon electricity reduced the demand for EUAs and led to an estimated forecast fall in the 2020 EUA price from €60 to €50/EUA (CCC, 2009). The financial crisis and the predicted fall in future carbon intensive industry and electricity demand further undermined the 2020 forecast price to €20/EUA, at which level low-carbon generation was not profitable without support. This demonstrates the fundamental flaw in the design of the ETS, by setting a quota on EU emissions rather than on price, the resulting carbon price signals that were intended to guide long-term investment decisions are unstable, and lacked credibility, being vulnerable to changing political decisions, in this case setting a higher renewable energy target. If the ETS had migrated from a cap-and-trade system to a carbon price support system (Grubb and Newbery, 2008), then additional renewable energy would have reduced total emissions without affecting the carbon price.

3.1 Supporting Renewable Electricity

At privatization, the lack of decommissioning fund for the existing nuclear fleet so concerned the Treasury that it imposed a fossil fuel levy to raise the necessary money to fund future decommissioning, and it also imposed a nonfossil fuel obligation (NFFO) on distribution companies to buy nuclear power, which was subsequently extended to include renewable power. Renewable generation was secured by tender auctions that initially paid an average price of 7.51 p/kWh, which fell with subsequent rounds so that by the last round (i.e., the fifth) in 1998, the average was only 2.71 p/kWh. This combination of long-term contracts, paying a fixed feed-in tariff (FiT), determined by competitive bidding appeared as an ideal market-guided way of stimulating renewables development. According to Connor (2003, p.76), onshore wind costs fell from 10 p/kWh in 1990 to (rather hypothetical) 2.88 p/kWh in 1998. There was one serious flaw, in that there were no penalties for bidding and then not delivering. Optimistic wind developers frequently found that they were unable to buy the site or install the turbines at an acceptable cost, or failed to secure planning consent, or needed to relocate slightly, but were unable to under the agreed contract terms. Pollitt (2010), citing evidence from Wong (2005), notes that between 1990 and 1998 only 75 or 25% of the 302 awarded wind projects were ever built and these amounted to 391 MW or only 15% of the tendered 2659 MW. In the last round none of the 33 wind projects awarded was ever contracted.

NFFO was replaced by the Renewables Obligation (RO) Scheme in 2002, under which suppliers are required to secure a specified share of electricity from renewable electricity or pay a penalty price (whose revenue is recycled back to the renewable generators). Different renewables receive different levels of support, so that until 2013 onshore wind receives 1 RO certificate (ROC) per megawatt hour generated, while offshore wind receives 2 ROCs until 2015.

Figure 1.5 shows that the value of an ROC, determined by their supply and mandated demand, has been fairly stable at around £50/MWh, but as the wholesale price has been very volatile, the revenue enjoyed by renewables has also been very volatile. Clearly wind farms that were happy with prices in the period before 2004 would have been substantially overrewarded thereafter, while those tempted in at the recent peak in 2009 might be regretting their decision. Such price volatility must surely raise the cost of capital by reducing the level of debt that is prudent for financing wind, in comparison with an FiT which guarantees future revenues, subject to the modest and predictable volatility in annual wind output.

The *Renewables Directive* raised the required level of UK renewables to a level that most observers considered would be infeasible (or excessively costly) under the current system of support. Figure 1.6 shows that, despite having a better wind resource than Germany, the United Kingdom has been lagging far behind but has ambitions to deliver 27,000 MW by 2020. This should be feasible if the United Kingdom replicates the German rate of roll out from 2000 to 2010. That might require replicating their FiTs and perhaps also reforming the planning system, which has been argued to

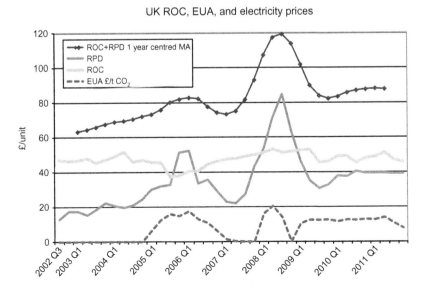

FIGURE 1.5   The wholesale spot price, the support for renewables, and their revenue.

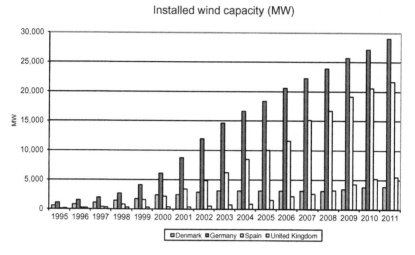

FIGURE 1.6   Evolution of wind capacity installed in various EU countries. *Source: IEA, EWEA.*

be the major reason why Britain has been so slow in developing its wind resource.

3.2 The Pressure for Reform

Figure 1.4 shows that by 2009 the EUA price had fallen below €15 and as such it was quite inadequate to justify mature zero-carbon generation from nuclear power (or the near mature onshore wind power). In addition, the EU Large Combustion Plant Directive (LCPD) imposes tight emissions standards that would require major refits of coal-fired generation or massive plant closure before 2016. In the United Kingdom, some 12 GW of the older coal-fired plant (about 20% of peak demand) will close by the end of 2015, while an additional 6.3 GW of aging nuclear plant will also close by 2016. Despite the need for new investment, uncertainty over future energy policy has encouraged utilities to delay new build, further raising concerns over security of supply.

In response to these concerns, Ofgem launched *Project Discovery* in June 2009 and reported on February 3, 2010[5] recommending "far reaching energy market reforms to consumers, industry and government" and concluded that "The unprecedented combination of the global financial crisis, tough environmental targets, increasing gas import dependency and the closure of ageing power stations has combined to cast reasonable doubt over whether

---

5.   At   http://www.ofgem.gov.uk/Media/PressRel/Documents1/Ofgem%20-%20Discovery% 20phase%20II%20Draft%20v15.pdf.

the current energy arrangements will deliver secure and sustainable energy supplies."

After a pause and a change of government to the Conservative and Liberal Democrat coalition on May 6, 2010, DECC launched its consultation on *EMR* in December 2010 (DECC, 2010). Its diagnosis was similar to that of *Project Discovery* (and various others, including the CCC)—the carbon price was now too low to support unsubsidized nuclear power, while the wholesale electricity price was set by fossil fuel prices (and the ETS). Fossil generators thus had a natural hedge as the difference between the electricity sales price and the cost of fuel is reasonably stable, while that for nonfossil generation is very volatile as their variable costs are low and constant. Looking forward, nonfossil generation faced volatile EUA prices that are too low and sensitive to political intervention, thus undermining their future credibility.

The consultation estimated that the cost of meeting the Government's environmental targets by 2020 in electricity alone amounted to £120 billion, or over £12 billion per year compared with less than £5 billion in 2008 (which itself is nearly 80% above the previous decade's average).[6] This is considerably above financial analysts' estimates of the capacity of the major incumbent generation companies to finance on their own, indicating the need to access new sources of finance. Given the high capital cost of most low-carbon options, anything to derisk investments and lower the weighted average cost of capital (WACC) would have significant benefits in terms of lower costs and prices. A reduction in the equity risk premium and an increase in the debt share might reduce the WACC by 1% (or even more for smaller entrants), which would reduce the capital cost by £1.2 billion each year by 2020, or nearly £45/year per household, compared with current electricity bills of £450/year (although domestic consumers consume about 40% of the total, electricity prices fed through other goods that are ultimately consumed).

4 REFORMING THE ELECTRICITY MARKET

The consultation proposed a carbon price floor (CPF) to ensure that the carbon price moved on a trajectory that would ensure the commercial viability of nuclear power without further support, and this was the subject of a separate and rather hasty consultation by HM Treasury, with draft legislation published on January 11, 2011. The levels announced in the budget in March 2011 would support the price of CO_2 starting at £16/t in April 2013, rising to £30/t (€35/t) in 2020, and projected to rise to £70/t by 2030, all at 2009 prices (HMT, 2011a). By itself, any tax, and particularly a carbon tax that might adversely impact British competitiveness, would not be credible,

---

6. £4.3 billion at 2005 prices (Office of National Statistics).

as it could be reversed in any subsequent budget. Indeed, past protests have reversed a similar road fuel tax escalator that was intended to steadily increase the real tax on motor fuel.

After the consultation, in July 2011, the Government published a *White Paper* (DECC, 2011), which is the normal precursor to proposed legislation, and in due course the Government set out its proposals for EMR in a *Draft Energy Bill* (HMG, 2012). Both the *White Paper* and now the *Draft Energy Bill* have been criticized, most cogently by witnesses who gave evidence to the Parliamentary Energy and Climate Change Committee (HC, 2011, 2012).

The conclusions drawn in the *Draft Energy Bill* are that the effective carbon price needs to be raised for the electricity sector through a CPF and that price risk should be reduced by long-term contracts for differences (CfDs),[7] disingenuously called FiTs. In addition to these two central policies, EMR also proposes the future option of a capacity mechanism to allay security of supply fears and an Emissions Performance Standard (EPS) stated to be 450 gm CO_2/kWh,[8] primarily designed to rule out the construction of any unabated coal-fired power stations (but not gas-fired CCGT, needed for balancing intermittent/unreliable renewables).

4.1 The Carbon Price Floor

If the CfDs fix the price for generating low-carbon electricity, then the CPF would seem to be redundant, and arguably perverse in that it raises UK electricity prices relative to the Continent because of the likely considerably higher carbon price. Given the fixed and unaltered cap on total emissions in the EU ETS, the UK CPF would not reduce total EU CO_2 emissions at all. Further, it provides windfall profits to current heavily subsidized renewables as well as the existing nuclear power stations, while disadvantaging electricity production in the United Kingdom relative to our neighbors who can export to us, and also disadvantaging electricity-intensive

---

7. A classic two-side CfD sets a strike price, s, and a reference market price, p, for an amount M MW, and pays the generator $(s-p)M$ (or charges him or her if this is negative). The generator has an obligation to deliver M MW to the counterparty via the wholesale market, where if he or she is successful at trading he or she will receive price p and the counterparty is deemed to have received the delivery by buying in the wholesale market at that price. If successful, the generator will thus earn $pM + (s-p)M = sM$. If the generator fails to deliver, he or she earns only $(s-p)M$ which could involve a high penalty in peak hours when $p > s$, thus encouraging availability at times of system stress.

8. More correctly described as a limit on emissions of 3150–3942 t CO_2/MW of transmission entry capacity, if as the wording suggests it is based on an assumed base-load (which might be somewhere between 7000 and 8760 h operation) at that intensity per year. If a coal plant with 900 gm/kWh were to run only 50% of base-load hours, it should meet the EPS.

production that can migrate abroad (aluminum, steel, etc.), again with no global climate change benefit.

In defense of the CPF, it has a number of considerable countervailing benefits:

- As a fiscally generative new tax instrument, it ought to be attractive to cash-strapped EU governments in the current financial crisis, and if widely adopted by EU Member States, would go a considerable way to addressing the failure of the EU ETS.
- It reduces the apparent fiscal cost of supporting the long-term contracts, as well as meeting a coalition pledge that nuclear power would not be subsidized (arguably also an EU requirement under state aids). Unlike many other EU governments, the United Kingdom is held to account by the Office of National Statistics that deems any levies on electricity bills to finance renewables, efficiency measures, and the like as taxes, so there is no perverse inducement to choose one form of levy over another in terms of budgetary accounting.
- It discourages overhasty investment in fossil generation that would lock in carbon emissions for the life of the plant. It does so by clearly signaling the risks to any investor who anticipates high-capacity factors using conventional generation, whether coal or gas.

The Government has consulted on the role of gas in the electricity market (DECC, 2012b) and planned to produce a strategy statement in the autumn 2012. There were concerns that the Government (under pressure from the Treasury) prefers to meet its near-term climate change targets by increasing the share of gas generation.[9] The fuel cost of coal-fired generation (including the cost of purchasing EUAs) was £32/MWh and for gas-fired generation was £41/MWh in early 2012.[10] At these fuel prices but with the mid-2020s forecast CPF of £50/t, their fuel costs alone would be £72/MWh and £61/MWh, respectively (to which would have to be added a margin to cover the capital and other fixed costs). As such they were likely to be at a level at which they would only operate at low-load factors. The CPF therefore signals to generators wishing to build fossil generation that they cannot be assured of adequate returns after 2020 and they should therefore consider carefully and probably delay making such investments. In particular, unabated coal is at risk from the EPS, and new gas-fired generation would likely be delayed until the details of any proposed capacity mechanisms are published—making it more likely that such a mechanism will be needed.

---

9. *Financial Times*, July 19, 2012 "Reality tests coalition's energy reforms."
10. First quarter, 2012, DECC *Energy Price Statistics*, EUA price £5/t, same period, assuming 38% efficient coal plant and 55% efficient gas plant.

5 CRITICISMS OF THE EMR

As the CPF was introduced in the budget of 2011, subsequent criticisms of the EMR were concentrated on the remaining elements. In the course of extensive consultation (DECC, 2010, 2011) and enquiries by the Parliamentary Committee on Energy and Climate Change (HC, 2011), several points were repeatedly stressed but did not gain much traction in the *Draft Energy Bill* or the amplifying *Policy Overview* document (DECC, 2012a). The Committee expressed three major concerns:

1. Different technologies require different contract designs.
2. The proposed contracts would not have creditworthy counterparties.
3. The Treasury by limiting consumer expenditure on supporting these contracts would make their future availability uncertain.

These three concerns will be considered in order.

5.1 Contract Design

The first criticism was that as various low-carbon technologies differ considerably (as between controllable nuclear power and unpredictable wind and solar photo-voltaic — PV), so should their contract design, instead of all being straightjacketed by a single CfD. This places the responsibility on the generator to market his or her power and pay any imbalance charges. While a CfD is entirely suitable for controllable nuclear power, it unnecessarily raises risks and costs for unpredictable renewables such as wind and solar PV (Newbery, 2012a,b). The Parliamentary Committee worried that all contracts were being forced into a common format so that they could collectively seek approval as legitimate state aid, whereas if nuclear power were to be treated differently from renewables (which can already be state-supported) they would be subject to objections (HC, 2012, ¶142, p. 30). If that is indeed the defense, then it is a very indirect argument for choosing a single-contract design for all technologies.

The intention, clearly stated in the consultation documents, is to encourage the entry of new capital into the industry (as the balance sheets of the incumbents fall far short of the capacity to finance the estimated £85 billion of new generation investment needed). It is therefore crucial to reduce entry barriers and hence unnecessary complexity. For wind farms, that means paying a fixed FiT or contract price per megawatt hour of electricity generated (or per megawatt per hour if available for dispatch but constrained off) and ensuring a guaranteed off-take. The contract price should properly reflect the true cost of transmission and balancing and the benefits of meeting the renewables and carbon targets at least cost (paying less per megawatt hour in more windy than less windy locations to extract the site rent). This design avoids the need either to continually contract to sell power and then deal with the balancing mechanism or subcontract on less advantageous terms to an incumbent (whose risk and credit would therefore be unnecessarily further strained). Removing the obligation for suppliers to contract for renewable energy further undermines the ability of new entrants to be sure that they can sell their power (HC, 2012, ¶59).

The *Draft Energy Bill* (at ¶54) makes some allowances in this direction:

54 In the short term, however, variation in CfDs may be needed for some technologies—within intermittent (i.e. generation that is inherently variable and dependent on primary power sources outside the control of generators, e.g. wind, wave, and solar) and baseload (i.e. generation that generally operates continuously to serve the minimum electricity demand over a given period of time ("baseload")) classes—in recognition of their different risk profiles (for example early stage CCS projects, due to their demonstration status), to ensure they come forward at a reasonable cost. Any variations agreed will have to represent value for money and maintain a level playing field in line with our approach to securing state aid clearance.

Further details are provided in the *Policy Review* annex B (DECC, 2012a). It reiterates the principle that "the aim is to deliver CfD terms that are largely standardised across technologies" (*ibid* p. 35) but with some technology specific adjustments. Thus, the strike price for wind is the *hourly* day-ahead auction price[11] for the GB zone (which raises the question whether there will a single price for GB, or logically, at least one for England and Wales and a separate zonal price for Scotland). More sensibly, the proposal is "minded to pay the CfD on the basis of metered output unless the reference price is negative, in which case to pay on a measure of availability" (*ibid* p. 33). This appears to avoid the problem of exposure to imbalance risk and the need to find a buyer for the power, but the Parliamentary Committee criticized the *Draft Bill* for failing to clarify who would actually buy the contracted renewable power (HC, 2012, pp. 33–35) and considered that this would create a considerable barrier to entry and hence raise its cost.

The Parliamentary Committee did not, however, comment on the way in which the strike price for renewables will be set, which appears to imply paying the same price per megawatt hour regardless of the location of the wind power (although with a difference onshore and offshore). The whole advantage of setting a contract price for each wind farm is that it allows site-specific characteristics to be taken into account without providing windfall profits to wind-favored (but possibly expensive to connect) locations, a problem exacerbated by the massive regulatory failure engineered through the transmission *Connect and Manage* regime (Newbery, 2011).

---

11. Newbery (2012b) argued that the original proposal of a day-ahead base-load reference price for wind at least provided some positive basis uplift as wind is currently positively correlated with price, although this basis was risky and might reverse with massive wind.

On the positive side, the proposals urge a move to tenders or auctions as soon as possible. Elsewhere, Newbery (2012b) has argued that this would be best achieved if the sites were separately secured through an intelligently designed planning process, possibly with tenders by landowners to secure them at least cost, and then auctioned through a process that allowed the SO to determine the least total system cost offers, taking into account the costs of balancing (which would fall on the SO) and transmission reinforcement (borne by the same parent body, NGET) and also the local wind resource. The result would be that the prices paid per megawatt hour for wind would vary by location and transfer the site rents to customers, upon whom the extra costs of supporting renewables will fall under current arrangements.

As an aside, the United Kingdom should follow good Continental practice and exempt commerce and industry from levies to fund the contracts, which are purely designed to raise revenues to provide the public good of renewables support and which should not, under good fiscal practice, distort production decisions. This would not stop generators offering long-term contracts to industries that would like to lock in a real price. A managed transition from the present array of consumer levies to imposing the standard rate of value-added tax (VAT) on energy would help achieve this, and a transition date could be chosen so that this would be roughly revenue neutral, and although it would exacerbate the Treasury's problem with the resulting expenditure commitment it would make explicit the nature of the Government's commitment in supporting the contracts.

Most of these problems of marketing wind power would be avoided if Britain reintroduced a voluntary pool for wholesale electricity with central dispatch for contracted plant, ideally with nodal pricing as in the US standard market design.

5.1.1 ROC Rebanding and Future Wind CfDs

Meanwhile, the Government has to reassure renewables investors that they can continue to invest while the details of the new contract regime is clarified. As noted earlier, renewable electricity is currently supported by ROCs, with different types of renewables enjoying different levels of support—1 ROC/MWh for onshore wind and 2 ROCs/MWh for offshore wind. The Government reviews the level of support every few years and published its new ROC banding proposals in July 2012 (DECC, 2012a) confirming (at ¶3.7) its intention to reduce the level of support for onshore wind to 0.9 ROCs/MWh for new accreditations and additional capacity added in the banding review period (April 1, 2013 to March 31, 2017). For offshore wind, the proposals are to continue the current 2 ROCs/MWh for additional capacity added in 2013/14 and 2014/15, reducing to 1.9 ROCs/MWh in 2015/16 and 1.8 ROCs/MWh in 2016/17.

As is standard, the Government is required to publish an *Impact Assessment* (IA, DECC, 2012b, effectively a social cost−benefit analysis) of proposed policy changes, and these reveal some disturbing implications of the apparent intention of continuing to set a single support scheme for each technology regardless of location. Appendix F table F1 is partly (just onshore wind) reproduced below.

Table F1: ROCs required for investment for each technology in each year based on DECC in-house analysis and rationales for proposed bands

	Country	Cost tranche	2013/14	2014/15	2015/16	2016/17	Rationale for bands
Onshore >5MW	E&W	Low	0.5	0.4	0.4	0.5	No changes are proposed to the consultation proposals
		Low-medium	0.7	0.7	0.7	0.7	
		Medium	1.0	0.9	1.0	1.0	
		Medium-high	1.3	1.2	1.2	1.3	
		High	1.5	1.5	1.5	1.5	
Onshore >5MW	Scotland	Low	0.2	0.2	0.2	0.2	Incentivises the more cost-effective onshore deployment.
		Low-medium	0.5	0.4	0.4	0.5	
		Medium	0.7	0.6	0.7	0.7	
		Medium-high	0.9	0.9	0.9	0.9	
		High	1.2	1.1	1.1	1.2	
Offshore Round 2	UK	Low	2.0	1.8	1.4	1.5	No changes are proposed to the consultation proposals. Analysis shows that if offshore wind is to make a cost-effective contribution to the 2020 target it is necessary to encourage some
		Low-medium	2.3	2.0	1.7	1.7	
		Medium	2.5	2.3	1.9	1.9	
		Medium-high	2.8	2.6	2.1	2.1	
		High	3.1	2.8	2.4	2.4	
Offshore Round 3		Low	2.6	2.5	2.5	2.5	
		Low-medium	2.9	2.8	2.8	2.8	

What this implies is that in England and Wales 0.4 ROCs would have been adequate for low-cost areas and 0.7 for low−medium-cost areas, while in Scotland the number of ROCs needed is 0.2 less for each cost band. This demonstrates the poor design of the ROCs which pay the same regardless of location and by implication the danger of replicating that for the proposed CfDs.

It is a simple matter to compute a very rough estimate of the cost saving of having locationally differentiated ROCs and CfDs. For example, if the shares of each band are the same the saving would be very roughly calculated as: $0.5 \times 33\% + 0.2 \times 33\%$ ROCs @ £50/ROC = £12/MWh for England and Wales and $0.7 \times 25\% + 0.46 \times 25\% + 0.23 \times 25\%$ ROCs @ £50/ROC = £10/MWh for Scotland, or a simple average of £11/MWh. The new ROC banding period is supposed to deliver 6.4 TWh/year of onshore wind in GB (¶41, p. 18), so we are talking of £70 million per year which could have been clawed back to consumers. One might also note that onshore wind is assumed to have a hurdle rate of 9.6% (presumably real but not stated in table A.3) which only falls to 8.6% with CfDs. Given a sensible FiT and minimizing the transaction costs, an 80:20 debt:equity ratio should be plausible and with the proposed Government's Green Investment Bank[12] lending at say 3% real with equity at say 8% this should give a WACC of 4% instead of 8.6%. If onshore wind turbines cost roughly £1 million/MW and there will be an extra 2.6 GW, costing £2.6 billion, the extra annual saving might be £120 million per year. Altogether a better designed support scheme sensibly financed by low-risk public funds would save consumers nearly £200 million per year (and that is just for the new build onshore in the 4-year period, not even for all renewable electricity).

The conclusion is that for onshore wind, the planned new ROC subsidy is likely to be overgenerous, and the sooner it is replaced by a market-determined locationally specific FiT the better.

5.2 Credit Risk

The second concern strongly expressed by the Parliamentary Committee (HC, 2012, pp. 21−28) is that the Government (more specifically the Treasury) appears unwilling to act as the underwriter of, or counterparty to, these long-term contracts. Despite an invitation from the Parliamentary Committee, the Treasury declined to appear and be questioned on this critical issue, leaving the more pessimistic observers to conclude that if the counterparties to the contracts were asset-poor supply companies,[13] their creditworthiness would not reduce the cost of capital to finance the contract-holding investors one iota, thereby totally undermining the *raison d'etre* of the EMR. Other critics went as far as to suggest that the proposed synthetic counterparty would not be legally enforceable (HC, 2012, ¶81).

Optimists argued that a mechanism overseen by Ofgem and the SO for passing through the contract costs to final bills, as with the current ROC Scheme, would suffice and be adequately bankable. The *Policy Overview* (annex B, p. 68) argues that these arrangements should allay concerns but recognizes that not everyone is convinced and proposes to "continue to actively consider the merits of alternative models that use a single counterparty."

If these perceptions of credit risk do in fact reduce financial institutions' appetite to fund a large debt share at competitive rates of interest, as pessimists claim, then the obvious solution is for the Government to be the

---

12. "The UK Green Investment Bank, one of the major policies of the coalition government, will begin operating in April 2012, according to Nick Clegg. He said that some of the bank's early targets would be 'offshore wind, waste and non-domestic energy efficiency'" (BBC News, May 23, 2011, available at: http://www.bbc.co.uk/news/science-environment-13502121.
13. DECC (2012b, p. 33) is minded to place the obligation on suppliers in proportion to their market share.

counterparty for all such long-term contracts (as investors thought was promised in the *White Paper* (HC, 2012, ¶76)) and/or be willing to lend on them through a properly funded Green Investment Bank. This would also address the much delayed ambition to stimulate growth in the economy through leveraging investment. There is a growing if much belated awareness that this is necessary and the Government announced in July 2012 that it would underwrite up to £40 billion of infrastructure investment.[14]

Part of the concern appears to be that the Government is afraid of a suit brought against any contracts for nuclear power that has a Government guarantee as breaching state-aid conditions (HC, 2012, ¶78), which might take years to resolve through the EU courts. If so, that would seem to strengthen the case for separating out renewables support, which has already been exempted from state-aid rules, and for which proper FiTs are in common use across the Continent, and which would in any case be the correct choice for the United Kingdom. However, it would expose nuclear to a possible failure to secure exemption.

Nuclear power is inevitably a special case, not least as there may be only one remaining viable company, EdF, willing to build at least the first few plants, making any auction process moot. It should be recognized as special, with the first few plants treated, as in the United States, as first-of-a-kind projects worthy of additional subsidy to deliver the cost reductions needed for a wider scale roll-out. It is very hard to believe that either offshore wind or CCS plant (coal or gas) will be competitive with nuclear power within 20 years, even including all the extra social costs and benefits of each, and consequently, if we are serious about meeting our medium-term carbon targets, considerable nuclear investment will be needed from 2020 onward. The sooner any state-aid issues are resolved with the EU the better. The EU has given its commitment to its *2050 Energy Road Map* that stresses the need for urgent electricity decarbonization to avoid locking in fossil fuels for the long lives of any new fossil plant. It ought to accept the logic of not ruling out nuclear power in countries willing to support it by clarifying its view on state aids.

5.3 Funding Risk

The third concern expressed by the Committee is that the Government has made clear that it wishes to limit its levy expenditure to support low-carbon contracts and to that end may restrict either the volume of CfDs or the funds available at some future date, up to the amount set out in the Levy Control Framework (DECC, 2012a, annex B, pp. 25–26; HC, 2012, pp. 30–32; HMT, 2011b). At first glance this seems reasonable, given the financial havoc wrecked by open-ended and *ex post* excessive subsidies to PV in

14. http://www.bbc.co.uk/news/business-18880354 accessed on 19 July.

Germany, Spain, and Italy, given the speed with which PV can be ordered and connected, but is surely less of a problem with other renewables that have a considerably longer gestation. The logic can also be criticized as the claim is that it is designed to minimize the impact on consumer bills, but if gas prices fall and as a result the levy required to fund the CfDs rises, consumer bills will not have risen (although they might otherwise have fallen). The effect is therefore for the Government, in control of the policy and the risks arising from it, attempting to shuffle off these risks onto those for whom the costs of bearing that risk will be higher.

If it takes investors several years and considerable up-front costs to secure sites, planning permission, and grid connections, they will be more reluctant to do so without an assurance that there will be contracts available at the end of the process. Keith Anderson of Scottish Power expressed this cogently, arguing that it might cost £100−150 million to get to the final investment decision (FID), and if he did not know before that whether he would get a contract it would represent "an unacceptable risk" (HC, 2012, ¶106).

To partially allay this concern, the proposal (DECC, 2012a, annex B, p. 12, ¶18) is to publish strike prices for renewables for 5 years from 2013 to provide some comfort. (These may and presumably will vary by year and technology, see p. 22, ¶18, but not as they should, by location.) This again implies a countrywide wind power price, for example, which will lead to windy places being developed first regardless whether they are the cheapest in total transmission and generation cost. This problem would be avoided if the tender auction were run for the site at the start of the process, as suggested earlier.

There is some reassurance that CfDs should be available before financial close so that developers can present the financiers with the necessary risk assurances, and the claim is that this will reduce risk compared to the current RO scheme (although whether that has attracted adequate outside entry is itself not clear, given the high wind resource in the United Kingdom and the disappointing amount installed).

5.4 Bureaucracy Versus the Market

The consultation over EMR reawakened the debate about whether a liberalized market could deliver a low-carbon power sector, particularly as the necessary investments would not be commercially viable at market-determined prices. Two views have emerged: that if the market is provided with suitable price signals (for carbon, for the external benefits of developing renewables) and offered suitable risk-reducing instruments, then the benefits of competition can be retained. The other view is that some form of central planning, perhaps the single buyer model that was entertained but then rejected in the 1990s by the EU in successive Electricity Directives, would be necessary to deliver the required decarbonization at an acceptable pace and cost. This debate has widened to consider how best to intervene in the market if that model is preferred, and whether the proposals in the EMR meet these requirements.

Helm (2012) has been a cogent critic of the Government's EMR proposals and in particular for its lack of stability and growing complexity. His main criticism is that of its "Gosplan approach versus market-based interventions." He raises legitimate concerns about how the Government will set and justify the nuclear CfD strike price, other than basing it on the cost and then "doctoring" forecast electricity prices to justify that the contract is in the money. It seems inevitable that at least for the first few reactors the appropriate contract will involve some cost sharing between EdF and the contract counterparty, given the considerable uncertainty and ignorance about what a reactor would actually cost to build in the United Kingdom. The *Policy Review* expects the strike price to be based on the FID enabling process (annex B, table 1).

Helm's point seems to be that the Government, in setting the strike price, is placing an unknown obligation on consumers (the ultimate counterparty to the contracts) as the future electricity price is so uncertain. But it is hard to see how this invalidates setting any strike price, nor why it is so different from the old CEGB world in which consumers paid the cost of whatever the CEGB decided to build (nor indeed the obvious point that taxpayers have to pay for whatever the Government chooses to do, whether to invade Iraq or hugely expand the health service). An inability to forecast future electricity prices is hardly a reason for not granting a fixed (real) price for a long-term contract if that is the cheapest way to deliver investments that are widely agreed to be necessary. Putting the risk elsewhere would merely increase the cost of capital and the cost to consumers or taxpayers, or delay such investments.[15] Helm's claim (¶24) that "forecasting future gas prices is hazardous, ... Yet this is what setting the strike price requires" seems incorrect, except to the extent of being able to prove without doubt that the contract is "unsubsidized." The strike price is ideally set at the efficient cost of the capacity to be secured, independently of the market price. How much of it should be procured is largely determined by the carbon and renewables targets that the Government has already signed up to. Arguably the issue is not that setting the strike price requires a forecast of future electricity prices but determining the volume

---

15. There is a common but fallacious argument that the allocation of the risk only affects incentives not the cost of that risk. It is standard to note that sharing the risk equally between two parties halves the total cost, as the risk cost rises as the *square* of the exposure, and also that if different risks are not perfectly correlated then sharing them lowers the total cost—in the extreme case in which they are perfectly negatively correlated aggregating the two risks eliminates the cost, as when an upstream generator and a downstream supplier each with fixed contracts for inputs and outputs hedges the intermediate wholesale market risk via contracts or vertical integration. Allocating the risk cost to all consumers, for whom the exposure is 1−2% of expenditure, is thus more cost effective than allocating it to a generator whose net revenue exposure may be more than 100%.

of CfDs that can be met by the levy control requires such a forecast; the lower the future electricity price, the higher will be the levy required to fund existing contracts. Again this seems to be an abrogation of the Treasury's willingness to take on risk that it can bear at far lower cost than anyone else.

The more interesting question that seems largely to have been overlooked is what will happen to electricity prices as the CPF rises above the projected £30/t CO_2 in 2020 to DECC's forecast of £70/t in 2030. If a reasonable amount of fossil generation remains on the system in the period 2020–2030, it will be price setting some considerable fraction of the time, and at £70/t the price would almost certainly make nuclear contracts in the money if the cost of capital is really brought down as EMR proposes. For example, a CCGT would have a fuel cost of £70/MWh at 2012 fuel prices, to which would have to be added a markup to recover fixed and capital costs.

The average base-load price will then depend on how prices are set when either interconnectors or renewables/nuclear are price setting, and this is very hard to predict, depending on how much interconnector capacity is built, how much wind and PV is on the system at each end of these interconnectors, and whether the interconnectors are constrained or not. It is, however, likely that if the high carbon prices are maintained then the nuclear CfDs (and onshore wind) will be in the money.

If carbon (and thus to a considerable extent electricity) prices are set to rise in real terms, then should the electricity strike price also rise but from a lower level? The counter-argument favoring a fixed (real) price is that it front-end loads the subsidy and that reduces the credibility risk and hence the cost of capital. Indeed, for CCS there is a similarly strong case for subsidizing the initial capital cost directly to even more front-end load the support (Newbery et al., 2009). However, it may be worth exploring the relative cost to consumers of a lower initial strike price that escalates at above the rate of inflation versus a stable real price (much as water prices were regulated to rise in real terms at a K-factor to finance investment).

Helm's other major criticism is that the Government will be "picking winners" in determining the mix of technologies to support through tailored CfDs, and thus will be prone to lobbying, when it should just allow the market to choose the least cost mature technology choice in the light of a suitable carbon price. Most renewables (but perhaps not onshore wind) are immature and support under the Renewables Directive is best justified as a means of lowering their cost to become competitive with properly carbon-priced generation. Helm recognizes that supporting immature renewables should be considered under a carefully designed research and development (R&D) policy, and this is surely a strong argument to put to the EU as outlined below. Unfortunately, we are meantime stuck with the targets that we perhaps imprudently agreed.

6 CONCLUSIONS

The United Kingdom's misleadingly named EMR has the sensible objective of lowering the cost of reaching its ambitious and legally binding carbon and renewable targets. This aims to do by reducing risk via credible long-term contracts. This chapter accepts that the diagnosis in the EMR *White Paper* and underlying the *Draft Energy Bill* is largely sound, in contrast to many who have claimed that it is "focused on simply shifting risk around" (Yarrow, 2011). The cost of risk depends on how it is allocated, and risk sharing (through contracts passed through to consumers or taxpayers) can greatly reduce the cost of risk. Instead, the real problem is the well-known principal-agent problem of retaining incentives while reducing the risk, where markets, auctions, or benchmarking can all play their part.

One of the main concerns with the proposed EMR, widely expressed in the EU, is whether it represents a retrograde step, replacing market-driven investment decisions with a single, possibly state controlled, buyer model. The deeper concern is whether liberalized electricity markets are compatible with a low-carbon electricity industry. The argument of this chapter is that careful design of the policy instruments needed to support currently uncommercial zero-carbon generation ought to be able to retain the benefits of liberalized markets, namely, competition to drive down costs and stimulate innovation, but that indeed poorly designed interventions could easily undermine the liberalized market and raise the costs substantially higher than needed.

If that is appreciated, it ought to be possible to reduce risk and the resulting cost of capital while retaining and possibly even improving the incentive properties of the current electricity market. The intensity of disagreement with this will doubtless be high, as many benefit from the lack of contestability in the present opaque and illiquid British market, and enjoy the rents that derive from a mismatch between targets, reflected in ROC premia, and the ability of the current system to deliver on those targets.

The Parliamentary Committee (HC, 2012), building on a large number of submissions and witnesses, has justifiably criticized the details set out in the *Draft Energy Bill* as failing to ensure that these contracts will be credible and hence financeable by cheap debt without government underwriting or guarantees. They further criticize the barriers to entry created by a lack of an assured off-take of renewable power, and the "one-size fits all" approach to contract design.

The *Energy Bill* could still be rescued and meet its stated objectives with some moderate but important changes but that would still leave the electricity market poorly designed to handle the growing volume of low-carbon generation anticipated—clearly a task for a future occasion, as is tidying up and making more effective measures to improve energy efficiency and hence balancing demand and carbon reduction. At the European level, Britain should continue to press for reforms to the EU ETS, failing which it should encourage each Member State to embrace a CPF, to build up a coalition in favor of at least reducing the ETS caps to reflect higher planned renewable energy and lower demand because of the financial crisis. The test for an adequate carbon price is whether it would be able to support moderately mature zero-carbon technologies such as nuclear power and onshore wind. It would also be desirable to begin a debate on how to avoid a *2030 Renewables Directive* for what by then should be largely mature low-carbon options that should be commercial given a solution to the more pressing need for an adequate carbon price.

REFERENCES

Bower, J., 2002. Why Did Electricity Prices Fall in England and Wales?: Market Mechanism of Market Structure? Oxford Institute for Energy Studies Working Paper, EL O2.

CCC, 2009. Meeting Carbon Budgets—the Need for a Step Change. Committee on Climate Change. Available from: <http://www.theccc.org.uk/reports/1st-progress-report> (accessed 22.03.2013).

Connor, P.M., 2003. UK renewable energy policy: a review. Renew. Sustainable Energy Rev. 7 (1), 65–82.

DECC, 2010. Electricity Market Reform: A Consultation Document, December. Available from: <http://www.decc.gov.uk/en/content/cms/consultations/emr/emr.aspx/> (accessed 22.03.2013).

DECC, 2011. Planning our Electric Future: A White Paper for Secure, Affordable and Low-Carbon Electricity. Available from: <http://www.decc.gov.uk/en/content/cms/legislation/white_papers/emr_wp_2011/emr_wp_2011.aspx/> (accessed 12.07.11).

DECC, 2012a. Electricity Market Reform: Policy Overview, May. Available from: <https://www.gov.uk/government/uploads/system/uploads/attachment_data/file/48371/5349-electricity-market-reform-policy-overview.pdf> (accessed 22.03.2013).

DECC, 2012b. A Call for Evidence on the Role of Gas in the Electricity Market, 17 March. Available from: <https://www.gov.uk/government/publications/a-call-for-evidence-on-the-role-of-gas-in-the-electricity-market> (accessed 22.03.13).

DG Comp (2007). DG Competition Report on Energy Sector Inquiry, SEC(2006) 1724, Brussels, 10 Jan. Available from: <http://ec.europa.eu/competition/sectors/energy/inquiry/full_report_part1.pdf>.

Evans, J., Green, R., 2005. Why did British Electricity Prices Fall After 1998? Department of Economics, University of Birmingham, Discussion Paper 05–13. Available from: <http://ideas.repec.org/p/bir/birmec/05-13.html> (accessed 22.03.13).

Grubb, M.G., Newbery, D.M., 2008. Pricing carbon for electricity generation: national and international dimensions. In: Grubb, M., Jamasb, T., Pollitt, M. (Eds.), Delivering a Low Carbon Electricity System: Technologies, Economics and Policy. Cambridge University Press, Cambridge, UK, pp. 278–313. (Chapter 11).

HC, 2011. Electricity Market Reform. House of Commons Energy and Climate Change Committee Fourth Report of Session 2010–12, HC 742, 16 May, Vol. 1.

HC, 2012. Draft Energy Bill: Pre-legislative Scrutiny. House of Commons Energy and Climate Change Committee First Report of Session 2012–13, HC 275-1, 23 July.

Helm, D., 2012. EMR and the Energy Bill: A Critique. Available from: <http://www.dieterhelm.co.uk/node/1330> (accessed 22.03.13).

HM Government, 2012. Draft Energy Bill, CM 8362, May.

HMT, 2011a. Budget 2011, HC 836, March 2011.

HMT, 2011b. Control Framework for DECC Levy-Funded Spending. Available from: <http://hm-treasury.gov.uk/psr_controlframework_decc.htm> (accessed 22.03.13).

Newbery, D.M., 1998. The regulator's review of the English electricity pool. Utilities Policy 7 (3), 129−141.

Newbery, D.M., 2005. Electricity liberalisation in Britain: the quest for a satisfactory wholesale market design. Energy J. Special Issue on European Electricity Liberalisation, D. Newbery (Ed.), pp. 43−70.

Newbery, D.M., 2006. Electricity liberalization in Britain and the evolution of market design. In: Sioshansi, F.P., Pfaffenberger, W. (Eds.), Electricity Market Reform: An International Perspective. Elsevier, Amsterdam, the Netherlands, pp. 109−144, ISBN13 978-0-08045-030-8 (Chapter 4).

Newbery, D.M., 2011. High Level Principles for Guiding GB Transmission Charging and Some of the Practical Problems of Transition to an Enduring Regime. Available from: <http://www.ofgem.gov.uk/Pages/MoreInformation.aspx?docid=93&refer=Networks/Trans/PT> (accessed 22.03.13).

Newbery, D.M., 2012a. Reforming competitive electricity markets to meet environmental targets. Econ. Energy Environ. Policy 1 (1), 69−82.

Newbery, D.M., 2012b. Contracting for wind generation. Econ. Energy Environ. Policy 1 (2), 19−36.

Newbery, D.M., Pollitt, M.G., 1997. The restructuring and privatisation of the CEGB—was it worth it. J. Ind. Econ. XLV (3), 269−303.

Newbery, D.M., Reiner, D., Jamasb, T., Steinberg, R., Toxvaerd, F., Noel, P., 2009. "Carbon Capture & Storage (CCS): Analysis of Incentives and Rules in a European Repeated Game Situation" Report for DECC. Available from: <http://cleanenergysolutions.org/content/carbon-capture-and-storage-ccs-analysis-incentives-and-rules-european-repeated-game-situatio> (accessed 22.03.13).

Pollitt, M., 2010. UK renewable energy policy since privatisation. In: Moselle, B., Padilla, J., Schmalensee, R. (Eds.), Harnessing Renewable Energy in Electric Power Systems: Theory, Practice, Policy. RFF Press, Washington, DC, pp. 253−283.

Wong, S.-F., 2005. Obliging institutions and industry evolution: a comparative study of the German and UK wind energy industries. Ind. Innov. 12 (1), 117−145.

Yarrow, G., 2011. Response to the Electricity Market Reform Consultation. Available from: <http://www.rpieurope.org/publications/2011/G_Yarrow_Response_to_the_Electricity_Market_Reform_Consultation.pdf> (accessed 20.03.13).

Chapter 2

Electricity Market Reform in Britain: Central Planning Versus Free Markets ☆

Malcolm Keay, John Rhys and David Robinson
Oxford Institute for Energy Studies

1 INTRODUCTION

In Chapter 1 of this book, Newbery looks at the evolution of the UK electricity market.[1] Market reforms are currently under way there in response to the challenge of meeting ambitious targets for renewable energy and decarbonization. Newbery concludes that the UK approach is largely sound, though concerns remain about whether it represents a step back from market liberalization; indeed, he suggests that there could be a fundamental concern about whether liberalized markets are compatible with a low-carbon electricity industry.

This chapter takes forward and generalizes the analysis of Chapter 1, looking in particular at that fundamental question: whether and to what extent are strict environmental (greenhouse gas) targets incompatible with liberalization? Many countries, particularly in Europe, are facing challenges similar to those of the United Kingdom in relation to renewables and emissions targets—is it inevitable that reforms to meet these targets will result in a reduction in the role of market forces?

The chapter consists of five sections in addition to this introduction:

- Section 2 looks at the nature of the challenge and the reasons why existing markets, even supported by carbon pricing, are unlikely to deliver governments' objectives.

☆ This chapter draws in part on a working paper, *Decarbonisation of the electricity industry—is there still a place for markets?*, by the same authors, published on the Web site of the Oxford Institute for Energy Studies in 2012.
1. In fact, most of the discussion relates to the Great Britain market—Northern Ireland has its separate arrangements. However, these complications do not affect the main themes of this chapter.

- Sections 3–5 look at alternative approaches: improving existing markets, a carbon-intensity target, and a single-buyer approach which might in practice leave room for some competition.
- Section 6 summarizes and concludes.

There are of course many other structures (such as vertically integrated local or national monopolies) that might provide possible models for decarbonization but they are not considered here—the aim is to explore how competition and strict emissions goals can be reconciled, especially in the European context.[2]

2 THE CHALLENGE

2.1 Policy Targets

The United Kingdom is in a special position by virtue of its legally binding target of an 80% reduction in greenhouse gas emissions by 2050 under the Climate Change Act 2008, but it is not alone in having stringent emissions goals. Many other countries are committed to the same ultimate aim. The European Union (EU) has a target of at least an 80% reduction by 2050, and individual EU countries have their own targets, many comparable to those of the United Kingdom (Germany and France, for instance, also have an 80% emissions reduction target for 2050). The G8 also adopted an 80% target at its summit in Italy in 2009.[3] So while not all industrial countries have binding goals for 2050, all are committed to moving in the same direction as the United Kingdom in reducing emissions and all have made commitments under the Copenhagen Accords to meet CO_2 reduction targets for 2020; the EU, for instance, has committed to a 20% reduction in CO_2 emissions by 2020, compared to 1990.

In most cases, industrial countries have adopted renewables targets. All European countries currently have binding targets under the Renewables Directive; the majority of US states have renewables portfolio standards[4]; Japan is strengthening its renewables targets to enable a reduction in dependence on nuclear power (Miyamoto et al., 2012).

The overall strategy of concentrating on electricity for the deepest reductions is also a common feature. In the United Kingdom, the Climate Change Committee saw decarbonization of electricity as one of the first and most important areas—it concluded that "Any path to an 80% reduction by 2050

2. See Fox-Penner (2010) for other possible models for the future that build on the two main US regulatory approaches: vertically integrated and regulated utilities and the deintegrated structure with retail choice.
3. G8 Leaders Declaration: *Responsible Leadership for a Sustainable Future*, paragraph 65, on Summit Web site.
4. US EIA Web site *Today in Energy*, February 3, 2012.

requires that electricity generation is almost totally decarbonised by 2030"
(CCC, 2008, p. 173), and this has been at the heart of the government's
strategy.

In Europe, the European Commission's *Roadmap for moving to a
competitive low-carbon economy in 2050* (EC, 2011) sets out a plan to meet
the long-term target of reducing domestic emissions by 80–95% by mid-
century. The strategy in the roadmap is broadly similar to that adopted in the
United Kingdom. The Commission notes that:

*Electricity will play a central role in the low carbon economy. The analysis shows
that it can almost totally eliminate CO_2 emissions by 2050, and offers the prospect
of partially replacing fossil fuels in transport and heating . . .. The share of low
carbon technologies in the electricity mix is estimated to increase from around 45%
today to around 60% in 2020, including through meeting the renewable energy
target, to 75 to 80% in 2030, and nearly 100% in 2050.*

EC (2011, p. 6)

As Boltz (Chapter 8) and Riesz et al. (Chapter 25) point out, this move
toward greater use of renewable and other low-carbon sources is taking place
across Europe and is posing challenges for market operation.

Of course, there are still many differences in national circumstances.
The route the United Kingdom has chosen reflects its rigorous emissions
targets, the relatively high (in European terms) carbon intensity of its electricity system, the significant generation investment urgently required due
to the imminent closure of many of its conventional plants, and the fact
that it is an island system. Some countries (e.g., the United States) may
share a similar point of departure with the United Kingdom in terms of carbon intensity but have less stringent targets and be happier to rely on the
expected switch to gas. Others (like Germany) face special problems
because of the nuclear phaseout. Others again (specifically France) begin
with relatively low levels of carbon intensity—and a long and successful
tradition of national planning—while others suffer from limited integration
and little investor confidence in the stability of regulation (e.g., Spain).
Each of these different characteristics will influence the choice of the decarbonization route.

At first sight it might appear that a number of other countries have more
successful decarbonization programs than the United Kingdom; even if they
face pressure to meet increasingly ambitious targets, they may therefore be
able to rely essentially on business as usual (BAU). But this cannot be taken
for granted. Germany, for instance, is widely recognized as having an effective program of encouraging investment in renewable sources via feed-in
tariffs (FiTs). However, meeting its decarbonization target would require a
step change in effort. Germany remains a relatively carbon-intensive economy with CO_2 emissions per head above the European average (Table 2.1).
The carbon intensity of its electricity generation is likely to increase rather

TABLE 2.1 Measures of Carbon Intensity for Selected European Countries

Country	Carbon Intensity of GDP (ppp) (kg CO_2/US$ 2000)	CO_2 per Head (t/cap)	Carbon Intensity of Electricity (g CO_2/kWh)
France	0.21	5.49	90
Germany	0.33	9.16	430
Italy	0.26	6.47	384
Spain	0.27	6.17	299
United Kingdom	0.27	7.54	450

Source: IEA (2011)

than fall over the next decade or so, as the nuclear phaseout is implemented (Buchan, 2012). So it would be difficult to argue that Germany is at present on track for the decarbonization of its electricity sector. In any event, if the end result were a system consisting almost entirely of generators with FiTs, even in the form of market premium FiTs, the role of markets would be very limited. Like other countries with ambitious emissions goals, Germany will face major challenges in meeting its targets while leaving space for competition.

The main difference between the United Kingdom and other countries is that most of the latter are not subject to the same degree of urgency—just as the United Kingdom was a pioneer in liberalization, so it is in market reform for decarbonization. As with liberalization,[5] other countries may have the good fortune of learning from the UK's successes and being able to avoid the UK's mistakes. This chapter considers whether there are ways of achieving national goals for renewable energy and emissions reduction that leave a bigger role for markets and competition.

There are inherent difficulties in meeting this challenge. Low-carbon generation investments present special problems for markets, in particular because of the inflexibility and capital-intensive nature of these sources.[6] A key issue is investment certainty—perhaps the most important issue for any investor in a capital-intensive, long-lived, immobile, and highly use-specific asset (which describes most low-carbon generation investments). Investors seek adequate security in the revenue stream and protection from regulatory or other risks outside their control, such as the risk that the

---

5. A general critique of some of the shortcomings of the UK liberalization model, in the context of the decarbonization agenda, is given in Rhys (2009).
6. See Keay (2006, 2011).

government will in future introduce measures which undermine the wholesale price assumptions, and hence the projected revenues, on which the investment was predicated.

This is a longstanding issue for the power sector and helps explain why the sector developed in so many countries through vertically integrated monopoly. The problem is exacerbated in the context of low-carbon investment because of the difficulties in making that investment attractive relative to coal- or gas-based generation, whose levelized costs are currently lower than most low-carbon alternatives.

These difficulties include "incumbent advantage"; volatility in fossil fuel prices induces corresponding volatility in the wholesale electricity price, when the system is dominated by fossil plant (Anderson, 2007). Fossil generators' profit streams are to some degree insulated from this volatility since their input costs are highly correlated to the prices achieved for their output. Low-carbon generators do not enjoy this advantage; in addition to requiring larger amounts of capital, they suffer more volatile revenues unless there is a regulatory or contractual remedy.

A further layer of uncertainty for investors is that in most countries the existing wholesale market structures were designed to suit the technical, cost, and operating characteristics of fossil plant. These market arrangements will require very substantial revision with the advent of large volumes of zero marginal cost plant, a significant part of which may be intermittent or inflexible or have other characteristics very different from conventional thermal plant. The processes of short-term system control and optimization and of wholesale price formation, as they have operated for the last two decades, may be unsustainable beyond the short to medium term. These future changes in market structure and operation are not directly considered here (though see Chapter 25); however, they should be borne in mind in assessing the options. The likelihood that future market changes will be needed may act in the short term as an additional disincentive to investment because of the uncertainty it creates. In the longer term, it is important that the measures introduced to encourage low-carbon investment are flexible enough to accommodate future market changes and new technologies.

A final problem is that the huge transformation of the power sector to achieve decarbonization, and its possible expansion to accommodate transport sector load, will pose very demanding issues for the coordination of generation and infrastructure investment, including conventional transmission infrastructure, CO_2 gathering networks for carbon capture and storage (CCS) plant, and the development of smart grids to accommodate demand-side measures.

2.2 Market Signals for Low-Carbon Investment

The core of the problem is the need for low-carbon investment. In the United Kingdom, for instance, new generation and transmission capacity will

TABLE 2.2 UK Generation Investment to 2020 (GW)

Technology	Conventional	Renewable Scenarios		
		Low	Middle	High
Nonrenewable generation capacity				
Coal	3.9	3.9	3.9	3.9
CCGT	7.5	5.2	4.5	4.2
Nuclear	3.5	3.5	3.5	3.9
Total nonrenewable	14.9	12.6	12.3	12.0
Renewable generation capacity				
Onshore wind	0.4	7.1	8.5	9.6
Offshore wind	0.9	38.4	45.8	61.2
Biomass	0.0	3.6	4.9	5.6
Other	1.0	1.0	1.0	1.0
Total renewable	2.3	50.1	60.2	77.4
Total generating capacity investment	17.2	62.7	72.5	89.4

Source: DECC (2009)

have to be built at a rate unprecedented in UK history—the investment required is estimated by the government at around £110 billion of which about £75 billion would be generation investment.[7] Table 2.2 gives one estimate of the scale of investment required for various renewables scenarios,[8] showing the significant increase in the total investment needed as compared with a BAU case. It should be noted that the amount of fossil investment required is only a little lower on a high renewables scenario than on a BAU basis because of the need for backup.

---

7. HMG July 2011. This estimate has recently been challenged—the Chief Executive of SSE is quoted in *Power in Europe* June 25, 2012, as saying that only £70–75 billion would be needed and that existing mechanisms could deliver this. But this only adds to the uncertainties policy-makers are facing. £70 billion would still be a significant increase on current investment levels. Furthermore, the factors that led to the lower investment estimate (lower electricity demand, shale gas potential, nuclear extensions, etc.) are likely to encourage investors to hold back or invest in gas-fired generation, rather than investing in expensive low carbon capacity, accentuating what is already a problem with liberalized markets, as discussed in Keay (2006).

8. The table comes from one of the background papers prepared for the UK Renewables Strategy (DECC, 2009).

The pace of investment required will vary between countries, but in nearly all cases it will need to be well above historic levels if it is to achieve the goal of decarbonizing the electricity sector. Bauknecht et al. (Chapter 7), for instance, indicate the investment required in Germany up to 2050, which will include significant amounts of fossil generation (55 GW) in addition to over 100 GW of renewables. Pfaffenberger (Chapter 4) emphasizes the investment implications, for instance in transmission networks, of the government decision to shut down nuclear and to expand renewable energy capacity. The problem is in giving the appropriate signals for generation investment within a liberalized market, where no generator is required to invest at all, much less invest in a particular type of plant, especially when the plants are required for policy reasons and are intrinsically unattractive to investors.

One alternative for governments is to go for a centralized system, which directly guarantees remuneration of the investment needed. This is the route chosen by the UK government in its Electricity Market Reform (EMR) proposals. However, that approach involves a considerable degree of government decision making: on the form and content of the FiT contracts for difference (FiT CfD), on the capacity required, on pricing and technology choices, and many other details, thus giving rise to the problems inherent in centralized approaches—of risk-averse and conservative decision making, lack of innovation, and so on. It is therefore worth investigating whether there are approaches to the decarbonization goal that leaves more room for the operation of market forces.

2.3 Stronger Carbon Price Signals

A textbook "first-best" approach to the issue would indicate one of the following approaches as providing a theoretically optimal approach to policy:

- A carbon tax that correctly reflects the "social cost of emissions," that is, the estimated cost of the damage that they cause through their impact on future climate or
- A quantity limit for cumulative emissions that arguably should be weighted according to the date of the emission and then shared out through a market in tradable permits. The limit in principle represents a calculated optimal feasible level of emissions.

Estimates of either the social cost or the optimal quantity limit are both difficult and contentious, but, subject to a number of qualifying assumptions, one would expect the approaches to be broadly equivalent, at least in a world of certainty and predictability, and to provide an optimal policy outcome. The right level of carbon tax/price would result in the optimum quantity of cumulative emissions and vice versa.

Quite apart from the substantial practical issues of resolving what is the right estimate of social cost or the optimal quantity target, there are major problems for governments in either version of these "market-driven" textbook approaches to policy. Even if there were universal acceptance of a "moderately high" cost of carbon, such as the figures around £56 per tonne of CO_2 that are to be found in the current Treasury/DECC guidance for policy evaluation,[9] the imposition of such a figure as a tax on emissions would have immediate political, sectoral, and distributional consequences that most governments are not currently prepared to contemplate. The same qualifications apply to any quantity limit scheme that imposes severe limits and has a similar effect on consumer prices.

The most significant of these adverse features of the purely market-driven approach, from any government perspective, are the following.

First, it will typically be the case that many of the necessary investments can be induced by offering a much lower price than the £56 per tonne and are indeed profitable at much lower prices (even if these are a little higher than current prices). The higher price, if available to everyone, will benefit investments that have already been made, such as those in existing nuclear plant and some less carbon-intensive gas plant. This will mean high levels of windfall profits to some producers; these profits will derive from much higher revenues paid by consumers. On the other hand, many of the investments seen by governments as necessary (e.g., offshore wind generation) are unlikely to be profitable, even with a £56 carbon price, and will not therefore take place, so the government cannot rely on the price alone to deliver either its emissions or its technology targets.

By contrast, direct government intervention can be used to discriminate along the supply curve and minimize the extra profits accruing to low-cost producers, including existing nuclear and gas producers, while enabling higher cost investment to take place. Discrimination can be achieved, for example, by limiting the benefits to new plant and providing financial or other support only for quotas of particular technology types, while still in principle encouraging competition among suppliers to meet those quotas. This may be construed as unhelpful intervention in technology choice, but it is a tempting option for governments. It provides a significant means of limiting windfall profits and the overall cost to government or consumer, while ensuring that targets are met. An alternative approach which can help avoid direct government intervention is to mandate suppliers or producers to deliver low-carbon investment or to allow a single buyer or vertically integrated utility to contract long term for supplies.

The second general concern for government is the political impact of the significantly higher consumer prices that would result from the textbook

---

9. See Treasury (2011).

approach and would run the risk of undermining political support for the whole emissions reduction program, especially where they resulted in big windfall profits. Governments are already having trouble justifying the increase in costs associated with the early phases of decarbonization, and they are naturally seeking ways to limit the substantial price increases expected.

The third problem is the intrinsically global nature of the issue. The cost–benefit trade-offs implied in the textbook, market-driven, approach outlined above are by their nature global in scope. The estimated social cost in this context is an aggregate global cost. But this is not a benefit that accrues to, for example, UK citizens as compensation for foregoing their own consumption or for investing in emissions reduction. Costs and benefits, from a UK perspective, quickly become asymmetric, as consumers bear the whole of the cost but perhaps get back (even in the long term) only 2% of the benefits. This, particularly in the absence of effective global agreements, strongly reinforces an unwillingness to use either a social cost or a purely market-driven approach as a basis for pricing emissions.

Finally, there is the question of whether a long-term price or quantity-based approach can be made credible. Given the various political risks mentioned above, how can any government make a credible commitment that a price or quantity limit will be maintained not only by itself but also by its successors, in the face of uncertain global developments? If it cannot, investors will give the price or quantity signal declining weight over time, and this would significantly increase the risk-weighted cost.

Investment, particularly if it is to be obtained with a modest cost of capital, depends not just on an assumed revenue stream but on the degree of confidence that can be attached to those revenues. Taxes are unlikely to provide sufficient long-term credibility—tax policies are vulnerable to a wide range of economic and fiscal uncertainties and government statements of intent are unlikely to be bankable. Variants of the carbon-price approach, which are based on a rising price over time, are particularly vulnerable to this problem— they suggest that the government lacks the will to take difficult short-term decisions and is therefore deferring the pain of meeting the climate targets. As they look into the future, and see the risk of increasing political pain, the more difficult it, therefore, is for investors to be sure that the expected carbon price will be delivered.

Credibility can be improved in various ways, for instance, via international treaties to underpin the degree of political commitment, but the most readily acceptable forms of commitment seem to be long-term contracts (which are effectively the basis of the UK's EMR approach) or vertically integrated monopoly. It may well be that one of the key challenges facing governments wishing to reconcile liberalization and decarbonization will be to find new instruments, which provide a similar degree of reassurance to investors, while preserving the flexibility necessary to allow competitive markets to reach the most efficient solutions.

These considerations broadly explain a strong political preference for direct interventions, as witnessed in the UK's EMR. These may be "second best" in a theoretical sense, but the government expects them to have the desired effect of reducing emissions without some of the pain of major price adjustments.

One can categorize the range of plausible interventions, including both the UK's current EMR proposals and the variants and alternatives we explore in this chapter, in terms of the ways in which they attempt to deal with the inability to allow an "unconstrained" purely market-driven approach. The categories are not necessarily mutually exclusive, and it is possible that the best policies may require a combination of approaches. They are as follows:

- *Limited adoption of the pure market or CO_2 price-driven approach*, typically through much more limited CO_2 price rises or less-demanding quantity restrictions. Price should always be seen as having an important role to play, even if, as this chapter argues, the very high price level necessary to achieve the policy aims on its own is not likely to be a feasible option. The EU emissions trading scheme (ETS) is an example of limited adoption of a market-driven approach, but it has substantial practical weaknesses. First, it has been set with very limited ambition on carbon emission limits, and in consequence has failed to deliver carbon prices that are sufficiently high to provide a commercial case for the level of low-carbon investment that is acknowledged as necessary. Second, both the volatility of prices and the relatively short duration of the scheme undermine confidence in its ability to support revenues over the long term.[10]

- *Mandating* low-carbon outturns by imposition of low-carbon technology requirements on the supply businesses, on generators, or on vertically integrated utilities. This includes policy interventions that still leave individual market players to work out how best to achieve the mandated outcome. In our chapter, we discuss two particular interventions that both work essentially by mandating an outcome. These are the carbon-intensity obligation and a development from the EMR proposals that would mandate a single buyer. The Californian approach is based on mandating vertically integrated utilities. Designing a mandatory approach to be effective will imply considering what market and incentive structures the approach creates.

- *Reform of wholesale market structures* through specific measures such as capacity payments, markets for ancillary services, and better functioning carbon markets. This is explored in the next section of this chapter and has to be viewed in the general context of efficient market operation. Wholesale electricity market structures can have a significant effect on

---

10. See Newbery (Chapter 1), the figure entitled "EUA price October 2004–December 2011" for the current low level of EUA prices and their historic volatility.

prices and on investment prospects even in the absence of low-carbon policy imperatives. But they will also need to change to reflect the technical realities and opportunities for power systems with low-carbon generation. This section identifies some of these issues in relation to capacity payments and ancillary services, along with some proposed reforms of the EU ETS.

3 IMPROVED MARKETS

Most EU countries and many US states have competitive wholesale energy markets that work reasonably well in the sense of short-term operations. However, the design of markets could be improved in many cases to encourage more effective competition and provide better signals for efficient investment, including by new entrants in conventional and low-carbon generation. Improvements in the world's large carbon market, the EU ETS, are also under consideration. The discussion below looks at

- Short-term energy markets (Section 3.1)
- Capacity markets and incentives for reliability (Section 3.2)
- Incentives for low-carbon generation (Section 3.3)
- Organized ancillary markets (Section 3.4)
- Improved carbon markets (Section 3.5).

3.1 Short-Term Energy Markets

We have identified four specific areas for improving short-term energy markets: greater liquidity in wholesale electricity markets, better balancing markets, nodal prices, and integration of continental electricity markets.

- Liquidity: In most countries, wholesale electricity markets are relatively illiquid, compared with other commodity markets like oil. This tends to discourage an entry for both independent generators and specialized retail companies. The former find it difficult to lock in revenue streams, especially for longer periods; the latter to find suitable financial contracts to hedge the price that they offer to their customers. In the United Kingdom, Ofgem[11] have considered two specific proposals to address these problems: to oblige the main vertically integrated companies to act as "market makers," guaranteeing a supply of particular financial products and to require mandatory auctions of a proportion of their generation in the form of forward financial contracts of different lengths. Ofgem have decided to pursue the idea of mandatory auctions. Other regulators might wish to consider these and other measures to improve liquidity.

---

11. See Ofgem (2012a).

- Better balancing markets: In some systems (like Great Britain), balancing markets effectively penalize companies that are out of balance, so as to encourage companies to balance their operations in advance of despatch. Apart from encouraging the vertical integration that has reduced liquidity in the market, this approach imposes high costs on intermittent renewables (whose output is less predictable). Redesign of balancing markets could reduce these disadvantages, while maintaining system security.[12]
- Nodal pricing: Currently, wholesale prices in many EU countries do not always reflect transmission considerations. This distorts investment decisions, in particular by ignoring the additional transmission costs required to integrate new sources of generation. Nodal pricing would give better signals for the location of new plants,[13] including renewable generation.
- Integration of EU markets: In spite of the commitments and legislation to create single European markets for gas and for electricity, the EU is still made up of regional electricity markets that are poorly interconnected due to transmission constraints; this can make it more difficult to develop and integrate large volumes of renewable generation. While the EU has progressed quite substantially in the integration of European electricity markets, the process of full integration will require substantial investment in interconnection and the political will to support these investments.[14]

3.2 Capacity Payments, Contracts, and Markets

Most EU countries have energy-only markets, as illustrated in Figure 2.1.

However, there is a growing consensus among academics and industry analysts and participants that the energy-only structure is not well adapted to remunerating firm capacity in systems with large amounts of intermittent and capital-intensive generation sources.[15] A number of countries that do not already have them are therefore now considering the introduction of capacity remuneration mechanisms. The models can be separated into two broad categories: payment-based approaches and quantity-based approaches. The payment-based approach involves providing capacity payments—set by an administrative decision or a mechanism—to supplement the payment in the energy market. The quantity-based approach involves determining the

---

12. See Ofgem (2012b).
13. See Newbery (2011).
14. See Jacottet (2012) for an overview of some of the issues.
15. See in particular, Cramton and Ockenfels (2012). See also Robinson et al. (2010). Hereafter, we refer to this paper as The Brattle Group (2010).

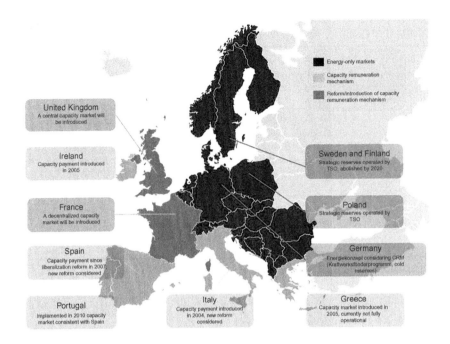

FIGURE 2.1   Capacity remuneration mechanisms in Europe. *Source: Based on IHS CERA March 2011, with update from Cailliau (2012).*

quantity of resource required and then allowing capacity prices to be set by competitive mechanisms.

Table 2.3 summarizes the arguments for and against the payment-based approach, referred to here as "capacity payment" models. This approach has been adopted in a number of countries, including Spain and much of Latin America. The greatest problems with this approach are the significant discretion given to the regulator that can create regulatory uncertainty for investors, and the fact that administrative prices will be either too high or too low—leading either to excess investment and cost or to inadequate investment.

Quantity-based models impose an obligation, typically on retail or distribution/retail companies or on the system operator, to purchase a sufficient quantity of capacity to meet the demand of their customers, including a target reserve margin. Government may itself take on the obligation or transfer it to a central purchasing agency. Table 2.4 summarizes the advantages and disadvantages of the quantity-based models. Their main advantage is that they can ensure resource adequacy, including reserve capacity to back up renewable power. They also provide revenue security

TABLE 2.3 Capacity Payments Models

What Payments?	Energy Markets Payments and Administratively Determined Capacity Payments	
Resource adequacy	*Advantages*	*Disadvantages*
	• Energy price spikes can be mitigated • Greater revenue certainty • Provides regulator with flexible tool	• No guarantee that the desired level of capacity is reached • Can distort market prices • Potential to undermine demand-side response • Increases regulatory risk
Low-carbon objectives	• Less risk than for energy only of problems over resource adequacy	• Heavy burden on regulators • Increased regulatory risk, particularly if there is political involvement

TABLE 2.4 Quantity-Based Models

What Payments?	Energy Market Prices and Capacity Market Prices	
Resource adequacy	*Advantages*	*Disadvantages*
	• Should achieve the desired level of reliability • Can provide revenue security • Encourages innovation • No distinction between new and existing capacity	• If based only on bilateral contracts, can create market power and raise transaction costs for small players • Complex if organized capacity market is included
Low-carbon objectives	• Helps create sufficient operating reserve capacity	• Low-carbon quantity requirements can be complex • FiTs can overcompensate (or be insufficient)

for investors and can thereby help to lower costs. Their main disadvantage is that they may be complex to manage, especially if they include organized capacity markets.

The quantity-based model has an increasing number of adherents: it is the basis for most of the capacity remuneration mechanisms in the United States and is being considered in a number of EU countries.

One specific variant of the quantity-based approach that seems to be gaining support in the EU is the use of "reliability" option contracts,[16] usually with the following characteristics:

- A quantity-based obligation placed on the buyer of capacity—which in the EU could be a retail company, a large customer, the system operator, or a single buyer established by the government.
- Sellers of capacity are required to provide evidence that they can meet their commitments to provide "reliable" power.
- A long-term option contract enables the seller to earn a stream of income (i.e., the option fee), subject to the obligation to supply electricity at the strike price when the option is called, for example, when market prices exceed the strike price.
- An auction or other competitive process is used to select the plants that win the long-term contracts and to determine the price of the option.

3.3 Incentives for Low-Carbon Generation

Even with the changes suggested above, the incentives to build low-carbon generation would need strengthening, since the primary objective of a capacity market is to ensure resource adequacy, not decarbonization.

There are three standard approaches to encourage investment in low-carbon sources: FiTs set by government or regulators, obligations imposed on buyers or generators, and tax credits. Table 2.5 summarizes some of the pros and cons of the main methods.

TABLE 2.5 Different Methods to Promote Low-Carbon Generation

Advantages	Disadvantages
FiTs and tax credits	
1. Flexibility in the pace of growth of renewable installed capacity 2. Certainty about cash flow	1. Little control over capacity 2. Can be very expensive 3. No incentive to run when needed 4. Not technology neutral
Quantity-based approaches	
1. No expensive overshooting 2. Encourage innovation (if no or limited technology carve out)	1. Only work if no barriers to building plants 2. Revenue uncertainty can complicate financing 3. Often higher transaction costs

16. See Cramton and Ockenfels (2012) and EWI (2012). This approach has been adopted in Colombia and in the New England Power Pool. In Great Britain, the DECC is considering a version of this model, as is the Italian regulator.

Many countries (e.g., Germany, Spain) have successfully built up the share of renewable energy through FiTs and tax credits. They have the advantage of flexibility and provide some certainty over revenue streams—thereby lowering the cost of debt financing. However, FiTs are increasingly being questioned: by governments because of the difficulty of setting prices administratively and the risk of overpayment when prices are set too high and by investors because of the lack of legal guarantees in the face of retroactive changes to the FiTs.

Obligation-based systems have been adopted in other countries, including the United Kingdom (the Renewable Obligation Certificates) and the United States (Renewable Portfolio Standards). In these systems, load-serving entities, like retail supply companies, are obliged to purchase a certain share of their energy from renewable sources. The main attractions are that these systems do not encourage expensive overshooting and they can encourage innovation, provided there are very limited technology "carve outs" that favor specific technologies. Some countries combine quantity-based obligations with FiTs. For instance, China combines renewable obligations on the main coal-based generators, with auctions to determine FiTs; the result is that almost all of the wind parks in China are owned by the large coal-based generators.

The specific proposal here is to build on the quantity-based approach outlined above to meet reliability (i.e., resource adequacy) requirements. Whichever organization has the obligation to meet quantified reliability standards would be required also to obtain a rising percentage of contracted power from near zero-carbon sources. The potential near zero-carbon alternatives would include renewable power, nuclear power, demand reduction, and (possibly) CCS-based coal and gas. The goal is to select the least cost combination of supplies, not to promote specific technologies, so banding would not be part of the original proposal. This does not rule out subsidies or special support but does require that this happen outside the electricity and carbon markets (e.g., through taxes).

3.4 Organized Ancillary Service Markets

With the growth of intermittent renewables, the economic importance of back-up reserve has grown.

Ancillary services are at present normally the subject of regulation or bilateral contracts with the system operator, in some cases involving an auction procedure. The proposal here is to create more comprehensive organized ancillary service markets in which all potential suppliers of back-up reserve are able to compete for contracts. As with the reliability contracts, the system has four key features:

- An obligation on the system operator to contract for ancillary services, to provide a particular level of reliability to the system.

- Bidders for the service (e.g., reserves) must provide evidence that they are able to meet the technical requirements for providing the service, either physical or contracted.
- Contracts have a similar form to the reliability contracts, in that they guarantee an option fee to the chosen sellers and specify a strike price at which the services will be guaranteed (indexed to the underlying fuel, where applicable).
- A competitive procedure, probably an auction, would be used to select the companies to provide the service and to set the market-clearing price.

3.5 Improved Carbon Markets

Finally, an important part of the new picture would be more effective carbon markets. Although the authors are skeptical about the ability of carbon markets to act as the sole instrument for bringing forward low-carbon investment, it is possible that improved carbon markets, along with the other market changes suggested above, could combine to create the necessary framework conditions within which such investment would be more attractive. This is in addition to the benefits that an efficient carbon market would have in the short term, in particular encouraging the replacement of coal by gas-fired plants.

The first step might be a more general one to reduce, eliminate, or "neutralize" technology-specific quotas in policy making. No one doubts that renewable energy (and energy efficiency or at least energy conservation) will play an important role in the reduction of CO_2 emissions. However, if quotas for renewable energy are introduced without tightening the CO_2 emission target, this reduces the demand for emission allowances, muffling the price signals and discouraging investment in all low-carbon technologies that do not receive special subsidies, thereby discouraging innovation.

There are different ways to remedy this distortion. One is to reduce or even eliminate technology-specific quotas for future investment, while leaving intact any obligations already entered into. An alternative would be to neutralize the impact of quotas on ETS prices through an adjustment (i.e., a tightening) of the EU's own emission targets.

A second element might be to manage carbon prices directly, for instance by introducing a floor price as in the United Kingdom, managing the price through a central banking arrangement or through automatic "adjustments" that would change the supply of allowances to reflect prices—in other words to tighten supply when prices were too low.

The problem with these ideas is that they involve a fundamental rethink of the EU ETS, which is very difficult politically to achieve. The debate currently revolves around operational options within the current structure: "setting aside" allowances to reflect the current recession or tightening the emissions target to 25%—instead of 20%—below 1990 levels. Both of these would have the effect of raising the CO_2 price, but neither of them would address the fundamental problem of uncertainty about future price levels.

Overall, while a package including the sorts of improvements listed above would undoubtedly help improve market signals and give better incentives for low-carbon generation, not all the reforms are equally realistic. Most of the identified reforms to the short-term markets are relatively straightforward or have been implemented in one or more jurisdictions, even if they are controversial. Introducing quantity-based obligations and markets for capacity and ancillary services may be more challenging, but these have been introduced already in some EU countries and are under consideration in others. However, the prospects for improving long-term signals for low-carbon investments in the EU ETS are arguably less favorable in current circumstances.

Furthermore, there is no guarantee that the changes suggested above, even if all were implemented together, would bring forward the required amounts of carbon generation at a fast enough rate; apart from anything else, the uncertainties produced by the scale of the changes are likely both to create a hiatus in investment as the implications are absorbed and add to future uncertainty—how would an investor know that a government (or its successor) might not decide on further tinkering with the markets?

4 AN ALTERNATIVE MARKET-BASED APPROACH: TRADABLE CARBON-INTENSITY TARGETS

4.1 Tradable Carbon-Intensity Targets

A more radical approach would be the use of *tradable carbon-intensity targets* to promote low-carbon generation.

The basic idea is that there would be a cap, expressed in terms of carbon intensity (g/kWh), applying to all electricity generation within a given system and set in advance over a long time period. The obligation would be tradable. There are broad precedents for such a scheme, in particular in the United States where the idea of carbon-intensity limits has some traction; there have been proposals to apply similar limits to electricity (the "Clean Energy Standard") and for trading schemes in the transport sector.[17] The discussion below focuses on the United Kingdom but, *mutatis mutandis*, could be applied across many systems.

---

17. See, for instance, EIA (2011), which looks at the implications of an intensity-based approach for electricity and CSEM (2009), which looks mainly at motor fuel.

TABLE 2.6 CO_2 Emissions from Electricity and Heat Generation by Particular Fuels (2007–2009 Average)

Fuel Source	Range of Emissions Within OECD Europe (g CO_2/kWh)	Average Emissions (g CO_2/kWh)
Solid fuel	554–1290	830
Oil	333–749	591
Natural gas	218–308	328

Source: IEA (2011)

The carbon intensity of electricity generation varies substantially between different systems; it depends on a number of factors including the age, efficiency, and mode of operation of the plants in place, but the principal driver is the technology and fuel source concerned. Table 2.6 shows the range of emissions from various sorts of fossil plant in Europe.

It will be seen that for each fuel source there is a considerable range of emissions, reflecting the age of the plants, their efficiency and the precise technology in use—for example, combined cycle plants and combined heat and power plants have higher efficiencies and hence lower emissions per kilowatt-hour. Overall coal plants have higher emissions than oil, and oil than gas, but even natural gas plants do not offer high enough efficiencies to meet governments' ultimate decarbonization objectives. However, with significant investment in low-carbon options such as nuclear or renewables those objectives are within reach—as shown in Table 2.1, the carbon intensity of generation in France is only 20% of the UK level.

The United Kingdom already has an informal intensity cap—the government's implicit aim (HMG, 2011) is to get carbon intensity below 100 g/kWh by 2030, from around 480 g today. This could be translated into a formal arrangement by setting a cap for each year up to 2030 on a declining trend. For the purposes of illustration this could be a steady reduction of 20 g/kWh per year (i.e., 460 g in 2012, 440 g in 2013, and so on), though in practice the trajectory would need to take account of the potential pace of investment. Further targets for the 2030s could then be set in, say, 2020, by which time more information should be available about the viability of CCS, new nuclear, etc., but the government could indicate at an earlier stage its expectation that the cap would fall and be no higher than, say, 50 g/kWh.

The obligation would give a clear signal to generators about the nature of the future capacity needed; because of its tradability, it would give flexibility in operation and strong incentives for cost minimization.

4.2 How the System Might Work

The requirement would be imposed via an obligation on all generators, whether new or incumbent, to meet the carbon-intensity cap. They could comply with this obligation by one of the following methods:

- Keeping the carbon intensity of their own generation within the cap;
- Buying carbon-intensity reduction certificates (CIRCs) from other generators to bring their intensity down to within the cap;
- Paying a penalty (or buying reserve CIRCs from the government) to make up any shortfall.

The obligation would apply across a generator's entire fleet (which could be an individual plant only) and would be calculated by taking total carbon emissions during the year (or other chosen period) divided by total electricity generation from that fleet or plant. Both figures are in principle easily obtainable and unambiguous.

The example in Table 2.7 shows how the scheme might operate. Say the target for 2020 is 300 g/kWh and there are three generators: A has a mainly coal fleet and an intensity of 600 g/kWh; B has a mixed fleet and an intensity of 400 g/kWh; C has a fleet composed entirely of zero-carbon generation. In total they produce 1 TWh and 300,000 tonnes of CO_2:

> To comply with the target, both generator A and generator B would buy credits from generator C until they were within the limit.
> The units in the example are actual or virtual electricity output at a given carbon intensity. In principle, it would be possible to trade carbon instead (as has been proposed for one variant of the US Clean Energy Standard). Indeed, in terms of trading within the year, the proposed scheme is effectively an ETS. The key difference between it and a conventional trading scheme is that it gives more straightforward, credible, predictable, and long-term signals at the point of investment.
> The obligation could be placed on either suppliers or generators; the option illustrated here is a generator obligation. The proposed US Clean

TABLE 2.7 Example of Carbon Intensity

	Output	Carbon Intensity (g/kWh)	Carbon Emissions (tonnes)
Generator A	300 GWh	600	180,000
Generator B	300 GWh	400	120,000
Generator C	400 GWh	0	0
Total	1 TWh	300	300,000

Energy Standard by contrast would be a supplier obligation (like, say, the renewables obligation in the United Kingdom). There are arguments for both approaches and the choice would depend on the nature of the system concerned and the specific objectives and orientation of government policy.

Another major issue is the nature of enforcement—what happens if targets are missed? There are two main options:

1. Punitive: The principle here is to set a high penalty for any failure to meet the target so as to provide strong incentives for compliance. This is more likely to ensure that the targets are met; but it adds to risk and could push up costs.
2. Cap on cost: The aim would be to set a reserve price; the government would then issue as many extra CIRCs as needed at that price. The price could be set at the upper end of the range of the expected costs of low-carbon generation; this would provide an incentive to reduce costs and cap the burden on consumers but would not ensure that the target was met.

Although the approaches are different in principle they could be combined in practice—for example, a contingency reserve of CIRCs could be established of, say, 5–10% of total generation; anything in excess of that would bring the penalty system into operation. (The threshold could be set individually for each generator or could be a "double threshold"—that is, it would apply only when the system as a whole exceeded the threshold but would then apply to each individual generator's own excess figure.)

In principle, the boundaries for trading could be set as widely as desired and the scheme could operate at, say, a US or European level. In practice, in such a situation, transitional arrangements might be needed, given the differences in carbon intensity between, say, Greece and Estonia on the one hand (both more than 700 g/kWh) and Sweden and France on the other hand (both less than 100 g), or the similar differences in carbon intensity between US states. It should be possible to develop a scheme which required all countries or states to move over a (long) period of time toward a common intensity target, but meanwhile recognized different targets in different countries as a basis for trading.

4.3 Main Benefits of Scheme

The scheme should provide the following benefits:

- Simplicity, credibility, and predictability for producers: Generators would know a long time in advance what their targets would be and could plan their investment programs around the targets. The problem of making a carbon price credible, would also largely be overcome—an intensity target embodies the decarbonization objective directly (at least in countries with relatively low demand growth) so is less subject to uncertainty.

- Business friendly: This form of target is familiar to generators and easy to implement. Indeed, many generators have used such a target—for example, E.ON had the aim of reducing carbon intensity by 10% between 2005 and 2012[18]; Vattenfall wants to reduce by 50% by 2030[19]; RWE set a target of 50% reduction between 1990 and 2015.[20] A statutory target would give a clear frame of reference for investment planning.
- Transparency for investors: Investors would be able to assess generators on the basis of the carbon intensity of their existing fleet, and their plans for reducing carbon intensity. This would create investment pressure for emissions reductions and increase the attractiveness of low-carbon generation projects.
- Efficiency and cost-effectiveness: The target would provide strong incentives for cost reduction, which the EMR proposals and other European renewables support mechanisms arguably fail to do. It would encourage entrepreneurs to find both the lowest cost forms of low-carbon investment and the most efficient trajectory for reduction.
- Technology neutral: The proposal is technology neutral, allowing markets to discover the best mix of technologies. This has many advantages, in particular, it would encourage innovation and force governments to clarify their objectives if they decide to give additional technology-specific support not based on emissions reduction benefits. The UK government has made clear its preference to move toward a more technology neutral scheme after 2020.
- Market neutral: The arrangements could in principle be combined with almost any market structure. Market reforms need not be determined by the current urgency of investment support and could be developed in line with the development of the low-carbon system; in the future situation, when the objective (getting below, say, 50 g/kWh) has been reached, the intensity target would be no more than a general background framework. By contrast, the EMR proposals are closely tied up with existing market structures, making it difficult to see any exit strategy for the government.
- Good for the demand side: It should be relatively easy to accommodate the demand side in the new arrangements. In this, they differ from the EMR capacity proposals. These are based on the model of the future provision of large increments to supply and are not well suited to demand-side measures, which tend to develop more gradually and less predictably. By contrast, under the intensity scheme, provided they were verified, demand-side measure would have a high market value and

18. E.ON (2006). Corporate Social Responsibility Report.
19. Vattenfall (2011). Corporate Social Responsibility Report.
20. RWE npower Corporate Responsibility Report.

tradability; long-term advance commitments and investments would not be needed. The incentives for developing such measures would be high.

The intensity proposal therefore seems to have many potential benefits. Against that must be set the key questions of credibility and uncertainty— the proposal would involve a major change in approach for most governments and would not directly guarantee investor revenues. Would it provide a strong enough basis for investment and be credible over the long term? Given the novelty of the approach, there can be no definitive answer to these questions but it is arguable that a commitment to an intensity target would be more credible than a commitment to a future carbon price or emissions cap. It is simpler in form, more closely related to the emissions objective, more likely to deliver a least cost solution and does not depend on (or constrain) assumptions about the growth in Gross Domestic Product (GDP) or electricity demand, so there is less reason to expect the government to change the target.

5 A CENTRAL BUYER

5.1 Background and Objectives

If certainty of investment is taken as the overriding criterion, an alternative radical approach would be to take the logic of the EMR to its ultimate conclusion and adopt a fully developed central-buyer approach designed to address the interacting market failures implicit in meeting the low-carbon policy objective:

- the difficulties in internalizing the costs of emissions in decisions made by the main parties in the power sector, either via the cap and trade or an estimated social cost approach and
- issues intrinsic to the power sector, in particular those of coordination, and of regulatory certainty as a basis for investment.

This could be argued to imply the need for a high degree of central direction and coordination, as an alternative to increasingly complex and frustrating attempts to manipulate the market to achieve a policy determined outcome.

The proposition[21] is for a central agency to take responsibility for delivering both the low-carbon emissions target and a sufficiency of total generating capacity and associated transmission, purchasing from multiple generators and selling on power to suppliers.

The concept is not new; it is essentially the same as that of the single buyer, which at one stage was widely supported as an alternative model for the development of markets in the EU. Nor need it mean an abandonment of markets: a central purchasing approach can be designed to include

---

21. Both the diagnosis of this issue in the context of UK liberalized markets and UK policy and a proposal for a central buyer approach are set out in more detail in Rhys (2010a).

most if not all the beneficial competitive pressures and market disciplines associated with the best features of the UK model. Early versions of the 1990 England and Wales privatization model were in effect based on the notion of a single-buyer function exercised jointly by the 12 distribution companies. Under this scheme, the 12 would forecast their own requirements and make separate contracting choices before pooling their contracts for operational purposes. One proposed scheme, the "distributors' pool," would then have had the National Grid dispatching generation plant under contract.

An alternative, much closer to the solution finally adopted, was based on a "generators' pool"; this was intended to allow generators to collectively meet their scheduling and dispatch arrangements by trading through an actual- or bid-based (as opposed to contractual) merit order so as to maximize efficiency. The latter was eventually modified in relatively minor ways to create the actual 1990 market structure, the main modification being to ensure that the pool was open to a wider range of participants.

As these examples demonstrate, the concepts of central purchasing and organization have not been totally foreign to the development of the UK or EU models of competitive industry structure.

5.2 Advantages

This proposal would deal with the problems associated with investment uncertainty and coordination:

- The central agency would enjoy the downstream security of revenues, ultimately from consumers, which would enable it to offer bankable long-term contracts to low-carbon generation investors; the viability of these contracts would not be dependent on a highly uncertain carbon price. In effect the agency would perform the traditional role of a vertically integrated utility, although it might also be described as a single buyer. The outcomes include a lower cost of capital, and hence potentially lower prices as well as greater attractiveness to the community of infrastructure investors.
- Since the contracts could if necessary include provision for the central agency to dispatch output as required, subject to the contract terms, it could, when the occasion arose, seek to optimize generation over much longer periods, for example, a month or a year. This would provide a structure within which the optimization of more complex low-carbon system operations can take place.
- It can deal with broader coordination issues in a much more straightforward way, negotiating directly with the Grid as well as with suppliers and local networks. The model also offers a variety of ways in which the central agency/supplier/customer interface can be managed to accommodate demand response, distributed generation, and local networks.

There are many options for the precise form of a central buyer: a range of possibilities has been explored for the United Kingdom, and the reader wishing to explore this proposal further is referred to Rhys (2010b).[22]

6 CONCLUSIONS

The main conclusion of this chapter is that the scale of the investment required to decarbonize power generation will pose major challenges for liberalized electricity systems. Left to themselves, existing markets will not generate the level and types of investment which many governments are aiming at. It is difficult for those governments to find market-friendly measures to guide investment; low-carbon generation tends to be unattractive to investors because of its inflexibility, high levelized costs, and capital intensity. Against this background, the United Kingdom, facing rigorous emissions targets and an inflexible legal framework, has chosen an approach designed to reduce risk for investors by providing revenue security, underwritten by a legal and regulatory framework determined directly by the government rather than relying on markets.

The move to central direction seen in the United Kingdom, if replicated more widely, would significantly reduce the role of markets in electricity. It is for debate how problematic that outcome would be—in many countries the degree of electricity competition (at least as revealed, for instance, in the level of customer switching) is currently limited; furthermore, there is no clear consensus about the benefits of full electricity liberalization—many countries have stopped short of this goal. Nonetheless, advances in information and communication technology, such as smart grids and smart meters, may be opening up significant new possibilities for the extension of market forces and consumer choice; it would be unfortunate if the liberalization experiment were abandoned before it had really been tried.

Governments with ambitious targets for electricity decarbonization should therefore give full consideration to whether there are market-oriented ways of achieving that objective—other countries may have more time than the United Kingdom to explore alternative options.

Some possible approaches are discussed above. It is possible that the more radical options might be seen as unproven and potentially risky—for instance, It would in principle be possible to combine aspects of the different approaches—for example, giving a single-buyer carbon-intensity targets but leaving it free to adopt the most cost-effective route, or introducing a carbon-intensity target along with market improvements. Each government will of course want to take account of its own national circumstances and goals. This chapter has argued that, in doing so, they should not assume that

---

22. See also the following Web site:http://ccpolicygroup.blogspot.com/search/label/Power%20sector.

the only way forward is to restrict the role of markets by providing detailed central direction themselves; they should be prepared to explore options which are designed to meet strict environmental targets while leaving scope for the free operation of market forces and consumer choice.

REFERENCES

Anderson, D., 2007. Electricity Generation Costs and Investment Decisions. UKERC Working Paper, February 2007.

BERR, 2008. Growth Scenarios for UK Renewable Generation Report by Sinclair Knight Merz BERR Publication URN 08/1021 London, June 2008.

Buchan, D., 2012. The Energiewende—Germany's Gamble, OIES SP26, June.

Cailliau, M., 2012. Marcel Cailliau, Chair of Eurelectric TF Market Design for RES Integration. The role of energy-only markets in the transition phase towards building a low carbon economy. Electricity Markets at the Crossroads: Which Market Design for the Future, January 19, 2012, p. 7.

CCC, 2008. Climate Change Committee. Building a Low Carbon Economy UK, 2008.

Cramton, P., Ockenfels, A., 2012. Economics and design of capacity markets for the power sector. Z. Energiewirtsch 36, 113–134.

CSEM, 2009. Taxes and Trading Versus Intensity Standards. Centre for the Study of Energy Markets, University of California, CSEM WP 190.

DECC, 2009. The UK Renewable Energy Strategy Cm 7686 July 2009.

EC, 2011. European Commission COM/2011/0112 final.

EIA, 2011. Energy Information Administration. Analysis of the Impact of a Clean Energy Standard, November 2011.

EWI (Institute of Energy Economics) at the University of Cologne, 2012. Investigation into a Sustainable Electricity Market Design for Germany. Summary, April.

Fox-Penner, P., 2010. Smart Power: Climate Change, the Smart Grid and the Future of Electric Utilities. Island Press.

HMG, 2011. Planning our Electric Future: A White Paper for Secure, Affordable and Low-Carbon Electricity.

HM Treasury, 2011. Valuation of Energy Use and Greenhouse Gas Emissions for Appraisal and Evaluation. DECC, London, October.

IEA, 2011. CO_2 Emissions from Fuel Combustion—2011 Edition. OECD/IEA, Paris.

Jacottet, A., 2012. Cross-Border Electricity Interconnections for a Well-Functioning EU Internal Electricity Market. OIES, Oxford.

Keay, M., 2006. The Dynamics of Power. OIES, Oxford.

Keay, M., 2011. Can the market deliver security and environmental protection in electricity generation? In: Rutledge, I., Wrights, P. (Eds.), UK Energy Policy and the End of Market Fundamentalism. Oxford University Press, Oxford.

Mitchell, C., Woodman, B., 2011. Learning from experience? The development of the renewables obligation in England and Wales 2002–2010. Energy Policy 39, 3914–3921.

Mitchell, C., Bauknecht, D., Connor, P., 2006. Effectiveness through risk reduction: a comparison of the renewable obligation in England and Wales and the feed-in system in Germany. Energy Policy 34, 297–305.

Miyamoto, A., Ishiguro, C., Nakamura, M., 2012. A Realistic Perspective on Japan's LNG Demand After Fukushima, June 2012, NG62, OIES, Oxford.

Newbery, D., 2011. High Level Principles for Guiding GB Transmission Charging and some of the Practical Problems of Transition to an Enduring Regime. EPRG, University of Cambridge, Cambridge, Revised April 22, 2011.

Ofgem, 2012a. Retail Market Review: Intervention to Enhance Liquidity in the GB Power Market, February 22, 2012.

Ofgem, 2012b Open Letter: Ofgem Decision to Launch a Significant Code Review (SCR) of the Electricity Cash-out Arrangements, March 28, 2012.

Rhys, J., 2009. Will Markets Deliver Low Carbon Generation? SPRU Electronic Working Paper Series No.17, 2009.

Rhys, J., 2010a. Reforming UK Electricity Markets. A Purchasing Agency for Power. Oxford Energy Forum, July 2010.

Rhys, J., 2010b Can UK Electricity Markets Deliver a Low Carbon Future? Finding the Way Forward. BIEE Academic Conference Paper.

Robinson, D., Pfeifenberger, H., Hesmondhalgh, S., 2010. Resource Adequacy and Renewable Energy in Competitive Wholesale Electricity Markets. The Brattle Group Discussion Paper, September.

Chapter 3

The French Paradox: Competition, Nuclear Rent, and Price Regulation

Jacques Percebois
University of Montpellier I Head of CREDEN

1 INTRODUCTION

The French power sector has a special status in Europe for two reasons: its proportion of nuclear energy is very high, around 75%, and the price of electricity is low by European standards and remains largely controlled by the government—at least when it comes to households (Percebois and Wright, 2001). The prices for industrial customers are even more strongly controlled by government, but this should be partially phased out by 2015. As described by Boltz in Chapter 8, recent directives to open national markets to competition, imposed by European Commission (EC), have already changed many things: the incumbent utility, Electricité de France (EDF), no longer has a monopoly and consumers, households included, can freely choose their retail supplier. In practice, however, the incumbent's market share remains very high and the network operators—Réseau de Transport de l'Electricité (RTE) for transport and Electricité Réseau de Distribution France (ERDF) for power distribution—are subsidiaries of the incumbent, which remains 85% government-owned. This explains why, at the EC's request, a number of arrangements have been introduced to provide balance between competition and the government intervention within the French electricity market.

The choice of nuclear allows France to benefit from a relatively low price for electricity, certainly lower than the average European price, an advantage French consumers do not intend to relinquish (Figure 3.1). In this context, the word "competition" for the average European is synonymous with lower prices. For the French consumer, faced with the prospect of domestic prices rising to average European market levels, the word "competition" is synonymous with higher energy prices. This peculiar state of affairs explains one of the paradoxes of the current French electricity market. The EC mandates are essentially requiring domestic French prices to rise to increase competition

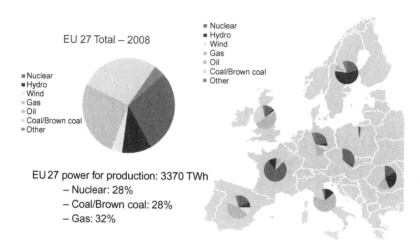

EU 27 Total – 2008

- Nuclear
- Hydro
- Wind
- Gas
- Oil
- Coal/Brown coal
- Other

- Nuclear
- Hydro
- Wind
- Gas
- Oil
- Coal/Brown coal
- Other

EU 27 power for production: 3370 TWh
- Nuclear: 28%
- Coal/Brown coal: 28%
- Gas: 32%

FIGURE 3.1   European electricity generation mix.

while to many observers, competition is actually known for driving prices down. This chapter describes some peculiarities of the French electricity market, including the mechanism introduced to balance liberalization, competition, and regulated electricity prices. At the heart of the problem is the sharing of the "nuclear rent". However, it is necessary to arrive at a proper definition of the "nuclear rent" and to clarify various related concepts including differential rent, monopoly rent, and scarcity rent.

The chapter is organized as follows: Section 2 provides a brief history of the evolution of electricity market in France. Section 3 describes different stages of electricity market liberalization in Europe and their implementation in France. Section 4 introduces the various concepts associated with the "nuclear rent" previously mentioned, which are critical to understanding the paradoxes of French electricity market. Sections 5 and 6 describe the so-called Regulated Access to Incumbent Nuclear Electricity (ARENH) mechanism, designed to share the "nuclear rent" to encourage more competition with the incumbent dominant utility in the French market. Section 7 discusses the current energy debate in France including a discussion of nuclear energy's future prospects followed by the chapter's conclusions.

2 ABBREVIATED HISTORY OF FRENCH ELECTRICITY SECTOR

The evolution of the French energy and electricity sector is very closely tied to French government policies since World War II. In France, the government has always played an important role in energy decisions because of the strategic importance of energy for a country, which has very few natural resources. The period covering 1946 to present may be subdivided into four distinct periods further described in Percebois (1989) and Hansen and Percebois (2010) corresponding to four different strategies depending on whether priority was placed on addressing shortages, ensuring the security of supplies, or minimizing the cost of access to energy to the consumers—particularly industrial consumers—as presented in Table 3.1 and described below.

During the first period from 1946 to 1960, priority was given to fighting energy shortages. The reconstruction of the French economy following World War II increased the country's energy needs, which were met by domestic, and some imported coal, hydroelectricity and relatively small amount of imported oil, mostly for use in the transportation sector. At the time, the French economy was a virtually closed economy relying primarily on domestic resources. During this period, French coal production was ramped up reaching its peak in 1959. Coal, which in 1960 represented almost 55% of primary energy consumption, was extensively used for heating to produce steam, to generate electricity, and even used to produce synthetic gas. After the Lacq gas field was discovered in 1951 and placed into operational use in 1955, manufactured gas was gradually overtaken by natural gas, from domestic sources at first, and then imported gas from the Netherlands, Algeria, Russia, and Norway. Large investments were made in hydroelectricity, which accounted for more than 10% of French primary energy in 1960 and has remained a significant source of electricity production to present.

TABLE 3.1 Four Stages of the Evolution of French Energy Policy, 1960–2010

French Primary Energy Balance (%)

Sources	1960	1973	1990	2010
Coal	54.5	15.6	8.3	4.0
Oil	31.6	67.3	38.7	32.0
Natural gas	3.4	7.2	11.3	15.0
Nuclear	–	2.2	34	40.0
Hydro	10.5	5.4	5.5	6.0
Renewables	–	2.3	2.2	3.0
Total (%)	100	100	100	100
Total (Mtep)	85	180	230	276
Independency rate (%)	62.1	24.4	48.7	49.1

The main means of achieving this "nationalist" energy policy was to have the public sector in a monopoly position. During this period, coal, gas, and electricity were nationalized with the creation of Charbonnages de France, Gaz de France, and EDF in 1946, which were granted EPIC (Etablissement Public à caractère Industriel et Commercial) status (public industrial and commercial undertakings) and endowed with strong and independent management. Only the oil sector was spared nationalization and that goes back to prior laws passed in 1928, which gave the government a monopoly role in oil importation. These privileges were subsequently delegated to foreign oil companies including multinationals, such as Shell, ESSO, BP, and Caltex, or to domestic ones, particularly the Compagnie Française des Pétroles—the predecessor to Total, in which government held a third of the capital. The cost of energy in France was relatively high during this period because energy prices were largely dependent on the price of coal, the dominant fuel. French coal mines are deep and costly to operate.

During the second period between 1960 and 1973, government priority shifted to holding down the cost of supplying France with energy. The growing internationalization of the French economy, following the passage of the Treaty of Rome in 1958, required that to be competitive, the national industry should benefit from production costs substantially equivalent to those of its main foreign competitors. Among the implications of this treaty was that it no longer made sense to rely on expensive domestic coal when abundant oil was available on the international market at low cost. In 1960, the French government established a plan—the J.M. Jeanneney Plan—to gradually phase out expensive French coal and encouraged the substitution of imported oil in all economic sectors—including in electricity generation. This explains why by 1973 the coal share had dropped to 15.6% of the primary energy consumption in France and why the oil share rose from 31.6% in 1960 to 67.3% in 1973 as shown in Table 3.1. Not surprisingly, the French energy independence rate—defined as the ratio of domestic energy production to total consumption—fell from 62.1% in 1960 to 24.4% in 1973. Clearly, increased reliance on imported oil had its consequences.

During the third period between 1973 and 1990, government priority radically shifted away from overreliance on imported oil and to enhancing the security of the French energy supply. The first oil shock of 1973 revealed how vulnerable France had become to imported oil. In 1974, the government adopted a new energy policy—the Messmer Plan—named after the Prime Minister during Georges Pompidou's presidency, with the following three objectives:

- rational use of energy leading to energy savings in all sectors of the economy;
- geographical diversification of imported energy sources including oil, natural gas, and coal;

- strong emphasis on nuclear-generated power, the only large-scale domestic source of energy for France (Figure 3.2).

The French nuclear *turnaround* was hugely successful.[1] By 1990, the nuclear share in the primary balance had reached 34% compared to 2.2% in 1973, whereas the share of oil dropped to 38.7% (Table 3.2). During this period, the coal share continued to decrease, that of natural gas, mostly imported, continued to rise and the country's energy independence rate rose to 48.7%.

The fourth phase, which covers the period from 1990 to present, coincides with the gradual liberalization of network industries, especially electricity and natural gas, by the EC, as described by Boltz in Chapter 8. The introduction of competition, encouraged by the European directives, gradually modified the organization of the French energy industry, with the abolition of the import and production monopolies and the partial opening of the public industries' capital, while adhering to the three main objectives of the 1974 Messmer Plan. Another major emerging concern during this period was the threat of climate change and new emphasis on sustainable development. These concerns, in turn, intensified government's focus on low-carbon nuclear energy, energy efficiency, and sustainability.

It is often said that there is no common energy policy in Europe, when in fact there is a Pan-European competition policy that is increasingly applied to the energy field with three main objectives:

- first is energy competitiveness, which includes access to energy at the lowest cost;
- second is security of the energy supplies;
- third is the fight against global warming, which includes support of renewables and low-carbon energy resources described in Chapters 5 and 7.

In France, the government endorses these principles and the current structure of the primary energy balance reflects this policy: the coal share—entirely imported since the last French coal mine was closed in 2004—dropped to 4%, that of oil stabilized around 35%, that of natural gas around 15%, whereas nuclear energy represents about 40% of primary consumption and 75% of the electricity production, among the highest in the world. The hydroelectric share is 3.5% and that of renewable energy, except hydroelectricity, remains moderate (2.5%). The rate of energy independence, as defined earlier, has also stabilized at around 49%.

As can be seen from the preceding discussion, since 1945, the French government has and continues to play a central role in the field of energy. Moreover, it has been able to adopt important policy choices and implement them in response to domestic and international developments or constraints.

---

1. As described in Chapter 4, it is rather ironic that in 2011, Germany embarks on a *turnaround* to phase out the German nuclear sector by 2022.

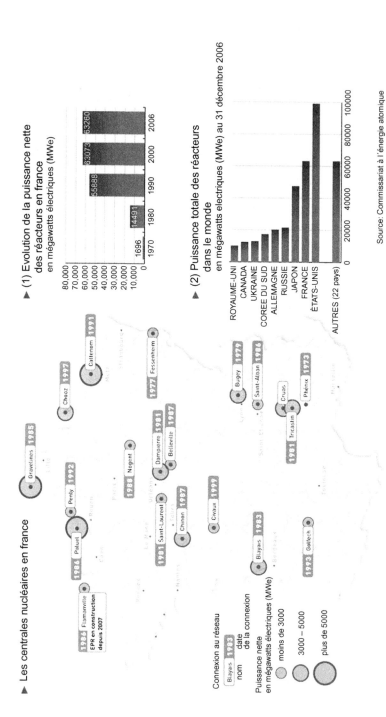

FIGURE 3.2 Location of nuclear power stations in France with date of connection to the network. (1) Evolution of nuclear power in France (MWe), (2) nuclear power in the world. *Source: Commissariat a l'Energie Atomique.*

TABLE 3.2 Evolution of the French Electricity Production

TWh	1960	1970	1980	1990	2000	2010	2010 Installed Capacity (GW)
Conventional thermal (coal, oil, gas)	32	79	119	45	50	59	**26.4**
Hydro	41	57	70	57	72	68	**25.3**
Nuclear	0	5	58	298	395	408	**63.1**
Other renewables	0	0	0	0	0	15	**5.7**
Total	73	141	247	400	517	550	**120.5**

It can be said that the energy policy in France has been among a few areas where there has been strong political consensus. The consensus in favor of a dominant nuclear component, however, may not be as strong as it used to be following the Fukushima accident in March 2011, as further described in this chapter.

Aside from the changing public and political sentiment, the predominance of nuclear power in the French electric power sector, and overall energy mix, is likely to face a number of challenges due to the European directives that aim to increasingly expose the network industries—electricity, gas, railway, and telecommunications—to competition (e.g., see Bouneau et al., 2007).

Under the Treaty of Rome, which was signed in 1957, the energy policy was first and foremost a matter to be decided by each member state. One of the cornerstones of the European Union (EU) is the promotion of competition within a "common" market, eventually leading to a "single" market. This goal, according to the final stages of the current EU directives, is expected to be substantially achieved by 2014. The EC is not empowered to dictate what a country's energy choices must be but it has the authority to make sure that the market competition remains the cornerstone of all energy undertakings, except in justified or exceptional cases. Germany's recent decision to unilaterally phase out its nuclear fleet, without even taking the trouble to notify its partners and the Brussels Commission, illustrates that each country is in charge of its own affairs in the energy domain.

Today, energy policy in Europe is composed of several national policies, which are the result of history and geography. It is through gas, oil, and electricity transport systems that the "Energy Europe" is gradually taking shape but it is far from complete.

Still, the market should play its role and hold down prices for the end user. The producers who, like EDF, own large nuclear plants—63 GW out of approximately 120 GW installed capacity—which produce low-cost electricity, certainly cheaper than electricity produced with gas or coal, naturally enjoy a comparative advantage when competition is introduced. As will be described later, this "nuclear rent" could be compared to a "scarcity rent" if the competing producers cannot—on a long-term basis—get access to the low-cost nuclear energy, if they did not make the choice before the opening of the market or they cannot do it for legal or political reasons. If the price of electricity is not held down by the government of the net exporting country and reaches the European equilibrium price, the incumbent in the exporting country will make excess profits: this is the price to pay for past decisions. But all the providers will sell at the same price on the final market and the competition will allow entrants to win market share by accepting lower profits than those earned by the exporting country. If the net exporting country's prices are held down by the government at the nuclear level for political or social reasons, then domestic and foreign competitors will not be able to compete with the incumbent since the prevailing selling prices in that country will necessarily be lower than the higher costs of nonnuclear electricity. In this case, the new entrants will not be able to win market share in that country, which may distort competition. This is the core of the problem facing policy makers in EC and in France.

The existence of the nuclear "scarcity rent" is a real problem in view of the European policy makers, which led to the New Organization of the Electricity Market (NOME) law in France in 2010 and the establishment of the ARENH mechanism by mid-May 2011. These two decisions take up the recommendations of the reports drawn up by the Champsaur I (NOME) and Champsaur II (ARENH) Commissions (Champsaur, 2011) further described below.

3 THE DIFFERENT STAGES OF THE LIBERALIZATION OF ELECTRICITY IN EUROPE

The main differences among the EU countries are manifested in the fuel mix used in the production of electricity (Table 3.3). At the EU level, most electricity is generated with natural gas, roughly 32%; nuclear and coal-fired plants represent roughly 28% of total generation each, with the nuclear's share expected to decline over time due to recent decisions made in some countries, notably Germany, as described in Chapter 4. Only 6% of the European electricity is currently produced by oil. The remaining 14% is generated by hydropower, wind, solar power, or biomass. These are averages for the EU as a whole, but vary significantly from country to country.

In stark contrast to the EU-wide averages, in France nuclear power accounts for 75% of generated electricity, far ahead of Belgium at 53%, Germany at 28%—before roughly half of the reactors were shut down in 2011—the United Kingdom at 19%, and 0% in Italy. Table 3.3 compares the prevailing fuel mix in EU 27, France, and Germany prior to 2011.

TABLE 3.3 Electricity Generation Mix in EU 27, France, and Germany in 2010[a]

Sources	EU 27 (%)	France (%)	Germany[a] (%)
Nuclear	28	75	28
Hydro	7	12	3
Coal	28	5	51
Gas	32	4	9
Renewables	5	4	9
Total	100	100	100

[a]*Prior to 2011 decision to shut down half of the operating fleet.*

The difference among EU member states is quite noticeable. The percentage of electricity generated by natural gas is more than 60% in the Netherlands and 40% in Italy, whereas in France it hardly reaches 4%. Almost all Poland's electricity is produced from coal, roughly 92%, whereas in France it is around 5%. In the aftermath of the Fukushima disaster, some countries, notably Germany and Belgium, have decided to phase out nuclear energy while others, including Italy, renounced it, whereas others, including France, the UK and a few others, continue to officially support the nuclear option. It is clear that this is far from a consensus among EU countries in terms of the future role of nuclear energy in the power generation mix and this has serious consequences when national borders are fully opened to competition and trade since the cost of electricity production varies significantly from one country to another within the Union.

Against this background, the encouragement of competition in the energy field has been achieved in several stages via a number of directives, primarily relating to gas and electricity, since oil is already a sector wide open to global competition, as is coal. The general idea is that all the monopolies must disappear except the "natural" ones, those that control the transport and distribution networks. The implementation of this idea has been attempted in several major steps, also described by Boltz:

- The first "electricity" directive, issued in 1996, which imposed the opening-up to competition for electricity production and commercialization by introducing progressive eligibility of end users in stages over time.
- The second directive, issued in 1998, which applied the same process to the natural gas industry.

- The second "Energy Package", issued in 2003, which strengthened the provisions of the first directive to gas and electricity markets and required the legal unbundling of network activities.
- The third "Energy Package", issued in 2009, which imposed the ownership unbundling of network activities and includes strong environmental provisions.

This progressive opening of electricity and gas markets to competition is based on a number of legal and procedural measures further described in Percebois (2008) and highlighted below:

1. First, the suppression of legal monopolies on import, export, gas and electricity production, and marketing. These activities are part of what is called the "deregulated" sector and the competition authorities are taking steps to make sure that there is no collusion—for example, in price setting—nor any predatory strategy—for example, by offering prices below cost, i.e., dumping.

2. Second, the expanding eligibility of consumers to select competing retailers, which means that consumers can henceforth choose their supplier. Eligibility has been gradually introduced since 2000, first reserved to industrial consumers and now available to virtually all consumers as of July 1, 2007. Like an industrial consumer, households can now select to stay with the incumbent or choose an alternative supplier. In some EU countries, all the gas and electricity rates are "market offer" prices, or OM for *Offre de Marché*, which means that the supplier can freely negotiate with the customer and offers prices which are generally aligned with spot prices, with a partial or complete indexation more or less in real time. In a few other countries, as is the case in France, the customer may choose between two options:
   - Either to stay with the regulated selling price, *Tarif Réglementé de Vente* or TRV, which is established by the government. In the case of electricity, this rate will be phased out in 2015 for industrial customers but should remain in effect for households. The regulated prices for gas are also expected to be phased out,
   - Or to opt for market offer (OM) prices, which are freely negotiated with the supplier. Since prevailing electricity and gas prices on the "spot" markets are relatively volatile, especially for electricity, which cannot be stored, these negotiated rates are sometimes higher and sometimes lower than the regulated prices. The EC hopes that in the end, only OM prices will remain.

3. Third, the third party access (TPA) to electricity and gas transportation and distribution considers these activities as "natural monopolies," meaning that the duplication of networks is viewed as unnecessary and costly. Moreover, due to the inherent *economies of scale* of networks, economists believe that such monopolies must be considered as "essential facilities"

opened to all operators. Incumbent suppliers, such as EDF, for electricity, and GDF-SUEZ, for gas, as well as other new entrants, must be able to use these facilities to serve their customers via the payment of reasonable tariffs set according to prevailing rules—for example, covering legitimate costs—transparent rules—for example, perfect information on the available capacities of networks—and above all, applied in a nondiscriminatory fashion to all competing suppliers. Under such a scheme, the incumbent, who initially owned the network, should *not* be allowed to block entry, which would be tantamount to a denial of rights to new entrants. These network services must consequently be "regulated," in other words, the tariffs must be set by an independent regulatory commission, which ensures that the necessary investments are made, that system access is neutral, transparent, and nondiscriminatory, and that the unused capacities are available for the market—a rule known as "UIOLI," for "use it or lose it." In France, the Commission for Energy Regulation (CRE) proposes rates to the Minister in charge of Energy, who has 2 months to issue an opinion. These tariffs must allow the recovery of costs incurred including the investment costs and the operating costs, using a "cost-plus" approach, plus the interest on the invested capital.[2]

4. Fourth, the development of a gas and electricity "spot" market, allowing exchanges between producers and suppliers, some of whom are seeking to buy gas and electricity to supply their customers, while others want to sell depending on the prevailing circumstances. For example, a "day-ahead" spot market has been established in which blocs of electricity are exchanged "overnight" and prices are negotiated for each hour of day and night.

As a result of these initiatives, the price paid by a consumer is made up of several components with parts of the price being *regulated* while others are determined by the *unregulated* components. For those who chose the TRV option, where available, the price paid is relatively stable but rate increases are regularly scheduled to take into account the prices reported on the market, after the CRE has issued its opinion. For those who chose the OM option, the price paid is more volatile but not necessarily higher than the TRV since the price is negotiated with the supplier, who can make attractive commercial offers.

In France, households have generally opted for the TRV option because the regulated prices are often lower than the market ones. This is due to the

---

2. In theory, the third directive imposes a system of ownership unbundling but exceptions are possible and France was able to implement an ITO (independent transmission operator) system where transmission and distribution networks remain the property of EDF but they have to show proof of their independence by adhering to particularly strict "code of good conduct." In the long run, there is no doubt the ownership unbundling will emerge as the ultimate solution to the ownership issue.

fact that the regulated prices are frequently set based on the historical cost of nuclear energy, which tends to be low. Currently, only the incumbent EDF can offer the TRV. The OM rates are negotiable both by the incumbent and competing suppliers. The proportion of clients who have switched suppliers in Europe varies from one country to another; it is high when the prevailing prices before the opening of the market were high, and low when the opposite was the case. In France, the proportion of customers who have switched is particularly low as shown in Table 3.4, for the same reason.

Clearly, the EC expects a gradual migration toward price *convergence* over time among the European countries as a result of the imposed directives. In practice, however, this price convergence is hard to observe to date, at least for electricity, firstly because there is still transmission congestion at borders and also because the electricity generation mix and costs are vastly different from one country to another for such convergence to take place. The cost and price disparities in terms of kilowatt-hour production costs explain why the price convergence has not taken place at the consumer level.

In the case of electricity, if we take into account the TRV option, the per kilowatt hour generation cost measured at the output terminals of power stations and the marketing cost together represent about 40% of the price including all taxes for a French household. The transmission network access tariff represents about 10%, the distribution network access tariff about 25%,

TABLE 3.4 Market Shares of Electricity Suppliers in France at March 31, 2012

End Users	Residential End Users	Nonresidential End Users
Sites number	30,677,000	4,921,000
Regulated tariffs	28,756,000	4,235,000
Market prices (suppliers)	1,921,000	686,000
(EDF)	(11,000)	(316,000)
(Entrants)	(1,910,000)	(370,000)
Market share of entrants (%)	6.2%	7.5%
Consumption TWh (per year)	144.3 TWh	294.8 TWh
Regulated tariffs	135.3 TWh	161.2 TWh
Market prices	9.0 TWh	133.6 TWh
Market share of entrants (%)	6.1%	21.8%

Source: CRE.

and the remainder, approximately 25%, is made up of taxes, which include local taxes, value-added tax (VAT), and CSPE, which is a "contribution to the public electrical utility" designed to cover the extra costs linked to the spatial equalization of rates and the subsidies provided to renewable energies in the form of feed-in tariffs (FITs). For gas, the components of the price paid by a domestic consumer (TRV) are as follows: 50% for the gas, which is mostly imported, 35% to cover access to distribution and transmission network, and 15% for taxes. The price for gas delivered to the end user does not vary much from one European country to another since all the countries are supplied under similar conditions, and imported gas prices tend to be set at EU-wide levels. Prices for electricity delivered to the end user, on the other hand, vary significantly from country to country since the production costs vary noticeably, depending on whether the power is generated by nuclear power plants, from hydroelectric, or by coal or gas-fired power stations. Presently, the cost of nuclear and hydroelectric generated power is considerably lower than the cost of fossil units, especially if the price per ton of CO_2 is included.

For instance, in 2010 the electricity price in France, which is comprised 75% nuclear and 12% hydropower, was much lower than the prevailing average price in the EU. This cost disparity results in retail tariffs for industries to be 27% and for households 39% less than the EU average as shown in Figure 3.3.

In reality, the goal of opening the network industry to competition is not to align 27 competitive markets but to create a single market for gas and electricity in the EU. This obviously requires removing the remaining

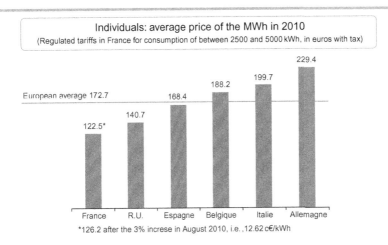

FIGURE 3.3 Average price for the domestic sector in selected European countries in 2010, euros per megawatt-hour.

transmission bottlenecks among the member states as well as implementing the EC directives. Efforts to strengthen the cross-border interconnections, without jeopardizing the reliability of the network within each member state, were initiated by the European electrical system operators well before the enactment of the European directives on opening the market to competition.

4 NUCLEAR DIFFERENTIAL RENT AND SCARCITY RENT MUST NOT BE CONFUSED

In a competitive market, assuming the power generation fleet is optimal, the selling price allows the recovery of the full fixed and variable costs of the infrastructure used if the pricing is based for each period—off-peak hours, full hours, peak hours—on the fleet's marginal costs. At peak hours, variable and fixed costs of peaking plants, such as gas turbines, must be covered. At off-peak hours, only the variable costs of the marginal plants used must be covered; sometimes this will be a coal-fired plant, and at other times it may be a nuclear plant. This requires using the plants in order of increasing marginal costs, that is, adhering to the "merit order" rule. In theory, this is what a perfect market or a central planner does, as explained in Hansen and Percebois (2010).[3] The fixed costs of baseload plants, e.g., nuclear, are covered thanks to the high selling prices set at peak hours. Therefore, selling a nuclear kilowatt-hour based on the gas turbine's marginal cost at peak hours is not an unjustified *rent* but rather a legitimate "differential rent." It is the means to retrieve a markup for covering the nuclear fixed costs because nuclear plants are typically price takers in markets where marginal prices are set by more expensive peaking units. The same applies when selling a nuclear kilowatt-hour based on a coal-fired plant's variable cost during off-peak hours; in this case, the prevailing price contributes to retrieving some of the fixed costs of nuclear power. Thus, the rule is simple: charge the variable cost of the basic nuclear plant during the off-peak hours, charge the variable cost of the coal plant during intermediate hours, and charge the variable and the fixed cost of the peaking plants, typically gas turbines, during the peak hours of the year.

The problem with the elegant theory is that the market price is often too low during peak hours, usually below the level necessary to recover the full fixed and variable costs of the gas turbine. As a result, not all fixed costs of the various plant components are recovered, and more seriously, this frequently leads to inadequate investment in peaking plants. This is the

---

3. Several other chapters cover this topic, including Chapter 9.

"missing money" problem referred to by Stoft (2002). The missing money problem has led to calls for the introduction of a "capacity market" to guarantee sufficient return on peak investments.[4]

Setting these arguments aside, by selling the nuclear output based on the nuclear variable cost at off-peak hours of the year, based on the variable cost of coal during intermediate hours, and on the basis of fixed and variable costs of gas turbines during peak hours, allows nuclear plants to recover their variable and fixed costs. But for this theory to work, the power generation fleet must be optimal, which is rarely the case. In practice, sometimes it operates at overcapacity, sometimes at undercapacity. Moreover, as indicated above, the price is sometimes too low during peak hours to cover the full cost of peaking gas turbines.

If, due to the market supply scarcities or because of high cost of fossil fuels, the equilibrium price is, on average, higher than what would be necessary to cover the complete cost of nuclear plants, then the generator benefits from a *rent*, that is, either a "monopoly rent"—if the high price results from the producer's market power due to a "capacity withdrawal"—or a "scarcity rent"—if the high price is the consequence of the unbalanced nature of the power generation fleet, for example, too many expensive thermal plants where it would be preferable to have more nuclear or hydro plants.

In practical terms, the observed high prices during certain periods on the European market do not seem linked to a producers' market power. It is the high cost of fossil energy, exacerbated during certain periods by the high cost of CO_2, which explains why nuclear operators can recover more revenues than would be necessary to cover the full costs of their operation while selling at the same price as the conventional thermal units. As it turns out, this is the situation observed today in Europe, especially in France and Belgium, two countries where the proportion of nuclear in electricity production is particularly high, 75% and 53%, respectively, as previously mentioned.

Thus, the question is who deserves to benefit from this scarcity rent: the incumbent, the consumer, or the government on behalf of its citizens? In France, it was essentially the consumer who benefited from the nuclear rent in the past thanks to low retail electricity price and who benefits from it today thanks to the ARENH mechanism described below. Figure 3.4 shows the link between the intensity of electricity investments over the 1955—2010 period and the electricity tariffs paid by the French end user, since the first oil shock. As illustrated in the graph, the electricity tariff paid by French consumers increased during periods when investments in nuclear plants were

---

4. The issue of energy-only versus energy and capacity markets is controversial and extensively covered in the literature. Chapter 19, for example, suggests that energy-only markets are adequate; as does Chapter 10, which describes another successful energy-only market in Texas. Chapter 9, on the other hand, describes the capacity mechanism in the context of the PJM market.

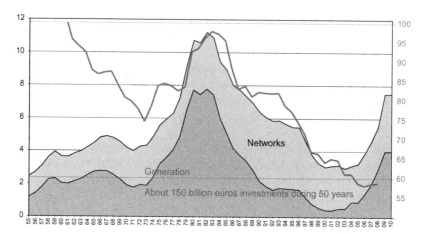

FIGURE 3.4   Evolution of electricity investments in France (1955–2011). Power generation and network investment in constant euros 2007; billion euros left axis. Tariffs in red (industrial tariffs in real terms, euros/MWh right axis) are linked to investment effort since the first oil shock in 1974 (start of nuclear program). (For interpretation of the references to color in this figure legend, the reader is referred to the web version of this book.) *Source: EDF.*

made—the consumers had to pay for it—and decreased when new investments were no longer necessary. The French nuclear investment has been mainly concentrated during the 1974–1990 period while the benefits of these investments are observed today.

5 SHARING THE "NUCLEAR SCARCITY RENT": THE ARENH MECHANISM

When French industrial consumers gained eligibility in 2000, abandoning the regulated selling prices (TRV) in favor of the OM price was highly profitable since the market price was roughly 25% lower than the regulated rate. This was the result of two factors: first, the existence of an electric supply overcapacity in Europe and second, because of prevailing low oil prices. At the time, the price of a barrel of oil was around $20 compared to $40 in 1981, and more than $110 in 2011. As a result, the price of natural gas, which in Europe is indexed to the price of oil, was also relatively low in long-term supply contracts. The price of coal on the international market was also low and the EC had not yet established the CO_2 quota system—it was established in 2005.

As a result, the price of thermal electricity generated from gas, coal, and even fuel oil was low. With the electricity produced in the EU being 60% thermal-based, the market price was relatively low most of the time. Things, however, changed in 2004 when oil prices began to increase (Chevalier and

Percebois, 2008). It exceeded $100 a barrel at the beginning of 2008 before reaching the peak of $147 in July 2008, then dropped to less than $40 at the end of 2008 and climbed to around $100/$110 after early 2009, and remains high by historical standards.

Consequently, the electricity prices prevailing on the spot market, which had historically correlated with price of oil, exceeded the regulated selling price. The industrial customers, who had opted for the OM price, sought to return to the regulated price. This, however, was neither consistent with the law nor with the EU directives or the spirit of market liberalization. Thus, they lobbied the French Parliament, which passed a law setting the *transitory tariff system* or TARTAM. This tariff provided for the possibility, for a limited period, to return to a regulated price based on the former-regulated price with a positive differential of 23% to take into account the advantage granted to these industrials for 4 years. Not surprisingly, this scheme has been criticized by the EC, which started two legal proceedings against France for failing to follow EC directives and for what it considers "state-sponsored assistance" to French industrial consumers. With EDF being a mostly state-owned public corporation, the TARTAM scheme was deemed comparable to a public subsidy.

5.1 Subsidy or No Subsidy?

At the heart of the debate is that the regulated selling prices of electricity in France, the TRV, is set by the government and is based on the production costs of the French power fleet, which is nearly 90% determined by the low-cost nuclear and hydro component—as previously noted,[5] whereas the OM prices paid by the customers who took advantage of eligibility are based on prices observed in the wholesale spot European electricity market, which for the most part are based on the coal-fired or gas power stations' production cost, which is noticeably higher. Moreover, German thermal power plants are often on the margin on wholesale European market, which determine the prevailing market prices in France, Germany, and Benelux. During the intermediate and peak periods, all suppliers are granted the same costs to encourage competition. In the base period, when nuclear and/or hydropower is on the margin, new entrants are challenged on price. The dilemma is how best to reconcile the EC's two contradictory goals:

1. On the one hand, allow French consumers to benefit from the low-cost nuclear power, which they financed with their prior investments in retail rates. As shown in Figure 3.4, French consumers paid substantially higher retail rates between 1974 and 1985 when the bulk of the French nuclear

---

5. Depending on the year, some 75−78% of French generation is from nuclear power and 12−13% from hydro.

fleet was financed, 50% was paid by self-funding and 50% by borrowing on the financial markets.

2. On the other hand, complying with the European Directives and the "spirit" of liberalization of the electricity market, which implicitly requires "entrants" to win market shares at the expense of the incumbent? This vision relies on the conventional view of competition (e.g., the Harvard Business School variety), which requires the incumbent to give up its excessively dominant position, as reflected by examining the HHI (Hirschman−Herfindahl Index). Yet, the fact remains that the new entrants cannot effectively compete with EDF since the French tariffs are capped at the French electricity cost level, which is much lower than the production cost of competing generators who did not opt for nuclear power or were not within the capacity of the French incumbent.

The first Champsaur Commission was established in 2009 to examine and resolve this contradiction. This *ad hoc* Commission nominated by the French government included four members of the Parliament and four energy experts including Jacques Percebois, the author of this chapter. Three potential options were identified by the Commission and presented to the government:

1. The first option was to completely *deregulate* all prices, thereby doing away with the regulated selling prices for both the industrial and the domestic sectors. In this case, the price paid by the French consumer would be on par with the European market price, which means that it would have to *rise* noticeably. This would be the ultimate "liberal" solution advocated by Brussels and many economists. But how would one explain this option to French consumers, who actually *paid* for the nuclear energy in higher prior tariffs. How would anyone be able to rationally justify that they are going to lose their comparative cost advantage in a competitive market—where most consumers expect lower prices? Such a solution would have important ramifications in purchasing power of domestic consumers and in competitiveness of industrial consumers. Moreover, it would be ruinous to the social acceptance of nuclear energy in France, killing prospects for future investments.

2. The second option was to assign the "nuclear rent" to the government, owner of EDF. This rent could be considered and treated as a common tax revenue, in virtue of the principle of nondedication of government income, or it could be applied against particular expenses in the energy sector—for example, to modernize the electricity distribution networks—or redistributed to the French consumer as a "negative CSPE." In the latter case, the French consumer would retrieve a portion of the prior nuclear investment. In this case, the government should only recover the "scarcity rent" and

not the "differential rent" described above.[6] This scheme would increase the incumbent's cost to the same level as the "entrants" in the off-peak period.

3. The third option was to allow EDF's competitors partial access to EDF's historical nuclear power, at least during off-peak periods by essentially lowering the supply cost of the "entrants" to that of the incumbent. Under this scheme, instead of levying the "scarcity nuclear rent," it is shared with the competitors. The "differential rent," however, would not be shared since it applies to the recovery of the fixed cost of nuclear power—only the "scarcity rent" would be shared. The distinction between the two schemes is not necessarily clear, as graphically illustrated in Figure 3.5.

After examining the options, the third option was selected by the Champsaur I Report and this is what ended up in the NOME law, further described below. The first option was essentially ruled out from the beginning based on the government's mission statement. The regulated price for access to nuclear energy during the off-peak period became embodied into ARENH.[7]

In this context, one must wonder if it is logical to ask EDF to share part of the "scarcity rent" with its competitors. EDF owns the nuclear plants and the ARENH mechanism considers them as a kind of "essential facility." But, if the government were to reject the recommendation of the commission in option 3, it would either have to abolish TRV or risk retaliation from the

---

6. As it turns out, the Belgian government chose this solution to secure the nuclear rent of GDF-SUEZ, the private owner of seven nuclear reactors. The tax was subsequently added to the tax revenues, but the producer strongly contested its excessive amount. It must be noted that there is no longer any regulated selling price in Belgium since all prices are market based for both the industrial and the domestic clients. However, in the government's view, the profits realized by the incumbent are excessive and it wants to levy all or part of this "nuclear rent." The problem, in this dispute, may be that the government is confusing "scarcity rent" with "differential rent." In some ways, this situation can be compared to the one observed with the "sunk costs" that some incumbents highlighted in the context of the introduction of competition. At the request of the incumbent, the government must take charge of the additional costs linked to decisions taken before the liberalization, on the grounds that these decisions are a handicap in a competitive market, for example, overcapacity investments to prevent from any risk of failure. By taxing the nuclear rent, the government requires the firm to pay part of the excess profits linked to decisions, sometimes guaranteed by the government, made in a protected market, which provides the firm with excess profits due to the introduction of competition. The argument is that at least *part* of these excess profits must be given back to the community who paid for them in the first place.
7. Another possibility could have been the concept of virtual power plants (VPPs), which would have required EDF to auction off part of its nuclear-generated electricity to its competitors. But under the existing system, the auctioning price would have been close to the spot market price, hence would have not addressed the fundamental problem. Stated differently, the "sourcing" cost of acquiring EDF's nuclear power to competitors would not have been significantly reduced with such a system.

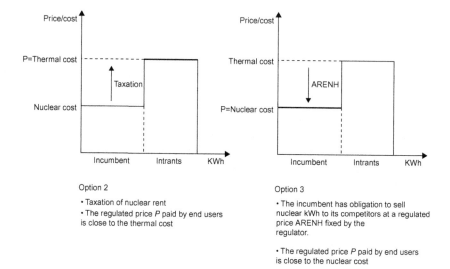

Option 2
- Taxation of nuclear rent
- The regulated price *P* paid by end users is close to the thermal cost

Option 3
- The incumbent has obligation to sell nuclear kWh to its competitors at a regulated price ARENH fixed by the regulator.
- The regulated price *P* paid by end users is close to the nuclear cost

FIGURE 3.5   Solutions to implement fair competition during off-peak period.

European authorities. The threat brandished by Brussels, which demanded an immediate phaseout of TARTAM, was either to impose the principle of reciprocity regarding market shares won by each incumbent outside its national base—which would have been tantamount to preventing EDF from competing in other European countries—or to institute a market share ceiling or a limited installed electricity capacity owned by the incumbent—as was done in Italy, imposed by the Italian Parliament. In any case, the EC threat against France was imminent and "credible," resulting in the adoption of ARENH.

The Champsaur I Report recommended that CRE implements the ARENH scheme and the government decided to gradually introduce it during a transitional period from 2011 to 2013. The government asked a new Commission, Champsaur II, composed of three members, Paul Champsaur, Bruno Durieux, and Jacques Percebois, to suggest a level of compromise between what EDF was asking and what the other parties wanted. This new commission delivered its report in March 2011 further detailed in Champsaur (2011). The government subsequently established the ARENH scheme in accordance with the report's recommendations and the new system went into effect on July 1, 2011, which also was the date for the abolition of TARTAM proposed under the NOME law.

5.2 Implementation of ARENH Scheme

As described in the preceding section, the scheme introduced by the NOME law is a compromise between what is economically optimal, politically acceptable, and operationally feasible. The following principles were used to reconcile between maintaining the regulation of tariffs, as required by the European Directives, while transposing them into French law. The main features of ARENH mechanism are summarized as follows:

- To allow French consumers to continue to benefit from the competitiveness of the French nuclear fleet while at the same time allowing competition to develop across all customer segments, EDF has to sell part of its nuclear energy to its competitors at the same price it sells to its own customers. This price must correspond to the historical cost of the existing fleet of the 58 operating nuclear reactors, including an allowance for a normal margin. The ARENH-regulated price is set according to a "cost-plus" principle. It is initially set by the government to be adjusted by the CRE starting in 2013. It only takes into account the historical costs of the existing nuclear plants, not any new reactors to be built in the future or those currently under construction.
- The volume of sales under ARENH is capped at 100 TWh, which corresponds to about 25% of the current nuclear production in France. Only the French customers residing in France can have access to this allocation, the so-called ARENH volume, which is available to EDF's competitors in accordance with their client portfolio residing in France. The scheme allows an alternative supplier to pass the ARENH price, which is similar to the wholesale price, to clients. A supplier who loses clients loses the corresponding ARENH rights.
- The law specifically *excludes* the renewal cost of the nuclear plants in the ARENH price. The regulation concerns the *existing* assets and their possible life extension, which will imply rejuvenation costs, including upgrading and safety investments. These costs, incidentally, are estimated at 55 billion euros for the 58 reactors, including safety investments imposed by the Nuclear Safety Authority after Fukushima, but investments in new reactors are excluded.[8]
- The renewal price of the nuclear power plant will be included in the electricity price paid by the final consumers and not in the ARENH price, which is only a component of the final price.

---

8. The decision on whether to extend the existing reactors' lifetime beyond 30–40 years requires the approval of the ASN, the French Nuclear Safety Authority. The authorization is given to each reactor separately and for only 10 years at a time. The former government wanted to extend the lifetime of the existing 58 reactors to 60 years. As of 2012, the average age of French nuclear plants is 27 years and all the reactors currently in use should reach the age of 40 between 2020 and 2025. The new French government wants to reduce the share of nuclear power to 50% of electricity production in 2025, which will imply stopping one reactor out of two when they reach their 40th birthday. As proposed, one reactor out of two will be maintained in operation, the other one will be replaced by gas- or coal-fired thermal power plants or by renewable energies. The decision was recently made by the new French President.

- The ARENH measure is expected to last until 2025, with the understanding that the law may be amended and/or possibly extended in the interim years.
- The assets' base taken into account for the ARENH calculation is that of the accounting cost of historical nuclear estimated at 22 billion euros in 2010: it is the residual book value registered in EDF's accounts, roughly 15 billion euros, plus a share of the assets dedicated to cover the nuclear "downstream" charges on a long-term basis, roughly 7 billion euros, which corresponds to 15/40th of the dedicated assets since, of a 40-year-old lifetime, 15 years will apply to the ARENH regulation.[9]
- The ARENH price gradually increases over time at the rate of inflation to account for the rise in operational expenses.[10]
- The ARENH price resulting from the work of the Champsaur II Commission has been set at 39 euros/MWh, with a 38−40 euros price range. The government opted for 40 euros/MWh for 2011—that is, to say the last 6 months since the measure took effect in July 2011—and 42 euros/MWh for 2012. The subcomponents of this price are as

---

9. This measure does not meet with an average economic cost and it is intentional since the investment has already been paid by the consumer. In 2011, about 75% of the nuclear power investment has already been paid off by the French consumer. If a nuclear power plant secondary market were to exist, the current power generation would be, without any doubt, estimated at more than 110 billion euros. It is on this level that EDF "promotes" its power plant fleet. But such a market does not exist. The "overnight" cost estimated by aggregating the investments made, without taking into account what has been paid off (as if the power plant fleet had been built overnight), would be in the order of 83 billion euros. But it is fair to take into account what has been paid off in the ARENH calculation so that the consumer does not pay the nuclear power plant fleet twice. This is why only the book value has been accepted, in accordance with the NOME law provisions. But it is logical to consider the return on the capital invested and the Champsaur Report accepts the WACC (weighted average cost of capital) suggested by EDF, that is, 8.4% before tax. However, all operational charges, including the accounted "pass-through" rejuvenation investments, are taken into account in the ARENH calculation. The advantage of a "pass-through" logic is that the firm gradually retrieves the maintenance and power plant extension investments, without going into debt to do it. Moreover, this method is neutral regarding the lifetime of a power plant fleet. Otherwise, a hypothesis should be made on the renewal period, which is hard considering uncertainties since the decision will depend on the ASN, so as to be able to choose the pace of the payoffs. Note that the nuclear power activity invests with a political risk that other industrial activities do not usually have to confront: the "nuclear power plant shutdown" political risk. This means that the investments can at any time become "sunk costs." Hence, the "pass-through" system limits certain drawbacks. Only the oil industry is familiar with the same type of risk during the nationalizations without compensation, as has sometimes happened in some producing countries.

10. As for the price of electricity paid by the final consumer, it will probably increase faster than inflation because both the investments to be made to modernize and develop the electricity transmission and distribution networks and the continuously increasing CSPE cost will have to be taken into account.

TABLE 3.5 Breakdown of Cost Components in ARENH[a]

Capital return on past nonamortized nuclear investments (dismantling included)	6 euros/MWh
New investments for extension of life expectancy of the existing nuclear park (from 40 to 60 years)	8 euros/MWh
Operating costs of the existing nuclear park (58 reactors) (fuel, salaries, taxes)	25 euros/MWh
Total	39 euros/MWh

[a]Actual cost of the existing nuclear fleet supported by the incumbent, this fleet having been largely amortized over time.

follows: 25 euros for operating expenses, 8 euros for maintenance and new investments for the extension of power plant fleet lifespan, which are recovered by the pass through mechanism, and 6 euros for the remuneration of the past investments on the basis of the WACC (Table 3.5). Following the Fukushima disaster, the government considered opting for the highest price range insofar as security investments should be higher in the future.

It must be noted that currently, the price of fossil fuel-generated electricity is higher than the nuclear-generated price in Europe—at least higher than historical nuclear costs but not necessarily the cost of third-generation European pressurized reactor (EPR)-type nuclear power plant—because the price of the gas imported to Europe is indexed to oil, which is presently high. But if the EU, and France in particular, decided to aggressively exploit the continent's nonconventional gas reserves, things could change and the same phenomenon seen in the United States could happen in Europe. If such a scenario were to happen, the low price of shale gas would diminish the cost advantage of nuclear and even the competitiveness of coal-fired power plants.[11]

---

11. In the case of the United States, the low price of shale gas has led to an increase in American coal exports, which has resulted in lower coal prices in international market, compromising the profitability of the European combined cycle gas plants. The situation is further exacerbated because these gas power plants are often used as "backups" for intermittent renewable energies reducing the low load factor. The net result is higher generation cost of gas plants because the price of imported gas is indexed to oil price and at the same time, the load factor of these power plants decreases because of the priority given to wind and photovoltaic energies. In this case, coal-fired power stations once again become competitive, especially as the cost for a ton of CO_2 is currently low—approximately 7 euros—on the European market.

6 ARENH AND CURRENT COST OF NUCLEAR NOT TO BE CONFUSED

The Court of Auditors delivered its report on the nuclear power cost in France in early 2012 (Cour des Comptes, 2012), and some observers immediately pointed out that the value assigned by the Court in determining the unit cost of nuclear is noticeably higher than the level set by the ARENH, a regulated price at which EDF must sell its nuclear output to its competitors under the terms of the NOME law, already described.

The "Current Economic Cost" of nuclear generation, as calculated by the Court of Auditors, is the price that an entrant on the nuclear power market would be willing to pay to "rent" the current power plant fleet, rather than build one. This price reflects the economic value of an asset built over the years, based on the observed investments assessed in 2010 currency, including provisions for the end of the cycle costs. Based on this interpretation, the cost was estimated at 49.5 euros/MWh by the Court as illustrated in Table 3.6.

TABLE 3.6 Average Historical Cost of Nuclear Generation for the Existing Fleet of 58 Reactors

	Billion Euros 2010		Billion Euros 2010
I. Past investments		II. Investments due to life expectancy of the existing park (PWR)	
UNGG (first generation)	6	Life expectancy to 60 years (+ safety costs due to Fukushima)—**Subtotal**	**55**
PWR (second generation)	96	III. Charges for the future	
Fuel cycle (AREVA)	19	Dismantling	36
Research expenses (EDF, CEA, AREVA)	55	Used uranium management	15
Superphenix	12	Waste management	28
Subtotal	**188**	**Subtotal**	**79**
Annual operating costs (for 410 TWh)	9	*Average cost/MWH (investments + operating costs, life-expectancy investments excluded)*	€49.5

Source: Cour des Comptes (2012) Report.

TABLE 3.7 Various Costs of Nuclear (MWh)

Approaches	Euros/MWh
Cost of new nuclear (generation 3, EPR) (estimate)	75
Current economic cost (with extension of life expectancy investments) (generation 2, PWR)	54 (Cour des Comptes)
Current economic cost (without extension of life expectancy investments) (generation 2, PWR)	49.5 (Cour des Comptes)
ARENH (cost for the incumbent EDF taking into account amortized capital and extension of life expectancy investments) (generation 2, PWR)	39–42 (Champsaur)

This cost estimate does not include the upgrading and safety investments imposed on the existing fleet following the Fukushima accident, which are estimated around 55 billion euros, as previously mentioned. If these required investments are included, the price rises to 54 euros/MWh. If the capital investments already paid by consumers are deducted from this[12] the price falls to the same level as the ARENH, that is, 39 euros/MWh. The results of the various cost options are summarized in Table 3.7.

The 49.5 euros/MWh "Current Economic Cost" shown in the accompanying table captures the current economic cost of nuclear generation taking into account the "overnight" cost of nuclear, which consists of 83 billion euros, the accumulated interest during construction, 13 billion euros, the front and back end fuel cycle costs the research expenses, and the cost of first-generation reactors, as summarized in Table 3.6.[12]

The total cost of capital estimated by the Court, defined as "user cost of the installed nuclear assets," is 9.104 billion euros per year, and the power plant fleet operational cost is 11.043 billion euros a year, including 8.024 billion euros for the operating expenses and 1.076 for spent fuels and waste treatment management, which totals 20.147 billion euros in 2010 currency.

---

12. The "overnight" cost accounts for aggregate investment to build the current 58 plant fleet representing about 62,510 MW of installed capacity generating 407.9 TWh in 2010. As the name implies, "overnight cost" assumes that these reactors could be built "overnight," i.e., without incurring any interest. Since this is not the case, interests accumulated during construction, 13 billion euros, must be added to get a more accurate picture. In case of the French nuclear investments, roughly one half of the financing was realized by self-financing, while the other half was borrowed, including on the American bond market.

Considering the 2010 production of 407.9 TWh, the "Current Economic Cost" is estimated at 49.5 euros/MWh as reflected in the accompanying table.[13]

In summary, the Court of Auditors' approach is a *retrospective* technique calculating the cost of the existing nuclear fleet to the French consumers. The ARENH approach is a *marginal* one, which calculates how much the nuclear fleet costs to EDF today, knowing that the incumbent has already recovered a good portion of the investment cost from the consumers. If EDF were to sell nuclear-generated output to its competitors, which ultimately reaches French consumers, on the basis of the cost calculated by the Court of Auditors, EDF would over-recover a part of the capital costs because the current cost of nuclear-generated power for the firm is lower than the average cost borne by consumers. However, when the nuclear power plant fleet is to be renewed, the tariffs will have to be increased to finance the new investments.

7 THE FRENCH NUCLEAR DEBATE

Having described the various schemes to share the historical bounty of inexpensive nuclear power with new suppliers so that they can effectively compete with the incumbent EDF in its home turf, this section explores the current debate on future of nuclear energy in France. In this context, three main issues are being debated in France concerning the future of the electricity sector, and nuclear's role therein, further explored below:

1. First, what is the optimal structure of the electricity mix for France going forward?
2. Second, is it justified to subsidize renewable energy resources in the context of a market where prices are supposed to be cost-reflective?
3. Third, do current market price signals provide sufficient incentive to invest in peaking capacity to prevent future power failures?

---

13. Note that this cost is independent of the lifetime of the power plants. If the new expenses of rejuvenation and safety measures following the Fukushima disaster are taken into account, the cost is estimated at 54 euros/MWh. The "Energies 2050" Commission, nominated by the French government in October 2011 and chaired by Jacques Percebois and Claude Mandil, with the mandate to analyze the various options for nuclear power in the future, draws on these figures to conduct its analysis. Four scenarios were to be explored: (1) extension scenario up to 60 years of the lifetime of the current power plants, (2) acceleration scenario of the transition to the third generation (EPR) and even the fourth generation (breeders), (3) scenario of partial nuclear phase-out to 50% of the nuclear share, (4) total nuclear phase-out scenario. The anticipated and provisional cost of the third-generation reactor (EPR) is today about €75 but it is not yet possible to anticipate the cost of megawatt-hour produced with breeders. The main recommendation made by the Commission was to opt for the extension up to 60 years of lifetime of the current nuclear park for economic reasons (see "Energies 2050" Report, 2012).

7.1 The Debate About the Electricity Mix

There is little doubt that nuclear will remain the main component of the French electricity mix at least through 2030. Deciding on a long-term policy beyond 2030s, however, is an open-ended question. Presently, more than 90% of the French electricity is produced without any CO_2 emissions.[14] It is a great advantage in the context where priority is given to low-carbon energy resources, as further described in Chapters 1 and 2 on the United Kingdom, Chapter 5 for the EU, and others in this volume.

As already described, the French decision to follow the nuclear option was made in the early 1970s to reduce French dependence on expensive and insecure imported oil while allowing the French end users to benefit from low and stable electricity prices. Some people would like to see the share of nuclear to remain at its current levels while others say it should be reduced to around 50% by 2030. The accident at Fukushima has brought the safety of nuclear energy into the debate, as described in Chapter 4 in the case of Germany and Chapter 23 in the case of Japan.

Fortunately, new and safer reactors like the EPR or breeder reactors like those developed under the Astrid program[15] are promised to become operational and hopefully competitive by 2040, in which case they can replace the existing reactors over time as they retire. This explains why an extension of life expectancy of the existing reactors from 40 to 60 years is the best solution, but only if safety investments are made to increase the confidence of the public.

In the nuclear field, France is, and is expected to remain, a leader in the world. Other sources of electricity generation should be used in small proportions: hydroelectricity, gas, coal, wind, and solar. The obligation imposed by the EC is to produce at least 23% of the French electricity from renewables by 2020 as described in Chapter 5. Since hydro already contributes at least 12%, efforts should be made to develop wind and solar to meet the 23% target by 2020, even if renewables must be largely subsidized.

The long-term fate of the French NOME law is also an open-ended question in the context where the consensus in favor of nuclear and electricity liberalization are now broken. For some people, future energy policy must, for environmental reasons, move away from fossil fuels, which would favor renewables and nuclear energy. Others favor moving away from fossil fuels *and* nuclear energy—in the case of Germany—which implies heavy reliance on renewable energies.

In the mean time, the prospect for fossil fuels to remain competitive must be considered, particularly if shale gas and shale oil are considered.

---

14. As already mentioned, roughly 75% nuclear, 12−13% hydro, and 3−4% wind.
15. Astrid is a breeder prototype, which should be operational by 2025; such a reactor is able to recycle plutonium produced by the existing PWR fleet and at the same time to use uranium 238.

Uncertainties remain concerning the competitiveness of gas and even coal power stations in France. Nearly 98% of the gas currently consumed in France is imported through long-term contracts with oil-indexation pricing. According to the International Energy Agency (IEA, 2011), shale gas reserves may amount to 5000 billion cubic meters in France. France's trade deficit amounts to 73 billion euros and energy imports account for 63 billion euros in 2011 currency. Therefore, the potential for exploiting a domestic energy resource should be closely considered.

Shale gas exploitation is a sensitive political issue. Currently, shale gas exploration is a taboo in France and hydraulic fracturing is banned because of its alleged negative effects on the environment, especially on underground water sources. It is, however, necessary to be less emotional and answer several key questions rationally, namely:

- Are the current estimates of recoverable shale gas reserves justified?
- At what cost can shale gas be profitably/commercially exploited?
- What would be the environmental impact of current and future shale gas drilling techniques?

In this context, a future scenario with a large part of electricity produced with domestic shale gas must be explored. A fall in the gas price due to a large production of shale gas in France, as well as a fall in the coal price on the international market, as it is now the case due to large exports coming from the United States, could make the competitiveness of nuclear energy questionable.[16] Under a low fossil fuel and carbon price scenario, nuclear loses much of its attractiveness.

The report published in January 2012 by the French Cour des Comptes has concluded that the current price of the French nuclear generation, including reactor dismantling and nuclear waste management costs, remains competitive with the alternative options, at least at currently high international oil and gas prices. However, the cost of nuclear energy is expected to rise in the future because new investments will be required. Consequently, the regulated cost-reflective price set by the government would also increase. This will justify efforts to engage in energy efficiency and social measures to help people said to be suffering from fuel poverty. Whatever the option chosen, the future price of electricity will rise for the end user, because all segments of the electricity chain will become more costly: investments will be required in the generation sector to improve the safety of the existing reactors or replace them, investments will also be necessary to develop transmission and distribution infrastructures and to develop European interconnections and new subsidies will be necessary to promote wind and solar energies, topics

---

16. The price of coal has fallen from \$120 to \$83 per ton between January and June 2012 on the international market, and the price of CO_2 remains low, currently around €7 per ton.

TABLE 3.8 French Electricity Costs (Euros/MWh, 2012—networks tolls and taxes excluded)

Power Station Types	Euros/MWh
Hydroelectricity	30–40
Nuclear (PWR) (existing fleet)	40–50 (75 for new nuclear EPR)
Coal ($85/ton)	55–70
Natural gas (combined cycles) ($12/MBtu)	70–80
Onshore wind	80 (FIT)
Offshore wind	200 (FIT)
Solar (photovoltaic)	250–400 (FIT)

Source: "Energies 2050" Report (2012).

extensively described in companion chapters in this volume, including Bauknecht et al.

By all indications, however, France should be able to maintain its current comparative advantage in the nuclear field option relative to other European countries as illustrated in Table 3.8 unless oil and natural gas prices collapse. In June 2012, the cost of a megawatt-hour produced with a gas combined cycle power station was about 70 euros compared to 50 euros with a pressurized water reactor (PWR) nuclear power station; the gas was bought at around $12/MBtu. With a coal power station, the cost was 55–60 euros on the basis coal at $90 per ton. Table 3.8 provides a comparison of alternatives and their currently estimated costs.

7.2 The Debate About the Subsidization of Renewable Electricity

As described earlier, in a competitive market, power plants are normally dispatched in merit order, with the least cost options first, the most expensive last. A notable exception to this principle, however, applies to renewable energies. According to prevailing European law, renewable energies are given a priority on the grid and must be paid high FITs set by the government. These high FITs are justified in view of the high cost of noncompetitive new technologies, which would otherwise not be able to compete with the prevailing market prices. In case of France, the following FITs apply in 2012:

- €82/MWh for onshore wind,
- €200/MWh for offshore wind,
- €250–400/MWh for photovoltaics.

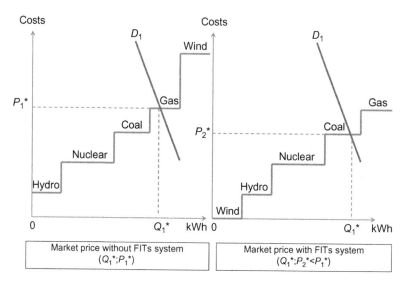

FIGURE 3.6 Illustration of perverse effects due to the implementation of a FITs system for renewables. The market price is no longer cost-reflective as the supply curve shifts to the right.

The priority given to renewables has a perverse effect because on the spot market the marginal cost of renewables is zero and consequently the merit order curve shifts to the right (Figure 3.6), which reduces the market clearing price and jeopardizes the profitability of natural gas power stations. This phenomenon is described in more detail in Chapters 5 and 7.

Needless to say, renewables are expensive compared to the average spot market prices which tend to be around €55/MWh during off-peak and €75/MWh during peak periods.[17] It is also necessary to take into account the costs associated with providing the back-up power to intermittent renewables. During periods without any wind or sun, it is necessary to keep gas power stations in operation and consequently the load factor of these gases, and sometimes coal, power stations remains relatively low, which is costly. The extra cost of the wind and solar generation is shared among all customers via a surcharge.

As described in several companion chapters in this volume including Chapter 25, a system with large percentage of intermittent renewable output may also experience periods of negative prices on the spot market. The emergence of negative prices for electricity on the European spot market is

---

17. Of course, electricity prices on the European day-ahead spot market are highly volatile: the market anticipates the failure costs when electricity demand is high and there is a risk of capacity shortages. During heat waves or extremely cold periods, market prices can spike to reach €3000/MWh, and occasionally more.

the consequence of the rapid development of wind power, notably in Germany, as a result of:

- low demand for electricity during off-peak periods, typically nights and early morning hours and
- strong winds which run the wind turbines at full capacity.

It is important to remember that wind power has priority on the network with purchase obligation at lucrative-guaranteed FITs. Electricity is not a product like any other since it cannot be easily stored and because supply must constantly match the changing demand, so as not to risk power failure. This has resulted in a windfall gain for dam owners with large reservoirs, particularly in Switzerland, who can store the excess electricity produced by wind. The hydro operators are paid to store the surplus electricity, which they can use to fill the reservoirs. The stored energy can then produce electricity at peak times when the spot price is high, selling it to Italian consumers, among others. Curtailing wind-generated power is prohibited by law; stopping some thermal power plants for short periods is too costly and consequently it is better to sell this renewable electricity at negative prices during certain periods. This phenomenon has been observed on numerous occasions since 2009 in Germany and once in France at the beginning of 2012. During these episodes, the customer who can store energy is paid to "buy" electricity.

7.3 The Debate About Peaking Capacity

Investing in peaking plants is costly because the capacity is infrequently used, typically for a small number of hours a year.[18] Peaking units are usually combustion turbines running on gas or oil, making them environmentally detrimental because they emit high levels of CO_2 relative to baseload plants, especially nuclear with zero emissions.

The problem has become more pronounced in liberalized markets due to the lack of sufficient incentives to invest in peak capacity since the profitability of such investments is deemed to be risky, yet is critical to the stability and reliability of the grid. Previously, the incumbent, fearing power failure, tended to *overinvest* knowing that the costs would be recovered in regulated prices. In a liberalized market, it is the market that sets the price during peak times and the operators take a risk investing in costly equipment, which may not be required. Consequently, the investors under-invest in peaking capacity in the hope that the deficiency in supply will increase the market prices. They will thus be rewarded for their unscrupulous behavior. In the best case, there is a problem of "free

---

18. For a discussion, refer to Chapter 9.

riding," where each operator relies on its competitors to ensure that the peak demand is met.[19]

Today the regulator has to implement a system, which correctly remunerates the peak capacity, even if this capacity is not required on the grid, regardless of the equilibrium market price. It then raises the problem of "missing revenue," which is explained by the fact that pricing based on the marginal costs or the prices observed on the market may be insufficient to ensure correct remuneration of the peak capacity. Some believe that peak management must be considered as a public service and that the system operator, RTE, who is in charge of transmission network in France, should undertake such investments. Others think that each supplier should be obliged to contribute to the balancing of supply and demand by providing either additional physical capacities or a cancellation portfolio. In case of breach of such an obligation, penalties could be imposed. The French legislation has opted for a capacity obligation imposed on each supplier.[20]

8 CONCLUSIONS

As described in the body of the chapter, ARENH is the consequence of policy decisions seeking to promote market competition while recognizing that prices remain differentiated based on national cost differentials, in this case, low-cost French nuclear power. Had such a solution been ruled out, it would have been necessary to deregulate all electricity prices in France, which would have resulted in French prices *rising* to the average European levels. This is why the NOME law allows the competitors to buy a part of the off-peak electricity generation from EDF at a regulated price set by the government and correlated to the historical cost of the French nuclear fleet.

This compromise allows the present dual system for electricity pricing for French consumers to remain: regulated tariffs fixed by the government and set according to the production cost of the nuclear energy, and market prices indexed to the prices observed on the European wholesale electricity market. The domestic end users have the option to choose the one that best suits them. But for how long would this compromise solution last?

It is necessary to dissociate ARENH and the average cost of nuclear energy as recently calculated by the French Court of Auditors. ARENH is the actual cost for EDF today. The average cost calculated by the Court is the cost estimated for the long run for the French end user, taking into account the amortized capital costs paid through past electricity tariffs.

---

19. The 2010 NOME law, already mentioned, offers EDF's competitors the right to buy part of the nuclear electricity at a regulated low price, and has, in turn, obliged these competitors to provide the capacities necessary to meet the demand of their customers.
20. Bowring (Chapter 9) describes how PJM manages this issue through a capacity market.

And this explains why in France it is necessary to accept the "distorted" market rules, at least for the time being. Other European countries, however, have also introduced some distortions on the energy market. The unilateral decision by Germany to phase out their nuclear fleet, for example, is a big blow to the establishment of a common energy market in Europe. This move will make it even more difficult to achieve price convergence for all final consumers in Europe. *It is a contradiction to ask the market to set the price for the end user and at the same time to leave national policy makers to choose the energy mix.*

The future of nuclear energy should not be questioned in France and no one expects a total phase-out of nuclear power. Unfortunately, a reduction of the proportion of nuclear energy in the French electricity mix should be expected. How much and when is hard to tell since the final decision is essentially a political one. It is, however, important to point out that the relatively low electricity prices offer a competitive advantage to the French industry.

REFERENCES

Bouneau, C., Derdevet, M., Percebois, J., 2007. Les Reseaux Electriques au Coeur de la Civilisation Industrielle. Préface Andris Piebalgs, Timée Editions, Paris, pp. 175.

Champsaur, P., 2011. Rapport de la Commission sur le Prix de l'ARENH. DGEC, Ministry of Energy, Paris (March).

Chevalier, J.M., Percebois, J., 2008. Gaz et Electricite: Un Enjeu pour l'Europe et Pour la France. Report for the Prime Minister (C.A.E.), La Documentation Française, Paris.

Cour des Comptes, 2012. Les Coûts de la Filière Electronucléaire. Rapport Public, Paris, <www.ccomptes.fr>.

Energies 2050 Report, 2012. President: Jacques Percebois, Vice-President: Claude Mandil, C.A.S. and D.G.E.C, Ministry of Energy, Paris, February, pp. 550.

Hansen, J.P., Percebois, J., 2010. Energie: Economie et Politiques. Editions de Boeck, Préface de Marcel Boiteux et Avant-propos de Jean Tirole, pp. 780.

IEA, 2011. World Energy Outlook 2011. OECD/IEA, Paris.

Percebois, J., 1989. Economie de l'Energie. Préface de Yves Mainguy, (Ed.), Economica, Paris, pp. 689.

Percebois, J., 2008. Electricity liberalization in the European Union: balancing benefits and risks. Energy J. 29 (1), 1–19.

Percebois, J., Wright, P., 2001. Electricity consumer under the state and the market: French and U.K. electricity supply industries. Utility Policy 10 (3), 167–179.

Stoft, S., 2002. Power System Economics: Designing Markets for Electricity. MIT Press, Cambridge, MA.

Chapter 4

Turnaround in Rough Sea—Electricity Market in Germany

Wolfgang Pfaffenberger[a] and Esther Chrischilles[b]
[a]Professor of Economics, Jacobs University, [b]Research assistant, Cologne Institute for Economic Research

1 INTRODUCTION

With the passage of the energy law of 2005, the electricity market in Germany underwent an additional reform after the reform of 1998 initiated by the European Union (EU) as described by Brunekreeft and Bauknecht (2006). Network access and pricing were subjected to the new regulation; the wholesale market developed strongly and the prices on the European Energy Exchange (EEX) took on a leading role. Renewable energy was promoted by a special feed in law that guaranteed high prices for qualifying renewable energies while at the same time guaranteed the sale of renewable electricity produced. Figure 4.1 shows that about 80% of electricity generated in Germany in 2010 came from fossil and nuclear sources.

Prior to the enactment of the 2005 reforms, the Bundestag had decided in 2000 to gradually abandon nuclear power with all nuclear plants scheduled to close by 2022. In 2010, concerns about meeting the country's greenhouse gas emission targets, the high cost of renewable energy, and the problems associated with integrating intermittent renewable sources into the network led the government to decide to prolong the operating time of several nuclear plants. With the introduction of a new tax on nuclear fuel, the government wanted to channel some part of the expected windfall profits from nuclear generation into the federal budget.

After the nuclear accident in Japan in 2011, however, the government decided to immediately close down all eight older reactors without a proper legal basis, and later in the same year, legislation was adjusted to close these reactors down permanently. For the remaining nine reactors, the closure date adopted was similar as in the 2000 policy decision and depends on production relative to a maximum defined in the law.

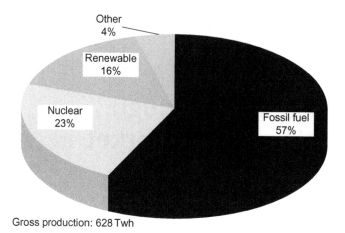

Gross production: 628 Twh

FIGURE 4.1 German electricity generation mix in 2010, in percentage. *Source: BMWI, Energiedaten (2012).*

A short-term decision on a long-term issue may be "politically expedient;" however, it could create follow-up problems that were not carefully considered at the time of the decision making in March 2011.

This chapter covers the ramifications of the so-called *Energiewende* or *turnaround*, which has become the dominant issue in German energy sector eclipsing all else. Section 2 gives an overview of the electricity market in Germany; Section 3 explains the Energiewende in more detail. Nuclear energy is the most important carbon-free baseload electricity available at relatively low cost; therefore, the question arises of how this resource can be replaced within a relatively short time frame and what this implies for the electricity system as a whole as well as the impact of the nuclear phase out on German economy.

Section 4 explains the implications of this *turnaround* for generation and transmission of electricity, security of supply, competition, as well as financial issues. Location of generation and consumption will be further apart in the future. The implications of this change are also dealt with by Knieps in Chapter 6. The *turnaround* to a considerably higher share of often intermittent renewables leads to a number of additional problems and is described in Chapters 7 and 25, covered by Bauknecht et al. and Riesz et al., respectively, and others.

Section 5 looks at implications for the German economy. It is expected that electricity prices will rise due to the higher share of renewables and/or the higher cost of CO_2 certificates if more fossil fuel has to be used for electricity generation.

Section 6 explains the policy implications of the government's political decision. As described by Boltz in Chapter 8, within the EU, there is an agreement on a number of policy objectives, and the question is whether the German economy can reach the agreed targets. The lessons that can be learned from the German experience are discussed in the conclusion.

TABLE 4.1 Basic Data on the German Electricity Market, 2010

Basic Data for Germany		2010
Population	Million	81.8
GDP (2005 prices)	Billion €	2369
Primary energy consumption	Petajoule	14,044
Electricity		
Installed capacity gross	GW	170
Therein renewable	GW	60
Therein nuclear	GW	21.5
Production gross	TWh	628
Therein renewable	TWh	101
Therein nuclear	TWh	140
Consumption gross	TWh	610
Consumption per capita	kWh/person	7457
Consumption per GDP	kWh/1000€	257

Source: BMWi (2012)

2 THE ELECTRICITY MARKET IN GERMANY

2.1 Basic Information

The German electricity market is the largest within the EU (Table 4.1). It is connected to all neighboring countries and there are considerable power flows between countries. The technical integration of the central European network was achieved by the Transmission Service Organizations (TSOs) long before the market opening in Europe. The market integration, of course, began with the introduction of the European reforms.

Figure 4.2 shows the map of Germany with the location of major power stations and the transmission grid.[1]

Figure 4.3 shows the footprint of the four transmission grid operators or TSOs in Germany with the length of the transmission grid indicated in Table 4.2.

---

1. The size of the circles in the map indicates the size of generation units. Only units ≥100 MW are included. The map does not show small renewable units. Further information on the location of renewable plants is provided in Section 4.

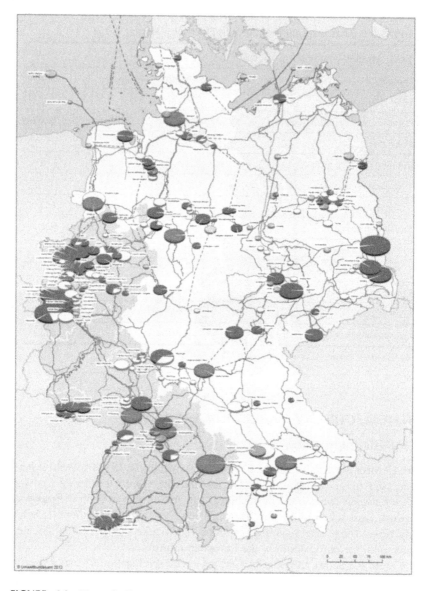

FIGURE 4.2 Map of German power stations and high-voltage transmission system. *Source: Umweltbundesamt (2012).*

The first phase of market opening in the German power market may be found in Brunekreeft and Bauknech (2006). After the reform of 2005, many additional changes took place. Below, an overview of the state of the industry before the *turnaround* is presented with the implications of the *turnaround* for the industry covered in Sections 3 and 4.

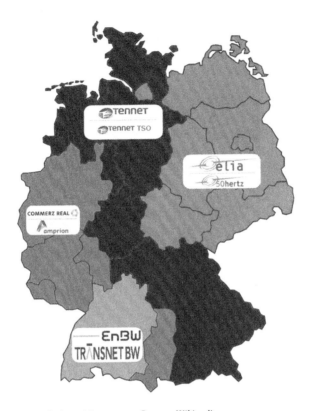

FIGURE 4.3   Transmission grid operators. *Source: Wikipedia.*

TABLE 4.2 Length of 220 and 380 kV Network

Amprion	EnBW	Tennet	50 Hz
11.000 km	3.674 km	10.667 km	9.750 km

Source: Wikipedia

2.2 Legal Framework

The passage of the 2005 law introduced a new regulatory agency as well as new regulation covering network pricing. Legal unbundling of transport and distribution networks from other activities of integrated companies and the regulation of network access conditions including network pricing created the conditions for competition in the wholesale as well as the retail market. There are no network prices for power fed into the network; therefore, the location of plants is not influenced by network prices.

The preliberalized market had considerable surplus of network capacity, so that the allocation of plants was, to some extent, independent of network considerations. This has changed due to the increase in the share of renewables and the influence of transport cost of fuels on the location of fossil fuel plants in an unbundled system. Presently, TSOs have to connect new plants without any discrimination, but at the same time cannot guarantee that new plants can feed in their production when the network is congested.

The changes in location and structure of production need to be mirrored in the development of the transport grid. The legal framework obliges the TSOs to cooperate and present a long-term development plan to be agreed upon by the regulatory agency. The main mechanism for the coordination between production and transport infrastructure is a political/administrative one not necessarily driven by economic considerations, as proposed by Knieps in Chapter 6.

2.3 Generation

Figure 4.4A and B shows the development of capacity and production of electricity for the period 2005–2010. It can be seen that capacity in renewable generation has increased considerably—from 32 to 60 GW within 5 years—whereas total production remained almost the same.

Table 4.3 shows that in 2010, 44% of total capacity in the renewable segment produced only 17% of total electricity.[2] The load factor in the renewable segment fell due to wind and photovoltaic sources increasing their share in capacity and due to climatic conditions that resulted in relatively low contributions to production. The capacity share of nuclear and fossil plants fell from 87% to about 80% in 2010. During this period, fossil fuels and nuclear generation dominated German power generation. These plants serve as a backbone to keep the system stable.

The traditional segment of generation is dominated by four large companies with a balanced portfolio of plants—Eon, RWE, Vattenfall Europe, and EnBW. In addition, a number of larger municipal plants produce part of their electricity need in local plants, often using heat from thermal plants—combined heat and power or CHP—for their district heating systems. A small new segment of small-scale decentralized plants, outside the renewable segment, is also emerging. However, decentralized plants face economic difficulties unless they can make use of financial support by producing heat and electricity in CHP mode.

The renewable segment produces at fixed prices and has priority for feeding renewable power into the network whenever it is available. The nonrenewable segment has to bid its output in the open market, at variable prices, and has to adjust its production according to the quantities produced by renewables.

---

2. This is among the characteristics of many renewable technologies, that is, a relatively low capacity factor.

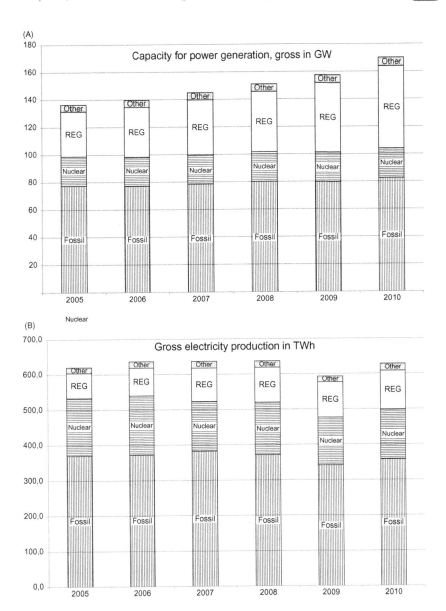

FIGURE 4.4 Capacity (A) and production (B) in Germany, 2005–2010, in GW and TWh, respectively (REG = renewables hydro, wind, biomass, and photovoltaic). *Source: BMWi, Energiedaten (2012).*

The European Energy Exchange (EEX) has assumed a leading role in setting wholesale prices for spot and futures transactions. Also the over-the-counter transactions follow EEX wholesale prices, as further described in Chapter 8 by Boltz, describing the preferential treatment of renewable generation.

TABLE 4.3 Shares of Renewables in Capacity and Production and Capacity Use of Renewables in h/Year

	2005	2006	2007	2008	2009	2010
Capacity (%)	24	26	29	32	37	44
Production (%)	11	12	15	15	17	17
Capacity use (h/year)	2164	2164	2373	2245	1980	1825

Source: BMWi Energiedaten (2012); own calculation

2.4 Transmission

The German high-voltage grid was originally operated by the four large integrated utility companies, Eon, RWE, Vattenfall Europe, and EnBW. Currently three of these are partly or fully unbundled from their original parent companies as described in Section 2.1. Network prices are regulated by the federal agency. The obstacles to competition in the first phase of market opening of 1998 are thus substantially removed.

The main concerns regarding the German transmission infrastructure now lie in two areas:

1. Strengthening the transmission infrastructure to better cope with power flows between European countries and improving the possibilities of trade.
2. Improving and extending the network to deal with the rising share of renewable energy and the fluctuations caused by intermittent renewable resources, especially wind.

In both areas, a considerable amount of investment will be needed in the future and the regulator is under pressure to allow rates of return that are attractive to the capital market.

2.5 Distribution and Retail

A large share of the distribution networks in Germany is operated by local companies, which often also serve as retailers. Legal unbundling of retail and distribution is only required for larger companies. The network prices of all distributors are regulated. Currently, consumers can make use of independent retailers or buy power from other suppliers.

In the first phase of market opening, municipalities often sold at least part of their utility companies to large integrated utilities. In contrast, there is now a tendency to try to buy back the local networks to gain influence in the local energy market. This local influence, however, may be mostly an illusion because of the high degree of regulation applied to unbundled networks (Menges and Müller-Kirchenbauer, 2012).

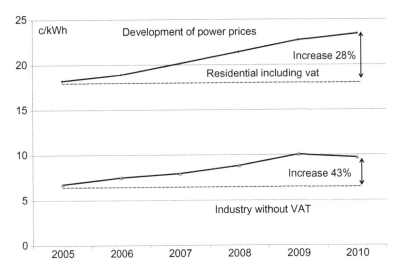

FIGURE 4.5   Development of German power prices in the residential and industrial sector, 2005–2010, cents/kWh. *Source: BMWi (2012).*

2.6 Performance: Prices, Environment, Competition

Figure 4.5 shows the development of power prices for residential and industrial customers for the period 2005–2010. The price increase during this period was considerably larger than the moderate rate of inflation. Prices increased because of the introduction of the European CO_2 certificate scheme in 2005, the general increase in the price of fuels and, most notably, the increase in renewable generation. The additional cost of renewables is financed by a surcharge levied by the TSOs.

The relative price increase was higher for industrial customers because the high taxes for residential customers and the higher network prices for distribution in the low voltage grid reduce the relative importance of the power price for residential customers. Industrial consumers with high specific electricity cost are partly exempt from part of the surcharge for renewables; but because of the small number of companies eligible for this exemption, this does not significantly influence the average figures presented in the diagram.

Table 4.4 shows the emission of CO_2 associated with the production of electricity, currently the most important source of these emissions in Germany. Due to the age distribution of fossil fuel plants, the efficiency of these plants does not match the modern standards, which keeps emissions high. This was partly compensated by the carbon-free production from nuclear and renewable plants. The original government policy was to reduce future carbon emissions by substituting renewables for coal while keeping nuclear generation for a longer time. This idea was abandoned in 2011 with the *turnaround* policy as further described in Section 4.

TABLE 4.4 CO_2 Emission of Power Production in Million ton CO_2

2005	2006	2007	2008	2009	2010
325	330	340	319	294	302

Source: Umweltbundesamt (2012).

Due to the high industrial concentration in generation according to Monopolkommission (2011)—an independent advisory body on competition policy—competition has only gradually improved in the wholesale and retail segment of the market. The commission has proposed a number of additional measures especially relating to regulation and the activities of the cartel office to improve the conditions of competition. The commission also has criticized the forms of promotion of renewables, demanding a more market-oriented approach. The arguments regarding industrial concentration lost a lot of relevance after the decision in 2011 to abandon nuclear energy because the market power of companies with a relevant nuclear portfolio has been drastically reduced. The argument regarding the market orientation of renewables has become even more important for the future as discussed in Section 4 and also in Chapter 7 by Bauknecht et al.

3 THE ENERGIEWENDE 2011

The capacity of German nuclear plants in 2011 was about 20 GW. In March 2011, the federal government decided to close 8 of the 17 plants with a total capacity of about 8 GW. The remaining nine plants will be closed step by step from 2015 to 2022.

This move was motivated by a strong antinuclear sentiment after the Fukushima accident in Japan and is further described in Chapter 23 by Asano and Goto. The German nuclear law allows direct intervention by the federal government in case of imminent danger from any nuclear plant. However, the accident in Japan caused by an earthquake and a tsunami did not change the risk of operation in any German nuclear plant. All German plants at that time had a valid license as long as they conformed to the prevailing safety standards. To suspend the operating licenses of the eight affected plants was definitely outside of the legal framework in place in March 2011. Later in the year, the legal framework was amended. Three of the four affected companies are now claiming damages; the fourth company EnBW, which is 100% state owned, probably hopes to profit from the legal claims of the others if they prevail in the courts.

Figure 4.6 shows the effect of the nuclear phase out compared to the production of 2010. In 2011, due to the mild climate, electricity demand fell relative to 2010. The considerable loss of nuclear potential production could

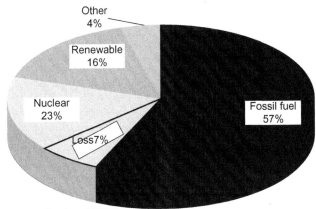

Gross production: 628 Twh

FIGURE 4.6 Production of 2010 with closure of eight nuclear plants. *Source: Own calculation on the basis of 2010 data by BMWi (2012).*

also be compensated by an increase in the production of renewables and a reduction of exports. Due to its central geographic location within Europe, a lot of electricity flows through Germany. In recent years, Germany had an *export surplus* of between 15 and 20 TWh; however, in 2011 this was reduced to 6 TWh and the country is likely to become a *net importer* as a result of the nuclear phase out.

Moreover, due to the concentration of heavy industry with a high dependence on nuclear electricity in some southern regions of Germany, there were fears that a shortage of electricity, together with a lack of network capacity, could lead to critical situations from time to time. Clearly, the present grid was not built for large amounts to be transported from the north to the south of Germany. According to information from TenneT—one of the four German TSOs—the number of critical situations experienced in their grid alone was more than 1000 in 2011, nearly three episodes per day, and three times higher than in 2010. Due to special efforts undertaken by the TSOs, including the delay of necessary maintenance and other operational measures, no breakdowns have occurred in the high-voltage grid to date. Yet, it is clear that the grid will experience high stress as a result of the new requirements necessitated by the *turnaround*.

Figure 4.7 shows the effect of phasing out the remaining nuclear plants on electricity production. By 2022, about 100 TWh of nuclear electricity will have to be replaced by other generation sources, imports, or by a reduction in demand through energy efficiency gains.[3]

---

3. For each plant, a maximum production for the remaining time has been defined, the exact time pattern depends on the performance of the plants.

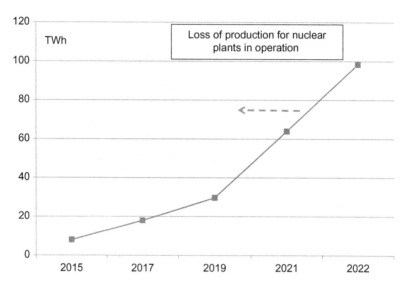

FIGURE 4.7 Loss of German electricity production due to the nuclear phase out, 2015–2022, in TWh. *Source: Own calculation on the basis of 2010 data and the amendment of the nuclear law.*

4 FUTURE IMPLICATIONS OF THE *TURNAROUND*

As already mentioned, in 2000, the German parliament decided to gradually abandon nuclear energy by 2022 by setting a ceiling on the production of each nuclear plant. At the time, the government negotiated an agreement with the industry so that there would be no legal claim for damages from revoking the original unlimited license for operating the plants. The plan called for plant closures to begin around 2012 and end about 2022. At the time, it was assumed that there would be sufficient time to make a smooth transformation away from nuclear power.

Until 2005, the government target for the reduction of CO_2 in Germany was 25% relative to 1990 (DPG, 2005). This target, never realistic, was silently abandoned in 2005 in favor of the Kyoto target, which required a 21% reduction for all greenhouse gas emissions by 2012. This target has been achieved; but it is not clear what will happen after the nuclear phase out. In this context, finding a generation mix that meets the country's environmental goals without stifling the German economy, and acceptable to the public, have been drastically reduced as shown in Figure 4.8. A similar debate is taking place in Japan, where the future of nuclear energy is being decided as described in Chapter 23 by Asano and Goto.

4.1 Generation

At the time of the original nuclear phase out in 2000, there may have been lingering hopes within the industry that a future parliament might revise the

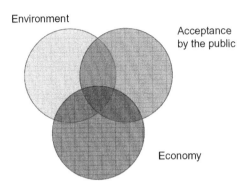

FIGURE 4.8 Scope for decision on future energy mix. *Source: Pfaffenberger and Hille (2004).*

decision on nuclear energy. This wishful thinking resulted in reluctance on the part of the industry to invest in replacement capacity and lost time. In 2000, a revised renewable energy law was passed and later revised several times (EEG, 2009). Investment in renewable energy plants was boosted and there were hopes for a fast replacement of nuclear energy by biomass and offshore wind energy.

There are, however, serious limits to increased reliance on biomass due to competition of land use for food production. Similarly, offshore wind has proved to be much more technically and financially challenging than originally expected, leading to delays and rising costs. The long distance to shore, depth of the water, environmental restrictions, capacity restrictions for special equipment, high network cost and high production cost for electricity are among the main reasons for the slow development of offshore wind despite the potential size of this option, including higher load factors compared to onshore wind.

As further described in other chapters, the growing share of intermittent renewable sources changes the role of thermal power plants. They must compensate for the fluctuations in renewable generation to keep the system stable. From a long-term perspective, conventional plants must be very flexible and at certain times, when renewable production overshoots demand, electricity must be stored.

The transformation of the system thus does not only have to increase the share of renewables but also have to completely change the structure of conventional generation. An early analysis of the impacts may be found in Krämer (2003).

To encourage further growth of renewables, special provisions have been made to minimize the risk to investors, including priority to be fed into the grid. Conventional thermal plants, on the other hand, have to adjust their output continuously to allow virtually all available renewable generation to be absorbed. This reduces incentive to invest in conventional plants, whose value is determined by market-determined prices on the spot and futures market. Under a scenario of continuous increase in renewable generation, market price signals do not support investment in conventional plant because market price is determined by marginal cost of existing plants. These do not support the full cost of new plants.

One of the problems to be solved in the course of the *turnaround* is to change the market rules such that system stability is maintained through appropriate price incentives. Different proposals are currently being discussed. In this context, it is important to remember that the EU has created an internal market for goods and services. Therefore, the question arises whether the market for system stability should be organized on a national or a European scale (Maurer et al., 2012), as also described in Chapter 8 by Boltz.

On a European scale, the need for reserves will probably be smaller because of time-dependent demand differences among countries as well as the differences in the portfolio of generation units and their availability. This requires sufficient transborder capacity in the high-voltage grid, which must be further enhanced. A pan-European approach is probably the more economic solution in the long run, but it requires a lot of institutional changes, not easily or quickly implemented. It is also doubtful whether national governments are willing to risk security of supply problems by giving up the possibility of intervention in the national electricity system.

It is, however, fair to say that due to the scale of the German power sector, in the long run, the *German turnaround* is likely to become a *European turnaround*. Currently, due to overproduction of renewable energy at certain times that lead to unplanned flows into neighboring countries, neighboring countries must take actions to protect their grids, a reality reflected in observation below:

Germany's nuclear shutdown aggravates the grid spill-over problem for its neighbours. Several of the closed reactors are in southern and south-west Germany which is now more dependent on increasing amounts of wind power from northern and eastern Germany. But there is congestion on internal German transmission lines. Because electrons follow the path of least resistance, wind power has been spilling over in 'loop flows' into Poland and the Czech republic in the east and the Netherlands in the west. These countries have now placed so-called phase shifting transformers on their borders with Germany to prevent their grids being disrupted.

(Buchan, 2012, p. 5)

4.2 Transmission

Within Germany, the transport grid is one of the relevant bottlenecks for a successful *turnaround*. Due to natural geographical factors, wind energy is concentrated in the northern part of Germany, especially near the coast of the North Sea and the Baltic Sea. Figure 4.9 shows the density of current installed capacity in MW/km^2 clearly indicating that renewable capacity is much more concentrated in the north.[4]

---

4. The map does not include offshore plants. It also does not indicate the amount of electricity produced, which is higher in coastal areas due to higher load factors and stronger wind regimes.

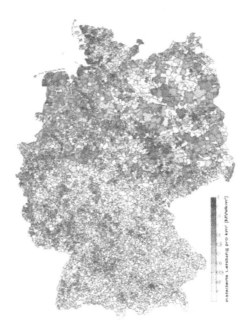

FIGURE 4.9   Wind energy installations in Germany in 2009, in MW/km². *Source: dena/IWES.*

Moving forward, the distance between electricity producing and consuming regions increases with the rising share of renewables produced in more distant regions and also increases considerably in regions where large nuclear installations have been closed.

Figure 4.10 shows the regional imbalances that are expected for 2020 due to the increase of renewable production and the expected closure of nuclear plants as originally planned in 2000.[5]

Given these realities, the four TSOs have proposed to make substantial adjustments in the following areas (the dotted lines in Figure 4.2 show some of the planned extensions):

- Add four high-voltage DC lines for long-distance transport from North to South (between 2000 and 3000 km depending on time horizon and demand development).
- Add about 1500 km of new high-voltage AC lines.
- Improve and extend about 4000 km of existing lines (Netzentwicklungsplan, 2012).

---

5. The additional effects of the decision taken in March 2011 to close the eight older nuclear plants are not considered (dena, 2010).

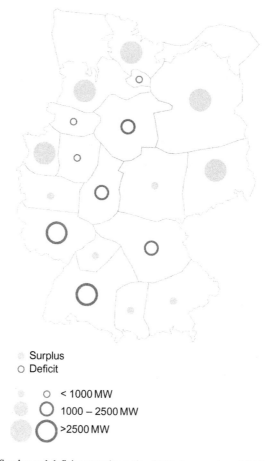

Surplus
○ Deficit

○ < 1000 MW
○ 1000 – 2500 MW
○ >2500 MW

FIGURE 4.10   Surplus and deficit power by region 2020. *Source: dena (2010).*

The investment required for the next 10 years is estimated to be approximately €20 Billion.[6] Clearly, limitations in the transmission grid could become a serious bottleneck to the successful implementation of the *turnaround* as observed below:

Extending the grid is an essential precondition for the success of the turnaround. The speed of improvement of the grid determines the speed of the turnaround. If renewable energy continues to grow faster than the grid develops, the targets of the energy turnaround and the security of supply are endangered.

(Netzentwicklungsplan, 2012, p. 149) (Own translation).

---

6. This does not include the investments to connect offshore wind converters. The estimated investment required for offshore connections, according to reports in the media, is of the same order of magnitude.

Whether the plans for the improvement and extension of the grid can be implemented and whether the offshore connections can be installed in time depends on:

- The compatibility with other legal requirements regarding infrastructure projects, protection of nature and environment, as well as other issues.
- The time-consuming procedures for getting various approval, permits, and for allowing the participation of local population.
- Competence and cooperation of authorities on three levels of government, federal, state and local government, and their coordination.
- Ability to finance the necessary investments in a timely manner.

To expedite things, changes to the legal framework have been introduced including a new law passed in 2011 to change the authority for building long-distance power lines (NABEG, 2011). For interstate power lines, the procedures for approval are now with the federal agency for network regulation (Bundesnetzagentur). It is hoped that the time-consuming approval procedures, currently taking as long as 10 years, can be shortened and streamlined.[7]

4.3 Distribution

Traditionally, the distribution system was built in a hierarchical way: from source, typically a transformer connecting to the grid, to point of destination, industrial or residential consumer. With a rising share of distributed generation, this historical paradigm is obviously changing. Instead of the hierarchy with a definite direction of flow, power flows both ways. The control of such networks needs information systems and control devices that allow decentralized units to feed into the system, while at the same time, keep the system stable. This raises a number of technical, institutional, and economic questions. For example, the rights and duties of independent producers in

---

7. For large infrastructure projects, environmental impact analysis must be done and the input of local residents considered. The network development plan of the TSOs gives an outline of the necessary improvements and extensions but does not go into the details of the exact routes for new power lines. Once details of the routes are made public, significant opposition can be expected. Acceptance for power lines in the public is very low due to widespread fear of electromagnetic radiation. It is now also popular to propose the use of high-voltage subterrestrial cables rather than overhead wires, which are considerably more expensive.

For offshore wind, the main change was to oblige the TSO adjacent to offshore wind and to connect all offshore plants (§17 Abs. 2a EnWG). This has increased the incentive for investors of wind energy converters because the high cost of network connections will have to be paid by the TSO and thus the network users, not plant owners. This puts a large financial burden on the TSO (see Section 4.5) and creates complex liability problems. Legislation is now proposed to remunerate plant owners if electricity cannot be transported from offshore plants due to disturbances on the offshore network. In plain language, that means that all electricity consumers will pay for wind energy *not* produced through the network charges.

relation to the operator of the distribution network must be defined and the present legal framework needs to be adjusted.

At the distribution level, significant investment in hardware, software, and education is necessary to change from the hierarchical, one-way flow to network with bidirectional flows. The investment required to upgrade the more than 1 million km German distribution network is estimated to be around €21−27 billion (Küffner, 2012).

Economic questions also relate to appropriate price incentives for local generators. At present, the promotion of renewables and CHP is based on fixed prices for the former and a premium payment scheme for the latter, which are independent of location. For a distributor, the proper incentives must encourage optimization of the intake of power from external sources coupled with those from distributed sources while recognizing the cost of network investments and operations. For such an optimization to take place, the TSO needs instruments that also relate to the activity of independent producers or virtual power stations. Fixed feed in prices or premiums cannot accomplish this task under a scenario where the share of distributed generation is increasing, as is likely to be the case. It may be that the distribution network has to be extended to cope with the task. The present market regime does not lead to an economic optimum, which balances network costs against production costs. In the course of the *turnaround*, these questions need to be addressed so that potential investors receive clear and effective incentives.

4.4 Competition and Markets

The present system for promoting renewables in Germany is challenged on at least three arguments:

1. The first challenge is on legal grounds: The present system of promotion is based on fixed prices for renewables combined with an absolute merit order priority, which effectively reduces the market risk for producers of renewables. The extra cost is financed by a levy charged by the TSOs to the consumers. This is an interesting form of direct government intervention without the use of government funds. This, however, violates the general rule implicit in the German constitution, which broadly states that activities promoted by a specific law and benefiting the general public should be financed by the public budget and not by levies to specific groups. Financing renewables from public budgets would also have the advantage that the distribution effects would be alleviated. Public financing of renewables also contradicts European rules against the use of subsidies and is therefore unlikely to be the solution for the future.

2. The second challenge has to do with European market rules on trade: The present system constitutes a barrier for trade and thus violates the rules for the internal European market. This is well known, but as of yet, no

political agreement has been reached between the member countries for a harmonized scheme for the promotion of renewables.
3. The third challenge has to do with market regime: As Kopp et al. (2012) have shown, it is unlikely that many renewables can be integrated into the market regime by letting them supply to the wholesale spot and futures market. Because marginal cost of many renewables is almost zero, whenever renewables set the market price, the price will be essentially zero. In a regime with large volumes of renewables, the wholesale price would tend to be close to zero for considerable hours of the year. Under such circumstances, there would be no incentives to invest in plants, a problem also described in Chapters 5 and 25.

The solution to a market-oriented scheme for promotion of renewables is a combination of a quota system with elements of a capacity market. This could also solve the legal problem. In such a system, the government establishes the rules, but no levies are necessary. Renewables earn the market price plus a premium. The premium itself depends on the obligation set by government rules as further described in Chapter 7. For a discussion of capacity markets, also refer to Cramton and Ockenfels (2012) and Maurer et al. (2012).

The transformation of the subsidy scheme for promotion of renewables will take time to implement. The present system has the strong support of a powerful lobby because it benefits many investors.

4.5 Finance

It goes without saying that the *turnaround* also requires large amounts of investment in generation, transmission, and the network. For renewable energy plants, typically large investment is needed in the beginning followed by low operation costs because the energy source is free. This increases the relative share of capital expenditure and thus increases investment risks (Figure 4.11). It remains to be seen under what conditions the capital market is willing to finance large-scale investment in renewable energy and for how long.

Additionally, the financing of grid extensions, particularly grid connections for offshore wind parks, may delay the *turnaround*. The large amount of investment required in the near future will probably not be possible without a considerable share of the risk secured by government guarantees. The political decision to implement the *turnaround* puts the capital market in a strong position. It is rather ironic that, in the midst of the current financial crisis, when capital is abundant and interest rates are at historically lows levels seldom seen, the financing of politically desired infrastructure investments should pose a problem. However, investors also look at the technical risks that are gradually becoming apparent, and thus ask for risk premiums

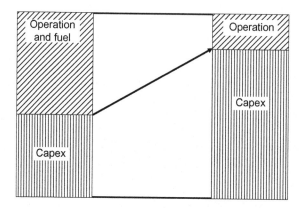

FIGURE 4.11 Relative change in cost structure of German power sector investments as it moves toward more renewable resources.

and require that the cost of offshore transmission failures be covered by the customers. Currently, pending legislation requires the TSOs to reimburse the producers in case of network failures and have the right to pass these costs to the network users.

5 WHAT DOES THE *TURNAROUND* MEAN FOR THE GERMAN ECONOMY?

Not surprisingly, the question whether and how the transformation of electricity supply in Germany affects the economy has caused a lively and ongoing debate. The *turnaround* comes with significant risks and some opportunities. In the long run, the *turnaround* can reduce price risks and import dependency on fossil fuels and, if successful, may serve as a model for an environmentally sustainable electricity supply system unlike any other in the world. In the short run, industries engaged in renewable energy or the so-called green economy can be expected to benefit from political support and the subsidies. At the same time, there are considerable risks concerning network stability and rapidly rising retail electricity prices, which affect the competitiveness of the German industrial sector, briefly described in this section.

5.1 Industry, Trade, and Commerce

The economic impacts of the *turnaround* are particularly relevant for the industrial sector. Industry is a fundamental pillar of income and employment in Germany. With over 28% of value added and over one-quarter of jobs in 2010, it is more important than in the average OECD or EU. The German industry plays a major role as energy consumer, using over 28% of the final energy. Industrial consumption of electricity is even more pronounced accounting for 44% of total consumption. The commercial sector consumes

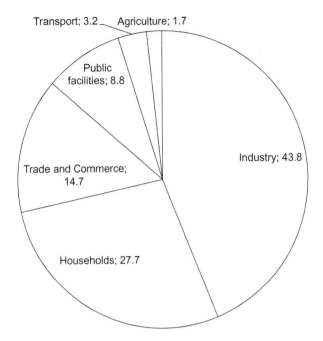

Transport; 3.2   Agriculture; 1.7
Public facilities; 8.8
Trade and Commerce; 14.7
Households; 27.7
Industry; 43.8

FIGURE 4.12 Share of final electricity consumption by consumer groups. *Source: BMWi (2012).*

roughly 14% with the balance going to the residential and other sectors as shown in Figure 4.12. Within the industrial sector, electricity is the second component with a 31% share (Figure 4.13).

The main concerns about the *turnaround* are focused on affordability and reliability of electricity. Figure 4.14 summarizes the responses of business community to a survey regarding the expected impact of the *turnaround* on business. Answers are differentiated by the share of energy in total costs of a given business. The survey suggests that the industry believes that the *turnaround* is likely to increase electricity prices and lead to lower power quality and reliability.

Concerns about power quality appear warranted with rising frequency of critical situations following the shutdown of the eight nuclear plants as described in Section 3. Short-time interruptions or voltage fluctuations can affect a wide range of industrial processes.

The same applies to concerns about rising electricity prices for industrial consumers, which have already risen considerably over the past few years and are progressively moving away from European averages. An average industrial plant using about 5000 MWh per year paid 8.7 cent/kWh at the end of 2007 compared to 11.3 cent at the end of 2011, all taxes included, except the value added tax (VAT). This is a 30% rise within 4 years, while prices in the EU rose by only 17%. Part of the increase is caused by politically induced price elements, which almost tripled since 2007, while prices without taxes nearly remained constant. Figure 4.15 shows the electricity prices for different industrial consumers compared to the average of the

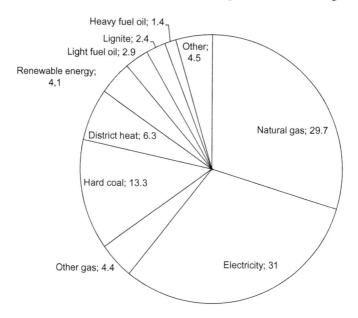

FIGURE 4.13   Share of industrial final energy by source. *Source: BMWi (2012).*

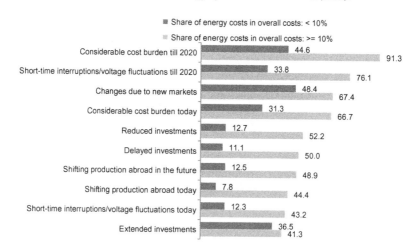

FIGURE 4.14   Perceived and expected consequences of the *turnaround* for the German economy. *Source: IW-Panel of Environmental Experts (IW-Umweltexpertenpanel) 3/2012, Cologne Institute of Economic Research. Survey among 111 company representatives responsible for environmental issues.*

EU-27, including all taxes and levies. Figure 4.16 shows the electricity prices of a medium industrial consumer within the EU.

Prices are expected to rise in the future mainly because of the following three effects:

1. First, the variable cost of coal and gas plants, which are needed to compensate for fluctuations in intermittent renewables, exceeds the variable cost of nuclear power, which is being phased out (Wissel et al., 2008), thus increasing wholesale prices. At times, when renewable energies provide a large amount of electricity, the wholesale price drops, reducing the incentive for investment and could lead to higher prices during peak hours (Erdmann, 2011).

2. Second, electricity suppliers have to buy CO_2 certificates within the European Emission Trading System (EU-ETS) if they want to be allowed to burn fossil fuels. In the absence of carbon-free nuclear generation, conventional thermal plants have to buy ever-larger number of CO_2 certificates through 2022 resulting in rising prices. Additionally, emission certificates will be further limited by the EU from 2013 onwards as described in Section 6.1, resulting in higher variable cost of fossil fuel

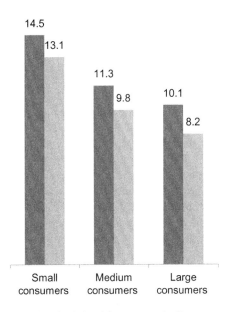

■ Germany   ▦ EU-27

FIGURE 4.15   Electricity prices for industrial consumers in Germany compared to the EU-27 average in 2011; all taxes and levies included (without VAT) in cent/kWh.

Small: 20 MWh < consumption < 500 MWh.
Medium: 2000 MWh < consumption < 20,000 MWh.
Large: 70,000 MWh < consumption < 150,000 MWh. *Source: Eurostat.*

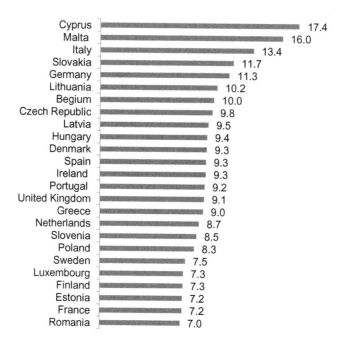

FIGURE 4.16 Electricity prices for industrial consumers in the EU in 2011; taxes and levies included (without VAT) in cent/kWh.
2000 MWh < consumption < 20,000 MWh *Source: Eurostat, Austria: no data available.*

plants and higher prices on the wholesale market, which will be passed through to consumers.

3. Third, the additional cost of renewable electricity is financed by a surcharge, which is passed on to consumers. The surcharge is adjusted every year on the basis of the amount of renewable energy fed into the network and the feed in price for the various renewables.[8] Increasing the volume of renewables and the share of more expensive renewables will increase the surcharge. As shown in Figure 4.17, the renewable energy surcharge has already risen considerably since it was established. The surcharge paid by consumers will soon exceed the average wholesale price. This means that the price for electricity paid by consumers is likely to double. Due to a significant expansion of photovoltaic and offshore energy production, the subsidy cost is expected to triple by 2025 (Erdmann, 2012). Additionally, there will be the indirect costs including the network expansion costs and the costs associated with offshore wind parks.

The surcharge for renewables will rise by around 50% from 2012 to 2013 and from 3.6 to 5.3 c/kWh (FAZ, 2012) with further increases expected in the

---

8. The surcharge is not the feed in price. For details on the renewable energy law, see Chapter 7.

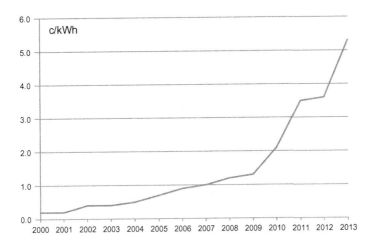

FIGURE 4.17   Development of the renewable energy surcharge in cent/KWh. *Source: BdEW (2012) and TSOs.*

following years depending on the portfolio of renewables and offshore wind generation. A conservative estimate of the costs, assuming that the capital expenditure for network connections and extension will be about €6 billion per year and assuming 5 c/kWh surcharge for renewables comes out to around €36 billion per year—roughly 1.4% of the German GDP. This direct effect, one can argue, is *affordable* for a rich society to achieve higher environmental quality. Problems are rather caused by the indirect effects on the economy due to a loss of competitiveness of the industrial sector. By some estimates, the total costs of the *turnaround* could be as high as €335 billion by 2030 including network extension and additional cost of renewables (Erdmann, 2011).

The price effects of the *turnaround* affect different industries differently. Electricity-intensive sectors such as paper, metal, chemical, glass, textile, rubber, plastic, food, and feed industries will be more vulnerable to lose their competitiveness as shown in Figure 4.12. Metal industries are over four times and paper industries almost five times more electricity intensive than average industry. Electricity intensive industries in Germany make up almost 80% of the electricity consumption in the manufacturing sector as described by Petrick et al. (2012) and Eichhammer et al. (2011).[9]

Aside from these substantial risks, the *turnaround* implies some opportunities for the German economy. As energy efficiency gains importance as a competitive factor, the demand for related technologies and services increases,

9. To sustain competitiveness and prevent a shifting of energy and carbon intensive productions abroad, several exceptions from electricity levies and taxes for industrial consumers have been introduced. The definition of whether a company is energy intensive or not, however, is inconsistent and trade intensity is not considered at all.

leading to growth of the so-called green technologies. Moreover, opportunities arise for marketing and exporting renewable energy hardware and related products, sectors which have experienced substantial growth in the past few years. Investment in facilities using renewable energies almost tripled between 2004 and 2009, thus allowing the sector to employ almost 340,000 people by 2009— almost twice as many as in 2004. Even accounting for sectors with declining employment, there has been net gains in overall employment (BMU, 2011).[10]

More sustainable benefits of the *turnaround* might lie in reduced import dependency of fossil fuels and related price risks in the long run. This is relevant as fossil fuels are still used for the bulk of electricity generation. Germany imports almost 90% of its natural gas and 80% of its coal. Prices for gas and coal grew by 72% and 84% between 2005 and 2011, respectively. With declining production within the EU, the dependency on imports is expected to increase.

5.2 Private Households

Private households will experience higher retail prices as a consequence of the *turnaround*, and because demand for electricity is relatively inelastic, the loss of purchasing power is relevant. Moreover, the impact on the residential sector will be disproportionately higher because a wide range of electricity-intensive industrial consumers are excluded from higher prices, which means the burden will fall heavily on small- and medium-sized enterprises as well as private households.

Currently, retail electricity price for private households is 25.3 cent/kWh, including VAT. It is more than twice as high as for an average industrial consumer. Residential prices have increased by almost 40% within the last 6 years, primarily due to the politically induced levies and taxes, which make up approximately half of the electricity tariff. This has had a noticeable impact on low-income households, since electricity bills constitute a relatively large share of their income. Moreover, high-income households often install photovoltaic modules and thereby gain from green power subsidies (Techert et al., 2012).

The extent to which the additional costs of the *turnaround* are passed on to residential consumers will depend on the evolution of the renewable subsidy schemes as well as other adjustments in the electricity market. Without substantial modifications, the current provisions will result in significant increase in direct and indirect costs, potentially reaching over €300 billion within the next 20 years (Erdmann, 2011). To ensure the acceptance and economic viability of the *turnaround*, more cost-efficient systems of financing are already under consideration as discussed in Section 4.4. Furthermore, the increasing revenues from future auctions of higher-priced CO_2 certificates may be used to reduce the burden on retail consumers.

---

10. These gains, however, have been made because of the heavy renewable subsidies and it is not clear if the gains will be sustainable once subsidies are removed.

6 EUROPEAN POLICY TARGETS

Energy policy in the member countries of the EU is strongly influenced by common European regulation, as described in Chapter 8. The implementation of the policy is shared between the union and the member states according to the Treaty of Lisbon Art. 4 (EU, 2009). Needless to say, Germany's unilateral decision on *turnaround* will have consequences for the EU, and for Germany's ability to reach the targets set by the Union for greenhouse gas reduction, the share of renewables and energy efficiency further described below.

6.1 Reduction of Greenhouse Gas Emissions

When the original 15 member states ratified the Kyoto Protocols, the target was to reduce greenhouse gas emissions to 8% by 2012. Moreover, the EU has set a target to reduce its emissions by at least 20% by 2020. Thus far, Germany has fulfilled its obligations, but the future course of emissions depends on what replaces carbon-free nuclear energy. In this context, Germany could experience potential increases in emissions, which would influence the overall EU outcome due to the size of the German emissions.

The main instrument for greenhouse gas reduction for the electricity sector is the tradable certificate scheme, introduced in 2005 to meet the Kyoto target. Originally, free certificates were allocated to member states on the basis of past emissions and had limited impact.

Only large emitters (≥ 20 MW) are included in the system. National governments keep a register of all emitters, who have to balance their emissions with a corresponding amount of certificates. Because certificates are tradable, the price depends on supply and demand. In the future, certificates will be auctioned by national governments and electricity producers will be required to buy their certificates from 2013 onward. CO_2 extracted and stored according to the rules of carbon capture and storage does not require CO_2 certificates.

The EC plans to reduce the total number of certificates by 1.74% per year annually, which is expected to stiffen the price over time. Since the carbon market is a European market, additional CO_2 emitted in one country, for example, Germany, will drive the price up for all member states, potentially leading to windfalls profits for countries with a high share of carbon-free electricity, for example, France.

At present, there is no international agreement for climate policy, while the EU has traditionally assumed a leading role in such agreements. It remains to be seen how European climate policy will be adjusted in the context of future international negotiations and whether the German *turnaround* will result in similar policies in other countries, possibly including Japan.

6.2 Reduction of Primary Energy Consumption

The EU has also set a target to reduce primary energy consumption by 20% by 2020. Germany has also set ambitious targets for itself for the reduction of primary energy demand.[11] To meet these targets, significant improvements in utilization of energy across all economic sectors are required, something that has been difficult to achieve. The German *turnaround*, due to the statistical effects, may result in symbolic reductions even if final energy consumption remains the same.[12] Reduction of primary energy consumption in the past was often accompanied by an increase in electricity consumption, something that is likely to be true in the future.

6.3 Share of Renewables

The third part of EU's 20-20-20 by 2020 is a target to increase the share of renewable energy in *total*—not electricity—energy consumption to 20% by 2020 to improve security of supply and protect the climate. For each member country specific targets have been defined depending on the level of GDP and the potential for renewable energy production. The 2020 target for Germany is 18%. The EU has introduced a mechanism by which every second year the progress achieved is compared to the target. Sanctions have been defined in case the share of renewable energy lags behind the defined target.

Fundamentally, there is no contradiction between the European target and the German *turnaround*. Germany's efforts to increase electricity production from renewables will help reach the European target. As previously described, the present system of promoting renewables, however, is in conflict with the European directives for an open internal market for electricity. The German push for more renewable generation increases the pressure on the EU to finally develop an acceptable scheme to subsidize renewables compatible with EU's open market principles, something that have not been successful in the past.

At the present, there is an intensive debate on the introduction of elements of a capacity market in the market rules in Germany. Proponents of this approach suggest that it will be necessary to create incentives for flexible power stations to stabilize the fluctuations from renewables. If such market rules are introduced on a national level in Germany, it may interfere with the operations of the EU-wide exchanges. The German *turnaround* therefore

---

11. If nuclear energy is replaced by imported electricity or renewables, about one-third of the reduction required will result automatically due to the statistical valuation of these different energies. For imported electricity and renewables, the primary energy equivalent of 1 kWh is 1, whereas for nuclear or fossil energy it depends on the efficiency of transformation.

12. As described in footnote 11, for nuclear energy, the efficiency used is 33%; hence, the nuclear phase out counts as a reduction of primary energy use.

may create additional pressure on the EU to redevelop their energy policy in this area. Which is another way to repeat what was said earlier, that because of the size and central location of Germany, the German *turnaround* may lead to an EU-wide *turnaround*.

7 CONCLUSION

Germany has embarked on a risky path to phase out its nuclear fleet under the government's *turnaround* strategy. The original decision in 2000 to abandon nuclear power had reasonable lead time to implement but this opportunity was squandered. Now, there may not be enough time to successfully implement the *turnaround*.

The lessons to be learned from the German experience, summarized below, could benefit other countries wishing to transform their electricity industry to rely on a higher share of renewables and no nuclear power.

- Implementing major changes in energy policy requires a long time horizon. In a democracy with changing priorities and policies, it is not always easy to give the industry sufficient lead time to achieve the desired objectives. The German experience shows how the *turnaround* could have been handled smoother, at lower costs and with fewer dislocations.
- Decisions regarding the long-term development of renewables must be integrated with corresponding expansion of the transmission network.
- Electricity producers need long-term incentives to restructure their generation portfolio to cope with a higher share of mostly intermittent renewables.
- Promotion of renewables will be necessary for a long time span and require compatible electricity market rules.
- For a successful transformation of the industry, it seems advisable to establish a round table with all affected players so that coordination between TSOs, conventional generators, renewable developers, the regulator, government, and customers can be discussed and agreed.
- In the case of German *turnaround*, the increased share of renewables plus the necessary infrastructure investments are likely to result in significant increases in overall cost. How these additional costs are allocated and shared is an important issue to gain public acceptance of the scheme in the long term. This issue, initially neglected, may emerge as an obstacle to the success of the transformation.

The success or failure of the German experience will ultimately depend on its indirect effects and the competitiveness of electricity-intensive industrial sectors of the economy. The *turnaround* may become a success story if can be successfully implemented and leads to major gains in productivity and efficiency and to creation of a thriving green economy. At the moment, it is hard to imagine how this can happen. Time will tell.

REFERENCES

BdEW, 2012. Erneuerbare Energien und das EEG: Zahlen, Fakten, Grafiken (2011). Berlin.

BMU, 2011. Erneuerbar beschäftigt! Kurz- und langfristige Wirkungen des Ausbaus erneuerbarer Energien auf den deutschen Arbeitsmarkt. Berlin.

BMWi, 2012. Energiedaten.

Brunekreeft, G., Bauknecht, D., 2006. Energy policy and investment in the German power market. In: Sioshansi, F., Pfaffenberger, W. (Eds.), Electricity Market Reform: An International Perspective. Elsevier, Amsterdam, pp. 235–263.

Buchan, D., 2012. The Energiewende—Germany's Gamble. The Oxford Institute for Energy Studies, Oxford.

Cramton, P., Ockenfels, A., 2012. Economics and deisgn of capacity markets for the power sector. Z. Energiewirtschaft 36, 113–134.

dena, 2010. Integration erneuerbarer Energien in die deutsche Stromversorgung im Zeitraum 2015–2020 mit Ausblick auf 2025. Berlin.

DPG, 2005. Climate Protection and Energy Supply in Germany 1990–2020. DPG (German Physical Society), Berlin.

EEG, 2009. Gesetz für den Vorrang Erneuerbarer Energien (Erneuerbare-Energien-Gesetz—EEG).

Eichhammer, W., Kohlhaas, M., Neuhoff, K., Rohde, C., Rosenberg, A., Schlomann, B., 2011. Untersuchung des Energiesparpotentials für das Nachfolge-modell ab dem Jahr 2013ff zu den Steuerbegünstigungen für Unternehmen des Produzierenden Gewerbes sowie der Land- und Forstwirtschaft bei der Energie- und Stromsteuer. Berlin.

Erdmann, G., 2011. Kosten des Ausbaus der erneuerbaren Energien. vbw.

EU, 2009. Treaty on the functioning of the European Union.

FAZ, October 8, 2012. Der Ökostrom wird für die Verbraucher immer teurer. FAZ.

Kopp, O., Eßer-Frey, A., Engelhorn, T., 2012. Können sich erneuerbare Energien langfristig auf wettbewerblich organsieierten Strommärkten finanzieren? Z. Energiewirtschaft 36, 243–255.

Krämer, M., 2003. Modellanalyse zur Optimierung der Stromerzeugung bei hoher Einspeisung von Windenergie. vdi, Düsseldorf.

Küffner, G. (2012). Gegenverkehr im Stromnetz. Frankfurter Allgemeine Zeitung, T1., 18 September 2012.

Maurer, C., Tersteegen, B., Zimmer, C., 2012. Anforderungen an den konventionellen Kraftwerkspark—wieviel und welche Kraftwerkskapazität wird benötigt. Z. Energiewirtschaft 36, 147–154.

Menges, R., Müller-Kirchenbauer, J., 2012. Rekommunalisierung versus neukonzessionierung in der Energiewirtschaft. Z. Energiewirtschaft 36, 51–67.

Monopolkommission, 2011. Energie 2011: Wettbewerbsentwicklung mit Licht und Schatten, Sondergutachten 59.

NABEG, 2011. Netzausbaubeschleunigungsgesetz Übertragungsnetz. Berlin.

Netzentwicklungsplan, 2012. Netzentwicklungsplan Strom 2012—Entwurf derÜbertragungsnetz betreiber. Berlin.

Ockenfels, A., Grimm, V., Zoettl, G., 2008. Strommarktdesign—Preisbildungsmechanismus im Auktionsverfahren für Stromstundenkontrakte an der EEX. Gutachten im Auftrag der European Energy Exchange AG.

Petrick, S., Rehdanz, K., Wagner, U., 2012. Energy use pattern in German industry: evidence from plant-level data. Jahrbücher Nationalökonomie 231, 379–414.

Pfaffenberger, W., Hille, M., 2004. Investitionen im Liberalisierten Energiemarkt: Optionen, Marktmechanismen, Rahmenbedingungen. Bremen.

Techert, H., Bardt, H., Niehues, J., 2012. Ungleiche Belastung durch die Energiewende: vor allem einkommensstarke Haushalte profitieren. Wirtschaftsdienst 92, 507–511.

Umweltbundesamt, 2012. Entwicklung der spezifischen Kohlendioxid-Emissionen des deutschenh Strommix 1990–2010 und erste Schätzung für 2011.

Wissel, S., Rath-Nagel, S., Blesl, M., Fahl, U., Voß, A., 2008. Stromerzeugungskosten im Vergleich. IER Arbeitsbericht 4.

Chapter 5

The Growing Impact of Renewable Energy in European Electricity Markets

Reinhard Haas, Hans Auer, Gustav Resch and Georg Lettner
Energy Economics Group, Vienna University of Technology, Vienna, Austria

1 INTRODUCTION

Despite the high costs, generating electricity from renewable energy sources is considered as environmentally benign (Resch et al., 2008). In recent years, due to generous support schemes in a number of countries, electricity generation from renewables is growing at a remarkable rate as illustrated in Figure 5.1 for EU-27 countries between 1990 and 2010. The growth of "new" renewables excluding hydro is even more impressive over the same period from less than 1% to about 9%, mainly from wind and biomass (Figure 5.2).

The rapid growth of renewable generation is mostly attributed to schemes[1] like Feed-in Tariffs and Tradable Green Certificates encouraged by EU policies (e.g., European Parliament and Council, 2001). These generous subsidies are expected to continue in one form or another in the coming years resulting in continued growth of renewable generation. The Directive 2009/28/EC (EC, 2009), for example, requires moving from 11% reliance on renewables in overall energy supply in 2010 to 20% by 2020. Various estimates (Resch et al., 2012), suggest that renewables' contribution as a percentage of electricity generation within the EU will rise from 21% in 2010 to 34−36% by 2020. Beyond 2020, the policy framework is less certain but a strong renewable energy component for the period up to 2050 is among the central features of the current EU policy debate.

Germany plays a significant role in this context due to the large size of its economy and the political decision to phase out its nuclear fleet by 2022, mostly to be replaced by renewable generation, as further described in Chapters 4 (Pfaffenberger and Chrischilles, 2013) and 7 (Bauknecht et al., 2013). The

---

1. For further details on the support schemes, refer to Auer et al. (2009), Haas et al. (2008, 2011a,b) and Jacobsson et al. (2009).

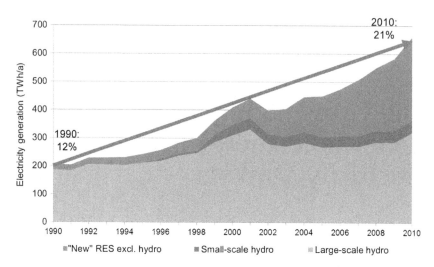

FIGURE 5.1 Historical development of electricity generation from all renewables including hydro in EU-27 between 1990 and 2010 (TWh). *Source: EUROSTAT (EU, 2012).*

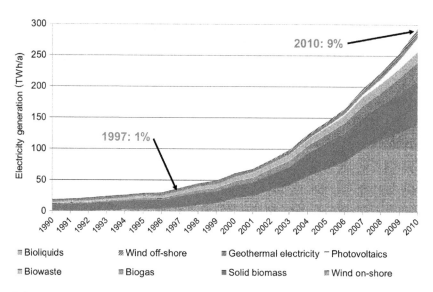

FIGURE 5.2 Development of electricity from "new" renewables (excluding hydro) in EU-27 between 1990 and 2010 (TWh). *Source: EUROSTAT (EU, 2012).*

impact of the German "Energiewende" or turnaround, plus similar developments in other EU countries, is likely to fundamentally change the electricity supply system in Europe. Already, the impact of large amounts of renewable generation is being felt on the spot market prices at the German electricity exchange, EEX (www.eex.com), as shown in Figure 5.3.

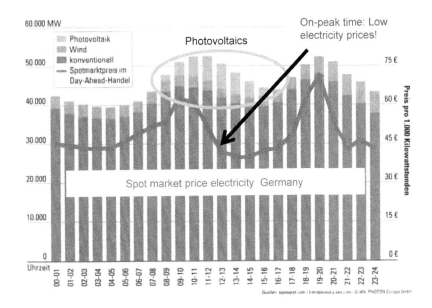

FIGURE 5.3 Example for the impact of PV capacities on the price developments in the German electricity market on October 22, 2011.

The growth of renewables, especially photovoltaics (PV), is expected to become even more pronounced by 2020. In Germany alone, total installed PV capacity is projected to increase from about 20 GW installed by the end of 2011 to at least 50 GW by 2020. This is roughly half of total fossil and nuclear capacity in Germany in 2011. The continued growth has increased the need to address a number of critical issues including:

- Suggestions for the implementation of capacity markets to ensure supply security (see also Chapter 7);
- Calls to reexamine unsustainable subsidies which encourages more renewable generation to be added to the network while bypassing the normal market incentives that applies to conventional technologies;
- Reexamining the long-term impact of renewables on retail tariffs on households and industry;
- High expected additional costs for grid extension and storage, which are necessary to compensate for the intermittency and unpredictability of renewables.

This chapter examines the historical growth, the current situation, and the future prospects of renewables for electricity generation in Europe. Similar arguments, of course, apply to other regions of the world with ambitious renewable targets. The main objective is to examine the possible effects of further uptake of renewables on prices in European electricity markets including:

- The impact of renewables at specific times of the year when they shift the supply curve of conventional generators in wholesale markets leading to low or even negative prices;
- The impact of intermittent renewables on the cost of fossil-fueled plants, mainly natural gas, which are needed for reliability reasons;
- Change of spreads between high- and low-price levels.

The chapter is organized as follows: Section 2 looks at history and lessons learned from the past and at future prospects of intermittent renewables in EU-27. Section 3 presents the basics on how prices come about in liberalized electricity markets and the impact of intermittent renewables on wholesale electricity prices. The impact of larger shares of intermittent renewables on the prices in electricity markets is examined in Section 4. Section 5 asks whether new market rules are necessary followed by the chapter's conclusions.

2 HISTORICAL TRENDS AND FUTURE PROSPECTS

The rise of nonhydro renewables is a relatively recent phenomenon. Prior to the 1990s, hydropower was virtually the only form of renewable energy in Europe—as in many other countries. Historically, hydro generation was encouraged through government intervention in the form of state-owned companies who enjoyed low risk, low interest rates, and long depreciation periods, which enabled them to finance the construction of hydro plants. Moreover, these companies had the security of captured customers who had to pay back the investment through their electricity tariffs.[2]

The seasonal variability of hydro resources was managed by reliance on thermal plants as schematically shown in Figure 5.4 for Austria for a typical year. In summer, hydropower from run-of-river plants, as well as from hydro storage, could meet virtually the whole load while in winter thermal plants are needed to provide about half of the electricity. The excess capacity could be exported, or conversely, any deficits could be met by imports. As illustrated, exports tend to be higher in summer and imports in winter for Austria.

This basic scheme that relied on significant investment in hydro capacity in numerous European countries—e.g., Norway, Sweden, Spain, France, Austria, and Switzerland—was augmented with sufficient amounts of thermal, fossil-fueled, or nuclear plants. Over time, countries learned to rely not only on their internal resources but also on those of their neighbors through

2. The same applied to nuclear investments, as described by Percebois (Chapter 3) in France, for example.

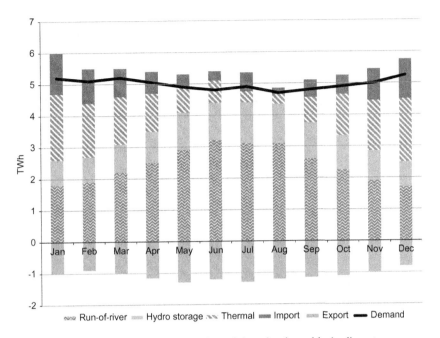

FIGURE 5.4   Historical pattern of meeting demand through a thermal-hydraulic system.

transmission interconnections. Denmark, for example, was—and still is—the thermal backup system for Norway and Sweden, and vice versa. Similarly, Austria and Switzerland, countries with large hydro resources, developed strong ties with Germany to exchange electricity as needed to balance generation and load. Austria and Switzerland, for example, routinely meet Germany's peak demand from hydro storage while Germany exports—until 2010 before the shift in its nuclear policy—baseload electricity.

The renewable component of electricity generation in EU-27 has continually grown from 12% to 21% between 1990 and 2010 as illustrated in Figure 5.5. Even more impressive is the growth in electricity generation from natural gas from 9% to 24% over the same period while the share of coal and petroleum has decreased. Nuclear electricity reached its peak of about 33% in the mid-1990s, dropping slightly to 28% by 2010, a trend that is expected to continue with the German nuclear phaseout.

The composition of the renewable electricity mix between 1990 and 2010 changed remarkably with hydropower dominating the picture in 1990 (Figure 5.6A), while wind, biomass, and solar have become noticeable by 2010 (Figure 5.6B).

In terms of installed capacity, the growth of renewables is impressive in selected European countries as shown in Figure 5.7 with major differences among countries especially in the case of Germany, Italy, and Spain where significant amounts of wind and PV have been added during the past decade.

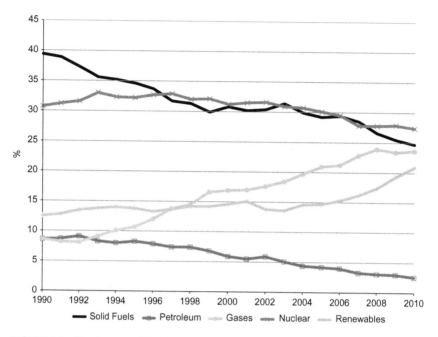

FIGURE 5.5 Development of fuel shares of electricity generation in EU-27 from 1990 to 2010. *Source: EU (2012).*

These large capacity additions, of course, do not translate directly into generation due to the relatively low capacity factor for wind and solar PV: a typical wind farm may operate some 1800–2300 h/year; the corresponding figure for PV may be in the range 800–1200 h/year depending on the location. Consequently, despite the growth of new renewables, fossil fuels—and nuclear in the case of France—remain dominant in UK, Italy, and even in Germany.

A key question in this context is the impact of the continued rise of intermittent renewables as required under EU directives as well as national policies such as Germany's nuclear phaseout. These issues were the subject of Re-Shaping, a research project examining the consequences of policy choices on future deployment of renewables and their costs (Resch et al., 2012). The study examined two scenarios: a business-as-usual (BAU) case assuming that all relevant energy policies and market rules will remain unchanged and a scenario of "strengthened national policies" (SNP), considering improved financial support as well as the mitigation of noneconomic barriers that hinder further deployment of renewables.

Under the BAU scenario, moderate deployment of renewables can be expected as illustrated in Figure 5.8, with the total share of renewables, including hydro, in gross electricity generation growing from 21% in 2010 to 29% by 2030. In this case, the share of intermittent renewables, in particular

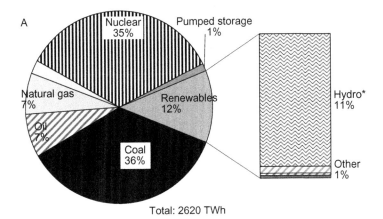

FIGURE 5.6   (A) Electricity generation in EU-27 by fuel in 1990. (B) Electricity generation in EU-27 by fuel in 2010.

of wind and PV, will rise from 5% to 12% over the same period. Under the SNP scenario, the share of total renewables is expected to rise to 54% by 2030, and intermittent renewables such as wind and PV could reach 29%.

The share of total and of intermittent renewables (wind and PV only) in gross electricity demand varies considerably from one country to another by 2030 as shown in Figure 5.9 for the two scenarios. Aside from larger total renewable penetration levels under the BAU case, the growth of intermittent generation under the SNP scenario could reach significant levels in some countries. For example, intermittent renewables are expected to account for roughly 94% of total domestic consumption in Denmark, 63% in Ireland, and 53% in the United Kingdom by 2030. Clearly, this would require adequate

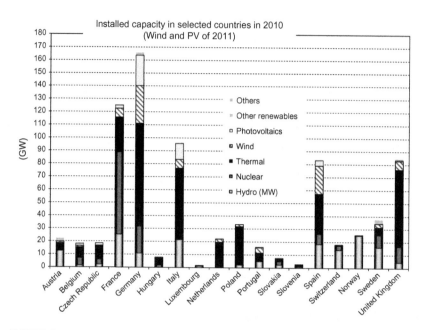

FIGURE 5.7 Power plant capacities in selected European countries in 2010 (for PV and wind in 2011). *Sources: UCTE, EWEA, and EPIA.*

investment in flexible backup generation and storage for system stability and reliability. In some regions, such as the Iberian peninsula, with limited transmission connections, this may pose special challenges. In Spain, the share of intermittent renewables as a percentage of total consumption is expected to increase to more than 24% while for Portugal an increase to 32% is expected.

As illustrated in Figure 5.10, the overall growth of electricity generation from renewables within the EU-27 block by 2030 is impressive, especially under the SNP scenario with potentially large increases in onshore and off-shore wind generation during the 2020–2030 period. But even under the BAU case, renewable generation is projected to increase from 650 TWh in 2010 to 1200 by 2030. With strengthened policies, and in particular in the case where noneconomic barriers are removed, a deployment of 2150 TWh appears feasible by 2030—that represents more than a doubling of renewable generation compared to BAU case.

Other key technologies are PV, offshore wind, solid biomass, biogas, and solar thermal electricity. Offshore wind generation, for example, is projected to grow from 6 TWh in 2010 to 76 TWh by 2030 under BAU case; the corresponding number under the SNP scenario is 417 TWh for 2030, an astonishing number. Other renewable options such as hydropower (both large and small scales), biowaste, and geothermal appear less significant and their deployment is only marginally influenced by renewable support schemes.

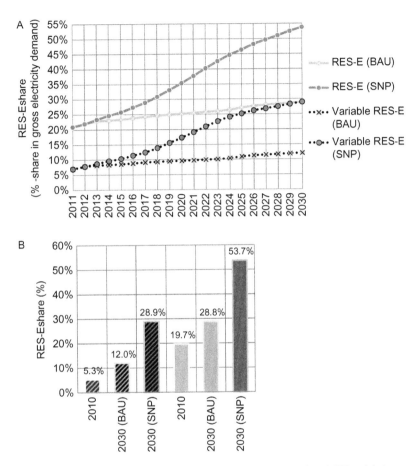

FIGURE 5.8   Share of total and of intermittent renewables (wind and PV only) in gross electricity demand in the period 2011–2030 (A) and by 2010 and 2030 (B) in the EU-27 according to selected scenarios: BAU and SNP. *Source: Re-Shaping study (Resch et al., 2012).*

3 THE CHANGES AFTER THE LIBERALIZATION OF THE ELECTRICITY MARKETS

This section discusses how the liberalization of the electricity markets in Europe changed the formation of prices in wholesale markets, and—as further described in Section 4—the impact of rising shares of renewables on spot market electricity prices.

The liberalization process in Europe started in the late 1980s in the United Kingdom and gradually migrated to continental Europe with the 1999 EU directive (EC, 1997).[3] One of the major features of the liberalized

3. For further details, see for example, Boltz (Chapter 8) or Haas et al. (2006).

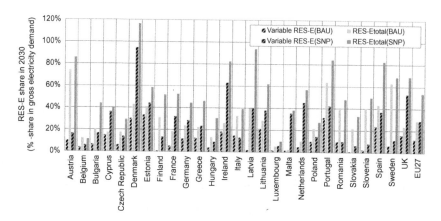

FIGURE 5.9 Country-specific share of total and of intermittent renewables (wind and PV only) in gross electricity demand by 2030 according to selected scenarios: BAU and SNP. *Source: Re-Shaping study (Resch et al., 2012).*

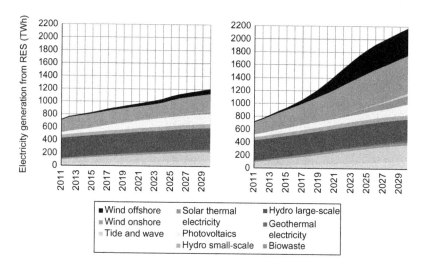

FIGURE 5.10 Development of electricity generation from renewables by technology in the EU-27 up to 2030 according to the BAU case (left) and the case of SNP (right). *Source: Re-Shaping study (Resch et al., 2012).*

electricity markets was that the pricing regimes changed. In former regulated markets, prices were established by setting a regulated tariff, which was calculated by dividing the total costs of supplying service by the number of kilowatt-hours sold—with some differences between different groups of customers. The major change that took place after the liberalization was that prices were now expected to reflect the marginal costs of electricity generation (Stoft,

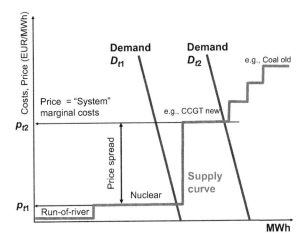

FIGURE 5.11   How prices come about in markets with conventional capacities including large run-of-river hydro.

2002; Boltz, 2013). At the time when liberalization started, considerable excess capacities existed in Europe. This led to the expectation that prices will (always) reflect the short-term marginal costs (STMCs) as illustrated in Figure 5.11. The graph shows a typical merit order supply curve with conventional capacities including large hydro. The typical historical pattern of electricity generation in the European electricity markets consisted of conventional fossil, nuclear, and hydro capacities. Since the late 1990s, most of the time nuclear contributed the largest share, followed by fossil and hydro.[4] Nonhydro renewables were not a significant factor until recent times.

As shown in Figure 5.11, the intersection of the supply curve with demand determines the market clearing price at the system marginal costs. The curve D_{t1} shows the demand curve at times of low demand, e.g., at night, and p_{t1} is the resulting (low) electricity price. D_{t2} shows high demand times, e.g., at noon, and p_{t2} is the resulting (high) electricity price. The difference between p_{t2} and p_{t1} is the so-called *price spread* further described below. It provides useful information, for example, on the economic attractiveness of storage, which will be of high relevance in markets with large share of renewables. Until recently, the price spread has been of interest mainly with respect to pumped storage. That is to say, during periods when prices are low, water can be pumped into reservoirs; while generating electricity when the opposite is true.

The price patterns in different European electricity markets are shown in Figure 5.12 for the period from 2000 to 2012 where price

---

4. In principle, this pattern can be found in every market, also in the NORDPOOL, where Denmark provides the fossil backup capacities.

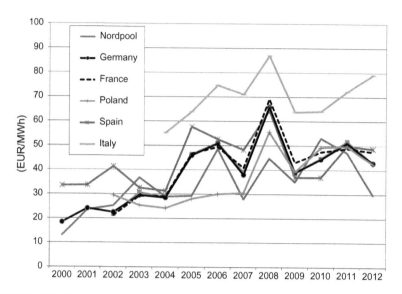

FIGURE 5.12   Development of spot market prices in different European electricity markets 2000–2012 (2012 preliminary).

volatility and considerable differences between various submarkets are observed. Italy tends to experience higher prices and more volatile due to its overreliance on imported electricity, congested transmission lines, and heavy reliance on expensive natural gas. In the case of the NORDPOOL, which includes Sweden, Norway, Finland, and portions of Denmark, the pattern is different due to heavy reliance on hydro and lack of strong interconnection with continental Europe. Despite these differences, a remarkable convergence of prices has taken place even in the case of the isolated Iberian peninsula, which is not yet fully integrated into the European network due to transmission limitations. The reason for high prices in 2008 in continental Europe was the low hydro availability while the falling prices since 2008 may be attributed to the current European economic crisis.

The STMC price regime, illustrated in Figure 5.11, of course, will not be permanent nor always apply. Once excess generation capacity is exhausted, there will be a shift toward long-term marginal costs (LTMC). Similarly, generators are likely to behave strategically during high demand periods in markets with limited peaking capacity. Moving forward, one can expect deviations from the STMC price regime, as illustrated in Figure 5.13. Moreover, as described below, the introduction of large amounts of renewables with essentially zero marginal cost will further affect the principles behind STMC, a feature described in a number of chapters in this volume.

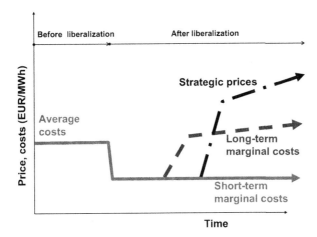

FIGURE 5.13 Changes of pricing electricity before and after liberalization of electricity markets.

4 THE IMPACT OF HIGHER SHARES OF RENEWABLES IN ELECTRICITY MARKETS

Against the backdrop of rapidly rising renewable generation across the EU, notably in countries like Germany, Denmark, Spain, and others, three key questions arise.

1. First, what is the impact of large amounts of renewable generation feeding the grid especially during low demand periods when renewables shift the supply curve of conventional electricity virtually "out of the market."
2. Second, what is the impact of intermittent renewable generation on the costs at which fossil—especially natural gas—capacities are offered.
3. Third, what is the effect of renewables on the *price spreads*, already defined, over time.

4.1 The Impact of Renewable Generation on Market Price

With a few exceptions, such as hydro with large reservoirs and geothermal, renewable technologies by their nature tend to be intermittent, not entirely predictable, nor dispatchable. These are familiar characteristics, which have been recognized in networks with large amounts of run-of-river hydro and wind for some time, for example, in Denmark (Lund, 2005) or Spain (Zubi, 2011). As noted by Riesz et al. (Chapter 25) and Bauknecht et al. (Chapter 7), large amounts of wind generation during low demand periods result in lowering market clearing price, occasionally leading to negative prices (Nicolosi, 2010). However, the wind-driven effects mostly happen during off-peak hours when prices are already low, causing them to became even lower. This

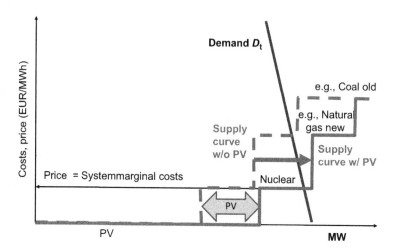

FIGURE 5.14 Merit order supply curve with and without additional PV capacities at on-peak time of a bright summer day with STMCs for conventional capacities.

phenomenon has led to increased volume of intraday trading, where traders attempt to take advantage of the price differentials in different markets.

The pattern of wind-driven prices is now being experienced in countries like Germany with the rapid growth of PV generation. Figure 5.3 shows an example of price formation on the German electricity market on October 22, 2011, a sunny day. The PV generation during midday hours not only displaces virtually all hydro generation but also results in lower spot prices during a period when they tend to be high. This is an example of how the rise of renewables will impact spot prices, trading patterns, and dispatching of conventional generation. Similar patterns are experienced in other sunny regions such as Italy and Spain where solar generation has a significant impact on midday prices.

The explanation is simple. On a sunny day with ample solar generation, the supply curve is shifted to the right as schematically shown in Figure 5.14, which essentially pushes nuclear and fossil-fueled generation "out of the market." If the impact of PV on a sunny day in October, not a peak period for solar generation in Germany, can be as dramatic as is shown in Figure 5.3, one can expect much more dramatic impacts on market prices during future summer months.

4.2 The Impact of Renewable Generation on Fossil Plants[5]

Aside from the above-described effects, intermittent renewables will also influence the costs at which fossil generation—especially natural gas—are offered. The illustration in Figure 5.11 is based on the STMCs of

---

5. The following analysis draws on Haas et al. (2013).

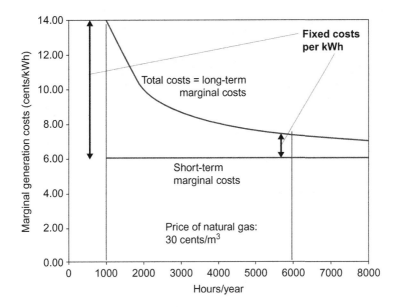

FIGURE 5.15 STMCs (variable) and LTMCs (total) of electricity generation in a CCGT plant depending on yearly full-load hours.

conventional generation, which may correspond to some 6000 full-load hours per year.[6] The revenues derived from these hours must cover both the fixed and variable costs, as illustrated in Figure 5.15. The graph schematically shows the total and variable (short-term) electricity generation costs of a new combined-cycled gas turbine (CCGT) based on its annual full-load generation hours. As can be seen, the share of fixed costs is considerably higher when the plant operates at full load for a minimal number of hours, say, 1000 h/year[7] as opposed to a high number of hours, say 6000 h/year.

Historically, different types of fossil plants were dispatched to meet the load over the course of all hours in a year and it was generally possible to recover the fixed costs when more expensive plants, usually gas-fired peaking units, set the price. In recent years, frequently old and mostly depreciated coal plants determined the STMC, allowing new peaking units such as CCGTs to recover their fixed costs. In the future, where renewables with virtually zero marginal costs set market clearing prices at least at some points of time, this may no longer be the case. This problem, sometimes referred to as the "missing money" problem, is also described by others including Percebois (2013, Chapter 3) and Bowring (Chapter 9) or Walls et al. (2007).

---

6. That assumes roughly 70% capacity factor.
7. Of course, full-load hours vary year by year depending on demand, hydropower, and other factors.

It leads to lack of sufficient investment in peaking capacity and storage, which in fact will be sorely needed to deal with the intermittency of renewables. Adib et al., for example in Chapter 10, describe the resource adequacy issues confronting the Texas market, which will only become more pronounced in markets with growing renewable share.

As the preceding discussion illustrates, the issue of missing money is likely to become more common and serious as renewables begin to dominate many European markets, as is expected for Germany by 2050. Under such a scenario, only highly flexible CCGT plants can remain viable as described by Auer (2011) or Carraretto and Zigante (2006). But as the number of hours such plants are needed to operate drops, say to 1000−2000 h/year, different pricing strategies including the implementation of capacity markets may become more relevant. Regardless of what such schemes may be called, pricing based on long-term marginal costs (LTMC), which include capacity costs, are likely to become more prevalent than today, further discussed in Section 5.

How the growth of renewables might impact future pricing strategies of fossil or biomass power plants over time is a subject of speculation. As schematically shown in Figure 5.16, the merit order supply curve and the high- and low-demand curves are affected based on the availability of renewables, which tend to be intermittent and not entirely predictable. The illustration shows three examples for supply curves: merit order supply curves for

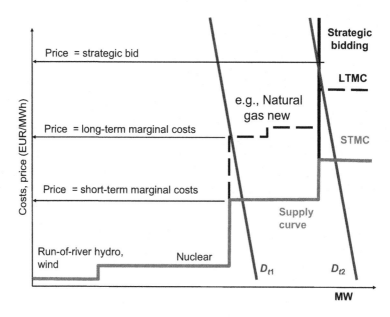

FIGURE 5.16 Merit order supply curve for STMC versus LTMC of CCGT plants or strategic bidding at times with low-intermittent renewable.

STMC versus LTMC of CCGT plants and a supply curve for strategic bidding, which is shown as a vertical line.

Depending on where the supply and demand curves intersect, different prices would prevail. The price may be extremely high in the case of strategic bidding—not sustainable in the long run—or lower when long-term marginal costs are included, or even lower if STMCs prevail.

4.3 Changes in Price Spreads

Of further relevance in this context is how changes in spot market prices and bidding patterns is likely to impact future "price spread," the difference between the off-peak and the on-peak price for electricity. Figure 5.17 shows

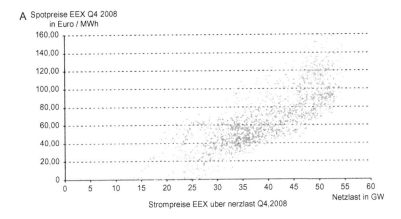

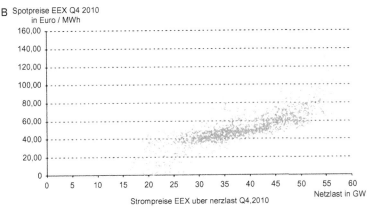

FIGURE 5.17   Comparison of price spreads on the German spot market in 2008 (A) and 2010 (B). Development of price spreads on the German spot market—prices (vertical axis) over electricity demand (horizontal axis)—(A) fourth quarter 2008 and (B) fourth quarter 2010. *Source: LBD (2011).*

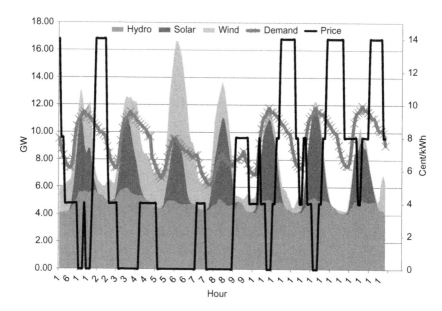

FIGURE 5.18 Development of intermittent renewables from wind, PV, and run-of-river hydro plants over a week in summer on an hourly basis in comparison to demand and resulting electricity market prices with total costs charged for conventional capacities.

the spot market prices—on the vertical axis—and the corresponding electricity demand—on the horizontal axis—for the fourth quarter of 2008 and 2010. Clearly, the difference between high prices and low prices is lower in 2010 (Figure 5.17B) compared to 2008 (Figure 5.17A). The major reason was that hydro availability was rather low in 2008 leading to comparably higher prices than in 2010, as illustrated in Figure 5.12. This suggests that the change in price spreads between 2008 and 2010 can be attributed—at least to some extent—to the availability of hydro. As more intermittent and variable renewables are added to the network, this trend can be expected to increase.

How will the price spread in European markets evolve in the future as larger amounts of PV, solar thermal, and wind generation are added to the network? The consequence for electricity prices are shown in Figure 5.18 where a hypothetical scenario with high levels of generation from wind, PV, and run-of-river hydro plants over a week in summer are depicted using synthetic hourly data for an average year in Germany. The graph shows significant volatilities in electricity market prices with total costs charged for conventional capacities—black solid line—ranging from 0 to 14 cents/kWh[8] within very short-term time intervals. In practice, of course, the prices may not go to zero but would be rather low.

---

8. The 14 cents/kWh in this example results from Figure 5.15 with full-load hours of about 1000 h/year.

The longer-term impact of variable and intermittent renewables on price spread on the grid is subject to speculation. The intuitive explanation is that when renewables are plentiful, say during windy or sunny periods, the prices will be extremely low, approaching zero or possibly going negative, while at other times—when demand is high and renewables are scarce—prices can be much higher due to strategic bidding by fossil generators exercising market power. This is graphically shown in Figure 5.18.

A more important development may be that future high prices will not necessarily appear at peak-demand times but at times of low renewables availability as also described by Riesz et al. (Chapter 25) and Bauknecht et al. (Chapter 7). This will also change the operation of pumped hydro facilities and lead to new investment in energy storage technologies to take advantage of significant price differentials. Over time, the familiar patterns of the night-to-day shift of generation will change in response to the unpredictable and variable rhythm of renewable generation. The most likely consequence of increased price volatility will be to make storage and flexible peaking units much more valuable than they currently are.

5 NEW MARKET STRUCTURES

As already discussed, the current electricity spot markets rely mainly on the basic principle that at every point-in-time prices are equal to the system marginal costs. This, of course, allows the coexistence of different market segments to take advantage of arbitrage opportunities, which explains futures and forwards markets, day-ahead, intraday, and balancing markets. The question is whether there is a need for fundamental changes in market mechanisms to accommodate large amounts of new renewable generation as also described by Bauknecht et al. (Chapter 7).

This has raised interest in so-called capacity markets in some circles. The major argument of the *apologists* of this idea is that in the absence of some sort of fixed "stand-by fee" paid to fossil-fueled plants, the owners/operators will shut them down as they become marginally profitable or not profitable at all.[9] Ironically, these plants will increasingly be needed to maintain system reliability and serve as backup for intermittent renewables. In some markets, as in Texas, offer cap prices have been raised partially to allow peaking units and those needed for resource adequacy to gain more revenues during periods of supply scarcity and high prices. In this context, if the regulators are willing to accept occasional high spikes in spot prices that are significantly above the STMC—without accusing the stakeholders for

---

9. Investors may, of course, behave strategically by simply threatening to shut down plants, thus resulting in scarcity and higher prices.

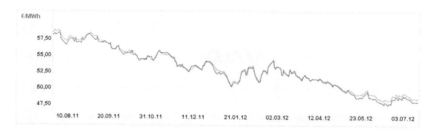

FIGURE 5.19 Prices of futures in Germany (price baseload year futures for 2013 (red) and for 2014 (gray)). (For interpretation of the references to color in this figure legend, the reader is referred to the web version of this book.). *Source: EEX (July 17, 2012).*

abuse of market power—there might be no need for additional capacity markets.[10]

With respect to time-dependent market structures, different new patterns are likely to emerge. Regarding the role of hedging and future contracts, an argument recently raised is that in markets with high shares of intermittent renewables no hedging is possible and future markets will break down. Ironically, as described in Section 4.3, the very opposite may be true. Hedging and tradable long-term contracts (LTC), to a large extent, will assume the role of capacity markets. For example, LTC traded years ahead on an annual basis will serve to secure the long-term provision of capacity. If LTC prices increase, it is a sign that capacity is becoming scarce. This, for example, may provide an incentive to "mothball" old fossil plants rather than shutting them down permanently. If LTC prices are decreasing, the opposite may happen. The closer the delivery date of contracts, the more fine-tuned will be the capacity reservation due to LTC purchasing.[11] The appealing aspect of this solution is that it works as a "voluntary capacity market."

The prices of future contracts in Germany for the years 2013 and 2014 have dropped continuously from August 2011 to July 2012 as shown in Figure 5.19. This can be seen as a sign that currently no capacity shortages are expected for these years.

In addition to these conventional future markets, short-term markets like intraday and balancing markets are showing growing relevance. In this context, it is likely that also *long*-term markets for these products will emerge. Another effect will be a continuing opening and extension of the balancing markets for electricity. The geographical areas for these products will become larger and, hence, more competition will take place. Such competition is likely to happen rather at the level of decentralized balancing organizations than at the current spot markets.

---

10. As described in Chapter 10, this appears to be the current thinking in Texas.
11. For instance, if good hydropower conditions are observed, less capacity will be hedged than vice versa.

As noted by a number of others in this volume, including Reeves in the *Preface*, eventually, it will be necessary to accept a new paradigm shift in our understanding of the whole electricity system, including switching from an inflexible and one way grid where variable load is met with changes in generation to a more flexible and smarter system that allows two-way flows, including greater scope for demand participation by consumers

6 CONCLUSIONS

The major conclusion of this chapter is that the electricity market and the electricity supply system of the future will look quite different than today while many of the fundamentals will remain. By and large, most of the effects of renewables are already known, what is new is that the variability of their generation will further increase if much higher quantities of wind and PV are fed into the grid, as appear to be the case for the EU.

The effects of these developments on the prices in electricity markets will be:

- Much more price volatility from hour to hour and day to day,
- Increasing relevance of intraday markets,
- Higher prices for fossil capacities and storage technologies for balancing the intermittent renewable generation,
- Growth of balancing markets and intensified competition at the level of decentralized balancing organizations.

In this context, a key question is whether and when an electricity market fully or mostly supplied by renewables can be brought about? While this issue has not been part of the analysis in this chapter, it is not difficult to predict that an electricity market fully supplied or mostly by renewables can only be achieved at an abnormally high cost. The simple explanation is that such a system will need large amounts of flexible generation and/or storage or equally large demand-side resources.

REFERENCES

Auer, J., 2011. Grid Regulation in Competitive Electricity Markets (Habilitation thesis). EEG, Vienna University of Technology, Vienna.
Auer, J., Resch, G., Haas, R., Held, A., Ragwitz, M., 2009. Regulatory instruments to deliver the full potential of renewable energy sources efficiently. Eur. Rev. Energy Markets 3 (2), 91–124 (special issue on "Incentives for a low-carbon energy future," ISSN 1782-1029).
Bauknecht, D., Brunekreeft, G., Meyer, R., 2013. From niche to mainstream: The evolution of renewable energy in the German electricity market, this volume.
Boltz, W., 2013. The challenges of electricity market regulation in the European Union, E-Control Austria, this volume.
Carraretto, C., Zigante, A., 2006. Interaction among competitive producers in the electricity market: an iterative market model for the strategic management of thermal power plants. Energy 31 (15), 3145–3158.

EC, 1997. Directive 96/92EC of the European Parliament and of the Council Concerning the Common Rules for the Internal Electricity Market. Official Journal L27 of 1/30/1997, Luxemburg.

EC, 2009. Directive on the Promotion of the Use of Energy from Renewable Sources, Brussels.

EU, 2012. EU Energy in Figures, Brussels.

European Parliament and Council, 2001. Directive of the European Parliament and of the Council on the Promotion of Electricity Produced from Renewable Energy Sources in the Internal Electricity Market. Directive 2001/77/EC—September 27, 2001, Brussels.

Haas, R., Auer, J., Glachant, J.-M., Keseric, N., Perez, Y., 2006. The liberalisation of the continental European electricity market—lessons learned. Energy Stud. Rev. 14 (2), 1–29.

Haas, R., Meyer, N.I., Held, A., Finon, D., Lorenzoni, A., Wiser, R., et al., 2008. Promoting electricity from renewable energy sources—lessons learned from the EU, U.S. and Japan. In: Sioshansi, F.P. (Ed.), Competitive Electricity Market. Elsevier, Oxford, Amsterdam.

Haas, R., Resch, G., Panzer, Ch., Busch, S., Ragwitz, M., Held, A., 2011a. Efficiency and effectiveness of promotion systems for electricity generation from renewable energy sources—Lessons learned from EU countries. Energy 36, 2186–2193.

Haas, R., Panzer, Ch., Resch, G., Ragwitz, M., Rice, G., Held, A., 2011b. A historical review of promotion strategies for electricity from renewable energy sources in EU countries. Renew. Sustain. Energy Rev. 15, 1003–1034.

Haas, R., Lettner, G, Auer, J., Duic, N., 2013. The looming revolution: how photovoltaics will change electricity markets in Europe fundamentally. Energy, forthcoming.

Jacobsson, S., Bergek, A., Finon, D., Lauber, V., Mitchell, C., Toke, D., et al., 2009. EU renewable energy support policy: Faith or facts? Energy Policy 37 (6), 2143–2146.

LBD-Beratungsgesellschaft (Hrsg.), 2011. Energiewirtschaftliche Erfordernisse zur Ausgestaltung des Marktdesigns für einen Kapazitätsmarkt Strom. Abschlussbericht.

Lund, H., 2005. Large-scale integration of wind power into different energy systems. Energy 30 (13), 2402–2412.

Nicolosi, M., 2010. Wind power integration and power system flexibility—An empirical analysis of extreme events in Germany under the new negative price regime. Energy Policy 38 (11), 7257–7268.

Percebois, J., 2013. The French paradox: Competition, nuclear rent and price regulation, this volume.

Pfaffenberger, W., Chrischilles, E., 2013. Turnaround in rough sea: The German electricity market, this volume.

Resch, G., Held, A., Faber, Th., Panzer, C., Toro, F., Haas, R., 2008. Potentials and prospects for renewable energies at global scale. Energy Policy 36 (11), 4048–4056.

Resch, G., Panzer, C., Ortner, A., Busch, S., Hoefnagels, R., Junginger, M., et al., 2012. Renewable energies in Europe—Scenarios on future European policies for RES. Reshaping project report D22, Vienna. Available from: <www.reshaping-res-policy.eu> (accessed March 2013).

Riesz, J., Gilmore, J., Hindsberger, M., 2013. Market design for variable generation, this volume.

Stoft, S., 2002. Power System Economics. IEEE Press, Piscataway, NJ.

Walls, W.D., Rusco, F.W., Ludwigson, J., 2007. Power plant investment in restructured markets. Energy 32 (8), 1403–1413.

Zubi, G., 2011. Technology mix alternatives with high shares of wind power and photovoltaics—case study for Spain. Energy Policy 39 (12), 8070–8077.

Chapter 6

Renewable Energy, Efficient Electricity Networks, and Sector-Specific Market Power Regulation [☆]

Günter Knieps
Institute of Transport Economics and Regional Policy, Albert-Ludwigs-Universität Freiburg, Gemany

1 INTRODUCTION

The political goal in European countries is to increase significantly the share of renewable energy in electricity generation during the next decades. The European Council of March 2007 decided on a mandatory target of a 20% share of energy from renewable resources in overall community energy consumption by 2020 (Council of the European Union, 2007, p. 21). Since the renewable energy potentials vary strongly between Member States, national overall targets for the share of energy from renewable sources in gross final consumption of energy in 2020 vary between 10% for Malta and 49% for Sweden, with 18% for Germany.[1] According to the Energy Roadmap 2050 of the European Commission (2011, p. 11) wind energy from the Northern Seas and the Atlantic sea basin as well as wind and solar power from the Mediterranean countries can provide important contributions to electricity supply. In Germany wind energy is considered to be of particular importance. The recent developments regarding increasing generation of renewable

---

[☆]Helpful comments by participants of research seminars at the University of Freiburg and the Karlsruhe Institute of Technology, in particular, by Gert Brunekreeft, Martin Keller, Kay Mitusch, Birgit Rosalowsky and Hans-Jörg Weiß are gratefully acknowledged. Special thanks are due to the editor of this volume for his ongoing encouragement and his helpful comments and suggestions.
1. Directive 2009/28/EC of the European Parliament and of the Council of 23 April 2009 on the promotion of the use of energy from renewable sources and amending and subsequently repealing Directives 2001/77/EC and 2003/30/EC, OJ L 140/16, Annex I.

energy create challenges for electricity networks. Whereas electricity genera-tion with conventional energy sources like coal, gas, and oil can be located rather close to demand, electricity generation with renewable energy natu-rally depends on the location of its source. For example, in Germany wind energy is largely generated in the north of the country and transmission to demand centers located in the south requires large transmission capacities.[2] This results in increasing scarceness of network capacities and, due to strongly fluctuating wind conditions, an increasing need for efficient alloca-tion of electricity transmission capacities.

The recent laws and regulations regarding electricity transmission in Germany prohibit transmission pricing for injection of electricity at genera-tion nodes and only allow transmission prices at customer nodes. In addition, according to the renewable energy law there are obligations to guarantee prioritized access to renewable energy, irrespective of the capacity con-straints and scarcities of the electricity networks. Moreover, network carriers may be obliged to extend their infrastructure capacities, irrespective of whether economic incentives are available. Under these currently existing institutional constraints economically unfavorable developments can already be observed: excessive demand for transmission capacities and resulting negative electricity prices during periods of strong wind energy injection. As a consequence, a large requirement for investments in additional trans-mission capacities is considered unavoidable (dena, 2010, p. 16). Although potentials of demand-side management for households and industry are also taken into account (dena, 2010, p. 409), a consistent economic analysis of optimal transmission prices and resulting incentives for capacity extensions is missing.

Priority access and guaranteed access for electricity from renewable energy sources are also considered to be important for the integration of renewable energy sources into the European markets for electricity. Member States, however, have the flexibility to implement different schemes of sup-port for energy from renewable sources at the national level. These may include investment aid, tax exemptions, renewable energy obligation support schemes as well as direct price support schemes including feed-in tariffs and premium payment levels.[3] Subsidies for renewable energy plants and connec-tion lines to the grid are very common in European countries. Focusing on electricity networks, it is interesting to note that Germany is not nearly the only country favoring the transmission of electricity from renewable

2. See Pfaffenberger and Chrischilles (2013) in this volume, Chapter 4, Figure 4.2.
3. Directive 2009/28/EC of the European Parliament and of the Council of April 23, 2009 on the promotion of the use of energy from renewable sources and amending and subsequently repeal-ing Directives 2001/77/EC and 2003/30/EC, OJ L 140/16, recitals 60 and 25, Article 2(k).

resources.[4] Priority usage of the grid is granted in 15 member countries. Obligations for grid expansions, in particular required for renewable energy transmission, are also implemented in several Member States.

Given the institutional background of liberalized European energy markets, electricity generation and transmission cannot be determined jointly by a central agency.[5] A precondition for exploiting the potentials of competitive decentralized electricity generators is that each electricity generator has the competency for making individual investment in power plants as well as (short-run) generation decisions.

The goal of this chapter is to develop an analytical framework for efficient transmission charging structures and network infrastructure investments from a disaggregated point of view. The opportunity costs of usage of electricity networks have to be properly reflected by electricity transmission prices in order to provide the economically efficient incentives for electricity injection and extraction at different locations (nodes). Moreover, the incentives for investments in transmission networks are derived simultaneously based on revenues from efficient transmission charges.

The resulting disaggregated nodal pricing framework provides valuable insights for the current reform debate in European electricity markets, in particular concerning the necessity to provide proper economic incentives for independent providers of renewable energy as well as the proper economic incentives for the customers, including the development of smart grids. The term smart grid emphasizes the increasing role of adding information and communication technologies to the electricity networks and complementary parts (e.g., smart metering at customer side). A particular focus is on medium voltage distribution networks with increasing relevance of distributed generation of renewable energy sources. Whereas in traditional electricity networks power was only injected at nodes located at (high voltage) transmission networks, decentralized injection of renewable energy in (medium voltage) distribution networks gains increasing importance (Brandstätt et al., 2011, pp. 244f.). As a consequence power flows are changing, even flows from distribution networks to transmission networks may become relevant. Nevertheless, a phasing out of transmission networks cannot be expected (Haber and Bliem, 2010, pp. 4f.).

---

4. Based on RES LEGAL, legal sources on renewable energy (www.res-legal.de), a collection of legal sources on support schemes for renewable energy and grid issues in the EU's 27 Member States. The subsections "overview of support system" and "overview of grid issues" include extensive information on support schemes and grid regulatory issues focusing on connection to the grid, use of the grid and grid expansion for each member country. In addition, a comparison of grid regulations in different member countries differentiating between connection to the grid, usage of the grid and grid expansion is provided.
5. Directive 96/92/EC of the European Parliament and of the Council of December 19, 1996 concerning common rules for the internal market in electricity, OJ L 27/20.

The chapter is organized as follows: In Section 2, electricity systems are introduced from a disaggregated perspective. Due to liberalized electricity markets competitive independent generators arise which take disaggregate transmission pricing signals of network providers as given. In Section 3, the social welfare maximizing benchmark of disaggregated nodal pricing under variable investments is developed. In contrast to integrated nodal pricing reflecting the total value of electricity consisting of generation and transmission costs at different nodes, disaggregated nodal pricing consists of three separate elements: First, electricity transmission prices raised by the network carrier, consisting of node-dependent injection and extraction prices based on system externalities. System externalities changing power loss and network scarcity are the opportunity costs of electricity injection or extraction depending on the node (location) where generation or extraction takes place.[6] Second, the generalized merit order indicating at which nodes injection is worthwhile so that generation costs and injection price do not exceed marginal willingness to pay on the wholesale market. Third, nodal prices at extraction nodes reflecting the sum of the (uniform) wholesale price and node-dependent extraction price.

In Section 4, disaggregated nodal pricing and market power regulation are analyzed. First the minimal regulatory basis is derived consisting of monopolistic electricity networks. Second the incentives of unregulated electricity network carriers are analyzed. Compared to social welfare maximization the pricing rule for electricity injection and the investment rule remain unchanged, whereas the optimal consumption rule changes due to the marginalization of the network monopolist. It can be shown that regulation of network-specific market power due to the monopolistic bottleneck of electricity networks should not result in regulation of network pricing structures or regulatory obligations to invest in network capacities. Instead, incentive regulation and subsequent price cap regulation should be applied. Finally, lessons for regulatory reform of European electricity networks are drawn, taking into account the increasing role of renewable energy and smart grids.

2 DISAGGREGATED NODAL PRICING

2.1 Electricity Systems: Integrated Versus Disaggregated Perspective

The traditional nodal pricing approach has been developed from a vertically integrated utility perspective. Before market liberalization electricity

---

6. In order to simplify notation, the term electricity transmission networks shall be applied, without differentiating between (high-voltage) transmission networks and (medium-voltage) distribution networks. Since all nodes are connected, the principle of system externalities holds globally throughout transmission and distribution networks.

systems consisting of electricity generation units and electricity transmission networks were typically organized as vertically integrated monopolies. The starting point was the peak load pricing model for electricity generation for heterogeneous plant types for a single welfare maximizing public utility reflecting the merit order of different plant types, whereby the output of each plant is valued at the last most expensive unit applied (Crew and Kleindorfer, 1976). The extension of this model to spatially located demand and supply over a fixed network with a given number of physical locations (nodes) became known as nodal pricing. Optimal spatial spot pricing for electricity indicates the value of energy at a given time and location. Demand and generator availability is considered to be stochastic, taking into account unanticipated demand fluctuations as well as probabilities of generator outages. Prices at each node reflect the integrated locational value of electricity including the costs of electricity generation and the transmission costs within the network (Bohn et al., 1984, p. 364).

Decision-making of integrated generation and transmission network providers has been considered as a complex operational research problem[7] focusing on optimization problems over which a unique decision-maker has unlimited control.[8] The focus has been on global optimization of an integrated generation and transmission network provider resulting in integrated centralized socially optimal investment and production decisions of electricity generation as well as transmission (Bohn et al., 1984; Caramanis et al., 1982; Schweppe et al., 1988).

The focus of this chapter, however, is on the incentives of electricity network owners to provide the economically efficient pricing signals for the usage of transmission networks to the decentralized and competitive electricity generators as well as the electricity consumers. In this context a disaggregated market framework is considered focusing on independent generation and demand nodes under the energy balance requirements from an *ex ante* (day ahead) perspective leaving the handling of real-time energy balance requirements to network carriers and markets for operating reserves. Only real power flow is considered.[9]

To localize the true opportunity costs of electricity transmission as well as the proper incentives for investments in electricity networks, simultaneous

---

7. "Traditional OR activity has focused on "optimization" or (primal) problems over which the decision-maker has direct control. As technology has developed, there has been a clear trend toward including more areas, more factors, more detail, more complexity, and more objectives within the scope of our analyses." (Read, 1996, p. 131).

8. It is interesting to note that one of the founders of the basic concepts of linear programming and related duality theory, A.W. Tucker, strongly motivated his research on optimization modeling with the relations between linear programming and the theory of electrical networks as modeled by Kirchhoff and Maxwell (Kuhn, 2002, p. 132).

9. For the differentiation of a decision-relevant "core framework" and the consideration of reactive power and similar ancillary services, see Gribik et al. (2007).

optimization of generators' investment and production decisions not only results in unnecessary complexities but also seems incompatible with the market framework of liberalized electricity markets. In order to take into account the role of markets for electricity generation and the competitive search process for locations, technologies, etc., free entry of electricity generators based on decentralized entrepreneurial decisions under competition should be reflected in the framework of the analysis.[10] Decentralized independent generator decisions are of particular relevance in the future world of strongly increasing generation of renewable energy taking place at different locations throughout Europe, which cannot be assumed to be in the hand of an integrated network owner.

It is important to distinguish between the possibilities of decentralized decision-making by application of dual programming approaches and the introduction of competition at least in some parts of the value chain. According to Read (1996, p. 131) decentralization of decision-making requires a demand for simpler analyses, taking into account the concerns and responsibilities of different decision-makers. This would imply a greater role for dual analyses and for the design and management of markets in the operational research literature. However, as long as one integrated electricity company has final control and complete information over all relevant variables, it becomes irrelevant whether the solutions (within the company) are found by a primal- or dual-programming approach (using shadow prices).[11] Moreover, it seems impossible to imitate the dynamic effects of competition on markets by operational research optimization procedures.

The aim of the disaggregated regulatory approach is to exploit the potentials of competition in network industries. This requires identifying those subparts of an electricity sector with network-specific market power and those subparts with competitive potentials. Since electricity networks are monopolistic bottlenecks (characterized as natural monopolies with irreversible costs) adequate access regulation to electricity networks is required to allow competition on the markets for electricity generation (Knieps, 2006, pp. 61ff.). The market power of generators and the role of extending or restricting this market power by alternative organizations of transmission rights (Gilbert et al., 2004) may be of particular relevance in the context of interconnector auctions (Höffler and Wittmann, 2007).

---

10. Although the costs and benefits of the competitive market framing by its very nature cannot be quantified *ex ante*, central planning cannot be a superior alternative. For the important role of competition as a search procedure for new information on production technologies and demand conditions on markets, see von Hayek, 1948.

11. According to the duality theorem for linear programming, for each (primal) maximization problem, there is an associated equivalent (dual) minimization problem. For example, decentralized decision-making within a firm with a large number of plants each using some overall company resources is optimized by application of dual accounting prices (Baumol, 1977, pp. 108, 117; Dantzig, 2002).

2.2 The Disaggregated Model Framework

In contrast to the integrated approach of Bohn et al. (1984) and Schweppe et al. (1988), in the following generator investment and generator locations are taken as exogenous and not optimized endogenously. Competitive generators make their generation and investment decisions independently, taking the charges for power injection as given. The opportunity costs of electricity injection and extraction give the decisive pricing signals; the network provider has to provide to the independent generators and consumers. The individual decisions of generators to inject electricity depend on the opportunity costs of injection and the marginal generation costs at their individual node (location).

The underlying market framework is assumed to be as follows. Generators inject electricity at different nodes depending on their location buying the right and obligation to inject on a day-ahead market. Netting is endogenously carried out at the different nodes, so that (net) generator as well as (net) extraction nodes can be differentiated. It is assumed that generators as well as consumers do not have market power in their demand for transmission capacities and take the network prices of the network provider as given. Within the transmission network the number of generators and consumers at the different nodes is large enough, so that individual contracts for net transmission of electricity from a generator at node i to a consumer at node j are not incentive compatible, because the consumer at node j is only interested in the wholesale market price for electricity, irrespective of whether electricity is generated at high marginal costs of generation and low opportunity costs of injection or vice versa. In particular, the bilateral case of interconnector capacity markets (Höffler and Wittmann, 2007) is not considered.

It is well known from network economics that congestion pricing (short-run problem) and investment decision in network capacities (long-run problem) should be considered simultaneously (MacKie-Mason and Varian, 1995, pp. 301ff.; Mohring and Harwitz, 1962). Network capacity should be extended to the point where the marginal costs of an extra unit of capacity are equal to its marginal benefits of reduced congestion. The extension of network capacities to such a degree that congestion disappears would result in overinvestment and cannot be considered a socially optimal solution.

In the following the goal is to derive an analytical framework for pricing and investment decisions of electricity network providers. Investment decisions and production decisions of electricity generators are not under the decision authority of the network operator, but taken in a decentralized manner by competing generators. Neither the marginal costs of generation nor the investment problem of the generators at different nodes are considered to be under the control of the network capacity provider.

The focus is on the different nature of congestion externalities in electricity networks compared to other network industries (e.g., gas transportation and railways). The transmission of electricity within a multinode network is subject to physical—technical characteristics (Ohm's law and Kirchhoff's law), resulting in loop flows. If injection or extraction at a node takes place, the usage of all lines between all nodes is affected, changing power loss and network scarcity and therefore causing system externalities. In contrast, single line (local) congestion externalities arise on a gas pipeline (Cremer and Laffont, 2002), a highway (Mohring and Harwitz, 1962), a railroad track or airport, where the opportunity costs of infrastructure usage are limited to the specific individual infrastructure.[12]

It is assumed that the electricity transmission network of a typical network provider consists of an exogenously given number of H lines. All lines are connected, building a transmission network between all nodes (net injection nodes and net extraction nodes).

Marginal generation costs at net injection node j are assumed to be constant.

λ_j $j = 1,\ldots,J$ exogenously given
w_i infrastructure size (capacity) of each line i, $i = 1,\ldots, H$
$w = (w_1,\ldots, w_H)$ vector of transmission network infrastructure investment
$g_t = (g_{1t},\ldots, g_{Jt})$ net injection in node $j = 1,\ldots, J$
$d_t = (d_{1t},\ldots, d_{Kt})$ net extraction in node $k = 1,\ldots, K$
$z_{it}(g_t, d_t, w)$ power flow on line i in period t, $i = 1,\ldots, H$

Electricity injection in nodes j may increase or decrease the power flows on different lines, because opposing flows are canceling each other out, the same holds also for electricity extraction in nodes k:

$\partial z_{it}/\partial g_{jt} > 0$ increasing injection results in increased power flow on line i
$\partial z_{it}/\partial g_{jt} < 0$ increasing injection results in decreased power flow on line i
$\partial z_{it}/\partial d_{kt} > 0$ increasing extraction results in increased power flow on line i
$\partial z_{it}/\partial d_{kt} < 0$ increasing extraction results in decreased power flow on line i
$z_t = (z_{1t},\ldots, z_{Ht})$ vector of power flows along each line in period t
w_i is assumed to be a continuous variable (scalar), with no indivisibilities
$\rho(w_i)$ investment costs of w_i
T life time of investment in lines, $t = 1,\ldots, T$ periods
$\hat{Z}_i(w_i)$ $i = 1,\ldots, H$ maximal capacity of line i depending on the size of the line
$\partial \hat{z}_i/\partial w_i > 0$ maximal power flow increases with the size of the line (infrastructure capacity)

---

12. Within the Internet architecture of differentiated services networks data packet transmission, the application of strict priority scheduling may provide a structure for congestion externalities between different service classes, such that lower priority class data packets have to bear higher externality costs than higher priority class data packets (Knieps, 2011). Whereas in the Internet traffic network the externalities within and between traffic classes are the result of entrepreneurial decisions by the traffic providers, system externalities of injection or extraction in an electricity transmission network with a given number of interconnected nodes are the result of physical laws and cannot be influenced by organizational design.

A change in the power flow on line i does not only depend on a change in investments on line i, but also on a change in investments on all other lines. It is assumed that an increase in investment increases network capacity.[13] The underlying assumption typically applied in network industries is that additional incremental investment is beneficial by reducing congestion externalities and capacity constraints (Schweppe et al., 1988, p. 247).

$$\frac{\partial z_{it}}{\partial w_i} < 0; \quad \frac{\partial z_{it}}{\partial w_j} \neq 0 \quad i = 1, \ldots, H$$

$V_t(z_{it}(g_t, d_t, w))$ power loss on line i in period t
$V_t(z_t) = \sum_{i=1}^{H} V_t(z_{it}(g_t, d_t, w))$ power loss of flow Z_t

$$\frac{\partial V_t(z_t)}{\partial w_i} = \sum_{i=1}^{H} \frac{\partial V_t(z_{it}(g_t, d_t, w))}{\partial w_i} < 0$$

A marginal increase of investments on line i reduces power loss on all lines (electricity always searches the way of lowest resistance).

p_{kt} (d_{kt}) denotes the inverse demand function for electricity extraction at node k in period t; p_{kt} is the ("gross") price (total user costs) for electricity at node k in period t. The consumer price should include generation costs and opportunity costs for transmission through the electricity network (injection and extraction price). Assume that there are zero-income effects associated with the demand functions p_{kt}, so that consumer surplus of total demand for electricity at extraction node k in period t is

$$S(d_{kt}) = \int_0^{d_{kt}} p_{kt} \, d\tilde{d}_{kt} \quad \text{thus,} \quad \frac{\partial S}{\partial d_{kt}} = p_{kt}, \quad k = 1, \ldots, K \text{ (exogenously given)}$$

and total consumer surplus of demand on different extraction nodes in period t is

$$S(d_t) = \sum_{k=1}^{K} S(d_{kt}), \quad t = 1, \ldots, T$$

---

13. There are counterexamples, where strengthening an arbitrary line may even decrease capacity (see the example in Wu et al., 1996, p. 16); however, under regular conditions (e.g., a proportionate increase of injection and line resistance) this assumption seems justified (Schweppe et al., 1988, p. 247, Appendix p. 319). In transportation economics there is a similar example, the so-called Braess paradox, which points to the possibility that the building of an additional road may even increase congestion problems. However, Samuelson has already shown that adequate congestion prices will prevent the occurrence of this paradox (Samuelson, 1992, p. 7).

3 THE SOCIAL WELFARE MAXIMIZING BENCHMARK UNDER VARIABLE INVESTMENTS

In contrast to integrated nodal pricing reflecting the total value of electricity consisting of generation and transmission costs at different nodes, disaggregated nodal pricing consists of three separate elements: (1) electricity transmission prices raised by the network carrier, consisting of node-dependent injection and extraction prices, (2) the generalized merit order indicating at which nodes injection is worthwhile so that generation costs and injection price do not exceed marginal willingness to pay on wholesale market, and (3) nodal prices at extraction nodes reflecting the sum of the (uniform) wholesale price and node-dependent extraction price. Disaggregated nodal prices can be derived for a transmission network of a historically grown size $\overline{w}$ exogenously given infrastructure (short-run) or alternatively for a variable infrastructure size w. In the following the disaggregated nodal price under variable infrastructure shall be derived simultaneously. We shall see that the structure of the disaggregated nodal price is identical for both cases but the system externalities and thereby the optimal injection and extraction charges will differ.

3.1 Socially Optimal Network Investments and Disaggregated Nodal Pricing

In the following the first best social welfare maximizing benchmark of optimal network investment and disaggregated nodal pricing decisions are derived. The investment and production decisions of the generators are taken independently, taking the node-dependent injection prices of the network carrier as given.[14] Due to this entrepreneurial independence of decentralized competitive generators this welfare maximization problem is not equivalent to a primal optimization problem of vertically integrated electricity firms.

$$\max_{g_{kt},d_{jt},w_i} \sum_{t=1}^{T} \left(S(d_t) - \sum_{j=1}^{K} \lambda_j g_{jt} \right) - \sum_{i=1}^{L} \rho_i(w_i) \tag{6.1}$$

$$z_{it}(g_t, d_t, w) \leq \hat{Z}_{it}(w) \quad i = 1, \ldots L, \quad t = 1, \ldots, \overline{1} \quad | \mu_{z_i}(w) \tag{6.2}$$

---

14. It should be noted that under the assumption of a static environment where all decisions on generation are in the hands of the network owner, the relevant system externalities as well as the socially optimal network investment decisions can be derived in an integrated primal optimized model context, see Bohn et al. (1984) for given investment level and Schweppe et al. (1988), Chapter 10, pp. 237ff. for variable infrastructure level. However, the disaggregated nodal pricing approach reflects the entrepreneurial flexibility of generators on European electricity markets; moreover, the model framework gains from significant analytical simplifications.

$\mu_{z_i}(w)$: shadow value of line constraints

$$g_t = d_t + V_t(z_t) \quad | \; \theta_t \tag{6.3}$$

θ_t: shadow value of system balance equation

$$L = \sum_{t=1}^{T} \left(S(d_t) - \sum_{j=1}^{K} \lambda_j g_{jt} \right) - \sum_{i=1}^{H} \rho_i(w_i) -$$
$$\sum_{t=1}^{T} \left(\sum_{i=1}^{H} \mu_{z_{it}} (z_{it}(g_t, d_t, w) - \hat{z}_{it}(w)) - \theta_t(d_t + V_t(z_t) - g_t) \right) \tag{6.4}$$

$$\frac{\partial L}{\partial g_{jt}} = -\lambda_j - \sum_{i=1}^{H} \mu_{z_{it}} \frac{\partial z_{it}(\cdot, w)}{\partial g_{jt}} - \theta_t \left(-1 + \frac{\partial V_t(z_t)}{\partial g_{jt}} \right) = 0, \quad j = 1,\dots,J, \; t = 1,\dots,T \tag{6.5}$$

$$\frac{\partial L}{\partial d_{kt}} = \frac{\partial S}{\partial d_{kt}} - \sum_{i=1}^{H} \mu_{z_{it}} \frac{\partial z_{it}(\cdot, w)}{\partial d_{kt}} - \theta_t \left(+1 + \frac{\partial V_t(z_t)}{\partial d_{kt}} \right) = 0 \quad k = 1,\dots,K, \; t = 1,\dots,T \tag{6.6}$$

$$\frac{\partial L}{\partial w_i} = -\frac{\partial \rho_i}{\partial w_i} - \sum_{t=1}^{T} \left[\sum_{j=1}^{H} \left(\mu_{z_{it}} \left(\frac{\partial z_{jt}}{\partial w_i} - \frac{\partial \hat{z}_{jt}}{\partial w_i} \right) + \theta_t \frac{\partial V_t(z_t)}{\partial w_i} \right) \right] = 0 \quad i = 1,\dots,H \tag{6.7}$$

By means of Eqs. (6.5)–(6.7) optimal decisions for network carrier (network prices and optimal network investment levels) as well as the decentralized decision-making of competitive generators resulting in the generalized merit order and the optimal consumption rule can be derived.

From simultaneous solution of Eqs. (6.5)–(6.7) follows:

$g_t^* = (g_{1t}^*, \dots, g_{Kt}^*)$ socially optimal net injection in node j, $j = 1, \dots, J$
$d_t^* = (d_{1t}^*, \dots, d_{Jt}^*)$ socially optimal net extraction in node k, $k = 1, \dots, K$
$w^* = (w_1^*, \dots, w_L^*)$ socially optimal infrastructure level in line i, $i = 1, \dots, H$

3.1.1 Socially Optimal Investment Rule in Electricity Networks

From (Eq. 6.7) follows:

$$\frac{\partial \rho_i}{\partial w_i} = - \sum_{t=1}^{T} \left[\sum_{j=1}^{H} \left(\mu_{z_{it}} \left(\frac{\partial z_{jt}}{\partial w_i} - \frac{\partial \hat{z}_{jt}}{\partial w_i} \right) + \theta_t \frac{\partial V_t(z_t)}{\partial w_i} \right) \right] \quad i = 1,\dots,H \tag{6.8}$$

The infrastructure capacity of each line should be extended to the point where the marginal costs of an extra unit of capacity are equal to the marginal benefit through reductions of system externalities (reductions of power loss and alleviations of line flow constraints) on all lines.

3.1.2 Socially Optimal Injection and Extraction Pricing Rules at Optimal Investment Level

From Eqs. (6.5) and (6.6) system externalities according to opportunity costs of injection and extraction at different nodes and subsequently the optimal network injection and extraction pricing rules (at optimal investment level) can be derived:

System externality of injection at node j:

$\sum_{i=1}^{H} \mu_{z_{it}} \partial z_{it}/\partial g_{jt}$ impact of marginal increase of injection at node j on scarce transmission capacity plus
$\partial \theta_t V_t(z)/\partial g_{jt}$ impact of marginal increase of injection at node j on power loss.

System externalities of extraction at node k:

$\sum_{i=1}^{H} \mu_{z_{it}} \partial z_{it}/\partial d_{kt}$ impact of marginal increase of extraction at node k on scarce transmission capacity plus
$\partial \theta_t V_t(z)/\partial d_{kt}$ impact of marginal increase of extraction at node d on power loss.

Socially optimal injection pricing rule at optimal investment level:

$$\tau_{jt} = \sum_{i=1}^{H} \mu_{z_{it}} \frac{\partial z_{it}(\cdot, w^*)}{\partial g_{jt}} + \theta_t \left(\frac{\partial V_t(z_t)}{\partial g_{jt}} \right) \quad j = 1, \ldots, J, \quad t = 1, \ldots, T \quad (6.9)$$

Socially optimal extraction pricing rule at optimal investment level:

$$\tau_{kt} = \sum_{i=1}^{H} \mu_{z_{it}} \frac{\partial z_{it}(\cdot, w^*)}{\partial d_{kt}} + \theta_t \left(\frac{\partial V_t(z)}{\partial d_{kt}} \right) \quad k = 1, \ldots, K, \quad t = 1, \ldots, T \quad (6.10)$$

Since system externalities at an injection node may be positive or negative, the optimal network injection prices may be positive or negative. The same holds for extraction nodes, so that optimal network extraction prices may also be positive or negative.

Opportunity costs of network usage (injection or extraction) and subsequently the optimal (injection or extraction) network usage prices are different at optimal investment level w^* compared to a short-run given investment level $\overline{w}$.

3.2 The Generalized Merit Order Rule and the Optimal Consumption Rule

It is well known that the merit order of electricity generation without networks is constituted according to increasing marginal generation costs until the most expensive unit matches demand, determining the so-called system lambda (Crew and Kleindorfer, 1976). In contrast, taking into account the opportunity costs of electricity transmission, the decision-relevant costs for electricity generators not only consists of the marginal costs of generation, but also of the opportunity costs of electricity injection. The original merit order according to marginal generation costs (without network transmission costs) becomes irrelevant. Instead the generalized merit order according to the increasing sum of marginal generation costs and opportunity costs of injection becomes relevant.[15]

It is assumed that a competitive wholesale market for electricity exists, so that a wholesale price for electricity already injected into the network exists reflecting a market value according to the shadow value of the system balance equation θ_t (the opportunity costs of system balance equation). These opportunity costs of the last unit of demand which is still served are knowable to the different generators at the different nodes (locations).

The network carrier requires injection prices according to opportunity costs of injection

$$\tau_{jt}, \ j = 1, \ldots, J, \ t = 1, \ldots, T$$

The wholesale market sends θ_t: Marginal willingness to pay at the system limit (last marginal unit which is served (or just not served)).

From Eqs. (6.5) and (6.9) follows for the marginal most expensive generator:

$$-\lambda_j - \tau_{jt} + \theta_t = 0$$
$$\Rightarrow \quad \theta_t = \lambda_j + \tau_{jt}$$

The generators at a net injection node make a decentralized decision according to the generalized merit order rule, whether to inject or not inject.

$$\lambda_1 + \tau_{1t}, \lambda_2 + \tau_{2t}, \ldots, \lambda_J + \tau_{Jt}$$

If $\lambda_j + \tau_{jt}, \ \leq \theta_t$ inject
If $\lambda_j + \tau_{jt}, \ > \theta_t$ do not inject

Generators sell electricity to a wholesale market, if the individual marginal costs of generation and the opportunity costs of injection into the transmission network at its node (location) are competitive and do not exceed the marginal willingness to pay in the wholesale market. This may result in a reverse of the (hypothetical) original merit order without network considerations.

---

15. Thus, in our model framework only decision-relevant injection and extraction nodes are analyzed.

Consumers at extraction nodes have to pay the uniform wholesale price and the node-dependent opportunity costs of extraction. The network carrier requires extraction prices according to the opportunity costs of extraction

$$\tau_{kt}, \quad k = 1, \ldots, K, \quad t = 1, \ldots, T$$

From Eq. (6.6) follows:

$$\frac{\partial S(d_t)}{\partial d_{kt}} - \tau_{kt} - \theta_t = 0 \qquad \frac{\partial S}{\partial d_{kt}} \text{ exogenously given } k = 1, \ldots, K$$

$$\Rightarrow p_{kt} \qquad\qquad = \theta_t \qquad\qquad\qquad + \tau_{kt}$$

price for demand　　　wholesale price
in node k in period t　（irrespective at which node　extraction price at node k
　　　　　　　　　　generation takes place)

From transportation economics the arbitrage argument is well known that in a social welfare optimum the price differences at different locations reflect the marginal transportation costs between these two locations. Due to system externalities this conclusion does not hold for electricity transmission.[16] Differences between optimal injection prices at two injection nodes j_1 and j_2 reflect the fact that electricity injection at node j_1 and node j_2 causes different system externalities, irrespective of whether the transmission line between both injection nodes is congested or not. Possibilities for arbitrage of competitive generators at the different locations do not exist. Only the sum of generation costs and injection price is relevant for determining whether a generator is successful in matching the wholesale price, irrespective of whether a generator has high marginal costs of generation and low opportunity costs of injection or vice versa. Similarly, differences between optimal extraction prices at two extraction nodes k_1 and k_2 reflect the fact that electricity extraction at node k_1 and node k_2 causes different system externalities, irrespective of whether the transmission line between both extraction nodes is congested or not. Possibilities for arbitrage between consumers at the different locations do not exist. Thus the price differences for demand at the different nodes are stable and are the result of different system externalities at the different locations.

3.3 The Financial Viability Problem

The purpose of first best optimal electricity injection and extraction prices is to provide incentives for efficient usage of electricity transmission capacities.

---

16. Wu et al., 1996, p. 15, already conclude that (integrated) nodal price differences do not equate to marginal transportation costs between those nodes.

Nevertheless optimal network usage fees also contribute to the financing of electricity network infrastructures. The net revenues to finance infrastructure by social welfare maximizing injection and extraction prices derived in Eqs. (6.9) and (6.10) are:

Revenues from the socially optimal injection price at optimal investment level:

$$\sum_{j=1}^{J} R(\tau_{jt}) = \sum_{j=1}^{J} \tau_{jt} g_{jt} = \sum_{j=1}^{J} \left[\sum_{i=1}^{H} \mu_{z_{it}} \frac{\partial z_{it}(g_t, d_t, w^*)}{\partial g_{kt}} g_{jt} + \theta_t \left(\frac{\partial V_t(z_t)}{\partial g_{jt}} \right) g_{jt} \right] \quad t = 1, \ldots, T$$

Revenues from the socially optimal extraction price at optimal investment level:

$$\sum_{k=1}^{K} R(\tau_{kt}) = \sum_{k=1}^{K} \tau_{kt} d_{kt} = \sum_{k=1}^{K} \left[\sum_{i=1}^{H} \mu_{z_{it}} \frac{\partial z_{it}(g_t, d_t, w^*)}{\partial d_{kt}} d_{kt} + \theta_t \left(\frac{\partial V_t(z_t)}{\partial d_{kt}} \right) d_{kt} \right] \quad t = 1, \ldots, T$$

In the following we assume that z_{it} is a homogenous function of degree 0, such that scarcity on lines only depends on the relation between power flow and dimension and not on the absolute size of the infrastructure. In other words: z_{it} is homogeneous of degree zero if for all $\alpha > 0$, $z_{it}(\alpha g_t, \alpha d_t, \alpha w^*) = \alpha^0 z_{it}(g_t, d_t, w^*)$. Thus, V_t is also homogeneous of degree zero, such that power loss only depends on the relation between power flow and dimension and not on the absolute size of the infrastructure.[17] In other words: if for all $\alpha > 0$, $z_{it}(\alpha g_t, \alpha d_t, \alpha w^*) = \alpha^0 z_{it}(g_t, d_t, w^*)$, then $V_t(z_{it}(\alpha g_t, \alpha d_t, \alpha w^*)) = \alpha^0 V_t(z_{it}(g_t, d_t, w^*))$.

Under the assumption that z_{it} and subsequently V_t are homogeneous of degree zero, such that the Euler theorem can be applied, and taking into account the optimal investment rule (6.8), we obtain:

$$\sum_{t=1}^{T} \sum_{j=1}^{J} \left[\sum_{i=1}^{H} \mu_{z_{it}} \frac{\partial z_{it}(g_t, d_t, w^*)}{\partial g_{jt}} g_{jt} + \theta_t \left(\frac{\partial V_t(z_t)}{\partial g_{jt}} \right) g_{jt} \right] +$$

$$\sum_{t=1}^{T} \sum_{k=1}^{K} \left[\sum_{i=1}^{H} \mu_{z_{it}} \frac{\partial z_{it}(g_t, d_t, w^*)}{\partial d_{kt}} d_{kt} + \theta_t \left(\frac{\partial V_t(z_t)}{\partial d_{kt}} \right) d_{kt} \right]$$

$$= -\sum_{t=1}^{T} \left[w_i^* \sum_{j=1}^{H} \left(\mu_{z_{it}} \left(\frac{\partial z_{jt}}{\partial w_i} - \frac{\partial \hat{z}_{jt}}{\partial w_i} \right) - \theta_t \frac{\partial V_t(z_i)}{\partial w_i} \right) \right] = \sum_{i=1}^{H} \frac{\partial \rho_i}{\partial w_i} w_i^*$$

Only in the special case, if there are constant returns to scale in infrastructure construction, then $\sum_{i=1}^{H}(\partial \rho_i / \partial w_i) w_i^* = \sum_{i=1}^{H} \rho_i(w_i^*)$ and total

---

17. This assumption is typically made in network economic literature, if infrastructure size is rather large (Keeler and Small, 1977, p. 2; Mohring and Harwitz, 1962).

investment costs are covered by revenues from socially optimal injection and extraction prices. With decreasing returns in infrastructure construction, $\sum_{i=1}^{H}(\partial \rho_i/\partial w_i)w_i^* > \sum_{i=1}^{H} \rho_i(w_i^*)$, and socially optimal injection and extraction charges exceed total investment costs. However, for the relevant case of increasing returns in infrastructure construction of electricity transmission networks, $\sum_{i=1}^{H}(\partial \rho_i/\partial w_i)w_i^* < \sum_{i=1}^{H} \rho_i(w_i^*)$ holds and socially optimal first best injection and extraction charges do not cover total investment costs. Thus, a regulator enforcing first best injection and extraction charges might jeopardize the financial viability of the electricity transmission network.

It would be detrimental for the regulatory agency to prescribe first best welfare maximizing disaggregated nodal pricing. For the relevant case of economies of scale in electricity transmission networks this would result in a regulatory caused deficit and a subsequent distortion of the financial viability of the transmission network carrier.

4 DISAGGREGATED NODAL PRICING AND MARKET POWER REGULATION

4.1 Minimal Regulatory Basis

In the disaggregated regulatory framework the smallest possible regulatory basis has to be chosen, applying regulatory instruments strictly to those parts of a network where network-specific market power exists. In particular, the cost covering of the different electricity generators and the cost covering of the transmission network provider are to be differentiated. Generation capacity at different nodes is chosen endogenously and the cost covering problem is solved endogenously on the generator side. Depending on the type of generation units economies of scale in the size of generation units are different and will be exhausted. Free market entry results in the emergence of competing generators. Market power regulation of electricity generators remains superfluous. Under decentralized production decisions of electricity generators incentives for injection (without subsidy) will arise as long as the price θ_t on the wholesale market covers the node-dependent injection price and the variable generation costs.

In contrast, electricity transmission network carriers are neither confronted with active network competition nor with potential competition. Electricity networks have the characteristics of a natural monopoly. For a particular geographic area the interlinking of all points of generation and consumption provided by one network carrier is most cost-efficient. Furthermore, the setup of electricity transmission networks involves geographically sunk irreversible investments. Thus electricity transmission networks are monopolistic bottlenecks. This results in the necessity to regulate the network-specific market power of electricity network carriers (Knieps, 2006, pp. 53, 61ff.).

4.2 The Profit Maximizing Transmission Network Carrier

A natural behavioral assumption is that network carriers do not maximize social welfare but maximize their monopolistic profit. Thus in the following disaggregated nodal pricing shall be considered under the profit maximizing objective function. The network owner provides disaggregated nodal pricing signals to two sides: to the different generators located at the (net) injection nodes and to the different consumers located at the (net) extraction nodes. Since the consumers, via the wholesale market price, (implicitly) also pay the injection price, the network monopolist will avoid the fallacy of double marginalization. It is well known that incentives for double marginalization only exist in the institutional framework of independent upstream and downstream monopolists (Machlup and Taber, 1960; Spengler, 1950). Since monopolies always want to minimize production costs, marginalization only takes place at the extraction nodes (demand side):

$$\max_{g_{kt},d_{kt},w_i} \sum_{t=1}^{T}\left(\sum_{1}^{K} p_{kt}(d_{kt})d_{kt} - \sum_{j=1}^{J}\lambda_j g_{jt}\right) - \sum_{i=1}^{H}\rho_i(w_i) \tag{6.11}$$

$$z_{it}(g_t,d_t,w) \le \hat{z}_{it}(w) \quad i=1,\ldots H, \quad t=1,\ldots,T, \quad |\mu_{z_i}(w) \tag{6.12}$$

$$g_t = d_t + V_t(z) \quad |\theta \text{ but only } w_i \text{ can be influenced} \tag{6.13}$$

$$L = \sum_{t=1}^{T}\left(\sum_{1}^{K} p_{kt}(d_{kt})d_{kt} - \sum_{j=1}^{J}\lambda_j g_{jt}\right) - \sum_{i=1}^{H}\rho_i(w_i) -$$
$$\sum_{t=1}^{T}\sum_{i=1}^{H}\mu_{z_{it}}(z_{it}(g_t,d_t,w) - \hat{z}_{it}(w)) - \theta_t(d_t + V_t(z_t) - g_t) \quad t=1,\ldots,T \tag{6.14}$$

$$\frac{\partial L}{\partial g_{jt}} = \lambda_j - \sum_{i=1}^{H}\mu_{z_i}\frac{\partial z_{it}(\cdot,w)}{\partial g_{jt}} - \theta_t\left(-1 + \frac{\partial V_t(z_t)}{\partial g_{jt}}\right) = 0 \quad j=1,\ldots,J, \quad t=1,\ldots,T \tag{6.15}$$

$$\frac{\partial L}{\partial d_{kt}} = p_{kt} + \frac{\partial p_{kt}}{\partial d_{kt}}d_{kt} - \sum_{i=1}^{H}\mu_{z_{it}}\frac{\partial z_{it}(\cdot,w)}{\partial d_{kt}} - \theta_t\left(+1 + \frac{\partial V_t(z_t)}{\partial d_{kt}}\right) = 0 \tag{6.16}$$
$$k=1,\ldots,K, \quad t=1,\ldots,T$$

$$\frac{\partial L}{\partial w_i} = -\frac{\partial \rho_i}{\partial w_i} - \sum_{t=1}^{T}\left[\sum_{j=1}^{H}\left(\mu_{z_{it}}\left(\frac{\partial z_{jt}}{\partial w_i} - \frac{\partial \hat{Z}_{jt}}{\partial w_i}\right) + \theta_t\frac{\partial V_t(z_t)}{\partial w_i}\right)\right] = 0 \quad i=1,\ldots,H \tag{6.17}$$

Compared to social welfare maximization, the optimal consumption rule changes due to the marginalization of the network monopolist, leading to a monopolistic markup $a_{kt}^m = -(\partial p_{kt}/\partial d_{kt})d_{kt}$ in prices for electricity extraction

(Eq. 6.16).The gross monopoly price p_k^m at the extraction nodes k at time t can be derived from Eq. (6.16) as follows:

$$MR_{kt} = p_{kt} + \frac{\partial p_{kt}}{\partial d_{kt}} d_{kt} = \theta_t + \tau_{kt} = MC_{kt}$$

Thus:

$$\frac{p_{kt}^m - MC_{kt}}{p_{kt}^m} = -\frac{1}{\varepsilon_{kt}} \text{ with } \varepsilon_{kt} = \frac{\partial d_{kt}}{\partial p_{kt}} \frac{p_{kt}}{d_{kt}}$$

$$p_k^m = \theta_t + \tau_{kt} + a_{kt}^m = \theta_t + \tau_{kt}^m$$

$\tau_{kt}^m = \tau_{kt} + a_{kt}^m$ is the monopolistic extraction charge at node k in period t, characterized by a monopolistic markup.

The markup at the different extraction nodes will vary according to price elasticities of demand, so that extraction nodes with inelastic demand will have to bear a higher markup compared to extraction nodes with more elastic demand.

Compared to social welfare maximization the pricing rule for electricity injection (Eq. 6.15) and the investment rule (Eq. 6.17) remain unchanged.

From simultaneous solution of Eqs. (6.15)−(6.17) follows:

$g_t^m = (g_{1t}^m, \ldots, g_{Kt}^m)$ profit maximizing net injection in node $j, j = 1, \ldots, J$
$d_t^m = (d_{1t}^m, \ldots, d_{Jt}^m)$ profit maximizing net extraction in node $k, k = 1, \ldots, K$
$w^m = (w_1^m, \ldots, w_H^m)$ profit maximizing infrastructure level in line $i, i = 1, \ldots, H$

Although the pricing rule at injection nodes does not change because no markup is raised at injection nodes, system externalities at injection nodes change due to a changed power flow and thereby injection prices τ_{jt}^m differ from socially optimal injection prices τ_{jt}.

4.3 The Fallacies of Price Structure Regulation

Market power regulation of electricity transmission networks should not intervene in disaggregated nodal pricing structures. Instead price cap regulation of transmission price levels should be applied, with entrepreneurial flexibility regarding injection and extraction pricing structures (Knieps, 2006, p. 65f.).

As has already been shown incentives to take into account the opportunity costs of capacity usage and subsequent system externalities at injection nodes as well as extraction nodes exist not only under welfare maximizing conditions but also for an unregulated transmission network monopolist. Regulatory interventions forbidding the raising of injection charges conflict with an efficient allocation of transmission capacities.

Price differentiation based on system externalities should allow financial viability of the transmission network owner. Since economies of scale of capacity expansion are considered to be of particular relevance in electricity transmission networks (Brunekreeft et al., 2005, p. 75) an entrepreneurial search for cost covering pricing structures becomes relevant. Similar to the unregulated transmission network monopolist price cap regulation leaves entrepreneurial flexibility for choosing pricing structure. Thus a price cap-regulated monopolist will again not have incentives for double marginalization and thereby only raising a markup on the system externalities of electricity extraction nodes, raising injection prices according to system externalities at injection charges. Similar to the unregulated monopoly case, the markup at the different extraction nodes will vary according to differences in the price elasticities of demand, so that at extraction nodes with inelastic demand consumers will have to bear a higher markup compared to extraction nodes with more elastic demand.

As an alternative to linear extraction prices optional two-part tariffs can be applied, with an option for households to choose between a fixed network access charge combined with a lower variable extraction price versus the linear variable extraction price.

4.4 The Fallacies of Investment Obligations

Although the investment rule under profit maximization does not change compared to the socially optimal investment rule, the benefit of extending the size of capacity by reducing system externalities is calculated on the base of power flows resulting under profit maximization. Due to the markup on socially optimal extraction prices the demand at profit maximizing extraction prices is lower than the demand at socially optimal extraction prices. Therefore, the socially optimal investment level given the lower monopolistic traffic flow would lead to overinvestment. Instead the profit maximizing investment level taking into account profit maximizing injection and extraction charges is smaller than the socially optimal investment level taking into account socially optimal injection and extraction charges.

Assuming again that the functions z_t (and therefore V_t) are homogeneous of degree zero, such that Euler's theorem can be applied, it follows:

$$\sum_{t=1}^{T}\sum_{j=1}^{J}\left[\sum_{i=1}^{H}\mu_{z_{it}}\frac{\partial z_{it}(g_t,d_t,w^*)}{\partial g_{jt}}g_{jt}+\theta_t\left(\frac{\partial V_t(z_t)}{\partial g_{jt}}\right)g_{jt}\right]+$$

$$\sum_{t=1}^{T}\sum_{k=1}^{K}\left[\sum_{i=1}^{H}\mu_{z_{it}}\frac{\partial z_{it}(g_t,d_t,w^*)}{\partial d_{kt}}d_{kt}+\theta_t\left(\frac{\partial V_t(z_t)}{\partial d_{kt}}\right)d_{kt}\right]=\sum_{i=1}^{H}\frac{\partial \rho_i}{\partial w_i}w_i^*$$

$$>$$

$$\sum_{t=1}^{T}\sum_{j=1}^{J}\left[\sum_{i=1}^{H}\mu_{z_{it}}\frac{\partial z_{it}(g_t,d_t,w^m)}{\partial g_{jt}}g_{jt}+\theta_t\left(\frac{\partial V_t(z_t)}{\partial g_{jt}}\right)g_{jt}\right]+$$

$$\sum_{t=1}^{T}\sum_{k=1}^{K}\left[\sum_{i=1}^{H}\mu_{z_{it}}\frac{\partial z_{it}(g_t,d_t,w^m)}{\partial d_{kt}}d_{kt}+\theta_t\left(\frac{\partial V_t(z_t)}{\partial d_{kt}}\right)d_{kt}\right]=\sum_{i=1}^{H}\frac{\partial \rho_i}{\partial w_i}w_i^m$$

Investment obligations forcing the monopolistic network transmission carrier to conform to a socially optimal investment level (given socially optimal transmission flow) would therefore result in socially inefficient over-investment. This argument also holds for price cap-regulated transmission networks as long as the resulting price level allows a positive markup on extraction charges under social welfare maximization.

5 CONCLUSIONS

Two lines of institutional reforms can be observed in European energy policy: market liberalization and the political support of electricity generation by means of renewable energy sources. Market liberalization opens up possibilities for decentralized, competitive generators of renewable energy. The latter increases the need for time-sensitive and locational differentiated congestion management in electricity networks which is systematically disturbed by network priority rules for renewable energy. Moreover, the prohibition of node-dependent transmission charges is detrimental. The opportunity costs of capacity usage for the whole network (system externalities) not only result at customer nodes but also at generation nodes. The decision-relevant costs for electricity injection are therefore not only the marginal costs of generation but also the opportunity costs of network usage at the generation node. For a proper application of disaggregated nodal pricing, the system externalities of injection by renewable energy sources are to be taken into account, irrespective of whether injection takes place in the transmission or distribution network. Since injection charges vary according to the location of the renewable energy plant, incentives arise for the generators of renewable energy not only to focus on generation costs but also to choose the proper location of electricity generators taking into account locational differentiated injection charges.

This generalized merit order rule is strongly ignored in the current energy policy reform process in Europe. As a consequence wind energy is injected, even if the opportunity costs of network usage are extremely high, leading to inefficient power flows. Although the marginal costs of regenerative energy generation are typically rather low, the opportunity costs of injection of a specific regenerative energy (e.g., wind energy) may (depending on the specific point in time) be very high. Without efficient congestion management this may even lead to the necessity for inefficient injection of energy at other nodes in order to avoid the breakdown of electricity transmission. By means of smart metering and subsequent peak load and responsive pricing the demand of households and firms can be influenced, reducing the demand at times when disaggregated nodal extraction prices are high.

Lump sum subsidies of investment for regenerative energy generation do not create inefficient generation incentives. In contrast, subsidies for regenerative energy generation depend on the amount of injected regenerative

energy and ignoring the opportunity costs of network usage leads to inefficient incentives to inject large amounts of energy. Producers of regenerative
energy only have incentives to sell their energy at market conditions if
they are confronted not only with generation costs but also with the system
externalities of injection.

REFERENCES

Baumol, W.J., 1977. Economic Theory and Operations Analysis. Prentice-Hall, London.

Bohn, R.E., Caramanis, M.C., Schweppe, F.C., 1984. Optimal pricing in electrical networks over space and time. Rand J. Econ. 15 (3), 360–376.

Brandstätt, C., Brunekreeft, G., Friedrichsen, N., 2011. Locational signals to reduce network investments in smart distribution grids: what works and what not? Util. Policy 19, 244–254.

Brunekreeft, G., Neuhoff, K., Newbery, D., 2005. Electricity transmission: an overview of the current debate. Util. Policy 13, 73–93.

Caramanis, M.C., Bohn, R.E., Schweppe, F.C., 1982. Optimal spot pricing: practice and theory. IEEE Trans. Power Apparatus Syst. PAS-101 (9), 3234–3245.

Council of the European Union, 2007. Brussels European Council 8/9 March, Presidency Conclusions, Brussels, May 2.

Cremer, H., Laffont, J.-J., 2002. Competition in gas markets. Eur. Econ. Rev. 46, 928–935.

Crew, M.A., Kleindorfer, P.R., 1976. Peak load pricing with diverse technology. Bell J. Econ. 7, 207–231.

Dantzig, G.B., 2002. Linear programming. Oper. Res. 50 (1), 42–47.

dena (Deutsche Energie-Agentur), 2010. dena-Netzstudie II, Integration erneuerbarer Energien in die deutsche Stromversorgung im Zeitraum 2015–2020 mit Ausblick 2025. Berlin, November.

European Commission, 2011. Energy Roadmap 2050, Communication from the Commission to the European Parliament, the Council, the European Economic and Social Committee and the Committee of the Regions, Brussels, December 15, 2011, COM(2011) 885 final.

Gilbert, R., Neuhoff, K., Newbery, D., 2004. Allocating transmission to mitigate market power in electricity networks. Rand J. Econ. 35 (4), 691–709.

Gribik, P.R., Hogan, W.W., Pape, S.L., 2007. Market-Clearing Electricity Prices and Energy Uplift, Harvard Environmental Economics Program, Working Paper.

Haber, A., Bliem, M.G., 2010. Smart grids—auswirkungen auf die netzentgelte. Energiewirtschaftliche Tagesfragen 60, 2–5.

Höffler, F., Wittmann, T., 2007. Netting of capacity in interconnector auctions. Energy J. 28 (1), 113–144.

Keeler, Th.E., Small, K., 1977. Optimal peak-load pricing, investment and service levels on urban expressways. J. Polit. Econ. 85 (1), 1–25.

Knieps, G., 2006. Sector-specific market power regulation versus general competition law: criteria for judging competitive versus regulated markets. In: Sioshansi, F.P., Pfaffenberger, W. (Eds.), Electricity Market Reform: An International Perspective. Elsevier, Amsterdam, pp. 49–74.

Knieps, G., 2011. Network neutrality and the evolution of the Internet. Int. J. Manage. Netw. Econ. 2 (1), 24–38.

Kuhn, H.W., 2002. Being in the right place at the right time. Oper. Res. 50 (1), 132–134.

MacKie-Mason, J.K., Varian, H.R., 1995. Pricing the Internet. In: Sichel, W., Alexander, D.L. (Eds.), Public Access to the Internet. MIT Press, Cambridge, MA, pp. 269–314.

Machlup, F., Taber, M., 1960. Bilateral monopoly, successive monopoly, and vertical integration. Economica 27, 101–119.

Mohring, H., Harwitz, M., 1962. Highway Benefits: An Analytical Framework. Northwestern University Press, Evanston, IL.

Pfaffenberger, W., Chrischilles, E., 2013. Turnaround in rough sea—electricity market in Germany. In: Sioshansi, F.P. (Ed.), Evolution of Global Electricity Markets: New Paradigms, New Challenges, New Approaches. Elsevier, Amsterdam.

Read, E.G., 1996. OR modelling for a deregulated electricity sector. Int. Trans. Oper. Res. 3 (2), 129–137.

Samuelson, P.A., 1992. Tragedy of the open road: avoiding paradox by use of regulated public utilities that charge corrected Knightian tolls. J. Int. Comp. Econ. 1, 3–12.

Schweppe, F.C., Caramanis, M.C., Tabors, R.D., Bohn, R.E., 1988. Spot Pricing of Electricity. Kluwer Academic Publishers, Boston, Dordrecht, London.

Spengler, J.J., 1950. Vertical integration and antitrust policy. J. Polit. Econ. 58, 347–352.

von Hayek, F.A., 1948. The meaning of competition. In: von Hayek, F.A. (Ed.), Individualism and Economic Order. George Routledge & Sons, London, pp. 92–106.

Wu, F., Varaiya, P., Spiller, P., Oren, S., 1996. Folk theorems on transmission access: proofs and counterexamples. J. Regul. Econ. 10, 5–23.

Chapter 7

From Niche to Mainstream: The Evolution of Renewable Energy in the German Electricity Market

Dierk Bauknecht[a], Gert Brunekreeft[b] and Roland Meyer[c]

[a]*Oeko-Institut - Institute for Applied Ecology, Postfach 17 71, D-79017 Freiburg, Germany;*

[b]*Jacobs University Bremen, Bremer Energie Institut, College Ring 2, D-28759 Bremen, Germany;*

[c]*Postdoctoral Research Fellow, Jacobs University Bremen, College Ring 2, 28759 Bremen, Germany*

1 INTRODUCTION

In many countries, electricity generation from renewables is taking over a relevant share of the market, and political objectives aim at further increases. As renewables are no longer a niche technology, the interaction between renewables and the electricity system can no longer be neglected.

This has two dimensions: first, in operational terms, renewables affect the supply–demand balance, and electricity prices and renewables operation should therefore react to market signals to the extent possible. Second, the question arises as to whether renewables with their generally low marginal costs undermines the marginal cost-based market model and thus exacerbates the investment problem that potentially exists in electricity markets.

Looking at the first dimension, the chapter provides an overview of different market integration strategies and their pros and cons. While market integration is relatively straightforward for controllable renewables, it is more challenging for intermittent renewables from wind and solar. At first sight, full market integration can provide the most efficient operation of renewable plants. However, a closer and more differentiated evaluation of the potential to integrate renewables into market is required that moves beyond the debate on feed-in tariff (FiT) with no market integration versus quota with full market integration. It is not just that full market integration neglects the trade-off between investment and market integration and can undermine support mechanisms in place. Rather, full market integration only addresses part of the problem and can even be counterproductive in the case of system balancing.

The second dimension relates to the remuneration of generators and its impact on generation investments and reliability. The increasing share of intermittent renewables with low marginal costs raises the question of whether tradition energy-only markets provide sufficient investment incentives for conventional generation to maintain an adequate level of reliability in the long run. The concern of a potential "missing-money problem" that may suppress efficient generation investment has led to the discussion of (potentially) necessary changes in the market design. Notably in Europe, capacity markets have recently moved into the focus of discussion since the remuneration of capacity, in addition to energy sold, seem to be more adequate for traditional generators which are facing lower utilization but are needed to back up electricity supply. The chapter analyzes the political and economic perspectives of capacity markets and related remuneration mechanisms as part of the market design issues raised by large-scale integration of renewables.

This chapter analyzes the issues with respect to the support and market effects of increasing renewables for Germany. Due to its remarkable development of renewables in combination with the phaseout of nuclear power production, Germany makes a compelling case for a structural (r)evolution of its electricity market.

This chapter consists of four sections in addition to the introduction. Section 2 provides an overview of the current development of the German electricity sector. Section 3 examines the evolution of the renewables support schemes against the background of the growing shares of renewables and the scopes and limits to expose renewables to market incentives. Section 4 examines the market impact of the green transition and discusses options for a stronger capacity-based remuneration of generators. Section 5 concludes with an outlook on the main conceivable market developments that a future market design should account for.

2 MARKET INTEGRATION OF RENEWABLE ELECTRICITY IN GERMANY

Over the last decade, Germany has experienced a remarkable growth in electricity generation from renewables. While renewables contributed only 7% in 2000, they accounted for approximately 22% of overall production in the first half of 2012 (BDEW, 2012). As a consequence, renewables have overtaken both hard coal and nuclear power in terms of generation, only lignite plants generate more.

What is more, there are ambitious political objectives to further expand renewables capacity. According to the new German energy policy, further described in Chapter 4 by Pfaffenberger and Chrischilles, under "Energiekonzept der Bundesregierung," the CO_2 emissions shall be reduced by 40% by 2020 and 80% by 2050, compared to 1990, respectively

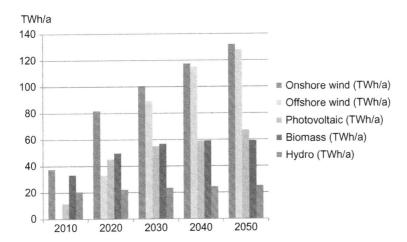

FIGURE 7.1   Development of renewables supply in Germany by 2050. *Source: Based on BMU (2012); base scenario.*

(Bundesregierung, 2010). In achieving these ambitious policy goals, the decarbonization of the energy sector plays an important role.

Against this background, the share of renewables is expected to increase to 42% by 2020. According to the German Ministry of the Environment, the German energy market will eventually reach an 85% renewable share by 2050 (BMU, 2012). Figure 7.1 shows the expected development for different types of renewables. The largest increase is expected from onshore and offshore wind capacity, while photovoltaic (PV) and biomass will also provide a significant contribution to electricity supply. The overall development of renewables and their share in total electricity production is shown in Figure 7.2. A detailed analysis of developments in the German electricity system is provided by Pfaffenberger and Chrischilles in Chapter 4.

As long as renewables played only a minor role, the main concern has been to design a support scheme in such a way that it is both effective and efficient in promoting renewables plants. The focus was on the economics of individual plants. Due to the growth of renewables, the focus has shifted from individual plants to the electricity system. This involves the following questions: what are the effects of renewables on the system, how can they be integrated into the system, and how can the electricity system indeed be turned into one that can accommodate a dominant share of renewables.

One important aspect of this debate is the network integration of renewables. This includes network connection, network operation, and network development (Bauknecht, 2012; Eclareon and Öko-Institut, 2012). For a detailed analysis of the specific challenges of increasing renewables for electricity networks and a discussion of solutions, see Chapter 6. Besides network integration, the other main element of the integration of renewables in the system is market

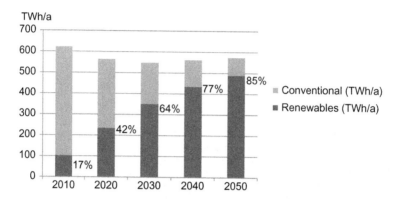

FIGURE 7.2 Development of renewables supply and shares in total electricity production in Germany by 2050. *Source: Based on BMU (2012); base scenario.*

integration. While network integration in terms of network access is important so that every single plant is able to sell its electricity, market integration becomes particularly relevant with a rising share of renewables, especially fluctuating renewables. Once these technologies reach a critical mass, it becomes ever more problematic to operate them in a niche market separate from the wholesale and balancing service markets and neglect their impact on these markets.

In Germany, the development of renewables has triggered a discussion as to whether the current fixed price system can be maintained, and more market-based approaches have already been introduced in the Renewable Energy Act (Erneuerbare Energien Gesetz – EEG). The main support for renewables in Germany has been an FiT scheme that came in force in 2000. Under this scheme, tariffs are fixed for typically 20 years, with a predefined tariff reduction for new installations. Under such a scheme, renewables generators obviously do not have any incentive to react to market signals, and their sole objective is to maximize their output, irrespective of market conditions and demand changes.

On top of the feed-in scheme, an optional premium scheme has been introduced by the amended EEG that came into force at the beginning of 2012. The new option is a sliding premium scheme, which on the one hand introduces an element of market risk, and on the other hand compensates plant operators for the additional costs they incur when managing their plants in the market rather than selling their output at a fixed tariff.

The market premium is basically defined as the difference between the technology-specific FiT and a reference market price. This reference price is defined as the price that an average plant could have earned. If plants manage to outperform the average electricity price, they can increase their revenues. Similarly, there is downside risk if plants generate in such a way that they earn less than the "average" plant. This mechanism is designed in such a way that the characteristics and generation profiles of different technologies are taken into account.

Although the new scheme is still geared toward the FiT levels, plants now have to sell their generation directly in the market. Compared to the previous fixed premium, renewables generators are exposed to some level of upside and downside risk. As a consequence, producers in principle have an incentive to optimize operation of their plants and marketing strategies in order to increase market revenues compared to the reference market price.

To compensate generators for the costs they incur due to market participation, the market premium scheme includes the so-called "management premium." The management premium should cover the pioneering effort of administration, improved production forecast, and marketing.

The premium scheme has significantly stimulated direct marketing of renewables by plant operators, particularly in the case of wind power. In August 2012, 25 GW of renewables operated under the premium scheme. Most of this is onshore wind with more than 21 GW so that roughly two-thirds of the onshore wind capacity has opted for the market premium (EEG/KWK-G, 2012). The main reason for this development is that the revenue from the management premium exceeds the downside risk. It is doubtful whether the market participation of renewables plants has significantly affected their actual operation, which would be necessary to exploit their flexibility and thus reduce the flexibility that is required from the system.

The transition from a fixed FiT to a premium scheme in Germany is an example of the evolution of support schemes for renewables toward more market-based approaches, further discussed in Section 3. There is also a discussion emerging about replacing the FiT with a quota scheme.

Apart from the shift toward renewables, the green energy transition in Germany is driven by the phaseout of nuclear power production, further described in Chapter 4 in this volume. Figure 7.3 shows the development of generation capacities (in gigawatt and shares in total installed capacity) as forecasted by BMU (2012).

As a result of increasing renewables, the amount of fossil generation capacities is expected to decrease by almost 50% by 2050. Over and above the decommissioning of capacity caused by the nuclear phaseout, there will be a significant reduction of coal and lignite plants. Despite this reduction in fossil generation capacity, conventional generators will still be needed, especially in a transitional phase as backup capacity, to compensate for the fluctuating supply from renewables. Given the expected retirement of power plants, Germany will face additional investment needs for conventional plant. According to EWI (2012), 55 GW of new conventional generation capacity will be needed by 2030, of which 44 GW are gas plants. The growing importance of gas plants to provide flexible capacity reserves explains why the share of gas capacity is not expected to decrease over time. Moreover, a growing demand for storage is expected especially after 2030 when renewables increasingly lead to negative residual load.

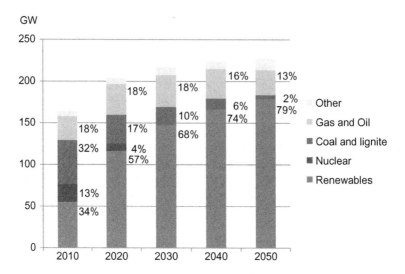

FIGURE 7.3 Development of gross generation capacity until 2050. *Source: Based on BMU (2012); base scenario.*

These large and rapid structural changes will put the electricity market under enormous pressure and require a rethinking of the current market design with respect to the support and integration of renewables in the market and the remuneration of generation capacity. The following section analyzes the evolution of support schemes for renewables, while Section 4 discusses the market effects of growing renewables that drive the debate on necessary adjustments of the current market design.

3 THE EVOLUTION OF SUPPORT SCHEMES FOR RENEWABLES

3.1 Rationale for Market Integration

Since the introduction of support policies for renewables, there has been an ongoing debate about the best approach to promoting these technologies. On the one hand, there are price-based approaches whereby policymakers define a certain payment above market prices to make up for the costs of renewables above market levels. On the other hand, there are quantity-based quota systems with green certificates in which policymakers fix a certain volume of renewables and the level of support that results from the certificate prices will in principle be determined by the market.

Table 7.1 provides an overview of various support schemes; these are illustrated in the figure below the table. More than a decade of practical experience with these schemes has shown that feed-in schemes are typically both more effective and efficient in providing a stable investment

TABLE 7.1 Overview of Support Schemes for Renewables

	Fixed FiTs	Variable FiTs	Sliding Premium	Premium with Cap and Floor	Fixed Premium	Quota with Green Certificates
Direct market participation	No	No	Yes	Yes	Yes	Yes
Revenue	Fixed FiT	FiT depending on, for example, market price	Difference between reference price and market price	Market price plus premium. Premium levels depend on cap and floor and market price	Market price plus fixed premium	Market price plus certificate price

Low price risk   High price risk

Revenue under the various support schemes are shown.
Blue, market price; green, certificate price; red, overall remuneration.
Source: Authors' own illustration.

environment and promoting renewables technologies (Cory et al., 2009; Rathmann et al., 2011). One reason for the inefficiency of quota systems is that, in their original design, they do not distinguish between different technologies and generators, so that all generators receive the same certificate price. Although quota schemes mainly benefit low-cost technologies, the lack of differentiation typically leads to high producer rents. If technology banding is introduced, this problem can be overcome to some extent, yet at the expense of the original simplicity of the approach. Moreover, as a result of both electricity and certificate price risk, generators are typically faced with high-risk premiums when financing their plants. In the United Kingdom, it has been decided that the scheme will be abolished by 2017. More details on the UK quota system are provided by Newbery in Chapter 1.

Nevertheless, with an increasing share of renewables, countries with feed-in schemes come under pressure to revise their approach and make generation from renewables more responsive to market signals. This has triggered the development of support mechanisms that are located between the two extreme cases, as shown in Table 7.1. The development in Germany that was briefly described in the previous section is a case in point. Market integration of renewables has become a more pressing issue on the agenda for a number of reasons, including:

- First, in many countries renewables have increased significantly in the last decade. Both technical and economical characteristics of renewables have an impact on electricity markets, including lower market prices or even negative prices that have already been observed, for example, in Germany, and that may occur more frequently with an increasing share of renewables (Andor et al., 2010). Figure 7.4 shows the minimum hourly spot prices for the years 2010/2011 and illustrates a couple of examples where negative price spikes occurred at the German spot market. In order to reduce this impact of renewables on the market and other market players, it may be argued that renewables should be integrated into electricity markets so that they can react to market signals. For example, renewables would stop generating in times of oversupply, thus reducing negative prices in the market (Nicolosi, 2011).

- Second, support and protection from the market mechanisms is normally granted only for a limited period of time, for example, 20 years in Germany. Afterward renewables producers have to participate in the electricity markets, even in countries that offer fixed FiTs. One can therefore argue that support instruments should be revised so that renewables

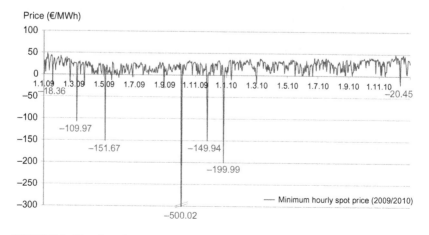

FIGURE 7.4 Negative price spikes at the German spot market. *Source: EEX.*

producers can gain market experience before they are obliged to participate directly in the wholesale markets.

- Third, it can be argued that shielding renewables from market pressure
  have been appropriate as long as these technologies have been in a development phase. In that period, support schemes separated from the markets
  can provide the necessary investment security. However, once they
  become more mature, they should be treated more like other generators
  and integrated into the competitive market.

The following sections discuss market integration of renewables in more
detail. The focus of this section is on variable renewables like wind and solar
power. For dispatchable renewables like biomass, it seems more straightforward to provide flexibility and react to market signals. However, even for
these plants investment in storage is needed to increase their flexibility.

3.2 Flexibility of Renewables and Market Risk

Two characteristics of fluctuating renewables are important to bear in mind
when discussing its potential for market integration (IEA, 2011):

1. the generation of these plants does not necessarily match with the fluctuating demand (*variability*) and
2. the generation profile of fluctuating renewables plants is typically only to
   some extent predictable (*uncertainty*).

For example, Figure 7.5 illustrates the variability problem for 1 week in
March if today's generation profile of variable renewables in Germany is

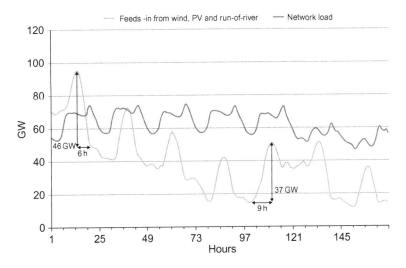

FIGURE 7.5   Illustration of the variability of renewables feed-in. Authors' own illustration.

scaled up to a 50% share. In some hours, there will be a negative residual load, that is, an oversupply of renewables, while most of the time they do not generate enough to match demand. Variability also implies that there are steep downward and upward ramps where renewables output changes quite rapidly, which require other generators, storage or demand to follow this profile.

Uncertainty implies that renewables generation does not match forecasted generation so that other generators, storage or demand need to be available to reduce or increase their output or their consumption in order to fill the gap between forecasted and actual renewables generation. In the case of conventional generation, this is particularly problematic if due to their minimum stable generation, providing capacity implies that they also need to generate electricity while providing balancing capacity. In this case, providing capacity reduces the capacity of the system to absorb renewables.

Neither is a new phenomenon as power plants have always had to react to uncertain demand variations. However, with the increase of fluctuating renewables, these two issues become more prominent on the supply side. While demand variability typically follows certain patterns, the combination of both demand and supply variability leads to more irregular residual load that has to be covered by the rest of the system.

Due to the specific characteristics of variable renewables plants, their ability to react to market signals in a flexible way is limited. Despite significant technological progress that has reduced the costs of renewables, there is only a limited potential to increase the flexibility of these technologies, which makes it difficult for renewables plants to become "normal" market participants. Before discussing market integration strategies that aim at exploiting this flexibility, Table 7.2 shows how renewables plants can offer flexibility.

If renewables plants are integrated into the market, their revenue risk will increase. This can be divided into three risk elements (Mitchell et al., 2006) further described below:

1. price risk
2. volume risk
3. balancing risk.

This differentiation is important for the further analysis because the extent to which generators are exposed to these risk dimensions should match their ability to manage these risks, described further below, by providing the flexibility described in Table 7.2.

First, in liberalized markets prices are generally volatile, and market integration exposes generators to market price fluctuations. This includes both a short-term and a long-term dimension. Short-term price risk refers to a setting where generators are incentivized to adapt their production profile to the market situation, that is, generate when prices are high, thereby providing the flexibility potential outlined in the previous section.

TABLE 7.2 Renewables Characteristics and Flexibility Potential

Renewables Characteristics	Possible Responses	Potential
Variability	Turn down renewable plants	In principle full downward flexibility. But this option is at the expense of overall renewable generation and should be limited to critical situations, for example, negative residual load where no alternative flexibility is available, rather than being used as a fulltime flexibility option.
	Manage maintenance periods taking into account market needs	Potential limited, wind and solar have low maintenance requirements.
	Choose locations that provide favorable generation profile	Potential limited and choice of locations driven by many other factors, for example, overall wind yield, infrastructure, acceptance.
Uncertainty	Improve forecasts	Very relevant potential, but forecast innovations not driven by individual renewable generators.
	Provide efficient balancing	Variable renewables can participate in balancing markets. But balancing on a microlevel (e.g., one storage unit per renewables plant) likely to be inefficient.

A fixed FiT does not entail any price risk whereas both quota and premium support schemes (with a premium payment above market prices) involve a market price risk since renewables generators have to sell their output directly on the market. However, even under a quota or premium scheme it is not guaranteed that short-term price signals feed-through to a generator's production decision if generators sell their output based on long-term contracts.

Second, a generator's revenue is always the function of both price and volume. With the FiT which is mostly combined with priority dispatch, there is no market-related volume risk involved since the whole production can always be sold (there is always the risk that variable generators cannot produce due to a lack of wind, solar, etc.). As opposed to a priority dispatch regime, both under the quota and the premium scheme, operators have to find counterparties to sell their production. Hence, there is always the risk that a plant cannot sell its total production and thus loses revenue. However, in the case of renewables with low marginal costs, it is rather unlikely that plants will not be dispatched.

Third, market integration also increases the balancing risk for generators. If they sell their output on the market, they are typically not only exposed to fluctuating market prices, but they also become responsible for imbalances. This balancing responsibility generally entails the additional risk of imbalance penalties to renewables producers. The imbalance risk is similar to the general market price risk, in that it is driven by imbalance prices on the market. However, in addition it also depends on a generator's imbalances. Clearly, balancing always puts intermittent renewables producers at a relative disadvantage to conventional, more predictable generators due to their limited forecasting capabilities.

3.3 Market Integration and Market Design

Importantly, market integration is not just about support instruments for renewables but also about market design and the interaction between the two (Hiroux and Saguan, 2009; see Chapter 25). If renewables plants are exposed to the market risks outlined in the previous section, there should be no market risk due to markets that are not functioning properly. This means that market should be transparent and liquid (Weber, 2010).

Moreover, if renewables are required to participate in the market, the market design should not only be geared toward large-scale controllable plants but should also enable renewables to manage their specific risks. The design of electricity markets should enable plants to participate in wholesale and possibly even balancing markets and provide the electricity products they can generate (Barth et al., 2008). If renewable generators are required to take on balancing responsibilities and are exposed to balancing market prices, these markets should be designed so that generators are not unduly penalized for their generation characteristics, especially in the case of fluctuating generation (Vandezande et al., 2010). In that sense, market design is complementary to support instruments that promote market integration.

A market design that accommodates the specific characteristics of variable and uncertain renewables can to some extent also be an alternative to integrating renewables into markets. This means that the market should be designed in such a way as to ensure that the required flexibility to manage variability and uncertainty of renewables is provided in the most efficient way. Gearing the market toward the requirements of renewables is likely to solve a larger part of the problem than requiring renewables to provide that flexibility and manage market risks themselves, which they can do only to a very limited extent.

In terms of uncertainty, trading close to real time is particularly important for fluctuating renewables, because generation forecasts significantly improve closer to real time. Intraday markets are therefore particularly important as the share of renewables increases. Of the similar importance if the gate-closure time that regulates how much ahead of real time, the system operator takes over with real-time balancing. There is a positive relation between short gate-closure time and accurate forecasting (Barth et al., 2008; Müsgens and Neuhoff, 2006).

3.4 Where Should Renewable Plants be Exposed to Risk?

At first sight, full market integration can provide the most efficient operation of renewable plants. In this view, the competitive market is the best way to exploit flexibility, and market integration provides renewable plants with complete transparency about market conditions and incentives to react to these conditions.

However, there are a number of arguments against full market integration. First of all, there is a trade-off between full market integration and the risk it entails on the one hand and the need to provide a stable investment environment for renewables in order to enable further renewables expansion (Mitchell et al., 2006). Market integration may help exploit flexibility of renewables but can undermine support schemes that are in place.

In addition to this argument, a second issue is that renewables can offer only limited flexibility, as was shown above. From an individual generator's perspective, this implies that it has only a limited capability to manage market risks. From a system perspective, this means that exposing renewables to market prices is an insufficient approach to provide the required flexibility. Hence, renewable plants should only be exposed to the type and level of risk they can manage whereas other measures are required to deal with the remaining variability and uncertainty in the system. These measures include an appropriate market design as well as measures that help to develop other flexibility options, such as storage and demand-side management; see the following section.

The differentiation between different market risks that was put forward above also enables a differentiated view on market integration. If the analysis of the flexibility potential of renewables is combined with the differentiated risk analysis, it provides the basis for evaluating the potential to integrate renewables into markets that moves beyond simply juxtaposing "feed-in with no market integration" on the one hand and "quota with full market integration" on the other hand.

Table 7.3 provides an overview of possible responses to the variability and uncertainty of renewables as well as possible policy instruments to promote these responses. In terms of variability, fluctuating renewables plants can offer some level of flexibility that can be exploited by exposing renewables to price risk and volume risk. This requires functioning markets. However, the flexibility potential that can be exploited is rather limited, especially when certain renewables production targets are to be reached. In order to tap this limited flexibility potential, renewables generators do not need to be fully integrated into market, that is, sell their electricity directly on the

TABLE 7.3 How to Deal with Renewables Variability and Uncertainty

Renewables Characteristics	Possible Responses	Policy Instruments
Variability	Turn down renewables	Expose plants to some price risk.
	Manage maintenance periods taking into account market needs	Require support mechanisms that expose renewables to market signals and functioning markets.
	Choose locations that provide favorable generation profile	Exposure to price risk does not necessarily require direct market participation of renewables but can also be achieved by FiTs that are linked to market conditions. Curtailment agreements between renewables and network operators.
Uncertainty	Improve forecasts	But not necessarily by exposing individual renewables generators to balancing risk.
		Smaller systems are more difficult to forecast.
		Other factors may be better positioned to provide efficient forecasts.
	Provide efficient balancing	Critically depends on a competitive market and flexible market design (intraday market, etc.) rather than balancing by individual renewables.
		Balancing incentives in the support scheme only address the smaller part of the problem.

competitive market with full risk exposure. Rather, there are examples of feed-in schemes that introduce an element of short-term price risk through varying FiTs.

Such a scheme is already in place, for example, in Hungary where tariffs vary by time and weekday depending on the demand profile—peak, valley, and deep valley periods—but also vary between summer and winter (Couturen and Gagnon, 2010; Rathmann et al., 2009). Weather-dependent renewables are exempted from these tariff variations since they cannot adjust their production like other generators. Whereas tariffs are not directly linked to the short-term market situation in this mechanism, but follow a predefined pattern; with a higher share of renewables the tariff could be partly linked to short-term market signals. While feed-in schemes with fixed tariffs in principle represent capacity-based support, such varying FiTs would introduce an energy-based element.

There are also schemes in which there are no support payments in the case of negative market prices, for example, in Denmark (Eclareon and Öko-Institut, 2012). Another option to enhance the use of flexibility potential would be to allow for voluntary curtailment agreements between renewables generators and network operators, as further described in Brandstätt et al. (2011).

In terms of uncertainty, renewables would need to be exposed to balancing risk to be incentivized to reduce uncertainty. The key measures for dealing with uncertainty are to improve forecasts on the one hand and to provide efficient balancing to manage remaining imbalances on the other hand. In both cases, it is doubtful whether individual renewables plants are in the best position to implement these responses.

In the case of improving forecasts, this can in principle be carried out by individual generators, but may as well be conducted on a more aggregated level, especially as small systems are inherently more difficult to forecast.

As for balancing remaining uncertainties, the key question is also whether this should be done on an individual or on a system level. Providing balancing capacity on an individual level (e.g., a battery next to each wind generator) requires a higher overall balancing capacity and leads to higher costs than balancing on a system level where individual imbalances at least partly offset each other. Exposing renewables generators to balancing risk may incentivize them to utilize balancing capacity in an inefficient way if deviations from forecasts are managed for individual generators. In Germany, renewables plants are not exposed to any balancing risk, unless they participate in the market premium scheme with full balancing risk. In the European Union (EU), there is a wide variety of mechanisms to expose renewables plants only to some extent to the balancing risk, thus better reflecting their technical capabilities (Eclareon and Öko-Institut, 2012).

The effect of exposing renewables generators to imbalance risk is mainly to shift the costs of imbalances to these generators. However, this happens without exploiting much additional potential to manage this uncertainty. Therefore, when it comes to dealing with imbalances of renewables, the key instrument is to set up functioning intraday and balancing markets that can provide the required flexibility at least cost. Renewables plants should participate in these markets wherever possible. However, exposing variable renewables plants to balancing risk may even be counterproductive if it leads to local balancing activities that increase the overall demand for balancing capacity. For example, if every domestic PV plant would be balanced by a domestic battery, the overall storage demand would increase significantly.

3.5 Guidelines for the Market Integration of Renewables

The previous sections have shown that there are a number of reasons why renewables support schemes are broadly migrating from fixed feed-in schemes toward schemes that allow for a higher degree of market integration. At the same time, it was argued that the benefits of market integration in terms of exploiting flexibility are rather limited, especially in the case of variable renewables. Market design is the more relevant issue when it comes to providing the required flexibility from various options, including storage, demand-side flexibility, and renewables flexibility. In the German example, two-thirds of wind plants have voluntarily moved from the fixed feed-in scheme to the market-based premium scheme. However, this has been mainly triggered by the additional management premium that plants can receive in Germany which is described in Section 2 and that leads to additional revenue for renewables plants and thus additional total support costs. It is doubtful whether this market integration has triggered any changes in the plants' operation.

Nevertheless, it is unlikely that support schemes that were designed to promote renewables in a niche market can survive when renewables become the dominant source of generation. The following guidelines should be considered when pursuing market integration of renewables:

- A key principle for designing markets and support schemes for high shares of renewables and promoting market integration of renewables should be that the main objective for the future market is not to select between investment in conventional generation on the one hand and renewable generation on the other hand. Rather, it should support the implementation of the renewable targets (see Section 2 for the German targets) at least cost. This is different from the previous notion that renewables require temporary support and can then start competing against conventional generation in the market.
- The main rationale for integrating renewables into electricity markets is to exploit their flexibility potential. This requires a clear understanding of the—limited—flexibility potential of renewables and what this flexibility can contribute to solve the overall system challenges. More work needs to be done in that area.
- Market integration of renewables is a matter of both adapting support schemes and setting up adequate markets. Renewables should not be exposed to market risk when markets are not ready yet.
- At the same time, some of the system challenges that result from renewables should be tackled by providing adequate markets rather than exposing renewables to market risks. For example, in order to provide efficient balancing, functioning short-term and balancing markets are arguably more important than exposing renewables to balancing risk.

- In terms of market design, renewables integration requires functioning markets in general as well as more specific mechanisms to deal with the uncertainty of renewables, namely intraday markets and short gate-closure times.
- In terms of support scheme design, any revision of existing schemes should ensure that it does not undermine investments in renewables.
- For evaluating different support scheme designs, it is proposed to differentiate between price, volume, and balancing risk. Renewables generators should only be exposed to market risk they can manage and where they can provide flexibility to the system. Especially in the case of fluctuating renewables, market integration has to be in line with the variability and uncertainty of their generation profile.
- Exposing renewables to balancing risk should therefore not be a priority. Mechanisms that incentivize renewables to use flexibility options on a microlevel, for example, to balance individual wind plants, should be avoided.
- Exposing renewables to price risk does not necessarily imply that these plants should directly participate in the market. Feed-in schemes can also be made more responsive to market signals if the FiT is no longer fixed but partly varies depending, for example, on market prices. Another option is voluntary curtailment agreements.

Overall the question is not how renewables can be integrated into the existing market that was designed for a different setting. Rather, as renewables become the major player, the main challenge is to design the market around the specific characteristics of renewables. One of the reasons why fully integrating renewables into the current market is not an option is that the current market may not even be able to enable investment in conventional generation. This is further discussed in Section 4.

4 THE EVOLUTION OF ELECTRICITY MARKET DESIGN

4.1 Energy-Only Markets and the "Missing-Money Problem"

The current transformation of the German electricity sector toward a large-scale renewables supply has recently put the discussion on a future market design on the top of the political agenda (Brunekreeft and Meyer, 2011). Looking abroad, however, shows that the general debate on market design is much broader and started more than a decade ago. In their seminal papers, Cramton and Stoft (2006) and Joskow (2006) provide a theoretical analyses of the so-called "missing-money problem" which is also the main driver of the German debate. They argue that the traditional remuneration of generators on basis of their electricity sales may not provide sufficient revenues to incentivize the investments needed to maintain an adequate level of generation capacity in the long run. Cramton and Ockenfels (2012) give a more recent analysis of the problems of traditional spot markets and describe how a change of market design may successfully address the issues of generation adequacy.

As in most European countries, the German electricity market is organized as an *energy-only* market. This means that generators only receive revenues for their electricity sales (in megawatt hour) to the market.[1] With these revenues both variable (mainly fuel) and fixed (capacity) costs need to be recovered. Moreover, all investment incentives result from the energy-based market revenues.

Market prices and revenues result from matching demand and supply bids. All generators bidding into the market are ranked according to their marginal cost bids. The resulting *merit order* determines the order in which generators are scheduled to serve demand at least cost. Accordingly, the merit order represents a variety of different generation technologies since different load profiles favor different cost characteristics. *Base load generators*, like nuclear of coal plants, are characterized by high capacity costs and low fuel costs. Due to their low marginal bids, they are placed at the left-hand side of the merit order, with a high utilization that allows them to recover their high fixed costs. In case of higher demand, occurring less frequently, *peaking units*, like gas-fired plants, facing higher marginal costs are scheduled. Figure 7.6 shows the merit order of conventional generators (excluding renewables) in Germany for 2011.

Since the market price is determined by a *uniform price auction*, all generators receive the price which is set by the marginal unit, that is, the most costly unit in the merit order which is needed to serve current load. Hence, all generators benefit from high scarcity prices in cases of peak demand as these provide additional (inframarginal) rents needed to recover total costs. A well-functioning energy-only market should be able to provide cost recovery for all (cost) types of generators to ensure an adequate level of supply security.

However, there are reasons to doubt that scarcity prices will actually be high enough to sufficiently recover the necessary investment. The main concerns relate to peaking units providing necessary reserves in cases of high demand or generation outages. Due to limited possibilities of electricity storage and demand response, at least in the medium term, these reserves capacities are essential for supply security. Given their low level of utilization, however, investment risks are high, and any suppression of efficient scarcity prices will most likely result in underinvestment.

---

1. The only exceptions are the markets for operating reserves organized by the transmission system operators. Here the participating generators typically receive both a capacity payment (per megawatt of reserves provided) and an energy payment (per megawatt hour of electricity actually dispatched).

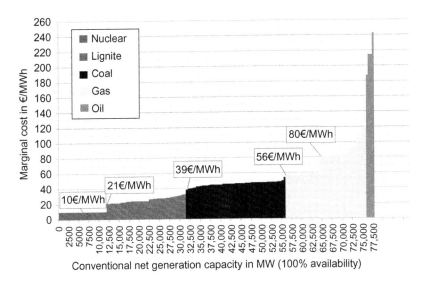

FIGURE 7.6   Merit order of conventional generation in Germany 2011. *Source: Retrieved from www.gruene.at; calculations based on EEX, UBA, BMU (2010), and statistics of the coal industry.*

There are two main lines of arguments that support the concerns of insufficient investment incentives:

1. First, a shortfall of investment may occur as a direct result of political or regulatory *price intervention*. The main problem of competitive oversight is to distinguish whether high market prices result from scarcity or an abuse of market power (Joskow, 2006). In the former case, high prices are economically justified since they act as a price signal to attract new investments in case they are needed to restore an efficient market equilibrium. In case of market power abuse, however, a part of the capacity is artificially withheld from the market by dominant market players aiming to increase profits by influencing the market price for their remaining electricity sales. Regulators or policymakers may use several instruments to intervene in the market with the aim to protect consumers from extreme price spikes. One way to do so is to implement a price cap on the market. The side effect of such a measure is that it may suppress real scarcity prices, and hence, the market fails to incentivize new investments. The same effect may result from introducing a bid cap at the power exchange. At the European Power Exchange (EEX) that covers the German market, price bids of generators are limited to −3000 €/MWh and 3000 €/MWh, respectively. By now, this bid cap does not appear to be binding since previous price caps were significantly below the cap (Brunekreeft and Meyer, 2012a). Given that backup capacities and

storage facilities crucially depend on price spikes, the price cap argument may become an important issue in the future.

2. A second line of arguments refers to the increasing *investment risks* as a result of current market developments. For reserve capacities, the market environment changes toward a lower level of utilization in exchange for earning higher price spikes when they are actually scheduled. Given that scarcity prices may be limited to a few hours, investment in backup units is getting increasingly risky. On the one hand, there is good reason to argue that private investors are and should be able to manage *internal market risks* on their own since those risks occur in other industries as well. On the other hand, the electricity supply industry differs from most other industries in that it faces significant *external market risks* caused by uncertain changes of future regulation and market design. An important aspect of regulatory risk is the so-called "commitment problem" which occurs if the regulator or the legislative authority is not able to credibly commit to a certain long-term policy.[2] If an investor anticipates the risk of unfavorable market interventions, for example, introducing a price cap to address market power problems, he or she might hesitate to invest in the first place. In this sense, the current market design discussion could be a kind of self-fulfilling prophecy if investors decide to postpone their investments until policymakers decide how to proceed with the market design.

4.2 The Impact of Increasing Renewables on Market Prices and Investment Incentives

Against the general background of potential investment problems in liberalized electricity markets, the German discussion on a future market design is specifically driven by its challenge of integrating large amounts of renewables into the electricity market. The central concern is that scarcity prices may not be high enough to ensure an adequate level of reserve capacity in the long run. However, even if the strong increase of renewables is not considered a separate driver of the missing-money discussion, it will most likely exacerbate the problem. The reason is that the dynamic development toward renewable supply leads to the so-called "merit order effect" which denotes the rightward shift of conventional generators in the order of dispatch.

The merit order effect results from the fact that renewables, notably wind and solar power, have zero or low marginal costs and are hence placed on

---

2. Regulatory commitment problems may also play an important role for network investment incentives (Brunekreeft and McDaniel, 2005).

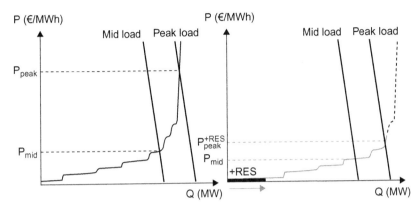

FIGURE 7.7   Price decrease due to increasing renewable energy sources (RES). Authors' own illustration.

the left-hand side of the merit order. Figure 7.7 illustrates this effect for different demand levels. Two related effects can be observed:

- First, the *utilization* of generators decreases. For conventional utilities, an increasing renewables supply is equivalent to a demand reduction. As the right-hand side of Figure 7.7 shows, some peaking generators (indicated by the dotted line) may not be scheduled anymore in "regular" peak situations, that is, in absence of unexpected outages of generation capacity.
- Second, the *average system price* decreases since for most load situations the relevant part of the merit order is the flat area between base and mid load, while peak situations with high scarcity prices become rare. As mentioned before, however, scarcity prices are important for fixed cost recovery of all generators, not only the peaking units. A permanent lack of these inframarginal rents lowers investment incentives and may endanger generation adequacy in the long run.

The decreasing system price does not only affect investment in conventional generation but also makes it impossible to phase out renewables support and integrate these plants into the current market.

Figure 7.8 illustrates the development of gross margins for different coal plants and gas and steam turbines from 2004 to 2012. These numbers show a constant decrease of the margins since 2009 which can be attributed to the increasing competition notably by renewables. This development has recently been (temporarily) reversed by the abrupt start of the nuclear phaseout with the shutdown of eight nuclear plants in 2011. On the other hand, however, due to the need for backup reserves, the reduction of nuclear capacity increases rather than decreases the need for investment in conventional reserve capacities, as the investment forecasts in Section 2. Hence, the nuclear phaseout

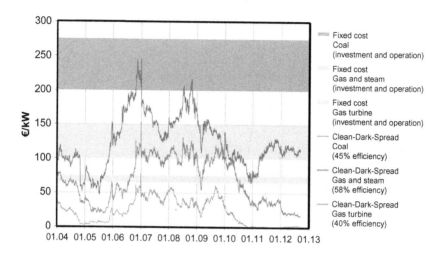

FIGURE 7.8 Development of margins for coal plants and gas and steam turbines. *Source: Matthes et al. (2012, p.19).*

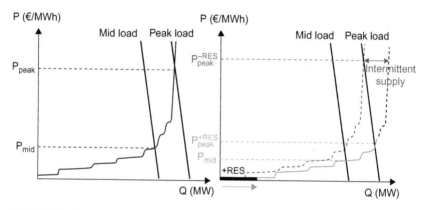

FIGURE 7.9 Intermittency of renewable energy sources (RES). Authors' own illustration.

does not eliminate the missing-money problem but rather tightens the problem if market prices fail to deliver efficient investment signals.

An important aspect of the "German type" of the missing-money problem is the *intermittency* of renewables supply. As shown in Section 2, Germany is aiming to provide more than 40% of electricity supply from renewables by 2020 and more than 60% by 2030. Among these, wind and solar power are going to make up the dominant part. The main problem is that supply from wind and solar power is highly fluctuating and, hence, not reliable enough to ensure supply security in the absence of conventional reserve or storage capacity. Figure 7.9 illustrates the effects of intermittency on the market outcome.

Obviously, the intermittency problem exacerbates the missing-money risk in two ways:

1. First, the dominance of renewables capacity in the market increases the variability of supply. The higher the share of renewables, the more important it is to provide for sufficient amounts of reliable backup or storage capacity to accommodate for the risk of a shortfall of renewables supply. This backup need explains why in spite of the decreasing share of conventional generation capacity, the investment incentives for these plants are essential for supply security. In fact, given the expected retirement of conventional power plants, EWI (2012) even forecasts additional investment needs of 55 GW of conventional generation capacity by 2030, of which 44 GW are gas plants.

2. Second, given the large amounts of renewables capacity, scarcity situations occur rarely. As a result, reserve units face a level of utilization which is both low and difficult to predict in the long run. Hence, they require even higher peak prices to recover their fixed costs than they would in a world with lower share of renewables capacities. According to the previous market design studies, the most recent of which is EWI (2012), it appears to be unlikely that scarcity prices will suffice to recover these investment costs. Hence, it would be risky to leave the issue of generation adequacy to the energy-only market alone.

4.3 Evolution or Revolution of the Market Design with Respect to Capacity Mechanisms?

To address the issue of generation adequacy, several countries have implemented different forms of *capacity mechanisms* in addition to the energy-only market. The idea of such mechanisms is to provide additional and more stable revenues for generators which are based on capacity instead of actual generation. Such a change in the revenue driver seems to be perfectly in line with the changing role of conventional generators. In an electricity market dominated by renewables, their task is no longer the large-scale supply of electricity but the provision of capacity reserves for situations in which they are needed due to insufficient supply from renewables. In the following, four basic forms of capacity mechanisms are described and examined in the context of possible implementation in Germany (Brunekreeft et al., 2011; de Vries, 2007).

1. The simplest form of capacity mechanisms consists of administratively set *capacity payments* that aim to prevent existing generators from leaving the market or to incentivize new investments. A more elaborate version of capacity payments is established in Spain, where a capacity-based investment incentive is granted that decreases with the reserve margin, that is, the more excess capacity there is in the market, the lower is the

investment payment. At the time the investment incentive is granted, the payment is fixed to a period of 10 years. A second component of the Spanish system is an availability payment for existing generators that are not able to earn adequate returns on the market (Federico and Vives, 2008).

2. *Strategic reserves* are implemented in Sweden and Finland. A central authority, typically the transmission system operator (TSO), auctions a certain amount of reserve capacity.[3] These capacities are withdrawn from the regular energy market and are centrally dispatched only in specific scarcity situations. A possible trigger for reserves may be a certain level of day-ahead price that is reached in a peak situation. The auction price determines the capacity revenues for participating generators. The effect of a strategic reserve depends strongly on the dispatch price and the amount of acquired reserves. The minimum goal is to prevent those generators from leaving the market which are not utilized on the day-ahead market to recover their costs. The higher the dispatch price or the size of the reserve market, the higher will be the scarcity on the regular market as a result of withholding capacity. Higher scarcity prices may trigger additional investments so that the overall amount of available capacity may increase (EWI, 2012).

3. *Capacity credits* are a more complex model known from the US market PJM, where a separate *capacity market* in addition to the energy market has been established. To ensure an adequate level of generation capacity, suppliers have to purchase the so-called "capacity credits" to cover their forecasted peak demand—plus a reserve margin of 15%—up to 3 years in advance. The generators' revenues result from the capacity price. The advance of 3 years aims to signal scarcity early enough to incentivize sufficient investments in time (see Chapter 9).

4. *Reliability options* can be described as a "financial version" of capacity credits. Instead of physical capacity, suppliers are obliged to acquire option contracts from generators selling the required amount of electricity for a predefined price. This can be organized as follows: a central auctioneer, for example, the TSO, acquires the option contracts covering peak demand plus reserve margin. The option premium defines the generators' capacity revenues. The options are centrally called as soon as a certain trigger price on the day-ahead market, as defined in the contracts, is reached. Generators have to pay the difference between the spot price and the option price to the auctioneer who passes these payments through to consumers. While the auction guarantees the availability of sufficient generation capacity, similar to the capacity credits model, there are two

---

3. For an analysis of different auction designs for a strategic reserve see Brunekreeft and Meyer (2012b).

positive side effects. First, consumers are protected against price spikes as they only pay the electricity price defined in the option contract.[4] Second, the model addresses market power issues in scarcity situations, since it incentivizes generators to make their capacity available on the market in times of high prices. Since producers have to pay the price difference—spot price minus option price—to consumers anyway, they do not gain from price manipulation by withholding capacity from the market.

All these capacity mechanisms have their respective pros and cons which should be considered and weighted in order to figure out which model is most suitable to address the specific market design issue in Germany. Several studies have been conducted or are currently underway to evaluate the necessity and the possible design of a capacity mechanism for Germany and/or its neighboring countries in the EU.[5] The most important evaluation criteria used in the analysis of models are efficiency, implementation costs, and compatibility with market developments.

The efficiency criterion refers to the question of whether a capacity mechanism is able to achieve the *optimal level* of generation capacity, and whether it provides those reserves at *least cost*.

With regard to the *optimal volume of capacity*, the main distinction of capacity mechanisms is between selective and comprehensive models. In case of *selective* models, only a part of the market is covered by the capacity mechanism, *while comprehensive* mechanisms cover the whole market. Selective mechanisms are considered less efficient, since they do not control the overall amount of generation capacity but only part of it. Examples are capacity payments and strategic reserves which address those generation units that fail to cover their full costs in the regular market. By restricting the capacity mechanisms to those units, the remaining amount of generation capacity depends as before on the energy-only market prices. Any price distortion of the market caused by the capacity mechanism or any prediction error concerning the market-based investment incentives may result in a suboptimal level of overall generation capacity. This is different in the case of a comprehensive capacity mechanism since it directly steers the optimal reserve margin by controlling the overall amount of capacity in the market (Brunekreeft et al., 2011).

With respect to the *least-cost procurement* of reserves, the main difference is between *administrative* and *auction-based* capacity mechanisms. Administratively determined capacity payments are typically considered less

---

4. If the spot price is high, consumers receive the price difference and are therefore hedged against higher prices.
5. See, for instance, Brunekreeft et al. (2011), BNE (2011), Ecofys (2012), EWI (2012), LBD (2011), and Matthes et al. (2012).

efficient, especially since the most likely create windfall profits for those generators that are able to recover their costs from market revenues anyway. Auction-based models, which are all models described above except for the capacity payments, provide cost efficiency by applying a competitive bidding process among generators. Given an adequate auction design, these models appear to be more suitable for providing generation capacity at least cost.

Implementation costs mainly refer to the strength of market intervention or complexity of changes to the current market design that are required to implement the model. From this point of view, capacity payments and strategic reserves appear to be easily implemented since the energy-only market remains the main pillar of the market design. In contrast, capacity credits require the establishment of a new separate market which may cause further market design problems like market power and regulatory control of availability of capacity. Related to the implementation costs, the choice of model depends on the available *time frame*. While capacity payments and a strategic reserve may be implemented within a relatively short time period, the more advanced capacity market models may require 5−7 years to be implemented, and even more time until investment incentives finally turn into available capacity.

Finally, *compatibility* with market developments means that a capacity mechanism should not hamper but foster current and future market developments. Notably, the evolution of the electricity sector toward more active demand-side participation, and more broadly the development of smart grids, as well as the integration of storage facilities to increase system flexibility, will play an important role in a future market design. All these developments will contribute to a stable and secure energy supply and should therefore be carefully considered when implementing a new market design. In particular, demand response and storage may become important factors in reducing the needs for conventional reserve capacity. Hence, it is advisable for these elements to be integrated into a capacity mechanism. The strategic reserve model implemented in Sweden provides an example of how demand-side reserves can participate in a reserve auction as a complement to generation reserves (Brunekreeft et al., 2011). Also capacity credits and reliability option models can be designed to include demand-side participation.

Due to its simplicity, a strategic reserve is one of the options currently discussed as a possible short-term solution for Germany. Such a mechanism could be implemented quickly and does not require a large intervention into the market design. As mentioned above, however, the disadvantage of a strategic reserve is its lack of efficiency, since it only covers a selective part of the market. There are also concerns that a strategic reserve leads to higher market prices and thus windfall profits for existing generators as part of the generation capacity are withdrawn from the market and held in the strategic reserve (LBD, 2011). Furthermore, implementing a reserve market alone does not address the source of the problem which is the lack of market-based incentives for renewables. Instead, it creates another submarket in which real competition is replaced by auction-based "as-if" competition.

The choice of an adequate market design for the long run depends on a number of specific market characteristics. In the case of Germany, two aspects seem to be of particular importance.

1. First, the central location of Germany and its strong integration with neighboring countries require considering the possible cross-border effects when implementing a new market design. By now, capacity mechanisms in Europe were mainly implemented on a national level, while energy companies think and act globally. A stronger integration of Europe's electricity markets requires a more integrated and coordinated approach for the market design as well. To give an example, introducing capacity payments in one country may attract investments that would otherwise have occurred in neighboring countries and exacerbate the problem of generation adequacy abroad. Before deciding about a permanent market design adjustment on the national level, it may be advisable to take account of the European developments to avoid an inefficient allocation of resources. At the same time, however, as long as system security is mainly a national responsibility and there are only limited interconnector capacities, national responses are to some extent unavoidable.

2. Second, electricity markets are currently subject to significant changes with respect to demand-side participation and the integration of storage facilities. Both aim to increase market flexibility by enhancing the responsiveness of electricity demand to market prices. A main part of the development is the implementation of smart grids. Increasing demand response and storage alleviates the problem of generation adequacy since it reduces the need for conventional backup capacities. Furthermore, as the discussion in Section 3 shows, the support scheme for renewables is an elementary part of the development and, hence, the market design discussion. All these market developments should be considered when deciding on a future market design.

5 CONCLUSION

The fundamental changes that the German electricity market is currently facing suggest that the challenge of a future market design goes beyond the issue of integrating more renewables into the current market. It is rather a question of redesigning the market to fit in with a world with renewable-dominated electricity supply and a more active demand side. The political vision of a low-carbon energy sector requires an integrated approach to account for the ongoing structural changes that alter the roles and responsibilities of all market participants.

This implies that the two developments that have been presented in this chapter and that are currently taking place in Germany—reform of support instruments for renewables and market redesign to enable power plant investment—should increasingly converge.

On the one hand, the support for renewables will increasingly move from feed-in schemes that were essentially capacity based to more energy-related elements. They should be adjusted to allow for market-based incentives within the limits of renewables flexibility.

On the other hand, for conventional and storage capacities needed to back up the system in cases of a shortfall of renewables supply, the market design adjustment should go in the opposite direction by supplementing the energy-based payments by a more stable capacity-based remuneration.

The support schemes for renewables should in the long run be integrated into an adequate market design that allows for market-based investment signals for all kinds of capacity. The critical task is to find an optimal trade-off between increasing market incentives and reducing investment risks.

A future market design should foster demand-side participation, smart grids, and the integration of storage facilities including new developments like power-to-gas. A more flexible demand side reduces the need for conventional generation capacity and helps to enhance supply security.

However, one problem of choosing an optimal market design for the future is that many market developments are at an early stage. The technological and economic perspectives of smart grids or the implementation and regulatory treatment of storage facilities are still under discussion. Furthermore, the growing integration of the European markets may require an integrated European approach to a future market design. Such an approach is still lacking. Hence, the current market environments rather ask for an evolution than a revolution of the market design.

The question of which variant of the available market design models would be the best solution for Germany, or whether a new one is needed, and how the future market should be designed in detail, is currently in the focus in the academic and political discussion.

REFERENCES

Andor, M., Flinkerbusch, K., Janssen, M., Liebau, B., Wobben, M., 2010. Negative strompreise und der vorrang erneuerbarer energien. Z. Energiewirtschaft 34, 91−99.

Barth, R., Christoph Weber, C., Swider, D., 2008. Distribution of costs induced by the integration of renewables power. Energ. Policy 36 (8), 3107−3115.

Bauknecht, D., 2012. Transforming the Grid—Electricity System Governance and Network Integration of Distributed Generation. Baden-Baden.

BDEW, 2012. Brutto-Stromerzeugung nach Energieträgern 2012, Bundesverband der Energie- und Wasserwirtschaft e.V. (BDEW), May 2012.

BMU, 2010. Leitstudie 2010. Langfristszenarien und Strategien für den Ausbau der erneuerbaren Energien in Deutschland bei Berücksichtigung der Entwicklung in Europa und global, Study conducted by Deutsches Zentrum für Luft- und Raumfahrt (DLR), Fraunhofer Institut für Windenergie und Energiesystemtechnik (IWES), Ingenieurbüro für neue Energien (IFNE), December 2010.

BMU, 2012. Leitstudie 2011. Langfristszenarien und Strategien für den Ausbau der erneuerbaren Energien in Deutschland bei Berücksichtigung der Entwicklung in Europa und global, Study conducted by Deutsches Zentrum für Luft- und Raumfahrt (DLR), Fraunhofer Institut für Windenergie und Energiesystemtechnik (IWES), Ingenieurbüro für neue Energien (IFNE), March 2012.

BNE, 2011. Kapazitätsmarkt. Rahmenbedingungen, Notwendigkeit und Eckpunkte einer Ausgestaltung, Bundesverband Neuer Energieanbieter e.V. (bne), September 2, 2011.

Brandstätt, Chr., Brunekreeft, G., Jahnke, K., 2011. Large-scale integration of wind: flexible voluntary curtailment agreements under a feed-in system with renewables priority to avoid inefficiently negative power prices. Energ. Policy 39 (6), 3732−3740.

Brunekreeft, G., Damsgaard, N., De Vries, L., Fritz, P., Meyer, R., 2011. A Raw Model for a North European Capacity Market—A Discussion Paper, Final Report, Elforsk.

Brunekreeft, G., McDaniel, T., 2005. Policy uncertainty and supply adequacy in electric power markets. Oxf. Rev. Econ. Policy 21 (1), 111−127.

Brunekreeft, G., Meyer, R., 2011. Kapitalkosten und Kraftwerksinvestitionen bei zunehmender Einspeisung aus erneuerbaren Energien − Die Diskussion um Kapazitätsmärkte. Z. Wirtschaftspolitik 60 (1), 62−73.

Brunekreeft, G., Meyer, R., 2012a. Auction Design for a Strategic Reserve Market for Generation Adequacy: On the Incentives Under different Auction Scoring Rules. Mimeo, Bremer Energie Institut/Jacobs University Bremen, Bremen.

Brunekreeft, G., Meyer, R., 2012b. Preisspitzen und Investitionsanreize im deutschen Strommarkt: Die Bedeutung der EEX-Gebotsgrenze für die Versorgungssicherheit. Mimeo, Bremer Energie Institut/Jacobs University Bremen, Bremen.

Bundesregierung, 2010. Energiekonzept für eine umweltschonende, zuverlässige und bezahlbare Energieversorgung, BMWi und BMU, September 2010.

Cory, K., Couture, T., Kreycik, C., 2009. Feed-in Tariff Policy: Design, Implementation, and RPS Policy Interactions. Golden, CO. National Renewable Energy Laboratory Technical Report No. TP-6A2-45549. March 2009. <http://www.nrel.gov/docs/fy09osti/45549.pdf/> (accessed 2.11.2012).

Couturen, T., Gagnon, Y., 2010. An analysis of feed in tariff remuneration models: implications for renewable energy investment. Energ. Policy 38 (2), 55−96.

Cramton, P., Ockenfels, A., 2012. Economics and design of capacity markets for the power sector. Z. Energiewirtschaft 36 (2), 113−134.

Cramton, P., Stoft, S., 2006. The Convergence of Market Designs for Adequate Generating Capacity Manuscript, 25 April 2006.

De Vries, L.J., 2007. Generation adequacy: helping the market do its job. Utilities Policy 15, 20−35.

Eclareon and Öko-Institut, 2012. Integration of Electricity from Renewables to the Electricity Grid and to the Electricity Market—RES-INTEGRATION. Final Report. March 2012.

Ecofys, 2012. Notwendigkeit und Ausgestaltungsmöglichkeiten eines Kapazitätsmechanismus für Deutschland. Zwischenbericht. Studie im Auftrag des Umweltbundesamtes, June 2012.

EEG/KWK-G, 2012, Informationsplattform der Übertragungsnetzbetreiber, http://www.eeg-kwk.net. (accessed 2.11.2012).

EWI, 2012. Untersuchungen zu einem zukunftsfähigen Strommarktdesign. Gutachten von dem Energiewirtschaftlichen Institut an der Universität zu Köln im Auftrag von dem Bundesministerium für Wirtschaft und Technologie (BMWi). March 2012.

Federico, G., Vives, X., 2008. Competition and Regulation in the Spanish Gas and Electricity Markets. Reports of the Public-Private Sector Research Center.

Hiroux, C., Saguan, M., 2009. Large-scale wind power in European electricity markets: time for revisiting support schemes and market designs? Energ. Policy 38 (7), 3135–3145.

IEA, 2011. Harnessing Variable Renewables—A Guide to the Balancing Challenge, Paris.

Joskow, P.L., 2006. Competitive Electricity Markets and Investment in New Generating Capacity. Center for Energy and Environmental Policy Research (CEEPR), Cambridge, MA.

LBD, 2011. Energiewirtschaftliche Erfordernisse zur Ausgestaltung des Marktdesigns für einen Kapazitätsmarkt Strom, Study conducted for Umweltministerium des Landes Baden-Württemberg, LBD-Beratungsgesellschaft, December 2011.

Matthes, F., Schlemmermeier, B., Diermann, C., Hermann, H., von Hammerstein, Ch., 2012., Fokussierte Kapazitätsmärkte. Ein neues Marktdesign für den Übergang zu einem neuen Energiesystem. Study for WWF Germany.

Mitchell, C., Bauknecht, D., Connor, P.M., 2006. Effectiveness through risk reduction: a comparison of the renewable obligation in England and Wales and the feed-in system in Germany. Energ. Policy 34 (3), 297–305.

Müsgens, F., Neuhoff, K., 2006. Modelling Dynamic Constraints in Electricity Markets and the Costs of Uncertain Wind Output. <http://www.dspace.cam.ac.uk/bitstream/1810/131648/1/eprg0514.pdf/> (accessed 02.11.2012).

Nicolosi, M., 2011. The Impact of Renewables Policy Setting on Integration Effects—A Detailed Analysis of Capacity Expansion and Dispatch Results. MPRA Paper, No. 31835.

Rathmann, M., et al., 2009. Renewable Energy Policy Country Profiles, Re-Shaping Project Report D3. Available from: <http://www.reshaping-res-policy.eu/> (accessed 02.11.2012).

Rathmann, M., et al., 2011. Towards Triple-A Policies: More Renewable Energy at Lower Cost, Re-Shaping Project Report D16. <http://www.reshaping-res-policy.eu/downloads/Towards-triple-A-policies_RE-ShapingD16.pdf/> (accessed November 2012).

Vandezande, L., Meeus, L., Belmans, R., Saguan, M., Glachant, J-M., 2010. Well-functioning balancing markets: a prerequisite for wind power integration. Energ. Policy 38, 3146–3154.

Weber, C., 2010. Adequate intraday market design to enable the integration of wind energy into the European power systems. Energ. Policy 38, 3155–3163.

Chapter 8

The Challenges of Electricity Market Regulation in the European Union

Walter Boltz
Executive Director, E-Control Austria

1 INTRODUCTION

Fifteen years after the first European electricity directive came into force,[1] liberalization of the European electricity market is still work in progress. The challenge has been to expand markets beyond national borders to regional or European dimensions.

The process has gained quite some momentum. Diverse market coupling models in several regions have contributed to better market integration. The elaboration of a target model outlining a common view on what the future wholesale market should look like in general terms, and how available interconnection capacity should be calculated and made available to the market has been an important step. In February 2011, the European Union's (EU's) Council of Ministers decided that the EU internal energy market should be completed by 2014, that is, all new market rules should be in place by that time. This common market should even extend to balancing energy and intraday bilateral trading—a national prerogative until recently, despite some bilateral integration.

From a political perspective, the European challenge also stems from the close relationship of many governments with "their" incumbent energy companies. Energy companies are often seen as national champions in industrial policy and being responsible for securing the necessary primary energy for the future. As primary energy from national sources is becoming scarce in Europe, this role is likely to gain further importance.

While at national level, this kind of "protectionist" aspect ought to be less important, the European road to a common market is paved with

1. Directive 96/92/EC.

Evolution of Global Electricity Markets.

TABLE 8.1 Generation Capacity in the EU (in megawatts)

EU 27	2000	2008	2009	2010
Nuclear	136,347	132,842	132,861	130,538
Fossil fuel fired	391,306	445,428	454,155	462,173
Hydro	135,636	141,694	142,905	142,726
Other renewables	21,942	94,748	111,561	133,940
Solar	82	10,102	15,244	22,981
Wind	12,808	64,034	74,614	83,819
Biomass	3940	9352	10,019	10,071
Biogas	975	3799	3092	3891
Not specified	440	1198	1143	1144
Total installed capacity	636,161	315,910	842,624	870,521

Source: Eurelectric (2011)

national hurdles, national politics,[2] and the inability of many governments to make their markets more competitive. It is this quite deplorable but fundamental starting point of liberalization that one must take into account when analyzing future challenges.

Indeed, new challenges loom ahead, such as the implications of the political decision to switch from a dispatchable electricity system to a less dispatchable intermittent one, which is often highly concentrated locally. Table 8.1 shows the development of generation capacity in the EU since 2000, and more specifically the increase in intermittent generation capacity from solar and wind.

This chapter comprises three sections. Section 2 describes the past and ongoing bottom-up development of cross-border European market liberalization in electricity. Section 3 analyzes the challenges ahead, and discusses how these might undermine results achieved so far in terms of market integration and liberalization. Section 4 draws a number of conclusions.

2 THE ROAD TO INTEGRATED MARKETS IN THE EUROPEAN UNION

The European regulatory framework has evolved from mere principles toward more detailed regulation, advancing through the three periods shown

---

2. For a more detailed discussion of national policies, see the specific chapters on United Kingdom, France, and Germany in this book, Chapters 2–4.

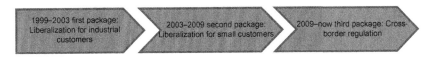

FIGURE 8.1   Stages of European liberalization.

in Figure 8.1 and further described below. During this development, the design of many detailed rules was initially left to Member States; national implementation was to follow the general principles of nondiscrimination and a stepwise opening of the market to competition. As explained in this chapter, this approach has not proven effective.

2.1 The Initial Setup: 1999–2003

The electricity (and gas) markets have been one of the last remains of former product markets separated by different regulation, product definition, abundant barriers to entry, and so on. The process of market integration started in 1985 with the EU Commission's White Paper on Completing the Internal Market.[3]

An internal document[4] highlighted the most important barriers to an integrated electricity market. As long as specific legal provisions concerning electricity markets were missing, only general rules on the free movement of goods, which are part of the "European Constitution," were applicable in order to open up markets to competition.

The resistance of governments to further market opening can be explained by their inherent interests in this area. The low-price elasticity allows governments to use the industry as a vehicle to subsidize primary energy production by forcing purchase of local primary energy instead of foreign sources. In some cases, it also allows the sector to act as a stabilizer as part of general economic policy (companies in some cases had to invest or had to hire people in order to alleviate pressure in labor markets or to produce countercyclical investment).

In 1991, the European Commission proposed the first directive "concerning common rules for the internal market in electricity."[5] The aim of this directive was to guarantee the free movement of capital, that is, the right to freely establish companies, construct and run power stations in the whole Community, and the free movement of goods to serve any final customer. Unbundling of transmission system operators (TSOs) was perceived to be of vital importance already at that time, and the draft directive foresaw unbundling of commercial operations from generation and distribution. As a concession to some coal-producing Member States, a possibility for them to direct

---

3. COM(85) 310.
4. COM(88) 238 final.
5. COM(91) 548 final; together with the respective directive on the internal market in natural gas, this is called the first package.

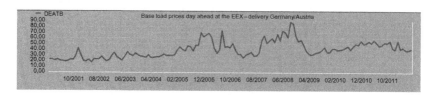

FIGURE 8.2   Base load prices (in €/MWh). *Source: EEX.*

TSOs to give priority of up to 20% of dispatched energy to generating units using indigenous primary fuel sources was introduced. Member States were to comply with the directive by the end of 1992.

However, during negotiations between Council, Parliament, and the Commission, the liberalization of retail markets was watered down, and only liberalization of industrial customers was retained. In addition, the negotiations took so long that the liberalization was effectively delayed by about 7 years.[6] For those companies, which had "suffered" public intervention in their investment policy or other economic decisions before liberalization, the directive allowed for a transitional regime under the headline of "stranded costs." It implied not implementing free access to the grid for producers, which in turn excluded the producers of a Member State under a transitional regime from potential export markets via reciprocity rules. As it turned out, only few Member States were able or willing to use this provision and instead preferred financial subsidies following general state-aid rules.

Although from a regulatory point of view many issues were still open during this first phase of market liberalization, the experience of liberalized industrial customers (eligible customers) was quite positive. The German Dow Jones VIK index, collecting price information from big industrial energy consumers, decreased by some 25% between 1998 and 2001.[7] These low industrial prices were made possible by wholesale prices as low as 20 €/MWh in 2001 (Figure 8.2). Despite the fact that primary energy prices were at very low levels (about 10 €/MWh for natural gas and $39/t for coal)[8] electricity prices did not allow any capital amortization in this period. At the very early stage, prices did not even cover short-run marginal cost, indicating that the price was not a reliable indicator, probably due to the fact that in 2001, monthly traded volumes were only about 1 TWh for German delivery (Figure 8.3).

Countries suffering from grid congestion to low-price areas, such as the Netherlands, had considerably higher prices in this first phase of liberalization (Figure 8.4).

---

6. By February 19, 1999.
7. Madlener and Jochem (2001).
8. BP (2012), Heren NBP Index, BAFA German Border Prices for natural gas and McCloskey Northwest Europe marker prices for coal.

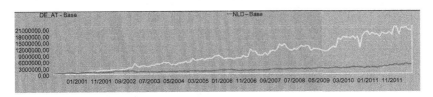

FIGURE 8.3 Traded volumes in Germany and the Netherlands (in megawatt hour). *Source: EEX, APX.*

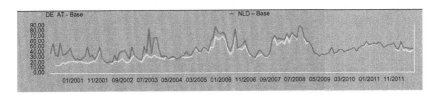

FIGURE 8.4 Base load prices in Germany and the Netherlands (in €/MWh). *Source: EEX, APX.*

It also became apparent that achieving a level playing field between the main energy companies in Europe was not possible. The vague directives relying on national implementation had brought about a patchwork of highly diverse systems, from markets fully liberalizing the whole value chain to markets that were only opened to competition to the absolutely necessary extent. In addition, the diversity of production technologies also had, and still has, important financial ramifications. Base load companies such as lignite and nuclear producers generate a high annual net cash flow in order to amortize the initial investment, used to create reserves. The financial power of base load companies in comparison to peak load producers was often perceived as unfair competition after liberalization.[9]

This raised the fear among many Member States that by opening the markets "their" incumbents might fall prey to major foreign market players such as the French EdF or German E.ON and RWE. And indeed the United Kingdom, the first country to liberalize its market (except the Nordic countries), saw major electricity companies bought by foreign producers E.ON and EdF in 2001.[10]

In addition, many regulatory deficiencies were identified, *inter alia*:

- The role of the TSO as the main operator in most markets was not defined thoroughly. The lack of market transparency led to a situation

---

9. This of course has to be distinguished from future investment decisions where financial risks are quite different.
10. The biggest European generators are EdF (Electricité de France SA, F), E.ON (E.ON AG, Ger), GdF-Suez (GDF Suez SA, F/Bel), RWE (RWE AG, Ger), and Enel (Enel S.p.A, It), providing almost 50% of EU supply.

where vertically integrated companies enjoyed a high level of inside information and were able to profit from often opaque transport capacity allocation procedures. These procedures very often allowed for long-term contracts in combination with first-come-first-served principles or to a lesser degree pro rata, which effectively excluded new market players from getting network access and entering markets.

- There was no cross-border coordination regarding general market design or daily business issues. This means that roles and responsibilities as well as rights and obligations of different parties in different markets were hugely inconsistent and this diversity turned out to be a serious obstacle to market integration.
- Transmission congestion revenues were sometimes used as an "extra" on top of the recovery of the regulatory asset base, that is they constituted additional profit for TSOs. Such a situation, of course, reduced incentives to alleviate system congestion.
- Regulation was performed by authorities independent from the industry but not from politics. In most Member States specific regulatory authorities were established, but for instance, in Germany the ministry for economy was responsible for regulation during this time. In such a system, it was quite frequent that politics could intervene even in day-to-day decisions of regulators. Regulators were seen as "network regulators" in many jurisdictions, leaving market design and competition issues to politics and/or competition authorities. This patchwork of responsibilities did not allow for a consistent regulatory approach across borders. By that token it was competition policy in many cases which, in the course of mergers, improved market rules as concessions (remedies) from major energy companies.
- Transposition of the directive by Member States into national law was typically not comprehensive or satisfactory. Infringement procedures were opened against eight Member States.

2.2 First Steps Toward Deepening Regulation: 2003–2009

In March 2000, the European Council, that is the heads of state, asked the Commission to complete market liberalization and to speedup the process. In 2003, the so-called second package came into force and had to be transposed into national law by July 2004. By that time, all nonhousehold customers were eligible for choosing their supplier and by July 2007, all customers were to be liberalized. For the first time, the directive stipulated the establishment of sector-specific regulatory authorities in all Member States, but there was no requirement for independence from "politics" in their decisions.

In many Member States, this did not change the former situation much and ministries kept the power to take final decisions on issues such as network tariffs. Still, the provision guaranteed some kind of transparency as regulators had to send their draft decisions to ministries, thereby forcing the latter to provide some well-founded reasoning in case of deviation from the draft. One of the most important tasks of EU regulators during this period was to enforce the national provisions that had resulted from transposition of the directive. This was specifically the case for enforcing unbundling of network operators from competitive parts of vertically integrated companies. For TSOs and distribution system operators (DSOs), the directive demanded at least unbundling the operator in legal terms, which means establishing at least a separate legal entity and making sure that the management is not part of the structures of the integrated company. To ensure nondiscrimination in day-to-day business, network operators had to establish compliance programs including legal and management unbundling. The box below provides a snapshot of the European legal framework, which is critically pertinent to the chapter's focus.

The European Legal Framework:

In the context of European Union 27 (EU 27), legal acts have to get approval from the Council of Ministers, that is, the 27 national ministers who are responsible for energy affairs in Member States, and the European Parliament. Proposals for legislative acts may only be put forward by the EU Commission, which handles the EU's central administration.

Effectively, the Council position is a compromise between different national interests, the Parliament is the democratic European chamber and the EU Commission serves as the technocratic European element in this process.

If no initial agreement can be found, legal acts are often finally negotiated between Council and Parliament in the so-called conciliation committees, in which the EU Commission acts as a mediator.

There are two main forms of legal acts: the "directive" is an act that stipulates goals and general principles. Member States have to transpose these into national law so that they become legally applicable. The "regulation" is directly applicable after coming into force. In the case of directives, the EU Commission monitors transposition and can launch cases against Member States at the European Court of Justice if transposition is not in line with the objectives of the directive.

Apart from the apparent modification, in terms of the regulatory regime, a more detailed design was implemented. The box above describes the main types of European legal acts as well as the decision-making process in the area of energy law in the EU. The second package consisted of a directive, comparable to the 1996 directive in scope, and a regulation,[11] which laid down the design for cross-border markets. Several new issues were addressed in this regulation:

---

11. 1228/2003/EC, Regulation on conditions for access to the network for cross-border exchanges in electricity; please note that as opposed to directives, regulations are directly applicable and do not have to be transposed into national law in Member States.

- the "Inter TSO (transmission system operator) Compensation Mechanism" (ITC), which compensates for costs incurred by hosting cross-border flows;
- clarification of nondiscrimination in network charges with respect to the origin or destination of flows;
- transparency requirements for TSOs;
- market-based congestion management;
- clarification on how congestion revenues of TSOs might be used;
- provision for new interconnections to be able to receive an exemption from third-party access under specific conditions;
- enforcement and penalties of up to 1% of annual turnover of a company.

It is quite evident that the first pillar of the second package was to increase the independence of "system relevant" players, that is, TSOs in wholesale markets, DSOs in retail markets, and regulators for oversight and enforcement. The second pillar was to achieve a situation in which "incentives" were consistent with market liberalization.

However, national transposition and the possibilities for enforcing regulation were both disappointing. The former did not lead to a uniform market model in Europe and especially congested transmission lines proved to be barriers to market integration. Available capacity was sometimes allocated using nonmarket-based methodologies such as first come first served, giving advantage to incumbent traders. In addition, existing rules were quite often not accompanied by proportionate enforcement powers for regulators. Where this happened, the economic benefit of noncompliance was higher than the potential penalties. Regulators started a project called "regional initiatives" in 2006, which put in place road maps toward electricity market integration in seven overlapping European regions. Geographically, they reflected the existing network congestions.

Essentially, the objective of a transparent process was to put pressure on all players and not to delay integration. At the same time, market players profiting from integration (mainly traders) were able to identify more clearly those regulatory elements that prevented them from entering into beneficial contracts.

Price correlation at that time, depicted in Table 8.2, is quite revealing.[12] Price dynamics were not in line with each other, as shown by low correlation coefficients between most markets. In such a situation, many chances for profitable arbitrage trading arise. This is why power exchanges and traders were very active in pushing for more integration.

---

12. Using correlation coefficients as indicators of market integration is often criticized and rightfully so: there are many factors influencing correlation (e.g., correlated primary energy for marginal capacity) and it does not include price levels at all. Still, it indicates common price dynamics, which, from an economic policy perspective, at least partly covers the main interest of industrial customers in terms of energy cost.

TABLE 8.2 Correlation Coefficients of Day-Ahead Base Load Prices at European Exchanges in the Netherlands (APX), Germany (EEX), Nordic Region (Nord Pool), Spain (OMEL), France (Powernext)

		APX	EEX	Nord Pool	OMEL	Powernext
2002	APX	1				
	EEX	0.67	1			
	Nord Pool	0.23	0.04	1		
	OMEL	0.26	0.52	−0.50	1	
	Powernext	0.65	0.83	0.08	0.55	1
2003	APX	1				
	EEX	0.68	1			
	Nord Pool	0.24	0.20	1		
	OMEL	0.37	0.48	−0.33	1	
	Powernext	0.69	0.88	0.15	0.52	1
2004	APX	1				
	EEX	0.71	1			
	Nord Pool	0.12	0.35.	1		
	OMEL	0.41	0.47	−0.03	1	
	Powernext	0.72	0.91	0.29	0.50	1
2005	APX	1				
	EEX	0.83	1			
	Nord Pool	0.50	0.53	1		
	OMEl	0.60	0.56	0.30	1	

Source: CEER (2006)

The model region where market integration was, and still is, most advanced in Europe is the Nordic region, comprised of Finland, Sweden, Norway, and Denmark. The Norwegian power exchange, Nord Pool Spot, runs implicit auctions to allocate network capacity of the entire region and only in times of congestion, the algorithm calculates partial optima for the price zones in the Nordic area (market splitting).

The exchange has implemented a high degree of transparency in the market, where fundamental information on planned and unplanned outages of

TABLE 8.3 Percentage of Hours with Different Day-Ahead Prices

2005		NO1	NO2	SE	FIN	DK1	DK2
				Lower spot price than:			
NO1	Higher spot price than:		17	3	4	4	2
NO2		17		2	3	8	3
SE		18	17		1	6	0
FIN		24	23	8		9	7
DK1		51	53	47	47		42
DK2		28	29	17	17	15	

NO1, South Norway; NO2, North Norway; SE, Sweden; DK1, Denmark (Jutland); DK2, Denmark (Zealand); FIN, Finland.
Source: CEER (2006)

generation capacity, for instance, is made available to the market participants.

This market model has attracted much liquidity. Consequently, today about three-quarters of consumption in the Nordic countries is traded day ahead on Nord Pool Spot.

The success of this region in terms of efficient allocation of transmission capacity and liquid wholesale trading still influences the European vision of a target electricity model,[13] which favors bilateral trading and does not include central dispatch of electricity like in a pool model. Table 8.3 shows the number of hours in which day-ahead prices differed between the market areas in the Nordic market. For instance, day-ahead prices in south Norway differed from those in Sweden in 21% of hours, so during 79% of the time prices in these two bidding areas the same. Compared to other EU markets, such a dynamic combination of bidding areas was very innovative.

In 2006, power exchanges already existed in many markets on the European continent. The concept of market splitting with one exchange only was therefore not conceivable anymore. However, in November 2006, trilateral market coupling between the Belgian Belpex, the French EPEX (European Power Exchange), and the Dutch APX-Endex power exchanges was established. For this purpose, a new power exchange had to be installed in Belgium (Belpex). Today, almost all European Member States have their own exchanges in place, or trading on one of the existing exchanges for their respective bidding areas is possible at least for day-ahead products. The most

13. See footnote 26.

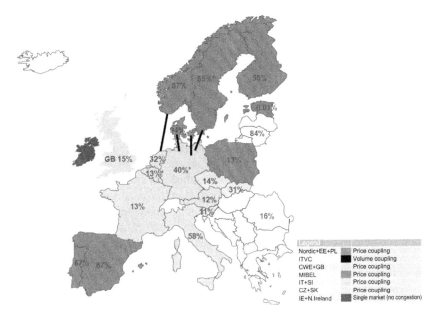

FIGURE 8.5   Exchange-traded volumes as percentages of national demand. *Source: ACER (2012).*

important exchanges with cross-border significance are OMEL in Spain for the Spanish and Portuguese market, EPEX Spot in France for the French, German, Swiss, and Austrian markets, and Nord Pool Spot in Norway for the Nordic market. Figure 8.5 depicts the day-ahead trading volume in relation to national consumption of electricity on power exchanges.

Despite this encouraging example of regional cooperation, for instance, the Council of European Energy Regulators (CEER)[14] asked for "political commitment to the [...] Regional Initiative," while at the same time expressing skepticism as "Member States still tend to protect national champions."[15] A prominent example of this was the German merger case of E. ON-Ruhrgas, where the competition authority prohibited the merger whereas the Minister of Economic Affairs overruled the Cartel Office.[16]

The above example of market coupling should not be mistaken for an indication of a general development in this direction in Europe. Many Member States, especially in eastern Europe, did not have wholesale trading at all, with regulated consumer prices and publicly managed long-term contracts between suppliers and producers; trading did not make sense and was

---

14. On CEER see for instance: Vasconcelos (2001).
15. Op.cit.
16. Newbery (2009).

not even endorsed in some cases. Starting in the mid-1990s, many eastern European countries privatized their infrastructure industry. Investors sometimes paid prices beyond the economic value of companies. In return, some countries guaranteed a stable and high cash flow for the investors. E.ON, RWE (Germany) and EdF (France) were the most important western investors[17] in the 1990s, which did not have any interest in competition in these countries after the enlargement of the EU in 2004.[18]

Over time, as individual contracts between suppliers and customers expired, more and more industrial and commercial customers were able to choose from among different suppliers. Pressure was building up on regulated prices for these customers, which were high, in comparison to potential import prices. This development was supported by more and more network capacity being allocated in nondiscriminatory procedures. As mentioned above, the final step of liberalizing the retail market was taken in 2007, when all households became "eligible customers." However, the generally low level of switching in the household sector at that time[19] did not contribute to the pressure to open wholesale markets.

In general, the main area for improvement was found to be related to investment in interconnection capacity. Examples of permission procedures taking 10 years and more for simple high-voltage overhead lines pointed at insufficiency in this respect. However, analysis of the reasons for such lengthy procedures yielded quite a range of answers. Dealing with vertical foreclosure in its sector inquiry,[20] the EU Commission found that the "current level of unbundling of network and supply interests" would have "negative repercussions on market functioning and on incentives to invest in networks [...]." Figure 8.6 shows existing network congestions in the EU. Apart from internal bottlenecks in many Member States, congestions mainly occur around the North Sea, where most wind generation is being installed. Therefore, the Nordic, Great Britain, and Irish markets must be better interconnected with the European mainland. In addition, the northern part of the mainland needs reinforced transmission to the south.

Independence of TSOs was also one of the key areas of necessary improvement suggested by Glachant et al. (2006). Of course, it is quite difficult to prove that the lack of investment is due to vertical integration, or the extent to which this is the case.

Other reasons, frequently put forward by TSOs, are quite burdensome permission procedures for new transmission lines and the lack of economic

---

17. On mergers in eastern Europe: LaBelle, Kaderjak (2008).
18. On energy company strategies in Europe in general: Midttun, Omland (2004).
19. Exceptions were in Great Britain with an annual switching rate of almost 20% and the Nordic market with typically some 8% annual switching rate. All other countries had switching rates below 5%.
20. COM(2006) 851 final.

FIGURE 8.6   Congested network areas in the EU. *Source: ENTSO-E (10-year network development plan 2012).*

incentives.[21] A look at the situation in Europe and the incentives for vertically integrated companies reveals that in some cases, all arguments are correct. In export-oriented countries, a vertically integrated company's economic incentive not to invest in is certainly quite low, so permission procedures or inadequate return on investment are more important. However, in import countries or countries with old and inefficient production facilities there is a clear economic argument for a vertically integrated company not to invest in additional interconnection capacity as this would lead to lower wholesale prices and reduce profits of the vertically integrated company.

21. Dobbeni (2007).

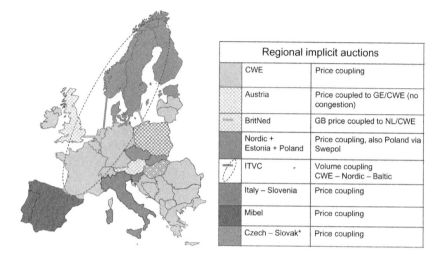

Regional implicit auctions	
CWE	Price coupling
Austria	Price coupled to GE/CWE (no congestion)
BritNed	GB price coupled to NL/CWE
Nordic + Estonia + Poland	Price coupling, also Poland via Swepol
ITVC	Volume coupling CWE – Nordic – Baltic
Italy – Slovenia	Price coupling
Mibel	Price coupling
Czech – Slovak*	Price coupling

FIGURE 8.7   Coupled markets in the EU. *Czech-Slovak coupling is going to be extended to Hungary. *Source: Europex.*

Even if none of the arguments above was valid, investment in networks might still be hampered. Given that TSOs are subject to national law, cross-border projects might fall into a regulatory gap. When deciding whether to integrate an investment into a TSO's regulatory asset base, national authorities had to assess the investment's cost and benefit within the perimeter of the national boundaries. So, if there was no immediate benefit for the local population or economy, for instance, in the case of envisaged facilitation of transit flows, typically such an investment was not to be accepted by the regulator on the basis of national law.

2.3 A System of Cross-Border Regulation: 2009 to Present

Three years after the trilateral market coupling,[22] the spot markets between Denmark and Germany were coupled with the so-called interim tight volume coupling (ITVC) mechanism. Sweden followed suit in 2010 (via the Baltic Cable). By November 2010, the two areas were combined and enlarged, so that, via the German market, the coupled area comprised eight countries, that is France, Belgium, the Netherlands, Luxemburg, Germany, Austria, Denmark, and Sweden (Figure 8.7). Since then (January 2011), also Norway has joined via a cable to the Netherlands, and Great Britain joins via BritNed, a cable between Great Britain and the Netherlands. The European experience seems compatible to PJM (Pennsylvania-New Jersey-Maryland) to a certain degree, where successful market integration was a major goal,

---

22. See above.

and more and more electricity systems joined this market. In terms of physical interconnection and integration, the European development is quite unique.

Even so, the EU agreed on a third package of liberalization.[23] Backed by rising prices and a common suspicion that incumbents might collude and manipulate prices by withholding capacity—although only minor cases of price manipulation were ever proved—both politics and the electricity industry in Europe by and large supported further regional integration efforts. The core objectives were structural unbundling of TSOs from vertically integrated utilities, a governance process for a future market design based on independent TSOs and regulators, and the establishment of umbrella institutions for network and regulation at European level. European energy regulators also received powers to closely monitor electricity markets and to enforce rules. Until then monitoring and enforcement was quite differently organized in all Member States and in many cases, the lack of powers played an important role in the sluggish development. The third package also stipulates that national energy regulators must be independent in their day-to-day decisions even from politics.

As Glachant (2006) pointed out, however, the industrial reference model shifted from disintegrated to vertically integrated companies already in the early 2000s. Finally, the compromise found is to allow alternative unbundling models, that is, an independent system operator (ISO) model, where investment decisions and operational decisions have to be outsourced to fully independent units, and the independent transmission operator (ITO) model, which is a model of pure legal and functional unbundling with a high level of regulatory oversight. In the end, the third package has a bearing on industrial structures in some cases only. When the package was agreed, the European Commission forced some integrated companies, especially in Germany, to divest transmission companies as a remedy in competition cases. However, Vattenfall decided to sell the formerly vertically integrated TSO voluntarily.

Also in terms of national compliance with the third package, the picture is a quite disappointing one today. It seems that eight Member States have not yet fully transposed the directive, and only a few TSOs have been certified as being in line with the unbundling rules, despite the March 2012 deadline. In 2011, the European Council called for complete market integration by 2014, that is, all new network codes and market rules should be in place by that time.

Moving forward, the governance process for developing the future market design is quite complex. First, the European Commission defines priorities with regard to Framework Guidelines for 14 areas where harmonization is necessary; these include network security and reliability, connection, access,

---

23. For electricity: Directive 2009/72/EC; Regulation EC/714/2009; and Regulation EC/713/2009

data exchange, interoperability, emergency procedures, capacity allocation and congestion management, technical trading rules, transparency rules, balancing rules, transmission tariffs, and finally energy efficiency regarding electricity networks. These are areas which were identified as important for enhancing secure network operation or integrating wholesale markets in electricity. Put differently, these are the areas on which market foreclosure had been based so far.

For these priority areas, the Agency for the Cooperation of Energy Regulators (ACER) publishes legally nonbinding Framework Guidelines in which they define objectives and principles of final Network Codes. In the end, total of all codes will constitute the market design in Europe.

Based on the Framework Guidelines, the European Network of TSOs for electricity (ENTSO-E) produces Network Codes. After an ACER opinion, these are sent to the European Commission to be made legally binding via a process involving Member States (comitology). No code has run through the full process yet. One of the main reasons for this multistage process is the resistance of national governments to shifting power from national to Community level.

ACER's Framework Guidelines on Capacity Allocation and Congestion Management[24] implement the so-called target model which has been prepared since 2008. These are the main market design documents for the future. The main principles laid down are:

- Coupling of bidding zones in day-ahead markets; there is one algorithm which determines volumes and prices simultaneously (marginal prices);
- Bidding zones are combinations or parts of control areas and can be changed if network structure changes; they are defined and valid for all time horizons, that is, also for forward markets;
- In forward markets, physical and financial transmission rights may coexist (but not on the same border);
- Intraday markets should be served by a pan-European platform (continuous trading including capacity pricing) for capacity allocation—but more details are still to be developed.

Furthermore, ENTSO-E must develop a Community wide 10-year network development plan, which is nonbinding but serves as a reference for national network plans to be developed by TSOs and approved by national regulators. Regulators have the task to finally enforce those national plans, if necessary by open tendering.

In comparison to the second package, the third one provides for a much more centralized system for cross-border issues and comprises European cooperative organizations for TSOs and regulators. There is also an already quite advanced discussion on how to speedup permitting procedures for

---

24. ACER (2011).

transmission projects.[25] Its core is to define projects of European interest, which enjoy fast-track procedures and potentially financial support.

Even though the new arrangements have not yet produced results, we might again face challenges that demand new solutions and mechanisms for the electricity system to be put in place before the existing system can prove that it works. Unresolved issues of today's markets still stem from national energy policies that distort competition in integrated markets. While the past liberalization packages were to overcome national protectionist strategies for the first time, national policies threaten to undermine the market mechanism itself.

3 A NEW AGE OF REGULATORY INTERVENTION

This section describes the main future challenges for the EU market, all of them in one way or the other related to the transformation of the system to integrate more renewable production and specifically more intermittent generation. Even so, we can distinguish between supply-side aspects (Section 3.1) and demand-side aspects (Section 3.2).

3.1 Integrating Renewable Energy: A Difficult Transition Ahead

An aspect of the third package not yet covered in this chapter are Europe's environmental goals, further explained in Chapter 5 by Haas et al. as well as chapters on United Kingdom and Germany. The EU's renewable energy share in total energy consumption is to reach 20% by 2020. For many Member States, this means that renewable electricity production must drastically increase—on an average from 21% to some 35% in 2020, as renewable sources in other areas are more limited (Table 8.4).

The most productive areas in Europe in terms of renewable generation are concentrated in specific locations, with high solar radiation around the Mediterranean Sea and strong winds at the Atlantic Coast and North Sea.

Estimates suggest that a share of some 35% in total generation requires installation of about 200 GW of additional renewable capacity.[26]

As renewables shares in electricity production increase, state-aid considerations will become more important. The wide variety of support mechanisms currently in place has already drawn criticism for giving inefficient signals to investors and allocating money to projects in highly subsidized areas instead of most productive wind and solar regions.[27]

However, this is not the main challenge. If, after this process, the market will indeed consist of one-third or more of subsidized assets, which in accordance with European law enjoy priority access to the grid and a guarantee

---

25. The European Infrastructure Package.
26. Neuhoff et al. (2011).
27. See Chapters 6 and 7 for further details.

TABLE 8.4 National Targets of Renewable Energy as Part of Energy Consumption in the EU

EU Targets in Percentage	2005	2020
Sweden	39.80	49.00
Latvia	32.60	40.00
Finland	28.50	38.00
Austria	25.80	34.00
Portugal	20.50	31.00
Estonia	18.00	25.00
Rumania	17.80	24.00
Denmark	17.00	30.00
Slovenia	16.00	25.00
Lithuania	15.00	23.00
France	10.30	23.00
Bulgaria	9.40	16.00
Spain	8.70	20.00
EU	*8.50*	*20.00*
Poland	7.20	15.00
Greece	6.90	18.00
Slovakia	6.70	14.00
Czech Republic	6.10	13.00
Germany	5.80	18.00
Italy	5.20	17.00
Hungary	4.30	13.00
Ireland	3.10	16.00
Cyprus	2.90	13.00
The Netherlands	2.40	14.00
Belgium	2.20	13.00
United Kingdom	1.30	15.00
Luxemburg	0.90	11.00
Malta	0.00	10.00

Source: Directive 2009/28/EC

that costs are covered for a certain period of time, this will have major ramifications for the whole market mechanism.[28] Installed capacity in Europe
currently amounts to about 100 GW for wind[29] and 50 GW for solar,[30] which
is about 17% of total installed capacity. In terms of energy covered by a support scheme, shares peak at almost 30% (such as in Spain and Denmark),
and in Germany the share is about 15% of produced energy. The consequences of further investment in intermittent generation can be gauged from
the experience in Germany which today has a share of almost 50% in PV
(Photovoltaics) and 30% in wind capacity[31] in Europe and about 55 GW out
of 170 GW of total installed capacity. As Neuhoff et al. (2011) points out in
an analysis of different congestion management mechanisms, even today
wind can dramatically change the regional scope and price levels of markets
in an appropriate short-term, that is, nodal pricing system.[32]

Taking into account that in the future a much higher share of electricity
might be produced by nondispatchable but locally concentrated sources, the
extent to which an old-fashioned wholesale electricity market will still be
manageable seems quite uncertain. Network flows will get less predictable.
The main challenges in this respect are:

- Congestions within bidding zones;
- Term shift in wholesale trading;
- Curtailment of renewable energy;
- Marketing of renewable energy.

3.1.1 Existing Bidding Zones May Have More Internal Congestion

The European target model is based on the assumption that at least in continental Europe control areas or countries are the usual units of markets, with
internal congestion in exceptional cases only. Coupling is achieved via flow-
based capacity calculation and implicit auctions using one single optimization algorithm run by power exchanges or other entities.[33] These assumptions
were derived from historical experience. However, as network flows are
increasing, even old control areas might not withstand future flows. Bidding
zones are conglomerates of control areas. If there is more internal congestion, the market size might shrink by decoupling control areas—which at the
same time implies a reduction of liquidity. This is the very reason why most
regulators today support measures to stabilize bidding zones. As shown
above, markets are still very illiquid in many Member States so that market

---

28. A majority of Member States uses FiTs to support renewable electricity production. For a
more detailed analysis of the German situation, see Pfaffenberger/Chrischilles in this volume.
29. EWEA (2012).
30. EurObserv'ER (2012).
31. See footnotes 21 and 22.
32. See Chapter 2.
33. European Commission (2011).

fragmentation in the existing bilateral trading market with vertical integration between production and supply might have noticeable adverse competitive effects.

In addition, this is an issue of national industrial policy. If national electricity markets were broken up into bidding zones, the expansion of renewable energy in the system might create a competitive disadvantage for some higher parts of the Member States concerned. Energy-intensive industry would either ask for counterbalancing measures, for instance, reduced network tariffs, or heavily oppose any policy aimed at increasing the share of renewables.

One measure to overcome market splitting is to use redispatching. Redispatching means using capacity which is not in the money in order to reduce network flows. Today, redispatch measures are typically of minor scale (below 1% of total network cost), but with an upward trend. In the future, they are likely to significantly increase as more and more intermittent generation comes online.

In combination with other constraints such as must-run conditions or priority dispatch of renewable generation, this will make for a reduction of the size of the freely traded market, and efficiency in the system will be jeopardized.

3.1.2 Intermittency Might Cause a Term Shift

Intermittent generation implies uncertain supply. Adequate production forecasts are only possible a few hours before real time, that is, it is only then that the energy becomes marketable. This might lead to a shift from long-term trading to short-term or even close to real-time physical trading.[34] In Germany, for instance, the volume in traded forward contracts has decreased by more than 50% during the last 4 years (Figure 8.8).

In the European context, where market concentration is high, this might increase the profitability of withholding capacity.[35] One of the most important arguments in favor of the bilateral Nordic model in Europe was its higher resilience to market manipulation as most of the prices are locked in via forward markets, where the possibility of capacity withholding does not play an important role.[36] If short-term day-ahead and intraday markets or even balancing energy markets with specific technical requirements for producers become more important, this will facilitate collusion. Fewer suppliers might be able to actively participate in these markets. Higher frequency of trading transactions will further facilitate collusive behavior. However, Europe's

---

34. Please note that it is not clear which part of the evident reduction in trading volumes is due to this fact and which part has to be attributed to the current economic environment.
35. For the British pool model see for instance: Newbery (1997); Newbery (2005); Green (2006).
36. For counterarguments see for instance: Chloé Le Coq (2004).

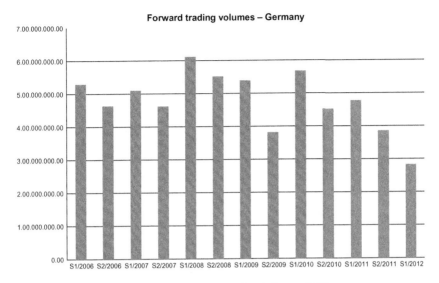

FIGURE 8.8   Traded volumes at EEX (in megawatt hour). *Source: EEX.*

current efforts to achieve cross-border integration of balancing energy markets and intraday markets will hopefully balance this. Even so, Europe certainly faces a transition phase in terms of time scales, where markets are not yet integrated and short-term physical markets are gaining in importance.

An immediate consequence of this development is that market monitoring will have a more important role to play as the competitive range further decreases.

In a market with a high share of intermittent generation, the role for traditional base load generation is unclear if not precarious. It does not easily react to short-term price signals or fluctuations in demand and supply and so contributes to a situation that even results in negative wholesale prices in individual hours.

Recently, many regulators and government policy makers are discussing and/or proposing the implementation of capacity markets at wholesale level.[37] The underlying intentions are diverse: in France, for example, it is to reduce peak load in winter; in Great Britain, the aim is to support investment in new conventional generation capacity. In this context, energy-only markets are put into question in general,[38] even more so as simulations show that the average load factor of a mid-merit and peak power station will drop further as renewables continue to grow. Some studies have concluded that

37. France and Great Britain already have developed plans of capacity mechanisms; in Germany there is a broad discussion about the necessity of installing such mechanisms ongoing.
38. For many: Joskow (2008).

there will be a lack of investment in the future in the absence of capacity payments for traditional generation.[39]

To answer this question we must draw on empirical evidence. Today, there are sufficient investment projects in the pipeline in Europe. In addition, generic capacity payments may not be the right instruments to get the necessary highly flexible generation units.

The basic question is whether the actual discussion is window dressing for another issue. Companies with fossil fuel production, which may be driven out of the market by subsidized renewable electricity, might be forced to shut down at least older parts of their production fleet. Theoretically, this is an intended consequence of subsidizing renewable producers in the first place. At the same time, the high gas prices in Europe typically place mid-merit combined cycle gas turbines (CCGT) out of the market, which adds to load factors and frustrates new investment. In such a situation, companies are quite likely to seek compensation for keeping plants in operation, for instance, via capacity payments. If we were to take a national approach here, the early experience in PJM suggests that there would be negative effects. Any general change of European market design must be careful in this respect. However, the probability of agreeing on a European system is very small, so that monitoring the effects on competition and state-aid rules will be quite important.

3.1.3 Subsidized Renewable Energy Will Be Curtailed More Often

If redispatch measures become more expensive or if network stability is in danger, intermittent generation might need to be curtailed. This happens quite frequently, although on a small-scale so far. Extensive and frequent curtailment of renewable generation, however, is certainly not in line with the objective of reaching a 20% renewables share in total energy consumption in 2020. In addition, this means that under a feed-in tariff (FiT) system producers would have to get paid even when they are not generating. One of the reasons for curtailment is that in many jurisdictions within Europe renewable energy, if technically possible, is connected to the grid.

3.1.4 Marketing Subsidized Renewable Energy

As more and more energy is produced under FiTs, where the energy is normally sold centrally by the TSO or some other body, the market power of this body will exceed any market power of a competitive company today. (With decentralized selling of renewable energy, most producers would face problems in terms of the necessary economies of scale.) The regime under which any central marketing body acts is crucial. Usually, one of the objectives is to sell the energy to the market at prices as high as possible in order

---

39. For the French discussion/solution see Chapter 3.

to reduce the necessary amount of subsidies that have to be collected from final customers in some way or another. This objective is not different from objectives of normal traders. What is different is that if the seller is a TSO, it certainly has access to a lot of information about the system as a whole. We must not underestimate the potential for insider trading[40] in this area. In terms of political economy, the cost of integrating renewable energy can either be attributed to the transition toward a low-carbon energy mix or can be hidden somewhere in the system and incorrectly be presented as a consequence of the liberalized market. The decision taken on this issue will influence the extent to which further liberalization or more state intervention will happen in Europe.[41] Prices will increase during the transition anyway.

3.2 Opening Markets to Demand Participation[42]

The third package foresees general rollout of smart meters.[43] Member States have to make sure that load profile meters are in place in 80% of households by 2020. In most Member States, there is only a limited number of households with smart meters today, one notable exception being Italy. Still, the third-package requirement is mandatory unless a Member State provides a negative cost-benefit analysis to show that smart metering would not be feasible.

The expected result of smart metering is that consumers will have better information about their consumption, and suppliers could start offering new time-dependent rates; the new metering infrastructure might even pave the way for services related to home automation. Although it is too early to predict the main actual benefits, one potentially very attractive feature is that via these meters consumers could (e.g., via aggregators) jointly participate in wholesale markets, especially in short-term markets such as intraday and balancing. This would create an additional means to balance supply and demand and thereby contribute to stabilizing the system. And yet, this is only part of the story. The subsidies needed for renewable energy will drive retail prices up. The pure energy price will play a less important role and account for a smaller share in the total bill of households or even commercial customers. As systemic costs (ancillary services, network, fees to finance renewables' support, etc.) increase, competition and the gain from being an active customer in the market will edge downward. Providing a service to the system might be a more lucrative possibility for households to decrease their energy

---

40. Insider trading in spot markets has explicitly been prohibited in Europe since the end of 2011.
41. Pollitt (2012).
42. In the book's preface, Reeves refers to consumer participation as a critical issue.
43. In 2009, the European Commission also gave a mandate to the European standardization organizations CEN/CENELEC (M/441) to develop an open architecture for utility meters.

bill than switching to another supplier in the future. Such a situation will have various ramifications particularly in the following two areas.

3.2.1 Aggregators Are Needed

Who should take on the role of aggregator for small consumers to participate in the market? In principle, network operators and suppliers come to mind. Starting in the Nordic region, the so-called supplier-centric model is gaining support. It implies that, wherever possible, services should be provided by suppliers. It seems quite obvious that the suppliers will be the aggregators of demand as they participate in the wholesale markets. However, most local retail markets are currently dominated by incumbent suppliers, which could constitute an entry barrier for new market players as it raises the minimum market size, where a typical retailer can perform its business efficiently.

3.2.2 Providing Services Might Need Some Initial Investment

In a system as described above, the distribution of opportunities among households might be uneven. Participation might need initial private investment and consumption levels suitable for management according to short-term (price) signals. This means that the spread of household bills might increase. Regulatory intervention might follow.

4 CONCLUSION

After 15 years, the European development toward integrated cross-border wholesale markets has acquired critical momentum. During the next 2 years, large parts of the European market design should become binding including rules and timelines for implementation.

However, the challenges for the next decade might put into question the market design that is developed now. The transition from a dispatchable system with a high share of fossil fuels toward a system where the latter are replaced by renewables might demand other markets and mechanisms, closer to real time and network physics. In principle, the Nordic market model, based on bilateral contracts without central dispatch, has been the blueprint for most of continental Europe. As shown above, the high variability in the system caused by intermittent generation does not sit well with this market design. Some elements of central dispatch (like redispatch) or even installing mandatory pools are therefore very likely to be on the agenda of energy regulators and politics in the future. The lack of a harmonized policy toward renewable energy could lead to more centralized market models.

In such market models, monitoring will be even more important than in the past as remuneration of producers does not depend on bilateral contracts but on bidding behavior and regulatory rules.

As prices increase, securing public support for the transition to a low-carbon system is essential. This is the second main challenge in Europe's electricity market: to guarantee an acceptable price level for final customers. One way to do so is to facilitate participation in the market for demand response or ancillary services. But whether this will be sufficient remains yet to be seen.

The core tendency has been a development toward more centralization or better cooperation at European level. Member States, which often used electricity companies as national champions in their industrial policy, have slowly transferred competences to entities that are independent from industry and politics. There is a need for coordinated planning, decision-making, and operation, aggravated by the increasing volatility in the system. As the focus of decisions shifts from regional or national considerations toward achieving an overall optimum, it remains to be seen whether European regulation will take the next step and succeed in delivering efficient market solutions in time.

REFERENCES

ACER–Agency for the Cooperation of Energy Regulators, 2011. Framework Guidelines on Capacity Allocation and Congestion Management for Electricity. Available from: <http://acernet.acer.europa.eu/portal/page/portal/ACER_HOME/Public_Docs/Acts%20of%20the%20Agency/Framework%20Guideline/Framework_Guidelines_on_Capacity_Allocation_and_Congestion_M/FG-2011-E-002%20(Final).pdf/> (accessed 09.08.12).

ACER–Agency for the Cooperation of Energy Regulators, 2012. ACER/CEER Annual Report on the Results of the Internal Electricity and Natural Gas Markets in 2011, Luxembourg: Publications Office of the European Union.

BP, 2012. BP Statistical Review of World Energy. Available from: <http://www.bp.com/sectionbodycopy.do?categoryId=7500&contentId = 7068481> (accessed 08.08.12).

CEER, 2006. Council of European Energy Regulators, ERGEG's Assessment of the Development of the European Energy Markets, Brussels. Available from: <http://www.energy-regulators.eu/portal/page/portal/EER_HOME/EER_PUBLICATIONS/NATIONAL_REPORTS/NR_2006/E06 MOR-02-03_AssessmentReport_Draft-V9.doc/> (accessed 08.08.12).

Dobbeni, D., 2007. Regulatory and technical challenges for the European electricity market. Eur. Rev. Energy Markets 1 (2), 77–107 (Claeys & Casteels, Leuven, Belgium).

Eurelectric, 2011. Power Statistics & Trends 2011, Brussels. Available from: <http://www.eurelectric.org/powerstats2011/> (accessed 15.10.12).

EurObserv'ER. 2012. Photovoltaic Barometer, No 7. Available from: <http://www.eurobserv-er.org/pdf/baro208.asp> (accessed 08.08.12).

European Commission, 2011. Public Consultation on the Governance Framework of the European Day-Ahead Market Coupling, D(2011) 1176339.

EWEA–European Wind Energy Association, 2012. Wind in Power: 2011 European Statistics. Available from: <http://www.ewea.org/fileadmin/ewea_documents/documents/publications/statistics/Stats_2011.pdf/> (accessed 08.08.12).

Glachant, J.-M., Belmanns, R., Meeus, L., 2006. Implementing the European energy market in 2005–2009. Eur. Rev. Energy Markets 1 (3), 51–85 (Claeys & Casteels, Leuven, Belgium).

Green, R., 2006. Market power mitigation in the UK power market. Util. Policy 14, 76–89.

Joskow, P.L., 2008. Capacity payments in imperfect electricity markets: need and design. Util. Policy 16, 159–170.

LaBelle, M., Kaderjak, P., 2008. Impact of the 2004 Enlargement on the EU Energy Sector. Corvinus University of Budapest, Budapest, Hungary.

Chloé Le Coq, 2004. Long-Term Supply Contracts and Collusion in the Electricity Market. SSE/EFI Working Paper Series in Economics and Finance No. 552. Available from: <http://swopec.hhs.se/hastef/abs/hastef0552.htm> (accessed 09.08.12).

Madlener, R., Jochem, E., 2001. Impacts of Market Liberalisation on the Electricity Supply Sector: A Comparison on the Experience of Austria and Germany, vol. 74. Forschung im Verbund, Vienna, Austria.

Midttun, A., Omland, T., 2004. Configuration and performance of large European companies: a statistical analysis. In: Finon, D., Midttun, A. (Eds.), Reshaping European Gas and Electricity Industries. Elsevier, Oxford, UK.

K. Neuhoff, R. Boyd, T. Grau, J. Barquin, F. Echavarren, J. Bialek, et al., 2011. Renewable Electric Energy Integration, Quantifying the Value of Design of Markets for International Transmission Capacity, DIW Discussion Papers 1166, Berlin. Available from: <http://www.diw.de/documents/publikationen/73/diw_01.c.387959.de/dp1166.pdf/> (accessed 09.08.12).

Newbery D., 1997. Pool Reform and Competition in Electricity. Available from: <http://www.econ.cam.ac.uk/emeritus/newbery/files/LBS.PDF> (accessed 09.08.12).

Newbery, D., 2009. Refining market design. In: Glachant, J.-M., Lévêque, F. (Eds.), Electricity Reform in Europe. Edward Elgar, Cheltenham, UK.

Newbery, D., 2005. Electricity liberalization in Britain: the quest for a satisfactory wholesale market design. Energy J.26, 43–70 (special issue on European Electricity Liberalisation).

Pollitt M.G., 2012. The role of policy in energy transitions: lessons from the energy liberalisation era. EPRG Working Paper 1208. Available from: <http://www.eprg.group.cam.ac.uk/wp-content/uploads/2012/03/EPRG1208_Complete.pdf/> (accessed 10.08.12).

Vasconcelos, J., 2001. Cooperation between energy regulators in the European Union. In: Henry, C., Matheu, M., Jeunemaître, A. (Eds.), Regulation in Network Industries, The European Experience. Oxford University Press, Oxford, UK.

Part II

The Evolution of Electricity Markets in The Americas

The Evolution of the PJM Capacity Market: Does It Address the Revenue Sufficiency Problem?

Joseph E. Bowring
Monitoring Analytics, LLC[1]

1 INTRODUCTION

There is continuing and evolving interest in whether wholesale power markets should include a capacity market in addition to an energy market. Several chapters in this book address issues related to capacity markets and other chapters address issues related to markets without capacity markets.[2] The discussion sometimes misses the essential point about the role of capacity markets in wholesale market design. Capacity markets were added to solve a defined market design problem that can also be solved in other ways. Capacity markets are not the only solution to the problem. The history of PJM (PJM Interconnection, LLC) demonstrates that some capacity market designs are not a solution to the problem. The goal of the discussion should be to weigh the attributes of capacity markets and the alternatives to capacity markets and make a rational policy decision that is consistent with competitive markets.

When wholesale power markets must meet defined reliability standards, a market design issue is created. How can the market provide guaranteed excess supply to ensure reliability while also providing sufficient revenue to support the ongoing operation of generating units and the incentives to invest in new and existing generation? Solutions to this revenue sufficiency

---

1. Monitoring Analytics, LLC is the Independent Market Monitor for the PJM wholesale power markets. Monitoring Analytics also provides consulting services related to market monitoring to those responsible for the design, oversight, and monitoring of other markets. Available at: <http://www.monitoringanalytics.com/home/index.shtml>.
2. Of particular relevance are chapters on United Kingdom, Germany, France, Australia, and Electric Reliability Council of Texas (ERCOT) in this volume, but capacity markets and the missing money problem are also mentioned in a number of other chapters.

problem include relying on cost of service regulation, relying on bilateral contracts with regulatory approval, permitting the exercise of market power, administratively setting high prices at times of relative scarcity, and creating capacity markets. All of these approaches have been implemented and all result in higher revenues, but each has different attributes and market consequences. All of these approaches are administrative, but some incorporate market mechanisms more successfully than others. Capacity markets, especially when tightly integrated with energy markets, are a demonstrated, if still imperfectly implemented, market-based solution to the revenue sufficiency problem whereas scarcity pricing and reliance on market power have not yet been fully tested. But, one of the important lessons from the history of capacity markets in PJM is that the design of the capacity market matters and that it is possible to have a capacity market that does not resolve the revenue sufficiency problem.[3]

This chapter discusses how these issues have been addressed in the PJM markets. Section 2 covers the basic wholesale power market design issues and the market dynamics that led to the development of capacity markets. Section 3 is an overview of the PJM markets. Section 4 is a brief history of PJM capacity markets including the reasons for the original development of capacity markets and subsequent developments in PJM. Section 5 presents the actual market outcomes for PJM and implications for the market design. Section 6 explains some of the key issues with the current PJM capacity market design and suggests solutions. Section 7 is the conclusion.

2 MARKET ESSENTIALS

Despite substantial restructuring, wholesale power markets in the United States, and throughout the world, remain subject to extensive regulation, focused both on reliability and economics. Reliability regulations require that wholesale power markets result in the reliable provision of power, without either local or widespread blackouts, whereas economic regulation addresses revenue adequacy, incentives, and competitiveness issues.[4]

3. These papers provide theoretical approaches to market design issues and capacity markets: Cramton, P., Stoft, S. The Convergence of Market Designs for Adequate Generating Capacity with Special Attention to the CAISO's Resource Adequacy Problem. Available at: http://stoft.com/ (April 2006); Cramton, P., Stoft, S., 2005. A capacity market that makes sense. Electr. J. 18, 43–54; Joskow, P., Tirole, J., 2007. Reliability and competitive electricity markets. Rand J. Econ. 38 (1), 60–84; Hogan, W.W., 2005. On an 'Energy-Only' Electricity Market Design for Resource Adequacy; Joskow, P., 2008. Capacity payments in imperfect electricity markets: need and design. Util. Policy 16 (3), 159–170.
4. This chapter draws on related work of the author and on the work of Monitoring Analytics, LLC, the Independent Market Monitoring Unit for PJM, all cited herein. The work of Monitoring Analytics is based on contributions by all members of the MMU (PJM Independent Market Monitoring Unit). In particular, this chapter draws on analysis by Alexandra Salaneck on capacity markets, Keri Dorko on net revenues, and William Dugan on demand side resources.

As a result of these reliability requirements, PJM operates to a reliability standard of one loss of load event in 10 years.[5] To meet this target, PJM maintains a reserve margin, the level of which results from planning studies incorporating expected forced outage rates and expected load growth. An essential goal of the PJM market design is to maintain this reserve margin.[6] A reserve margin is equivalent to maintaining excess capacity to provide energy. Maintaining excess capacity means that the actual level of capacity generally exceeds the level that would result from the functioning of the energy market alone. Maintaining excess capacity means that the prices in the energy market are lower and less volatile than they would be otherwise and that the frequency and duration of high prices is less than it would be otherwise. Lower and less volatile energy prices mean lower revenues for all generating units. Lower revenues mean lower profits and lower profits mean reduced incentives to reinvest in existing plants or to invest in new plants. This is the revenue sufficiency problem which affects all wholesale power markets that are required to maintain a defined level of reliability. The revenue sufficiency problem is sometimes referred to as the missing money problem.[7]

The solutions to the revenue sufficiency problem are not limited to capacity markets and energy markets. The three basic approaches are the regulatory approach, the energy market approach, and the capacity market approach, but the energy market approach and the capacity market approach are not mutually exclusive. Solutions to the revenue sufficiency problem include relying on cost of service regulation to cover fixed costs, relying on bilateral contracts with regulatory approval to cover fixed costs, permitting the exercise of market power in the energy market, administratively setting high energy prices at times of relative scarcity, and creating capacity markets. All of these approaches have been implemented and all result in higher revenues and higher profits, but each has different attributes and market consequences.

The PJM market design relies on tightly integrated energy and capacity markets to solve the revenue sufficiency problem. The PJM market design, like all the organized wholesale power markets in the United States, includes an energy market with locational marginal prices (LMPs). The energy market accounted for 72.5% of the total cost of wholesale power in PJM in 2011. The PJM capacity market accounted for 18.2% of the total cost of wholesale

5. The one in ten standard is the result of North American Electric Reliability Council (NERC) decisions and is the generally accepted standard in US wholesale power markets <www.nerc.com/files/BAL-502-rfc-02.pdf>.

6. The Installed reserve margin (IRM) for 2015/2016 is 15.4%. The 2015/2016 planning parameters can be found at <http://www.pjm.com/markets-and-operations/rpm/ ~ /media/markets-ops/rpm/rpm-auction-info/2015-2016-planning-period-parameters.ashx > (accessed 26.10.12).

7. Other chapters refer to this problem, notably Chapters 3 and 5 by Percebois and Haas et al, respectively.

power in PJM in 2011. Transmission service charges were 7.1% of the total cost of wholesale power.[8]

Of the six organized wholesale power markets in the United States, PJM, New York Independent System Operator (NYISO), and ISO New England (ISO-NE) have capacity markets, California Independent System Operator (CAISO) requires utilities to purchase capacity through bilateral contracts subject to regulatory review, Midwest Independent System Operator (MISO) relies on state-level cost of service regulation to ensure adequate revenues and adequate capacity while including a voluntary capacity market, and ERCOT relies on a form of scarcity pricing.[9]

Relying on cost of service regulation, whether in the traditional rate base, rate of return sense, or via bilateral contracts, is a simple, direct way to ensure revenue adequacy for generating units. By definition, this approach is not a market approach and it does not create market incentives for entry from new participants, for investment in existing units, or for retirement of uneconomic units. The traditional state regulatory approach provides revenues only to the vertically integrated utilities that own generating plants and is inconsistent with the market restructuring that has occurred in large parts of the United States. The bilateral contract approach permits new entry, but competition is limited to an administrative process which results in a guaranteed flow of revenues to selected units for a defined period of time, typically less than the asset life. Both approaches rely on regulatory decisions to determine the location and technology of new generation and the revenues to new capacity. There is no market clearing price and capacity resources are paid different prices. There is no opportunity for merchant generation to compete directly based on price. There is no capacity market signal for demand side resources. There is no integration with the energy market to permit adaptation to higher levels of renewable resources.

Permitting the exercise of market power can result in higher net revenues for generating units if the owners of generation have the structural ability to exercise market power on the required scale, at high enough prices, for the required duration, and the incentive to exercise such market power.[10] Reliance on market power does not permit regulatory authorities to specify and realize the desired level of reliability. Such an approach may come with a price cap, which limits prices and profitability during periods when market power would otherwise permit higher prices. The exercise of local market

8. Monitoring Analytics, LLC. Section 1: Introduction. 2011 State of the Market Report for PJM, vol. 2. <www.monitoringanalytics.com/reports/PJM_State_of_the_Market_Report/2011.shtml> (accessed 15.03.12).
9. ERCOT market issues are discussed in Chapter 10 by Adib et al.
10. When individual unit owners set high prices during actual system scarcity conditions, it is not necessarily the exercise of market power, as the marginal cost for units at the end of their cost curves is not well defined.

power may not result in the intended outcomes including aggregate market level incentives or even effective local incentives. Reliance on market power to meet reliability goals results in a relatively high level of uncertainty about expected revenue levels and about investment in capacity.

Administratively setting high prices at times of relative scarcity can result in higher net revenues for generating units if the definition of scarcity, the administratively determined prices and actual supply and demand conditions interact in the intended manner. Reliance on administratively determined scarcity prices can result in uncertainty because triggering scarcity is a function of both the definition of scarcity and supply and demand conditions, the duration and frequency of which are not predictable.

Creating capacity markets can result in higher net revenues for generating units if the definition of capacity, the definition of demand for capacity, market power mitigation rules, the integration with the energy market design, and other market rules interact in the intended manner. Capacity markets permit the definition of reliability targets which can be achieved directly through the market design, although achieving reliability targets depends on market response to the resultant price signals. Reliance on capacity markets also includes a level of uncertainty related to the details and implementation of the actual market design.

Ongoing developments in the ERCOT market illustrate many of these issues. ERCOT recently implemented locational marginal pricing in its energy market, with an energy offer cap and no capacity market. ERCOT has experienced declining reserve margins and is currently discussing options including raising offer caps and introducing a capacity market.[11]

Although implementing an energy-only market with no administrative interventions is unlikely to succeed at providing wholesale power market outcomes with the desired level of reliability at the lowest possible cost, the performance incentives associated with such a market are a strength of the energy-only approach and are a key benchmark for assessing performance incentives in any wholesale power market design. These performance incentives are simple. A resource is paid the market price, when it produces energy, for the amount of the energy produced and it is not paid when it does not produce energy. The more complex market designs have in many cases attenuated these fundamental incentives. It is essential that the market designs replicate these performance incentives.

3 OVERVIEW OF PJM

PJM Interconnection, LLC operates a centrally dispatched, competitive wholesale electric power market that, as of September 2012, had installed

---

11. The Brattle Group. ERCOT Investment Incentives and Resource Adequacy. <http://www.brattle.com/_documents/UploadLibrary/Upload1047.pdf> (accessed 01.06.12).

generating capacity of 182,873.5 MW and about 800 market buyers, sellers, and traders of electricity in a region including more than 60 million people in all or parts of Delaware, Illinois, Indiana, Kentucky, Maryland, Michigan, New Jersey, North Carolina, Ohio, Pennsylvania, Tennessee, Virginia, West Virginia, and the District of Columbia (Figure 9.1).[12] In 2011, PJM had total billings of $35.9 billion.[13]

PJM operates a day-ahead energy market, a real-time energy market, a capacity market, a regulation market, synchronized reserve markets, a day-ahead scheduling reserve market, and financial transmission rights markets.[14]

Of PJM's total installed capacity, 109,904.1 MW are coal plus nuclear, 60.1% of the total, 826.6 MW are wind plus solar, 0.5% of the total, and an additional 4.3% are hydro (Table 9.1).

Total generation output in PJM for January through September, 2012 was primarily from coal, nuclear, and gas, 95.3%, with 3.3% from wind, solar and hydro (Table 9.2).

The peak load in PJM was 154,344 MW in 2012 (Figure 9.2). PJM peak load has increased as a result of additions to the PJM footprint as well as internal load growth.[15]

4 HISTORY OF CAPACITY MARKETS IN PJM

A capacity market was not included when the PJM markets were created.[16] Nonetheless, PJM's members relied upon capacity obligations to ensure reliability and to allocate the costs of that reliability when PJM was structured as a power pool. That contractual system ultimately resulted in a capacity market as PJM evolved from a power pool into a market. In the power pool, PJM member utilities determined their capacity obligations based on forecast load plus an agreed upon reserve margin that was designed to achieve the desired reliability objective. PJM member utilities were required to build or buy capacity to meet their defined obligation or pay a capacity deficiency charge if they did not. The capacity deficiency charge was calculated in the same way that the net cost of new entry is calculated in the current PJM

12. See Company Overview. PJM.com. PJM Interconnection LLC. <http://pjm.com/about-pjm/who-we-are/company-overview.aspx> (accessed 12.07.12).
13. Monitoring Analytics, LLC. "PJM Geography" for maps showing the PJM footprint and its evolution prior to 2011. 2011 State of the Market Report for PJM, vol. 2, Appendix A. <www.monitoringanalytics.com/reports/PJM_State_of_the_Market/2011.shmtl> (accessed 15.03.12).
14. Monitoring Analytics, LLC. 2011 State of the Market Report for PJM. <www.monitoringanalytics.com/reports/PJM_State_of_the_Market/2011.shtml> (accessed 15.03.12).
15. For additional information on the "PJM Integration Period," see the "2011 State of the Market Report for PJM," vol. 2, Appendix A, "PJM Geography." <www.monitoringanalytics.com/reports/PJM_State_of_the_Market/2011.shtml> (accessed 15.03.12).
16. PJM Supporting Companies. Request for authorization to engage in market-based transactions. Docket No. ER97-3729-000 (July 14, 1997).

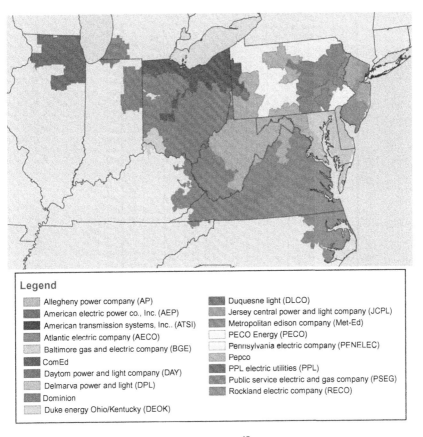

Legend

Allegheny power company (AP)	Duquesne light (DLCO)
American electric power co., Inc. (AEP)	Jersey central power and light company (JCPL)
American transmission systems, Inc.. (ATSI)	Metropolitan edison company (Met-Ed)
Atlantic electric company (AECO)	PECO Energy (PECO)
Baltimore gas and electric company (BGE)	Pennsylvania electric company (PENELEC)
ComEd	Pepco
Daytom power and light company (DAY)	PPL electric utilities (PPL)
Delmarva power and light (DPL)	Public service electric and gas company (PSEG)
Dominion	Rockland electric company (RECO)
Duke energy Ohio/Kentucky (DEOK)	

FIGURE 9.1   PJM's footprint and its 19 control zones.[17]

capacity market, equal to the annual levelized cost of a combustion turbine (CT) peaking unit less net energy market revenues. The associated bilateral market in which PJM members exchanged capacity obligations was the first semblance of a capacity market.

In the power pool, PJM capacity obligations combined with state cost of service regulation to maintain a reliable pool with capacity and energy adequate to serve load and revenues adequate to cover the costs of existing and new generation. The rudimentary capacity market that existed in the PJM pool prior to the creation of the PJM market was similar to the current reliance in some wholesale power markets on state cost of service regulation with a bilateral market for trading capacity.

17. Monitoring Analytics, LLC. 2012 Quarterly State of the Market Report for PJM: January through September. <http://www.monitoringanalytics.com/reports/PJM_State_of_the_Market/2012.shtml> (accessed 15.11.12).

TABLE 9.1 PJM Installed Capacity (By Fuel Source): January 1, May 31, June 1, and September 30, 2012[18]

	1-Jan-12		31-May-12		1-Jun-12		30-Sep-12	
	MW	Percent	MW	Percent	MW	Percent	MW	Percent
Coal	75,190.4	42.0	79,311.0	42.8	79,664.6	42.9	76,739.2	42.0
Gas	50,529.3	28.3	51,940.1	28.0	52,709.1	28.4	51,995.2	28.4
Hydroelectric	8047.0	4.5	8047.0	4.3	7879.8	4.2	7879.8	4.3
Nuclear	32,492.6	18.2	33,085.0	17.9	33,149.5	17.8	33,164.9	18.1
Oil	11,217.3	6.3	11,494.7	6.2	10,767.2	5.8	11,531.7	6.3
Solar	15.3	0.0	16.3	0.0	47.0	0.0	47.0	0.0
Solid waste	705.1	0.4%	689.1	0.4	736.1	0.4	736.1	0.4
Wind	657.1	0.4	660.1	0.4	779.6	0.4	779.6	0.4
Total	178,854.1	100.0	185,243.3	100.0	185,732.9	100.0	182,873.5	100.0

18. Monitoring Analytics, LLC. Section 6: Net revenue. 2011 State of the Market Report for PJM, vol. 2. <www.monitoringanalytics.com/reports/ PJM_State_of_the_Market/2011/shtml> (accessed 15.03.12).

TABLE 9.2 PJM Generation (By Fuel Source (GWh)): January Through September 2011 and 2012[19]

		Jan–Sep 2011 GWh	Percent	Jan–Sep 2012 GWh	Percent	Change in Output
Coal		279,501.2	48.0	252,201.1	41.7	(9.8%)
	Standard coal	270,273.8	46.4	244,255.4	40.3	(9.3%)
	Waste coal	9227.4	1.6	7945.7	1.3	(0.5%)
Nuclear		195,196.7	33.5	205,503.9	33.9	5.3%
Gas		82,130.5	14.1	119,274.3	19.7	45.2%
	Natural gas	80,774.5	13.9	117,560.4	19.4	45.5%
	Landfill gas	1355.9	0.2	1713.5	0.3	26.4%
	Biomass gas	0.1	0.0	0.4	0.0	175.0%
Wind		7924.5	1.4	9936.8	1.6	25.4%
Hydroelectric		11,379.8	2.0	9768.1	1.6	(14.2%)
Waste		4254.8	0.7	4199.9	0.7	(1.3%)
	Solid waste	3318.0	0.6	3462.3	0.6	4.3%
	Miscellaneous	936.8	0.2	737.6	0.1	(21.3%)
Oil		2207.7	0.4	4401.7	0.7	99.4%
	Heavy oil	1844.8	0.3	4122.7	0.7	123.5%
	Light oil	334.3	0.1	265.8	0.0	(20.5%)
	Diesel	15.9	0.0	8.2	0.0	(48.2%)
	Kerosene	12.7	0.0	4.9	0.0	(61.8%)
	Jet oil	0.1	0.0	0.0	0.0	(29.1%)
Solar		37.9	0.0	195.8	0.0	416.9%
Battery		0.2	0.0	0.4	0.0	131.1%
Total		582,633.3	100.0	605,482.0	100.0	3.9%

19. Monitoring Analytics, LLC. Section 6: Net revenue. 2011 State of the Market Report for PJM, vol. 2. <www.monitoringanalytics.com/reports/PJM_State_of_the_Market/2012.shtml> (accessed 15.03.12).

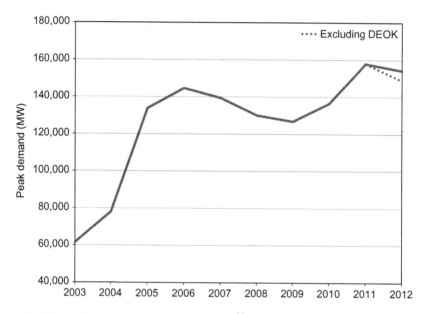

FIGURE 9.2 PJM footprint peak loads: 2003–2012.[20]

The daily capacity credit market (CCM), introduced in 1999, was the first formal capacity market in PJM. The current reliability pricing model (RPM) capacity market design was implemented in 2007, after 8 years of the daily CCM design, to resolve the revenue sufficiency problem in a manner consistent with other aspects of the PJM market design.

One of the important lessons from the history of capacity markets in PJM is that the design of the capacity market matters and that it is possible to have a capacity market that does not resolve the revenue sufficiency problem.

Table 9.3 shows PJM market milestones including the April 1, 1999, creation of the PJM market with LMPs and market-based generation offers, the 1999 introduction of the first PJM capacity market, and the 2007 introduction of the RPM capacity market design.

4.1 Implementation of a Capacity Market

A formal capacity market was implemented by PJM in response to the early stages of retail restructuring rather than in response to developments in the wholesale power market.[21] The development of the PJM capacity market did

---

20. The Duke Energy Ohio/Kentucky (DEOK) Transmission Zone joined PJM on January 1, 2012.

21. See Bowring, J., 2008. The evolution of PJM's capacity market. In: Sioshansi, F.P. (Ed.), Competitive Electricity Markets. Elsevier, for more complete history of the development of PJM capacity markets.

TABLE 9.3 PJM Market Milestones[22]

Year	Month	Event
1996	April	FERC Order 888, "Promoting Wholesale Competition Through Open Access Nondiscriminatory Transmission Services by Public Utilities; Recovery of Stranded Costs by Public Utilities and Transmitting Utilities"
1997	April	Energy Market with cost-based offers and market-clearing prices
	November	FERC approval of ISO status for PJM
1998	April	Cost-based Energy LMP Market
1999	January	Daily Capacity Market
	March	FERC approval of market-based rates for PJM
	March	Monthly and Multimonthly Capacity Market
	March	FERC approval of Market Monitoring Plan
	April	Offer-based Energy LMP Market
	April	FTR Market
2000	June	Regulation Market
	June	Day-Ahead Energy Market
	July	Customer Load-Reduction Pilot Program
2001	June	PJM Emergency and Economic Load-Response Programs
2002	April	Integration of AP Control Zone into PJM Western Region
	June	PJM Emergency and Economic Load-Response Programs
	December	Spinning Reserve Market
	December	FERC approval of RTO status for PJM
2003	May	Annual FTR Auction
2004	May	Integration of ComEd Control Area into PJM
	October	Integration of AEP Control Zone into PJM Western Region
	October	Integration of DAY Control Zone into PJM Western Region
2005	January	Integration of DLCO Control Zone into PJM
	May	Integration of Dominion Control Zone into PJM

(*Continued*)

---

22. Monitoring Analytics, LLC. Section 1: Introduction. 2011 State of the Market Report for PJM, vol. 2. <www.monitoringanalytics.com/reports/PJM_State_of_the_Market/2012.shtml> (accessed 15.03.12).

TABLE 9.3 (Continued)

Year	Month	Event
2006	May	Balance of Planning Period FTR Auction
2007	April	First RPM Auction
	June	Marginal loss component in LMPs
2008	June	Day-Ahead Scheduling Reserve (DASR) Market
	August	Independent, External MMU created as Monitoring Analytics, LLC
	October	Long-Term FTR Auction
	December	Modified Operating Reserve Accounting Rules
	December	Three Pivotal Supplier Test in Regulation Market
2011	June	Integration of ATSI Control Zone into PJM

not create the demand for capacity. The demand for capacity was created by the PJM power pool rules that predated the creation of the market and that were established to ensure reliability. The first PJM capacity market was intended to create a competitive mechanism to facilitate retail competition by permitting owners of capacity to trade capacity credits so that the rights to existing capacity could follow load.[23] The design of the first PJM capacity market did not contemplate or incorporate incentives to build new generation.

In the PJM footprint, retail restructuring created the opportunity for new entrants to compete with incumbent utilities to serve retail loads. The new entrants did not own capacity, but faced the same requirement as the incumbent, vertically integrated utilities to own capacity equal to their load plus a reserve margin. The relatively small new entrants were not in a position to build new capacity and therefore needed a way to acquire capacity to meet reliability obligations that was competitive and flexible enough to permit new entrants to buy and sell capacity in small increments as loads switched suppliers. As retail competition extended more widely over the PJM footprint, existing utilities also needed a way to sell capacity in excess of current requirements if load was lost to new competitors. Reliance on voluntary bilateral sales and purchases facilitated the exercise of market power by

---

23. This first PJM capacity market was a capacity credit market that permitted generation owners to retain full ownership and control of generating assets while selling the rights to capacity in desired increments, capacity credits, for purposes of meeting the reliability requirements of load serving entities.

incumbents and served as a barrier to entry, when the new entrant had to purchase capacity from the utilities with which it was competing for retail load.

As a result of regulatory actions by the Pennsylvania Public Utility Commission, PJM implemented a mandatory capacity market as of January 1, 1999, which was structured as a daily market to permit load to switch suppliers.[24] The full PJM energy market design became effective on April 1, 1999.[25]

4.2 The Design of the PJM RPM Capacity Market

The net revenue performance of PJM markets over the period from 1999 through 2005 led to stakeholder discussions which culminated in a redesign of the capacity market. In PJM, net revenue from the energy, capacity, and ancillary services markets was, in general, well below the replacement cost of capacity for all unit types over that period. In addition, the absence of a locational capacity market in the presence of substantial variations in local capacity market conditions meant that local market prices did not reflect local market conditions.[26]

The historical level of net revenues in PJM markets was not the result of the $1000 per MWh offer cap, of local market power mitigation, or of a basic incompatibility between wholesale electricity markets and competition. Competitive markets can, and do, signal scarcity and surplus conditions through market clearing prices. In PJM as in other wholesale electric power markets, the application of reliability standards meant that scarcity conditions in the energy market occurred with reduced frequency. In PJM, a combination of the energy market and the daily capacity market did not directly compensate the resources needed to provide for reliability.

The experience with a daily capacity market resulted in reconsidering the role of the capacity market in PJM. It became clear that a capacity market is a formal mechanism, with both administrative and market-based components, used to allocate the costs of maintaining the level of capacity required to maintain the reliability target. A capacity market is an explicit mechanism for valuing capacity and is preferable to nonmarket and nontransparent mechanisms for that reason.

---

24. Pennsylvania Public Utility Commission. Interim Order. Docket No. I-00980078 et al. (September 17, 1998).
25. PJM Interconnection, LLC. Filing of amendments to Amended & Restated Operating Agreement. Docket ER99-196-000 (October 14, 1998); PJM Interconnection, LLC. Amendment to Filing Regarding Capacity Credit Markets. Docket ER99-196-000 (November 19, 1998).
26. Monitoring Analytics, LLC. Section 4: Capacity Market. 2011 State of the Market Report for PJM, vol. 2. <http://www.monitoringanalytics.com/reports/PJM_State_of_the_Market/2011. shtml> (accessed 15.03.12).

PJM's RPM capacity market, implemented in 2007, is an explicit effort to incorporate these concepts. The RPM capacity market design is intended to send signals to the market based on the locational and forward-looking need for generation resources to maintain system reliability consistent with long-run competitive equilibrium in the energy market. The RPM capacity market is explicitly designed to provide the revenue adequacy required for system reliability.

RPM is a 3-year forward-looking, annual, locational market, with a must offer requirement for capacity resources and a must buy requirement for load, with specific obligations on capacity resources, with performance incentives, that includes clear market power mitigation rules and that permits the direct participation of demand side resources. Capacity resources in PJM are required to make offers in the day-ahead energy market every day, to invest in adequate transmission interconnection capability to be deliverable to load, and to have their energy subject to mandatory recall if needed in a PJM-defined emergency.[27]

In the PJM market design, capacity markets and energy markets are tightly integrated. Generating units do not have capacity components and energy components. A unit is viable based on its total revenues without regard to whether they come from the capacity market or the energy market. The capacity market is designed to ensure that capacity resources cover their fixed and variable costs from a combination of energy and ancillary market net revenues and capacity market revenues.

The RPM capacity market is an annual construct. Generation owners must sell capacity for a year and customers must buy capacity for a year. Net revenues and costs are evaluated for a year. The obligations of capacity resources are for a year. The relevant year in the capacity market is termed the delivery year, which is 3 years forward from the date of the capacity market base residual auctions (BRAs).

The PJM capacity market is also a locational construct. Locations in the capacity market are termed locational delivery areas (LDAs). LDAs are defined by the ability of the transmission system to permit the import of capacity into transmission zones and the supply and demand conditions in each LDA. While a relatively small number of LDAs have had separate prices, the price separation shows the significant difference in supply and demand conditions across the PJM footprint. Figure 9.3 shows the primary PJM LDAs that have cleared with separate prices.[28] The locational aspect of

---

27. Monitoring Analytics, LLC. 2011 State of the Market Report for PJM. Technical Reference for PJM Markets.

28. Rest of Regional Transmission Organization (RTO) is the RTO, excluding the nested LDAs that clear separately. The Mid-Atlantic Area Council (MAAC) area includes Western Mid-Atlantic Area Council (WMAAC), Eastern Mid-Atlantic Area Council (EMAAC), and Southwestern Mid-Atlantic Area Council (SWMAAC). WMAAC is analogous, for MAAC, to Rest of RTO, for RTO.

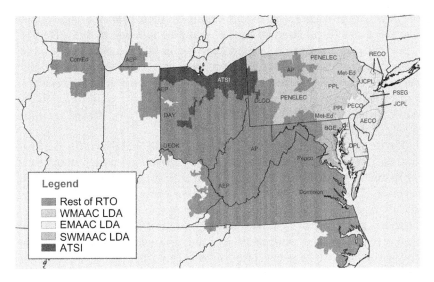

FIGURE 9.3   PJM locational deliverability areas.[29]

the RPM design is critical because it allows price signals consistent with actual locational capacity needs rather than providing a single and misleading clearing price across the entire footprint that would be too low to induce entry where needed and too high where entry is not needed.

Some market participants have asserted that new entry will occur only if new entrants receive long-term contracts for the sale of capacity at a guaranteed price. Rather than depending on long-term contracts, the PJM capacity market model depends on participant confidence that the market will remain in place for multiple years and that the market will reflect economic fundamentals as it clears each year. A number of important issues still need to be resolved to permit economic fundamentals to be reflected in capacity market outcomes.

The capacity market contributes to the revenue sufficiency of all unit types. Nuclear power plants, coal plants (CPs), combined cycle (CC) plants, and CT peaking plants all rely on the capacity market for revenue adequacy. Energy prices affect all unit types and create a revenue issue for all unit types.[30]

In return for revenue adequacy, generating units that are capacity resources are required, among other things, to provide a call on their energy at any time during the year, at the market price. Capacity resources are not

29. Monitoring Analytics, LLC. State of the Market Report for PJM: January through September. <http://www.monitoringanalytics.com/reports/PJM_State_of_the_Market/2012.shtml> (accessed 15.11.12).
30. Chapter 3 addresses this issue for the French market.

committed to providing energy only during system peaks or only during emergency conditions. Scarcity pricing would be a more cost-effective way of providing an incentive to provide energy for those limited hours. Base load plants are capacity resources but they are needed year round. The energy market would not function if capacity resources were only needed for short periods of peak demand. Customers would not receive the service they paid for when they paid for capacity if capacity was only needed for short periods of peak demand.

The demand for capacity includes expected peak load plus a reserve margin. The RPM demand curve, called the variable resource requirement (VRR) curve, is downward sloping rather than vertical and may result in the purchase of capacity in excess of the reserve margin. The total supply of capacity is generally only slightly larger than demand. Capacity in excess of demand is not sold and, if it does not earn or does not expect to earn adequate revenues in other markets or does not have value as a hedge, may be expected to retire. The actual level of purchased capacity under RPM has generally exceeded expected peak load plus the target reserve margin, resulting in reserve margins that exceed the target.[31]

The demand for capacity is almost entirely inelastic because the market rules require loads to purchase the system capacity requirement. The level of elasticity incorporated in the VRR curve is not adequate to modify this conclusion. The result is that any supplier that owns more capacity than the typically small difference between total supply and the defined demand is individually pivotal and therefore has structural market power.

In the capacity market, as in other markets, market power is the ability of a market participant to increase the market price above the competitive level or to decrease the market price below the competitive level. To evaluate whether actual prices reflect the exercise of market power, market offers must be evaluated for consistency with competitive offers. The market rules define the incremental cost of capacity for existing and new capacity resources, which define competitive offers.

Despite the fact that the market design for capacity leads to structural market power in the capacity market, competitive outcomes can be assured by appropriate market power mitigation rules. The RPM capacity market design includes detailed market power mitigation rules. Attenuation of the market power rules would mean that market participants would not be able to rely on the competitiveness of the market outcomes. However, the market power rules are not perfect and, as a result, competitive outcomes require continued improvement of the rules and ongoing monitoring of market participant behavior and market performance.

---

31. For more details, see Monitoring Analytics, LLC. Analysis of the 2014/2015 RPM Base Residual Auction. <http://www.monitoringanalytics.com/reports/Reports/2012/Analysis_of_2014_2015_RPM_Base_Residual_Auction_20120409.pdf> (accessed 09.04.12).

5 MARKET OUTCOMES IN PJM

The RPM design has been effective, using reliability, flexibility, and competitiveness as metrics. Capacity market prices increased and total market net revenues rose to levels consistent with new entry for CC units in eastern locations in PJM, where there was a need for capacity. The capacity market design has been flexible and incorporated the impacts of significant new environmental regulations, substantial changes in the relative costs of natural gas and coal, large additions of demand side resources and additions of renewable resources.

Market results have been competitive in the energy market and the capacity market from the beginning of PJM markets. While there have been issues associated with specific participant behaviors, this has not affected the overall conclusion that PJM markets have produced competitive outcomes.[32]

But the RPM design has not been as effective, using the level of total cost to customers as a metric. The RPM design continues to have issues that need to be resolved to ensure that reliability is provided at the lowest possible cost. Section 6 addresses current issues with the RPM design.

Section 5.1 reviews the history of capacity prices in PJM, Section 5.2 evaluates net revenue results, and Section 5.3 explains how the capacity market design has been flexible enough to respond to a series of exogenous changes.

5.1 Capacity Market Prices

Capacity market prices are a good indicator of the changes that resulted from the implementation of the RPM capacity market design in 2007. The capacity market prices illustrate the history of PJM capacity markets and the problems with the first, CCM, design. Figure 9.4 shows the history of PJM capacity prices since 1999. The data points (marked by triangles) for the RPM market are the capacity prices by location. Capacity market prices between 1999 and 2007 resulted from the CCM design and did not offset the net revenue shortfall from the energy and ancillary services markets. The capacity prices under the CCM design generally reflected the daily incremental cost of capacity, which was close to zero, in addition to some notable exercises of market power.[33,34] The relatively low prices demonstrate that the revenue sufficiency issue was not resolved by the CCM design. Capacity

---

32. Monitoring Analytics, LLC. 2011 State of the Market Report for PJM. <www.monitoringanalytics.com/reports/PJM_State_of_the_Market/2011.shtml> (accessed 15.03.12).

33. The market power mitigation rules under the CCM construct were limited and ineffective with the result that prices were at times affected by the exercise of market power.

34. Monitoring Analytics, LLC. Report to the Pennsylvania Public Utility Commission Capacity Market Questions. <www.monitoringanalytics.com/reports/Reports/2001.shtml> (accessed November 2001).

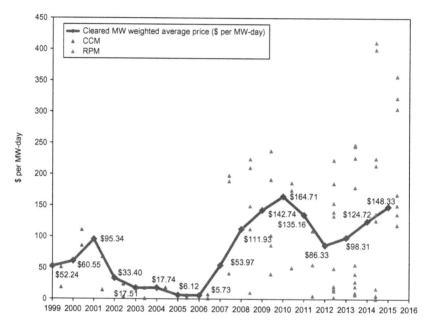

FIGURE 9.4   History of PJM capacity prices: calendar years 1999–2015.[35,36]

market prices increased substantially under the RPM design, but the increases varied significantly by location. The first full year of RPM capacity prices was 2008. Overall prices were higher under RPM than under CCM because the annual incremental cost of capacity is higher than the daily incremental cost of capacity. In addition, RPM prices at times reflected the full cost of new entry including annualized fixed costs as well as the annualized fixed costs of investments required to keep units operating. Capacity market prices were much higher in the eastern, more constrained parts of PJM, reflecting locational capacity supply and demand conditions consistent with the RPM design. The lower prices in the western part of the system, Rest of RTO, reflected the fact that supply exceeded demand by substantial amounts in that area.

35. Monitoring Analytics, LLC. Section 4: Capacity market. 2011 State of the Market Report for PJM, vol. 2. <http://www.monitoringanalytics.com/reports/PJM_State_of_the_Market/2011.shtml> (accessed 15.03.12).

36. 1999–2006 capacity prices are CCM combined market, weighted average prices. The 2007 capacity price is a combined CCM/RPM weighted average price. The 2008–2015 average capacity prices are RPM-weighted average prices. The CCM data points are cleared MW-weighted average prices for the daily and monthly markets by delivery year. The RPM data points are RPM resource clearing prices for individual LDAs.

5.2 PJM Market Net Revenue

While prices are a first indicator, net revenue is the key measure of the impact of PJM capacity markets. Net revenue is an indicator of generation investment profitability, and thus is a measure of overall market performance as well as a measure of the incentive to invest in new and existing generation to serve PJM markets. Net revenue is the contribution to total fixed costs received by generators from all PJM markets. Net revenue is the amount that remains, after short run variable costs have been subtracted from gross revenue, to cover total fixed costs which include a return on investment, depreciation, taxes and fixed operation and maintenance expenses.

The adequacy of net revenue can be assessed both by comparing annual net revenue to annualized total fixed costs and by comparing annual net revenue to annual avoidable costs. The comparison of net revenue to total fixed costs is an indicator of the incentive to invest in new and existing units. The comparison of net revenue to avoidable costs for potential new entrant units and for existing units is an indicator of the extent to which the revenues from PJM markets provide sufficient incentive for continued operation rather than retirement.

Although in the long run, net revenue from all sources is expected to cover the total fixed costs of investing in the generating resources needed to meet the reliability target, including a competitive return on investment, actual results are expected to vary from year to year. Wholesale markets, like other markets, are cyclical. When the markets are long, prices will be lower and when the markets are short, prices will be higher. Market conditions and market results can also vary by location.

The results for the period from 2009 through 2011 demonstrate the significance of the capacity market to net revenues for all three major categories of units.[37] These capacity market revenues would have been missing in the absence of a capacity market and, if missing, would have had a corresponding negative impact on the incentives to reinvest in units as well as to invest in new units. These capacity market revenues are a substantial addition to the revenues of all unit types, but only in the case of CC units in the eastern parts of PJM do the capacity market revenues fully resolve the revenue sufficiency problem. But, capacity markets are not a guarantee that all unit types will always receive 100% of full replacement costs, regardless of market conditions. Capacity market revenues comprise a larger share of total unit revenues for peakers (CTs) than for CC or base load coal units. CC and base load units run more hours under economic dispatch in the energy market, and thus energy market revenues are a larger share of total revenues for these unit categories.

---

37. Monitoring Analytics, LLC. Section 6: Net Revenue. 2011 State of the Market Report for PJM, vol. 2. <http://www.monitoringanalytics.com/reports/PJM_State_of_the_Market/2011.shmtl> (accessed 15.03.12).

TABLE 9.4 PJM Average Net Revenue for a CT Under Economic Dispatch by Market (Dollars per Installed MW-year)[38]

	Energy	Capacity	Synchronized	Regulation	Reactive	Total
2009	$8990	$47,188	$0	$0	$2384	$58,563
2010	$32,781	$55,186	$0	$0	$2384	$90,351
2011	$34,939	$45,972	$0	$0	$2384	$83,295

TABLE 9.5 PJM Average Net Revenue for a CC under Economic Dispatch by Market (Dollars per Installed MW-year)[39]

	Energy	Capacity	Synchronized	Regulation	Reactive	Total
2009	$44,553	$50,184	$0	$0	$3198	$97,936
2010	$89,027	$58,324	$0	$0	$3198	$150,549
2011	$103,726	$48,306	$0	$0	$3198	$155,230

Table 9.4 shows PJM average net revenue by market for a CT under economic dispatch. Capacity market revenues accounted for 66% of CT net revenue for the 3 years, with a maximum of 81% in the low energy market revenue year of 2009.[40] CT market revenues would have been substantially lower in the absence of a capacity market. For a new entrant CT in 2011, net revenues by zone ranged from 57% to 85% of annualized fixed costs, averaging 77% for PJM. While capacity market revenues have been substantial and served as incentive to reinvest in existing units, capacity market revenues have not, in general, been adequate to incent new investment in CTs in PJM, although they have been close in the eastern part of PJM.

Table 9.5 shows PJM average net revenue by market for a CC under economic dispatch. Capacity market revenues accounted for 40% of CC net revenue for the 3 years, with a maximum of 51%. CC market revenues would have been substantially lower in the absence of a capacity market. For a new entrant CC in 2011, total net revenues by zone ranged from 64% to 119% of

38. Monitoring Analytics, LLC. Section 6: Net revenue. 2011 State of the Market Report for PJM, vol. 2. <www.monitoringanalytics.com/reports/PJM_State_of_the_Market/2011.shtml> (accessed 15.03.12).

39. *Id.*

40. The net revenues are theoretical net revenues that would result for new entrant units based on their operating costs under actual market prices.

TABLE 9.6 PJM Average Net Revenue for a CP under Economic Dispatch by Market (Dollars per Installed MW-year)[41]

	Energy	Capacity	Synchronized	Regulation	Reactive	Total
2009	$47,467	$47,469	$0	$2051	$1783	$98,770
2010	$119,478	$54,670	$0	$898	$1783	$176,830
2011	$70,665	$44,282	$0	$1025	$1783	$117,754

annualized fixed costs, averaging 103% for PJM. Capacity market revenues have been substantial and served as incentive to reinvest in existing units and to invest in new CCs in the eastern part of PJM.

Table 9.6 shows PJM average net revenue by market for a CP under economic dispatch. Capacity market revenues accounted for 39% of CP net revenue for the 3 years, with a maximum of 48%. CP market revenues would have been substantially lower in the absence of a capacity market. For a new entrant CP in 2011, total net revenues by zone ranged from 20% to 31% of annualized fixed costs, averaging 26% for PJM. While capacity market revenues have been substantial and served as incentive to reinvest in existing units, capacity market revenues have fallen well short of providing an incentive to invest in new CPs anywhere in PJM.

The results from PJM markets not only illustrate the important role of capacity markets in providing revenues to generating units but also illustrate that market conditions play a critical role. PJM markets have been providing a signal for new CC entry and to a lesser extent new CT entry but have been providing a signal not to enter for new CPs. The net revenue sufficiency problem was fully solved for CC units, especially in the eastern part of PJM, was close to being solved for CT units, and was not close to being solved for CPs. This is a rational set of market price signals, consistent with underlying market conditions. Coal units have been less competitive in the energy market as coal prices moved much higher relative to gas prices and thus coal unit energy market net revenues declined. CC units have been more competitive in energy markets and thus CC energy market net revenues increased. Capacity markets do not guarantee revenue sufficiency regardless of market conditions.

In a perfectly competitive, energy-only market in long-run equilibrium, net revenue from the energy market would be expected to equal the total of all annualized fixed costs for the marginal unit, including a competitive return on investment, in the absence of externally imposed reliability requirements. The PJM market design includes other markets which contribute to

41. *Id.*

the payment of fixed costs. In a perfectly competitive market in long-run equilibrium, with tightly integrated energy, capacity and ancillary service payments, net revenue from all sources would be expected to equal the annualized fixed costs of generation for the marginal unit and the market design would meet externally imposed reliability requirements. In actual wholesale power markets, where equilibrium seldom occurs, net revenue is expected to fluctuate above and below the equilibrium level based on actual conditions in all relevant markets.

5.3 Flexibility

The capacity market in PJM provides a flexible market mechanism which permits responses to changes in market conditions. This flexibility has been demonstrated in the market response to significant modifications in environmental regulations affecting generating units and to fundamental changes in the markets for natural gas and coal. This flexibility has also been demonstrated by the successful integration of substantial amounts of demand side resources in the capacity market and, to a lesser extent, the integration of renewable resources.

Modified environmental regulations required significant new investment in some units in the PJM market, especially coal units.[42] Unit owners submitted offers into the next capacity market auction which incorporated the additional investment and avoidable costs associated with the modifications to the units required to comply with the environmental regulations. If the units cleared the market with the additional costs, the owners were given the signal to invest in the environment controls. If the units did not clear the market, the owners were given the signal to retire the units. Actual behavior was a function of individual assessments of the likelihood of the new environmental regulations actually being in place as well as the likelihood of retirement, regardless of market prices, given the investment costs faced by some units, and expectations about future energy and capacity market prices.

The environmental regulations had a significant impact on the capacity market clearing prices in some locations in the 2014/2015 annual capacity auction, as an example. As a result of the inclusion in market offers of the capital costs of complying with the new environmental regulations, prices in the rest of RTO were 34% higher, prices in MAAC were 30% higher, prices in EMAAC were 2% higher, and prices in Public Service Enterprise Group (PSEG) North were not affected. The overall impact of including

---

42. Monitoring Analytics, LLC. Section 7: Environmental and Renewable Energy Regulations. 2011 State of the Market Report for PJM, vol. 2. <http://www.monitoringanalytics.com/reports/PJM_State_of_the_Market/2011.shmtl> (accessed 15.03.12).

environmental requirements in capacity market offers was to increase total market revenues by $1.361 billion or 23%. The result was also to retain some capacity resources that depend on the capacity market for revenues adequate to cover the cost of investment to comply with new environmental regulations.[43]

The owners of generating units, especially coal units, reevaluated the economic viability of those units, given the combined impact of new environmental regulations and the reduction in the price of natural gas. As a result, owners retired a significant number of coal units. Actual and planned retirements from 2011 through 2019 are currently 19,142.8 MW in the PJM markets, about 11% of the total installed capacity on September 30, 2012.[44]

Planned generation in PJM enters generation request queues, which are used to manage the process of integrating new capacity into the markets. Capacity in generation request queues for construction during the 7-year period beginning in 2012 and ending in 2018 was 75,869 MW at September 30, 2012, or 42% of total installed capacity.[45]

From the implementation of the new PJM capacity market structure in 2007, there have been substantial additions of new resources, both generating units and demand side resources. Including new capacity imports induced by capacity market prices, a total of 27,379.6 MW of generation capacity were added, or 20,625 MW excluding new imports (Table 9.7).[46]

New demand side resources were also added. A total of 15,755.3 MW of demand side resources cleared in capacity auctions for the 2015/2016 delivery year, 9.6% of the total capacity cleared in that auction.[47] Table 9.8 shows the growth in the role of demand side resources in the RPM auctions. These demand side resources were added as a result of the incentives in the capacity market and not because of the energy market incentives to demand side resources, which were modified by Federal Energy Regulatory Commission (FERC) effective April 1, 2012.[48]

In 2011, capacity market payments were 97.7% of total payments to demand side resources in PJM markets.

Figure 9.5 shows all revenue from PJM demand side response programs by market for the period 2002–2011. In 2011, capacity market payments to

43. Monitoring Analytics, LLC. Analysis of the 2014/2015 RPM Base Residual Auction. <http://www.monitoringanalytics.com/reports/Reports/2012/Analysis_of_2014_2015_RPM_Base_Residual_Auction_20120409.pdf> (accessed 09.04.12).
44. Monitoring Analytics, LLC. 2012 Quarterly State of the Market Report for PJM: January through September. <http://www.monitoringanalytics.com/reports/PJM_State_of_the_Market/2012.shtml> (accessed 15.11.12).
45. Id.
46. Id.
47. Id.
48. 134 FERC 61,187 (Order No. 745).

TABLE 9.7 RPM Generation Capacity Additions: 2007/2008 through 2015/2016

ICAP (MW)

Delivery Year	New Generation Capacity Resources	Reactivated Generation Capacity Resources	Uprates to Existing Generation Capacity Resources	Net Increase in Capacity Imports	Total
2007/2008	19.0	47.0	536.0	1576.6	2178.6
2008/2009	145.1	131.0	438.1	107.7	821.9
2009/2010	476.3	0.0	793.3	105.0	1374.6
2010/2011	1031.5	170.7	876.3	24.1	2102.6
2011/2012	2332.5	501.0	896.8	672.6	4402.9
2012/2013	901.5	0.0	946.6	676.8	2524.9
2013/2014	1080.2	0.0	418.2	963.3	2461.7
2014/2015	1102.8	9.0	482.5	818.9	2413.2
2015/2016	6720.4	0.0	569.2	1809.6	9099.2
Total	13,809.3	858.7	5957.0	6754.6	27,379.6

demand side resources were $487 million, energy market payments were $2.0 million and synchronized reserve credits were $9.4 million.

If tightly integrated with the energy market as it is in PJM, the capacity market can also provide a flexible mechanism for addressing the impact of increased level of renewable resources as is occurring in other markets.[49] If the growth in renewables to meet aggressive targets results in lower energy prices, as it is expected to do, then energy market net revenues for traditional thermal resources will decline. In an energy-only market with scarcity pricing, the solution would be an increased frequency and duration of scarcity events. In a capacity market, as net revenue from the energy market declines, capacity prices increase, helping to ensure that the thermal units required for reliability continue to receive revenues adequate to support operations and ongoing investment in existing and new capacity.

---

49. Refer to Chapters 5 and 7 on European and German renewable issues and Chapter 25 for similar discussion.

TABLE 9.8 RPM Load Management Statistics: June 1, 2007 Through June 1, 2015[50]

	DR and EE Cleared Plus ILR		DR Net Replacements		EE Net Replacements		Total RPM LM	
	ICAP (MW)	UCAP (MW)	ICAP (MW)	UCAP (MW)	ICAP (MW)	UCAP (MW)	ICAP (MW)	UCAP (MW)
01-Jun-07	1708.1	1763.9	0.0	0.0	0.0	0.0	1708.1	1763.9
01-Jun-08	4029.4	4167.5	(38.7)	(40.0)	0.0	0.0	3990.7	4127.5
01-Jun-09	7138.3	7374.4	(459.5)	(474.7)	0.0	0.0	6678.8	6899.7
01-Jun-10	8892.2	9199.3	(499.1)	(516.3)	0.0	0.0	8393.1	8683.0
01-Jun-11	10,570.7	10,935.6	(1017.3)	(1052.4)	0.2	0.2	9553.6	9883.4
01-Jun-12	9073.1	9407.0	(2173.4)	(2253.6)	(33.7)	(34.9)	6866.0	7118.5
01-Jun-13	10,916.7	11,329.7	0.0	0.0	0.0	0.0	10,916.7	11,329.7
01-Jun-14	14,641.3	15,183.2	0.0	0.0	0.0	0.0	14,641.3	15,183.2
01-Jun-15	15,194.0	15,755.3	0.0	0.0	0.0	0.0	15,194.0	15,755.3

50. Monitoring Analytics, LLC. 2012 Quarterly State of the Market Report for PJM: January through September. <http://www.monitoringanalytics.com/reports/PJM_State_of_the_Market/2012.shtml> (accessed 15.11.12).

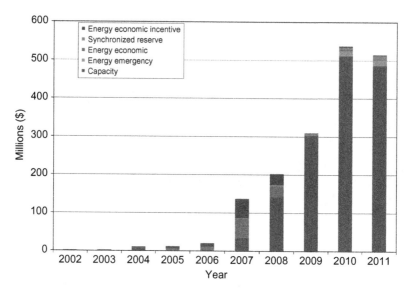

FIGURE 9.5   Demand response revenue by market: calendar years 2002–2011.[51]

Renewable energy sources in PJM remain at a modest 3.3% share of output (for the first three quarters of 2012), but the generation queues for planned resources include substantial amounts of renewables. Wind projects account for approximately 26,495 MW of capacity or 34.9% of the capacity in the queues, solar projects are 2675 MW, 3.5%, storage projects are 108 MW or 0.1%, for a total renewable level of 29,278, or 38.6% of all generation in the queues. Gas-fired combined-cycle projects account for 38,806 MW of capacity, 51.1%, or most of the balance of the capacity in the queues.

5.4 Summary

The PJM RPM capacity market design was implemented to address the revenue sufficiency problem that affected PJM as it affects all wholesale power markets. The design, while reliant on administrative rules, is market based, which results in the flexibility to respond to changes in market conditions including environmental regulations, large changes in the relative prices of input fuels, increased demand side participation, and increased renewable participation.

51. Monitoring Analytics, LLC. Section 5: Demand side response. 2011 State of the Market Report for PJM, vol. 2. <www.monitoringanalytics.com/reports/PJM_State_of_the_Market/2011.shtml> (accessed 15.03.12).

Of the other wholesale power markets in the world, only the ERCOT market in Texas has an energy-only design comparable to the PJM market or other Regional Transmission Organization (RTO)/Independent System Operator (ISO) markets in the US. ERCOT is currently considering whether to add a capacity market in response to the fact that, under an energy market, it is not possible to define and achieve a target reserve margin as it is with a capacity market.[52] Other markets frequently cited as energy only are the Australian National Energy Market (NEM) market and the Alberta market, neither of which are comparable to the wholesale power markets in the United States because they do not include locational marginal pricing, have highly concentrated generation ownership, and include substantial government ownership of both transmission and generation assets.[53] Those markets are so different in basic ways that they do not represent a useful point of comparison for US wholesale power markets. No wholesale power market in the world relies entirely on the energy market, without substantial administrative intervention, for reliable electric service.

6 CURRENT ISSUES WITH CAPACITY MARKET DESIGN

The introduction of the RPM capacity market design and subsequent modifications to that design have contributed to better functioning PJM markets. However, the current design of the PJM capacity market includes a number of significant, unresolved issues. The successful resolution of these issues is critical to the long-term success of the PJM capacity market.

The test of a market design is whether it can reproduce itself, that is, whether the incentives internal to the design will result in investment in the next generation of resources required for reliable operation at competitive prices. PJM markets have demonstrated flexibility in responding to changing market conditions and new generation has been added and more is planned. But it is not clear that incentives to invest are adequate or that the market design results in reliability at the lowest possible cost, consistent with a dynamic competitive outcome. Although PJM markets have continued to provide target levels of reliability, that outcome has depended on a significant role for an inferior DR product and benefited from low load growth. It is also not clear that the markets have resulted in the lowest possible price outcome. While the market results have been competitive, market design issues lead to questions about whether the capacity market can result in more efficient results over time, including the level and mix of resources which result from capacity market

---

52. See Chapter 10.
53. Refer to Chapter 19 on Australian market.

performance incentives. The RPM market design is working, but it has benefited from fortunate circumstances, including low load growth, which will not continue indefinitely.

The market design issues that have a significant impact on market outcomes in both the short and long terms include: the arbitrary 2.5% shift in the capacity market demand curve, the misdefinition of the demand response product, the lag in the net revenue offset, the design of performance incentives, and the role of state public utility commissions in the capacity market.

6.1 Demand Curve Shift

Under the current capacity market rules, PJM applies the "Short-Term Resource Procurement Target," which means that 2.5% of the reliability requirement is removed from the demand curve in the BRA in which total demand and supply are cleared. There is no supportable market rationale for this rule which inefficiently suppresses prices in the capacity market for all resource types including generation and demand side resources, and is the single largest design flaw in the PJM capacity market design.

The implementation of this arbitrary 2.5% reduction in demand significantly reduced the capacity market clearing prices and quantities in every capacity auction in which it was implemented.

Figure 9.3 shows the primary PJM LDAs that have cleared with separate prices. In the 2014/2015 BRA, the 2.5% demand reduction eliminated 3708.1 MW of demand and reduced prices by $16.70 per MW-day for the annual product in rest of RTO, and by $23.50 in rest of MAAC with no impact on prices in PSEG North. The reduction of demand for capacity by 2.5% in the 2014/2015 RPM BRA reduced total RPM market revenues by $1.2 billion, or 14.1%, compared to the total revenues that would have resulted if the market had cleared without the price suppressing impact of the rule.[54] In addition, cleared volumes of the annual product would have been higher in each LDA where prices would have been higher.[55]

The stated rationale for the 2.5% offset is that this provides for short lead time resource procurement in incremental auctions for the given delivery

54. Monitoring Analytics, LLC. Analysis of the 2014/2015 RPM Base Residual Auction. <http://www.monitoringanalytics.com/reports/Reports/2012/
Analysis_of_2014_2015_RPM_Base_Residual_Auction_
20120409.pdf> (accessed 09.04.12).
55. Id.

year.[56] But actual auction results show that the intended result has not occurred. In the BRA for the 2014/2015 delivery year, PJM procured Limited DR up to the limit for this product. As a result, no additional Limited DR could be procured in any subsequent incremental auctions.

The 2.5% demand reduction is not needed to facilitate participation from the resources that participate as Limited DR. There is no reason not to immediately end the use of the 2.5% demand reduction. The 2.5% demand reduction is a barrier to entry in the capacity market for both new generation capacity and new DR capacity. The reduction of demand in a market design that looks 3 years forward, to permit other resources to clear in incremental auctions, is not supportable and has no basis in economic logic. There are tradeoffs in using a 1-year-forward or a 3-year-forward design, but the design must be implemented on a consistent basis if it is to work. Removing a portion of demand affects prices at the margin, which is where the critical signal to the market is determined.

6.2 Limited DR

There are two categories of demand side products included in the RPM capacity market design. Energy efficiency (EE) resources are load resources that are offered in an RPM auction as capacity and receive the relevant resource clearing price. An EE resources are designed to achieve a continuous reduction in electric energy consumption during peak periods.[57] Demand resources (DRs) are interruptible load resources that are offered in an RPM auction as capacity and receive the relevant clearing price.

Effective with the 2014/2015 delivery year, the DR product was divided into three types. Annual DR is required to be available on any day for an unlimited number of interruptions for up to 10 h per interruption. Extended Summer DR is required to be available on any day from May through October for an unlimited number of interruptions for at least 10 h per interruption. Limited DR is required to be available during the period of June through September for defined peak hours for up to 10 interruptions for up

---

56. Monitoring Analytics, LLC and PJM Interconnection, LLC. Capacity in the PJM Market. <http://www.monitoringanalytics.com/reports/Reports/2012/ IMM_And_PJM_Capacity_White_Papers_On_OPSI_Issues_20120820.pdf> (accessed 20.08.12). PJM has also asserted that the 2.5% replaced the flawed Interruptible load for reliability (ILR) design, but there is no reason to extend the problems associated with the ILR design. In addition, PJM has indicated that the 2.5% is related to inaccurate demand forecasts by PJM, but it has no analysis to support the claim of systematic forecast bias, and PJM does not make this 2.5% adjustment in any other application of its forecasts, including most importantly, in transmission planning.
57. PJM Interconnection, LLC. Reliability Assurance Agreement among Load-Serving Entities in the PJM Region, Schedule 6, Section M.

to 6 h per interruption. Prior to the 2014/2015 delivery year, all DR was Limited DR.

The Limited DR and Extended Summer DR product types are both inferior to generation capacity resources because the obligation to deliver associated with both product types is inferior to the obligation to deliver associated with generation resources. The Annual DR product is almost comparable to generation, but is limited to 10 h per interruption and may only be called during defined hours each day. Generation resources are obligated to deliver energy every hour of the year if called.

The inclusion of the Limited DR product in the 2014/2015 BRA had a significant· negative impact on capacity prices. The clearing prices would have been substantially higher if all offers for Extended Summer and Limited DR products, including those within coupled DR offers, had been excluded, and only generation and annual DR had been offered in the 2014/2015 RPM BRA. The inclusion of the limited DR products reduced the RTO clearing price from $154.87 per MW-day to $125.99, 18.7%, reduced the MAAC clearing price from $202.80 per MW-day to $136.50, 32.7%, and had no effect on the PSEG North clearing price.[58] The inclusion of the limited DR products in the 2014/2015 RPM BRA reduced total RPM market revenues for the auction from $9.6 billion to $7.2 billion, 25%.[59]

While competition from demand side resources improves the functioning of the market, that is not the result if the demand side resources are inferior to other capacity resources. The purpose of demand side participation in RPM is to provide a mechanism for end-use customers to avoid paying the capacity market clearing price in return for agreeing to not use capacity when it is needed by customers who have paid for capacity. The fact that customers providing Limited DR only agree to interrupt 10 times per year for a maximum of 6 h per interruption represents a flaw in the design of the market. There is no reason to believe that the customers who pay for capacity will need the capacity used by Limited DR customers only 10 times per year. In fact, it can be expected that the probability of needing that capacity will increase with the amount of MW that Limited DR customers clear in the RPM auctions. This limitation means that such demand side resources sold in the RPM auctions are of less value in meeting market demand and cost less to provide than either annual DR or generation capacity. As a result, limited demand side resources could make lower offers than they would if they offered a resource with performance obligations comparable to those of generation or annual DR.

58. Monitoring Analytics, LLC. Analysis of the 2014/2015 RPM Base Residual Auction. <http://www.monitoringanalytics.com/reports/Reports/2012/ Analysis_of_2014_2015_RPM_Base_Residual_Auction_20120409.pdf> (accessed 09.04.12).
59. *Id.*

The capacity market design could be improved by modifying the definition of demand side resources to ensure that such resources provide the same value in the capacity market as generation resources. The elimination of both the Limited and Extended Summer DR products would help ensure that the DR product has the same unlimited obligation to provide capacity year round as generation capacity resources. Rather than creating a regulatory solution to a perceived issue with DR participation by creating an inferior product, a market solution is available. If a single demand side site cannot interrupt more than 10 times per year, a CSP could bundle multiple demand sites to provide unlimited interruptions.[60] The cost of providing bundled sites would be expected to be greater than a single site and the offer price of such resources would also be expected to be greater. Such a modification would help ensure that demand side resources contribute to the competitiveness of capacity markets rather than suppressing the price below the competitive level. Such a modification would permit the CSP to form its own view as to how often it expects to actually be interrupted and to construct a portfolio that permits such a response. The CSP would internalize the costs of providing a comparable resource rather than imposing that cost on other market participants.

6.3 Lag in Net Revenue Offset

The cost of capacity is a function of the net cost of new entry, which is the gross cost of new entry less the expected net revenues for a new entrant from all other PJM markets. The net revenue offset is the key link between the energy market and the capacity market. The goal of the net revenue calculation is not only to ensure that together they create the opportunity for full recovery of investment costs but also to ensure that there is not either systematic over or under payment.

The capacity market rules set the expected net revenues equal to the average of the net revenues for the prior 3 years for individual units, or for the reference technology in the case of new entry. The 3-year historical average will always be incorrect, but it will be a reasonable approximation of revenues during the delivery year if energy market conditions remain relatively constant over long periods. Unfortunately, this seldom occurs. The net revenue offset looks back 3 years at the time of the capacity auction, which looks forward 3 years, so the timing mismatch is substantial. If historical average net revenues include a very high net revenue year, the 3-year average may exceed the reasonably currently expected net revenues at the time of the auction or the actual net revenues during the delivery year. The result will be that capacity market prices are understated. The reverse can also occur.

---

60. A CSP is a curtailment services provider, an aggregator of individual DR providers.

This problem will be exacerbated under PJM's recently implemented energy market scarcity pricing rules.[61] To the extent that scarcity pricing results in substantially higher net revenues for a year, this will result in an increase in the 3-year historical average. If the scarcity pricing results are not repeated, the net revenue offset will be too high, potentially significantly, for 3 years with the result that capacity market prices will be too low. This could be further amplified if energy market prices and revenues trend downward for other reasons. The reverse could also happen if market participants expect scarcity conditions that are not reflected in the historical average net revenues.

While one way to address this issue would be to develop a forecast method for energy and ancillary service revenues that reasonably matched investor expectations at the time of the auction, doing so is virtually impossible. Another option would be to offset either scarcity revenues or all energy and ancillary service market net revenues in real time during the capacity market delivery year.[62] This would ensure that exactly the right offset is used and that the cost of capacity paid by customers and received by investors more accurately reflects the actual cost of capacity.

6.4 Performance Incentives

The goal of the capacity market performance incentives should be to match the incentives that would result from a competitive energy-only market. The performance incentives in the PJM capacity market fall well short of that objective. The most basic market incentive is that associated with the reduction of payments for a failure to perform. In any market, sellers are not paid when they do not provide a product. That is only partly true in the PJM capacity market. There are two areas where the performance incentives are inadequate, overpayment for underperformance and incorrect outage rate definition.[63]

A capacity resource will be paid 50% of its full capacity market revenues even in the case of complete nonperformance in the first year of such nonperformance. For example, a resource that sold 500 MW of unforced capacity at $150 per MW-day would be paid $75 per MW-day even if the resource did not produce energy when called during any of the PJM-defined approximately 500 RPM critical hours. That decreases to 25% in year two of sub-50% performance and to 0 in year three, but returns to 50% after 3 years of

---

61. 139 FERC 61,057 (April 19, 2012).
62. Monitoring Analytics, LLC. Protest and Compliance Proposal of the Independent Market Monitor for PJM. <www.monitoringanalytics.com/reports/Reports/2010/IMM_Protest_and_Compliance_Filing_ER09-1063-004_20100718.pdf> (accessed 18.07.10).
63. Monitoring Analytics, LLC and PJM Interconnection, LLC. Capacity in the PJM Market. <http://www.monitoringanalytics.com/reports/Reports/2012/IMM_And_PJM_Capacity_White_Papers_On_OPSI_Issues_20120820.pdf> (accessed 20.08.12).

better performance. Under some extreme circumstances, total nonperformance would result in total nonpayment as a result of penalties.

Not all unit types are subject to RPM performance incentives. Wind, solar, and hydro generation capacity resources are exempt from key performance incentives. Wind and solar generation capacity resources are not subject to peak hour availability incentives, to summer or winter capability testing or to peak season maintenance compliance rules. Hydro generation capacity resources are not subject to peak season maintenance compliance rules.[64]

Given that all generation is counted on for comparable contributions to system reliability, it would be efficient for all generation types to face the same performance incentives.

In the PJM capacity market, the forced outage rate is a performance incentive. Resource owners sell unforced capacity in the capacity market, which is installed capacity times one minus the forced outage rate for the resource. The higher the forced outage rate, the less capacity can be sold from a generating unit in the capacity market and the lower are the capacity market revenues for that unit. The capacity market creates an incentive to have low forced outage rates in this direct way. The forced outage rate also affects the level of payment actually received for the level of capacity sold in the RPM auctions. If the actual forced outage rate during critical hours is greater than the forced outage rate on which the sale of capacity is based, net payments to the capacity resource are correspondingly reduced subject to the floor of 50%. This is also a direct incentive, although attenuated, to have low forced outage rates during critical hours. The issue in the PJM capacity market is that the forced outage rates used to provide these incentives do not correctly measure actual forced outage performance.[65]

There is no reason not to reflect all outages in overall PJM system planning as well as in the economic fundamentals of the capacity market and the capacity market outcomes. Excluding some outages from the forced outage metric skews the results of the capacity market towards less reliable units and away from more reliable units both for existing units and potential new entrants. This incentive design is not consistent with an efficient outcome.

This issue has had a significant impact for some specific locational capacity markets in PJM. At an aggregate level, outages omitted from the outage rate calculation accounted for 11.6% of all forced outages in 2011. In 2011, for coal steam units, the difference was 995 MW of unforced capacity.

64. PJM Interconnection, LLC. Manual 18: PJM Capacity Market, Revision 15 (June 28, 2012), p. 98.
65. For a more complete discussion of this issue, see the IMM's White Paper included in: Monitoring Analytics, LLC and PJM Interconnection, LLC. Capacity in the PJM Market. <http://www.monitoringanalytics.com/reports/Reports/2012/ IMM_And_PJM_Capacity_White_Papers_On_OPSI_Issues_20120820.pdf> (accessed 20.08.12).

In the LDAs where this steam generation was located, PJM treated 995 MW as available when it was not available.

6.5 Reliability Issues for Individual States

The PJM footprint includes all or part of 13 states plus the District of Columbia. From the beginning of the RPM market, there has been tension between the perception of capacity market outcomes and the perceived needs of the individual states in the PJM footprint.[66] Each state public utility commission has the right and duty to ensure a reliable supply of electricity in their state, which is defined at a state level. The FERC has the responsibility to maintain just and reasonable rates for wholesale power, which it implements in large portions of the United States through the use of competitive markets operated by ISOs and RTOs. The challenge is to align the two objectives and to ensure that the wholesale market meets the reliability needs of the states.

A disconnect arose early in the implementation of the RPM capacity market for several states who were informed by PJM that although the capacity market was working, there were potential state-level reliability issues. This caused some states to consider whether they needed separate capacity procurement options to ensure state-level reliability. Despite the fact that there is no current evidence that there are reliability issues in any state that are not being adequately addressed via the capacity market, the efforts of states to create capacity procurement options have created potential issues for the wholesale market.

Regardless of other factors, states have a legitimate basis for concerns about reliability, including but not limited to those related to the construction of transmission lines; the existence of RMR contracts,[67] state and federal environmental requirements, potential unit retirements, the nature of competition for new entry, and siting issues for generation and transmission facilities.

With correct information inputs, markets are a flexible, least cost way to address these issues and uncertainties. Markets work best when the market design permits market outcomes to reflect the market fundamentals. Markets

---

66. Monitoring Analytics, LLC. Comments of the Market Monitor re In the Matter of the Board's Investigation of Capacity Procurement and Transmission Planning Docket No. EO 1105030 (October 14, 2011) and Comments of the Market Monitor re In the Matter of the Board's Investigation of Capacity Procurement and Transmission Planning Docket No. EO11050309. <http://www.monitoringanalytics.com/reports/Reports/2011/ IMM_Comments_NJ_EO_11050309_20110617.pdf> (accessed 17.06.11).
67. RMR contracts are required when a unit wishes to retire for economic reasons but PJM needs the unit to remain in service for reliability, typically for a period less than 18 months in duration while transmission upgrades are constructed. RMR contracts are an indication that the RPM market is not working well in the area where the RMR plants are located.

result in an appropriate assignment of risks and incentives between developers and customers. Markets may not work well if the market design does not permit market outcomes to reflect the market fundamentals.

If the wholesale market is to work effectively, it is essential that the inputs to the market are right, that states' reliability situation is fully and accurately reflected in market inputs and that the market design permits market outcomes to reflect these market fundamentals. For example, if a state decides that they want certain older units to shut down for environmental reasons, the state could take actions through environmental regulations which will directly produce that result. When the information about the reduced capacity which results is incorporated in the capacity market, the economic fundamentals will change correspondingly and the market will address any resultant shortfall in capacity.

The issue is that any capacity procured by a state must have a defined interaction with wholesale power markets. The state can procure capacity and pay the capacity on the side with no interaction with the wholesale capacity market. This option would have an impact on the wholesale energy markets if such units participated in PJM energy dispatch as energy-only units. Alternatively, a state can procure capacity, pay the capacity a guaranteed price, and require that the capacity offer into the wholesale capacity market. That capacity offer can be a competitive offer for new entry or the offer can reflect the side payments from the state and be set to zero. The option of offering at zero has created issues associated with potential impacts on the competitive wholesale power markets of contracts for capacity that are subsidized by payments through a state or other procurement process and that are not derived from the wholesale power markets.

In PJM, several states have procurement processes that are not consistent with the operation of a competitive capacity market. These processes, as currently implemented, would result in the procurement of capacity that is not needed for reliability; the procurement of capacity through a process that is discriminatory because it excludes existing generation; and the requirement to offer the procured capacity so that it clears in the PJM capacity auctions. The result of offering capacity purchased through such procurements in the PJM capacity market at prices less than cost would be to artificially depress prices in the PJM capacity market both inside and outside the individual states. These market impacts can be very large.[68]

One solution to this conundrum is for states to procure capacity through competitive nondiscriminatory auctions. Such an approach would permit states to enter into long-term contracts if they wish, while permitting the offer of the related capacity into the PJM capacity market at a zero offer price.

---

68. Monitoring Analytics, LLC. Impact of the New Jersey Assembly Bill 3442 on the PJM Capacity Market (January 6, 2011).

The Minimum Offer Price Rule (MOPR) was part of the initial RPM design and was intended to prevent the exercise of monopsony market power. The MOPR was modified in 2011 as a result of concerns about the states' procurement processes and associated plans to offer the procured capacity into the PJM capacity market.[69] The MOPR is currently undergoing another substantial revision in the PJM stakeholder process. The goal of the MOPR rule is to address the tensions between the role of the states and the role of the wholesale power markets as well as the role of competitive entrants, public power entities, and vertically integrated utilities.

6.6 Summary of Current Issues in the PJM RPM Market Design

The goal of any market design should be to get the prices right. Getting the prices right means that market clearing prices reflect the underlying supply and demand conditions in the market. Getting the prices right means that prices are not too low and that prices are not too high, but that prices induce entry when it is needed and that customers do not overpay for capacity. The PJM RPM capacity market design does, in general, get the prices right, but there are some key areas where capacity market prices do not reflect the underlying supply and demand conditions. Some features of the RPM design suppress the price, one feature results in a mismatch between capacity and energy market revenues, and some features result in customers overpaying for capacity.

The arbitrary 2.5% shift of the demand curve and the inclusion of the inferior limited demand side capacity product both contribute in a material way to price suppression in the RPM design as implemented. These features weaken the incentive to invest which is a strong point of the RPM design exactly when that incentive must be at full strength. The inadequate incentive to invest results not only in less than optimal new entry but also in misguided attempts to modify the capacity market design in other ways. For example, some market participants have suggested that load be required to enter into mandatory long-term contracts for at least a portion of total demand. In adequate incentives and the resultant less than optimal entry have created concerns about reliability in individual states in the PJM footprint that led to disputes about the MOPR.

The lag in the net revenue offset weakens the tight integration between the capacity market, and the energy and ancillary markets, which is a strong point of the RPM market design.

Some features of the capacity market design result in customers overpaying for capacity. A least cost market design would ensure that customers pay only for the most cost-effective set of generating units that provide a combination of energy and ancillary services when they are needed to meet

---

69. 135 FERC 61,022 (2011).

demand. To achieve the least cost design, generating units that receive capacity payments should meet correctly defined obligations, including making competitive offers every day in the energy market, correctly measuring their available capacity and receiving payment only when available to provide energy when needed to meet customers' demand.

Under the current PJM capacity market design, customers overpay for capacity when they pay capacity revenues even when units do not meet performance standards and when they pay for capacity that does not exist based on inaccurate outage rate metrics. While it is possible to maintain a reliable system with a capacity market design that overpays, that is, not in customers' best interests, is not in the interests of competitive generation suppliers and is not in the interests of a competitive market design.

7 CONCLUSION

It is essential that any approach to the design of wholesale power markets, and the design of capacity markets in particular, incorporate a consistent view of how the preferred market design is expected to work to provide competitive results in a sustainable market design over the long run. An efficient, sustainable market design means a market design that results in appropriate incentives to invest in new and existing units and to retire units such that reliability is ensured at the lowest possible cost as a result of the functioning of the market.

There are at least two broad paradigms that could result in such an outcome. The market paradigm includes a full set of markets, most importantly the energy market and capacity market, which together ensure that there are adequate revenues to incent new generation when it is needed and to incent retirement of units when appropriate. The market paradigm also includes the energy market only plus scarcity pricing approach, which has strong incentive characteristics. Under the market paradigm, there is competition to build new generation and to provide new demand side resources. Generation owners bear the risks associated with building and operating generating units. This approach will result in long-term reliability at the lowest possible cost if well designed. The market paradigm fits well with competitive retail markets.

The quasi-market paradigm includes an energy market based on LMP but addresses the need for investment incentives via long-term bilateral contracts or cost of service regulation. In the quasi-market paradigm, competition to build capacity is limited and does not include the entire market. In the quasi-market paradigm, there is no competition to provide demand side resources, which are mandated and provided under cost of service regulation. In the quasi-market paradigm, customers absorb the risks associated with new investment through guaranteed payments under either guaranteed long-term contracts or cost of service regulation. In the quasi-market paradigm, there is no market clearing price to incent investment in new or existing units or demand side resources. The quasi-market paradigm does not fit well with retail competition.

The RPM market has resulted in the addition of new generating capacity, of the retention of old generating capacity, and the addition of demand side resources. The flexibility of the RPM market design has permitted PJM markets to successfully adapt to significant new environmental regulations from the federal and state governments, to unprecedented levels of demand side resources, to growing levels of renewable resources, to dramatic changes in the relative prices of natural gas and coal, and to the retirement of generating units that resulted from the confluence of these factors.

An essential feature of the RPM capacity market design is its tight integration with the PJM energy and ancillary services markets. The level of offers in the capacity market and the shape and location of the demand curve in the capacity market are a function of the cost of capacity, net of the energy and ancillary services revenues earned by units in PJM markets. If net energy and ancillary services markets revenues decline, the price of capacity will rise, all else constant, and the reverse is also true. It is this tight integration between the capacity market, and the energy and ancillary services markets that contributes to the flexibility of the capacity market.

The PJM RPM market demonstrates how capacity markets can work successfully. The RPM capacity market design has succeeded to date in ensuring reliability in PJM. But the RPM capacity market design has some key flaws that it is essential to remedy if the RPM capacity market is to reach its potential and result in sustainable, reliable, competitive PJM wholesale power market outcomes at the lowest possible cost.

Chapter 10

Texas Electricity Market: Getting Better [☆]

Parviz Adib[a], Jay Zarnikau[b] and Ross Baldick[c]
[a]*Pionergy,* [b]*Frontier Associates and The University of Texas at Austin,*
[c]*The University of Texas at Austin*

1 INTRODUCTION

The Texas electricity market was previously described by Adib and Zarnikau (2006) in a preceding volume as "The Most Robust Competitive Market in North America." This market—specifically, the Electric Reliability Council of Texas or ERCOT—is frequently cited as North America's most successful in both generation and retail.[1] As indicated by Young Kim in Chapter 12 in this volume, the "differences in policies and market design explain why the state has become the most active and advanced competitive retail market in the nation."

Barriers to entry are relatively low, allowing new entrants to gain a foothold. Within most areas served by ERCOT, consumers enjoy the freedom to choose from a number of retailers offering different pricing plans and services.[2] The market rules are generally perceived to be fair and equitable. There is relatively limited regulatory oversight for retail pricing with market monitoring focused mainly on the operation of the wholesale market. Market forces are relied upon heavily to preserve reliability and resource adequacy. Among the states which have restructured their electricity markets to foster competition, Texas has the highest level of customer participation among loads eligible for customer choice and accounts for more than 32% of total electricity sales in the United States under competitive environments, as noted in Table 10.1 and Figure 10.1.

[☆]The authors would like to thank Ms. Margaret Marchant from Frontier Associates for her valuable editorial advice.

1. See, for example, Distributed Energy Financial Group (DEFG) (2011); Alliance for Retail Choice (2007); and Center for Advancement of Energy Markets (CAEM) (2003).

2. The principal exceptions are municipal utilities and cooperatives, which participate in the wholesale market but, with one exception, have not opened their retail markets to competition.

Evolution of Global Electricity Markets.

TABLE 10.1 US Eligible and Competitive Retail Power Sales by State (GWh)

Jurisdiction	Nonresidential Competitive Load			Residential Competitive Load		
	GWh	Eligible Percentage	Total Percentage	GWh	Eligible Percentage	Total Percentage
California	21,939	17.80	13.40	101	0.10	0.10
Connecticut	13,363	85.20	78.90	5583	45.60	43.00
Delaware	4068	75.00	62.00	115	3.80	2.50
DC	8318	83.90	87.50	127	6.30	6.20
Illinois	71,406	80.00	75.20	2211	5.40	4.70
Maine	4541	68.60	64.50	24	0.60	0.50
Maryland	28,514	81.70	78.50	5056	20.70	18.50
Massachusetts	23,094	76.20	64.50	2199	12.70	10.50
Michigan	7999	100.00	11.50	0	0	0
Montana	2453	100.00	27.60	1	0	0
New Hampshire	3380	57.20	52.80	9	0.20	0.20
New Jersey	33,325	79.40	70.40	3680	12.30	12.50
New York	54,795	68.20	59.30	8800	22.50	17.20
Ohio	52,746	58.60	52.30	14,872	33.10	27.90
Oregon	959	5.30	3.50	0	0	0
Pennsylvania	75,232	81.80	80.00	12,265	23.30	22.40
Rhode Island	2204	48.00	47.90	32	1.00	1.00
Texas	142,442	100.00	64.30	78,810	100.00	55.10
TOTAL	550,778	68.20	52.80	133,885	31.30	22.10

2011 GWh KEMA Reported Competitive Sales as Percent of 2011 Eligible Load and as Percent of EIA Reported Total Statewide C&I or Residential Load.
Source: Retail electric choice: proven, growing, sustainable, COMPETE, April 2012.

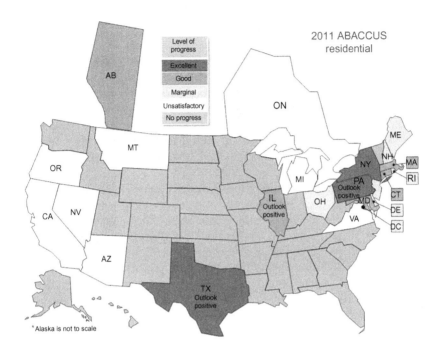

FIGURE 10.1 North American competitive residential electricity markets. *Source: The Distributed Energy Financial Group (2011) ABACCUS report.*

What useful lessons can be learned from ERCOT? How did this market achieve its success where others have not? Can Texas serve as a blueprint for restructuring initiatives elsewhere? Are the preconditions for ERCOT's success unique? Or will the long-term challenges associated with ensuring resource adequacy or the sensitivity of retail prices to volatile natural gas prices expose fatal flaws in a system that relies so heavily on market forces?

This chapter examines the transition by ERCOT to nodal pricing, steps taken to eliminate default retail prices, transmission expansion rules to accommodate renewable resources, policies to address market power, the introduction of advanced metering infrastructure, and efforts to integrate price-responsive loads in resource dispatch decisions. Despite many refinements, a number of challenges remain. Meeting these challenges may require better scarcity pricing, and a serious commitment to demand side resources and price-responsive loads in the market.

Following this introduction, Section 2 describes refinements to improve operation, market efficiency, and customer protection. Section 3 highlights emerging issues and challenges that must be met for Texas to maintain its leadership among restructured electricity markets, followed by chapter's conclusions.

2 RESTRUCTURING AND REFINEMENTS

Texas was the last state in the United States that brought its electric industry under regulation in 1975. The Public Utility Commission of Texas (PUCT or Commission), with three governor-appointed commissioners, took statewide responsibility to regulate rates and approve power plant and transmission expansion through administrative hearings and contested cases. Vertically integrated investor-owned utilities served most of the population centers and dominated the electric industry, along with a few major municipal utilities serving Austin and San Antonio and more than 70 rural electric cooperatives.

Customers experienced reasonable electricity rates under the regulatory regime. However, due to the competitive nature of their businesses and the need to further cut their production costs, industrial customers pushed for opening the electric sector to competition. The independent generators and power marketers, such as Enron, also applied pressure on legislators for further deregulation.[3] Texans enjoyed pursued relatively low rates, but because of relatively high consumption levels, had above average electricity bills. In 1997, Texas ranked 27th highest in the United States in electricity rates with an average residential rate of 7.82 cents/kWh, but had the 6th highest average annual residential bill of $1,066, largely due to heavy air-conditioning use (Adib and Zarnikau, 2006).

While the majority of Texas is served through ERCOT, three interstate regional reliability councils serve fringes of the state, as shown in Figure 10.2.[4]

Texas is the biggest of the lower 48 contiguous United States and has a population of 26 million. ERCOT operates as a single control area[5,6] and accounts for 75% of land and 85% of electricity load in Texas.[7] ERCOT serves about 23 million people accounting for more than 8 million electric meters. In 2011, ERCOT had a peak demand of 68,379 MW, a total electricity consumption of 335,000 gigawatt-hours (GWh), more than 40,000 circuit miles of high-voltage transmission, and more than 74,000 MW of capacity

---

3. See Wood and Gülen (2009).

4. In contrast to other states, most of Texas, located within ERCOT, is not under the jurisdiction of the Federal Energy Regulatory Commission (FERC). This is mainly because ERCOT is not electrically synchronized with either Eastern Interconnection or with the Western Interconnection transmission networks and its electricity transactions are not considered to result in interstate commerce. Hence, the PUCT solely regulates the electricity market within ERCOT. This has resulted in greater coherence in the retail and wholesale markets, more regulatory certainty and closer coordination between Legislative authorities and regulators involved in restructuring the electric industry within ERCOT.

5. ERCOT became the first US Independent System Operator (ISO) in 1997.

6. The electric system in ERCOT consisted of 10 control areas before restructuring in 2002.

7. The US electric system consists of three interconnections: the Eastern Interconnection, the Western Interconnection, and ERCOT.

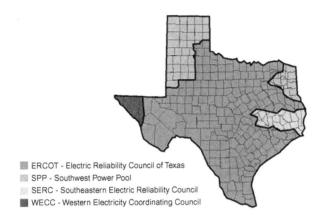

FIGURE 10.2   Regional reliability councils covering Texas. *Source: PUCT web site at: https://www.puc.state.tx.us/industry/maps/Electricity.aspx.*

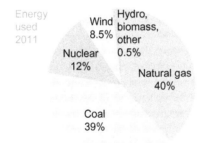

FIGURE 10.3   ERCOT electricity generation fuel mix. *Source: ERCOT (2012).*

provided by 550 generating units. Natural gas (57%) and coal (23%) account for 80% of total installed capacity (ICAP) (in MW). However, ERCOT still relies significantly on coal generation (in MWh) to meet its energy needs (ERCOT, 2012). Figure 10.3 reflects fuel mix in electricity generation.

In contrast to California, which opened wholesale and retail electricity competition simultaneously in early 1998, Texas took a staged approach.[8] Texas opened wholesale competition in September 1995 and allowed 7 years

8. For an excellent historical review of the legislative, regulatory, and market design aspects of electric restructuring and competition in Texas, please see Kiesling and Kleit (2009).

to pass before opening its retail electricity market to competition in January 2002.[9] This facilitated the transition from regulated rates to market-based retail pricing by first building a historical record of wholesale competitive prices. Independent generation companies built new capacity and established market positions to compete with incumbent generators such as TXU Electric, Houston Lighting & Power (HL&P), and American Electric Power (AEP). The PUCT finalized more than 20 rules governing the transition from regulated to competitive retail electricity service, including the establishment of requirements for retail electric providers (REPs) and the default service price, called "price-to-beat" (PTB) (Adib and Zarnikau, 2006).

Average residential prices in areas of Texas opened to retail choice increased at far greater rates than in areas of Texas not opened to competition (Zarnikau and Whitworth, 2005). This raised concerns that the market was not sufficiently competitive and that greater control over prices may be necessary. Nonetheless, the legislature and Commission opted to "stay the course." On January 1, 2007, the PTB fully expired and there was no longer regulatory oversight over the prices offered by the incumbent affiliated REPs (AREPs). The expiration of the PTB appears to have led to a reduction in the average prices charged by competitive REPs (CREPs) (Kang and Zarnikau, 2009; Swadley and Yucel, 2011).

Market restructuring has increased the sensitivity of retail electricity prices to changes in the price of natural gas. Under traditional regulation, price changes tend to reflect changes in the utilities' average fuel costs. Average fuel costs may represent the prices of a broad mix of fuel sources. In contrast, prices in the restructured ERCOT market closely follow marginal wholesale power costs, which are largely determined by natural gas prices (Woo and Zarnikau, 2009). Thus, in the local media restructuring is criticized as a failure when the price of natural gas is high and heralded as a success when gas prices drop.

Figure 10.4 contrasts the patterns in retail prices over time between the AREPs in their traditional service areas versus the prices charged by two sets of utilities which do not permit customer choice—investor-owned utilities serving outside of the ERCOT market and the state's two largest municipal systems (cities of Austin and San Antonio). The divergence in prices in competitive areas from other areas of the state was highest in mid-2006 and mid-2008. Recently, low natural gas prices have brought retail prices in

---

9. See Adib and Zarnikau (2006). The Law does not require municipalities and electric cooperatives, which roughly account for more than 20% of ERCOT load, to offer competition. However, they are allowed to opt in to retail competition. At the time of this writing, there has been a single cooperative to join retail competition, Nueces Electric Cooperative. In addition, retail competition has not been established in areas of Texas that remain outside ERCOT. Given that, less than 70% of the state's load enjoys retail competition.

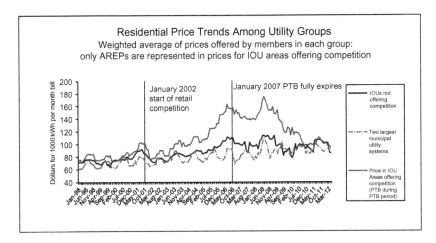

FIGURE 10.4 Historical retail prices offered by various load-serving entities 1998–2012 *Source: The* Monthly Retail Electric Service Bill Comparison for Residential Electric Service *data on the PUCT's web site available at: http://www.puc.texas.gov/industry/electric/rates/ RESbill.aspx (accessed 06.04.2013).*

competitive areas back into line with prices offered by the largest municipal systems and the vertically integrated utilities outside of ERCOT.

2.1 Legislative and Regulatory Activities

The Texas legislature has resisted temptations to overhaul the 1999 restructuring law. However, the legal frameworks through which the competitive markets operate continue to be updated.

The legislators established an Independent Market Monitor (IMM) in 2005, to be funded by an ERCOT fee and remain under Commission oversight.[10] In August 2006, the Commission selected a consulting firm through a competitive solicitation process to assume the duties of the ERCOT IMM.[11] Senate Bill 20 was also passed in 2005 to increase the state target for renewable resources to 10,000 MW by 2025.[12] In addition, the legislature

---

10. Senate Bill 408, *An Act Relating to the Continuation and Functions of the Public Utility Commission of Texas; Providing a Penalty,* was passed in 2005 during the 79th Session of the Texas Legislature.

11. The Commission's Market Oversight Division (also known as MOD Squad) was responsible for market monitoring duties from August 2000 to August 2006 when this responsibility was transferred to Potomac Economics.

12. Senate Bill 20, *An Act Relating to State's Goal for Renewable Energy,* was passed in 2005 during the 79th Session of the Texas Legislature. The 2025 target for renewable resources was exceeded by 2012.

passed Senate Bill 1125 in 2011 to increase the contribution of energy efficiency to meeting the state's growing electricity demand.[13]

In 2006, the Commission finalized two important market-related rules, SR 25.504 on market power and SR 25.505 on resource adequacy, establishing definitions for market power and market power abuse. In addition, SR 25.504 established a safe harbor whereby market participants with a smaller than 5% share of ERCOT-wide ICAP are deemed not to have market power. It also allowed market participants, including those with a larger share of ICAP, to file voluntary mitigation plans (VMPs) to ensure compliance with the rule.[14] To complement the market power rulemaking, the Commission finalized SR 25.505, setting an energy-only market system as the appropriate resource adequacy mechanism to address growing demand for electricity.[15] The rule tripled the existing wholesale market offer cap from $1000 to $3000 per MWh for energy or per MW per hour for ancillary service capacity.[16] Furthermore, market transparency was significantly enhanced by requiring most of the market and resource-specific information to be considered nonconfidential after 60 days.[17,18]

Meanwhile, the Commission also took steps to address retail electric services, enhance customer protection, and facilitate load response. SR 25.107, Certification of REPs, raised the financial and technical skills and requirements for REPs; SR 25.130 required the installation of advanced metering infrastructure for residential and small commercial customers that will be completed by the end of 2013; SR 25.475 enhanced information disclosure to small customers; SR 25.483 imposed specific requirements before a service disconnection could take place; SR 25.498 introduced prepaid services; and SR 25.507 provided opportunities for load to participate in emergency response service.[19]

---

13. Senate Bill 1125, *An Act Relating to Energy Efficiency Goals and Programs, Public Information Regarding Energy Efficiency Programs, and the Participation of Loads in Certain Energy Markets*, was passed in 2011 during the 82nd Session of the Texas Legislature.

14. See the Commission Substantive Rule 25.504, Wholesale Market Power in the Electric Reliability Council of Texas Power Region, available at: https://www.puc.state.tx.us/agency/rulesnlaws/subrules/electric/Electric.aspx (accessed 06.04.2013).

15. See Schubert et al. (2006).

16. As will be discussed in Section 3, the system-wide offer cap was raised again in June 2012 to $4500 effective on August 1, 2012.

17. See the Commission Substantive Rule 25.505, Resource Adequacy in the Electric Reliability Council of Texas Power Region, available at: https://www.puc.state.tx.us/agency/rulesnlaws/subrules/electric/Electric.aspx (accessed 06.04.2013).

18. See Adib et al. (2008) and Schubert et al. (2009).

19. For more details on each substantive rule, see the Commission Substantive Rules at: https://www.puc.state.tx.us/agency/rulesnlaws/subrules/electric/Electric.aspx (accessed 06.04.2013).

Since the early days of competition, the ERCOT market has not been immune from the concerns regarding market power abuse and manipulation.[20] Luminant—formerly known as TXU—and NRG Energy each account for about 15–20% of ERCOT's total generating capacity.[21] In 2004 and early 2005, the Commission Staff raised concerns that prices offered by Luminant were significantly above its marginal cost.[22] Similar behavior continued during 2005, particularly between June and September, resulting in higher than competitive prices in the ERCOT wholesale market.[23] The Commission Staff filed a Notice of Violation against Luminant for market manipulation and recommended an administrative penalty of $170 million.[24] The case was settled in December 2008, resulting in Luminant paying a $15 million penalty without acknowledging any wrongdoing. Luminant has since been operating under a VMP approved by the Commission.[25] In addition, the Commission recently approved two VMPs for NRG[26] and IPR-GDF Suez Energy[27] and there is currently one additional request for approval of VMP pending the Commission's approval.[28]

In response to Senate Bill 20 in 2005, the Commission passed SR 25.174, Competitive Renewable Energy Zone (CREZ) in 2009 facilitating expansion of transmission facilities.[29] The contentious proceeding that followed resulted in a final decision by the Commission to approve a plan to allow 10 Transmission and Distribution Service Providers (TDSPs) to construct transmission capable of transferring generation from a total of up to 18,456 MW of wind resources to major population centers, as identified in Figure 10.5. The original cost was estimated to be $4.93 billion,[30] but is now expected to

---

20. See Adib and Hurlbut (2008).

21. Texas law limits total installed capacity owned by each entity to only 20%.

22. See Hurlbut et al. (2004) and Potomac Economics (2005).

23. See Potomac Economics (2006).

24. See Potomac Economics (2007). "Investigation of the Wholesale Market Activities of TXU from June 1 to September 30, 2005," Potomac Economics, IMM, March. The report was further revised in September 2007 to reflect lower figures for profit for $19 million and rising prices in balancing energy market by $57 million. The Notice of Violation issued to TXU by the PUCT can be found on the Commission web site under Docket No. 34061.

25. PUCT, Docket No. 34480, *TXU Wholesale Companies' Request for Approval of a Voluntary Mitigation Plan Pursuant to Subst. R. §25.504(e)*.

26. PUCT, Docket No. 40488, *Request for Approval of Voluntary Mitigation Plan for NRG Companies Pursuant to PURA §15.023(F) and PUC Subst. R. §25.504(e)*.

27. PUCT, Docket No. 40503, *Request for Approval of Voluntary Mitigation Plan for IPR-GDF Suez Energy Marketing North America, Inc. Pursuant to PURA §15.023(F) and PUC Subst. R. §25.504(e)*.

28. PUCT, Docket No. 40545, Petition of Calpine Corporation for Approval of Voluntary Mitigation Plan.

29. For details, see the Commission Substantive Rules at: https://www.puc.state.tx.us/agency/rulesnlaws/subrules/electric/Electric.aspx (accessed 06.04.2013).

30. For more information, see the final order issued by the Public Utility Commission of Texas on August 15, 2008 at PUCT web site under Docket No. 33672.

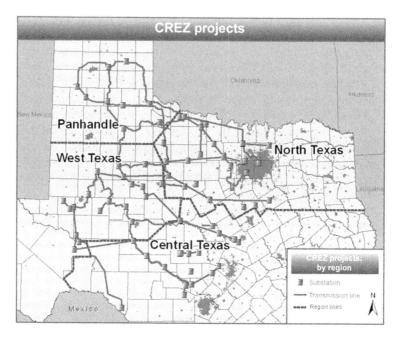

FIGURE 10.5   ERCOT competitive renewable energy zones transmission projects. *Source: PUCT web site at: http://www.texascrezprojects.com/projects.aspx (accessed 06.04.2013).*

approach $7 billion or more.[31] The TDSPs are expected to complete the new transmission lines by the end of 2013.[32]

The decision by the Texas legislation to increase the target for renewable resources and steps taken by the Commission to enhance renewable resource development, along with various federal incentives to increase renewable resources, have resulted in more than 10,000 MW of renewable resource capacity within ERCOT and is expected to increase as reflected in Figure 10.6.

Finally, in response to Senate Bill 943 in 2011,[33] the Commission has recently amended its SR 25.501 to enhance the diffusion and operation of energy storage facilities. This is an attempt to complement the operation

31. See PUCT Quarterly CREZ Progress Report No. 7, April 2012 update, prepared by RS&H. Also available at: http://www.texascrezprojects.com/quarterly_reports.aspx (accessed 06.04.2013).
32. For more information, see the final order issued by the PUCT on March 11, 2010 at PUCT web site under Docket No. 37902.
33. Senate Bill 943, An Act Relating to the Classification, Use, and Regulation of Electric Energy Storage Equipment or Facilities, was passed in 2011 during the 82nd Session of the Texas Legislature.

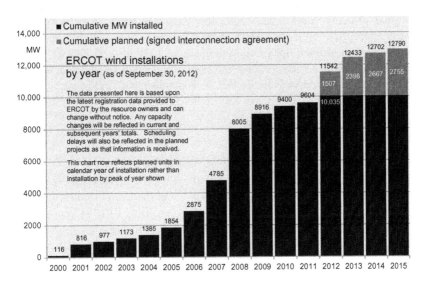

FIGURE 10.6 ERCOT wind capacity installed and future projection. *Source: ERCOT system planning monthly status report (September 2012).*

of an increasing amount of intermittent renewable resources capacity in ERCOT. To establish a level playing field, storage facilities are treated like generation facilities, and not subject to retail tariffs, rates, and charges or fees assessed in conjunction with the retail purchase of electricity.[34]

2.2 Efficiency and Operational Improvements

Since ERCOT began zonal market operation as a single control area on July 31, 2001, operational problems and high congestion costs were frequently observed. The Market Oversight Division (MOD) of the Commission, which was also responsible for ERCOT market monitoring activities, identified the shortcomings of the existing zonal design and recommended direct allocation of local congestion costs to address inadequacy of pricing signals in the ERCOT wholesale market. MOD summarized the only viable solutions as either directly "assigning local congestion costs within the existing ERCOT zonal model or conversion to nodal energy pricing LMP" (locational

---

34. See the Commission Substantive Rule 25.501, Wholesale Market Design for the Electric Reliability Council of Texas, available at: https://www.puc.state.tx.us/agency/rulesnlaws/subrules/electric/Electric.aspx (accessed 06.04.2013).

marginal pricing). Staff further indicated that "assigning local congestion fees is appropriate either as a permanent solution or as an interim approach should the Commission decide to implement an LMP at a later time."[35] The Commission finalized its decision for such a transition to nodal design by approving its Substantive Rule 25.501 in September 2003. However, it took a long time before ERCOT began full implementation of nodal operations on December 1, 2010.[36]

2.2.1 Transition to Nodal Market Operation

The transition to a nodal operation for the ERCOT market differed from other restructured wholesale electricity markets in the United States. In addition to the length of time that it took for such transition, ERCOT spent a significant amount of money for its market design. ERCOT first estimated a completion date of December 2008 and a project cost of $125 million.[37] In November 2008, ERCOT provided a revised plan to fully implement nodal operation by December 1, 2010 at an estimated project cost of $660 million.[38] Finally, the nodal operation began on December 1, 2010 at a cost around $550 million.[39]

ERCOT had the option to take off-the-shelf market design software in use in other markets for a much lower cost; however, to obtain consensus and satisfy some smaller electric cooperatives and municipal utilities, it decided to customize its market design and accept much higher total costs.[40] As a result, ERCOT's market design has noticeable differences from other markets in the United States.[41] For example, ERCOT's day-ahead market is voluntary and there are no negative consequences if some

---

35. Market Oversight Division, Staff Comments on Issues Related to the Transmission Congestion, Public Utility Commission of Texas, September 6, 2002.
36. See Schubert and Adib (2009) for a more detailed discussion on the evolution of market design in ERCOT.
37. See memo from Ron Hinsley, Chief Information Officer, to the ERCOT Board, listed under Item 6, Market Redesign Report, dated April 11, 2006. Also available at: http://www.ercot.com/calendar/2006/04/20060418-BOARD (accessed 06.04.2013).
38. See ERCOT November 26, 2008 News Release at: http://www.ercot.com/news/press_releases/show/281 (accessed 06.04.2013).
39. See presentation by Mike Petterson, ERCOT Controller, to the ERCOT Board, listed under Item 10a, dated January 18, 2011. Also available at: http://www.ercot.com/calendar/2011/01/20110118-BOD (accessed 06.04.2013).
40. There is still a question on whether it was a wise move to customize as opposed to taking an off-the-shelf software. The future performance and overall benefits may shed light on this question.
41. Some of the materials used in preparation of this discussion are based on personal communication by one of the authors with ERCOT Staff, Kenneth Ragsdale, dated May 24, 2012.

market participants decline to participate in that market besides the fore-gone opportunity costs. In contrast to other markets, ERCOT market design allows for:

- A unique market mitigation mechanism called Texas Two-Step, which is internalized in its security constrained economic dispatch (SCED), to address concerns regarding market power abuse.
- Unrestricted point-to-point (PTP) obligation bidding in the day-ahead market where market participants can bid between any two settlement points as long as they are not electrically similar.
- Unrestricted virtual offers/bids in the day-ahead market at any settlement point.
- Implementation of options for congestion revenue right (CRR) offers in the day-ahead market.
- High penetration of wind resources with the introduction of Voltage Stability Analysis Tool/Transient Security Analysis Tool (VSAT/TSAT) to set up the limits in real time to maximize utilization of the network and prevent artificial bottlenecks, such as conservative limit setting, in the transmission corridors from high concentration of wind in the west to load centers in north and south.

Furthermore, ERCOT enjoys state-of-the-art software unique to its mar-ket, such as the network model management system (NMMS).[42] NMMS, which is based on weekly model loads, provides a centralized common data-base to generate models for all processes including energy management sys-tem, market management system, system planning, outage scheduling, and congestion revenue rights auction.[43]

Some benefits have been reported since the beginning of nodal operation including immediate relief for ERCOT system operators and reduction in transmission-related curtailment of West Texas renewable resources.[44] As reported by the ERCOT IMM, "Overall pricing outcomes from the nodal real-time market have met expectations for improved efficiency." (Potomac Economics, 2012). A separate ERCOT back-cast analysis concluded that custo-mers could have saved \$90−180 million in their load charges if nodal operation

---

42. Trip Doggett, ERCOT President and CEO, CEO Update, A Presentation to the ERCOT Board, July 17, 2012, Slide 15. Also available under Item 4 at: http://www.ercot.com/calendar/2012/07/20120717-BOD (accessed 06.04.2013).

43. Some of the materials used in preparation of this discussion are based on personal communi-cation by one of the authors with ERCOT Staff, Kenneth Ragsdale, dated on May 24, 2012.

44. See presentation by Dan Jones, Director, ERCOT IMM, listed under Item 11, dated January 18, 2011. Also available at: http://www.ercot.com/content/meetings/board/keydocs/2011/0118/Item_11_-_IMM_Report.pdf (accessed 06.04.2013).

had been in place in 2008.[45] The nodal operation also allowed for better integration of intermittent resources and higher utilization of transmission facilities through more accurate calculation of their real time available transfer capabilities.[46] Other benefits are expected over time and it will be helpful to perform a detailed study after a few years to assess the operational and economic efficiencies realized from the transition to nodal operation.

2.2.2 Gradual Enhancement of Retail Operation

At the launch of retail customer choice, ERCOT experienced a great deal of difficulty in performing many of the retail customer service and billing activities that the vertically integrated utilities had performed so well. Some of the tasks that proved formidable to ERCOT in its role as Central Registration Agent included assigning account numbers Electric Service Identifier (ESI IDs) to new premises, customer switching, collection and distribution of billing data, and disconnection and reconnection of service. After a very difficult transition period, however, placing ERCOT into the role of Central Registration Agent appears to have succeeded in establishing trust in the system and reduced potential barriers to entry for REPs who are interested in competing in the ERCOT market.

As indicated by Young Kim in Chapter 12 in this volume, "Texas is the most active and competitive market in the United States as measured by a number of competitors, persistent migration levels, and total load migrated." As of August 10, 2012, ERCOT has completed about 7.7 million requests for customer switches to various REPs.[47] In 2010, 86 REPs were active.[48] Presently, about 40 REPs are in competition for residential loads, with 52 REPs serving at least 500 residential customers.[49,50] CREPs, or "nonlegacy retailers," serve over half of total sales in ERCOT.[51] Some REPs abandoned the retail market even before competition was fully initiated, including Enron, due to its bankruptcy, and Shell. American Electric Power Company

---

45. See a presentation by Mike Cleary, ERCOT Sr. Vice President and Chief Operating Officer, to the ERCOT Board, listed under Item 6, Market Benefits Observed: The First Six Months, dated June 21, 2011 (Revised July 5, 2011). In estimating savings in regulation services, ERCOT excluded the impact of an ice storm in the first few days of February 2011. The presentation shows that many expected benefits of nodal operation have already materialized, such as better locational pricing signals, convergence between day-ahead and real-time prices, and improved market transparencies. Also available at: http://www.ercot.com/calendar/2011/06/20110621-BOD.
46. See Potomac Economics (2012).
47. For more statistics on ERCOT retail electricity market, including Switch Status Trend Data, see: http://www.ercot.com/mktinfo/retail/ (accessed 06.04.2013).
48. PUCT (2011), Scope of Competition, p. 49.
49. See www.powertochoose.com (accessed 06.04.2013).
50. PUCT (2011), Scope of Competition, p. 49.
51. PUCT (2011), Scope of Competition, p. 53.

TABLE 10.2 Notable Bankruptcies of Texas REPs

Retailer	Date of Default on Obligations to ERCOT	Number of Affected Retail Accounts
Texas Commercial Energy	March 6, 2003	Unknown
Utility Choice Electric	2004	Unknown
Ampro	2006	Unknown
Buy Energy	2006	Unknown
Franklin Power Company	June 8, 2005	<3000
Blu Power of Texas	July 1, 2008	2092
Hwy 3 MHP	June 3, 2008	12,222
Sure Electric (Riverway Power Company)	June 10, 2008	6202
Pre-Buy Electric	May 16, 2008	8400
National Power Company	May 27, 2008	15,163
Abacus Resource Energy	February 9, 2011	7743
EPCOT Electric	July 12, 2012	5736

Source: ERCOT web site, various News Releases at: http://www.ercot.com/news/press_releases/ (accessed 06.04.2013).

also decided to transition out of the retail market before retail choice was fully initiated—most of its remaining retail customers were switched to Direct Energy. AES sold its retail operations to Constellation. Xcel Energy Retail Services left the market after selling its customer base to MPower. Some of the most recent merger and acquisition activities in the last few years included the purchase of Reliant Energy and Green Mountain Energy by NRG Energy. Also, Exelon, a major national player, bought Constellation New Energy, who in turn bought Star Tex. There have also been a few bankruptcies, as noted in Table 10.2.

Texas is among the leading states to implement smart meters for all residential and small commercial customers and is expected to have over 6.0 million smart meters by the end of 2013. More importantly, ERCOT is the only restructured market in the United States that has integrated smart meters into its wholesale market 15 min settlement process.[52] These new

52. According to an operations report by ERCOT CEO, Trip Doggett, to the Board of Directors on May 15, 2012, about 87.4% of ERCOT load is settled with 15 min interval data, representing about 5.3 million smart meters as of June 20, 2012. Also available at: http://www.ercot.com/calendar/2012/07/20120717-BOD, under Item 5, Operations Report (accessed 06.04.2013).

investments in advanced metering systems by the TDSPs provide a platform for future dynamic pricing programs and other value-added services, as well as better outage detection and customer switching capabilities.

While Texas enjoys the highest rate of customer switching among all restructured retail markets, the affiliate REPs still enjoy a sizable market share.[53] The top three retailers account for about 75% of total sales to residential customers in 2010.[54] The same three retailers dominate retail sales to nonresidential customers accounting for about 47% in 2010.[55]

Retailers associated with vertically integrated utilities which were unbundled as a result of restructuring still have close to half of the retail market share, if NRG Energy and Direct Energy—which acquired CenterPoint's and AEP's retail operations, respectively—are treated as an AREP. Yet, a variety of entities not present in this market prior to retail competition have gained a foothold, particularly, major national entities, such as GDF Suez, Constellation NewEnergy, EDF Industrial Power, Nobel Americas Energy, and NextEra, who have gained more than 30% of total retail sales to nonresidential customers.[56]

2.2.3 Utilizing Interruptible Loads as a Resource

Before restructuring, the ERCOT utilities reported over 3100 MW of interruptible load.[57] The PUCT recognized the need to retain the ability to use interruptible loads. In its order conditionally approving ERCOT's proposed protocols in 2001, the PUCT ordered ERCOT to "Develop additional measures and refine existing measures, to enable load resources a greater opportunity to participate in the ERCOT markets."[58]

The design of ERCOT's wholesale market permits "load resources" to compete "head-to-head" against generation resources to provide ancillary services, such as Responsive Reserve Service, which can be provided by interruptible loads with under-frequency relays, and Non-Spinning Reserve Service, which can be interrupted by the ISO with 30 min of notice. load resources that are not self-arranged can be offered into ERCOT's day-ahead

---

53. The percentage of competitive load served by nonaffiliated REPs in ERCOT as of the end of July 2012 was 70.4%, reflecting 58.6% for Residential, 82.7% for Small Nonresidential, and 80.0% for Large Nonresidential. Available at: http://www.ercot.com/content/mktinfo/retail/kd/ Switched_Premise_Report_EOM_2012_07.ppt (accessed 06.04.2013).

54. This represents 36.6% by TXU Energy, 26.9% by NRG Energy, and 11.3% by Direct Energy. For more details, see Kim in this volume.

55. This represents 18.5% by NRG Energy, 17.5% by TXU Energy, and 10.5% by Direct Energy. For more details, see Kim in this volume.

56. For more details, see Chapter 12 by Young Kim in this volume.

57. Project No. 22209: PUCT Market Oversight Division, *2000 Annual Update of Generating Electric Utility Data*, December 2000, p. 5.

58. PUCT, *Final Order in Docket No. 23220: Petition of the Electric Reliability Council of Texas for Approval of the ERCOT Protocols.*

market for ancillary services and, if selected, receive the market-clearing price for capacity.

The integration of load resources into the market for Responsive Reserves has been quite successful. As of May 2012, 207 load resources, working with over 10 scheduling entities, are qualified to provide ancillary services for a total capacity of 2500 MW.[59] The amount of load qualified to provide Responsive Reserves is well in excess of the limit on Load Resource participation in this market.[60] Load resources have been instrumental in preserving reliability, and typically two interruptions occur per year in response to under-frequency events. While there has been a good mix of load resources by size, it is noteworthy that about one-half of the total quantity of load resources is provided by five very large industrial loads.

After many years of debate, a new Emergency Interruptible Load Service (EILS) was launched in mid-2008. Under this program, interruptible loads, which are not otherwise providing an operating reserve, receive a bid-based payment for curtailing their purchases of electricity from the grid when ERCOT is in an emergency. The program was expanded to include small-scale distributed generation and aggregated load and was renamed Emergency Response Service in March 2012.[61]

3 THE REMAINING CHALLENGES

While the ERCOT market has achieved significant improvements during its first decade of operation, much work remains. The noticeable challenges include promoting demand response (DR), implementing effective scarcity pricing, improving incentives for investment, managing the integration of an increasing amount of wind resources, and maintaining consumer trust in the retail market, topics briefly described below.

3.1 Demand Response

DR is essential for the success of any competitive electricity market and ERCOT market has a lot to do in the next few years to improve its level of DR participation. Lack of adequate DR and of representation of demand willingness to pay into price formation should be considered the main problems in the ERCOT market. More effective participation by demand resources and price-responsive loads in the ERCOT wholesale electricity

---

59. Paul Wattles, Aggregations of Small Customers as Load Resources, Presentation by ERCOT Staff, May 24, 2012.
60. ERCOT procures 2,800 MW of Responsive Reserve Service for each hour and Load Resources can provide up to 50% of the total amount.
61. See the Commission Substantive Rule 25.507, Electric Reliability Council of Texas (ERCOT) Emergency Response Service (ERS), available at: https://www.puc.state.tx.us/agency/rulesnlaws/subrules/electric/Electric.aspx (accessed 06.04.2013).

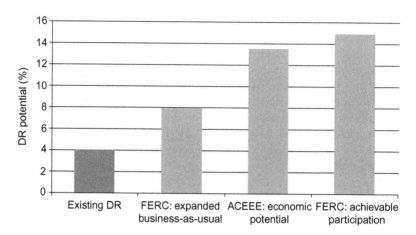

FIGURE 10.7 ERCOT DR peak demand reduction potential. Note: Existing DR estimated by Brattle to include large industrial price response. American Council for an Energy-Efficient Economy (ACEEE) economic potential is adjusted from the figure in ACEEE report to include only DR potential and exclude impacts from other sources. *Source: The Brattle Group (2012), Figure 30, p. 91.*

market could further enhance market operation, reduce the potential for market power abuse, and resolve the issues with scarcity pricing that have been blamed for discouraging capacity investment in ERCOT in recent years. It would also enhance the operation and effectiveness of ERCOT's energy-only resource adequacy mechanism that is under serious scrutiny in these days.[62] In a recent filing before the PUCT, the Brattle Group estimated the amount of "3-year Realistic Incremental DR Potential" to be between 1803 and 6177 MW.[63] Figure 10.7 shows potential DR for peak reduction in ERCOT market.

Under the zonal market structure, balancing energy prices were announced to the market roughly 10 min in advance of each 15 min settlement interval. This provided an opportunity for large industrial loads and load-serving entities at least partly exposed to balancing energy prices to reduce consumption during price spikes and for fast-start generators to come online and contribute to supply. The amount of DR to price spikes was approximately 600 MW—a modest, yet significant, response (Zarnikau, 2009; Zarnikau and Hallett, 2007).

The 10 min advance notice of prices was eliminated when the nodal market was established, in order to minimize short-term forecasting error and

62. The Brattle Group (2012) lists adequate and effective demand response as a prerequisite for success of an energy-only market to succeed in attracting enough investment to meet growing demand for electricity in ERCOT.

63. For more details, see the filing by the Brattle Group titled "Estimate of DR Potential in ERCOT," before the PUCT under Project No. 40000.

deter loads from deviating from scheduled or forecasted consumption levels (Zarnikau, 2009). Further, under nodal market structure, prices generally change every 5 min. Yet settlement intervals remain 15 min in duration and prices are announced at the start of each 5 min period. As a result, there is no advance notice of settlement prices. Therefore, price-responsive loads have no idea about the level of prices they are facing before the early minutes of the last 5 min interval within each 15 min settlement. This has significantly diminished the effectiveness of price-responsive loads and resulted in complaints by some industrial customers (Zarnikau and Adib, 2011).

As discussed further below, some important refinements in market design are in progress to enhance participation by demand resources and price-responsive load in the ERCOT market. ERCOT has refined its software to provide nonbinding price forecasts for each 5 min of the following 60 min.

Contrary to expectations, investment in advanced metering infrastructure has not yet led to large-scale residential sector DR. Extra steps have to be taken to facilitate aggregated load responses for customers with smart meters, which include residential and smaller commercial customers with loads smaller than 700 kW. REPs have been slow to offer retail dynamic pricing programs or other energy management services to these smaller customers, using the new metering infrastructure. While curtailable residential loads have the potential to provide ancillary services, ERCOT continues to grapple with how to reflect distributed demand-side resources in its network model and how to monitor and evaluate the performance of aggregations.

Both the Commission and ERCOT have been talking about the importance of DR since the opening of the competitive market in 2002. The challenge, however, is to make the transition from talking to taking actions that result in a clear plan with concrete and measurable results to achieve an adequate level of demand resources and load response in the next 3−5 years. Unfortunately, no specific plan is in place yet. However, the actions by the Commission and ERCOT in the next several months will demonstrate their commitment to achieving an adequate amount of DR for ERCOT market.

3.2 Resource Adequacy

Resource adequacy is another important and urgent concern facing the ERCOT competitive electricity market. The energy-only mechanism in place in the ERCOT market has not been able to attract adequate capacity expansion in recent years. But ERCOT is not alone. Most existing restructured electricity markets have experienced challenges in attracting sufficient capacity.[64]

---

64. For detailed discussion of energy-only and ICAP resource adequacy mechanisms see Adib et al. (2008) and Bowring (2008). Additional discussion could be found in Bowring in this volume.

Most restructured markets have an ICAP market where generators receive capacity payments for obligations to provide their capacity to serve demand for electricity. Some of these markets require retailers to contract for capacity to meet their demand and a mandated reserve margin to ensure resource adequacy. Markets operated by Pennsylvania–New Jersey–Maryland (PJM), New York Independent System Operator (NYISO), Independent System Operator in New England (ISO-NE), Midwest Independent System Operator (MISO), and California Independent System Operator (CAISO) in the United States and the Ontario market in Canada are among restructured markets with some kind of capacity market (Adib et al., 2008).

In contrast, ERCOT has an energy-only resource adequacy mechanism, along with Alberta in Canada, and Australia markets. An energy-only resource adequacy mechanism is an economically superior approach to the extent that it is able to elicit the true willingness to pay for electricity, rather than rely on administrative estimates of needed capacity, resulting in the most efficient market-based amount of reserve margin to ensure reliability.

In contrast to markets which provide capacity payments, generators in energy-only markets have to rely on energy payments from their customers to recover their original investments and generate a return.[65] For energy-only mechanisms to be successful, effective scarcity pricing is needed to send accurate price signals, and the demand side of the market must significantly alter consumption levels in response to prices. Neither of these two factors is currently present in the ERCOT market.

For an energy-only mechanism to be sustainable, actual prices have to be high enough above the cost of electricity generation to allow investors to make enough money to justify their investment. Table 10.3 shows average historical prices in the ERCOT market by load zones. Similarly, Figure 10.8 shows average monthly all-in prices, which include uplifts and ancillary services, for ERCOT.

Given low prices that have further declined in recent years due to much lower natural gas prices, the peaker net margins for natural gas units in recent years have not been high enough to attract new capacity expansion.[66] Figure 10.9 shows the peaker net margin for the ERCOT market.

Resource developers acknowledge that 2011 was an exception when the peaker net margin reached about $130,000 per MW. However, they still argue that prices are not high enough to attract new investment, and they

---

65. While ERCOT administered wholesale market relies on energy-only mechanism, majority of power in ERCOT region is handled through bilateral contracts where suppliers include premiums over energy prices to recover their investment and generate a return.

66. The peaker net margin shows extra revenue generated by a gas turbine after accounting for its operation. The ERCOT IMM assumes a heat rate of 10,000 BTU per each KWh of generation using Houston Ship Channel natural gas prices and a net revenue in 2011 for a new gas turbine ranging from $80 to $105 per KW per year. See Potomac Economics (2012).

TABLE 10.3 ERCOT Historical Real-Time Zonal Prices

	Average	Real Time	Electricity	Price
	2008	2009	2010	2011
ERCOT	$77.19	$34.03	$39.40	$53.23
Houston	$82.95	$34.76	$39.98	$52.40
North	$71.19	$32.28	$40.72	$54.24
South	$85.31	$37.13	$40.56	$54.32
West	$57.76	$27.18	$33.76	$46.87
Natural gas	$8.50	$3.74	$4.34	$3.94

Source: Potomac Economics (2012), p. ii.

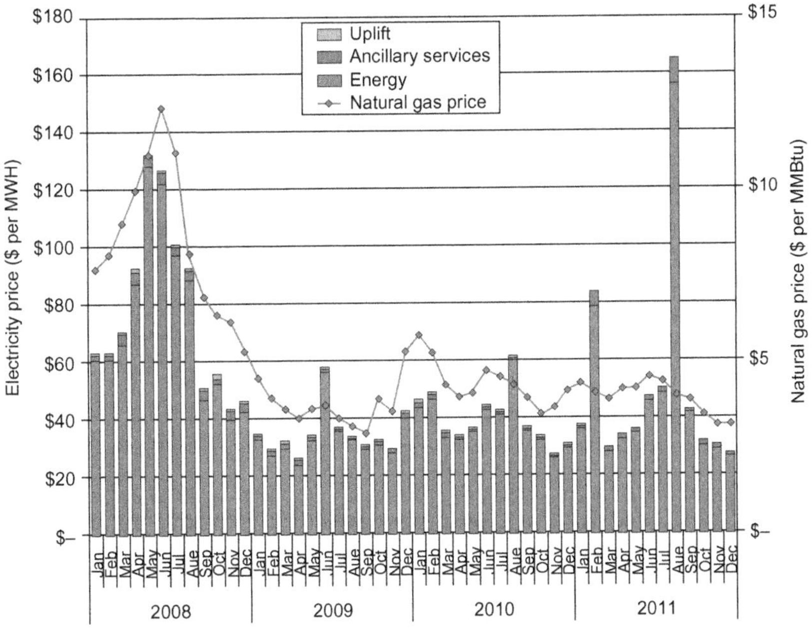

FIGURE 10.8   All-in price for electricity in ERCOT. *Source: Potomac Economics (2012), p. ii.*

question the effectiveness of ERCOT's scarcity pricing mechanism. In particular, resource developers refer to frequent out of market decisions by ERCOT operators to address reliability—including the frequent deployment of nonspinning reserves—which have suppressed real-time prices despite scarcity in the market justifying much higher prices.

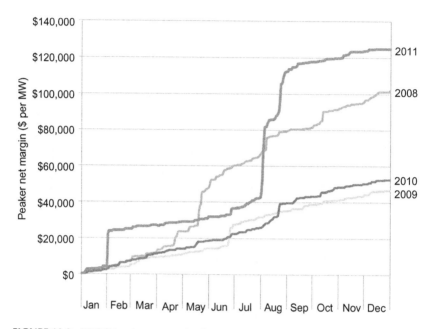

FIGURE 10.9   ERCOT peaker net margin. *Source: Potomac Economics (2012), p. xxii.*

Through the stakeholder process, ERCOT has taken steps to address some of the concerns by automatically setting prices at the system-wide offer cap, which was raised by the Commission to $4500[67] per MWh, during scarcity conditions effective on August 1, 2012. These occasions include circumstances when operating reserves—such as responsive reserves or regulation up—are deployed,[68] and energy is generated by units selected by reliability unit commitment (RUC) process[69] or for reliability must-run (RMR) services[70] that generate

---

67. See Rulemaking Project No. 37897, *PUC Proceeding Relating to Resource and Reserve Adequacy and Shortage Pricing*, Final Order on June 28, 2012.
68. The ERCOT Board approved NPRR 427 (*Energy Offer Curve Requirements for Generation Resources Assigned Regulation Up and Responsive Reserve Service*) in its December 2011 meeting. For more details, see: http://www.ercot.com/mktrules/issues/nprr/426-450/index (accessed 06.04.2013). Similarly, ERCOT protocols, Section 6, could be consulted for more details. Available at: http://www.ercot.com/mktrules/nprotocols/current.html (accessed 06.04.2013).
69. The ERCOT Board approved NPRR 435 Requirements for Energy Offer Curves in the Real-Time SCED for Generation Resources Committed in RUC in its February 21, 2012 meeting. For more details, see: http://www.ercot.com/mktrules/issues/nprr/426-450/index (accessed 06.04.2013). Similarly, ERCOT protocols, Section 6, could be consulted for more details. Available at: http://www.ercot.com/mktrules/nprotocols/current.html (accessed 06.04.2013).
70. The ERCOT Board approved NPRR 442 (Energy Offer Curve Requirement for Generation Resources Providing Reliability Must-Run Service) in its May 15, 2012 meeting. For more details, see: http://www.ercot.com/mktrules/issues/nprr/426-450/index (accessed 06.04.2013). Similarly, ERCOT protocols, Section 6, could be consulted for more details. Available at: http://www.ercot.com/mktrules/nprotocols/current.html (accessed 06.04.2013).

above their low sustainable limits. Similarly, minimum offer prices—$120 per MWh for online and $180 per MWh for off-line units—are set for energy deployed from units awarded for nonspinning reserves.[71]

As a longer term solution, Staff proposed raising the system-wide offer cap by 2015 up to $9000.[72] Unsurprisingly, generators supported an increase in the system-wide offer cap as early as possible. In contrast, nonaffiliated retailers, consumer advocates, and several legislators questioned the appropriateness of increasing the system-wide offer cap, particularly for the summer of 2012.[73]

Under ERCOT's sponsorship, the Brattle Group (2012) finalized a report in June 2012, providing options to address long-term concerns about resource adequacy. Later on, the Brattle Group tailored its policy recommendations into the following two "Composite Policy Options" filed on October 19, 2012[74]:

1. Energy-only market with support for DR
2. Texas capacity market

The first policy option maintains the current energy-only resource adequacy mechanism, further enhances and relies on auctions to procure demand resources, and, if needed, increases the amount of operating reserves to ensure the desired level of reliability sought by Texas regulators.[75] The second policy option closely resembles the ICAP mechanism in place in PJM market.

The ICAP resource adequacy mechanism is inconsistent with the current philosophy underpinning the ERCOT competitive market and ruins the "effective pricing signal" which is necessary for the energy market to work efficiently. This is because advanced payments to generators take a big portion of price and turn it into a fixed payment. The result will be a much

---

71. The ERCOT Board approved NPRR 428 (*Energy Offer Curve Requirements for Generation Resources Assigned Non-Spin Responsibility*) in its December 2011 meeting. For more details, see: http://www.ercot.com/mktrules/issues/nprr/426-450/index (accessed 06.04.2013). Similarly, ERCOT protocols, Section 6, could be consulted for more details. Available at: http://www.ercot.com/mktrules/nprotocols/current.html (accessed 06.04.2013).
72. See Project No. 40268, *PUC Rulemaking to Amend PUC Subst. R. §25.505, Relating to Resource Adequacy in the Electric Reliability Council of Texas (ERCOT) Power Region.*
73. See documents under PUCT rulemaking Project numbers 37897 and 40268. In particular, see comments by the Texas Competitive Power Advocates (TCPA) and Group of Competitive Texas Power Suppliers.
74. The Brattle Group filed a presentation for the October 25 Resource Adequacy Workshop under the PUCT Rulemaking Project No. 40000. The presentation begins by acknowledging that its recommendations are not based on an already approved "minimum acceptable reserve margin.
75. This Composite Policy Option reflects the first three options provided in the Brattle Group's June 2012 Report refined by some comments from market participants.

lower energy price which is not effective in attracting real market-based DRs.

Similarly, while there may be some desire to maintain a high-reserve margin to ensure a desired high level of reliability, with demand-side response setting the price occasionally, the competitive market will optimize the amount of reserve margins and avoid creating undesirable excess capacity. These results are market-based and economically efficient.

The ICAP mechanism may be a good transitional step from an existing regulatory regime to a real competitive market. However, for ERCOT, which has been operating a competitive market for more than 10 years, a move to an ICAP mechanism should be considered a backward move and should be avoided.

On October 25, 2012, the Commission finalized its decision to maintain energy-only resource adequacy mechanism. However, to enhance its effectiveness, the Commission approved increasing statewide offer cap to $5000 MWh or per MW per hour in June 2013, $7000 in June 2014, and $9000 in June 2015. In addition, the cap on peaker net margin was raised to $300,000 with the understanding that it will remain at three times of the cost of new entry for a gas turbine at 2014 and beyond.[76]

The authors agree that the most economically efficient approach for the ERCOT market is to maintain its energy-only resource adequacy mechanism and continue improving scarcity pricing. While the recent refinements will improve market operation and attract additional investment, the Commission and ERCOT must make a serious commitment to allowing market prices to reflect demand willingness-to-pay, which could result in occasional high prices but will significantly enhance load participation in this market. Such an approach would further support new investment when such investment is justified by market fundamentals. In particular, the following measures would be necessary:

- The Commission has to remain uncommitted to a particular level of reserve margin, including the current 13.75% set for ERCOT.[77] The appropriate level of reliability and the resulting optimal target reserve margin should be determined by balancing the marginal cost of outage to customers against the marginal value of additional capacity needed to eliminate the outage in question (Adib, 1991).
- It is prudent for the Commission to monitor market performance and refine the system-wide offer cap in the future if needed.

---

76. See Project No. 40268, *PUC Rulemaking to Amend PUC Subst. R. §25.505, Relating to Resource Adequacy in the Electric Reliability Council of Texas (ERCOT) Power Region.*
77. ERCOT operated under a 15% reserve margin for most of its existence since the 1970s. In 2002 its reserve margin was reduced to 12.5% where it remained until November 2010 when it was raised to 13.75%. The current reserve margin is set by ERCOT to maintain a desired level of reliability which reflects not more than 1 day of outage per 10 years.

- Encourage prices to approach the Value of Lost Load (VOLL) under conditions of curtailment of demand or declining reserves. ERCOT is expected to complete a VOLL study in the near future.
- Eliminate "free calls" on capacity utilized as ancillary services by ERCOT. It is a general experience to see some uncommitted capacity in the ERCOT market that is nevertheless ready to startup and begin operation. Historically, ERCOT has called on such uncommitted resources under emergency situations called Energy Emergency Alerts. Uncommitted resources called under such circumstances are not fully compensated for the capacity that they were effectively providing to the market.
- Allow full and unlimited participation by loads in the real-time market.

If the above market-based approach fails to attract adequate extra capacity, the Commission could then require load-serving entities to procure enough capacity to meet their load and required reserve margin.[78] This is the approach taken by some state regulators in other restructured markets such as CAISO, MISO, and PJM. Taking that approach may ultimately result in an ICAP market similar to other markets in the northeastern United States.[79]

3.3 Significant Increase in Wind Capacity[80]

The integration of more than 10,000 MW of intermittent renewable resources into the ERCOT grid has resulted in planning and operational difficulties, as well as economic challenges. Negative real-time market prices can result from wind resources competing with each other to maintain their electricity production in order to obtain a federally authorized Production Tax Credit (PTC). While such negative prices were very frequent in the spring of 2008, their recent occurrences are limited to late night and very early morning hours when load is at its lowest level and wind is at peak production. Negative pricing is a transitional issue that has lost its teeth and will be eliminated when federal incentives, such as the PTC and Investment Tax Credit (ITC) for each resource, expire.[81]

Wind generation-related planning challenges are focused mainly on the need for transmission upgrades and expansion to facilitate transfer of a significant amount of electricity generation from renewable resources in remote areas to population centers. Texas has been very successful in this area and could be seen as a model for others to follow, although systematic

---

78. This approach is consistent with recommendations by some interested parties in their filing under Project No. 40000, *Commission Proceeding to Ensure Resource Adequacy in Texas*.

79. Some form of capacity market (ICAP) is administered by market operators in three northeastern restructured markets: NYISO, ISO-NE, and PJM.

80. For further discussion of economic and operational challenges resulting from increasing integration of renewable resources into electric grids see Chapters 5, 7, and 25 by Haas et al., Bauknecht et al., and Riesz et al. in this volume.

81. These subsidies for new resources may be extended by the US Congress.

transmission planning optimization methods would improve the planning process compared to the *ad hoc* trial-and-error approach utilized by ERCOT.

Wind generation-related operational difficulties have included reactive power support, active power control, forecasting and scheduling, and accurate determination of required ancillary services to ensure reliable grid operation when dealing with intermittent resources (Reid et al., 2009). To address these and similar operational difficulties, ERCOT initiated and finalized several protocol revision requests (PRRs) and imposed new requirements to better integrate huge wind generation into its operation.[82]

Wind production forecasting is significantly improved and wind resources are now required to submit the most up to date meteorological data for their sites and use ERCOT's frequently provided forecast of wind generation potential as their planned operating levels (Saathoff, 2009). In addition, wind resources are required to follow performance metrics as expected from other generation resources, and have certain operational capabilities including voltage ride-through, reactive power, and ramp rate limitations (Bruce, 2012).

3.4 Market Design Improvements: Nodal 2.0

ERCOT nodal operation began in December 2010 with two known major shortcomings: (a) lack of co-optimization of energy and ancillary services in the real-time market and (b) inability to optimize the operation of less flexible units under the proposed 5 min dispatch intervals (Patton, 2005). The first one results in suboptimal resource utilization. The second one results in an improper treatment of "less flexible" units due to the fact that 5 min dispatch intervals would result in problems in dispatching and pricing of gas turbines and other units that cannot be dispatched on a 5 min basis. Such units should not be treated as price takers on occasions when they are truly reflecting the marginal units and should set market-clearing prices.[83]

In addition, ERCOT has identified several more refinements that will be implemented in the next few years. As indicated by ERCOT Staff in a white paper (ERCOT, 2011), the goal is to address several existing limitations with the following strategies:

- Improve economic efficiency and system reliability by providing a short-term look-ahead projection of system conditions.
- Provide more efficient commitment and dispatch of quick-start generation resources, load resources and storage resources for up to two hours.

---

82. The most important PRRs included PRR763, PRR771, PRR773, PRR794, PRR811, PRR812, PRR824, and PRR841. More details could be found at: http://www.ercot.com/mktrules/protocols/prr (accessed 06.04.2013).

83. See Direct Testimony of David Patton, *Proceeding to Consider Protocols to Implement a Nodal Market in the Electric Reliability Council of Texas Pursuant to P.U.C. Subst. R. § 25.501,* P.U.C.T. Docket No. 31540 (November 11, 2005).

- Allow load and storage resources to participate in the real-time market for energy and ancillary services.
- Allow real-time co-optimization of energy and ancillary services.
- Add capability to consider future topology changes, for example, outages.

ERCOT already has co-optimization of energy and ancillary services in its day-ahead market. The real-time co-optimization will allow consideration of the most recent information with regard to availability of resources, weather and fluctuations in demand for electricity, and the latest outages in transmission networks in order to optimize the use of available resources to meet electricity needs. This will result in more accurate pricing signals and better allocation of resources while minimizing total operating cost.

ERCOT has employed a look-ahead SCED mechanism similar to that currently in use by NYISO. The first phase was implemented in June 2012, which enables nonbinding price projections for each 5 min increment up to an hour into the future.[84]

Currently, load resources can participate in the day-ahead market. The improvement in market engine software could allow more active responses by loads during real-time operation. In fact, real-time load participation in SCED will allow loads to provide offers to curtail directly into ERCOT's dispatch model, which could result in better economic curtailments and price formation from explicit demand-side participation. If properly designed and implemented, this could provide opportunities for energy consumers who were not previously able to respond to prices to bid into SCED and curtail their loads at their preferred price levels.[85]

3.5 Other Wholesale Market Challenges

In spite of significant improvements, ERCOT has faced several other challenges, including market power over-mitigation, and inability to optimize the operation of less flexible gas units.

Through ERCOT's Texas Two-Step market power mitigation mechanism, constraints are divided into "competitive" and "noncompetitive" on annual and monthly bases using a procedure based on the Herfindahl–Hirschman index (HHI) and the shift factors of generators to constraints.[86] In step 1,

---

84. For more details see Nodal Protocol Revision Request 351 that was approved by the ERCOT Board in December 2011. Also available at: http://www.ercot.com/mktrules/issues/nprr/351-375/351/index (accessed 06.04.2013).

85. See Zarnikau and Adib (2011). Load participation in SCED was among five solutions recommended in this white paper.

86. As a threshold issue, and despite their pervasive use in assessments of market power in electricity markets, HHI-based metrics derive from antitrust law and have very little relevance to the normative assessment of competitiveness of electricity markets. See, for example, the discussion in Stoft (2002), Chapters 4–5, Lee et al. (2011a,b).

LMPs or reference prices for each resource node are determined subject to power balance and enforcement of the "competitive" constraints within the ERCOT transmission network.

In step 2, the higher of these reference prices and a mitigated offer cap (MOC) are used as a ceiling on each resource offer price to calculate the final LMPs subject to power balance, and both the "competitive" and "noncompetitive" constraints within the ERCOT transmission network.[87] The normative aim of such mechanisms should be to move prices toward competitive levels, including prices that reflect or approximate demand willingness-to-pay under shortage or scarcity conditions. Texas Two-Step mechanism has apparently resulted in occasions under certain circumstances when mitigation fails to produce competitive outcomes. The ERCOT IMM raised concerns and recommended solution with regard to unnecessary dispatch instructions issued to some units under certain circumstances that will result in oversupply of power and suppress prices.[88,89] As a solution, it is recommended to relax mitigation on resources which have no or minimal impact on noncompetitive constraints.[90]

More fundamentally, the Texas Two-Step will typically result in below-competitive prices in areas that are import constrained over "noncompetitive" constraints under the typical circumstance that the average cost to generate in such import-constrained areas is above the average of the mitigated prices. A move away from *ad hoc* market power measures toward assessment based on economic incentives would help to focus attention on the underlying issues of competition faced by market participants. Introduction of greater demand-side participation in SCED, including price setting by demand-side resources, will presumably reduce the need for market power mitigation and allow for demand and supply side resources to be relied upon more fully to set competitive prices.

3.6 Retail Market Challenges

In recent years, the challenges on the retail side have primarily focused on curbing deceptive trade practices among certain retailers, ensuring proper

---

87. ERCOT Protocols allow units that have previously agreed upon MOCs higher than their reference prices from the first phase to be paid their MOCs.
88. See IMM memo to the ERCOT Board of Directors Meeting on May 15, 2012, Item No. 7. Also available at: http://www.ercot.com/calendar/2012/05/20120515-BOD (accessed 06.04.2013).
89. See IMM presentation, over-mitigation in SCED, to the ERCOT Reliability Deployment Task Force (RDTF) on January 27, 2012, listed as IMM RTDF under Key Documents. Also available at: http://www.ercot.com/calendar/2012/01/20120127-RDTF (accessed 06.04.2013).
90. See related document presented to the ERCOT Congestion Management Working Group (CMWG) on May 8, 2012, Write Up on Over Mitigation under Key Documents. Also available at: http://www.ercot.com/calendar/2012/05/20120508-CMWG (accessed 06.04.2013).

consumer protections, and improving the accuracy of information provided to consumers. Credit requirements have been raised for REPs, posing a barrier to entry but providing greater assurance that poorly hedged retailers can weather spikes in wholesale prices. With new enforcement powers, the Commission has fined a number of retailers who failed to comply with Commission rules or engaged in deceptive trade practices.

While the spotlight is presently off of the retail market, further work remains. AREPs still account for a large share of the retail market, as noted above. Privacy concerns may intensify the backlash against the full dissemination of smart meter systems. Furthermore, it is to be seen whether steps will be taken to facilitate access to data by the third party to encourage innovative competitive services and reliance by REPs on smart meter data to offer value-added services to end use customers.

4 CONCLUSIONS

The Texas electricity market has successfully addressed challenges and introduced refinements in recent years to improve operational efficiency and transparency. Underpinning the success of the ERCOT market has been strong and unwavering belief among the state's policy makers that markets will lead to better outcomes than government regulation. Even when the 2006–2008 eras of high natural gas prices propelled electricity prices in areas open to retail competition to heights far in excess of the levels of electricity prices that could possibly have prevailed under traditional regulation, Texas "stayed the course."

Yet, somewhat ironically, maintaining a competitive market requires a great deal of attention and management from the state's legislature and regulatory agency. The market rules maintained by ERCOT continue to grow in scope and complexity. The state's legislature maintains a watchful eye and refines the market structure each time it meets. The state's regulatory commission vigilantly monitors the market, seeking to improve its operation. Texas is not completely immune to the political pressures reported in Chapter 1 by Newbery.

Despite the successful implementation of many refinements and recent steps taken by the Commission to address resource adequacy and scarcity pricing, a number of challenges remain to be addressed in the coming years, including:

- Lack of a level playing field for DR and price-responsive loads—a problem that is certainly not unique to ERCOT.
- Uncertainty regarding resource adequacy due to insufficient investment cost recovery for new resource developers. The recent decisions by the Commission to raise statewide offer cap may eliminate some of such uncertainty.

- Ineffective scarcity pricing which may work more effectively due to the recent decisions by the Commission to raise statewide offer cap.
- Market engine software updates to improve market efficiency by incorporating real-time co-optimization of energy and ancillary services and providing price projections for the next few hours.
- Over-mitigation of local market power through the existing Texas Two-Step mitigation mechanism.
- Lack of customer education regarding the potential benefits of the smart grid and advanced metering infrastructure.
- Slow progress by REPs in offering new and innovative products and services, which is anticipated to be addressed by full implementation of smart meters by 2013 and further reliance by REPs on advanced analytics to fully utilize the benefits of useful data generated by smart meters.

Chapter 2 by Keay, Rhys, and Robinson, as well as Chapter 7 by Bauknecht, Brunekreeft, and Meyer, in this volume voices concern that market forces in existing liberalized markets may not be able to achieve sufficient levels of investment in generating capacity. While these analyses focus on the investments necessary to decarbonize electricity generation in the United Kingdom and Germany, respectively, a similar concern is being voiced in Texas.

It may be noteworthy that the energy-only Australian market does not appear to be facing the same resource adequacy problems as ERCOT. As explained in Chapter 19 by Moran and Sood, higher wholesale price caps, spot price levels which better reflect the overall long-run marginal cost of supply, and recent milder summers in Australia may explain why that market is not feeling the strains now experienced in Texas.

Over the past 13 years, ERCOT has met its numerous challenges quite well. The degree to which it can rely upon market forces to meet this new set of challenges will determine whether it can retain its reputation as the best restructured market in North America.

REFERENCES

Adib, P., 1991. Direct Testimony in Docket No. 10473, Notice of Intent of Houston Lighting & Power Company for a Certificate of Convenience and Necessity for DuPont Project, Webster Units 1 & 2 Refurbishment Project and Greens Bayou Units 3 & 4 Refurbishment Project, Public Utility Commission of Texas, September, Austin, TX.

Adib, P., Hurlbut, D., 2008. Market power and market monitoring. In: Sioshansi, F.P. (Ed.), Competitive Electricity Markets: Design, Implementation and Performance. Elsevier, Oxford.

Adib, P., Zarnikau, J., 2006. Texas: the most robust competitive market in North America. In: Sioshansi, F.P., Pfaffenberger, W. (Eds.), Electricity Market Reform: An International Perspective. Elsevier, Oxford.

Adib, P., Schubert, E., Oren, S., 2008. Resource adequacy: alternate perspectives and divergent paths. In: Sioshansi, F.P. (Ed.), Competitive Electricity Markets: Design, Implementation and Performance. Elsevier, Oxford.

Alliance for Retail Choice, 2007. ARC's Baseline Assessment of Choice in the United States. May.

Bowring, J.E., 2008. Evolution of PJM's capacity market. In: Sioshansi, F.P. (Ed.), Competitive Electricity Markets: Design, Implementation and Performance. Elsevier, Oxford.

Bruce, M., 2012. Emerging Technologies Integration in ERCOT, A Presentation to the 2012 ERCOT Operations Seminar, January 12, Austin, TX. Also available at: <http://www.ercot.com/calendar/2012/01/20120127-ETWG>, Item 04 under Key Documents (accessed 05.04.2013).

Center for Advancement of Energy Markets (CAEM), 2003. Retail Energy Deregulation Index 2003, fourth ed.

Distributed Energy Financial Group (DEFG), 2011. The Annual Baseline Assessment of Choice in Canada and the United States (ABACCUS).

ERCOT, 2011. Functional Description of Core Market Management System (MMS) Applications for "Look-Ahead SCED," White Paper, V.0.1.2, presented in November 28, 2011 Workshop. Also available under LookAheadSced_1128workshop at: <http://www.ercot.com/calendar/2011/11/20111128-WMS> (accessed 05.04.2013).

ERCOT, 2012. ERCOT Quick Facts. July. Also available at: <http://www.ercot.com/about/profile/> (accessed 05.04.2013).

Hurlbut, D., Gauldin, J., Grasso, T., Greffe, R., Jaussaud, D., 2004. Staff Inquiry into Allegations Made by Texas Commercial Energy regarding ERCOT Market Manipulation, Project No. 27937, Public Utility Commission of Texas, January, Austin, TX.

Kang, L., Zarnikau, J., 2009. Did the expiration of price caps affect prices in the restructured Texas electricity market? Energ. Policy 37, 1713−1717.

Kiesling, L., Kleit, A.N., 2009. Electricity Restructuring: The Texas Story. American Enterprise Institute, Washington, DC (November).

Lee, Y.-Y., Baldick, R., Hur, J., 2011a. Firm-based measurements of market power in transmission-constrained electricity markets. IEEE Trans. Power Syst 26 (4), 1962−1970.

Lee, Y.-Y., Hur, J., Baldick, R., Pineda, S., 2011b. New indices of market power in transmission-constrained electricity markets. IEEE Trans. Power Syst. 26 (2), 681−689.

Patton, D., 2005. Direct Testimony in: Proceeding to Consider Protocols to Implement a Nodal Market in the Electric Reliability Council of Texas Pursuant to P.U.C. Subst. R. § 25.501. P.U.C.T. Docket No. 31540 (November 11).

Potomac Economics, 2005. Investigation into the Causes for the Shortages of Energy in the ERCOT Balancing Energy Market and into the Wholesale Market Activities of TXU from October 27 to December 8, 2004, Filing in the PUCT Docket No. 30513, April, Austin, TX.

Potomac Economics, 2006. 2005 State of the Market Report for the ERCOT Wholesale Electricity Markets, July, Austin, TX. Also available at: <http://www.potomaceconomics.com/index.php/documents/C6&C10>.

Potomac Economics, 2007. "Investigation of the Wholesale Market Activities of TXU from June 1 to September 30, 2005," Potomac Economics, Independent Market Monitor, March. The report was further revised in September 2007.

Potomac Economics, 2012. 2011 State of the Market Report for the ERCOT Wholesale Electricity Markets, July, Austin, TX. Also available at: <http://www.potomaceconomics.com/markets_monitored/ERCOT>.

Public Utility Commission of Texas, 2011. Scope of Competition in Electric Markets in Texas. Report to the 82nd Texas Legislature, January, Austin, TX. Also available at: <https://www.puc.state.tx.us/industry/electric/reports/Default.aspx>.

Reid, B., Thornton, A., Ngo, B., Adib, P., 2009. Wind Resource Integration Information for the Future. APX, San Jose, CA, Also available at: <http://www.apx.com/NewsAndEvents/Resources>, under Wind Resource Integration in Grid Operations (accessed 05.04.2013).

Saathoff, K., 2009. Wind Challenges & Integration Actions. A Presentation to the ERCOT Board of Directors, January 20, Austin, TX. <http://www.ercot.com/calendar/2009/01/20090120-BOD>, Item 11 under Key Documents.

Schubert, E., Adib, P., 2009. Evolution of wholesale market design in ERCOT. In: Kiesling, L., Kleit, A.N. (Eds.), Electricity Restructuring: The Texas Story. American Enterprise Institute, Washington, DC (November).

Schubert, E., Hurlbut, D., Adib, P., Oren, S., 2006. The Texas energy-only resource adequacy mechanism. Electricity J. 19 (10), 39–49 (Elsevier).

Schubert, E., Oren, S., Adib, P., 2009. Achieving resource adequacy in Texas via an energy-only electricity market. In: Kiesling, L., Kleit, A.N. (Eds.), Electricity Restructuring: The Texas Story. American Enterprise Institute, Washington, DC (November).

Stoft, S., 2002. Power System Economics: Designing Markets for Electricity. IEEE Press/Wiley-Interscience, Piscataway, NJ.

Swadley, A., Yucel, M., 2011. Did residential electricity rates fall after retail competition? A dynamic panel analysis. Energ. Policy 39, 7702–7711.

The Brattle Group, 2012. ERCOT Investment Incentive and Resource Adequacy, A Report Prepared for ERCOT, June 1, Austin, TX.

Woo, C.K., Zarnikau, J., 2009. Will electricity market reform likely reduce retail rates? Electricity J. 22 (2), 40–45.

Wood, P., Gülen, G., 2009. Laying the groundwork for power competition in Texas. In: Kiesling, L., Kleit, A. (Eds.), Electricity Restructuring: The Texas Story. American Enterprise Institute, Washington, DC (November).

Zarnikau, J., 2009. Demand participation in the restructured Electric Reliability Council of Texas market. Energy 35, 1536–1543.

Zarnikau, J., Adib, P., 2011. Advance Notice of Wholesale Electricity Prices: Recommended Solutions. White Paper presented to the ERCOT Demand Side Working Group, January, Austin, TX.

Zarnikau, J., Hallett, I., 2007. Aggregate industrial energy consumer response to wholesale prices in the restructured Texas electricity market. Energy Economics 30, 1798–1808.

Zarnikau, J., Whitworth, D., 2005. Has Electric utility restructuring led to lower electricity prices for residential consumers in Texas? Energ. Policy 34, 2191–2200.

Chapter 11

From the Brink of Abyss to a Green, Clean, and Smart Future: The Evolution of California's Electricity Market

Lorenzo Kristov and Stephen Keehn
Department of Market and Infrastructure Policy, California Independent System Operator

1 INTRODUCTION

California Independent System Operator (CAISO) is a balancing authority area and operator of wholesale spot markets that encompass most but not all of California. Figure 11.1 provides a schematic of the CAISO footprint, the major transmission lines, and key quantitative facts about CAISO markets and infrastructure. In January 2013, the Valley Electric Association in southwestern Nevada joined CAISO and became its first non-California participating transmission owner.

CAISO began operating in April 1998. Then in June 2000, there began a sequence of events that became known as California's electricity crisis of 2000–2001. As a result, California's electricity market evolution from 2001 to 2009 was mainly a crisis recovery period.[1] During the postcrisis years, CAISO and the state agencies—primarily the California Public Utilities Commission (CPUC)—redesigned the state's market structure to correct the flaws inherent in the original design that had set the stage for the crisis and allowed it to have such large impacts. The redesign was comprised of two major components: a new CAISO spot market based on the locational marginal pricing paradigm, and the CPUC resource adequacy program that established requirements for forward procurement of energy and capacity by the load-serving entities.

As these components were being put into place, two additional forces of change were gathering strength that would soon require additional reforms to

---

1. The crisis affected not only California but also had a chilling effect on restructuring efforts elsewhere, even outside of the United States. See Chapters 18 and 22.

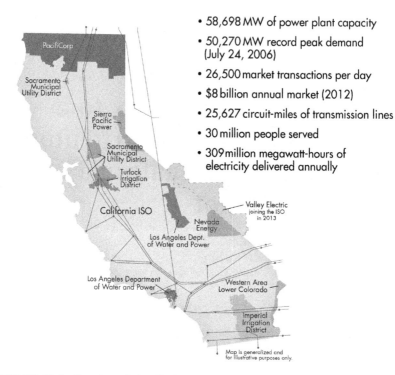

- 58,698 MW of power plant capacity
- 50,270 MW record peak demand (July 24, 2006)
- 26,500 market transactions per day
- $8 billion annual market (2012)
- 25,627 circuit-miles of transmission lines
- 30 million people served
- 309 million megawatt-hours of electricity delivered annually

FIGURE 11.1 Overview of the CAISO balancing authority area as of 2012 (© 2012, California ISO).

virtually every aspect of how the electricity sector functions. These two forces, whose impacts are yet to be fully realized, are new public policy mandates to substantially reduce the electricity sector's adverse environmental impacts, and the emergence to commercial viability of diverse new technologies entering the entire electricity supply chain from generation to end-use consumption. In particular, these forces are being felt in all three of the core functions that comprise the CAISO mission to provide open-access transmission service: operating a reliable grid, managing efficient spot markets, and planning transmission infrastructure expansion.

These new forces of change have revealed that the market structure designed and implemented following the crisis has not only corrected the conditions that allowed the crisis but it has also created a robust framework that will be readily adaptable to support the industry changes on the horizon. This chapter describes how industry changes are playing out in California, and how CAISO has been developing innovative responses to them.

Section 2 describes the electricity crisis of 2000–2001—the brink of abyss—and highlights the features of California's original market that contributed to the crisis. Section 3 explains how the subsequent market redesign has addressed the problems identified in the wake of the crisis and created a stable and flexible framework for future development. Section 4 describes the drivers of change that are mainly the state's environmental policy mandates and the rapid evolution of new technologies. Section 5 summarizes how these forces are affecting CAISO markets and operations. Section 6 provides a review of the initiatives CAISO has undertaken to become an innovative participant in accommodating the changes, supporting environmental policy objectives and facilitating beneficial technological advances. Section 7 then offers some conclusions.

2 THE BRINK OF ABYSS

Assembly Bill 1890, enacted by the California legislature in 1996, created the CAISO as part of a comprehensive restructuring of the electricity sector. The law was designed to create both a competitive wholesale spot market for energy and reserves and a program of retail competition or "direct access." The law combined the transmission systems of the three major investor-owned utilities—Pacific Gas & Electric (PG&E), Southern California Edison (SCE), and San Diego Gas & Electric (SDG&E)—into a single system under CAISO operational control. It also created the California Power Exchange as the central wholesale energy spot market, explicitly limiting CAISO's market role to functions essential to reliable grid operation, that is, spot procurement of reserve capacity and real-time balancing. CAISO began its operation on April 1, 1998.

In June 2000, barely 2 years after CAISO and the other components of California's restructured electricity sector started operations, spot energy prices in the new markets began to spike and, in one utility service territory, these spikes had alarming impacts on end-use customers' bills. Thus began California's electricity crisis of 2000−2001.[2] In fairly short order, the California Power Exchange, the central spot market, went out of business. One of the two large investor-owned utilities declared bankruptcy while the other teetered on the brink. CAISO's 25-member stakeholder governing board was dissolved by the legislature and replaced with a five-member board appointed by the Governor. And roughly nine billion dollars in wholesale energy charges by suppliers became the subject of prolonged litigation that remains pending against 15 suppliers more than 12 years later.

The crisis has been described as a confluence of perfect storm conditions, which included tight supply margins throughout the western region (creating opportunities for extensive exercise of supplier market power), lack of price responsive demand, and weak federal oversight of the prices charged by

---

2. For a thorough, detailed exposition of the crisis see Sweeney, J.L., 2002. The California Electricity Crisis, Hoover Institution Press, Stanford, CA and Sweeney, J.L., 2006. California electricity restructuring: the crisis and its aftermath. In: Sioshansi, F.P., Pfaffenberger, W. (Eds.), Electricity Market Reform, Elsevier, Waltham, MA.

wholesale sellers. These conditions in themselves would not have triggered such a crisis, however, had their impacts not been exacerbated by critical flaws in the original market design. A primary factor was the lack of forward energy contracting by the investor-owned utilities, to meet their load-serving requirements, which created vast exposure to high and volatile spot prices. This situation was a result of two prior design decisions. The first was a requirement that the utilities divest at least half of their thermal generation without any obligations on the facility buyers to sell back energy at pre-agreed prices. The second factor was a requirement that the utilities purchase through the Power Exchange all their energy needs above what their retained generation would provide.

Although divestiture of generation was already well accepted as a means to create a competitive generation sector, it was complicated in California by the need to recover stranded costs of utility-generating plants. The policy-makers who guided the restructuring effort expected generating resources would need to recover all their ongoing fixed costs, as well as their operating costs, through the spot markets and that competition would reduce spot market prices to the point where many utility resources would no longer be in the money. The policymakers therefore expected restructuring to create stranded utility-generating assets whose revenue gaps would need to be recovered from ratepayers, and divestiture became a key device for reducing the stranded cost burden. Moreover, divestiture without any sell-back obligations on the buyers was expected to help maximize the selling prices of these assets. Much to everyone's surprise, many of the divested thermal resources sold for more than their book values.

To create a deep and liquid spot energy market, state policymakers required the investor-owned utilities to bid nearly all their load and all of their retained generating assets into the day-ahead Power Exchange market. The initial market design had only two pricing zones—for northern and southern California—so that in most cases the price paid to the utility generation would offset the price paid by a comparable amount of their load, which left their net load exposed to Power Exchange prices. However, absent forward contracting to supply this net load, this degree of exposure turned out to be extremely costly once the Power Exchange prices increased in the crisis.[3]

Another factor that exacerbated the crisis impact was the retail rate freeze, which was intended to protect consumers in the transition to the new market structure. But when combined with high wholesale energy costs, it

---

3. In response to this situation, in 2001, the state entered long-term energy contracts with several suppliers on behalf of California load as a means of reducing supply uncertainty and exposure to volatile market prices. The need to recover the costs of these contracts then became the rationale for closing retail direct access to new subscribers. As discussed later in this chapter, the state reopened direct access to a limited extent only very recently.

caused the two largest utilities to suffer significant daily revenue losses. For the start of the new market, policymakers froze retail rates at 10% below their prerestructuring levels. At the same time, they established the competition transition charge (CTC), defined as the difference between the retail rate and the utility's total unit cost of providing electricity, as the mechanism to recover the stranded costs associated with utility-generation assets. The mechanism was designed so that, once a utility's stranded costs were fully recovered, both the rate freeze and the CTC would be removed. For the first 2 years, the margin was consistently positive so that by the end of May 2000, SDG&E had fully recovered its stranded costs and its rate freeze was lifted. Thus, when spot prices first began to spike in June 2000, the impacts only showed up on the bills of SDG&E customers. For PG&E and SCE, the price spikes caused the CTC to turn negative and the rate freeze prevented them from fully recovering their costs of serving customers.[4]

While many of these design features were beyond CAISO's control, there were two significant CAISO market features that contributed to the crisis. First, was the zonal market structure that consisted initially of only two large zones. This structure caused CAISO to accept day-ahead and hour-ahead energy schedules without regard to their impact on local or "intra-zonal" line limits. As a result, the combined energy flows from these schedules were often infeasible, requiring costly out-of-merit redispatches in real time to maintain reliable operations. Second, CAISO had relatively weak provisions for mitigating supplier market power, which suppliers exercised in supplying energy to meet system needs, as well as in offering decremental energy bids for redispatches to manage congestion caused by infeasible day-ahead schedules. The latter practice became known as the "DEC game."[5]

At the end of 2001, when the need to focus entirely on managing the crisis had subsided somewhat, CAISO initiated a comprehensive redesign of its markets which coincidently was about the same time that the Federal Energy Regulatory Commission (FERC) issued an order directing CAISO to fix its flawed zonal congestion management system.[6]

3 MARKET REDESIGN FOR EFFICIENCY AND FLEXIBILITY

The California electricity crisis provided some crucial insights that formed the basis for the CAISO's market redesign effort. The first insight was that reliable grid operations and well-functioning spot markets must be viewed as two

---

4. Sweeney, 2002 *op. cit.*, p. 172 et seq, gives a detailed explanation of how these features of the restructured market affected the financial situations of the investor-owned utilities.
5. Wolak, F.A., Bushnell J., Hobbs, B.F., Market Surveillance Committee of the California ISO, "Opinion on 'The DEC Bidding Activity Rule under MRTU.'" Available from: <http://www.caiso.com/1fbb/1fbbb95011100.pdf> (accessed 07.05.08).
6. *San Diego Gas & Electric v. Sellers of Energy and Ancillary Services*, 97 FERC 61,275 at 62,245 (2001).

sides of the same coin, not as separate functions. An original market design failing was that it placed all the centralized spot market functions within the Power Exchange, in which energy trading largely ignored the electrical properties and constraints of the grid. Meanwhile, it restricted CAISO to the bare minimum of market functions needed to maintain reliable operations, namely, real-time balancing and procuring reserves. The market redesign recognized that the markets must be aligned with the topology of the grid and the laws of physics. Moreover, the day-ahead and real-time energy markets cannot be abstracted from grid management requirements but must be the same mechanisms by which CAISO provides open-access transmission service, manages congestion to ensure feasible schedules, and dispatches resources to maintain real-time reliability. This key insight led CAISO to adopt the locational marginal pricing paradigm as its new market framework.[7]

The second crucial insight was that load-serving utilities must procure forward energy contracts to limit their exposure to spot prices. Moreover, the crisis clearly revealed the need to restore something equivalent to the traditional "obligation to serve" on load-serving entities, in the form of a requirement to procure sufficient generating capacity to meet peak system load plus a planning reserve margin and to make such capacity available to CAISO for commitment and dispatch on a day-to-day basis. This requirement was embodied in the Resource Adequacy Program, administered by the CPUC, which is an essential complement to CAISO's redesigned spot market. Taken together, the ability of load-serving entities to forward contract for energy and the requirement to procure resource adequacy capacity would serve not only to protect end-use customers from spot price volatility but also to provide the financial basis for independent investment in new generating facilities whose slow development under the restructuring had contributed to the tight supply conditions during the crisis.

The third crucial insight was that market design is not done once and then left alone. A new market structure must be flexible enough to adapt to evolving external conditions, new policy directions adopted by state and federal lawmakers, and technological advances without requiring fundamental changes. Although the potential impacts of state environmental policy goals and new technologies, such as smart grid, storage, and microgrids, were not yet on the horizon when CAISO began its market redesign effort, the need to design for flexibility became apparent within a few years as these developments were emerging. The next section will describe in detail how the flexibility of the new market structure has enabled CAISO to respond to changing

---

7. A full description of CAISO's comprehensive market redesign and the rationale behind the various design elements is provided in the direct testimony of Lorenzo Kristov, Attachment F to CAISO's February 9, 2006, filing of the market redesign tariff with the FERC, available from: <http://www.caiso.com/Documents/AttachmentF-DirectTestimony-LorenzoKristov_ExhibitNo_ISO-1_.pdf>.

conditions and new requirements. Whenever changes to the markets are considered, CAISO adheres to the first insight mentioned earlier, that markets must be aligned with the laws of physics and the physical characteristics and operating requirements of the grid.

CAISO locational marginal pricing-based markets launched on April 1, 2009, 7 years after the comprehensive redesign effort began, and have been operating efficiently ever since.[8] While the market redesign was still underway, the CPUC, in collaboration with CAISO, established the resource adequacy program, and began conducting regular long-term procurement proceedings to guide forward energy contracting by the investor-owned utilities. During this period, substantial new generation and transmission infrastructure was developed to relieve the tight supply conditions that had exacerbated price spikes and price volatility during the crisis. Also during the same period, FERC adopted stronger market oversight and enforcement capabilities. Together these changes addressed the conditions that allowed the crisis to occur, with the one exception that the amount of price responsive demand is still quite limited.

4 FORCES SHAPING THE TWENTY-FIRST CENTURY POWER SYSTEM

The electricity industry did not remain static while California redesigned the market structure to address the flaws that led to the crisis. Even before CAISO implemented its comprehensive redesign in 2009 it was clear that significant changes were on the horizon, with some already arriving, and that the markets would need to adapt. Although retail direct access had been suspended until the expiration of the long-term supply contracts the state had negotiated during the energy crisis,[9] some customers wanted more control of their own electricity use. During the redesign process, California adopted environmental policy directives that had implications for the composition of the supply fleet and would thus have large impacts on the spot markets, the real-time balancing functions, and transmission infrastructure needs. At the same time, new technologies have been maturing and increasing their presence on the supply and demand sides and in the retail and wholesale arenas. The sections below describe these factors in greater detail.

What is certain is that tomorrow's electric grid will not be our parents' electric grid. Balancing supply and demand used to be simply a matter of adjusting generators to balance relatively predictable changes in load. Load

---

8. Comprehensive documentation on the current CAISO market structure is provided in training materials, available from: <http://www.caiso.com/participate/Pages/Training/default.aspx>.

9. The California Public Utilities Commission suspended new direct access after September 20, 2001. See D.01-09-060 and California Public Utilities Code Sections 366 or 366.5, AB1X (2001).

varied based on well-established patterns and did not change much in response to system conditions. Power flowed from the generators to the load, generator responded to the directions from the system operator, and the system was balanced. But now the changes discussed earlier in this chapter are altering these basic premises. Not all generation can be counted on to respond to CAISO dispatch instructions. Indeed, not only is much of the new generation being connected to the grid non-responsive to operating instructions, its output also varies with the sun or the wind.

At the same time, demand is becoming a more active participant in the market through various demand response programs linked to spot prices and grid conditions. CAISO anticipates significant benefits to balancing the grid, as more customers take control of their energy use to manage their costs and their own environmental impacts and as the regulatory structure moves toward time-of-use and dynamic pricing. The smart grid offers the promise of even greater customer participation with its increased ways for CAISO operators to see current system conditions and dispatch demand- and distribution-connected resources.[10]

Although changes are occurring across the country, California is situated at the forefront and CAISO must deal with their impacts now. California has for many decades been a leader in environmental regulation. In recent years, California has taken the lead in mandating policies to increase renewable resources and reduce greenhouse gases. Further, California and its neighboring areas in the West are very well situated to develop renewable resources, including large-scale solar and wind installations as well as attaching small rooftop solar photovoltaic arrays to the distribution system. As the home of Silicon Valley and its capacity for technological innovation, California has a culture and history of developing and rapidly adopting new technologies. The payback from adopting new energy technologies is quicker for consumers in California, because of high energy prices, which are caused by numerous factors including environmental policies and lack of local inexpensive energy sources.

4.1 Desire of Customers to Self-Optimize Their Energy Use

One important initial restructuring feature in California was the introduction of retail direct access that allows electricity customers to choose their energy supplier. Retail direct access gained a strong foothold in California, as companies looked for ways to reduce their energy costs by purchasing power from suppliers who offered lower prices or reduce their environmental impacts by contracting with green energy suppliers. But when the energy crisis hit in 2000–2001, the retail suppliers faced the same wholesale market pressures that drove the

10. For a survey of developments in the area of smart grid, including a chapter on California, see Sioshansi, F.P. (Ed.), 2011. Smart Grid: Integrating Renewable, Distributed & Efficient Energy, Academic Press, Waltham, MA.

utilities into or to the brink of bankruptcy. Additionally, these retail suppliers faced competition from the frozen utility rates that did not reflect market forces and were kept artificially low. Added to that was the complication that their customers' energy payments flowed through billing systems of the virtually insolvent utilities. These pressures led many retail suppliers to struggle to stay afloat and in many cases return customers to the utilities.[11]

State policymakers suspended direct access in October 2001, allowing participating customers to remain signed up but prohibiting any further departures from utility retail service. This was motivated by the need to recover the costs of the long-term power contracts the state signed to bring an end to the crisis. The state contracts committed to pay high prices for energy that would supply the entire load in the state, not just the investor-owned utility customers. If direct access customers could avoid these costs they would have an unfair advantage. Thus, state policymakers decided not to allow new direct access to customers until the state contracts ended.

Customer desire for direct access did not disappear, however. For example, some commercial entities had direct access for some of their facilities but could not add other facilities to their contracts. Because some state contracts were for over 10 years, it became clear that something would have to be done to address these customers' concerns. The state legislature passed Senate Bill 695 in 2009, which ordered direct access to be opened to new customers, up to the highest level it had achieved before it was suspended. The CPUC adopted a 4-year phase-in mechanism that started in 2010, which released 35% of the available direct access capacity in each of the first 2 years, followed by 20% in the third year, and 10% in the fourth year.[12] Initial openings for new direct access capacity were quickly filled, and the percentage of direct access load in California increased from about 8.5% just before it was reopened to above 12% today (Figure 11.2).

In addition to direct access, customers are using other methods to manage their energy use and costs. With time-of-use and dynamic rate structures and new technologies providing greater visibility to and control over energy use and costs, customers are becoming more capable of optimizing their consumption decisions. The state's policies encouraging demand response and energy efficiency provide additional incentives for customers to optimize their energy use and for utilities and third parties to help customers achieve their goals. For example, all three investor-owned utilities in California are installing smart meters capable of providing customers with more detailed information about their energy use.[13]

---

11. See Chapter 12 by Kim on retail electricity markets in the United States.
12. See CPUC Decision (D.) 10-03-022, March 15, 2010.
13. See the CPUC web site on smart meters for links to decisions approving smart meter programs for the California utilities, utility web sites on smart meters, and information on opting-out and other issues. <http://PUC/PUC/energy/Demand+Response/ami.htm>.

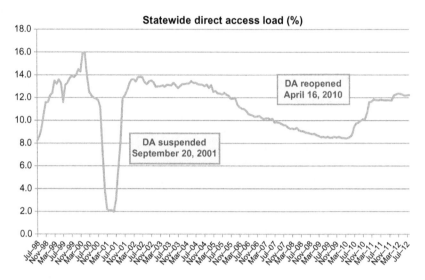

FIGURE 11.2 Participation in direct access, 1998–2012. *Source: CPUC web site, available at: http://PUC/NR/rdonlyres/5CF1FE63-1C5C-474F-A92B-C1067D1A21F9/0/Statewide DA_Aug2012.xls.*

State and national policies encouraging combined heat and power and distributed solar resources are also having an impact. California is now leading the nation in photovoltaic installations as a result of the California Solar Initiative,[14] in addition to recently setting a peak generation record of over 1000 MW for CAISO grid-connected resources alone.[15] Residential photovoltaic attractiveness is enhanced by a tiered rate structure that charges high prices for large monthly use that subsidize baseline rates, and a net energy metering policy that allows customers with behind-the-meter generation to pay only for their net monthly energy use.[16] These policies combine to enable customers with relatively high usage to avoid the high-price tiers entirely and end up with bills close to zero even though they may make extensive use of the distribution grid to take their excess power, and to provide them with power when their solar facilities are not generating.

14. The gosolarcalifornia web site (<http://www.gosolarcalifornia.ca.gov/>) stated, on October 24, 2012, that California leads the nation with 129,655 distribution-side solar projects accounting for 1362 MW of installed capacity. The web site also contains information on the various solar programs available to Californians. See also <http://www.californiasolarstatistics.ca.gov/> for California solar data through April 2013.

15. CAISO, 2012. California Independent System Operator, Sunny California Is America's #1 Solar Producer: Energy Officials Mark 1000 MW Solar Power Milestone at 2012 Stakeholder Symposium." Available from: <http://www.caiso.com/Documents/CaliforniaISOMediaAdvisory SunnyCalifornia-AmericasNumber1SolarProducer.pdf> (accessed 04.09.12).

16. See the CPUC web page on Net Energy Metering: <http://PUC/PUC/energy/DistGen/netmetering.htm>.

Commercial customers are generating their own electricity through distributed solar or other types of renewable resources, or with combined heat and power facilities that allow them to generate power for use in their own processes and sell the excess to their utility. California has a long history of encouraging these efficient energy uses, and recent decisions by the CPUC have made their adoption easier by instituting feed-in tariffs that take the excess power at fixed prices.

Some customers are taking the innovations even further. For example, the University of California at San Diego received a federal grant under the American Recovery and Reinvestment Act to implement a microgrid on its campus. The microgrid coordinates energy use across the campus with the generation from a combined heat and power facility and renewable resources to minimize the need to take power from the grid.[17]

4.2 State Environmental Policy Directives

Just as California led the nation in reducing auto emissions 40 years ago, California today is leading the way to a greener electricity grid and reduced greenhouse gas emissions. This includes the policies that have garnered the most attention, such as the renewables portfolio standard of 33% by 2020 and mechanisms to reduce greenhouse gas emissions to 1990 levels by 2020. Another policy is a preferential loading order for electricity supply, which requires investing in energy efficiency and demand response before investing in new generation, as well as preferring renewable to conventional thermal generation. The governor has called for installing 12,000 MW of distributed generation, which is stimulating innovative activities among state regulators, generation developers, and CAISO to achieve this target. California has also determined that power plants utilizing once-through cooling techniques must be retired or repowered before the end of this decade to avoid harmful environmental impacts.

4.2.1 Thirty-Three Percent Renewables Portfolio Standard

California initially adopted a renewables portfolio standard of 20% by 2010.[18] This applies to annual electricity consumption, so the standard means that 20% of the electricity consumed in the state on an annual basis must be provided from renewable resources. The state deferred the 2010 target date,

---

17. See, for example, James Newcomb, "The UCSD Microgrid – Showing the Future of Electricity ... Today," January 18, 2012. <http://blog.rmi.org/the_ucsd_microgrid_showing_the_future_of_electricity_today>.
18. The RPS program was initiated by SB 1078 (Sher), Stats. 2002, ch. 516. This bill set a goal for retail sellers to provide 20% of their retail sales from eligible renewable energy resources by 2017. Later SB 107 (Simitian), Stats. 2006, ch. 464, accelerated the 20% goal to 2010. See also California Public Utilities Code Section 399.

and the 20% standard was achieved by the end of 2011.[19] But the standard was increased in the meantime, first by a gubernatorial executive order[20] and then a legislated mandate for 33% by 2020 under Senate Bill X1-2. The new standard requires electric utilities to reach the 33% target in three compliance periods. Utilities must procure renewable energy equal to 20% of retail sales by the end of 2013 (which has already been met), 25% by the end of 2016, 33% by the end of 2020, and must maintain that percentage in subsequent years. SB X1-2 further applied the requirements not only just to those retail electricity suppliers regulated by the CPUC but also to the numerous municipal utilities in the state.

The impacts of the renewables portfolio standard on California's energy markets have already been significant and will increase as the level climbs to 33%.[21] As utilities and energy service providers comply with the mandates, CAISO planning and operations are affected in several ways. First, the need for renewable resources has led developers to propose large numbers of new generation facilities requesting interconnection to the grid. These facilities are often concentrated in areas that offer the best solar and wind conditions but tend to have very little load. This creates a need for new transmission infrastructure to connect the new resources and transmit their power to load centers.

Renewable resources are beginning to create significant operational issues for CAISO.[22] Unlike traditional generation, most renewables are neither able to forecast their output with great accuracy nor respond to the dispatch instructions. In contrast to the traditional situation where load is the primary source of output variation and generation responds accordingly to balance the system, renewable generation adds variability that also must be balanced. A related problem is overgeneration. In meeting the renewables portfolio standard, utility renewable energy contracts generally pay for as much energy as the resources can produce, even when the system does not need it.

---

19. This result follows reported renewable energy levels of 13% in 2006 and 17% in 2009. CPUC, 2012. Renewables Portfolio Standard Quarterly Report: 1st and 2nd Quarter 2012, p. 3. Available from: <http://PUC/NR/rdonlyres/2060A18B-CB42-4B4B-A426-E3BD C01BDCA2/0/2012_Q1Q2_RPSReport.pdf>.

20. Executive Order S-14-08 signed by Governor Arnold Schwarzenegger on November 17, 2008. Available from: <http://gov38.ca.gov/index.php?/executive-order/11072/>.

21. According to the CPUC, 2871 MW of new renewable generation capacity has come on line since 2003, and over 2800 MW more is scheduled to come on line by the end of 2012. CPUC, op. cit.

22. See Section 11.5 in this chapter for a description of and references to studies performed by the CAISO. While there are many reports on renewables integration, the following two cover the eastern and western interconnections of the United States: Enernex Corporation. January 2010. "Eastern Wind Integration and Transmission Study," NREL Report No. SR-550-47078. Available from: <http://www.nrel.gov/docs/fy11osti/47078.pdf>, and GE Energy. May 2010. "Western Wind and Solar Integration Study," NREL Report No. SR-550-47434. Available from: <http://www.nrel.gov/docs/fy12osti/54864.pdf>.

The absence of marginal fuel costs and the incentive structure of the production tax credits enable renewable generators to earn money even when spot market prices are negative.

The last issue has other serious market implications. Because renewable resources have negligible marginal cost, electricity spot prices will decrease as renewable volumes increase. Traditional thermal resources will see fewer energy sales and declining spot market revenues at the same time as their ability to respond to CAISO instructions will become more valuable in balancing the variability that renewable generation brings. The increasing need for flexibility in real-time balancing may also lead to more starts, stops, and output adjustments, as well as increasing the frequency with which units must operate at minimum load in order to maintain sufficient flexible capacity on line—all of which will likely increase costs for these generators at the same time their energy revenues are decreasing.[23]

4.2.2 Greenhouse Gas Regulation: California's Cap-and-Trade Program

One of the major environmental laws passed in California during the last decade was Assembly Bill 32 which established a goal of reducing greenhouse gas emissions to 1990 levels by 2020.[24] Executive Order S-3-05 then established a requirement to reduce emissions to 80% below the 1990 level by 2050. Figure 11.3 shows what these requirements would mean in comparison to a "business-as-usual" (BAU) scenario based on current expectations of growth in California's population and economy.

Assembly Bill 32 assigned responsibility for implementation of greenhouse gas regulation to the California Air Resources Board, which has developed a cap-and-trade program, to ensure that greenhouse gas emissions from large sources are actually reduced. Power plants in California and electricity imports into California fall under this mechanism, as do other large stationary sources of emissions. Covered entities will be required to obtain emission allowances equal to the amount of their greenhouse gas emissions. They may also use a limited number of emission offsets approved by the Board. Utilities in California and other covered entities will be allocated allowances initially but cannot simply apply them against their emissions. Instead, the investor-owned utilities will be required to offer their allocated allowances into an auction conducted by the Board, and then buy from the same auction the allowances they need to meet their requirements.

Cap and trade will increase the marginal costs of fossil fuel plants—even more so for the less efficient plants—thus decreasing their dispatch in the

---

23. See Chapters 5 and 7.
24. Global Warming Solutions Act of 2006 (Assembly Bill No.32). <http://www.leginfo.ca.gov/pub/05-06/bill/asm/ab_0001-0050/ab_32_bill_20060927_chaptered.pdf>.

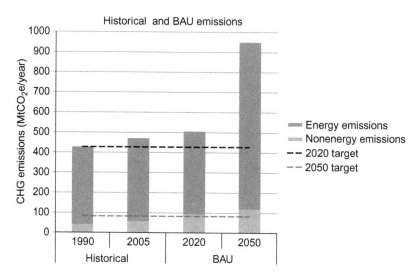

FIGURE 11.3 Implications of California's greenhouse gas reduction requirements. *Source: California Council on Science and Technology,* California's Energy Future—Portraits of Energy Systems for Meeting Greenhouse Gas Reduction Targets, *September 2012.*

spot market relative to renewable generation with lower marginal cost. The CAISO markets will explicitly consider the greenhouse gas allowance costs only to a limited extent, which is by expanding the allowable costs that a generator can include in its cost-based start-up and minimum load costs and in the default cost-based energy bids that are used for market power mitigation.[25] Beyond that, CAISO expects greenhouse gas allowance costs to be reflected in the energy bids of generators and parties importing energy into CAISO from out of state, which should serve to align the economic market dispatch with the state's environmental priorities.[26]

4.2.3 Once-Through Cooling Plant Retirements

Thirteen conventional thermal generators and California's nuclear generators (representing about 17,500 MW) within the CAISO balancing authority must retrofit, repower, or retire by 2020 and 2024, respectively, to comply with

25. CAISO. California Independent System Operator, Draft Final Proposal: Commitment Costs Refinements 2012. Available from: <http://www.caiso.com/Documents/DraftFinalProposal-CommitmentCostRefinements.pdf> (accessed 11.04.12).

26. The program is being administered by the Air Resources Board of the California Environmental Protection Agency. The Air Resources Board web site on the Cap-and-Trade Program provides the most up to date information about the program, which is scheduled for initial implementation on January 1, 2013. Available from: <http://www.arb.ca.gov/cc/capandtrade/capandtrade.htm>.

the state's once-through cooling mandate that restricts the use of coastal waters for power plant cooling.[27]

At the same time, the affected plants will become more valuable as more renewables come on line due to their locations and operational characteristics. These generators are mainly located close to load centers, in transmission-constrained areas, and as such are needed to maintain local reliability. Many of the once-through cooling generators are flexible, dispatchable, and can be started quickly. All but the nuclear plants are natural gas steam plants, which have relatively low minimum load levels and are able to ramp through a large region of their overall capacity. And all of the units feature large rotating generators whose spinning mass supplies inertia to the grid. To further complicate the matter, newer types of thermal resources that may replace these plants may not provide all the system benefits that the older types provide.

These issues mean that California must somehow reconcile the once-through cooling policy mandate with the renewables and greenhouse gas mandates to ensure continuing grid reliability. At the same time, as California is requiring once-through cooling plants to either retire or repower, these facilities are earning less revenues in the spot markets while becoming more valuable for their flexibility and local reliability benefits. One promising approach to this dilemma that CAISO is pursuing, discussed later in this chapter, is to incorporate a flexible capacity requirement into the Resource Adequacy Program. This would generate a stream of capacity payments to generators that can provide the flexibility CAISO needs for grid operations, while also creating a flexible ramping market product that will increase spot market revenues to flexible generators.

4.2.4 Distributed Generation

California's strategy to move to renewable generation does not rely solely on large-scale renewable generation, but includes 12,000 MW of distributed generation.[28] To help meet this goal, state policymakers have adopted policies and programs to encourage distributed generation development, and CAISO is supporting the effort with changes in its own market rules.

The California Solar Initiative[29] offers incentive payments at a rate that declines with resource size to encourage small solar deployment and enable the small solar industry to expand to achieve economies of scale and develop further innovations to drive down costs. Net energy metering allows the

---

27. In 2010, the State Water Resources Control Board adopted a Policy on the Use of Coastal and Estuarine Waters for Power Plant Cooling. For more information, see the State Water Resources Control Board Once-Through Cooling web site. Available from: <http://www.waterboards.ca.gov/water_issues/programs/ocean/cwa316/policy.shtml>.

28. See, for example, the California Governor's Conference on Local Renewable Energy Resources, July 25, 2011. Available at: <http://gov.ca.gov/s_energyconference.php>.

29. See the California Solar Initiative website for complete details. Available at: <http://www.gosolarcalifornia.org/csi/index.php>.

participating customers to have what amounts to free storage of their excess electricity production on the local distribution system, enabling customers with small solar installations to reduce their electricity bills to near zero.

Recent legislation has led to the development of feed-in tariffs for new, small (less than 20 MW), and efficient combined heat and power plants.[30] The ability to sell excess power to local utilities should help encourage the development of such resources. The CPUC also has a Self-Generation Incentive Program, which provides incentives for qualifying distributed energy systems, installed on the customer side of the meter. Technologies eligible for these incentives include microturbines, fuel cells, and advanced energy storage, in addition to more traditional technologies such as wind turbines, waste-heat-to-power technologies, and internal combustion engines.[31]

Distributed generation is often behind-the-meter generation that CAISO cannot dispatch and whose output CAISO cannot see. While distributed generation may help to decrease system peaks, it may also increase what appears as load variability on the grid. For example, much of this distributed generation is expected to be photovoltaic, which will vary with cloud cover, and will start and stop production in unison as the sun rises and sets, creating the need at these times for large amounts of ramping generation. Even with tools to improve the visibility of these resources, a large increase in distributed generation will likely increase the need for flexible generating capacity.

4.2.5 Loading Order

Utilities regulated by the CPUC have been instructed since 2003 to serve their electricity demand according to the loading order specified below, designed to reduce environmental impacts.[32]

1. Energy efficiency and conservation;
2. Demand response;
3. Renewable resources and distributed generation;
4. Clean conventional generation.

As with other state policies, this one is moving California toward a future with less conventional thermal generation capable of responding to dispatch

---

30. The CPUC website provides full details on the implementation of this legislation. Available at: <http://PUC/PUC/energy/Renewables/hot/feedintariffs.htm>.

31. The CPUC's Self-Generation Incentive Program is described at the following link: <http://PUC/PUC/energy/DistGen/sgip/>.

32. California Public Utilities Code Section 454.5(b)(9)(C) states: "The electrical corporation shall first meet its unmet resource needs through all available energy efficiency and demand reduction resources that are cost effective, reliable, and feasible." The Loading Order was first acknowledged in Energy Action Plan I, California Energy Commission, California Public Utilities Commission and Consumer Power, and Conservation Financing Authority, May 8, 2003. Available from: <http://docs.cpuc.ca.gov/word_pdf/REPORT/28715.pdf>.

instructions, more variable and distributed generation, and increased response and participation of customers in the electric markets.

To date, the loading order has not aligned well with forward planning needs. Utility-based energy efficiency and demand response programs are only approved on a short-term basis that looks ahead only a few years, while the time it takes to bring a major new generation plant on line can take five or more years. While the loading order requires load-serving entities to consider energy efficiency and demand response before other types of generation, these preferred resources do not participate in the long-term procurement processes as formal contractual commitments that are comparable to and compete against generation facilities. Until California adopts energy efficiency and demand response procurement policies, whereby suppliers commit to quantitative results for future time horizons comparable to the time it takes to develop generation resources, the loading order will be less fully realized than desired.

4.3 New Technologies

Technological innovation is a powerful force in its own right, apart from and parallel to any policy initiatives, that is driving changes to California's electricity production, transport, and use. The new innovation wave includes electric vehicles and smart end-use devices, generating and storage technologies, smart grid technologies, and other innovations such as fuel cells that are blurring the conventional boundaries between supply and demand.

4.3.1 Smart Grid

Smart grid innovations are permeating virtually the entire chain of electricity production, transport, and consumption.[33] At the consumption end, most consumers now have smart electricity meters installed that greatly increase customer access to a detailed daily and hourly electricity usage information. Today's capabilities are expected to be only the start of changes enabling customers to have more information and control of their energy use. Eventually, the smart grid will have dramatic implications for customer demand, distributed energy production, utility operations, wholesale spot markets, and the way the transmission grid is managed. The CPUC has instituted a smart grid proceeding, and the three investor-owned utilities have each filed smart grid deployment plans detailing how they envision implementing smart grid technologies over the next 10 years.

Although it may take many years to fully realize a world of smart electrical devices that receive dynamic price information from the grid and adjust the timing and duration of energy use, the initial elements of this vision are arriving now. Customers are purchasing electric vehicles and installing home charging stations that are typically configured to charge when prices are low.

---

33. See Sanders, Kristov and Rothleder, 2011. The smart grid vision and roadmap for California. In: Sioshansi (Ed.), 2011, *op. cit.*

It is now possible to purchase devices to control pool pumps and other appliances through the Internet, which allows customers to turn these appliances off remotely, if prices increase or there is call for conservation. CAISO has begun initial work on how to distribute prices and system condition information electronically, thus enabling devices that can receive the information to respond to changes in these factors.

A smart grid infrastructure will enable distributed energy resources and load, potentially aggregated by third parties, to participate directly in wholesale markets, and provide energy or ancillary services such as regulation or reserves. This will increase spot market participants and extend CAISO dispatch instructions to more parts of the electric grid, thus expanding the flexibility for real-time balancing.

A smart grid will provide utilities with more visibility and control over their distribution systems. Enhanced visibility and control will also extend to CAISO as the conventional functional distinctions between distribution and transmission become less relevant, and the utilities and CAISO collaborate to improve the operations of the whole system. Many distribution-connected energy resource developers are interested in participating directly in the wholesale markets, which will require utilities and CAISO to work, to ensure that the transmission-distribution interface does not present a roadblock.

A smart grid will have direct impacts on California's transmission network. CAISO is currently taking part in the west-wide syncrophasor project, which will provide significantly more detailed system conditions data and thereby improve both forecasting and operations.

4.3.2 Storage

The inability to store electricity in significant amounts has always been given in the electricity industry, but that limitation is now yielding to technological change. The term storage encompasses many possible technologies. Some, such as pumped storage hydro, have been around for years and are well understood, while others, such as flywheels and batteries, are relatively new and not yet commercially operating at large scales, while still others, such as railcar gravity systems, are developing utility-scale demonstration projects.[34]

Even so, while these storage types have different characteristics and behavior, it is clear the technologies offer the potential to bring about many changes to how the electricity markets and grids operate. For instance, storage devices will likely be able to provide faster ancillary services—particularly regulation, ramping, and spinning reserves—than most conventional resources. When teamed with variable renewable resources, storage can mitigate variability, increase capacity factors, and reduce overgeneration on the

34. The May 3, 2012 Economist Technology Quarterly recently offered a brief survey of the current development status of various storage technologies. Available from: <http://www.economist.com/node/21548495>.

grid. Expanded penetration of storage may also reduce needs for transmission upgrades in some situations.

4.3.3 Electric Vehicles

Independent studies show Californians are adopting electric vehicles at one of the fastest rates in the United States. This will have a large impact on electricity markets because of the load that these vehicles will create, as well as the ability for vehicles and charging stations to provide real-time balancing and regulation services. Such potential side benefits of changes to the transportation sector offer a prime example of how innovations on the horizon can provide solutions to the challenges presented by the greening of the electricity system.

5 HOW THESE CHANGES AFFECT THE ELECTRICITY MARKETS

Most forces and developments affecting California's electricity markets are interacting with each other rather than having discrete or independent impacts, and some changes driven by one factor may offer possible solutions to issues created by other forces. CAISO is therefore implementing a program of enhancements to different market elements and core functional areas in a manner that is logically planned out based on the expected path of future development, and flexible enough to adapt to surprises affecting the pace and nature of industry changes. CAISO continually monitors the impacts of these forces on real-time and day-ahead market performance and must address near-term needs expeditiously. Yet, it is also important to look at the longer-term implications, to ensure the right resource types are available in the right locations to maintain market stability and meet system operating needs. This section offers a brief survey of some key market and operational impacts from the changes described earlier.

CAISO anticipated significant changes were looming, and in 2007, began studying the potential grid impacts from renewable resources. This effort produced two reports: one on the operational impacts of 20% renewables; and another in 2011 concerning the 33% renewables portfolio standard.[35] The studies concluded that as more variable renewable resources are added to the system, increasing amounts of regulation and load following

---

35. Renewable integration studies performed by the CAISO. Available from: <http://www.caiso.com/informed/Pages/StakeholderProcesses/IntegrationRenewableResources.aspx>. See, especially: "Integration of Renewable Resources: Transmission and operating issues and recommendations for integrating renewable resources on the California ISO-controlled Grid" (November, 2007).
- "Integration of Renewable Resources: Operational Requirements and Generation Fleet Capability at 20% RPS" (August 2010).
- "Draft Technical Appendices for Renewable Integration Studies—Operational Requirements and Generation Fleet Capability" (October 2010).
- "Summary—Preliminary Results of 33% Renewables Integration Study—2010 CPUC Long-Term Procurement Plan Docket No. R.10-05-006" (May 2011).

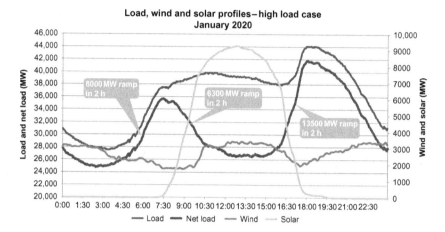

FIGURE 11.4 Net load ramping requirements expected in January 2020 (© 2012, California ISO).

capabilities are required to balance the grid. Somewhat surprisingly, however, the results show that the increased needs for ramping do not occur during the peak months, but are often projected to occur during the winter, spring, or fall periods.

Figure 11.4 provides an example of the situation CAISO is projecting. It shows the expected net load curve for a January day in 2020, assuming a high load forecast for the year. While the gross load curve is similar to what is seen today, with a morning ramp and an afternoon peak, the net load after accounting for wind and solar production shows a very different pattern. After a sharp upward morning ramp of about 8000 MW in 2 h, there is sharp downward ramp of about 6500 MW over 2 h that is followed by a fast evening upward ramp of 13,500 MW in 2 h. This pattern will require briefly turning on a large amount of flexible generation to meet the morning ramp, then turning it off, and then turning it on again for the evening ramp. This will require CAISO and generating plants to deal with more units cycling throughout the day.

Figure 11.4 also illustrates another potential issue for CAISO. The solar pattern shown in Figure 11.4 is very smooth in hitting a peak of around 9000 MW. Because the month in question is January, the possibility of cloud cover is a concern even though the facilities are located in areas with little rain. More troublesome would be a partial cloud cover situation that forces solar generation to quickly ramp up and down multiple times as the clouds come and go. Wind generation presents similar situations with rapid drops or increases in output that would lead to very steep and unpredictable ramps.

Figure 11.4 shows a forecasted winter day in California, and hints that there may be different ramping requirements during different months. Figures 11.5 and 11.6 look across the year to illustrate, respectively, the longest ramp and the maximum one-hour ramp expected in each month of

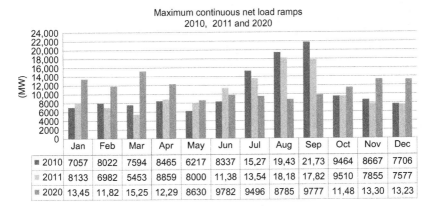

FIGURE 11.5 Maximum continuous ramps of net load expected in 2020 (© 2012, California ISO).

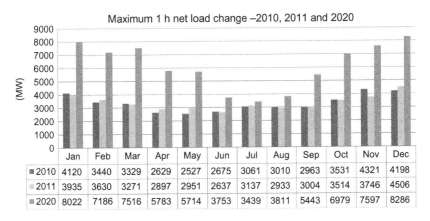

FIGURE 11.6 Maximum one-hour change in net load expected in 2020 (© 2012, California ISO).

the year. They show that the situation facing CAISO operations will likely be dramatically different in different months. Figure 11.5 shows that while the maximum continuous ramps during the summer months are predicted to be significantly lower than today, the amount of ramping required in other months will double today's requirements.

Figure 11.6 indicates that compared to today, the maximum 1 h change in net load expected in 2020 remains similar during the typical summer peak, but doubles during the other months. The operational challenge could be further complicated by once-through cooling plant retirements, which are typically large steam units that can provide large amounts of ramping capability. As these plants retire or are repowered and replaced with more efficient combined cycle plants having less flexible ramping capabilities, the challenge will be to ensure that sufficient resources are available on a daily basis to meet the ramping needs expected with 33% renewable energy in 2020.

The spring and fall increased ramping requirements concentration may also complicate generator maintenance scheduling. These are the months when generators typically perform annual maintenance, as they are usually not needed to meet load. In 2020, however, while the units may not be needed to meet peak loads, they may be needed to meet peak ramps. Hydro resources have traditionally provided another major flexible capacity source, but in the spring these resources may have limited ramping capability because of water flow conditions.

While Figures 11.5 and 11.6 indicate the expected operating requirements by 2020, Figure 11.7 offers an indication of how fast the changes will be occurring. It shows a typical March day, estimated for each of the years 2012 through 2020. What is striking is the change that occurs between 2014 and 2015. The patterns for the years 2012–2014 track fairly close together, but there is a pronounced shift in requirements in 2015, which increase each year through 2020.

Figure 11.7 also indicates a potential overgeneration problem. For the typical March day in 2020, the net daytime load decreases to just over 11,000 MW. Since this is close to the amount of hydro, nuclear, and contracted qualifying facilities on the system, none of which are flexible, there may not be enough flexible generation on line to provide downward ramping to manage overgeneration without reducing renewable generation. There also may not be enough flexible generation running to provide needed system inertia or to respond to sudden contingency events or even to respond to the anticipated upward afternoon ramp. This situation suggests a potential

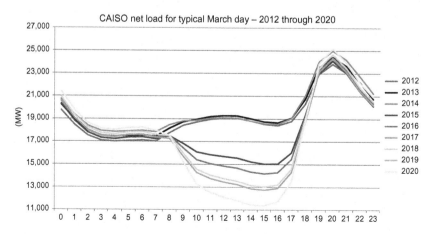

FIGURE 11.7 Expected annual changes in the pattern of spring net load (© 2012, California ISO).

operational benefit from facilities such as electric vehicle charging stations that can significantly increase midday demand and thereby help to mitigate the severity of these load-shape changes.

6 CAISO INITIATIVES AND INNOVATIONS

CAISO responses to the two major industry-transforming forces—policy mandates to reduce environmental impacts and advances in technology—are best described in terms of the three core areas of the its responsibility, all of which require innovative updating:

1. grid operations—reliable operation of the high-voltage transmission system;
2. spot markets—design and operation of wholesale spot markets both for trading energy and reserves and for allocating open-access transmission service; and
3. infrastructure development—planning the expansion of the transmission system and interconnection of generating facilities.

The CAISO is responding to the forces of change by anticipating and preparing for the new requirements they bring, and by proactively initiating market and operational changes to help facilitate the industry transformation.

6.1 Grid Operations

Ensuring sufficient operational capability must be viewed from a planning perspective, looking out 1−10 years to ensure the generation fleet will provide both the right performance capabilities and sufficient total installed capacity. In addition, a day-to-day real-time perspective is needed to ensure that these resources are available for unit commitment and dispatch instructions to maintain system balance. CAISO has performed several studies to estimate system needs under different renewable resource development levels and assess the current generation fleet capabilities. Also under development are new market constraints and market products to ensure efficient and operationally secure commitment and dispatch within the day-ahead and real-time markets structures.

6.1.1 Innovative Study Methodologies

CAISO has developed state-of-the-art study methodologies for evaluating system needs from a planning perspective in the context of integrating large amounts of renewable generating resources. The study proceeds in sequential steps illustrated in Figure 11.8. The first step is developing system-wide 1 min load, wind, and solar profiles for every minute of the year. CAISO uses operational data to describe current conditions and existing resources, and synthesizes wind and solar profiles for future resources based on resource locations and

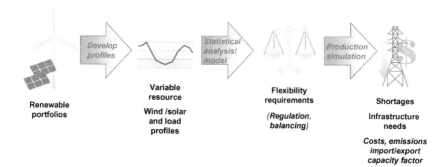

FIGURE 11.8 Renewable integration study process (© 2012, California ISO).

characteristics, day of year, time of day, and expected solar irradiance and wind conditions and their variability. The profiles are then used as inputs to a statistical analysis to calculate needed regulation and load following amounts. These requirements, along with hourly load and expected operating reserves, then become inputs to a production simulation that assesses the projected resource fleet's ability to meet the hourly load, maintain required operating reserves, and provide needed regulation and intra-hour balancing services.

Traditional planning assessments, such as loss-of-load probability, will need to be extended to assess fleet flexibility adequacy. CAISO will perform assessments to determine the impact potential flexibility shortages may have on operational performance criteria such as a Control Performance Standards (CPS1, CPS2, BAAL).[36] The assessment objective is to estimate the amount of flexible capacity needed to maintain a target reliability level. In addition to the flexibility studies, CAISO has performed other studies assessing the operational impact from high renewable penetration levels on frequency response.[37]

In addition to longer-term planning assessments, CAISO is developing shorter-term methodologies to quantify fleet capability needs to feed into annual resource adequacy procurement requirements for load-serving entities. By analyzing the net load range from trough to peak and the changes from 1 min to the next, CAISO can quantify the ramping needs over different time intervals (5 min, intra-hour, and the longest ramp period in a single direction) and use this analysis to specify needed amounts of flexible capacity.

6.1.2 Flexible Ramping Constraint and Product

"Flexi-ramp" is the CAISO term for a ramping capability procured through the markets. Besides the operational need to have sufficient ramping

---

36. Definitions of these terms and complete NERC standards documents are available from: <http://www.nerc.com/page.php?cid=2|20>.
37. Miller, N.W., 2011. General Electric, Report to California ISO (CAISO) Frequency Response Study. Available from: <http://www.uwig.org/Report-FrequencyResponseStudy.pdf>.

capability available, the increased system variability was creating instances of extremely high 5 min real-time prices when no energy supply shortage existed, rather only an inability for dispatched resources to ramp fast enough. These situations made it clear that the market procurement needed to focus on ramping capability as well as energy. CAISO initially implemented flexi-ramp through a constraint on the real-time market unit commitment and dispatch that ensures a sufficient amount of ramp-capable capacity is on line and available for dispatch, and is now in the process of designing a new flexi-ramp ancillary service that will be procured in the market optimization based on generator bids.

CAISO will procure the flexible ramping product in each 15 min real-time unit commitment procedure with the objective of providing sufficient ramping capability for the real-time 5 min dispatch to meet expected differences between the 15 min average net load and each 5 min interval net load. The flexible ramping product is similar to the load-following service referred to earlier, but focuses on deviations from the 15 min average net load rather than from the hourly average net load.

6.1.3 State-of-the-Art Control Room

To prepare for the evolving supply fleet operational impacts, CAISO has developed a state-of-the-art control room, including a new dispatch desk specifically dedicated to monitor and respond to changing renewable supply conditions. The control room incorporates advanced visualizations of the following:

- the current and expected production of all variable resources;
- changes in weather and other conditions that affect the output of these resources;
- expected impacts of projected renewable production on system imbalance conditions;
- updates on ancillary service availability and requirements; and
- impacts of the above factors on congestion.

The advanced control room integrates smart grid technologies such as phase angle measurement and visualization of grid security status. The state-of-the-art control room is critical to the CAISO mission of ensuring reliable grid operation under the anticipated changes to market and grid conditions discussed in this chapter.

6.2 Spot Markets

To align with the industry evolution in progress, CAISO has made several changes to its markets and initiated others. Some changes are designed to address specific near-term issues, while others will be more comprehensive.[38]

6.2.1 Near-Term Market Enhancements

CAISO is implementing changes to allow nongeneration resources, such as demand response and storage, to provide ancillary services, mainly by reducing the amount of time a resource must be able to provide the service. For regulation, CAISO reduced the requirement to 60 min for the day-ahead market procurement and 30 min for the real-time procurement. For spinning and nonspinning contingency reserves, the time that a unit must be able to provide energy was reduced from 2 h to 30 min. Also a program called Regulation Energy Management was adopted to enable resources, such as storage, to provide regulation service by adjusting their forecasted availability to reflect their charging state. For example, if a resource is dispatched in one interval to provide energy, for the next interval the algorithm would assume that the resource is consuming energy to recharge. If it is then needed in that next interval to provide energy for regulation, the algorithm would reduce the amount it could consume for charging. The new algorithm enables a storage device to provide as much regulation service as possible by intelligently monitoring and controlling its charging cycles.

To manage overgeneration through the market, CAISO will reduce the energy bid floor price in its markets, which ultimately limits market-clearing prices, from negative $30 to negative $150, and later to negative $300. At a negative $30 price, many wind generators still earn money because of production tax credits or bilateral contracts for renewable energy that compensate them for all output. Several years ago, CAISO created the Participating Intermittent Renewable Program (PIRP) that allows renewable generators to net their positive and negative real-time energy deviations from their hour-ahead schedules across the month and pay for the net at an average price. The program requires generators to self-schedule an independent hour-ahead forecast in each hour and prohibits them from submitting economic energy bids. CAISO is now modifying PIRP to allow these generators to submit decremental energy bids, to enable the real-time market to reduce their output economically to manage overgeneration.

In response to an order from FERC, CAISO is working (as are other organized markets) to modify its regulation market to incorporate a mileage payment. This will reward resources that are used more or can respond faster to signals from the automatic generation control system. The proposal includes a mechanism to account for and adjust payments based on the accuracy of the resource's response.

---

38. Chapter 25 by Riesz et al. discusses some market design considerations related to integration of variable renewable generation.

6.2.2 Design Changes in Progress

CAISO is revising some market rules to enable greater energy market partic-ipation by demand response and distributed generation resources. Traditionally, utility-triggered programs have been the primary mode of demand response participation. A recent FERC decision ordering demand response to receive the same payments as generation has created some yet-to-be-resolved issues because the required structure would allow demand response providers to receive double payments, first for the energy not used and then again at the market energy price.

Through demonstration projects CAISO is assessing the requirements for distributed generating resources to participate directly in the markets. Dispatchable, flexible distribution-connected generation, such as combined heat and power facilities, would be as valuable to grid operations as flexible transmission-connected resources, assuming the distributed resources have ade-quate visibility and respond to dispatch instructions. With Governor Jerry Brown's push to install 12,000 MW of distributed generation, flexible genera-tion on the distribution system could be extremely helpful in balancing the vari-able wind and solar resources expected to develop over the next few years.

Another factor motivating CAISO market participation is that load-serving entities procuring distributed resources want to count them toward their resource adequacy requirements. This highlights another way in which the Resource Adequacy Program will need to evolve in coordination with the CAISO markets to ensure that the electricity sector functions well as a whole system. A particu-lar concern is that if the distributed generation counting for resource adequacy is mostly solar and wind, it could crowd out flexible resource adequacy capacity and exacerbate the financial stresses on the flexible resources, thereby threaten-ing their financial viability and, in turn, further challenging grid operation.

CAISO tariff provisions known as "must-offer obligations" comprise a crucial link between resource adequacy capacity procurement by load-serving entities and its availability to CAISO for commitment and dispatch. Once a gen-erating plant is designated as providing resource adequacy capacity to a load-serving entity, the plant becomes subject to these provisions. For conventional dispatchable resources, CAISO enforces the must-offer obligations by inserting cost-based generated bids for a plant that fails to bid and has not reported an outage. For renewable generators, hydro resources subject to water management responsibilities and gas turbines limited by emissions permits, the tariff has a special category called "use-limited" resources that are not subject to the gener-ated bids provision. Also under exploration are revisions to the must-offer pro-visions to ensure that, as the amount of resource adequacy capacity coming from the use-limited category increases, CAISO will obtain the capacity's full resource adequacy value to meet operational needs.

A policy imperative implied by the above discussions of flexible resources is the need to develop a more robust economic framework to stimulate investment in, and ensure the commercial viability of, the flexible resources whose services will be more critically needed in the future. Many industry participants and policymakers recognize that the existing Resource Adequacy Program was not designed for the renewable resource future. The existing program looks at the system peak megawatt amount and adds a traditional planning reserve margin, looks only 1 year ahead in its annual procurement cycle, and does not distinguish among capacity types except to ensure sufficient capacity is procured in constrained load pockets. The graphs in the previous section indicate that this structure will not be sufficient to address flexible capacity needs, or accommodate the time it takes to permit and build a new generating plant.

To address flexibility issues, CAISO is working with the CPUC to include a flexible capacity procurement requirement in the Resource Adequacy Program. Still a work in progress, the initial CAISO proposal was to specify three aspects of flexibility: regulation capability, load following to meet 60 min ramping needs, and maximum continuous ramping capability. Load-serving entities objected saying it would be hard to meet procurement requirements because each generating plant might be capable of providing two or even all three of these capabilities. The CAISO and CPUC are exploring whether a simpler requirement with only one flexibility characteristic could work for an interim period.

To address the 1 year ahead limitation, CAISO has already modified its backstop capacity procurement rules to look at needs beyond the upcoming year, identify a generator that will be needed up to 5 years in the future, and procure that generator's capacity to keep it operating in the meantime. The need to look beyond the upcoming year became apparent recently when a newer flexible plant was not able to secure a resource adequacy contract and, faced with the prospect of earning only spot market revenues, was considering retiring. CAISO determined that the plant would be needed for its performance characteristics within the next 5 years and proposed offering a backstop capacity payment to forestall its retirement. General tariff provisions to support generators at risk of retirement that will be needed beyond 1 year in the future have now been developed.

The measures just described are only interim measures, however, until California implements some form of a multiyear forward resource adequacy mechanism that includes flexible capacity requirements. This may involve a centralized capacity market, and will need to be designed in collaboration with the CPUC and local regulatory authorities.

6.2.3 A Potential Real-Time Market Design

In response to FERC Order No. 764 on integration of variable energy resources,[39] the CAISO is considering a real-time market design concept that

---

39. "Integration of Variable Energy Resources," FERC Order No. 764 in Docket Nos. RM10-11-000, 139 FERC 61,246 (June 22, 2012).

could offer a solution to many of the operations-related market issues described earlier. The concept is to institute a 15 min interval real-time market in which energy is cooptimized with ancillary services. This market would be financially binding for settlement purposes and would establish the prices against which day-ahead virtual bids would be resolved. The existing 5 min dispatch would be retained as a balancing market; it would issue operationally binding dispatch instructions and set prices for settling deviations from the 15 min schedules.

This market design concept would align the CAISO real-time market with Order 764's requirement that balancing authorities offer 15 min interchange scheduling, yet would require only minimal changes to the existing real-time market infrastructure. Thus, the 15 min real-time market design could both address some major impacts of integrating renewable generation while simultaneously enhancing the coordination of markets and operations across the western region. The 15 min market structure would also provide opportunities for variable renewable energy resources to submit more granular schedules closer to real time and thus allow them to participate more fully in the CAISO energy markets.

6.3 Infrastructure Development

Transmission grid expansion and new generator interconnections were, until recently, driven by predictable load growth and incremental additions to or retirements from the supply fleet. The traditional procedures for performing these functions worked well in that context. The context has dramatically changed, however, which calls for innovative changes to these essential functions. On the positive side, California has a robust competitive marketplace for renewable generation projects with roughly four times as much generating capacity in the interconnection queue as will actually be needed. The challenge, however, is that efficient and timely infrastructure expansion decisions must be made under an unprecedented degree of uncertainty about which geographic areas will ultimately feature the greatest concentrations of new generation development. CAISO has developed a series of innovative reforms to shape infrastructure development accordingly.

6.3.1 Public Policy-Driven Transmission Planning

While it was becoming obvious that transmission expansion over the coming decade would be driven mainly by the need for access to areas with rich wind and solar energy resources, the traditional planning paradigm did not provide a basis for approving new transmission to meet such needs. Transmission upgrades to access renewable generation would not fit under the standard reliability or economic project categories. This meant that potentially costly transmission needed to access new renewable generation could not be developed through transmission planning. Instead, the separate interconnection process would approve upgrades to access new generation projects while transmission planning would address reliability and economic needs. Each process, independent and based on different criteria, would identify and approve potentially hundreds of millions of dollars of new transmission to address the different needs with no process for addressing all transmission needs holistically. CAISO responded by redesigning the transmission planning process to, among other things, include a new "public policy-driven" transmission category. The CAISO can now address all the drivers of grid expansion—reliability, economic, and now public policy—within the comprehensive annual transmission planning process.[40]

Following FERC approval of the revised planning process at the end of 2010, CAISO incorporated as a public policy planning objective the state's requirement to meet 33% of annual electricity demand from renewable resources by 2020. To implement this objective the planners expanded their studies to include an 8760 h production simulation to assess whether the renewable generation in each of the four CPUC-created resource portfolios could feasibly deliver 33% renewable energy on an annual basis.

In the 2011 planning cycle, CAISO added another factor to its public policy objective. Because load-serving entities must procure supply resources to meet their resource adequacy requirements in addition to meeting their renewable energy requirements under the 33% renewables portfolio standard, they generally want to obtain resource adequacy capacity from the same resources that supply their renewable energy. This procurement approach means that projects in the CAISO generation interconnection queue typically request more than simply the ability to connect reliably to the grid. They request "deliverability status" as well, to qualify to provide resource adequacy capacity to a load-serving entity. Absent such status, a project might either fail to obtain a power purchase agreement, or would have to accept a significantly lower contract price for its renewable energy.

What this additional factor meant for transmission planning was that CAISO had to add a deliverability study, to its assessment of public policy-driven transmission needs, which could indicate a need for additional additions or upgrades. The stage was set for CAISO's next major innovation, in infrastructure planning, which is described in the next section.

6.3.2 Integration of Generator Interconnection Procedures into the Transmission Planning Process

California's 33% renewables portfolio standard mandate triggered a tidal wave of renewable project proposals and generator interconnection requests.

---

40. Shortly after the CAISO filed its transmission planning reforms with the FERC in June 2010, the Commission issued a Notice of Proposed Rulemaking on transmission planning which included a requirement for a policy-driven transmission planning category. The Commission subsequently adopted this requirement in its Order No. 1000 in 2011.

The active queue contains roughly four times as much renewable generating capacity as will actually be needed to meet the mandate. As a result, there is great uncertainty as to which projects will actually achieve commercial operation. This creates a problem for the interconnection study process, because studies to identify network upgrades needed for each new annual cluster of requests start from the assumption that all active prior requests will ultimately become operational generating facilities. This assumption is necessary for accurately determining the incremental grid impacts from each new project cluster, and it has generally been valid when the amount of projects in the queue was in line with the amount of new generation that was needed and likely to be completed. But with today's queue volume, these assumptions are no longer valid and, consequently, the traditional study approach vastly overestimates the network upgrade requirements for each new cluster and leaves generation developers without realistic estimates of their transmission requirements. The result is that renewable project financing and bilateral energy contracting between project developers and load-serving entities have faced unmanageable uncertainties.

CAISO addressed these problems by integrating key aspects of the generator interconnection procedures into the transmission planning process. In July 2012, FERC approved the CAISO proposal, which is now being implemented for the latest interconnection cluster.[41] The new approach provides more realistic upgrade requirements for generation projects in the queue, even when the volume of requests is very large. It also addresses a need to limit ratepayer exposure to potentially high network upgrade costs resulting from generator requests to interconnect in inefficient locations.

The central new design concept is to use the transmission planning process, through the public policy-driven category, to identify and approve transmission upgrades to meet the interconnection needs of new generation facilities sufficient to achieve the state's 33% mandate. The new generation portfolio is developed by the CPUC with input from the California Energy Commission and municipal authorities. It reflects a highly probable geographic pattern of resource development based on the load-serving entities' procurement activities and the state's environmental assessment of preferred development areas.

The transmission approved in this manner is fully funded through transmission rates, and the new transfer capacity it creates is then allocated to meet the interconnection needs of generation projects that locate in areas supported by the transmission plan. CAISO allocation rules select the generation projects that will benefit from this transmission capacity based on meeting key development and financing milestones. If there are more project proposals in an area than the planned transmission can support, those

---

41. California Independent System Operator Corporation, *Order Conditionally Accepting Tariff Revisions*, FERC Docket No. ER12-1855 (July 24, 2012).

projects not successful in the allocation have to forego deliverability status and complete their projects as "energy only" facilities, pay for incremental upgrades without being eligible for cash reimbursement by ratepayers, or withdraw from the queue. Under this approach, the transmission planning process has become the venue for approving the most significant interconnection-driven upgrades as well as the vehicle for limiting ratepayer exposure to upgrade costs—all within a comprehensive planning process that also addresses reliability and economic project needs.

6.4 Remaining Challenges

Although CAISO has made great strides toward meeting the momentous changes occurring in the power industry, several key challenges remain.

- CAISO must develop a reliable methodology for determining the amounts of specific performance capabilities in specific locations that will be needed to ensure grid reliability in the future.
- Once the needs are determined, there must be a commercial and regulatory framework for making sure that the needed resources will be available.
- CAISO must work with distribution utilities, load-serving entities, and regulators to start thinking of transmission and distribution, and the wholesale and retail markets, as a single whole system. In the next few years, CAISO must expand its spot markets, resource visibility capabilities, load forecast methods, and grid operating practices to encompass the distribution level.
- CAISO must be sure its markets and operating structures are flexible and robust enough to adapt to the still uncertain path the industry evolution is taking. The strategy is to focus on needed operating characteristics and preferred market participant behaviors, rather than designing for specific types of technologies.

7 CONCLUSION

The electricity industry is going through an intense period of change, which requires organizations like CAISO that manage wholesale markets and transmission systems, to revise practically every aspect of their core functions. The main forces of change—energy and environmental policy mandates and rapidly evolving technologies—have already presented tough but surmountable challenges for market design, grid operations, and infrastructure planning. A common thread through all the present challenges is the need to make important decisions, ranging from very short-term unit commitment and dispatch decisions to longer-term market redesign and transmission upgrade decisions, in a context of much greater uncertainty than previously existed. The uncertainties cannot be eliminated, but they can be addressed and managed prudently and efficiently.

Within this context, and following the comprehensive market redesign of 2009, CAISO has fully transitioned from the crisis management imperative that dominated the post 2000–2001 period to take a leading role in designing and implementing initiatives to embrace and facilitate the ongoing green and technological transformation. CAISO foresees immense benefits accruing to electricity consumers, industry, and the environment and society as a whole from a power system that is smarter, greener, more distributed, and more efficient. Toward this end, CAISO will be continuing the program of change described in this chapter while ensuring that the spot markets remain stable and efficient and the grid remains reliable through the transition years.

Chapter 12

Unfinished Business: The Evolution of US Competitive Retail Electricity Markets

Young Kim
Former Associate Director, Retail Markets at DNV KEMA Energy & Sustainability

1 INTRODUCTION

One of the main reasons for restructuring electricity markets in the United States in the mid- to late 1990s was to give customers a choice. It was generally believed that the transmission and distribution of electricity was a natural monopoly, which required regulation, but the generation/supply of electricity should be deregulated and competitive. The states that were most willing to go down this path of restructuring the vertically integrated electric utilities were the ones that faced higher prices. One of the benefits of restructuring, early retail market advocates argued, was that competition would bring fairer and hopefully lower prices in the long run.

In the late 1990s and early 2000s, a total of 20 states plus District of Columbia (D.C.) passed laws to restructure their electricity markets to give some form of choice to customers (Figure 12.1).

Overall, the competitive retail market in the United States has enjoyed phenomenal growth. It has quietly become a $50–60 billion industry with more than 200 retail companies collectively serving approximately 14.5 million customers. The development of competitive retail markets in some states, however, has been more successful than in others. This chapter examines some of the reasons why certain markets flourished and will continue to do so, whereas other markets expanded in recent years but are doomed to give back their gains. Some of the key players and issues around market share and concentration will also be discussed, as these are measures of the health and vibrancy of competitive markets. The experience of competitive retail markets in the United States has an important bearing on similar developments in other parts of the world.

This chapter is organized as follows: Section 2 provides a global context for US competitive retail activity and market size; Section 3 covers the various retail choice models in the United States and provides views on the

Evolution of Global Electricity Markets.

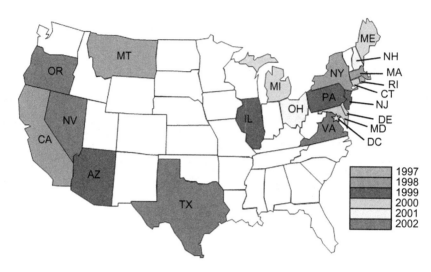

FIGURE 12.1 US states with retail choice and year opened. *Source: KEMA, EIA, 2012.*

possibility of expanding retail choice; and the chapter's conclusions are summarized in Section 4.

2 MARKET SIZE AND COMPETITIVE ACTIVITY

In the United States, total electricity consumption is estimated at 4238 TWh according to the Energy Information Agency (EIA).[1] Not all jurisdictions or electric utilities in the United States have been restructured to allow retail competition. A subset of total energy consumption in the United States is "eligible" to switch to a competitive supplier. Approximately 70 utility territories in the United States have competitive retail choice, and the eligible sales volume is an estimated 1305 annualized TWh.[2] The current overall switch rate is estimated at 54% of *eligible* sales, which is 708 annualized TWh of competitive retail sales. By customer segment, the nonresidential switch rate is 66% (562 annualized TWh) and the residential switch rate is 32% (146 annualized TWh) as shown in Figure 12.2.

Even though competitive sales represent just 17% of total electricity consumption in the United States, the figure of 708 annualized TWh would make it the sixth largest market in the world—behind India (755 TWh) and

---

1. Energy Information Agency's Annual Electric Power Industry Report, September 2012.
2. "Annualized" power sales estimates are for the current year. Annualized sales are an estimate of market size for the current period. To derive annualized estimates, the most recent switch rates (energy or adjusted peak load) are applied to the current year's estimate of eligible sales volume. Annualized sales are not directly comparable to calendar year sales estimates.

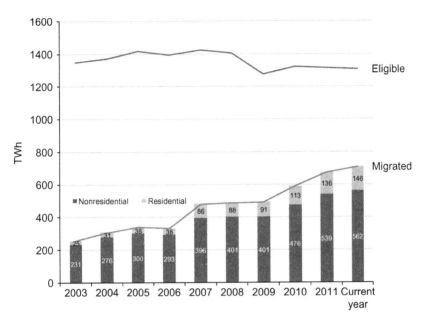

FIGURE 12.2   Eligible and competitive power sales, 2003 to current year (TWh). *Source: KEMA, EIA, State Commissions, 2012.*

ahead of Germany (590 TWh).[3] As a point of reference, total electricity consumption in the entire continent of Africa at 603 TWh is lower than the competitive retail market in the United States.

As described in other chapters (e.g., Chapters 1, 8 and 19) of this volume, the UK market (357 TWh) is fully competitive but the US competitive market is nearly double in size. In Australia (227 TWh), the three largest states—Queensland, New South Wales, and Victoria—are fully competitive. However, in both of these competitive markets, as critics point out, the vast majority of the market is served by the legacy or incumbent providers. In Germany, despite the opening of electricity markets to competition in 1998 for all customers, as of 2009 only 6% of residential customers and 14% of nonresidential customers have switched away from the utility. There has been relatively little movement in competitive retail markets in countries ranging from France to Japan, mostly for political reasons, as further explained in other chapters (e.g., Chapters 3, 4 and 23) of this volume.

Just as competitive activity in some international markets are more vibrant than in others, some states in the United States have strong competition whereas others have limited retail competition. Variations in policies across jurisdictions with open retail markets are reflected in the vast

---

3. IEA's Key World Energy Statistics, 2012. Annual electricity consumption data is from 2010.

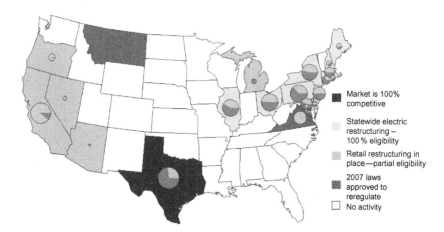

FIGURE 12.3 US states with retail choice—relative size of total and competitive markets, 2012. (For interpretation of the references to color in this figure legend, the reader is referred to the web version of this book.)[4] *Source: KEMA, EIA, State Commissions, 2012.*

differences in market activities. Figure 12.3 depicts states that allow customer choice regarding retail electric providers, the relative size of the eligible market, and the proportion of the market that is served competitively.

The main differences between the various retail choice models and default service structures in the United States can be categorized into three main segments further described in Section 3:

- No default service: Texas
- Market pass-through: New York
- Auction or request for proposal (RFP): Most PJM and New England states
- Hybrid or other: Ohio, California, and Michigan.

Despite the many differences in retail choice models, a common methodology is employed in this chapter to create comparable statistics across all choice markets. A few notes about the market sizing methodology:

- Historical, current, and projected size of the eligible, competitive, and default service markets are estimated for all restructured markets. The

---

4. In the map, the size of the pie associated with each state is the relative size of the total eligible electric market; the dark blue shaded area is the percentage of total eligible customers who have switched to a competitive supplier. Twenty states and D.C. have retail electric choice. Although two states passed reregulation laws in 2007, customers >5 MW are still allowed to shop in Virginia, and Montana customers can continue to take competitive service but cannot switch again if they return to utility service. Texas switching rates are for ERCOT only. The Texas market is fully competitive and therefore 100% of the market is served competitively. However, the total percent switched (to nonincumbent retailers) volumes are shown in this map.

eligible market is defined as the customers and corresponding energy sales that may choose a competitive supplier of electricity. The competitive market is defined as the customers and corresponding energy sales that are served by competitive or unregulated retail energy suppliers. The default service market refers to customers and sales that are served by the regulated utility and is calculated as the difference between the eligible and competitive markets.

• For historical eligible and competitive market size data (pre-2012), the source is the EIA Annual Electric Power Industry Report (EIA-861). For subsequent years, the size of the eligible market is estimated by applying growth rates from the EIA Short-Term Energy Outlook and Annual Energy Outlook. To estimate recent competitive migration, switching statistics from state commission reports are used.[5] For projections of future switching activity, forecasts are developed using an analysis of default service policies, recent switching trends, and competitive supplier activity.

3 RETAIL CHOICE MODELS

3.1 No Default Service—The Texas Model

Texas is generally regarded as a successful model in restructuring its electric power sector as further described in Chapter 10. Senate Bill 7, which restructured retail electric markets in the state, was signed into law on June 18, 1999.[6] In January 2000, utilities filed their business separation plans, which laid out how the utilities proposed to unbundle business activities into three units: a power generation company, a retail electric provider, and a transmission and distribution company. In some cases, the incumbent companies were completely separated (new owners). In September 2000, the Texas Public Utilities Commission (PUC) began certifying retail electric providers.

Base electric rates for most residential and small business customers were frozen until competition began in January 2002. Under competition, each customer except those in service territories of municipalities and cooperatives that have not opted for competition has the right to select an electricity provider. However, only territories within Electric Reliability Council of Texas (ERCOT) in Texas are open for competition, though retail access in the Panhandle and West Texas continue to be considered.

The Texas restructuring model is unique in the United States. for several important reasons. Differences in policies and market design explain why the state has become the most active and advanced competitive retail market in the nation.

---

5. The most recently available migration data as of this publication is the quarter ending June 30, 2012.
6. Senate Bill 7: http://www.legis.state.tx.us/billlookup/History.aspx?LegSess=76R&Bill=SB7.

The primary differences in Texas compared to other US retail markets include:

- Business separation. Incumbent, vertically integrated utilities were separated into three legally and structurally separate entities. In some cases, the incumbent companies were completely separated (new owners).
- Default service pricing. For the first 5 years, default service pricing was set through the Price to Beat (PTB) mechanism, which enabled competition. Upon termination of the PTB, the ERCOT retail market was fully competitive (meaning there is no default utility provider), which is unique in the United States.
- Public Utility Commission of Texas (PUCT) jurisdiction. The PUC in the only state commission that has primary jurisdiction over both the wholesale and retail electricity markets. In contrast, other states have to deal with federal regulators, namely the Federal Energy Regulatory Commission (FERC).
- Single retailer model. Texas law requires that the delivery companies effectively sell wire services to retailers that then resell the service to customers. Transmission and distribution companies in the Texas competitive market have very limited interaction with customers (mostly limited to receiving outage calls). Consequently, the wire companies do not create competitive barriers in Texas, unlike other US markets.

As mentioned in Chapter 10, Texas is the most active and competitive market in the United States as measured by number of competitors, persistent migration levels, and total load migrated.

Figure 12.4 shows the historical, current, and projected switch rates and competitive sales volume by customer segment (residential and nonresidential). Currently, the four largest utility distribution territories in Texas—Oncor, Centerpoint, American Electric Power (AEP) Texas, and Texas New Mexico Power—are 100% competitive. These four territories have a combined 226 annualized TWh. In terms of number of customers, these four territories comprise 6.4 million customers that are served competitively.

The 2002–2006 period, as described in Chapter 10, was a gradual transition to a retail market completely free of price controls. The expansion of the Texas competitive retail market bears this out. Total competitive sales volumes were only 78 TWh in 2003. By segment, the nonresidential market was 67 TWh or 51% switched, and the residential market was 11 TWh or 15% switched.

Until 2007, the PTB mechanism was in place, which allowed the competitive retail market to develop. PTB sets a price floor for the legacy or incumbent providers in the former vertically integrated utility territories.[7] By

---

7. The legacy or incumbent retailers (also known as affiliated retail electric providers) are as follows: TXU Energy in the Oncor distribution territory, Reliant Energy in the Centerpoint territory, Direct Energy in the AEP Texas territories (AEP North and AEP Central), and First Choice Power in the TNMP territories (FCP was acquired by Direct Energy in September 2011).

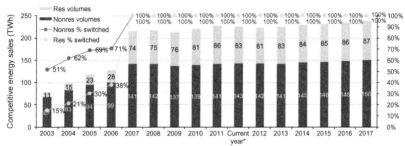

*Note that current year is an annualized estimate based on the most recent switch rates applied to the current year estimate of eligible sales volume or accounts.

FIGURE 12.4   Texas competitive power sales and switch rates, 2003−2017 (TWh, %), base case. *Source: KEMA, EIA, State Commissions, 2012.*

design, the PTB prices were higher than market rates, which created headroom for new entrants to compete against the affiliated retail electric providers (AREPs)—for example, Reliant in Centerpoint territory (Houston).

In 2007, the market development period (MDP) via PTB ended, and AREPs were free to compete without a regulated above-market rate. Total competitive sales jumped from 127 TWh in 2006 to 215 TWh in 2007. Today, just over 25% of total sales volume among the four distribution territories is served by the AREP in their respective territories.[8]

Since 2007, the size of the competitive retail market in Texas simply fluctuated with electricity usage patterns by customers in the four major distribution territories. There have been no regulatory changes that affected the size of the competitive market to the four distribution territories since 2007. In 2014, the competitive market is expected to grow by approximately 1.1 TWh with the expansion of choice in the Sharyland Utilities.

Market concentration in Texas is moderate, as the top five nonresidential retailers have a market share of 62%. The top firm in Texas (NRG) has 20% market share (Figure 12.5).

For the residential segment, there is more market concentration with 70% market share among the top three retailers. These three firms are all former incumbents, and the top firm in Texas Utilities (TXU) has 33% market share. Stream Energy is the only nonlegacy provider that is close in market share to the top three (Figure 12.6).

The key to the success of the Texas model—no default service—has led to the most dynamic and vibrant retail market in the country. There are 40−50 active retail companies that compete directly for residential or commercial customers in Texas. In each distribution service territory, a

---

8. The source is the most recent report card on retail competition from the Public Utility Commission of Texas, with data through March 31, 2012.

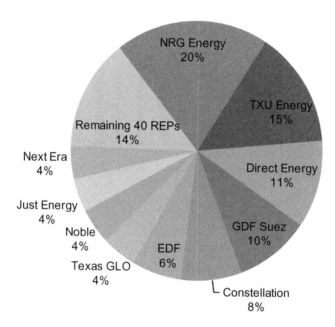

FIGURE 12.5 Texas nonresidential market share by retail company (2011).[9] *Source: KEMA, EIA, 2012.*

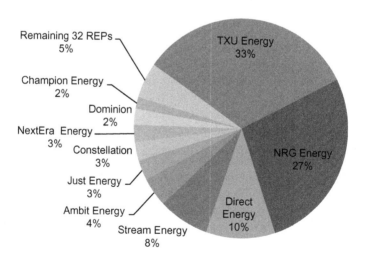

FIGURE 12.6 Texas residential market share by retail company (2011). *Source: KEMA, EIA, 2012.*

9. All market share data is sourced from the EIA (form EIA-861).

residential customer can choose from over 200 different offers. Retail offer innovation occurs in Texas with much more frequency than in other states, and the deployment of smart meters promises to accelerate this even further.

This model, where retailers are competing with each other, rather than an administratively determined default service price, requires firms to be nimble and reactive to competitors and customer needs. While the costs for operating in Texas are typically higher since retailers serve and bill customers directly, the margins for successful retailers are also higher. As long as there are no significant disruptions in the wholesale market (resource adequacy, scarcity pricing, etc.),[10] the retail market in Texas will continue to flourish.

3.2 Market Based on Cost Pass-Through—The New York Model

New York is unique because it does not have a statewide electric restructuring statute that requires customer choice. The New York Public Service Commission (PSC) established retail electric choice through individually negotiated utility settlement agreements in 1996–1997 and several other Commission developed policies.

Competition began in 1998 at different times in the utilities' service territories and was phased in fully in 2002, when all customers were able to choose. The retail rate structures vary significantly from utility to utility and are generally complex, although the PSC has increased the uniformity of rules and rate structures over time. In most cases, the utilities rely on a supply pricing mechanism that utilizes the New York Independent System Operator (NYISO) market clearing prices for energy, capacity, and ancillary services. However, certain key rate components and pricing mechanisms vary across New York utilities.

New York was one of the first states to base its default service rates on market pass-throughs and this has helped the state to avoid some of the "boom or bust" retail environments that characterize states with more static default service rates, resulting in headroom[11] that is relatively stable compared to other markets.

Pro-market retail policies established in the 2003–2006 period laid the foundation for persistent increases in residential and nonresidential migration levels. By August 2008, all six utilities had total retail switching rates (all customer classes by load) over 35%.

---

10. For more details, see Chapter 10.

11. Headroom is defined generally as the difference between (a) the bypassable components of the retail price of electricity that are provided by the default service utility should a customer switch away from bundled utility service and (b) the competitive retailer's cost of providing the equivalent set of products (also known as the retailer's cost of goods sold). Bypassable components are typically the utility's generation costs, capacity, transmission and ancillary services. In some cases, to encourage retail competition, utility commissions will include retail adders to increase headroom.

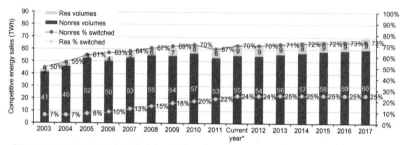

*Note that current year is an annualized estimate based on the most recent switch rates applied to the current year estimate of eligible sales volume or accounts.

FIGURE 12.7 New York competitive power sales and switch rates, 2003−2017 (TWh, %), base case. *Source: KEMA, EIA, State Commissions, 2012.*

The two major charges for energy supply from the utility are the Market Supply Charge (MSC) and the Merchant Function Charge (MFC). The MSC is the price the utility pays for the electricity that is consumed. The MFC covers other costs involved with serving default service, such as billing and collection charges. Combined, MSC and MFC dictate what customers pay for energy supply, and are the "price to compare."

Currently, New York has one of the most active retail electric markets, though most growth is now incremental, compared with prior rapid growth when the state's utilities implemented Purchase of Receivables (POR),[12] retailer or energy service company (ESCO) referral programs, and other measures. Transition periods (in which generation rate caps[13] were in place) have ended for all utilities. New York has adopted an ESCO Consumer Bill of Rights which limits termination fees for residential customers or customers solicited door-to-door. In addition, past practices that provided the utilities with incentives to move customers to retail choice providers (such as a $10 payment to the utilities for every account that moved to retail choice) have been terminated, as the PSC no longer considers such subsidies to be necessary given the level of retail choice activity in the state.

---

12. Purchase of Receivables is a policy that is designed to reduce the barriers of entry for retail competition. While implementation varies by state and/or utility territory, the fundamental concept is that utilities are responsible for calculating the percentage of bad debt or uncollectible expenses across their distribution territory and for spreading out those costs equally whether a customer switches to a competitor or stays with the default service utility. Since customer creditworthiness is a challenge in many markets, especially for mass markets, this policy encourages competitive retail entry by making retailers indifferent to a potential customer's bad debt risk.

13. In many developing retail markets, in order to avoid rate shock, state commissions capped the potential increases in the generation component of energy prices. With few exceptions, rate caps have been lifted and generation prices more closely follow competitive wholesale markets. In many cases, the lifting of these rates caps led to a very strong increase in competitive switching.

New York is one of the most vibrant competitive retail markets as measured by the number of competitors, switching levels, and competitive market size. Figure 12.7 shows the historical, current, and projected switch rates and competitive sales volume by customer segment (residential and nonresidential). Currently, New York total competitive sales are 64 annualized TWh, or 55% switched. The number of customers served by a competitive retail supplier is 1.6 million.

Due to a default service market design that is relatively stable, the New York market has shown a steady and sustainable development of competitive retail activity since 2003. Total competitive sales volumes were 43 TWh in 2003. By segment, the nonresidential market was 41 TWh or 50% switched, and the residential market was 2 TWh or 7% switched.

Switching in both segments expanded gradually since 2003 to their current levels. The New York nonresidential market is now 55 annualized TWh or 70% switched, and the residential market is 9 annualized TWh or 24% switched.

Over the next 5 years, the current switching levels in most utility territories in New York will continue to be supported and there will be incremental advances in some markets. Since New York has dynamic default service rates based on a pass-through of NYISO prices, the opportunity for competitive retailers will be consistent compared to some other markets.

In New York, nonresidential switch rates are expected to increase due to the lowering of the mandatory hourly pricing threshold by mid-2013. Typically, customers that are forced to take hourly spot pricing from the default utility seek out a fixed price arrangement from competitive suppliers to improve the predictability of their electricity costs. Therefore, lowering the hour price cutoff leads to higher switching. Niagara Mohawk (National Grid) will implement by June 2013 a cutoff of 250 kW for hourly pricing. Orange and Rockland will be lowering their threshold from 500 to 300 kW in May 2013, and possibly as low as 100 kW in the spring of 2015.

In the nonresidential segment for New York, the market share for the top five retailers is 70%. The top firm in New York (Hess) has 19% market share. Of the top five, only Consolidated Edison (ConEd) Solutions is affiliated with a regulated utility in the state (Figure 12.8).

In New York's residential segment, the top three command 59% market share. The top residential firm in New York (Direct Energy) has a market share of 28%. With the acquisition of New York State Electric and Gas (NYSEG) Solutions by Direct Energy in July 2012, there is no firm in the top five that is affiliated with a regulated utility in the state (Figure 12.9).

New York's default service model is the most attractive for retailers next to Texas. Since the regulated rate is a straight pass-through of costs from the wholesale market, retailers have less risk over the long run in terms of competing against the default service rates. This model is attractive for both retailers and customers.

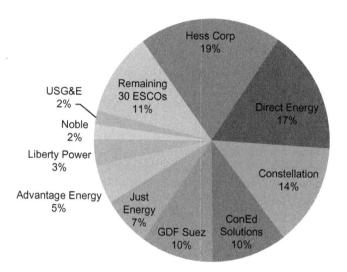

FIGURE 12.8 New York nonresidential market share by retail company (2011). *Source: KEMA, EIA, 2012.*

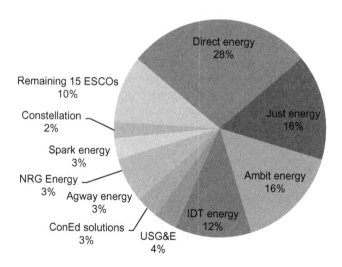

FIGURE 12.9 New York residential market share by retail company (2011). *Source: KEMA, EIA, 2012.*

In recent months, some of the policies that would keep the New York retail market on a growth path have been under some pressure. The retail enhancement of POR could be modified to make it less attractive for retailers to market to the credit-challenged. Hourly price thresholds could be maintained at current levels as opposed to being lowered for the smaller commercial customers. However, the foundation for retail markets remains intact, and that is the strong default service model. While New York did not go as far as Texas did in restructuring electricity markets, it is still second best.

3.3 Auction or RFP

Throughout many PJM and New England markets, default service procurement is conducted via auction or RFP. Although the method of procuring power can be similar across these markets, the exposure to wholesale market prices varies widely by state, utility territory, and customer class. In general, default service contracts are shorter and rate setting frequency is higher for larger customers. For example, customers in New Jersey with load >750 kW are on an hourly price if they receive supply from the default utility, which means that they have a very high level of exposure to the spot market. At the other end of the spectrum, residential customers in Illinois have their rates set once per year and default service is a blended rate based on procurements in the prior 3 years (Figure 12.10).

Compared to the Texas and New York models, auctions and RFPs are not as conducive to sustained levels of competitive market opportunity for all customer classes. Markets may advance or decline based on the timing of commodity prices. That is, in the current time period, competitive retailers enjoy a default service price that is above market in some cases due to long-term supply contracts (i.e., the inclusion of higher prices in prior years). When those high prices eventually roll off, the ability to undercut the default service price might disappear and retailers could exit the market.

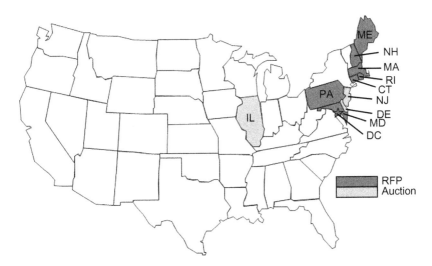

FIGURE 12.10 Map of US states with default service procurement via auction or RFP. *Source: KEMA, 2012.*

Within this type of default service model, the nuances are extremely important. These variations include rate setting frequency, length of default service contracts, laddered or multiyear procurements, and customer segmentation.

The three states within this default service category that merit special attention as being the most vibrant and therefore most useful for discussion purposes are Connecticut, Illinois, and Pennsylvania.

3.3.1 Connecticut

Utilities provide Standard Service (SS) to residential, small-, and medium-sized business customers (<500 kW) who do not receive power from competitive suppliers, and Last Resort Service (LRS) to larger (>500 kW) customers. SS and LRS rates are based on RFP solicitations for wholesale power supply with only minor adjustments by the Connecticut Department of Public Utility Control (DPUC). The bulk of SS is laddered over 3 years.

Prior to the expiration of generation rate caps at the end of 2006,[14] competitive retailers could not compete with the default service utility, and the retail market failed to develop.[15] In 2007, after rate caps were lifted, the Connecticut market expanded dramatically to 10 TWh or 33% switched (Figure 12.11). After the recession in the fall of 2008, and the sudden collapse of the wholesale power market, competitive retailers took advantage of the substantial headroom in the market. The Connecticut market is currently 19 annualized TWh or 69% switched. As in most jurisdictions, the Connecticut nonresidential market has more competitive activity with 13 annualized TWh or 86% switched, compared to the residential market with 6 annualized TWh or 47% switched. In terms of number of customers, 0.7 million are served competitively.

Connecticut is currently set up to be a "boom or bust" retail market because of the default service market design. Since smaller customers that

---

14. As noted above, generation rate caps were a common policy implemented throughout numerous jurisdictions in the early/transition stages of retail market development. Default service utilities were permitted to cap rates in order to avoid rate shock to their incumbent customer base.

15. Connecticut's initial price-to-compare was a capped rate Standard Offer (SO) that was in effect from 2000 through 2003. This was changed to a semi-capped rate Transitional Standard Offer (TSO) for the January 2004–January 2007 period. In theory, the TSO was capped to not exceed the utilities' 1996 bundled rates. However, the DPUC allowed utilities to recover energy and transmission costs through other charges. The generation portion of the TSO, unlike the SOS, was based on wholesale power costs resulting from periodic RFPs. Yet the TSO did not offer adequate headroom to competitive retail suppliers despite its market-based procurement mechanism and its ability to exceed rate caps. This limitation became especially apparent in the United Illuminating (UI) service territory, where most of the TSO supply had been procured in 2003 when wholesale electric prices were relatively low.

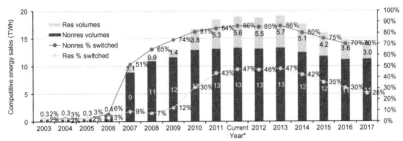

*Note that current year is an annualized estimate based on the most recent switch rates applied to the current year estimate of eligible sales volume or accounts.

FIGURE 12.11   Connecticut competitive power sales and switch rates, 2003–2017 (TWh, %), base case. *Source: KEMA, EIA, State Commissions, 2012.*

stay with the utility are served on a multiyear or laddered procurement schedule, current market prices are significantly lower than the default utility prices. In a declining price environment, the default utility price will include higher prices from prior years and will therefore be above market compared to competitive retailers, who generally procure short-term power under current market conditions.

The period following the recession of 2008 has been a "boom" market for Connecticut, but it will likely be followed by a "bust" market in the following years. Power prices are expected to rise due to economic recovery and costly environmental upgrades. Just as retailers had the benefit of expanding headroom in a declining price environment, they will experience shrinking headroom in future years in a rising price environment. By 2017, much of the gains in competitive retail markets after the recession will be given back.

The nonresidential segment is moderately concentrated for Connecticut, with market share for the top five retailers at 72%. The top firm in Connecticut (Constellation) has 28% market share (Figure 12.12).

In Connecticut's residential segment, the top five command 59% market share. The top residential firm in Connecticut (Public Power) has a market share of 19%. There are no retailers in Connecticut that are affiliated with the regulated utilities. Connecticut is the only state that reports up-to-date competitive market share information by customer counts[16] (Figure 12.13).

---

16. There is usually a time lag of at least 1 year for firm-specific market share information being made public. Utility commissions and federal agencies that collect this information deem this information too sensitive to proper functioning of competitive markets. This is why market share information presented here is from 2011 in most cases.

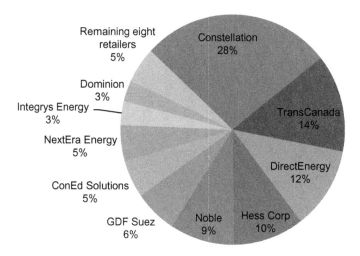

FIGURE 12.12 Connecticut nonresidential market share by retail company (2011). *Source: KEMA, EIA, 2012.*

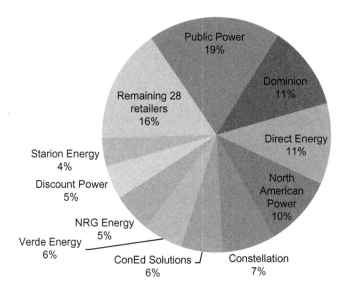

FIGURE 12.13 Connecticut residential market share by retail company (July 2012). *Source: KEMA, State Commission, 2012.*

3.3.2 Illinois

In the two largest utility territories, Commonwealth Edison and Ameren, default service supply is procured via auction through the Illinois Power Agency (IPA). Legislation from 2007 required swaps between the utilities

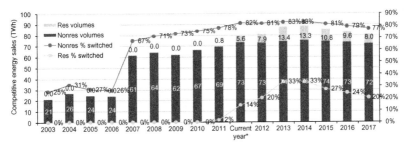

FIGURE 12.14   Illinois competitive power sales and switch rates, 2003−2017 (TWh, %), base case. *Source: KEMA, EIA, State Commissions, 2012.*

and their generation affiliates through 2012 in Ameren and 2013 in ComEd. Aside from these legacy swaps, the IPA procures volume for fixed price standard offer service (SOS) on a 3-year ladder (35% of supply procured 2 years in advance, 35% procured 1 year in advance, and 30% procured in the year of delivery). Customers <100 kW in ComEd are served on a fixed price SOS, whereas new customers <150 kW in Ameren are served on a fixed price SOS. Customers with higher load on default service are served on an hourly price, and these mandatory hourly cutoffs are the lowest in the United States.

Rate caps were lifted at the end of 2006, which led to the first step-change in the development of competitive retail markets in Illinois since it first opened. As shown in Figure 12.14, the nonresidential market expanded to 61 TWh or 67% switched (or 46% of the total Illinois market including residential). The legacy swaps in 2007, which were long-term contracts through 2012 and 2013 at the prevailing market prices, kept the small commercial market (<100 kW in ComEd, <150 kW in Ameren) from expanding significantly through 2009.

After 2009, the next period of market development in Illinois began. With declining wholesale power prices due to a slower economy, the legacy swaps were now above market, which created a "boom" scenario for competitive retailers. Nonresidential sales grew from 62 TWh in 2009 to 69 TWh in 2011, a growth of 11%. The current estimated size of the nonresidential market is 73 annualized TWh, which is 81% switched. In terms of customer counts, approximately 200,000 are currently served by competitive suppliers. Increased switching in the small commercial market is expected to drive the growth in the market through 2013. After 2014, the small commercial classes will begin to return to utility service as we enter into the "bust" phase resulting from rising wholesale prices and therefore declining headroom given this default service model. Hourly price customers, however, are expected to remain with competitive suppliers.

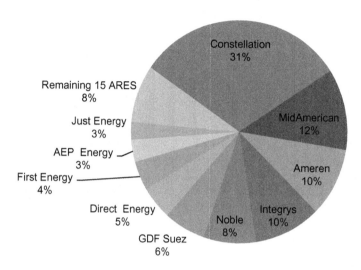

FIGURE 12.15 Illinois nonresidential market share by retail company (2011). *Source: KEMA, EIA, 2012.*

The residential market, which was insignificant in 2010, began to develop last year—and develop very rapidly, primarily because of municipal opt-out aggregation in Illinois.[17] Currently, 5.6 annualized TWh or 14% of the residential market is competitively served. Nearly 600,000 residential customers have currently switched away from their default utility. The combination of the legacy swaps and default service procurement timing (3-year laddering) have created a similar "boom" market for residential suppliers. In some cases, retailers are able to offer savings of 30% off the default utility rate.

Despite the swift pace of expansion in recent years, the residential market will retreat off their highs. The reason behind the success of aggregation was price savings, but it is not sustainable. When the favorable aggregation deals expire in 2014 and 2015, competitive retailers will not be able to offer the same savings levels. Nearly half of the aggregation deals will fail to renew, and these communities will return to utility default service. While municipal opt-out aggregation in Illinois has been deployed effectively to take

---

17. Municipal opt-out aggregation for electricity allows switching for an entire community—a township, city, or county—to a competitive retail supplier. Communities would vote to either pass or block an aggregation initiative through a referendum in a ballot of an election. If passed, a supplier is then selected by the community, typically through a competitive bidding process. For opt-out aggregations, all customers of a community that passed the referendum would be switched to the winning supplier automatically unless they specifically opted out. By contrast, opt-in aggregations would require customers in a community to send an affirmative request agreeing to the switch. Naturally, adoption rates for opt-out aggregations are considerably higher than opt-in.

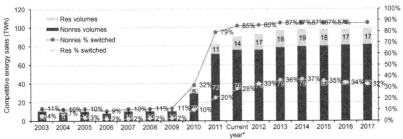

*Note that current year is an annualized estimate based on the most recent switch rates applied to the current year estimate of eligible sales volume or accounts

FIGURE 12.16   Pennsylvania competitive power sales and switch rates, 2003−2017 (TWh, %), base case. *Source: KEMA, EIA, State Commissions, 2012.*

advantage of the current "boom" market, it is not a policy that will form a foundation for long-term residential market development.

In the nonresidential segment for Illinois, market share for the top five retailers is 70%. The top firm in Illinois (Constellation) has 31% market share. Of the top five, three retailers (Constellation, MidAmerican Energy, and Ameren Energy Marketing) are affiliated with regulated utilities in the state (Figure 12.15).

Residential market share information in Illinois from 2011, given the low levels of switching in that year, would not be indicative of current market players. Since most of residential switching in Illinois has been through the aggregation channel, a summary of these deals would provide some visibility into Illinois residential market share. The retailers that currently supply aggregation contracts in Illinois number 7 in ComEd and 3 in Ameren. Retailers with the most number of aggregation deals are FirstEnergy Solutions (96), Homefield—an Ameren affiliate (58), Integrys (29), and MC Squared—a DPL affiliate (23).[18]

3.3.3 Pennsylvania

Default service varies by utility, but most procure supply use auctions and RFPs to achieve a mix of spot, short-term, and long-term purchases designed to achieve the least cost to customers. Most importantly, and in contrast to many of the other states in this auction/RFP category, most utilities in Pennsylvania have a relatively short length for default service procurement. With few exceptions, default service for smaller customers is laddered over 1−2 years, as opposed to 3.

18. The Illinois Commerce Commission (ICC) is the source of statistics on communities that have switched, with details on suppliers of the municipal opt-out aggregation programs.

Pennsylvania is a case study for the negative consequences of nonmarket-based generation rate caps: pent-up demand and lost customer value. In every Pennsylvania utility territory that lifted rate caps, switch rates rose dramatically in the following year. When rate caps were lifted, customers switched rapidly to take advantage of market-based rates. Large spikes in retail competitive activity occurred immediately following the expiration of rate caps. Duquesne caps were lifted in 2004, Penn Power in 2006, and Pennsylvania Power and Light (PPL) in 2009. Most recently Allegheny Power, Metropolitan Edison (Met-Ed)/ Pennsylvania Electric (Penelec), and Philadelphia Electric Company (PECO) caps ended in December 2010. The rate cap era, though recent, appears to be firmly in the past.

As shown in Figure 12.16, the competitive market expanded from 11 TWh (or 8% switched) in 2009 to 82 TWh (or 58%) in 2011. There are few precedents for this type of growth in competitive markets in such a short period of time. The Pennsylvania market is currently 90 annualized TWh or 64% switched. The nonresidential market has more competitive activity with 76 annualized TWh or 85% switched, compared to the residential market with 14 annualized TWh or 28% switched. In terms of number of customers, 1.6 million are served competitively.

In Pennsylvania, the residential market is making advances in this low-priced environment. One positive sign for retail markets was that the state's PUC decided in June 2012 to delay mailings to educate consumers about electric choice.[19] The PUC's primary reason for doing so was that customers were already receiving high volumes of mailings on shopping for power. Switch rates in all residential markets are up from Q1 to Q2 2012.

In addition to promoting competition and choice, the PUC is currently undertaking an investigation of possible changes to default service.[20] Pennsylvania is the only auction/RFP market that could potentially alter the structure of default service. In the most extreme scenario, the PUC may rule that default service should not continue in its current form, and the role of default service provider should be fulfilled by an entity other than the incumbent

19. In December 2011, the Pennsylvania PUC ordered all eight of the major electric distribution companies in the state to mail a postcard encouraging residential and small commercial customers to shop for competitive power supply. In August 2011, the PUC ordered two of the First Energy utilities, Met-Ed and Penelec, to mail 1 million letters to small volume customers informing them of available competitive supplier offers.
20. In August 2010, Direct Energy proposed that the default service role should be transitioned to competitive retailers as a condition of the First Energy–Allegheny merger. While the PUC ultimately decided to approve the merger without this condition (since it could not change the fundamental structure of default service for one set of utilities), the PUC opened a formal state-wide investigation into the matter in April 2011. Among the expected changes in default service is the introduction of retail opt-in auctions for customers not currently on competitive supply. The impact of opt-in auctions is expected to be minimal given the fairly low acceptance rates of opt-in programs.

electric distribution company. If this transition was to happen, Pennsylvania would follow the Texas model of no default service. The earliest any such changes would be implemented would be June 2015, after the current default service plans finish their terms.

While it is difficult to say for sure how far the retail market investigation will go, there is reasonable certainty that any changes to default service will be positive for the state's retail markets. Recent decisions by the PUC provide some guidance. A positive bellwether was the PUC's decision to allow Pike County, a small utility of approximately 5000 customers, to rely exclusively on the NYISO spot market for default service. The PUC determined that Pike County's approach satisfies the test of "least cost over time" and that they should not be required to procure default service with hedging. Though this is a unique case of a utility with a small customer base, it suggests that going forward the PUC will allow flexible interpretations of Act 129,[21] including less long-term default service procurement and possibly opening the door for retailers to provide default service.

Another informative decision was the PUC's response to First Energy's proposed default service plan. The combined First Energy utilities[22] proposed a 24-month default service procurement to fulfill all of their residential supply requirements. Approval of this plan would have likely resulted in a less vibrant retail market, and the PUC decided in August 2012 to reject this plan as proposed. The reason cited in the PUC's decision was the concern that a 2-year procurement would pose a risk that default service rates would not remain economical relative to other electric supply options. First Energy argued for rate stability in the near term, but the PUC considered the long-term view. Ultimately, the PUC adopted a mix of 12- and 24-month contracts for residential customers.

In the nonresidential segment for Pennsylvania, the market share for the top five retailers is 65%. Market concentration—compared to similarly sized markets—is low, which indicates vibrant competition. The top firm in Pennsylvania (First Energy Solutions) has 22% market share. Of the top five, two providers (First Energy Solutions and PPL EnergyPlus) are affiliated with regulated utilities in the state (Figure 12.17).

In Pennsylvania's residential segment, the top five command 79% market share. Dominion Retail has a dominant market share of 40%. Of the top five, two retailers (First Energy Solutions and Constellation) are affiliated with regulated utilities within Pennsylvania (Figure 12.18).

---

21. Act 129 is a revision of Pennsylvania's competition act passed in October 2008. Among other policy objectives, Act 129 required that default service providers procure a "prudent mix" of contracts to ensure "least cost to consumers over time." It was generally interpreted that the PUC would require a portfolio approach to default service procurement, but recent decisions challenge this assumption.
22. In Pennsylvania, the First Energy utilities are Met-Ed, Penelec, Pennsylvania Power, and West Pennsylvania Power.

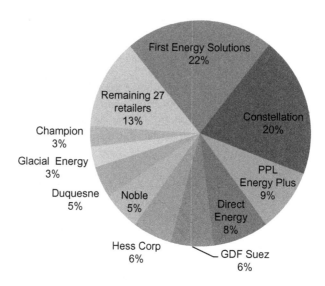

FIGURE 12.17   Pennsylvania nonresidential market share by retail company (2011). *Source: KEMA, EIA, 2012.*

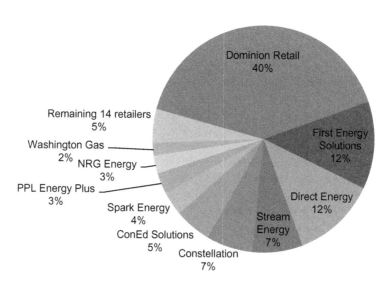

FIGURE 12.18   Pennsylvania residential market share by retail company (2011). *Source: KEMA, EIA, 2012.*

Others states/markets that secure default service through an auction or RFP model approach include:

Delaware: Prospects for future competitive activity continue to be mixed, as the state moves away from a pure auction-based standard offer supply (SOS) design. Although the majority of SOS power for Delaware customers is still procured via auction, policy makers have allowed utilities to develop integrated resource plans with carve-outs for other sources of power in attempts to mitigate prices. Removal of rate caps led to substantial increases in SOS power prices for residential and small commercial customers. Concerns about this increase led the Delaware legislature to pass the 2006 Retail Supply Act which allows Delmarva to enter into long- and short-term supply contracts, own/operate generation facilities, build generation and transmission facilities, invest in demand-side resources, and take any other PSC-approved action to diversify its SOS supply.

District of Columbia: The D.C. market continues to be in a holding pattern as the Commission considers the open question of SOS procurement. Pepco remains the SOS provider and procures power primarily via RFP. However, contracts comprising the portfolio for residential and small commercial customers may now include contracts greater than 3 years in length, and at least 40% of the SOS portfolio for residential and small commercial customers must be at least 3 years in length. Remaining load is procured via contracts lasting 1, 2, 3, or more years, with shorter contracts prohibited.

Maine: The SO price is set through an RFP process, and for the time being SO service is the only default service option available in Maine. SO contracts for residential and small commercial customers are about 3 years in length whereas those for the medium and large C&I range from 6 months to a year.

Maryland: Competitive RFPs are used to set SOS rates for customers under 600 kW. Customers with peak demand of over 600 kW are served on hourly prices. However, SOS procurement and the market structure in general are under review at the State PSC because of high prices from the 2006 move to market-based rates at the state's largest utility Baltimore Gas & Electric (BGE). In September 2011, the PSC issued an RFP for 1500 MW of new gas-fired generation capacity, but did not specify cost recovery. The RFP is mainly driven by the PSC's concern for reliability and as a means to lower capacity prices in the Reliability Pricing Model.

Massachusetts: Each utility solicits Default Service (referred to as "Basic Service") RFPs with separate bid prices for residential, commercial, and industrial customers.

New Hampshire: Default service varies by utility in New Hampshire. Public Service of New Hampshire (PSNH) continues to use its own generation for most of its supply, and PSNH bases its rates on estimates of short-term supply costs subject to NH PUC approval. Granite State Electric and Unitil, which are much more reliant than PSNH on outside generation, have based their DS rates on prices obtained through competitive RFPs.

New Jersey: A two-tiered default service structure generally referred to as Basic Generation Service (BGS) is in place. Auctions have been held each February since 2002 to secure supplies and to price two types of BGS: a Fixed Price service (BGS-FP) and an hourly, market-based Commercial and Industrial Energy Pricing service (BGS-CIEP).

Rhode Island: Narragansett Electric (National Grid) procures power via RFP for large customer classes.

3.4 Hybrid or Other—Ohio, California, and Michigan

Of the remaining open or quasi-open markets, Ohio, California, and Michigan have relatively high levels of retail activity but are subject to unique retail choice rules and market design elements, briefly outlined below.

3.4.1 Ohio

Often described as a "hybrid" market, Senate Bill 221 in 2008 required utilities to choose between an Electric Security Plan (ESP) and Market Rate Offer (MRO). Under ESPs, rates are set in a contested proceeding at Public Utilities Commission of Ohio (PUCO). To be approved, an ESP, in aggregate, must be more favorable than market pricing. Under the MRO, utilities would use a competitive bidding plan to acquire market supplies, though the PUCO declined to specify one method, and opened the door for portfolio approaches in addition to RFPs and auctions.

Prior to SB 221 in 2008, SB 3 in 1999 was enacted, which restructured the Ohio electric market. All customers of investor-owned utilities (IOUs) became eligible for retail electric choice at the start of 2001. Since then the market has had a phase of rapid growth, a period of retrenchment, followed by resumed growth due to above-market default service prices.

Soon after opening, the Ohio electric market became very active, mainly in the First Energy (FE) utility service territories of Ohio Edison (OE), Cleveland Electric Illuminating (CEI), and Toledo Edison (TE). In 2004, 43% of CEI's electric sales, 32% of OE's electric sales, and 25% of TE's electric sales were taking competitive supply. A number of factors contributed to these high switching rates. As part of a settlement agreement, FE offered power to competitive suppliers at below-market prices.

The restructuring act also allowed municipalities and other governmental entities to negotiate electric or gas supply deals for all customers within their boundaries except for those who affirmatively opted out. The largest of these governmental aggregation groups had 450,000 customers. High default service rates and shopping rate credits that increased over time also boosted switching in northern Ohio. The utility service territories of Cincinnati Gas and Electric (CG&E, now Duke Energy-Ohio) and Dayton Power and Light (DP&L) also experienced market activity during this period. In 2004, 25% of DP&L's and 18% of CG&E's electric sales were taking competitive supply.

During the 2001–2004 period, Ohio IOUs were in transitional phases called MDPs, during which default service rates were administratively set and frozen. The MDPs were scheduled to end in 2005 and 2006. Concerns that the post-MDP would expose customers to volatile market-based default service rates led the PUCO to request IOUs to propose plans for providing stable default service rates during the post-MDP. In 2004, the PUCO approved Rate Stabilization Plans (RSPs) for DP&L, CG&E, and the three FE utilities. In 2005, RSPs were also approved for the Ohio Power and Columbus Southern utility service territories—both owned by AEP. In January 2006, the PUCO approved a modified version of the RSP called the Rate Certainty Plan (RCP) for the three FE utilities. The adoption of the RSPs/RCPs had a chilling effect on the Ohio electric market. In 2005, even before most RSPs/RCPs went into effect, a number of Ohio's most active competitive electric suppliers exited the state. These retailers had difficulty offering savings off the RSP/RCP rates at a time when wholesale electric prices were relatively high. The end of FE's offer of subsidized power to retailers, costly transmission fees, and the failure of some retailers to lock-in enhanced shopping credits also contributed to the retail exodus.

The market rebounded due to ESP rates, which were set in 2008 before the fallout in the wholesale market. In many cases, these rates were above market prices. As shown in Figure 12.19, the Ohio market is currently 74 annualized TWh or 55% switched. The nonresidential market has more competitive activity with 59 annualized TWh or 65% switched, compared to the residential market with 15 annualized TWh or 35% switched. In terms of number of customers, 1.8 million are served competitively.

Looking forward, Ohio is very likely to experience the same "boom or bust" cycle because the default service mechanism of the ESP will remain for the foreseeable future. Although the option exists for an MRO, and several utilities have filed them, the PUCO has consistently rejected them and advised utilities to refile their default service plans under an ESP. In an important ruling, the PUCO in July 2012 approved FE's proposal to extend their current ESP through May 2016. FE's default service will continued to be laddered over 3 years across all customer classes (not just residential), and the steepest decline in switch rates will occur in the nonresidential segment.

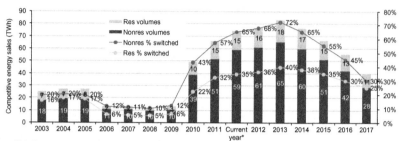

*Note that current year is an annualized estimate based on the most recent switch rates applied to the current year estimate of eligible sales volume or accounts.

FIGURE 12.19 Ohio competitive power sales and switch rates, 2003–2017 (TWh, %), base case. *Source: KEMA, EIA, State Commissions, 2012.*

Residential switch rates in Ohio will remain elevated because of long-term aggregation deals (9 years for the largest). Municipal aggregation, which has been successful at boosting residential switch rates in Ohio, has possibly been detrimental in some respects. Currently, 35% of the residential market has switched, primarily through aggregation. The downside of opt-out aggregation, however, is that many residential customers may not understand that they have switched. In a recent survey conducted in December 2011, only 35% of residential customers in Ohio were aware that they had a choice of retail electric supplier.[23] This remarkably low level of awareness suggests that customers of municipal opt-out aggregations do not make informed decisions on their provider of electricity.

Ohio's market can best be described as defense of incumbent territories by utility affiliates. Compared to state markets that have historically been able to sustain high diversity of competitive supplier offers, Ohio's competitive market is relatively concentrated by a few suppliers—notably retail affiliates of Ohio utilities. In the nonresidential segment, market share for the top five retailers is 94%. The top firm in Ohio (First Energy Solutions) is dominant with 58% market share. Of the top five, three providers (First Energy Solutions, Duke Energy Retail Sales, and DPL Energy Resources) command 87% of the market and are affiliated with regulated utilities in the state (Figure 12.20).

In 2011, there were five residential suppliers in Ohio, and First Energy Solutions commanded 86% of the market. The other retailers were Dominion Retail, Duke Energy Retail Sales, Direct Energy, and DPL Energy Resources.

23. Kim, Young and Dennis Zhang (DNV KEMA link to Power Choice Survey 2012).

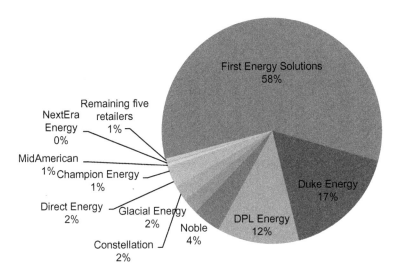

FIGURE 12.20 Ohio nonresidential market share by retail company (2011). *Source: KEMA, EIA, 2012.*

3.4.2 California

As described in Chapter 11, a poorly designed restructured market in California created the conditions for what became known as the California Energy Crisis of 2000–2001. During this period, there were severe price spikes in wholesale markets, which imposed large costs on California ratepayers. The negative experience in California set back electricity restructuring in this state and, as a consequence, in many other states in the United States.

In September 2001, Direct Access (DA) in California was officially suspended; however, customers who were served by ESPs prior to the closing date were allowed to continue to purchase electricity competitively, or could elect to return to utility service (after which they were subject to a 3-year minimum stay). In addition, legislation was adopted that put the utilities back into the business of supply procurement, which has led to new long-term PPA contracts and utility-owned generation, and the imposition of ongoing nonbypassable charges on retail choice customers.

Therefore, a retail market has continued to exist, although participation declined, as some customers chose to return to utility service, and no new customers were allowed to enter the market until 2009. In 2009, the legislature enacted SB 695, which allowed a limited reopening of DA for nonresidential customers. The law permits nonresidential DA, including customers

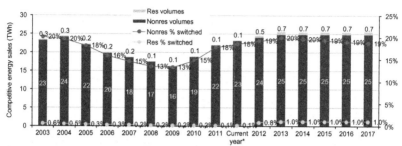

*Note that current year is an annualized estimate based on the most recent switch rates applied to the current year estimate of eligible sales volume or accounts.

FIGURE 12.21 California competitive power sales and switch rates, 2003–2017 (TWh, %), base case. *Source: KEMA, EIA, State Commissions, 2012.*

currently ineligible for DA,[24] up to a cap equal to the highest amount of historic competitive load in each service area during a 12-month period. Implementation was left to the California Public Utilities Commission (CPUC). The CPUC elected a 5-year transition for the new DA load. The shopping cap is effectively 18–20% of nonresidential sales volume (Figure 12.21).

Since the passage of SB 695, nonresidential switching has consistently and quickly filled the shopping cap. It is clear that current market conditions in California would support a considerably higher switch rate if not for the shopping cap. Since the shopping cap is filled by nonresidential customers, residential customers in California have few options to find a competitive supplier. Currently, state law permits residential switching through Community Choice Aggregation, which is similar to the municipal aggregation programs in Illinois and Ohio. Though these aggregations are fairly small and are not likely to have a large impact on overall switch rates, they do provide choice for residential customers.

3.4.3 Michigan

In 2008, Michigan retail choice was capped at 10% of utility average weather-adjusted retail sales for the preceding calendar year under HB 5524 (Figure 12.22).

Default service rate caps limited market activity during the 2000–2005 period. The caps lasted until December 2005 for residential customers;

---

24. California "suspended" competitive power market activity in 2001, except for customers who were grandfathered. Only customers (across all utilities and classes) who were taking competitive supply on or before September 2001 have been "grandfathered" and deemed eligible to continue to take competitive supply, but with onerous minimum stay requirements if they switch back to utility service.

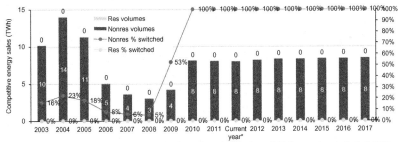

*Note that current year is an annualized estimate based on the most recent switch rates applied to the current year estimate of eligible sales volume or accounts.

FIGURE 12.22 Michigan competitive power sales and switch rates, 2003–2017 (TWh, %), base case. *Source: KEMA, EIA, State Commissions, 2012.*

December 2004 for small (<15 kW) commercial customers, and December 2003 for all other nonresidential customers.

Some switching occurred during the 2000–2005 period, helped by headroom that was created by securitization and rate-reduction credits, reduced distribution charges, and rate skewing that were only available to shopping customers. Yet this market activity was curtailed in 2004 when these credits were eliminated and new stranded cost charges were imposed. In the 2004–2005 period, shopping was also discouraged by new rules concerning how customers could return to full service and how much advanced notice they must give.

In 2009–2010, within a year of implementing the shopping cap, both Consumers Energy (CE) and Detroit Edison (DE) reached their 10% choice caps due to depressed wholesale prices. There is still a significant amount of load prepared to take competitive supply in the choice queue, should space become available as customers switch back to bundled service.

3.5 Potential Regulatory Expansion

With the recent trend of declining power prices, it has been a relatively favorable environment for advancing discussions on expanding capped or stalled markets and possibly opening up new states altogether. Expectations of progress on this front are fairly low given the regulatory hurdles and opposition from regulated utilities. For example, a push to reopen retail markets in Virginia is underway. Lobbying efforts have begun in Wisconsin and Florida, but it will be years before meaningful changes, if any, will be made in those states. A brief update of California, Michigan, and Arizona is provided below.

In California, any further expansion will require legislative approval. A bill to further expand retail choice (SB 855) was introduced in 2011, and may be acted on in 2012. Positive momentum in California exists since the expansion of the DA shopping caps was filled within seconds, and stakeholders have discussed using a lottery system for room that becomes available in the future.

The Michigan market will remain effectively closed to further choice absent some customer returns to bundled service which would free space under the choice cap. Although, several legislators have proposed legislation to raise the cap, prospects for passage are uphill, at best. Energy Choice Now, a coalition of retailers, customers, and other key stakeholders, was formed to help raise the 10% cap.

In Arizona, a settlement was passed in the Arizona Public Service (APS) territory to allow large customers (10 MW or higher) to shop, with migration capped at 200 MW. The settlement was approved at the end of May 2012 and no further market expansions in Arizona are expected in the near term.

Choice that is limited to a few is not really choice. It is important to provide all customers who want to switch from the regulated utility the option to choose. It has been argued by pro-competition advocates that choice in retail electricity has led to higher customer satisfaction due to price, product innovation, and greater flexibility and options. As a COMPETE Coalition report argues, customers in retail choice jurisdictions can benefit more quickly from falling wholesale prices whereas customers in regulated monopoly utility territories pay rates based on legacy cost structures.[25]

However, before the issue of expanding customer choice in stalled or closed markets can be confronted, the structural flaw in the current retail market design needs attention. It has been argued in this chapter that improvements to the alignment of wholesale and retail markets have enhanced retail competition. The path to full retail competition in the United States lies in making improvements first in markets that have already restructured.

4 CONCLUSION

There are numerous retail choice models in the United States, and the most important market design component is the default service procurement mechanism. The most durable choice model is full competition in Texas, in which there is no default service. The models in New York and Pennsylvania both have characteristics that are worthy of emulation. The history of retail choice in the United States has examples of policies that hampered market development with the most limiting being nonmarket-based price caps or a flat shopping percentage cap.

25. O'Connor, P. Retail Electric Choice: Proven, Growing, Sustainable. Prepared for the COMPETE Coalition. <http://www.competecoalition.com/files/COMPETE_Coalition_2012_Report.pdf> (accessed 03.04.12).

While some of these limits still exist in markets, such as Michigan and California, the primary barrier to sustaining momentum in competitive retail markets today is long-term default service procurement contracts. Nonresidential choice markets thrive when default service is based on the hourly spot price. Residential and small commercial customers will be given a wider array of retail electricity choices in the long run when default service market designs do not create a "boom or bust" cycle.

The overriding principle for retail market design should not be rate certainty (the regulated utility model). Rather, restructured markets should embrace the idea that competitive retailers know best how to serve their customers, and forcing most of them to compete only on price relative to the utility is the surest way to minimize the value creation possibilities of well-functioning retail markets.

Admirable progress has been made in competitive retail markets in the United States. Nearly 15 million customers are now being served by competitive suppliers. The switch rate within markets where customers can choose is now 54%, compared to just 19% in 2003. There is still plenty of room to grow in markets, particularly in the smaller commercial and residential customer segments. The key to strong and sustainable market development is a market design that more closely resembles free competition, not one that tries to retain the command-and-control aspects of traditional regulated electric monopolies.

Chapter 13

Fragmented Markets: Canadian Electricity Sectors' Underperformance

Pierre-Olivier Pineau

Associate Professor, HEC Montréal, 3000, chemin de la Côte-Sainte-Catherine, Montréal (Québec), Canada H3T 2A7

1 INTRODUCTION

Canada is a very important energy producer: in 2011, it was the sixth largest oil producer, third in natural gas, eighth coal exporter, and third hydropower producer (IEA, 2011), while it was the second world uranium producer (WNA, 2012). Overall, in 2009, Canada ranked fifth in terms of total primary energy production, only behind China, the United States, Russia, and Saudi Arabia (EIA, 2012). Given these statistics, one could expect that the energy sector in Canada is a defining sector, well structured and organized. In reality, the Canadian energy sector is broken into a multitude of uncoordinated subsectors, with a myriad of small players and no common voice, explaining why Canada has an inconsequential influence in world energy affairs. The electricity sector, on which this chapter focuses, is representative of this situation. Worst, it is currently not engaged in significant reforms that could generate additional economic and environmental benefits for all regions of the country.

In the next section, a description of the 10 provincial Canadian electricity markets is provided, with an analysis of their respective evolution over the last 15 years. Although each province has its own characteristics, they can be grouped in three categories:

1. Hydropower-dominated provinces: British Columbia (BC hereafter), Manitoba (MB), Quebec (QC), and Newfoundland and Labrador (NL), characterized by low production costs, a dynamic export orientation and public ownership.
2. Restructured provinces: Alberta (AB), Ontario (ON), and New Brunswick (NB) that moved away from the centrally managed model through the creation of an independent system operator (ISO) and more competitive

wholesale markets. These provinces struggle with investment challenges and high prices. Trebilcock and Hrab (2006), in a predecessor of the current volume, present the restructuring experiences of AB and ON. A more comparative perspective is provided in this chapter, with the extra benefit of additional years of reforms.

3. Traditional provinces: Saskatchewan (SK), Nova Scotia (NS), and Prince Edward Island (PEI) structured along vertically integrated utilities and highly dependent on fossil fuels, leading to high prices as in restructured provinces.

Figure 13.1 shows the geographic distribution of provinces and their installed generating capacity. It illustrates the fragmentation of the sector, with no province of the same group being neighbors, except in the smaller Eastern markets.

In Section 3, an argument is developed to make the case for greater electricity sector integration across provinces and with neighboring US states. Such harmonization would go much beyond the already in place transmission access policies, under which provincial transmission providers file Open Access Transmission Tariffs (OATT) with the US Federal Energy Regulatory Commission (FERC) in response to the 1996 order 888 *Promoting Wholesale*

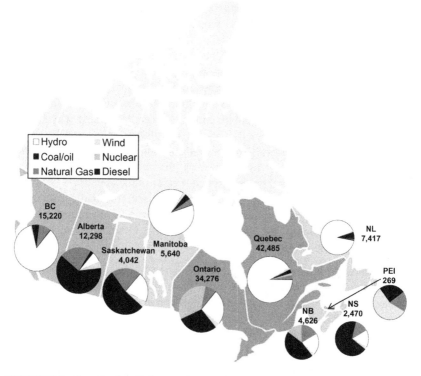

FIGURE 13.1   Canadian installed generating capacity, in megawatts, 2006 (NEB, 2008).

Competition Through Open Access Non-discriminatory Transmission Services by Public Utilities. It would involve more coordination in generation and transmission planning, discussions to implement standard market designs and the adoption of common greenhouse gases (GHG) emission targets, and renewable and efficiency goals. Even more critically, provinces would have to allow electricity prices to converge toward a market level better reflecting the value of energy and the state of the system. In short, Section 3 argues that the Canadian electricity sectors should embrace a new collaborative paradigm, in order to successfully meet the challenges of a competitive world under carbon constraints, followed by the chapter's conclusions in Section 4.

2 ELECTRICITY MARKETS IN CANADA

2.1 A Global View of the Canadian Electricity Sector

With its population of about 35 millions (0.5% of the world), the Canadian electricity production is impressive: 603 TWh in 2009 or 3% of world's total (IEA, 2011). This amounts to about 18,000 kWh per capita, six times the world average generation per capita. Furthermore, 75% of the Canadian electricity is produced with no direct GHG emission. In fact, 60% of the Canadian power production is coming from hydro (364 TWh in 2009) and 15% from nuclear sources or 90 TWh (IEA, 2011).

Paradoxically, GHG emissions from the electricity sector remain a major concern in Canada, as they represented 14% of the 690 million tons emitted in 2009 (Environment Canada, 2011), second only to the road transportation sector, responsible for 19% of all GHG emissions. This is explained by the important reliance on fossil fuels, especially coal, in many provinces (AB, SK, NB, and NS). Figures 13.1 and 13.2 illustrate the relative importance of the different sources of supply and trade in the 10 Canadian provinces. Table 13.1 presents the amount of generation capacity by type in Canada, with its distribution across classes of producers, government-owned, investor-owned, or industrial producers.

As shown in Table 13.1, provincial (and municipal, in some cases) governments are important owners of power plants in Canada, especially hydro, nuclear, and conventional steam ones (mostly burning coal). The private sector nevertheless owns an important share of the generation capacity, in all provinces—even if in some provinces the share of private ownership is as low as 2% (Manitoba).

This situation mostly results from the Canadian constitution, specifying that (article 92A.1):

in each province, the legislature may exclusively make laws in relation to [...]
development, conservation and management of sites and facilities in the province for
the generation and production of electrical energy.

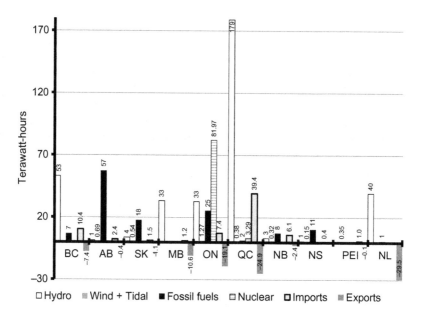

FIGURE 13.2 Provincial generation, imports and exports of electricity in 2010 (Statistics Canada, 2012a,b).

TABLE 13.1 Generation Capacity in Canada by Type and Class of Producer in 2009 (Statistics Canada, 2012c)

Type	Megawatt	Percentage	Government-Owned, %	Investor-Owned, %	Industry, %
Hydro	74,961	58	87	6	7
Wind	3,026	2	7	91	2
Tidal	4	0	0	100	0
Nuclear	13,345	10	62	38	0
Conventional steam	26,493	21	56	37	7
Combustion turbine	10,313	8	36	50	14
Internal combustion	601	0	40	22	38
Total installed capacity	128,743		72	21	7

A similar article specifies the same for nonrenewable natural resources, which leaves little room for the federal government in the energy sector. As a result, each province has its own electricity policy and regulatory agency, leading to disparate electricity tariffs, generation and transmission plans, and renewable/clean energy goals. This fragmented approach in almost all aspects of the sector[1] is the main cause of its underperformance, as detailed in Section 3.

The federal government, in Ottawa, is still involved in four energy areas (NRCan, 2012):

1. interprovincial and international energy trade and infrastructures, through the National Energy Board (NEB), an independent federal regulatory agency, located in Calgary, Alberta. The NEB is the Canadian equivalent of the FERC, albeit with less visibility, power, and drive to implement reforms;
2. regulation of nuclear power, through the Canadian Nuclear Safety Commission;
3. nuclear science and technology, through Atomic Energy of Canada Limited;
4. energy research and development, through different programs and research laboratories, such as CanmetENERGY, the principal federal performer of energy R&D.

In addition, the federal government supports renewable energy in various ways and has an impact on the energy sector through federal environmental legislation.

In this context, each province developed its electricity sector independently, leading to very different generation portfolios (Figures 13.1 and 13.2). Diverse market outcomes also result from this: Figure 13.3 illustrates the provincial per capita consumption in 2009 as a function of the average revenue per kilowatt-hour, in Can$ cents/kWh (¢/kWh), a proxy for the electricity price. It shows that per capita consumption varied from less than 10,000 kWh per year in some provinces including BC, ON, PEI, to more than 20,000 kWh in QC, while prices ranged from less than 6 ¢/kWh (MB) to almost 14 ¢/kWh (PEI). The per capita QC consumption is significantly higher than in other low-cost provinces (MB and BC) mostly because energy-intensive industries (aluminum, pulp, and paper) were attracted by the larger availability of hydropower. Electric heating, commonly used, also contributes to explain the high electricity consumption.

In the following sections, each province is reviewed and compared to other provinces of the same category (hydro, reformed, and traditional).

---

1. Reliability standards are however commonly established through the regional entities of the North American Electric Reliability Corporation (NERC).

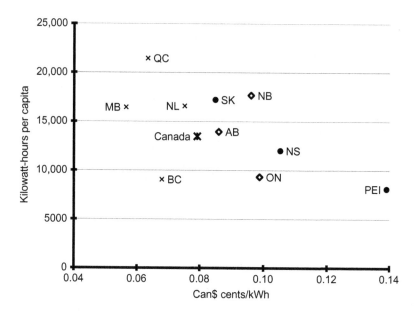

FIGURE 13.3 Provincial per capita electricity consumption and average revenue per kilowatt-hour in 2009 (Statistics Canada, 2012d).

2.2 Hydropower-Dominated Provinces: British Columbia (BC), Manitoba, Quebec and Newfoundland and Labrador (NL)

2.2.1 Overview

In these four provinces, producing 56% of the Canadian electricity, power is almost exclusively generated by hydropower plants: 89% in BC, 99% in MB, and 97% in QC and NL (Statistics Canada, 2012a). They all have a commercially oriented, provincial government-owned, integrated company dominating the market: BC Hydro, Manitoba Hydro, Hydro-Quebec, and NL Hydro. Given the historical low cost of their generation portfolio and their cost-of-service rate regulation, these provinces have the lowest electricity rates in Canada (Figure 13.3).

In Table 13.2, the main components of the market structure are presented, illustrating the importance of the provincial company in all subsectors of the market. These companies, however, have distinct generation, transmission and distribution business units, and file OATT with the US FERC.[2] This transmission open access allows the dominant company, private producers, and industrial ones to trade beyond their jurisdiction. The

---

2. NL does not file OATT because it is not interconnected with the United States (IEA, 2010).

TABLE 13.2 Market Structure and Main Players in Hydro Provinces

	BC	Manitoba	Quebec	NL
Generation	BC Hydro, FortisBC, Nelson Hydro, RTA, Teck + IPPs	Manitoba Hydro + small IPPs	Hydro-Québec (HQ) Production, RTA, TransCanada + other IPPs	NL Hydro (Nalcor) and Newfoundland Power (Fortis) + IPPs
Capacity in megawatts (2009)	15,220	5,640	42,485 (47,913)[a]	7,417 (1,989)
Transmission	BC Hydro[b]	Manitoba Hydro	HQ TransÉnergie	NL Hydro and Newfoundland Power
Distribution	BC Hydro, FortisBC, and some municipal utilities	Manitoba Hydro	HQ Distribution + 9 municipal distribution companies	NL Hydro and Newfoundland Power
Ministry responsible	Energy and Mines	Innovation, Energy and Mines	Natural Resources and Wildlife	Department of Natural Resources
Regulator	BC Utilities Commission	Public Utilities Board	Régie de l'énergie	Board of Commissioners of Public Utilities
System operator	BC Hydro	Manitoba Hydro	HQ	NL Hydro and Newfoundland Power
OATT since	2006	1997	1997	—
Market design	Centrally managed model with bilateral contracts			

[a]Hydro-Quebec has contractual rights over a 5,428 MW hydropower plant located in NL (Churchill Falls generating station) until 2041 (Hydro-Quebec, 2012a). In practice, this increases the QC generating capacity and decreases NL's one by 5428 MW. It corresponds to about 30 TWh per year transferred from NL to QC (Figure 13.1).
[b]Between 2003 and 2010, the independent, government-owned BC Transmission Corporation was in charge of transmission in BC (BC Hydro, 2012b).

wholesale market is also open. Independent power producers (IPPs) can therefore sign purchase power agreements (PPAs) with distribution companies or sell to large electricity consumers. The latter never happens,

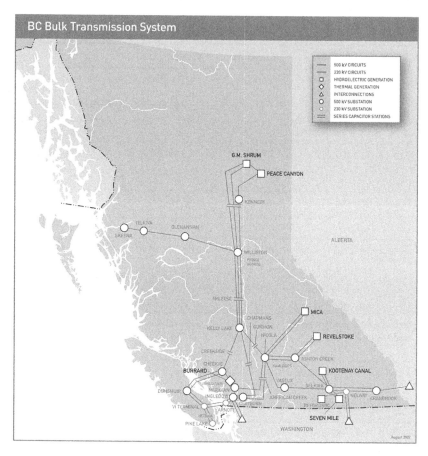

FIGURE 13.4　BC electricity sector (BC Hydro, 2012c).

however, as the distribution company offers low industrial rates, making it difficult for IPPs to compete.

The low cost of hydropower generation, leading to low electricity rates, never created an impetus for reform, at least from a consumer's perspective. From a shareholder's (government's) perspective, the incentives for reforms are divergent: on the one hand, maintaining low rates helps obtaining political support, but on the other hand, government's revenues could benefit more from the economic dividend resulting from the low-cost electricity production sold at higher prices (in export markets). These dual incentives led to a development strategy focusing on low provincial regulated rates, while at the time being active in export markets. BC Hydro indeed created as early as 1988 its trading subsidiary, Powerex, now active all across North America (Powerex, 2012). Figure 13.4 illustrates the BC system with its interties with

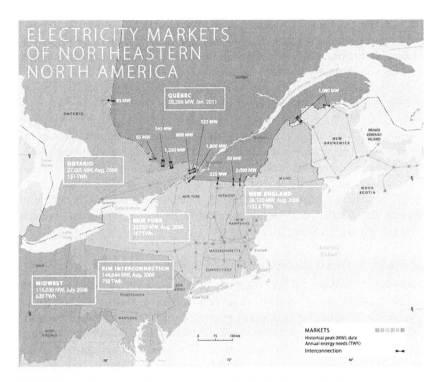

FIGURE 13.5 Quebec electricity sector and interties (Hydro-Quebec, 2011).

the United States (about 2,500 MW) and with Alberta (about 1,200 MW; Pineau, 2009).

HQ Energy Services (United States), a wholly owned subsidiary of Hydro-Quebec, obtained a license as a power marketer in the US wholesale market in 1997 (Hydro-Quebec, 2012b). This allows Hydro-Quebec to trade with all its neighbors, and beyond, through its multiple interties (Figure 13.5).

Manitoba Hydro became a full member of the Mid-Continent Area Power Pool (MAPP) in 1996, entering into power exchange agreements with US power marketers (Manitoba Hydro, 2012a). In 2005, Manitoba Hydro also became a full member of the Midwest Independent Transmission System Operator (MISO). The MB retail market remains nevertheless regulated and disconnected from wholesale price signals.

BC Hydro and Hydro-Quebec pay important yearly dividends to their provincial government: Can$230 million in 2012 for BC Hydro (BC Hydro, 2012a) and Can$1,958 million in 2011 for Hydro-Quebec (Hydro-Quebec, 2012a). They could generate these profits given the allowed rate of return on distribution and transmission activities and profits in the generation sector (partly relying on exports and energy marketing activities).[3] In MB, regulation does not allow Manitoba Hydro to include a rate of return on its cost of service.[4] Furthermore, it seldom pays a dividend to its shareholder, using net revenues to maintain low rates[5] and invest in new projects.

In NL, because of its remote location and sparse population, the context is slightly different. A provincial energy company, Nalcor Energy, was created in 2007 by the provincial government as a tool to implement its energy plan (NL, 2007). This plan aims at a better control over the development of the province's energy resources. Nalcor Energy is the parent company of NL Hydro and of other companies involved in hydropower and oil and gas development. Since 2009, Nalcor also has an energy marketing subsidiary to sell its increasing hydropower production in the Northeastern US electricity markets. As it is still in its development stage, Nalcor never paid a dividend to its shareholder (Nalcor, 2012).

2.2.2 Hydro Provinces Reform Assessment

Following the opening of the wholesale market in the US (FERC order 888 in 1996), hydro provinces have been very active to adjust their regulatory structure to optimize their external commercial activities. If some discussions to further reform these markets were occasionally held (see for instance Clark and Leach, 2007, for the case of QC), no major restructuring happened. Business units have been separated and OATT filed, to comply with US regulation, but competition remains marginal in the wholesale market: it is limited to a few PPAs with the main distribution company and absent in the retail market.

2.3 Restructured Provinces: Alberta, Ontario, and New Brunswick

The lack of hydropower potential limited the expansion of the generation segment in these provinces and made them rely much more on thermal generation (fossil fuel and nuclear), leading to higher production costs. Resulting high prices, together with a stronger appetite for competition in the organization of markets, opened the way to reforms in these provinces.

---

3. BC Hydro and Hydro-Quebec, in their generation activities, have to supply a "heritage" amount of electricity at historical low cost to their consumers: about 50 TWh in BC (BC, 2003) and 165 TWh in QC (Hydro-Quebec Act, article 22). Beyond these quantities, the cost of generation is not regulated in these two provinces, and distribution units have to sign competitive contracts with their generation sister unit or IPPs.
4. Only operating and capital costs can be charged to consumers (article 39.1 of The Manitoba Hydro Act, "Price of power sold by corporation").
5. Not only are electricity rates low in MB, but for commercial consumers (General Service Small and Medium), a *decreasing*-block tariff structure is adopted (Manitoba Hydro, 2012b), leading to little or no consumption efficiency incentives.

TABLE 13.3 Market Structure and Main Players in Reformed Provinces

	Alberta	Ontario	New Brunswick
Generation	ATCO, Enmax, Capital Power, TransAlta and TransCanada + IPPs	Ontario Power Generation, Bruce Power, TransCanada Energy, TransAlta + IPPs	NB Power Generation and Nuclear + 6 generation facilities owners
Capacity in megawatts (2009)	12,298	34,276	4,625
Transmission	ATCO Electric, AltaLink Management, EPCOR Utilities, ENMAX Power	Hydro One, Great Lakes Power, Canadian Niagara Power, Five Nations Energy and Cat Lake Power Utility	NB Power Transmission
Distribution	ENMAX, EPCOR, ATCO, FortisAlberta	More than 60	NB Power Distribution and Customer Service
Ministry responsible	Department of Energy	Energy	Department of Energy
Regulator	Alberta Utilities Commission	Ontario Utilities Board	Energy and Utilities Board
System operator	Alberta Electric System Operator	Independent Electricity System Operator	New Brunswick System Operator
OATT since	–	–	2004
Start of the market	January 1, 1996	May 1, 2002	December 1, 2004
Market design	Mandatory power pool	Power pool for real-time energy market with bilateral contracts, PPAs and regulated tariffs	Physical bilateral market with a redispatch market
Public reference price	Hourly pool price/system marginal price (SMP)	Hourly Ontario Energy Price (HOEP)	Final Hourly Marginal Cost (FHMC)

Indeed, impetus for reform came in these provinces from a desire to have less regulation and government's involvement (Daniel et al., 2007) and, at least in ON, to control cost (Trebilcock and Hrab, 2006). Table 13.3 provides an overview of these three restructured markets.

The key distinctive feature is the presence of an ISO setting a publicly available hourly reference price. Such ISO and reference price are absent in all other Canadian provinces. As shown in Table 13.3, AB was the first to open its market in 1996, followed by ON in 2002 and NB in 2004. NB, however, does not have a competitive power pool: its wholesale market is structured around physical bilateral contracts with a redispatch market, in which a "Final Hourly Marginal Cost" is defined.

NB implemented an OATT in 2004, when it restructured its sector, while AB and ON have put in place transmission service provisions and tariffs which are compatible with the OATT reciprocity and nondiscrimination, allowing them to not file OATT with the FERC (IEA, 2010).

In the next three sections, the restructuring process of each of these three provinces is reviewed. These sections do not enter into the details of the restructuring efforts. The goal is rather to provide an overview and to offer a global understanding of the Canadian electricity sectors. Daniel et al. (2007) and AESO (2006) offer a full focus on the AB experience, and Trebilcock and Hrab (2006) provides a complete ON review and analysis of the early reform experience in both AB and ON. Blakes (2008) and IEA (2010) also provide some additional details on the market structure of these two provinces. EDA (2006) presents the current ON electricity market in detail.

2.3.1 Alberta: Competition with Limited Interconnections

With the absence of provincial government-owned electricity company and already disintegrated electricity market structure (a combination of municipally owned and investor-owned utilities), AB had a natural predisposition for electricity market reform. The main driver for reform was a large support for less government involvement in the power sector. If it introduced a competitive power pool as soon as 1996 (Table 13.4), the regulation of wholesale electricity prices only stopped in 2001. Retail consumers have also been allowed to switch retailer from 2001 but could also remain with their distributor-retailer under the Regulated Rate Option (RRO). The RRO was regulated until 2006 and, from 2006 to 2010, it progressively evolved toward a monthly rate entirely based on projected market prices (month-ahead prices). As of May 2012, 35% of residential consumers had switched to a contract with a different electricity retailer (Alberta, 2012b).

The Alberta restructuring experience, while deployed in steps, still required adjustments: 2003 merger of the market and transmission operators into the AESO and addition, in 2008, of some long-term supply adequacy indicators, out of concerns that investment in generation would not be sufficient. Investment in transmission is also recognized to be an issue, not only within the province, but also with neighbors, as Alberta is "one of the least interconnected jurisdictions in Canada with only two transmission interties

TABLE 13.4 Timeline of Electricity Restructuring Events in AB (Alberta, 2012a; Daniel et al., 2007).

1996	Creation of the power pool and of the independent transmission administrator.
2001	Beginning of the current market structure and opening of retail competition (MSA, 2010).
2003	The Power Pool and the independent transmission administrator are merged into the Alberta Electricity System Operator (AESO).
2006	Regulated Rate Option—monthly price based on a mix of electricity purchased through long-term contracts and the following month's projected market price.
2008	Introduction of the long-term supply adequacy metrics (AESO, 2008).
2010	Regulated Rate Option—100% based on the following month's projected market price.

providing limited export and import capacity" (AESO, 2009; Alberta, 2008). Indeed, interties with BC and SK (Figure 13.6) can only import a maximum of 1200 and 150 MW, respectively, with the BC intertie not being able to reach that maximum because of internal network constraints in AB (AESO, 2009). This affects the ability to import cheaper electricity, mostly from BC, and to smooth out some price volatility.

The AB restructuring experience has been mostly an inward process, with the development of new Albertan institutions—AESO and the Market Surveillance Administrator (MSA)—which has not reduced electricity prices compared to jurisdictions with similar fossil fuel system (such as SK; Figure 13.2). It neither introduced greater confidence in the long-term adequacy of generation and transmission infrastructures.

2.3.2 Ontario: An 8-Month Experience

The ON electricity sector restructuring was also carefully prepared with a 4-year adjustment period (1998–2002). The main driver, in the ON case, was to try to infuse more rigor in power sector investment and expenditures, through competition and private ownership. If competition was indeed implemented, no privatization materialized in ON. The main reform mistake, however, was to let the retail price be entirely connected to the wholesale market price. This exposed unprepared and unhedged retail and commercial consumers, after May 1, 2002, to the exceptionally high price volatility of Summer 2002. It produced a political backlash that was fatal to the competitive system put in place. A rate freeze followed in December 2002, ending an 8-month full deregulation experience. Centralized planning and long-term supply contracts were reintroduced in 2004, with the creation of the Ontario Power Authority (OPA). See Table 13.5 for the timeline of

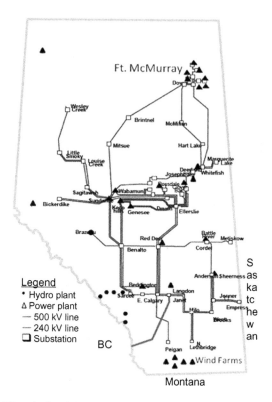

FIGURE 13.6 Alberta's electricity sector (E-T Energy, 2012).

the Ontario restructuring experience. Since 2005, with the introduction of the regulated price plan (ending the rate freeze) and of the smart meter initiative, the Ontario market remained mostly unchanged. Generators bid daily in a competitive markets, while being hedged against any risks with contracts for difference signed with the OPA.

The Ontario restructuring experience, while introducing more transparency in the Ontario electricity market, mostly through the IESO, has led to a hybrid system where competitive prices coexist with regulated ones, for both generators and consumers. With its smart metering initiative and time-of-use pricing, Ontario is the only jurisdiction in Canada that provides all consumers with more elaborated price information, beyond the monthly flat (or two tier) rate. This initiative is a success in terms of deployment, with 99% of residential and commercial consumers equipped with smart meters as of June 2012 (OEB, 2012c). Out of them, 89% are on the time-of-use regulated price plan.

More competition in the Ontario electricity system, however, is unlikely to develop under the current conditions, despite its various interties

TABLE 13.5 Timeline of Electricity Restructuring Events in ON (Canada Energy, 2012).

1998	Energy Competition Act, unbundling the integrated (generation and transmission) government-owned Ontario Hydro into different entities: (1) Ontario Power Generation (OPG) focused on generation, (2) Hydro One focused on transmission and distribution, and (3) the Independent Electricity System Operator (IESO) focused on electricity system dispatch (until 2004, the IESO was known as the Independent Market Operator). No privatization occurred.
2001	Bruce Power, a private company, assumes operational control of eight nuclear generating stations, a total of 6200 MW (Bruce Power, 2012). These units are leased from OPG.
May 1st, 2002	Opening of the wholesale market, with retail price based on wholesale market, without real-time metering and price information. Electricity bills double or triple during the summer due to exceptionally hot weather.
December 2002	The Electricity Pricing, Conservation and Supply Act, lowering and freezing the retail electricity price at 4.3 ¢/kWh (commodity charge, excluding all other charges such as network and other charges). This disconnected retail and wholesale markets.
2004	Electricity Restructuring Act, creating the Ontario Power Authority, to ensure an adequate, long-term supply of electricity (planning and long-term contracts with generators).
2005	Two major initiatives:
	Regulated Price Plan for residential and commercial consumers, "that provides stable and predictable electricity pricing, encourages conservation and ensures the price consumers pay for electricity better reflects the price paid to generators" (OEB, 2012a). Consumers without smart meters have an increasing block tariff, while consumers with smart meters have time-of-use tariffs. Both tariffs are adjusted every 6 months.
	Smart Metering Initiative, aiming at having all electricity consumers equipped with smart meters by 2010 (OEB, 2012b).

(Figure 13.7), offering many short-term trading opportunities. Indeed, the Integrated Power System Plan currently being prepared (OPA, 2011) requires the refurbishment of 10,000 MW of nuclear power, and the addition of 2000 MW of nuclear power by 2030, while shutting down all coal power plants by 2014. Such plans, because of the required financial guarantees, will heavily rely on government's involvement and will not be deployed through a competitive market.

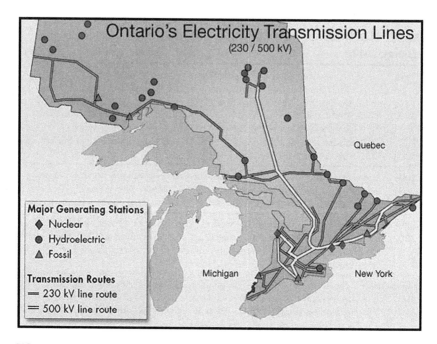

FIGURE 13.7   Ontario's electricity sector.

2.3.3 New Brunswick: Back to Vertical Integration

Restructuring in NB, implemented in 2004, was designed to avoid the price volatility experienced in AB and ON but also to foster competition. The creation of the New Brunswick System Operator (NBSO) guaranteed the-independence and transparency in the market, allowing long-term bilateral contracts to be signed between generators and both distributors and large consumers. The integrated New Brunswick Power Corporation (NB Power) became the New Brunswick Power Holding Corporation (NB Power Group) with subsidiary companies: Generation, Nuclear, Transmission, and Distribution.

In 2011, however, the government of NB announced the "Reintegration of NB Power" as a vertically integrated publicly owned utility (following the BC and QC model) and the dissolution of NBSO, with the system operation functions back to NB Power. These were actions 1 and 2 of the 2011 Energy Action Plan (New Brunswick, 2011). This move, back to a traditional struc-ture, comes after a series of energy events that have created high tensions in the NB electricity market:

- High cost of fossil fuels: With its high reliance on coal, oil, and natural gas power plants (more than 50% of its capacity), electricity production costs in NB have been hit by the price increase of these three commodities during the 2000–2008 period, especially since all of the coal, oil, and natural gas are imported.

- Problems with the nuclear refurbishment: The refurbishment work of the 680 MW Candu-6 nuclear power plant, commissioned in 1982, started in March 2008, and was scheduled to last 18 months and to cost Can$1.4 billion. Multiple delays have occurred, with overruns of Can$1 billion (CBC News, 2010) and return to service only planned for Fall 2012 (NB Power, 2011).
- Failed takeover of NB Power by Hydro-Quebec: In October 2009, the QC and NB governments unexpectedly announced the acquisition of most of NB Power assets by Hydro-Quebec. This would have been the first takeover of a provincially owned utility by another one. This plan would have provided price and supply security for NB, and a lucrative access to the NB and US market for Hydro-Quebec. However, NB and the transaction had to be canceled in March 2010 (Quebec, 2010).

These events led the province of NB to rethink its energy strategy and to focus on five key objectives (New Brunswick, 2011): low and stable energy prices, energy security, reliability of the electrical system, environmental responsibility, and effective regulation. In addition to a return to a vertically integrated electricity company, more emphasis will be put on regional electricity partnerships, such as the Atlantic Energy Gateway, an initiative "to promote and facilitate the development of clean and renewable energy sources in Atlantic Canada." (Canada, 2009).

2.3.4 Restructured Provinces Reform Assessment

None of the three reforms in AB, ON and NB can claim to be a success, at least using the six objectives of electric restructuring defined in AB (AESO, 2006):

1. A competitive, efficient, and innovative electricity marketplace.
2. New generators and many new service providers.
3. Informed consumers choosing from competitive, attractive options.
4. Continued downward pressure on rates.
5. Incentives for conservation and the wise use of energy.
6. Smart technologies and green power options that contribute to environmental goals.

If Alberta has a competitive marketplace (objective 1), no new generators made a difference (objective 2) and there is no significant innovation from service providers (objectives 1 and 2). In the only province with Smart technologies (ON), this development was a result of a government's decision—not of restructuring (objective 6). Only 35% of all consumers have changed provider in Alberta, illustrating that new options were not attractive or that consumers remain uninformed (objective 3). No price or consumption levels indication show progress on objectives 4 and 5, at least compared to other provinces (for instance, AB versus SK; Figure 13.3).

In all cases, significant market design changes have been implemented, adding to the cost of operating the market.[6] Only in AB was the reform consistently implemented, with no return to regulation (as in ON) or to a traditional market structure (as planned in NB). However, even if AB consumers now face fully competitive electricity prices, they have no opportunity to react to short-term price signals, as residential and commercial consumers are still billed at an average forward price.

Investment issues remain salient in all three provinces. The situation is well summarized by this observation, made by government of NB (New Brunswick, 2011):

New generation assets have not been built by private sector market participants to the extent predicted upon adopting a competitive electricity market model. Developers have found it very difficult to secure financing for new projects because financial institutions and other investors will not finance an independent power project without a long-term utility power purchase agreement in place as a secure source of future revenues.

2.4 Traditional Provinces: Saskatchewan, Nova Scotia and Prince Edward Island

In the three remaining Canadian provinces, no important change was brought to the electricity sector between 1995 and 2013. This is mostly due to the small size of their system (Table 13.6: generation capacity of 4,042 MW in Saskatchewan, and much less in the two other provinces) and to the lack of incentives to try to join a larger, more integrated, electricity system.

Table 13.6 summarizes the main features of the SK electricity sector and of the two other Atlantic provinces, NS and PEI. In all cases, an integrated utility is in charge of the sector, with some IPPs also generating power. Only in SK is the utility publicly owned.

In order to facilitate trade, these provinces have also started to file OATT, but relatively late compared to other provinces: 2001 in SK, 2005 in NS (Blakes, 2008), and in 2008 for PEI (IRAC, 2008).

---

6. In NB, part of the reasons for returning to a vertically integrated structure was the additional cost burden of the unbundled structure. This was explicitly recognized by the government in its Energy Action Plan (New Brunswick, 2011).

TABLE 13.6 Market Structure and Main Players in Traditional Provinces

	SK	Nova Scotia	PEI
Generation	SaskPower + about 10 small IPPs	NS Power (Emera) + IPPs	Maritime Electric (Fortis) + IPPs
Capacity in megawatts (2009)	4042	2470	268
Transmission	SaskPower	NS Power	Maritime Electric
Distribution	SaskPower	NS Power + six municipal utilities	Maritime Electric
Ministry responsible	Provincial Cabinet and Crown Investments Corporation of Saskatchewan	Department of Energy	Department of Finance, Energy and Municipal Affairs
Regulator	Saskatchewan Rate Review Panel	Utility Review Board	Regulatory and Appeals Commission
System operator	SaskPower	NS Power	Maritime Electric
OATT since	2001	2005	2008
Market design	Centrally managed model with bilateral conxtracts		

SK benefits from local coal supply, which partly explains the lower electricity price in this province, compared to NS and PEI (Figure 13.3). GHG emissions nevertheless remain a concern, which explains why SaskPower owns wind facilities and has signed PPAs with wind producers (SaskPower, 2012). Its limited interconnections with AB, MB and North Dakota, its US southern neighbor, allow for some trade, but market structures in these four jurisdictions remain designed independently.

In NS and PEI, the situation is slightly different. These provinces rely even more than SK on GHG-emitting thermal power plants, using increasingly expensive imported fossil fuels. Plans are therefore more actively made to increase the electricity supply from alternative sources. NS has indeed two tidal power plants and PEI elaborated a wind energy plan in 2008, aiming at 500 MW of wind capacity by 2013, up from only 111 MW of wind in 2009 (PEI, 2008). However, high cost and the absence of marketable carbon credits from wind-generated electricity have postponed the plan (Blackwell, 2012).

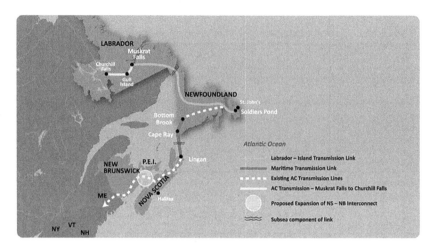

FIGURE 13.8   Transmission lines for the Lower Churchill projects (NL, 2010).

Hydropower supply from the Lower Churchill Project in NL is also seriously considered. Such hydropower supply would however require significant transmission investments, as illustrated by Figure 13.8.

Given the scope of such transmission lines and the limited size of the electricity market in Atlantic provinces, a joint ISO would make sense, as recommended to the NS government in a study it commissioned in 2009 (SNC-Lavalin, 2009). However, despite discussion around an Atlantic Energy Gateway (Canada, 2009), the Atlantic provinces have not taken steps to integrate their electricity market, in one way or another.

2.5 Insights from the Canadian Reform Experiences

Canadian electricity reforms have been only actively implemented in two provinces: AB and ON, even if NB did restructure more than any other provinces beyond these two. The paradox of these restructuring efforts is that while more competition was sought within provincial borders, no effort was made to increase competition from outside, including from the US. Such competition, in particular from hydropower provinces, BC in the case of AB, and MB and QC in the case of ON, would have raised the likeliness of decreasing prices, as cheaper sources of electricity would have entered the market.

In hydropower provinces, while minimal internal reforms were implemented to comply with US requirements, a very competitive and market-oriented behavior was adopted in export markets by provincially owned utilities. These provinces, and especially QC with huge energy storage capacity behind its dams, also benefit from cheap imports during off-peak hours, allowing them to do some intertemporal arbitrage.

Because of the lack of federal leadership in the electricity sector, and because provincial governments cherish their autonomy in managing their own electricity sector, no common Canadian reform ever attempted to create a common framework, such as in the European Union (EU, 2009) or as in the US, with the failed standard market design initiative led by the FERC (FERC, 2005). The next section presents the reasons for more integration among Canadian provinces and with US states, along with the main obstacles to such harmonization, and some possible paths that could be followed.

3 ELECTRICITY MARKET INTEGRATION FORCES AND OBSTACLES[7]

3.1 Electricity Market Integration[8]

Despite the fact that there are "few academic studies which have real theoretical depth (on regional power sector integration)" (ECA, 2010), there is an important consensus on the benefits of such electricity market integration. The United Nations has published many reports on the subject (see in particular UNECA, 2004 and UN, 2006) and so have the World Bank (ESMAP, 2010), the World Energy Council (WEC, 2010), the Organization of American States (OAS, 2007), and even the Commission for Environmental Cooperation (CEC, 2002). This latter organization is a North American organization established in 1994 along with the North American Free Trade Agreement (NAFTA). This literature identifies a series of potential technical benefits achievable through such initiative. Basically, benefits derived from efficiency gains are achievable through trade and increased productive efficiency. These benefits, in the context of electricity markets, are summarized in Table 13.7.

By increasing the regional scope of an electricity market, more of these benefits are achievable. In the Canadian context described previously, where there is a wide diversity of generation technologies and price levels, clearly pooling resources and harmonizing price levels could yield significant gains. There are four areas, in particular, from which efficiency gains would come in the Canadian case.

1. *Removing consumption subsidies in hydro provinces*: Through the low rates in hydropower-dominated provinces, a consumption subsidy is provided to electricity consumers, leading to poor incentives to implement

---

7. This section draws from Pineau (2009) and (2012).
8. See also Chapter 14, on the integration of Latin American energy markets, for more on integration issues.

TABLE 13.7 Potential Technical Benefits from Power Sector Integration
(CEC, 2002; ESMAP, 2010; Pineau, 2012; UN, 2006)

Improving reliability and pooling reserves	With access to the production facilities of its neighbors, each region gains access to much greater resources to meet the demand in the case of incident. This increases reliability and reduces the need for local reserves of production capacity.
Reduced investment in generating capacity	Thanks to pooling, each region can avoid costs of adding further capacity on its own.
Improving load factors and increasing demand diversity	Greater geographic reach often provides a more diverse demand, where peak periods do not coincide. This helps to avoid operating generating plants only for peak periods, and it uses the generator fleet in a more constant and efficient manner.
Economies of scale in new construction	With guaranteed access to a much larger market, larger generating stations can be installed, making some economies of scale accessible.
Diversity of generation mix and supply security	With more types of generation producing electricity, over a larger territory, the system is less exposed to events that affect a particular source of energy (low rainfall, lack of fuel, etc.). This increases the overall security of the integrated system.
Economic exchange	With a more diversified generating fleet and production costs, it is possible to use less costly technologies, situated in other regions, to meet various energy needs. It becomes possible to use lower cost, but distant, energy resources if equivalent local resources are not available. This reduces the overall operating costs of the system.
Environmental dispatch and new plant sitting	With a larger territory in which to choose the location of generation facilities, the best sites can be chosen (e.g., areas with less fragile ecosystems or zones with the most favorable winds for wind power).
Better coordination of maintenance schedules	Greater flexibility and reduced impact can be obtained with a more extensive production fleet.

energy efficiency. Opening provincial electricity markets and harmonizing them with their neighbors would raise prices in hydro provinces and lead to energy consumption reduction, which will be free for exports to "thermal" provinces or states—reducing price in these markets. See Pineau (2008) for more on such consumption subsidies and Billette de

Villemeur and Pineau (2012) for the conditions leading to gains from integration.

2. *Gains from less electrical heating*: In QC and some other provinces, space and water electrical heating is common due to the low price of (hydro)electricity. This is inefficient when, simultaneously, more provinces and states build natural gas-fired power plants. Indeed, the heating efficiency of natural gas (more than 85%) compared to producing electricity with natural gas (about 55% efficiency) justifies substituting natural gas to hydroelectricity for heating applications. Such substitution could only happen in a more integrated market, where incentives are aligned to make individual consumers make the correct choice.

3. *Optimize renewable energy location and production*: Rather than having every single province and state develop their own renewable incentive programs within their jurisdiction (where politically convenient), the best renewable opportunities should be developed through a regional process. With a "national electricity grid," as proposed by the province of NL (NL, 2007) and evoked in AB (AESO, 2009), more markets would have access to undeveloped hydropower. Such development could be cheaper and without intermittency issues, compared to local wind projects. Wind projects are often developed given the lack of local renewable energy alternatives.

4. *Maximizing hydro balancing for wind power*: While provinces with little hydro storage capacity develop significant wind capacity in Canada (AB and ON), real-time balancing opportunities with hydropower storage, available in BC, MB and QC, are impossible due to hour-ahead restrictions in interprovincial electricity trading, as opposed to 5-min dispatch schedules within provinces (IESO, 2012). This seriously limits the use of out-of-province hydro resources in balancing markets and leads to the use of more expensive in-province resources. Under a harmonized electricity market, resources would be used more optimally.

Although no estimate of gains resulting from integration has been computed in a Canadian context, the importance of hydropower in Canada is such that significant efficiency benefits would be obtained. Such belief is at least behind the main recommendation of the IEA concerning electricity markets in Canada (IEA, 2010):

The government of Canada should facilitate market opening and integration between provincial markets and with neighboring US markets to increase the transparency of generation investment signals, potential for competition in electricity wholesale and retail markets, and to simplify governance and oversight of reliability planning and system operation.

Within Canada, some voices also underscore, to various extent, benefits of integration: the Canadian Electricity Association (CEA, 2007), the

Canadian Academy of Engineering (CAE, 2009), and think tanks like the C.D. Howe Institute (Pierce et al., 2006) or the Canadian Center for Policy alternatives (Bigland-Pritchard, 2010).

3.2 Obstacles to Better Integration

There are three major obstacles hindering an integration process (Pineau, 2012):

1. the structure of political and electoral incentives at the provincial and federal levels;
2. the redistribution of the gains from a partial or complete integration; and
3. the lack of recognition of environmental benefits resulting from integration.

Because of the provincial nature of electricity systems and their historical role in bringing modern energy amenities to citizens, many consumers see their provincial utility as part of their culture and even, in some cases, of their identity (Froschauer, 1999). They resent seeing out-of-province interference in the electricity sector and consider, to some extent, local electricity production as being a "heritage" (such as in BC and QC—see footnote 5) that should be consumed locally. Given this, and the political convenience of announcing local "green energy projects and jobs," provincial political parties have little incentives to embark on a reform path that could result in a political backlash, such as experienced by the NB Premier after announcing the Hydro-Quebec takeover of NB Power (Section 2.3.3).

Beyond the provincial political sensitivities over electricity (and energy) issues, electricity integration would have some economic impact on wealth distribution. Consumers having access to low-rate electricity would face higher prices, and producers in higher-rate jurisdictions would lose market share when additional imports would flow into their markets. This contributes to explain why consumers in BC, MB and QC resist change, and why AB remains relatively isolated from its neighbors.

Finally, as hydropower exports do not get any recognition for GHG reductions induced in the market where they are sold, an important benefit remains unaccounted in such trade. This reduces the incentives to develop trade, as some gains are ignored. See Ben Amor et al. (2011) for an estimate of the GHG impacts of hydropower exports from the QC market.

3.3 Possible Approaches to Foster an Integration Agenda in Canada

There are many options to foster an integration agenda in Canada, and with the US, which are not mutually exclusive. They involve information, compensation, and regulatory/legal reforms.

First, not enough information is available on the gains resulting from integration. With more data on economic and environmental benefits of electricity market integration, the case could be better made that such reforms are worth it.

With such data at hand, some compensation schemes could be designed to overcome the resistance of consumers—voters in jurisdictions where price would increase. Because such trade reforms are welfare enhancing, such compensation could be made while leaving no player worse off after the change.

Finally, regulatory and legal changes could be made at different levels: trade agreements, system operations, energy regulatory boards, and possibly at other levels. Within Canada, free trade is not yet fully implemented—especially in energy. The Agreement on Internal Trade (AIT) is indeed being negotiated since 1995 with no agreement at all on energy, despite a common desire "to establish an open, efficient, and stable domestic [energy] market" (AIT, 2007). Through the AIT, and through similar free-trade agreement with the United States (such as the defunct Free Trade Area of the Americas; Pineau, 2004), some electricity market integration could happen. Regional transmission organization could be set up, as proposed for Atlantic Canada (SNC-Lavalin, 2009), with more compatible market designs than now. Provincial energy boards, regulating provincial energy sectors (Tables 13.2, 13.3 and 13.6) could also organize themselves as a unique organization with provincial branches—allowing for more coherence in regulatory decisions. The federal energy regulator (NEB) could play a coordination role in this harmonization of regulatory bodies across Canada.

All the approaches are of course difficult to initiate. However, there is an increasing political recognition in Canada and in the US that more cooperation among provinces and states is required in the energy sector. In July 2012, all provinces except BC agreed to work on a plan that would "ensure that the country has a strategic, forward thinking approach for sustainable energy development [...] [and] a more integrated approach to climate change, reducing greenhouse gas emissions and managing the transition to a lower carbon economy" (COF, 2012). This is often labeled as a Canadian National Energy Strategy. Furthermore, New England Governors and Eastern Canadian Premiers have also agreed in July 2012 to "begin work on creating a new regional process to identify longer-term opportunities in electricity markets, increase the flow of clean energy and associated infrastructure development to facilitate achievement of environmental objectives and ensure long-term security of supply through diversification and regional participation" ACNEG-ECP (2012). Although nothing more than working groups will result from these two political initiatives, they are signs that the integration issue is on the political agenda.

4 CONCLUSION

This chapter presented the fragmented Canadian electricity markets and the case for more integration. The 10 Canadian provinces can be grouped in three categories. First, the "hydropower-dominated provinces" (BC, MB, QC and NL), where a government-owned hydropower company is pervasive. Second, the "restructured provinces," where the electricity sector is designed around an ISO (AB, ON and NB). Finally, the "traditional provinces" (SK, NS, and PEI), which have relatively small, vertically integrated electricity sectors, dominated by thermal generation. If almost all provinces have implemented OATT and can therefore claim to have an open wholesale sector, only AB has a fully competitive electricity sector. ON has reintroduced regulation in 2002, 8 months only after its market liberalization, and NB now plans to revert to a fully integrated structure after 7 years of experience with its ISO.

The main paradox of electricity reforms in Canada is that the government-owned provincial hydropower companies are the most aggressive in competitive export markets, while by nature (and to follow their main mission) they should concentrate on their domestic market. This is due to the economic value their relatively cheap hydropower has in higher-cost neighboring markets. But despite this, no reform is implemented in these hydro provinces to build on their competitive advantage, such as Norway did with its electricity reforms and integration with other Nordic countries.

Gains from further integration, among Canadian provinces and with US states, would be coming from various sources. Of course, further trade benefits would contribute to these gains but also optimized investments and improved system operations. Political obstacles are however important and hinder the development of a more unified Canadian electricity market. But increasing political recognition that a national energy strategy is important, along with more data on economic and environmental benefits of regional power integration, may lead to positive developments. From the Canadian fragments, a harmonized electricity sector could deliver more than energy— increased wealth and lower environmental impacts. These are outcomes that can hardly be overlooked in our financially and environmentally constrained world.

REFERENCES

ACNEG-ECP, 2012. Resolution 36-1-Resolution Concerning Energy, 36th Annual Conference of New England Governors and Eastern Canadian Premiers. Canadian Intergovernmental Conference Secretariat, Ottawa.

AESO, 2006. The Path to Transformation—A Case Study of the Formation, Evolution and Performance of the Alberta Electric System Operator. Alberta Electricity System Operator, Calgary.

AESO, 2008. Long Term Adequacy Metrics, Threshold and Threshold Actions Recommendation Paper. Alberta Electricity System Operator, Calgary.

AESO, 2009. AESO Long-Term Transmission System Plan 2009. Alberta Electricity System Operator, Calgary.

AIT, 2007. Agreement on Internal Trade—Consolidated Version 2007. Internal Trade Secretariat, Winnipeg.

Alberta, 2008. Launching Alberta's Energy Future—Provincial Energy Strategy. Government of Alberta, Edmonton.

Alberta, 2012a. *Energy's History in Alberta*. Government of Alberta, Edmonton. Available from: <http://www.energy.gov.ab.ca/About_Us/1133.asp> (accessed 26.03.13.).

Alberta, 2012b. Switching Percentage by Group. Government of Alberta, Edmonton (Electricity Statistics Information System, 2012/07/16).

BC, 2003. Terms of Reference for Heritage Contract and Stepped Rates and Access Principles. Government of British Columbia, Victoria (Order in Council No. 0253).

BC Hydro, 2012a. BC Hydro Annual Report 2012. BC Hydro, Vancouver.

BC Hydro, 2012b. History. BC Hydro, Vancouver. Available from: <http://www.bchydro.com/about/who_we_are/history.html/> (accessed 31.06.12).

BC Hydro, 2012c. Maps—Bulk Provincial Transmission System. BC Hydro, Vancouver.

Ben Amor, M., Pineau, P.-O., Gaudreault, C., Samson, R., 2011. Electricity trade and GHG emissions: assessment of Quebec's hydropower in the Northeastern American market (2006–2008). Energ. Policy 39 (3), 1711–1721.

Bigland-Pritchard, M., 2010. Transforming Saskatchewan's Electrical Future—Part Three the Potential for Wind and Solar Power. Canadian Center for Policy Alternatives, Regina.

Billette de Villemeur, E., Pineau, P-O., 2012. Regulation and electricity market integration: when trade introduces inefficiencies. Energy Econ. 34 (2), 529–535.

Blackwell, R., 2012. Why PEI's wind plan is dying. Globe Mail (August 23). Available from: <http://www.theglobeandmail.com/report-on-business/industry-news/energy-and-resources/why-peis-wind-plan-is-dying/article1370245/> (accessed 26.03.13).

Blakes, 2008. Overview of Electricity Regulation in Canada. Blake, Cassels & Graydon LLP, Montreal.

Bruce Power, 2012. *A look back at the history of Bruce Power*. Bruce Power, Tiverton. Available from: <http://www.brucepower.com/about-us/guide-to-bruce-power/> (accessed 26.03.13).

CAE, 2009. Electricity: Interconnecting Canada a Strategic Advantage Report of the Canada Power Grid Task Force, Volume I—Findings, Conclusions and Recommendations. Canadian Academy of Engineering, Ottawa.

Canada, 2009. The Atlantic Energy Gateway—Backgrounder. Government of Canada, Ottawa. Available from: <http://www.ecoaction.gc.ca/news-nouvelles/20090329-1-eng.cfm/> (accessed 15.08.12).

Canada Energy, 2012. *Ontario Electricity Market Overview*. Canada Energy Wholesalers Ltd, Oakville. Available from: <http://www.canadaenergy.ca/index.php?hydro=oem> (accessed 15.08.12).

CBC News, 2010. Point Lepreau Delays Add $1B to Project, November 29, 2010. Available from: <http://www.cbc.ca/news/canada/new-brunswick/story/2010/11/29/nb-point-lepreau-refurbishment-update-211.html> (accessed 15.08.12).

CEA, 2007. The Integrated North American Electricity Market: Energy Security: A North American Concern. Canadian Electricity Association, Ottawa.

CEC, 2002. Environmental Challenges and Opportunities in the Evolving North American Electricity Market. Commission for Environmental Cooperation of North America, Montréal (Secretariat report to council under article 13 of the North American agreement on environmental cooperation).

Clark, C.R., Leach, A., 2007. The potential for electricity market restructuring in Quebec. Can. Public Policy 33 (1), 1–20.

COF, 2012. Premiers Guide Development of Canada's Energy Resources. Council of the Federation, Ottawa (Halifax, July 27, 2012).

Daniel, T., Doucet, J., Plourde, A., 2007. Electricity industry restructuring: the Alberta experience. In: Kleit, A.N. (Ed.), Electric Choices: Deregulation and the Future of Electric Power. Rowman & Littlefield, Lanham.

EIA, 2012. International Energy Statistics. U.S. Energy Information Administration, Washington, DC.

ECA, 2010. The Potential of Regional Power Sector Integration —Literature Review. Economic Consulting Associates, London.

EDA, 2006. Ontario Electricity Market Primer. Electricity Distributors Association, Vaughan.

Environment Canada, 2011. National Inventory Report 1990–2009: Greenhouse Gas Sources and Sinks in Canada—Part 1. Environment Canada, Gatineau (The Canadian Government's Submission to the UN Framework Convention on Climate Change).

ESMAP, 2010. Regional Power Sector Integration: Lessons from Global Case Studies and a Literature Review. The World Bank, Washington, DC (Regional Energy Integration Strategies Program—Solving Energy Challenges through Regional Cooperation, Briefing Note 004/10, Energy Sector Management Assistance Program (ESMAP)).

E-T Energy, 2012. *Available Electricity*. E-T Energy, Calgary. Available from: <http://www.e-tenergy.com/operations/available-electricity.html> (accessed 28.08.12.).

EU, 2009. *Internal Market in Electricity(from March 2011)*. European Union, Brussels. Available from: <http://europa.eu/legislation_summaries/energy/internal_energy_market/en0016_en.htm> (accessed 16.08.12).

FERC, 2005. Remedying Undue Discrimination Through Open Access Transmission Service and Standard Electricity Market Design—Order Terminating Proceeding. Federal Energy Regulatory Commission, Washington, DC (Docket No. RM01-12-000).

Froschauer, K., 1999. White Gold: Hydroelectric Development in Canada. UBC Press, Vancouver.

Hydro-Quebec, 2011. Developing Outside Markets. Hydro-Quebec, Montreal. Available from: <http://hydroforthefuture.com/projets/34/developing-outside-markets/> (accessed 28.08.12).

Hydro-Quebec, 2012a. Annual Report 2011. Hydro-Quebec, Montreal.

Hydro-Quebec, 2012b. History of Electricity in Québec—Hydro-Québec Highlights. Hydro-Quebec, Montreal. Available from: <http://www.hydroquebec.com/learning/histoire/faits_saillants.html/> (accessed 31.07.12).

IEA, 2010. Energy Policies of IEA Countries—Canada 2009 Review. International Energy Agency, Paris.

IEA, 2011. 2011 Key World Energy Statistics. International Energy Agency, Paris.

IESO, 2012. Introduction to Interjurisdictional Energy Trading—An IESO Training Publication. Independent Electricity System Operator, Toronto.

IRAC, 2008. In the Matter of an application by Maritime Electric Company, Limited for approval of an Open Access Transmission Tariff. Island Regulatory and Appeals Commission, Charlottetown, Docket UE20935, Order UE08-08.

Manitoba Hydro, 2012a. *History & Timeline*. Manitoba Hydro, Winnipeg. Available from: <https://www.hydro.mb.ca/corporate/history/hep_1990.html> (accessed 31.07.12).

Manitoba Hydro, 2012b. *Current Electricity Rates*. Manitoba Hydro, Winnipeg. Available from: <www.hydro.mb.ca/regulatory_affairs/energy_rates/electricity/current_rates.shtml>(assessed 31.07.12).

MSA, 2010. Alberta Wholesale Electricity Market. Alberta Market Surveillance Administrator, Calgary.

Nalcor, 2012. 2011 Business and Financial Report—Energy Comes With the Territory. Nalcor Energy, St. John's.

NB Power, 2011. Point Lepreau Generating Station Refurbishment Project Update—Fuel Channel Installation Completed. NB Power, Fredericton (November 14, 2011).

NEB, 2008. Coal-Fired Power Generation: A Perspective—Energy Briefing Note. National Energy Board, Calgary.

New Brunswick, 2011. The New Brunswick Energy Blueprint. Government of New Brunswick, St. John's (October 2011).

NL, 2007. Focusing Our Energy—Energy Plan. Government of Newfoundland and Labrador, St. John's.

NL, 2010. *Lower Churchill Project—Transmission Map*. Government of Newfoundland and Labrador, St. John's. Available from: <http://www.gov.nl.ca/lowerchurchillproject/map.htm> (accessed 10.01.12.).

NRCan, 2012. Overview of Canada's Energy Policy. Natural Resources Canada, Ottawa. Available from: <http://www.nrcan.gc.ca/energy/policy/1352/> (accessed 25.07.12.).

OAS, 2007. Regional Electricity Cooperation and Integration in the Americas: Potential Environmental, Social and Economic Benefits. Organization of American States, Washington, DC.

OEB, 2012a. Regulated Price Plan (RPP)—(RP-2004-0205). Ontario Energy Board, Toronto. Available from: <http://www.ontarioenergyboard.ca/OEB/Industry/Regulatory + Proceedings/ Policy+ Initiatives+ and + Consultations/Regulated + Price + Plan/> (accessed 09.08.12.).

OEB, 2012b. Smart Metering Initiative—History (RP-2004-0196). Ontario Energy Board, Toronto. Available from: <http://www.ontarioenergyboard.ca/OEB/Industry/Regulatory + Proceedings/ Policy + Initiatives + and + Consultations/Smart + Metering + Initiative + %28SMI%29/ Smart + Metering + Initiative + History/> (accessed 09.08.12.).

OEB, 2012c. Monitoring Report—Smart Meter Deployment and TOU Pricing—June 2012. Ontario Energy Board, Toronto.

OPA, 2011. IPSP Planning and Consultation Overview. Ontario Power Authority, Toronto.

PEI, 2008. Island Wind Energy—Securing our Future: The 10 Point Plan. Prince Edward Island, Charlottetown.

Pierce, R., Trebilcock, M., Thomas, E., 2006. Beyond Gridlock: The Case for Greater Integration of Regional Electricity Markets. C.D. Howe Institute, Toronto (C.D. Howe Institute Commentary No. 228).

Pineau, P.-O., 2004. Electricity services in the GATS and the FTAA. Energy Stud. Rev 12 (2), 258–283.

Pineau, P.-O., 2008. Electricity subsidies in low cost jurisdictions: the case of British Columbia (Canada). Can. Public Policy/Analyse de Politiques 34 (3), 379–394.

Pineau, P.-O., 2009. An integrated Canadian electricity market? The potential for further integration. In: Eberlein, B., Doern, B. (Eds.), Governing the Energy Challenge: Canada and Germany in a Multilevel Regional and Global Context. University of Toronto Press, Toronto.

Pineau, P.-O., 2012. Integrating Electricity Sectors in Canada: Good for the Environment and for the Economy. The Federal Idea, Montreal.

Powerex, 2012. About Us. Powerex, Vancouver. Available from: <http://www2.powerex.com/ AboutUs.aspx/> (accessed 31.07.12.).

Quebec, 2010. Energy Partnership with New Brunswick—The Government of Quebec Announces the End of Discussions. Government of Quebec, Quebec (March 24, 2010).

SaskPower, 2012. SaskPower Annual Report 2011—Energizing Growth. SaskPower, Regina.

SNC-Lavalin, 2009. Transmission and System Operator Options for Nova Scotia—Final Report. SNC-Lavalin, Montreal (Prepared for the Nova Scotia Department of Energy).

Statistics Canada, 2012a. Table 127-0002 Electric Power Generation, by Class of Electricity Producer. Statistics Canada, Ottawa.

Statistics Canada, 2012b. Table 127-00031 Electric Power Generation, Receipts, Deliveries and Availability of Electricity. Statistics Canada, Ottawa.

Statistics Canada, 2012c. Table 127-0009 Installed Generating Capacity, by Class of Electricity Producer. Statistics Canada, Ottawa.

Statistics Canada, 2012d. Table 127-0008—Supply and Disposition of Electric Power, Electric Utilities and Industry, Annual. Statistics Canada, Ottawa.

Trebilcock, M.J., Hrab, R., 2006. Chapter 12—Electricity restructuring in Canada. In: Sioshansi, F.P. (Ed.), Electricity Market Reform. Elsevier, Oxford, pp. 419–449.

UN, 2006. Multi Dimensional Issues in International Electric Power Grid Interconnections. United Nations, New York (Department of Economic and Social Affairs, Division for Sustainable Development).

UNECA, 2004. Assessment of Power Pooling Arrangement in Africa. United Nations Economic Commission for Africa, Addis Ababa.

WEC, 2010. Interconnectivity: Benefits and Challenges. World Energy Council, London.

WNA, 2012. *Uranium ProductionFigures, 2001–2011*. World Nuclear Association, London. Available from: July 24, 2012. <http://www.world-nuclear.org/info/Facts-and-Figures/Uranium-production-figures> (accessed 05.04.13).

Chapter 14

Latin America Energy Integration: An Outstanding Dilemma

Ricardo Raineri[a], Isaac Dyner[b], José Goñi[c], Nivalde Castro[d], Yris Olaya[e] and Carlos Franco[e]

[a]*The World Bank and Engineering School, Pontificia Universidad Católica de Chile;* [b]*CEIBA, CIIEN Universidad Nacional de Colombia;* [c]*Graduate School, Universidad de Concepción;* [d]*GESEL—Instituto de Economia—UFRJ;* [e]*Complexity Centre Universidad Nacional de Colombia*

1 INTRODUCTION

Latin America (Latam) is a region with a wide disparity in living standards, where poverty continues to be the region's main challenge. In some countries, per capita income reaches US$15,000 while in others it is barely above US$1000. Paradoxically, the region is rich in natural and energy resources that are not equally distributed among the countries. Some countries are abundantly endowed with oil, gas, and hydroelectric power, while others confront scarcity and a need for clean and competitive energy sources. In recent years, the economies of the region have been in a cycle of strong growth, low inflation rates, and unemployment. The regional GDP grew at 5.9% in 2010 and 4.4% in 2011, with 3.7% expected for 2012. Poverty and homelessness have fallen to their lowest levels in decades from 48.4% of the total population in 1990 to 30% in 2011 and, at the same time, the region has experienced significant improvements in productivity and real wages for those employed in urban areas. In 1990, there were 337 million workers in high and medium productivity segments, while in 2011, it increased to 565 million, expanding domestic markets and improving living conditions for millions of people.

In the context of the global economy, Latin-American countries are experiencing unprecedented changes with unpredictable consequences. The hegemony of some European countries is already in full retreat, as it is the relevance of Japan and, in a slowing but equally clear process, the loss of the relative importance of the US economy. At current growth rates, within the next decade, one should expect China to be the major economic power in the world, followed by the United States, while India, Brazil, and Mexico should be expected to be among the leading economies in the world. In this context, integration of energy markets in Latam is an opportunity for developing regional resources and increasing productivity in the region.

Worldwide market liberalization and integration of power supply are not new and started in the early 1980s (Newberry, 2005). Full integration of power supply markets between neighboring countries was led by the northern European nations (Jacobsen et al., 2005) and has been taking place in the European Union (EU) since the mid-1990s (Hass et al., 2006; Jamasb and Pollit, 2005). Integration may bring about benefits in terms of gains in competitive pricing and social welfare (Newberry, 2002). Other regions including North America, Latin America (Dyner et al., 2007), Australia, and Africa have shown inclinations toward integration with dissimilar results ranging from high success (e.g., PJM regional transmission organization in the US; Bowring, 2006 and in this volume) to a very slow start (Pineau et al., 2004 and also in this volume; Gnansounou et al., 2007).

As is argued in this chapter, the integration experience in Latam has had mixed results. On the one hand, there have been successful cases of bilateral agreements for energy trade but, on the other hand, there have been political and economic barriers to integration at the operating and market levels. This chapter is organized as follows: Section 2 briefly describes the market structure and technology mix for power generation in South America. Section 3 summarizes the main attempts at integration including their weaknesses and strengths before discussing the main lessons and work ahead for energy integration in Latam. Section 4 discusses energy integration and the geopolitics in Latam followed by conclusion.

2 SOUTH AMERICA'S ENERGY MARKET

The Latam market consists of almost 400 million citizens, more than 100 million customers, and significant opportunities to improve electrification rates (Table 14.1). According to IMF, regional growth is expected to double the rates projected for developed countries. In this context, as world energy consumption grew by 2.5% in 2011, in line with the historical average, and well below the 5.1% seen in 2010, Latam energy consumption has grown above world average, increasing 3.8% in 2011 and 6.4% in 2010, with a 3.2% annual average energy growth rate for the 2001−2011 period.[1] Energy consumption in Latam is expected to grow at 2% per year according to Energy Information Administration (EIA) projections, above developed countries but behind non-OECD Asia and Middle East. Although per capita energy consumption in Latam is considerably below the average of developed countries (Figure 14.1), it is projected to increase in proportion to the forecasted economic growth, requiring large investments in the energy

---

1. BP Statistical Review of World Energy, June 2012.

TABLE 14.1 Electrification Levels 2009

Country	Total Population (Thousands)	Total Customers (Thousands)	Population Served (Thousands)	Percent of Population with Electricity	Total Houses (Thousands)	Houses with Electricity (Thousands)	Percent of Households with Electricity
Argentina	40,134	13,730	n.d.	n.d.	n.d.	11,724	95.0
Bolivia	10,225	1634	7646	74.8	2502	1871	74.8
Brazil	193,934	65,722	190,575	98.3	57,037	56,140	98.4
Chile	16,970	n.d.	n.d.	n.d.	5133	4928	96.0
Colombia	44,460	10,441	38,620	86.9	10,571	9559	90.4
Ecuador	13,627	3745	n.d.	n.d.	n.d.	n.d.	n.d.
Paraguay	6,396	1279	6266	98.0	1204	1180	98.0
Peru	29,132	4879	22,443	77.0	7996	5917	74.0
Uruguay	3,345	1283	3271	97.8	1392	1361	97.8
Venezuela	28,584	5544	n.d.	n.d.	n.d.	n.d.	n.d.

Note: n.d.: no data.
Source: CIER (2011).

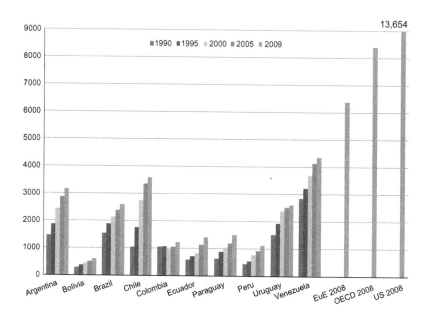

FIGURE 14.1 Per capita electricity consumption. *Source: CIER, May 2011.*

sector. Thus, in line with increases in electricity demand, more than 7000 MW of new generation capacity is required per year—more than half of this in Brazil.

As further described in Chapter 15 in this volume, Brazil is by far the country with the largest installed capacity in the region, and also the country which imposes the largest pressure on new investments to satisfy its growing energy needs. This also explains its active role in the international forum promoting deeper energy integration within the region, and on the search of successful bilateral integration agreements with neighboring and other regional countries.

2.1 The Energy Mix of the Region's Countries

There is a wide discrepancy in the energy mix of the countries in the region. Countries like Paraguay generate almost 100% of their electricity with hydropower, whereas fossil fuel capacity is above 40% in Argentina, Bolivia, Ecuador, and Peru. This fossil fuel generation base is large, considering the large hydropotential in some of these countries and in the region as a whole. Gas and oil supply for power generation is also varied and ranges from self-sufficiency in Colombia and Venezuela to dependence on imports as in Chile. The Latam energy mix is among the cleanest worldwide, with more than 30% of energy coming from hydroelectricity and other renewable energy sources, exceeding the clean energy mix of North America, Europe and Eurasia, Middle East, Africa, and Asia Pacific. Also, Latam is one of the regions with the greatest chances and potential to achieve a low-carbon intensive energy and electricity mix given the large and unexploited hydro-electric and other renewable energy sources available in the region (Table 14.2 and Figure 14.2).

Even though there is a huge potential for nonconventional and renewable energy sources, such as wind, solar, geothermal, and small hydro, they still represent a very small fraction of the regions' energy mix (Table 14.3). These technologies, however, are picking up fast as their prices become more competitive and as countries are diversifying their energy mix (Dyner et al., 2011). Brazil, Argentina, and Mexico are the only countries with nuclear power within the Latam region (Figure 14.3).

Also, depending on the industrial structure of the countries, electricity consumption can widely differ according to household and industrial consumption (Figure 14.4). For example, Chile, Peru, and Brazil are countries where industrial consumption represents a larger percentage of total consumption. Particularly, in the case of Chile, the large mining industry is an intensive energy user.[2]

Regarding electricity prices, there is a wide disparity in the electricity bill paid by final consumers in the region (Figure 14.5). Policies and poor regulations account for much of this, where extensive subsidies are not uncommon and trade barriers result in price disparities that cannot be explained by transaction or transportation costs. In some Latam countries, consumers pay more than twice the price paid in neighborhood countries. This suggests that big benefits can be achieved from electricity integration and energy exchange opportunities within the countries, the main focus of this chapter.

2.2 Electricity Exchanges

Electricity trade within the region is still modest (Table 14.4 and Figure 14.6). Setting aside a few exceptions such as electricity exports by Paraguay and Colombia, most countries are net importers of electricity, including Chile, Brazil, Argentina, Ecuador, and Uruguay.

Given the large price disparities within Latam, the potential for efficiency gains from greater integration and trade is significant. Countries like Brazil, Chile, Argentina, and Uruguay can benefit from the abundant hydro and hydrocarbon resources in countries like Colombia, Peru, Bolivia, Venezuela, and Uruguay.

---

2. In the case of Venezuela, the data is not disaggregated at the same level as for the other countries.

TABLE 14.2 Hydroelectric Potential in South America

| | Hydroelectric Potential in Late 2008 (MW) | | | Status of Hydropower Development in Late 2008 | | | |
| | | | | In Operation | | Under Construction | Planned |
	Gross Theoretical Capacity	Technically Exploitable Capacity	Economically Exploitable Capacity	Capacity (MW)	Generation (2008 in GWh)	Capacity (MW)	Capacity (MW)
Argentina	40,411	19,292	8904	9950	30,600	125	2800
Bolivia	20,320	14,384	5708	440	2310	88	2338–3064
Brazil	347,032	142,694	93,379	77,507	365,062	8580	68,000
Chile	25,913	18,493	11,073	5026	24,261	322	5800
Colombia	114,155	22,831	15,982	8996	43,020		429
Ecuador	19,292	15,297	12,100	2033	9040	1971	1706–6986
French Guyana	228	114	n.d.	116	512		
Guyana	9247	4224	2511	1	1		100
Paraguay	12,671	9703	7763	8130	53,710	395	200–1790
Peru	180,023	45,091	29,680	3242	19,040	220	98
Surinam	4452	1484	913	189	1360		0–732
Uruguay	3653	1142	685	1358	8070		70
Venezuela	83,447	29,795	11,416	14,567	86,700	2704	0–7250
Total South America	860,845	324,543	200,114	131,555			

Note: n.d.: no data.
Source: World Energy Council SER (2010). Used by permission of the World Energy Council, London, www.worldenergy.org.

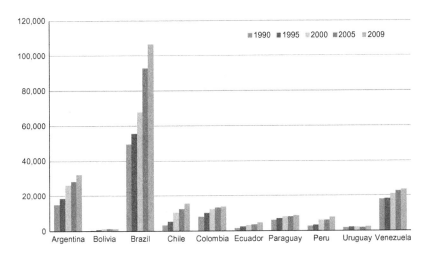

FIGURE 14.2   Installed power generation capacity (MW). *Source: CIER, May 2011.*

TABLE 14.3 Percentage of Renewable Energy Sources on Installed Capacity and Energy Generated in 2009

Nonrenewable or Thermal Energy Source			Renewable Energy Sources	
	Unit	Thermal (%)	Hydro (%)	Nonhydro (%)
Argentina	MW	68.4	31.5	0.1
Bolivia	MW	64.0	34.5	1.5
Brazil	MW	18.8	73.8	7.5
Chile	MW	63.6	34.8	1.6
Colombia	MW	33.3	66.5	0.1
Ecuador	MW	57.4	42.6	0.0
Paraguay	MW	0.0	100.0	0.0
Peru	MW	59.0	41.0	0.0
Uruguay	MW	31.7	61.4	6.9
Venezuela	MW	38.3	61.7	0.0

Source: CIER (2011).

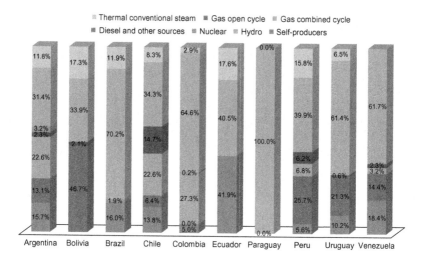

FIGURE 14.3   Generation capacity by as source percent of total generation capacity, 2009. *Source: CIER, May 2011.*

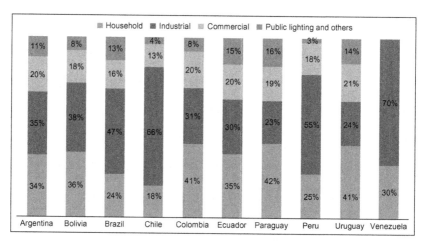

FIGURE 14.4   Energy consumption by sector as percent of final demand, 2009 (excluded network losses and self-consumption). *Source: CIER, May 2011.*

3 EXPERIENCES IN REGIONAL ENERGY INTEGRATION

This section briefly describes the power transmission and gas interconnection infrastructure that has been developed followed by examples of multilateral and some bilateral initiatives that have recently taken place.

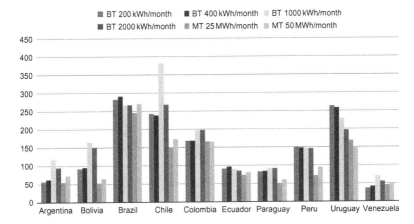

FIGURE 14.5   Final consumers' electricity price, 2010 (US$/MWh with value added tax, in effect for January 2010). *Source: CIER.*

3.1 Bilateral Initiatives

The bilateral Itaipu power plant project in the Paraná River, a joint venture between Paraguay and Brazil, is among the first and most important integration experiences in Latam. In this initiative, the talks between Brazil and Paraguay to exploit joint hydroelectric resources started in the early 1960s. The construction of the plant began in 1975 and ended in 2007 resulting in a capacity of 14,000 MW.[3] Other initiatives include the binational hydroelectric power plants of Salto Grande[4] (1890 MW), between Argentina and Uruguay, and Yaciretá[5] (3200 MW), between Argentina and Paraguay. After these first bilateral integration initiatives, other projects have taken place in the electricity and gas sectors. By 2011, CIER[6] reported 19 operating bilateral interconnections. Argentina is the most interconnected country with 10 bilateral interconnections with 5 neighboring countries; the least interconnected countries are Chile and Peru, with 1 bilateral connection for each (Table 14.5 and Figure 14.7).

---

3. On June 22, 1966, the Minister of Foreign Affairs of Brazil (Juracy Magalhães) and Paraguay (Sapena Pastor) signed the "Act of Iguazú," a joint statement that expressed the willingness to consider the use of water resources from both countries, in the stretch of the Paraná River "from and including the Falls of Guairá to the Iguazú River mouth." One of the clauses of the bill signed by both countries expected that the excess of energy that is not used by one of the countries would be sold exclusively to the other country participating in the project.
4. This is a project whose study began in 1938, and its construction started in 1974 and was completed in 1979.
5. This is a project whose study began in 1973, and its construction started in 1983 and was completed in 1989.
6. CIER: Comisión de Integración Energética Regional (Regional Energy Integration Commission), "Síntesis informativa energética de los países de la CIER" (*Brief Energy Background for CIER Member States*), May 2011.

TABLE 14.4 Electricity Exchanges Between Countries, 2009 (GWh)

		Export Country										
		Argentina	Bolivia	Brazil	Chile	Colombia	Ecuador	Paraguay	Peru	Uruguay	Venezuela	Total Imports
Import Country	Argentina	–	–	993	–	–	–	6831	–	251	–	8075
	Brazil	–	–	–	–	–	–	38,478	–	14	300	38,792
	Chile	1348	–	–	–	–	–	–	–	–	–	1348
	Colombia	–	–	–	–	–	21	–	–	–	–	21
	Ecuador	–	–	–	–	1077	–	–	63	–	–	1140
	Uruguay	963	–	505	–	–	–	–	–	–	–	1468
	Venezuela	–	–	–	–	282	–	–	–	–	–	282
	Total exports	2311	–	1498	–	1359	21	45,309	63	265	300	51,126

Exports to Brazil and Uruguay account for 402 GWh being passed in 2009 using the Argentinean Interconnection System (Sistema Argentino de Interconexión–SADI).
Source: CIER (2011).

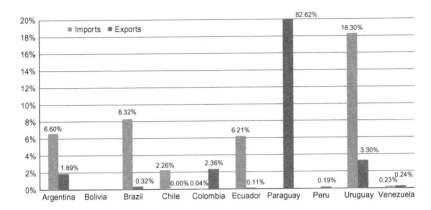

FIGURE 14.6   Electricity exchanges between countries as percent of local generation, 2009. *Source: CIER, May 2011.*

TABLE 14.5 Latam Electricity Interconnections

Interconnections	Number
Ecuador–Colombia	3
Colombia–Venezuela	3
Argentina–Paraguay	3
Argentina–Uruguay	3
Argentina–Brazil	2
Paraguay–Brazil	2
Argentina–Chile	1
Argentina–Uruguay	1
Brazil–Venezuela	1
Ecuador–Peru	1
Uruguay–Brazil	1 (under construction)
Colombia–Panamá	1 (under study)
Bolivia–Peru	1 (under study)

Source: CIER: Comisión de Integración Energética Regional (Regional Energy Integration Commission), "Síntesis informativa energética de los países de la CIER" (*Brief Energy Background for CIER Member States*), May 2011.

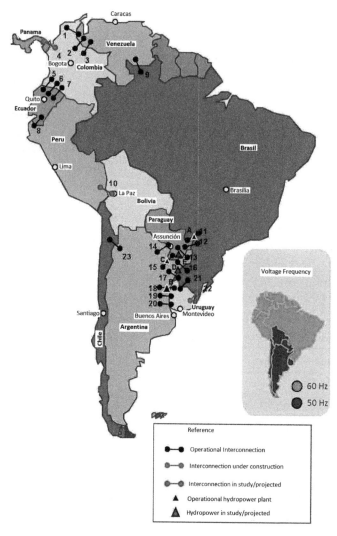

FIGURE 14.7   Major electricity interconnections in South America. *Source: CIER, May 2011.*

Latam integration arguably reduces the average costs of electricity supply, increases reliability and security, and promotes private investment in the sector (Ochoa et al., 2012), taking advantage of economies of scale and scope, the complementarities generated by the differences in time zones, peak load, and the complementarities in generation capacity technology and rainy seasons.

There are gas pipelines connecting Bolivia with Brazil and Argentina, Argentina with Paraguay and Chile, and Colombia with Venezuela

FIGURE 14.8   Major gas pipelines in South America. *Source: CIER, May 2011.*

(Figure 14.8). Gas trade along these pipelines has not been continuous, and has been interrupted between Argentina and Chile. Gas exports account for only 17% of the regional production. Nearly 43% of these exports, 16.41 billion cubic meters, are destined to North American, European, and Asian markets (British Petroleum − BP, 2011).

After recent major gas discoveries in Argentina, Brazil, Bolivia, Colombia, Mexico, Peru, and Venezuela, the only countries in Latam with almost no access to domestic gas are Chile, Paraguay, and Uruguay, along with Central American countries. Main natural gas reserves are estimated at 239 trillion cubic feet, allowing the region to supply 70 years of consumption, with an annual consumption level of nearly 3.4 trillion cubic feet in 2009. However, it is estimated that if discoveries of shale gas are confirmed, regional supply can feed more than 300 years at current levels of consumption.

3.2 Unasur

Unasur is the regional organization of the 12 South American states that aims to build a South American integrated area. It consists of a population of 400 million inhabitants, representing 68% of Latam's population. With the consolidation of Unasur, the 12 nations that make up this regional organization have decided to promote a participatory and consensual space for dialogue, and to undertake joint projects and lay the foundations to promote effective integration in the cultural, social, economic, policy, energy, and infrastructure projects. Energy integration is the key subject that acts as catalyst in Unasur, as this aspect of integration was stressed in 2008 at the inauguration of the Treaty. The political decision to use energy integration as an instrument for territorial transformation, reducing disparities and inequalities in welfare and quality of life over 17.7 million square kilometers is one of the core objectives of Unasur. Some modest achievements have already been attained.[7] However, in some schools of thought, integration is still seen with a strong ideological content, which limits subregional attempts, making it too dependent on the interest of the current governments to continue and back up the political views and practices that give sustainability to the energy integration idea. In this regard, the Free Trade of the Americas (FTAA) proposal provides some initial orientations.[8] Consequently, to succeed with the Unasur energy integration efforts, it is important to keep the discussion within a forum with a strong technical support. The strength of the integration process requires that all or at least most of the different political parties be convinced and committed with the process. The conviction must come from a broad political, social, and cultural base.

3.3 Central America Interconnection, SIEPAC

SIEPAC is the most successful multilateral attempt of regional electricity integration within Latam. To support the gradual consolidation of a Regional

---

7. Within the South American Energy Council of Unasur, which has defined an agenda to accomplish South American energy integration in a broad sense.
8. Free Trade Area of the Americas (Área de Libre Comercio de las Américas).

Electricity Market (MER)—Central America Market—SIEPAC created the legal, institutional, and technical mechanisms that promote private investment to add new power generation capacity. Power exchanges among the MER members are possible through the physical interconnection infrastructure. A new agency, the Central-American Electrification Council (CEAC) is responsible for establishing market rules and the institutions needed to enforce them.

The SIEPAC project includes the construction of a 1790 km long 230 kV transmission line, with an estimated transportation capacity of 300 MW (Figure 14.9). A private–public partnership—EPR (Empresa Propietaria de la Red)—builds the power transmission infrastructure with completion expected in 2013. The partners of this partnership own and operate the transmission networks of each one of the six countries. In addition, EPR has three partners from outside the region: ISA (Colombia), Endesa (Spain), and CFE (Mexico), where each partner owns 11.1% of the project's capital. The member countries provide funding through loans from multilateral banks such as the Inter-American Development Bank, IDB, and the Central American Bank for Economic Integration, who finance more than 70% of the total cost amounting to US$494 million.

FIGURE 14.9 SIEPAC transmission line. *Source: Empresa Propietaria de la Red (EPR), www.eprsiepac.com.*

The region is a net importer of electricity. In 2010, 354 GWh were exported from Mexico to Guatemala, and in 2010, Central America traded 1062 GWh—nearly 37% more than in 2009. These imports have substituted more than 700 GWh of electricity otherwise produced by inefficient and expensive plants, mostly oil generators (CEPAL, 2011a). Electricity trade is expected to increase after the SIEPAC infrastructure is fully operational and after the regulation for the regional power market is completed.

The goal of the SIEPAC project is to interconnect six Central American electricity markets: Guatemala, El Salvador, Honduras, Nicaragua, Costa Rica, and Panama, consolidating a regional energy market.

In May 2000, government representatives in the Steering Group approved the general design of the MER. MER is the seventh market that is superimposed on the six national markets in place. Its structure includes a regional regulatory commission and a Regional Operating Agency (EOR, 2010, 2011) in charge of electricity trade among Central American countries.

MER operates based on rules of the Framework Treaty of the Central American Electricity Market to ensure the free flow of electricity through the territories for themselves or for third countries in the region, subject only to the conditions set out in the Framework Agreement. The approval of the Framework Treaty of the Central American Electricity Market in December 2005 by the Regional Electric Interconnection (CRIE) was a milestone, allowing EPR to complete the conditions needed to start infrastructure expenditure as agreed with the IDB.[9]

The regulatory framework planned and executed the First Regional Transmission System and projected a second circuit. Additional funding which amounts to US$40 million for its accomplishment is expected through the Interconexión Eléctrica S.A. E.S.P. (ISA) Company.

Private utilities hold more than 60% of the total installed capacity in all of SIEPAC countries. In Costa Rica, the national power utility, ICE, owns more than 80% of the generation assets. Thermal plants account for nearly 40% of the total power generation, whereas hydropower generates over 50% of the total energy consumption; geothermal energy and wind power account for the remaining 10% of the total generation. Diesel and liquid fuels have the largest share of thermal generation—nearly 60% of the total (ECLAC, 2010).

3.4 Andean Markets (CAN)

The Andean community, CAN, includes Bolivia, Colombia, Ecuador, and Peru.[10] One of the goals of CAN is to integrate Andean Countries, promoting

---

9. http://www.proyectomesoamerica.org.
10. Argentina, Brazil, Chile, Paraguay, and Uruguay are associated members. Venezuela left the association in 2006 protesting against the free trade agreements executed by Colombia and Peru with the United States.

economic development, and energy integration is one of the strategies for achieving this goal. In addition to the benefit of improving efficiency in the use of generation capacity, integrating electricity systems allows CAN members to exploit the complementarities between their hydrological cycles, specifically, the complementarities of rainy seasons within Colombia, Ecuador, and Peru.

Venezuela was part of the Andean Community but left it in 2006. However, as trade is flowing again between the different countries, Venezuela is included as part of this block for energy integration purposes. In 2009, seven transmission lines connected Venezuela, Colombia, Ecuador, and Peru. Most of the interconnection lines in the region trade small amounts of energy and optimize energy resources in the borders. The most important interconnections are between Colombia and Ecuador and between Ecuador and Peru.

Ecuador, Peru, and Colombia can gain from electricity interconnection. Among these countries, power generation and electricity consumption have grown at rates above 2% for most of the first decade of the 2000s. Growth rates in Colombia have been mostly positive, between 1% and 6%, whereas in Peru and Ecuador, energy consumption has grown at rates between 2−10% and 2−15%, respectively.

In the same decade, generation capacity grew by 6.7% in Colombia, 38.3% in Ecuador, and 31.6% in Peru. Hydro accounts for more than 58% of the installed generating capacity, with Colombia having the largest hydro share (63%). Thermal generation fueled by natural gas, coal, and petroleum liquids accounts for nearly 40% of the total generating capacity. The share of other renewables is low. Each of these interconnected markets has an independent power dispatch, which allows each country to optimize its resources with independence from other markets.

Currently, projected interconnections between Colombia and Ecuador include a 213 km long, 230 kV double circuit line, which connects Jamondino (Colombia) and Pomasquí (Ecuador), and an 18 km, 138 kV single circuit line, connecting Panamericana (Colombia) and Tulcán (Ecuador). However, up until now, the bulk of electricity trade consisted of exports from Colombia to Ecuador, reflecting the lower marginal costs of the Colombian market, where exports/imports exchanges accounted for more than US$880 million between 2003 and 2011, and congestion fees exceeded US$300 million, with a positive trade for Colombia, the main exporting country.

Electricity trade between Ecuador and Peru has been negligible and mainly takes place for security of supply purposes, with a clear balance that favors Peru.

3.5 Andean Electricity Integration Initiative

In 2011, the Andean markets initiative (under CAN) started a wider plan where Bolivia and Chile became part of what is known as the Andean

Electricity Integration Initiative. In April 2011, ministers, deputy ministers, and senior officials of the Energy Sector of Bolivia, Chile, Colombia, Ecuador, and Peru met to discuss issues related to each country's electric infrastructure and the mechanisms to advance in a process of regional electricity integration. Resulting from this meeting, the Galapagos Declaration has agreed on creating the Council of Ministers for the Development of the Andean Electricity Corridor. Consequently, a request for technical and financial support was submitted to the IDB aiming at making progress toward the fulfillment of the goals defined in the work plan. The basic principles within this initiative aim to ensure the stability and legal and contractual assurance, free transit of electricity, together with free and nondiscriminatory access to the remaining capacity of the networks. Later, in a May 2012 meeting, The Council reviewed and approved the studies to be carried out with funding from the IDB in the areas of regulation and planning of interconnections. Also, a timetable and an institutional framework of the work to be conducted were agreed, expecting an approval in the next Council of Ministers meeting. The technical cooperation is financed by the IDB, through the "Support for Andean Studies of Electrical Interconnection, RG-T2056" Project.

The Andean Electricity Integration Initiative is so far one of the most comprehensive and sound multilateral integration initiatives in the region, taking clear steps to design the regulatory, commercial, and physical infrastructure needs to have a fully operational electricity corridor within the region.[11] If geopolitical barriers, trust, and beliefs are overcome, the next step to achieve electricity integration is to agree on a coordination scheme of the different ownership/regulatory/business models that exist within the region to support the huge investments needed for the countries' electricity interconnection.

3.6 Chile–Argentina Energy Integration

Energy integration between Argentina and Chile has gone through ups and downs (Raineri, 2006, 2007). Chile is a net energy importer, mostly of fossil fuels, which currently come from a wide range of countries and Argentina has become an importer of gas and other fuels.

One of the most recent and significant energy integration initiatives within the region, with deep effects on how the Latam energy integration is perceived, is the failed natural gas integration initiative between Argentina

---

11. Another project within the region is the Initiative for the Integration of Regional Infrastructure (IIRSA), which is a forum for dialogue between the authorities responsible for transport infrastructure, energy, and communications in the 12 South American countries. IIRSA's objective is to promote infrastructure development in a regional vision, ensuring the physical integration of South American countries and the achievement of an equitable and sustainable regional development.

and Chile, a case that dates back to 1995 when their governments signed a natural gas integration protocol, which served as the institutional framework to back up the companies' private contracts between natural gas producers in Argentina and consumers in Chile.

Under the protocol, private investors connected Chile and Argentina with seven gas pipelines, two in the north, two in the central zone of the country, and three in the south. At the time, in Chile, gas transportation infrastructure was complemented by large investments in natural gas power plants and natural gas distribution facilities, as well as infrastructure to substitute more expensive and less clean fuels in industrial processes and for residential consumption. In the power sector alone, between 1997 and 2004, 3316 MW of new capacity was built to run with imported natural gas from Argentina, and in 1999, an additional 643 MW natural gas power plant was built in the north of Argentina, close to the Chilean border, to export electricity to Chile through a 345 kV transmission line. Thus, between 1997 and 2004, 74% of the new power generation capacity built in Chile, to satisfy the 7.5% annual growth rate of electricity consumption, was based on Argentinean natural gas. Accordingly, electricity supply, and also industrial and residential natural gas consumption in Chile, became highly exposed to Argentinean natural gas supply.

Starting in 2004, Argentina's natural gas exports to Chile started to fall dramatically and have reached a standstill in recent years. This resulted in energy shortage which affected the Chilean economy and impacted the Argentinean economy, which collapsed in 2002 with the end of the fixed exchange rate regime,[12] which triggered a deep crisis in Argentina.[13] After the devaluation and as an attempt to control its adverse effects on the population and inflationary pressures, the Argentinean Government set price controls. Particularly, the natural gas price was set at low prices compared to other energy prices and world market fossil fuel prices. Domestic demand of natural gas in Argentina increased, sharply decreasing the surplus available for exports. On the supply side, and as a result of the low prices received by Argentinean natural gas producers, the incentives to invest in exploration and development of new gas fields and facilities were eroded. As price alterations continued, spare capacity and reserves dried out, and in an effort to assure the domestic supply of natural gas in 2004, the government imposed constraints on natural gas exports, demanding national gas

---

12. The scheme was established in 1991 to deal with hyperinflation, setting an exchange rate of one Argentinean peso for one American dollar.
13. According to World Bank studies, between October 2000 and October 2002, the percentage of the Argentinean population being declared under the poverty line increased from 33% to 58%.

producers to fulfill domestic consumption contracts, at low domestic prices, before satisfying their export contracts.

The resulting shortage of natural gas created serious adverse effects on the Chilean electric industry as well as on residential customers and industrial processes. At the time, about 35% of the installed Chilean capacity was natural gas fired plants. The shortages of natural gas combined with high fossil fuel prices resulted in near doubling electricity prices. To deal with the high dependence on Argentinean natural gas imports, Chile was forced to build Liquified Natural Gas (LNG) imports plants that were commissioned in 2009 and 2010. This marked the beginning of the end of the imported natural gas crisis that affected Chile with consequences that would take many years to overcome.

Currently, Chile has restarted its path for a new energy integration attempt within the region, but with the experience of this failed attempt in hindsight.

3.7 Energy Integration in Brazil and Neighboring Countries

Brazil has played an active role in the promotion of regional energy integration with bilateral agreements since the 1960s. In 1965, the Brazilian State electric company (Companhia Estadual de Energia Elétrica—CEEE) and the Uruguayan electric and telecom utility (Usinas Eléctricas y de Teléfonos del Estado—UTE) signed an agreement for exchanges in four different points of their borders. In July 1965, within the framework of this agreement, the first electricity interconnection in South America between Livramento and Rivera was inaugurated (Castro et al., 2012). After this first project, Brazil has continued supporting Latam energy integration, with the promotion of bilateral agreements, as the one with Paraguay, which allowed the construction of the binational Itaipú hydropower plant, with the support of multilateral energy integration agreements. More recently, in 2010, Brazilian President Luiz Inacio Lula da Silva signed an energy trade agreement with the President of Peru, Alan Garcia, between the two countries. This agreement provides support for the electricity interconnection of both countries, promoting energy exchanges between Peru and Brazil. The agreement also sets out rules for the participation of Brazil in the construction of hydropower projects in the Peruvian Amazonia, such as the 2200 MW Inambari project. In this context, Eletrobras[14] has also considered the construction of hydropower plants in Bolivia, Argentina, and Guyana.

---

14. Brazilian major electric utility company that is also Latam's largest power utility company. The Brazilian federal government owns 52% of Eletrobras stock.

4 ENERGY, INTEGRATION, AND GEOPOLITICS IN LATIN AMERICA

When it comes to energy integration, the European experience has been far more successful than the integration among Latam countries. There are many reasons for the European success which has made an important contribution to the sustainable development of the continent, elevating population living standards significantly.

The EU's success may be attributed to a strong and growing political block as defined in the Treaty on the Functioning of the European Union, in its Article 1941.[15] From an economic perspective, the EU considers that many energy system developments can be best achieved on an EU-wide basis while respecting the wishes of Member States. At the request of the European Commission in spring of 2006, regional initiatives were set up by the European Regulators' Group for Electricity and Gas (ERGEG). These initiatives were seen as a step to move from national electricity and gas markets to a single European energy market. The European Commission sees in the internal electricity market an opportunity to put the EU back on track for regional GDP growth, based on renewable energy sources with low CO_2 emissions and a lower dependence on foreign energy resources, topics further described in Chapter 8. The "Guidelines for trans-European Energy Infrastructure," issued by the European Commission in October 2011, unveiled a proposal to ensure that European energy networks and storage facilities are completed by 2020, where it identifies 12 priority corridors and areas covering electricity, gas, oil, and carbon dioxide transport networks. The Commission has proposed a regime of "common interest" for projects that contribute to implementation of these priorities.

A more successful model, summarized by Hope (2011) and Castro et al. (2012), exists within Nordic countries. In 1963, the transmission networks of Nordic countries consisting of Denmark, Finland, Norway, and Sweden created Nordel, an organization whose role is to integrate the transmission systems of these countries and create an MER. Iceland, despite having no electrical interconnection with the other Nordic countries, also was part of the organization. The Nord Pool emerged in 1996, under the initiative of Norway and Sweden, and quickly became a huge success and has been responsible for marketing significant part of the electricity consumed by the block. As an evidence of this, in 2008, the countries of the block had a power output of around 414 TWh, from an installed capacity of 97,199 MW. Of this total, 299 TWh were traded in the Nord Pool, which represents about

---

15. The EU capabilities related to combating climate change, greenhouse gas emission reductions in energy and other sectors are included in Articles 191–193.

70% of the entire production. The energy exchange in Nordel, including import/export to countries that are not part of the block as Russia, Germany, the Netherlands, Poland, and Estonia, was 66 TWh.

The countries that today are part of the EU could not have achieved the level of development, the strength of their democracy, and the life quality of its inhabitants without such successful integration. In the case of Latam, concerns about integration began more or less simultaneously, with an initial view of similar goals. European countries and North America are better integrated both politically and geographically, while the same issues become significant barriers to integration in Latam. Carvalho and Silva (2000) claimed that success integration processes in modern economies result from treaties signed among the governments, where the steps for the integration process should be discussed and analyzed by parliaments of the countries involved.

4.1 Integration Challenges

Why is it so difficult to achieve energy integration in regions such as Latam despite the obvious advantages? Mares (2004) pointed to political and economic instability, along with geopolitical interests and distrust, as some of the factors delaying energy integration projects in South America. There are many arguments or obstacles that obstruct energy integration including:

- Moving away from interdependency
- Political and economic risk
- Increased transaction costs
- Potential conflicts arising from the allocation of integration benefits such as congestion fees and ownership issues
- Asset ownership
- Allocation of resources under shortage scenarios
- Asymmetry regarding authority, influence in decision-making
- Loss of sovereignty
- The belief of unequal distribution of benefits among citizens/nationals and foreign investors/companies, which are behind profitable investments.

But beyond these arguments, two main difficulties stand out:

1. First is the weakness of the policymakers, political sector, business sector, social organizations, and the civil society beliefs with respect to the benefits of energy integration.
2. Second is the distrust between the states, especially neighbor states, sometimes nourished by regional or subregional leadership disputes among the largest countries in the region.

In general, the bilateral distrust has to do with unresolved border disputes after 200 years of independence. This is sometimes used as a factor of internal political support and national unity. Often, distrust has a real basis such as historical border attacks and border demarcation in dispute processes, and also failed energy integration projects, where energy integration agreements have not been respected. Such situations are also raised by some leaders or by certain institutions that look for support on the need to have full energy independence. The concept of full independency is contrary to the concept of energy integration and this idea is used as a synonym of national sovereignty and security.

Another ingredient that might also contribute against an integration process is the fact that international energy companies with regional interests are used to dealing with each country's operations in isolation. This may explain why such companies might not support regional integration.

Castro et al. (2012) argue that in Latam, energy markets liberalization have adversely affected the integration process, from a case where states were part of the process, in most cases through binational companies, to a case where private companies should lead the process requiring the support of national governments.

4.2 Integration Advantages

The lack of adequate energy integration results in inefficiencies that amounts to huge opportunity costs to improve the quality of life of Latin Americans. To move forward, there is the need to generate higher levels of mutual trust and transparency, for example, in areas such as the delivery of information in defense and military spending. On this, how Unasur is moving ahead in the South American Defense Council (CDS)[16] is essential, and this is one of the main regional integration drivers today. To the extent that the countries develop increased trust in each other, the armed forces increasingly know and trust each other in defense spending,[17] and to the extent that political leaders are convinced of the loss of opportunities for their nationals by maintaining, usually artificially, conflicts of various kinds between the nations.

---

16. In March 2009, the South American Defense Council (CDS) was established in Santiago, Chile, comprising the Ministers of Defense of the 12 Member States of UNASUR.
17. The first draft of the study "South American record in Defense Spending," was released in Quito. This is the result of the CSD work prepared by the Centre for Strategic Defense Studies (Centro de Estudios Estratégicos de la Defensa), another achievement of the CDS, based in Buenos Aires. This is the first regional attempt to create a "methodology to solve technical and design elements of the system on defense spending to create a common and generally accepted measuring system for our countries" (CDS resolution of May 2010).

Energy and geopolitics are very closely related. Integration is a factor in regional cohesion, based on an even geopolitical view: national security threats do not originate in neighboring countries, but they are outside the region, in international terrorism, in drug trafficking organizations, and the lack of social cohesion within our own countries.

How much political integration is required for electricity integration? Certainly, the efforts that have been seen within the last few years are a step ahead in the support of wider integration efforts within different fronts, on national defense and security, on road and rail infrastructure, on economic trade, and certainly on the increasingly political integration within the nations in the defense of international issues that are of common interest such as the promotion of democracy.

As it happens with the increasing population, the rising living standards, and the creation of wealth, energy integration also has benefits in terms of the following:

- Encouraging the energy market competition
- Taking advantage of larger economies of scale
- Reducing the reserve requirements.
- Optimizing the use of existing resources:
  - Resource dispersion
  - Complementarities
  - Time differences
- Improving technical reliability, safety, and quality of service.
- Reducing price distortions between different fuels and markets.
- Creating better market opportunities for large investments.
- Promoting long-term investments in infrastructure.
- Favoring the development of related industries in goods and services.
- Strengthening regional trade and increased productivity.
- Strengthening the region against the volatility of globalization.
- Facilitating a competitive insertion of the region in international markets.

In terms of the form that energy integration can adopt, and beyond the isolation of external electricity markets, countries can choose among the following different strategies:

- The corporate control model, where a country invests as a shareholder in other countries' energy companies, as it happens with Endesa, Suez, and other global energy companies with presence in Latam.
- The energy surplus export model, where minor infrastructure is built to ease the exports of energy surplus, as it happens with electricity exchanges between Ecuador and Peru or within Venezuela and Colombia.
- The infrastructure investment export model, where the countries explicitly build additional energy infrastructure to supply foreign markets, which

certainly is the most interesting, fruitful, and complex model, as is the case of the Itaipu power plant, where Paraguay entered in this joint project to export electricity to Brazil.

The strategy that each country follows to integrate with other countries in the region will define the size of the benefits it can expect from electricity and energy markets integration.

A 2009 UNDP study[18] examined the benefits of building and enforcing electricity interconnections within Bolivia, Chile, Colombia, Ecuador, and Peru. Among the benefits identified by the UNDP study is the greater convergence in electricity prices between the countries, and the economic benefits attained by the exporting countries, which can be used in social programs aimed at overcoming poverty. In the importing countries, the lower energy costs imply significant benefits in terms of productivity which release resources to be used in other initiatives. In addition, there are major environmental benefits in terms of the reduced emissions because of the replacement of more polluting thermal generation by cleaner generation as hydroelectricity. The UNDP study in electricity interconnections requires almost 1 billion US$ in investments in new or reinforced transmission lines, allowing up to 164,220 GWh of electricity exchanges that can lead to a reduction up to 123,165 tons of CO_2eq emissions, valued at US$1.232 millions in the period between 2014 and 2022.

Other studies have been led by the Regional Energy Integration Commission (CIER), who within its policy, supports American energy integration and take into account that the region needs to advance in the process of energy integration. The CIER study "Energy transactions among the systems of the Andean, Central America and the Southern Cone regions—Feasibility of their Integration" (CIER Project 15) shows that it is possible to optimize regional energy resources and make the region more competitive by the creation of interconnection schemes that respect the policies of each country, which do not require deep regulatory harmonization in the domestic markets, and also maximize the benefits for country consumers. With the participation of all countries in South America, Central America, and Mexico, the study was steered in two phases. Phase I was aimed to historically and critically analyze the gas and electricity interconnections, the energy markets, and their evolution in the three regions, from the technical and commercial operation, regulatory, and institutional point of

---

18. Prefeasibility Technical Economic Analysis of Electrical Interconnection between Bolivia, Chile, Colombia, Ecuador, and Peru ("Estudio para Análisis de Pre-factibilidad Técnico Económica de Interconexión Eléctrica entre Bolivia, Chile, Colombia, Ecuador y Perú," UNDP 2009).

view. Based on the above diagnosis, different electricity transaction scenarios were analysed in Phase II study, where 12 interconnection projects were evaluated, considering aspects such as economies of scale for hydro projects, operational safety and exchanges of opportunity, optimizing the use of existing infrastructure, operational safety, and energy exports. The favorable results obtained in the study indicate that there are significant opportunities for deepening regional energy integration.

The CIER study includes 15 countries: Argentina, Bolivia, Brazil, Chile, Colombia, Costa Rica, Ecuador, El Salvador, Guatemala, Honduras, Nicaragua, Panama, Paraguay, Peru, and Uruguay. It analyzes 12 regional interconnection projects, where 10 of them are profitable, taking advantage of the large complementarities of the energy mix within the region, with reduced CO_2 emissions, the use of hydroelectric plants' reservoirs as "energy reservoirs" that "store" not only water but also wind and biomass energy, and the large opportunities of wheeling, swaps, and opportunity energy exchanges taking advantage of large markets and scale economies of the energy projects.

5 CONCLUSION

The early result of the Latam integration process is in its earliest stages and only confirms the difficulties to build the appropriate institutional frameworks for integration. Few results are evident and the need of a political will to advance the process is clear. Simultaneously, there are significant serious differences and deficiencies in terms of the integration concepts and an important lack of complementation and consideration of several factors that characterize the integration processes on the political, economic, trade, culture, energy, infrastructure, education, and other areas.

It seems of utmost importance for the region to understand the importance of mutual trust between the nations, in issues like the national defense systems, the elimination of border conflicts, and overcoming poor educational training and cultural vestiges which have limited the energy integration progress.

Latam is at the beginning of a process of historical significance for the region, with huge opportunities. Progress has been made, and Unasur is a forum that's walking in a very interesting direction because, correctly enough, it has established the priorities in the areas of defense, energy, and infrastructure in the integration agenda, three areas that are entirely complementary for the Latam region.

6 ANNEX

Installed Capacity

Installed Capacity

Country	Argentina	Bolivia	Brazil	Chile	Colombia	Ecuador	Paraguay	Peru	Uruguay	Venezuela
MW										
1990	14,966	525	49,603	3372	8312	1717	6178	2842	1909	18,014
1995	18,511	709	55,497	5275	10,156	2465	6933	3196	2108	18,161
2000	26,357	1325	67,713	10,371	12,581	3348	8166	6070	2115	21,233
2005	28,292	1379	92,865	12,363	13,348	3567	8116	6200	2030	22,910
2009	32,144	1409	106,573	15,522	13,943	4838	8814	7982	2503	23,708
GWh										
1990	45,303	1901	211,328	13,851	33,877	6361	27,158	9558	7244	56,196
1995	62,809	2792	261,060	25,106	41,908	8405	41,607	13,080	6252	70,672
2000	88,965	3884	324,936	41,269	42,296	10,606	53,210	19,923	7365	89,488
2005	106,523	4908	405,100	52,479	50,430	13,404	51,047	25,510	7566	110,370
2009	122,286	6117	466,158	59,689	57,618	18,353	54,842	32,945	8022	124,823

(Continued)

(Continued)

Country	Argentina	Bolivia	Brazil	Chile	Colombia	Ecuador	Paraguay	Peru	Uruguay	Venezuela
kWh Per Capita
1990	1459	284	1554	1051	1058	589	641	444	1521	2837
1995	1882	378	1886	1763	1088	734	890	558	1934	3226
2000	2438	468	2142	2748	983	839	1044	776	2386	3697
2005	2871	521	2402	3358	1058	1147	1212	937	2518	4133
2009	3175	620	2600	3597	1233	1429	1501	1128	2625	4366

Source: CIER (2011).

Main Transmission Lines Length and Conversion Capacity, 2009

		Voltage				HVDC
		HVAC				
		100–150 kV	151–245 kV	246–480 kV	>480 kV	
Argentina	km	14,752	1578	1116	11,099	–
	MVA	n.d	n.d	n.d	n.d	–
Bolivia	km	774	1545	–	–	–
	MVA	420	575	–	–	–
Brazil	km	56,080	41,959	16,448	35,716	3224
	MVA	n.d	n.d	n.d	n.d	n.d
Chile	km	3594	8700	408	1002	–
	MVA	n.d	n.d	n.d	n.d	–
Colombia	km	10,090	11,647	–	2399	–
	MVA	11,439	13,393	–	7420	–
Ecuador	km	2392	2879	–	–	–

(Continued)

(Continued)

		Voltage				HVDC
		HVAC				
		100–150 kV	151–245 kV	246–480 kV	>480 kV	
	MVA	n.d	n.d	–	–	–
Paraguay	km	–	3672	–	16	–
	MVA	–	2837	–	2000	–
Peru	km	4057	5714	–	–	–
	MVA	6417	16,398	–	–	–
Uruguay	km	3554	11	–	771	–
	MVA	3177	70	–	1800	–
Venezuela	km	12,728				–
	MVA	38,553				–

Note: n.d.: no data.
Source: CIER (2011).

Resources

Estimated Resources of Technically Recoverable Shale Gas Basins for Selected 32 Countries, Compared with Existing Reserves, Production, and Consumption in 2009

	2009 Natural Gas Market		Proven Natural Gas Reserves	Technically Recoverable Shale Gas Resources
	(Trillions of cubic feet, in dry base)		(Trillions of cubic feet)	(Trillions of cubic feet)
	Production	Consumption		
Europe	10.8	14.6	186.2	639.0
North America	28.0	28.0	346.5	1931.0
Asia	5.7	6.3	174.6	1389.0
Australia	1.7	1.1	110.0	396.0
Africa	3.6	1.6	217.1	1042.0
South America	3.3	3.4	239.2	1225.0
Venezuela	0.7	0.7	178.9	11.0
Colombia	0.4	0.3	4.0	19.0
Argentina	1.5	1.5	13.4	774.0
Brazil	0.4	0.7	12.9	226.0
Chile	0.1	0.1	3.5	64.0
Uruguay	–	0.0		21.0
Paraguay	–	–		62.0
Bolivia	0.5	0.1	26.5	48.0
Total Previous Areas	53.1	55.0	1274.0	6622.0
Total World	106.5	106.7	6609.0	

Source: EIA (2011).

Interconnection Lines

Transmission

Countries	Location	Voltage (kV)	Comments
Ar–Bo	La Quiaca (Ar)–Villazón (Bo)	13.2	Existent
Ar–Bo	Pocitos (Ar)–Yacuiba (Bo)	33	Existent
Ar–Cl	Río Turbio (Ar)–Puerto Natales (Cl)	33	Existent
Ar–Py	Posadas (Ar)–Encarnación (Py)	33	Operating, 10 MW
Ar–Uy	Concordia (Ar)–Salto (Uy)	30	Not Operating
Bo–Br	Puerto Suárez (Bo)–Corumbá (Br)	13.8	Existent
Bo–Br	San Matías (Bo)–Corixa (Br)	35	Operating, 10 MW
Bo–Pe	Desaguadero (Bo)–Zepita (Pe)	24.9	Existent
Bo–Pe	Casani (Bo)–Yunguyo (Pe)	24.9	Existent
Br–Co	Tabatinga (Br)–Leticia (Co)	13.8	Existent
Br–Py	Ponta Pora (Br)–Pedro Caballero (Py)	22	Operating, 6 MW
Co–Ve	Arauca (Co)–Guasdualito (Ve)	34.5	Operating, 6 MW
Co–Ve	Pto. Carreño (Co)–Pto. Páez (Ve)	34.5	Operating, 7.5 MW

Ref.	Countries	Location	Voltage (kV)	Capacity (MW)	Comments
1	Co–Ve	Cuestecita (Co)–Cuatricentenario (Ve)	230	150	Operating (60 Hz)
2	Co–Ve	Tibú (Co)–La Fría (Ve)	115	36/80	Operating (60 Hz)
3	Co–Ve	San Mateo (Co)–El Corozo (Ve)	230	150	Operating (60 Hz)
4	Co–Pa	Cerromatoso (Co)–Panamá (Pa)	–	300	Under study
5	Co–Ec	Pasto (Co)–Quito (Ec)	230	200/250	Operating (60 Hz)
6	Co–Ec	Jamondino (Co)–Pomasqui(Ec)	230	250	Operating (60 Hz)
7	Co–Ec	Ipiales (Co)–Tulcán (Ec)	138	35	Operating (60 Hz)
8	Ec–Pe	Machala (Ec)–Zorritos (Pe)	230	110	Operating (60 Hz)

(Continued)

Ref.	Countries	Location	Voltage (kV)	Capacity (MW)	Comments
9	Br–Ve	Boa Vista (Br)–El Guri (Ve)	230/400	200	Operating (60 Hz)
10	Bo–Pe	La Paz (Bo)–Puno (Pe)	230/220	150	Project (50/60 Hz)
11	Br–Py	Salidas de Central Itaipú	500/220	14.000	Operating (60/50 Hz)
12	Br–Py	Foz de Iguazú (Br)–Acaray (Py)	220/138	50	Operating (60/50 Hz)
13	Ar–Py	El Dorado (Ar)–Mcal. A. López (Py)	220/132	30	Operating (50 Hz)
14	Ar–Py	Clorinda (Ar)–Guarambaré (Py)	132/220	150	Operating (50 Hz)
15	Ar–Py	Salidas de Central Yacyretá	500	3.200	Operating (50 Hz)
16	Ar–Br	Rincón S.M. (Ar)–Garabí (Br)	500	2.000/ 2.200	Operating (50/60 Hz)
17	Ar–Br	P. de los Libres (Ar)–Uruguayana (Br)	132/230	50	Operating (50/60 Hz)
18	Ar–Uy	Salto Grande (Ar)–Salto Grande (Uy)	500	1.890	Operating (50 Hz)
19	Ar–Uy	Concepción (Ar)–Paysandú (Uy)	132/150	100	Operating on emergency (50 Hz)
20	Ar–Uy	Colonia Elia (Ar)–San Javier (Uy)	500	1.386	Operating (50 Hz)
21	Br–Uy	Livramento (Br)–Rivera (Uy)	230/150	70	Operating (60/50 Hz)
22	Br–Uy	Pte. Médici (Br)–San Carlos (Uy)	500	500	Under construction (60/50 Hz)
23	Ar–Cl	C.T. TermoAndes (Ar)–Sub.Andes (Cl)	345	633	Operating (50 Hz)

Source: CIER (2011).

Power Plants

Ref.	Countries	Name	River	Installation Capacity (MW)	Comments
A	Br–Py	Itaipú	Paraná	14.000	Operating
B	Ar–Uy	Salto Grande	Uruguay	1.890	Operating
C	Ar–Py	Yacyretá	Paraná	3.200	Operating
D	Ar–Br	Garabí	Uruguay	1.500	In study
E	Ar–Py	Corpus	Paraná	3.400	In study

Source: CIER (2011).

Gas Pipelines Network and Natural Gas Proved Reserves

Ref.	Countries	Gas Pipeline	Diameter (inches)	Capacity (million m³/day)	Comments
1	Ar–Cl	San Sebastián (Ar)–Pta. Arenas (Cl) (Bandurria)	10	4	Operating
2	Ar–Cl	Batería de Recepción 7–T. del Fuego	6	1.5	Operating
3	Ar–Cl	Pta. Dungeness (Ar)–C. Negro (Cl) (Dungeness)	8	2	Operating
4	Ar–Cl	El Cóndor (Ar)–Posesión (Cl)	12	2	Operating
5	Ar–Cl	Pta. Magallanes (Ar)–Posesión (Cl)	18	1	Operating
6	Ar–Cl	L. La Lata (Ar)–Concepción (Cl) (Gas Pacífico)	24–20	3.5	Operating
7	Ar–Cl	La Mora (Ar)–Santiago (Cl) (Gasandes)	24	10	Operating
8	Ar–Cl	Cnel. Cornejo (Ar)–Mejillones (Cl) (Gasatacama)	20	9	Operating

(*Continued*)

(Continued)

Ref.	Countries	Gas Pipeline	Diameter (inches)	Capacity (million m³/day)	Comments
9	Ar−Cl	Gasod. Norte (Ar)−Tocopilla(Cl) (Norandino)	20	8.5	Operating
10	Ar−Bo	Ramos (Ar)−Bermejo (Bo)	88−13	1,2	Existent
11	Ar−Bo	Campo Durán (Ar)−Madrejones (Bo)	24	7	Existent
12	Ar−Bo	Miraflores (Ar)−Tupiza (Bo) (Puna)	−	−	Project
13	Ar−Br	Cnel. Cornejo (Ar)−S. Paulo(Br)	−	−	Project
14	Ar−Br	Aldea Brasilera (Ar)−Uruguayana (Br)	24	10/15	Operating
15	Ar−Uy	Gto. Entrerriano (Ar)−Paysandú (Uy) (Del Litoral)	10	1	Operating
16	Ar−Uy	Gto. Entrerriano (Ar)−Casa Blanca (Uy)	16	5−2	Existent
17	Ar−Uy	Bs. Aires (Ar)−Montevideo (Uy) (Cruz del Sur)	24	6	Operating
18	Bo−Br	Río Grande (Bo)−S. Paulo (Br)	32	30	Operating
19	Bo−Br	Río Grande (Bo)−Cuiabá (Br) (Gasbol)	18	2.8	Operating
20	Co−Ve	Est. Ballena (Co)−Maracaibo (Ve)	18	4.2	Operating

Source: CIER (2011)

Demand

Peak Load

		2005	2006	2007	2008	2009	Peak Load Month, 2009
Argentina		16,718	17,395	18,345	19,126	19,566	July
Bolivia		759	813	895	899	939	December
Brazil		60,918	61,782	62,895	65,586	69,193	November
Chile	SIC	1566	1676	1790	1897	1816	March
	SING	5764	6059	6313	6147	6139	December
Colombia		8639	8762	9093	9079	9290	December
Ecuador		2424	2642	2706	2785	2770	April
Paraguay		1354	1500	1521	1648	1810	November
Peru		3305	3580	3965	4199	4322	December
Uruguay		1485	1409	1654	1481	1684	July
Venezuela		14,687	15,945	15,551	16,351	17,337	September
South America		117.619	121,563	124,728	129,198	134,866	Not coincident demand

Source: CIER (2011).

Peak Load Growth Rate in Percent

	2005	2006	2007	2008	2009	Average 2005−2009
Argentina	7.2	4.0	5.5	4.3	2.3	4.7
Bolivia	7.7	7.1	10.1	0.4	4.4	5.9
Brazil	7.3	1.4	1.8	4.3	5.5	4.1
Chile	3.5	6.9	3.6	−0.7	−1.1	2.4
Colombia	3.7	1.4	3.8	0.0	2.1	2.2
Ecuador	2.7	9.0	2.4	2.9	−0.5	3.3
Paraguay	9.1	10.8	1.4	8.3	9.8	7.9
Peru	5.6	8.3	10.8	5.9	2.9	6.7
Uruguay	2.5	−5.1	17.4	−10.5	13.7	3.6
Venezuela	6.4	8.6	−2.5	5.1	6.0	4.7
South America	6.4	3.4	2.5	3.6	4.4	

Source: CIER (2011).

	Residential	Industrial	Commercial	Public Lighting and Others	Total Billed	Self-Consumption	Network Losses
Argentina	31,626	32,087	18,026	10,377	92,116	4400	17,104
Bolivia	1754	1863	869	380	4866	140	626
Brazil	100,638	199,505	65,981	54,520	420,644	5385	77,819
Chile	8901	33,400	6386	1994	50,681	1647	4990
Colombia	19,076	14,205	9384	3694	46,359	1081	7187
Ecuador	4687	3994	2581	1958	13,220	305	2722
Paraguay	2692	1517	1218	1036	6463	2	3069
Peru	6645	14,943	4815	684	27,087	488	3284
Uruguay	2946	1751	1528	1019	7244	154	1253
Venezuela	25,691	61,096			86,787	38,017	

REFERENCES

Bowring, J., 2006. The PJM market. In: Sioshansi, F., Pfafenberger, W. (Eds.), Electricity Market Reform—An International Perspective. Elsevier, Amsterdam, pp. 451−528.

British Petroleum, 2011. BP Energy Outlook 2030, London, January 2011. <http://www.bp.com/ liveassets/bp_internet/globalbp/globalbp_uk_english/reports_and_publications/statistical_ener-gy_review_2008/STAGING/local_assets/2010_downloads/2030_energy_outlook_booklet.pdf>.

Carvalho, M.A., Silva, C.L., 2000. Economia Internacional (International Economy). Saravia, Sao Paulo.

Castillo, D., 2006. Proyectos actuales y futuros en el marco de la decisión CAN 536 (Current and Future Projects within CAN 536 Decision Framework). <http://www.olade.org/FIER/ Documents/PDF-8.pdf/>, November 2012.

Castro, N.J., Silva, A.L., Rosental, R., 2012. Integração energética: uma análise comparative entre União Européia e America do Sul, TDSE No. 48 (Energy Integration: Comparative Analysis between European Union and South America, TDSE No. 48). GESEL, Rio de Janeiro.

CEPAL, 2011a. Centroamérica: Mercados mayoristas de electricidad y transacciones en el mer-cado eléctrico regional (Central America: Electricity Wholesale Markets and Transactions in the Regional Electric Market). D.F. ECLAC, México.

CEPAL, 2011b. Centroamérica: estadísticas del subsector eléctrico, 2010 (Central America: Statistics of the Electric Subsector, 2010). D.F. ECLAC, Mexico, Octubre de 2011.

CIER, 1997. Proyecto CIER 02. Mercados mayoristas e interconexiones de América del Sur. Fase II (CIER 02 Project, Wholesale Markets and South America Interconnections. Phase II).CIER, Montevideo.

CIER, 2006. Estudio de Transacciones de electricidad entre las regiones andina y américa central y mercosur, November 2006. Factibilidad de su integración, primera fase (Electricity Transactions Study for the Andean Regions and Centreal America and Mercosur. Integration Feasibility, Phase I). CIER. MH0938-PH057-06 Mercados energéticos consultores.

CIER, 2010. Proyecto CIER 15. Fase II, Informe Final − Energy transactions between the sys-tems of the Andean, Central America and the Southern Cone regions—Feasibility of their Integration.

CIER, 2011. Síntesis Informativa Energética de Los Países de La CIER. Comisión de Integración Energética Regional, Mayo de 2011 (Regional Energy Integration Commission, Brief Energy Background for CIER Member States, May 2011).

CONELEC, 2011. Estadística del Sector Eléctrico ecuatoriano Folleto Multianual. Quito, Agosto de 2011 (Statistics of the Ecuador Electric Sector Multi Annual Catalogue. Quito, August 2011).

Dyner, I., García, M., Rincón, J., 2007. Integration of power supply in Latin America. Working Paper, pp 20, Departamento de Ciencias de la Decisión y de la Computación, Universidad Nacional de Colombia, Carrera 80 No. 65 223, Medellín, Colombia.

Dyner, I., Olaya, Y., Franco, C.J., 2011. An enabling framework for wind power in Colombia: what are the lessons from Latin America? In: Haselip, J., Nygaard, I., Hansen, U., Ackom, E. (Eds.), Diffusion of Renewable Energy Technologies: Case Studies of Enabling Frameworks in Developing Countries, Technology Transfer Perspectives Series. Unep Risø Centre, Denmark.

EIA, 2011. World Shale Gas Resources: An Initial Assessment of 14 Regions Outside the United States. US Energy Information Administration. U.S. Department of Energy. <http:// www.eia.gov/analysis/studies/worldshalegas/pdf/fullreport.pdf>. November 2012.

EOR, 2010. Memoria EOR 2008−2009. Ente Operador Regional. San Salvador, Julio de 2010. (EOR 2008−2009 Annual Report. Regional Operational Entity. San Salvador, July 2010.)

EOR, 2011. Informe de Gestión Técnica y Comercial. Ente Operador Regional. San Salvador, Diciembre de 2011. (Technical and Commercial Management Report, Regional Operational Entity. San Salvador, December 2011.)

Gnansounou, E., Bayem, H., Bednyagin, D., Dong, J., 2007. Strategies for regional integration of electricity supply in West Africa. Energ. Policy 35 (8), 4142–4153.

Hass, R., Glachant, J., Keseric, N., Perez, Y., 2006. Competition in the continental European electricity market: despair or work in progress? In: Sioshansi, F., Pfafenberger, W. (Eds.), Electricity Market Reform—An International Perspective. Elsevier, Amsterdam, pp. 265–315.

Hope, E., 2011. 3er Latin American Meeting on Energy Economics ELAEE/IAEE. "Energy, Climate Change and Sustainable Development: Challenges for Latin America", April 2011, Buenos Aires, Argentina. Title of the Presentation: "Strengths and Threats for Regional Energy Integration: The Integrated Nordic Power Market—Integration with European Market".

Jacobsen, H.K., Fristrup, P., Munksgaard, J., 2005. Integrated energy markets and varying degrees of liberalisation: price links, bundled sales and CHP production exemplified by Northern European experiences. Energ. Policy 34 (18), 3527–3537.

Jamasb, T., Pollit, M., 2005. European electricity liberalisation. In: Newberry, D. (Ed.), The Energy Journal, pp. 11–41 (Special issue).

Mares, D.R., 2004. Natural gas pipelines in the southern cone. Working paper #29 Geopolitics of Natural Gas Study. Program on Energy and Sustainable Development. Stanford University, James A. Baker III Institute for Public Policy of Rice University. May 2004.

Newberry, D., 2002. Regulatory challenges to European electricity liberalisation. Swedish Econ. Policy Rev. 9, 9–43.

Newberry, D., 2005. European electricity liberalisation—introduction. In: Newberry, D. (Ed.), The Energy Journal, pp. 1–10 (Special issue).

Ochoa, M.C., Dyner, I., Franco, C.J., 2012. Simulating power integration in Latin America to assess challenges, opportunities and threats. Working paper, Universidad Nacional de Colombia. Working Paper, pp 32, Departamento de Ciencias de la Decisión y de la Computación, Universidad Nacional de Colombia, Carrera 80 No. 65 223, Medellín, Colombia.

Pineau, P.O., Hira, A., Froschauer, K., 2004. Measuring international electricity integration: a comparative study of the power systems under the Nordic Council, MERCOSUR, and NAFTA. Energ. Policy 32 (13), 1457–1475.

Raineri, R., 2006. Electricity market reform: an international perspective. In: Sioshansi, F.P., Pfaffenberger, W. (Eds.), Chile: Where It All Started. Elsevier, pp. 77–108 (Chapter 3: Series, Global Energy Policy and Economics).

Raineri, R., 2007. Chronicle of a crisis foretold: energy sources in Chile. IAEE Newsletter, pp. 27–30 (Fourth Quarter).

UN, ECLAC, 2010. Sede Subregional en México, Centroamérica: Estadísticas del subsector eléctrico, 2009. (Mexico Sub regional Seat: Central America: Statistics of the Electric Subsector, 2009) D.F. ECLAC, México (Editorial).

World Energy Council, 2010. Survey of Energy Resources. Regency House 1-4 Warwick Street, London W1B 5LT United Kingdom, ISBN: 978 0 946121 021.

XM, 2012. Informe de Intercambios Internacionales (Report on International Exchanges). XM Compañía de Expertos en Mercados. April 2012.

Part III

The Evolution of BRICs Electricity Markets

The Evolution of Brazilian Electricity Market

Luiz Pinguelli Rosa[a,b], Neilton Fidelis da Silva[a,b], Marcio Giannini Pereira[a] and Luciano Dias Losekann[c]

[a]*Energy Planning Program (PPE), Coordination of Postgraduate Programs in Engineering at the Federal University of Rio de Janeiro (COPPE/UFRJ), Building C, Room C-211, P.O. Box 68565, Cidade Universitária, Ilha do Fundão 21945-970, Rio de Janeiro, Brazil,* [b]*International Virtual Institute of Global Change—IVIG, Centro de Tecnologia, Building I—Room 129, P.O. Box 68501, Cidade Universitária 21945-970, Rio de Janeiro, Brazil,* [c]*Department of Economics, Fluminense Federal University, Rua Tiradentes, 17, Ingá 24.210-510, Niterói, Rio de Janeiro, Brazil*

1 INTRODUCTION

The Brazilian electricity market went through two institutional reforms in the last 20 years. The first reform in 1990s was inspired by the international power reform process and was gradually implemented. It included the usual reform measures, as unbundling, introduction of completion in generation and supply and creation of an independent system operator (ONS), an independent regulator (ANEEL) and a wholesale power market. Some power companies were privatized. However, a significant component remained under state ownership.

Since the institutional arrangements were not conducive to investment, installed capacity grew at a slower pace than power consumption. This imbalance resulted in a progressive depletion of Brazilian hydropower reservoirs, reaching alarming levels by 2001. The 2001/2002 power rationing had major economic and political impacts. It motivated a review of the institutional framework even before the transition to a truly competitive market was completed.

In 2004, the second reform process was implemented. The State regained its role in energy planning and long-term contracts became the only form to trade electricity. Distribution companies were obliged to acquire electricity through public auctions conducted by ANEEL and Energy Research Company (EPE),[1] an institution created to assist government on energy planning.

---

1. Aneel and EPE stand for Electricity Energy Agency and Energy Research Company, respectively.

Results of the second power reform process remain somewhat ambiguous to date. It virtually eliminated the risk of new power rationing, but generation expansion plans include expensive and high CO_2 emitting sources.

Today, the main challenges facing the Brazilian electricity sector are to provide low-carbon and affordable energy to a growing economy. Brazil enjoys a low-carbon footprint in the power sector resulting from high contribution of renewable sources, mostly hydro, in the generation mix. The challenge is to maintain this record as the share of hydropower in the generating mix decreases over time. The other challenge is to manage retail electricity prices, which have increased continuously in the last two decades. According to the International Energy Agency, electricity prices in Brazil are among the highest in the world (Firjan, 2011).

This chapter is organized as follows: Section 2 presents the main features of the Brazilian power system. Section 3 provides a description of the evolution of the Brazilian electricity sector, highlighting the two institutional reforms of the last 20 years. Section 4 presents future projections for the Brazilian electricity market followed by the conclusion of the chapter.

2 THE BRAZILIAN POWER SYSTEM

This section provides an overview of the Brazilian power system. It highlights the three characteristics that make the Brazilian case peculiar:

- A continental-sized transmission network;
- High growth of electricity consumption; and
- Predominance of hydropower generation.

Brazil has continental dimensions, with an area of 8.5 million square kilometers, population of 194 million people, and GDP of US$ 2.5 trillion in 2011. It is the most industrialized and diversified economy of Latin America, with a per capita GDP of US$ 12,788. For a discussion of Latin America's energy markets refer to Chapter 14 by Raineri et al.

The main electricity consumption centers (Figure 15.1) are interconnected in Brazil. The integrated power grid accounts for 98% of the national consumption of electricity,[2] 480 TWh in 2011 (EPE/MME, 2012). Interconnection has historically been focused on exploiting the country's vast hydroelectric potential, typically located far from consumption centers, and to take advantage of complementarities between regional hydrology. So, if there is a drought in one hydro generation region, the other regions can compensate for it (Table 15.1).

---

2. The rest is supplied by isolated systems, located mostly at the Amazon region. Those systems have progressively been interconnected to the national system and EPE predict they will respond for only 0.4% of total consumption.

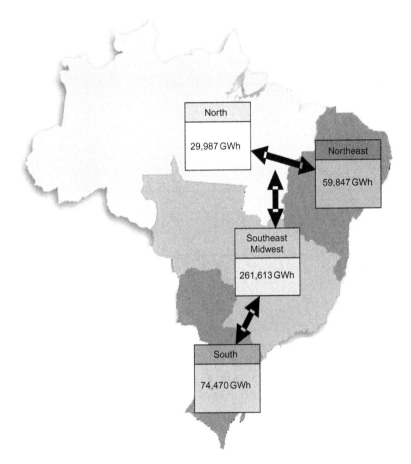

FIGURE 15.1 Electricity consumption by regional subsystem. *Source: Elaborated by the authors.*

The interconnected grid allows the operation of an integrated electricity wholesale market on a national level. Due to transmission constraints, the national system is divided in four subsystems: South, Southeast/Midwest, North and Northeast. The Southeast/Midwest is the most important one; it accounts for 60% of total Brazilian electricity consumption (Figure 15.1).

Prices in each area are set weekly by three load levels, heavy, medium, and light for each subsystem. In the last 5 years, weekly prices have been equal in the four areas 55% of the time. The major transmission constraint is between Southeast/Midwest and North/Northeast. Roughly 75% of the time prices are equal between South and Southeast/Midwest and between North and Northeast.

TABLE 15.1 Brazilian Power System in Numbers—2011

Installed capacity	123 GW
Hydropower	89 GW
Thermopower	31 GW
Consumption	433 TWh
Households	112 TWh
Industry	184 TWh
Commerce	73 TWh
Peak demand	71 GW
Hydro reservoirs maximum storage	202 TWh
CO_2 emissions	29.8 Mt

Source: EPE/MME (2012) and ONS (2012)

On average, Brazilian electricity consumption has increased 4.5% a year in the last 10 years. As per capita consumption, 2440 KWh/year, is still quite low by Organisation for Economic Co-operation and Development (OECD) standards, a sustained growth can be expected in the long term as national income rises. This growth creates a need for a continuous expansion of generating capacity.

Hydropower plants, with installed capacity of 123 GW, account for 71% of national installed capacity[3] (Figure 15.2). As many plants share the same river basin, most of the decisions on operating hydro plants are interdependent. The Brazilian hydroelectric plants count on large reservoirs that operate in a multiannual scheme[4] and can store half of the annual consumption of electricity in Brazil (Losekann et al., 2009). With an estimated potential of 260 GW, of which only one quarter is currently operating, hydropower should remain dominant in the national energy matrix in the coming decades. However, as most of the remaining hydro potential is located far off the consumption centers in environmental sensitive areas, such as in Amazonia, environmental and cost concerns suggest a progressive reduction in hydropower's share in the country's generating mix.

Other renewable sources are also relevant. Biomass represents 8% of total installed capacity, most of it in cogeneration units that use sugarcane bagasse

---

3. It considers the Paraguayan part of Itaipu power plant that is allocated to the Brazilian market and the small hydropower capacity.
4. Large reservoirs can store sufficient water to be used over multiple years.

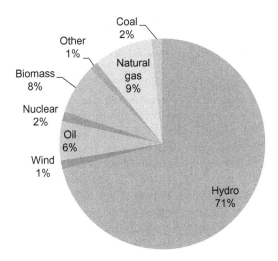

FIGURE 15.2   Brazilian electricity generation matrix. *Source: Elaborated by the authors. Data from ANEEL.*

and ethanol production. Wind power represents only 1% of total capacity but its share is rapidly increasing with about 7 GW under construction. With roughly 80% of the generation coming from renewable sources, the Brazilian power sector is not a significant source of CO_2 emissions in Brazil.

Natural gas is the most important fossil fuel, with 9% of total installed capacity. Its share increased sharply after the power rationing of 2001. This trend, however, stalled when political changes in Bolivia introduced uncertainty on Brazilian gas supply. Oil and coal account for only 8% of generating capacity with two nuclear plants accounting for 2% of the mix.

3 THE EVOLUTION OF BRAZIL'S ELECTRIC POWER SECTOR

This section covers the institutional evolution of the Brazilian power sector, focusing on the two phases of reform in the past two decades. The first phase, introduced in 1990s, ran into difficulties even before there was a chance to fully implement the scheme. This resulted in a return to state intervention in planning in the second phase of the reform, which was introduced in 2004.

3.1 Antecedents

Historically, the Brazilian electric sector has played a major role in the development of the country. From 1930 to late 1970, the energy sector contributed to the transformation of an agricultural exporting country into an industrialized one. This evolution has been driven mostly by State interventions and through funding and planning of power system expansion, a central part of the Brazilian Developmental State[5] policy.

Eletrobras was established to lead the technical and economical integration of the Brazilian power system, a goal set by the old National Electrification Plan (NEP) of 1954. Eletrobras holding company and its subsidiaries were in charge of generation and transmission. Distribution was done by companies owned by state-level governments.

This model was relatively successful until the 1980s. It was a period known as "lost decade," when Brazilian economic growth stalled and inflation accelerated. State-owned companies were used to avoid inflation, so electricity rates did not cover their costs. These policies created serious funding difficulties for Eletrobras and the entire power sector (Dias Leite, 2007).

In the 1980s, a neoliberal economic agenda was adopted to overcome the macroeconomic crisis. In the power sector, reform was initiated in 1996 with the assistance of Coopers & Lybrand in a study to:

- Develop a reform model for Brazilian power sector, appropriate to the privatizations process;
- Develop rules that could ensure free access to transmission network for any actor, with special attention to large consumers;
- Introduce new ways of marketing electricity among the utilities; and
- Set new regulatory requirements for the sector.

The restructuring of the electricity sector aimed to promote economic efficiency through competition and expansion through private investment.

3.2 The First Reform

The 1990s reform meant to foster private participation in the electricity sector and to introduce incentives for efficiency, mainly through liberalization of electricity generation. Investments in new projects were expected to soar through private sector initiatives. This was critical since at the time, power system expansion was depressed due to financial restraints on State-owned companies.

The scheme was gradually implemented (Table 15.2). First, it promoted the financial recovery of the sector. In 1993, law 8.631 abolished guaranteed remuneration and settled debts between utilities. As a result, the National Treasury absorbed US$23 billion of debts accumulated by the sector (Araujo, 2006).

Second, it eliminated barriers to private entrepreneurship. The sixth amendment to the Brazilian Constitution in 1995 allowed private

---

5. The governments of Getúlio Vargas (1951–54) and Juscelino Kubitschek (1956–61) were devoted to the great alliance that was kept, in one form or another, throughout the maintenance of the Developmental State, namely the tripod State/National Capital/Foreign Capital.

TABLE 15.2 Timetable of the First Phase of the Electric Sector Reform

Year	Event
1993	Law 8.631. Debt settling and end of guaranteed remuneration
1995	n 6 Constitution Amendment. Allowed private companies to exploit hydropower
1995	Law 8.987. Defined electricity supply licensing regime
1995	Law 9.074. Liberalized electricity supply to large consumers and established independent power production
1995	First privatization of distribution company
1997	Independent regulator (ANEEL) established
	First privatization of generation company (Cachoeira Dourada)
1998	Independent system operator (ONS) and wholesale electricity market (MAE) created
2001	Power rationing

Source: Elaborated by the authors

enterprises to participate in hydropower generation. The Concession Law (n. 8.987) defined the requirements that companies must follow to supply electricity. It introduced price-cap regulation to natural monopoly activities such as transmission and distribution, and sought competitive bidding for public concessions for hydropower plants, transmission lines, and distribution service areas. Law 9.074 liberalized electricity supply to large consumers, above 10 MW,[6] and created independent power producers (IPPs), which could compete with existing utilities to supply those consumers.

Even before the new reform framework was fully implemented, the first distribution company, Escelsa, was privatized in 1995. Since then, 23 state-owned companies have been privatized. The privatization process was intense in distribution but has faced many challenges in generation; only four generation companies, three of state-owned and one federal, were privatized.

A federal independent regulatory agency, ANEEL, was created in 1996 and started operating in 1997, when its directing board was nominated.[7] The agency, which is financed by a 0.5% tax in electricity tariffs, performs usual regulatory functions such as monitoring and approving tariffs

---

6. In 2000, this limit was reduced to 3 MW.
7. Aneel's board consists of five members, one is the general director, with noncoincident 4-year terms, nominated by the president, subject to the Senate approval.

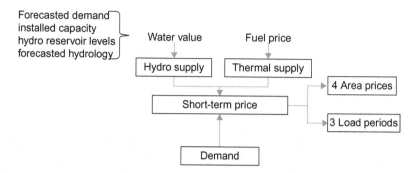

FIGURE 15.3   Schematic representation of short-term price formation. *Source: Elaborated by the authors.*

as well as granting concessions and coordinating power auctions (Araujo, 2006).

The initial reforms also promoted an industry redesign. Vertically integrated companies, for example, were obligated to ring-fence their businesses, forming new companies to operate each activity. The privatization was also intended to decrease the concentration ratio in generation.[8] However, as the privatizations did not go as far as originally envisioned, this objective did not reach its full intent.

In the transmission sector, private investments only applied to grid expansion. The new transmission lines, which can be constructed once recommended by ONS and approved by ANEEL, are sold in competitive auctions, allowing the winners to invest.

The short-term price of energy in the four subsystems, initially set on monthly basis is now set weekly. The dispatch is carried out by a stochastic computer model which calculates the short-term prices. The water value in the hydro reservoirs is the main variable in determining the dispatch order, and is set by the ONS by examining the hydrology and water levels in reservoirs and demand expectations as schematically shown in Figure 15.3.

The gradual implementation of the first reforms, leading to the competitive model, however, was interrupted by a major crisis in electricity supply that resulted in a power rationing. The original reform process did not motivate sufficient investments in new generation (Araujo, 2006; Losekann et al., 2009). Consequently, generating capacity grew at a slower pace than demand (Figure 15.4). This imbalance resulted in progressive drop in the level of

---

8. This happened to the generation company of São Paulo, CESP, which was divided in three generation companies (Paranapanema, Tietê, and CESP) and another one responsible for transmission, previous to the privatization.

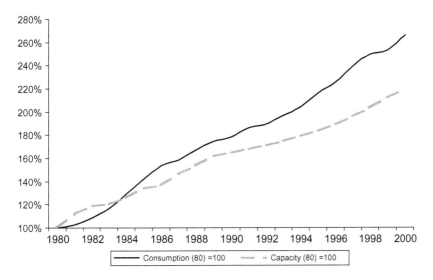

FIGURE 15.4 Installed capacity and electricity consumption evolution—1980/2000. *Source: Silva (2006).*

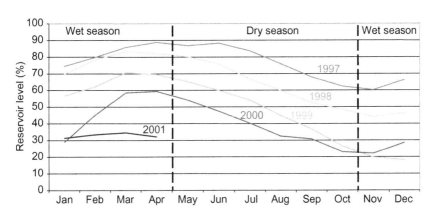

FIGURE 15.5 Reservoir levels (%) in Southeast/Midwest (1997–April 2001). *Source: Elaborated by the authors.*

storage in the hydroelectric reservoirs as shown in Figure 15.5[9] for the period 1997–2001.

A serious drought in 2001 depleted storage in hydro reservoirs, reaching alarming low levels during the normally wet season, November to April. To avoid a complete depletion of reservoirs during the following dry season,

9. Comissão de Análise do Sistema Hidrotérmico de Energia Elétrica (2001) offers a complete diagnosis of the rationing causes.

FIGURE 15.6   Institutional framework of the Brazilian Electricity Sector. *Source: Elaborated by the authors.*

May to October, which would be catastrophic, the Brazilian government implemented a mandatory consumption rationing in May 2001.[10] The mandatory rationing, which lasted until May 2002, resulted in 7% reduction in electricity consumption in 2001 with serious economic consequences. Estimated total cost of the rationing is close to 3% of the Brazilian GDP (Sauer, 2003), roughly US$ 40 billion. Economic consequences aside, this had serious political consequences. Opposition identified the power rationing as an evidence of bad government management. Indeed, President Cardoso was not able to assure election of his preferred successor, leading to the election of President Lula.

3.3 The Second Reform

The reform of Brazilian electric sector was an electoral commitment of President Lula in the election race of 2002. Once in office, the new model was debated during 2003 and a new regulatory framework was implemented in 2004.

The second reform aimed to provide security of supply and to avoid a rise of electricity prices. The government took back the planning role of the sector and significantly altered the wholesale market resulting in a new framework, schematically shown in Figure 15.6.

The National Council for Energy Policy (CNPE), created in the 1990s but not previously active[11] was given a central role in setting overall energy policy. It is composed of 14 members, of which 9 are ministers of State. The council set the main guidelines of Brazilian energy policy to be implemented by the Mining and Energy Ministry (MME).

Two new institutions were created, the EPE and the Electric Sector Monitoring Committee (CMSE). The first was created to assist the Energy

10. In general, consumers were required to reduce their consumption by 20%, relating to what was observed in 2000. Required reductions were different between consumption classes and regions.
11. CNPE was created in 1997, but its first meetings only took place in 2001 and were motivated by the power rationing.

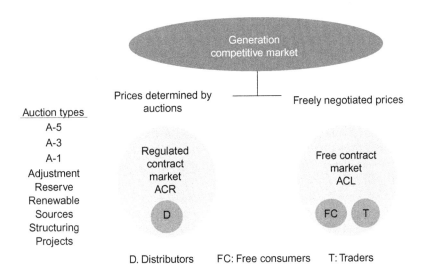

FIGURE 15.7   The Brazilian electricity wholesale market. *Source: Elaborated by the authors.*

Minister on sector planning, playing an important part at the expansion auctions. The second, CMSE—consisting of representatives of related institutions including Ministry of Energy, regulatory agencies, system operator, Chamber of Electric Energy Trade (CCEE) and EPE—is designed to coordinate and supervise the expansion process, identifying in advance any mismatch between supply and demand.

The Decree # 5163 of 2004 stated that all energy market trade must be carried out through contracts. The intention is to mitigate risks on power projects, making it easier to fund and finance. The CCEE, formerly Wholesale Power Market (MAE), is responsible for short-term market and adjusts any imbalances between effective and contracted volumes. Spot price is set roughly in the same way as before. Agents that are systematically exposed to spot transactions, for example, contract less than necessary, are subjected to penalties. Much of these requirements are to avoid future power shortages resulting from shortage of generation investment as occurred in 2000–01.

To facilitate trade in the wholesale market, two new markets were created as schematically shown in Figure 15.7: regulated contract market (ACR, in Portuguese) and free contract market (ACL). At the ACR, distribution companies buy energy in public auctions. They submit to the EPE their demand projections over a 5-year horizon. Based on those projections, the EPE sets the total market volume that will be offered in the auctions. These auctions, which are coordinated by CCEE and ANEEL, generators compete by making bids, in $/MWh and $/MW, to feed the distribution market. The winners sign contracts with the distribution companies that took part in the auction. In this way, the energy from each generator is divided among the distributors in the proportion represented by their market share in the total volume in the auction.[12] The wholesale selling price corresponds to the bids of generation companies (pay as bid) and the purchase price, paid by every distributor, is the average of sell prices.

The wholesale auction is designed to improve the efficiency and transparency of wholesale transactions by promoting the use of the best resources, sites and technologies to produce electricity at the least cost. The scheme makes electricity trade more transparent while stimulating entry of new competitors and preventing collusion among incumbent generators.

The Brazilian auction system adopts the lowest rate as the criteria to determine winners. It is similar to the German auction model (price descending). The auction is the reverse type, in which the bidding prices decrease. The MME and the EPE define the first bid, by setting the maximum electricity price.

The model distinguished energy generated by existing plants ("old energy") and energy that would be produced by new plants ("new energy"), being both negotiated in the ACR on different ways. The old energy is oriented to respond to the existing market when the model was created. In the auctions of "old energy," contract length is shorter, from 3–8 years, and they are usually carried out 1 year before delivery starts.

The "new energy" is focused on the growth of consumption over time. In the new energy auctions, hydro and thermal generations are auctioned under different rules. Hydropower is traded by 30 years concessions contracts, in delivered energy modality.[13]

Thermal power is contracted under the authorization regime for 20 years,[14] in available capacity modality where generators are paid by their capacity. When thermal power is dispatched, fuel costs are covered by side payments, which are passed through to final consumers. Therefore, thermal power plants are not exposed to dispatch risk, which is very high in a hydro-dominated system like Brazil.[15] Prior to the auction, thermal power projects

---

12. By the time of shaping this model, it was aimed to create a State company to buy all the energy and then sell it to the distributors, something similar to the idea of a single buyer. When the government concluded this possibility would increase the risk of contract break, the idea was left behind.

13. Hydro generators bid energy price ($/MWh) and are paid by their expected production. The bids must cover transmission costs relating to the delivery point, that is, the center of the submarket in which the generation project is located. In new energy auction where hydropower plants take part, there is a first phase where entrepreneurs compete to choose the hydropower project. Then, in the second phase those projects compete with others sources. Entrepreneurs cannot increase their bids from the first phase to the second.

14. In the first contracts to thermal power, their length was 15 years.

15. Brazilian natural gas power plants have a historical dispatch rate of 25% in the last 5 years. They can stay out of merit-order dispatch for long periods, sometime years, when hydrology is good. Energy-only payments imply a risky cash flow for thermal power plants.

must inform fuel prices. During the auction, they bid a required fixed income ($/KW) that would cover fixed costs.

To compare hydro and thermal projects, the auction system uses an index, cost benefit index or ICB, that considers expected dispatch and fixed and variable costs of each power plant. Each wining generator signs contracts with all participating distributors. Quantities are allocated in proportion to the forecasted demand of each distributor.

Usually, two "new energy" auctions are carried out each year to deliver electricity 3 and 5 years ahead (A-3 and A-5). The contracts from 20 to 30 years are negotiated in the auctions. In the A-3 auctions, projects with shorter construction time take part, such as thermal plants (except coal), wind, and Combined Heat and Power (CHP). The large hydro and coal-thermal plants generally participate in the A-5 auctions.

Auctions are typically designed to focus on one project, usually very large hydropower plants. Alternative renewable sources (wind, biomass, and small hydropower) are subject of specific auctions. Reserve auctions are designed to increase security of supply. As capacity traded in those auctions makes a reserve for the system, generators are paid by a charge that consists of final electricity prices.[16] Adjustment auctions are carried out to allow an intratrade between distribution companies, allowing overcontracted distributors to sell to the undercontracted ones, with a maximum length of 2 years.

At the ACL, large consumers[17] are free to choose their suppliers outside the centralized auctions. The energy is negotiated through bilateral contracts with generators and traders. The contracts last for different periods and short-term contracts are predominant. Large consumers can opt to continue to be supplied by a distribution company. They must inform 1−3 years in advance to switch to the ACL. Since 2004, the electricity trade has grown significantly in ACL, accounting for 25% of the electric energy market in Brazil.

After two decades of liberalization, the Brazilian electricity sector has experienced significant changes (Table 15.3). There are currently over 2000 traders in the wholesale market including 488 generators; most of them are independent producers (Table 15.4).

The Brazilian generation market is no longer concentrated. The largest generation company, Chesf, holds less than 10% of total installed capacity (Table 15.5). However, the three main generators are subsidiaries of Eletrobras Holding. Indeed, state-owned companies, at both the federal and state level, are still dominant in the generation business. Only 2 of the top 10 generators are private, Tractebel and AES Tiête.

---

16. The reserve auctions have been used to promote renewable (wind and biomass) in Brazil. It is odd, as those sources are intermittent and, therefore, inadequate to reserve purposes.

17. Consumption higher than 3 MW.

TABLE 15.3 Timetable of the Second Phase of the Electric Sector Reform

Year	Event
2002	Luiz Inácio Lula da Silva is elected with a reform mandate
2003	The new institutional model is announced
2004	The new regulatory framework passed
2004	First "old energy" auction. It negotiated the existing power capacity
2005	First "new energy" auction. It negotiated the expansion of power capacity

Source: Elaborated by the authors

TABLE 15.4 Agents Registered on the Wholesale Market (CCEE)

Agent Type	Number
Self producers	42
Generator (utility)	32
Distributor	46
Traders	141
IPPs	414
Free consumers	1427
Total	2102

Source: CCEE

In distribution, privatization has been more effective with 8 of the top 10 companies being private (Table 15.6). Eletropaulo is the largest distribution company in terms of total consumption, followed by Cemig and Copel, owned by the governments of Minas Gerais and Paraná, respectively.

3.4 Energy Auctions Results

The central theme of the second reform process is the energy auctions, particularly the ACR, which determines the future of generating expansion. The success of these auctions will determine the future of the energy mix and wholesale prices.

Since December 2004, when the first energy auction was held, 31 auctions have been carried out in Brazil: ten on existing energy, twelve on new

TABLE 15.5 Top 10 Brazilian Generation Companies

Company	Region	MW	Percentage of Brazilian Market	Ownership
Chesf	Northeast	10,615	8.9	Federal government
Furnas	South	9703	8.1	Federal government
Eletronorte	North	9131	7.6	Federal government
Cesp	Southeast	7455	6.2	State government
Tractebel	–	7145	6.0	Private
Itaipu	South/Southeast	7000	5.9	Binational
Cemig	Southeast	6782	5.7	State government
Petrobras	–	6,2	5.2	Federal government
Copel	South	4547	3.8	State government
AES Tiête	Southeast	2652	2.3	Private

Source: ANEEL

energy, three structuring, four reserve, and two on renewable energy.[18] As summarized in Table 15.7, most of the trading has been in existing and new energy auctions, with average price being lower for the former.

Figure 15.8 shows average prices by types of auctions and by energy sources. Belo Monte hydropower, auctioned in a structuring auction in 2010, has the lowest price (US$42/MWh). Wind price evolution is noteworthy, reaching US$152/MWh in the auction of PROINFA in 2005, but falling significantly in the more recent auctions. The same pattern applies to the reserve auctions.

The energy auctions are an effective instrument to promote investments in the Brazilian power sector. However energy prices varied. Prices of hydropower are close to wholesale price international experiences. Thermal power and some renewable energy are almost double that of hydropower.

18. As described in Section 3.3, reserve, structuring, and renewable energy auction types are oriented to specific energy sources.

TABLE 15.6 Top 10 Brazilian Distribution Companies

Company	State	GWh	Percentage of Brazilian Market	Ownership
Eletropaulo	São Paulo	2952	11.7	Private
Cemig	Minas Gerais	2003	7.9	State government
Copel	Paraná	1839	7.3	State government
CPFL	São Paulo	1651	6.5	Private
Light	Rio de Janeiro	1523	6.0	Private
Celesc	Santa Catarina	1185	4.7	Private
Coelba	Bahia	1179	4.7	Private
Elektro	São Paulo	942	3.7	Private
Celg	Goias	838	3.3	Private
Celpe	Pernambuco	787	3.1	Private

Source: ANEEL

TABLE 15.7 Traded Volume, Average Prices and Number of Contracts in the ACR Auctions (2004/2012)

Auction type	Traded Volume MW Avg	Average Price US $/MWh	Number of Contracts
Existing energy	19,987	45.46	1612
New energy	22,478	61.90	6728
Alternative renewable	900	74.05	1146
Reserve	2189	72.83	176
Total	45,554	59.17	9662

Source: Elaborated by the authors. Data from CCEE

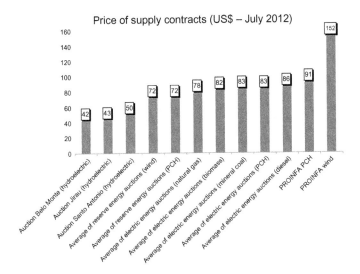

FIGURE 15.8   Price of supply contracts (US$/MWh). *Source: Elaborated by the authors, based on data from MME (2012) (b).*

4 FUTURE TRAJECTORIES

Based on the results of the expansion auctions and the Ten Year Energy Plan developed by EPE, this section presents a prospective analysis of future supply and demand, energy mix and forecasts of CO_2 emissions from the power sector for the next 10 years.

4.1 Consumption Forecast

EPE's Ten Year Energy Plan projects electricity demand for the period 2010−2020 in Brazil based on socioeconomic and demographic variables as well as analysis of energy-intensive industries. The economic scenario assumes sustained economic with average GDP growth rate of 5.0% per annum for the next decade.

According to EPE (2011) projections, electricity consumption will grow from 441 TWh in 2011 to 659 TWh in 2020, corresponding to an average growth rate of 4.7% per annum, with the commercial sector growing at 6.2% followed by the residential sector[19] as shown in Table 15.8.

---

19. The evolution of residential electricity consumption in Brazil results from an average growth of 2.5% of the number of consumers and a consumption per consumer expanding at a rate of 1.9% per year.

TABLE 15.8 Projection of Electricity Consumption 2011−2020 (GWh)

Year	Residential	Industrial	Commercial	Others	Total
2011	112,690	193,437	74,102	61,210	441,439
2015	135,682	229,870	93,495	70,723	529,769
2020	166,888	283,707	123,788	84,709	659,092
Period	Variation (percentage per year)				
2010−2015	4.8	4.6	6.2	3.7	4.8
2015−2020	4.2	4.3	5.8	3.7	4.5
2010−2020	4.5	4.4	6.0	3.7	4.6

Source: EPE (2011)

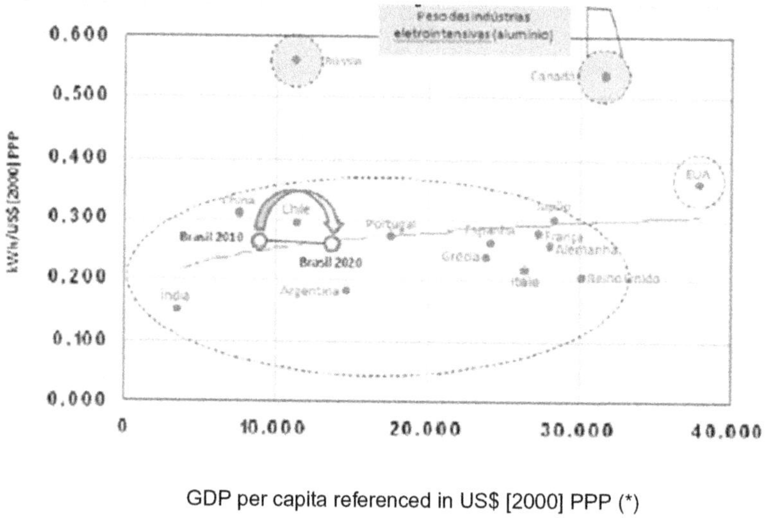

GDP per capita referenced in US$ [2000] PPP (*)

FIGURE 15.9   Energy intensity of the economy (2010−2020). (*)Data for the year 2007, for all countries except Brazil. Note: The power consumption includes autoproduction. *Source: IEA (2009): Key World Energy Statistics 2009.*

Brazilian *per capita* electricity consumption will increase from 2.4 MWh in 2011 to 3.5 MWh in 2020. However, the energy intensity of the economy will gradually decline over that period (Figure 15.9).

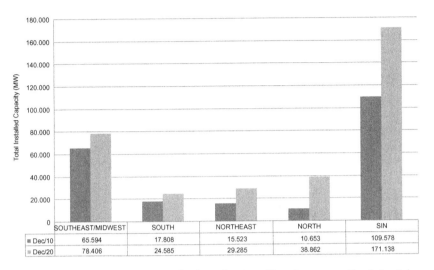

	SOUTHEAST/MIDWEST	SOUTH	NORTHEAST	NORTH	SIN
■ Dec/10	65.594	17.808	15.523	10.653	109.578
■ Dec/20	78.406	24.585	29.285	38.862	171.138

FIGURE 15.10 SIN installed capacity by subsystem. Note: It was considered the Itaipu import of power contracted to Paraguay. *Source: EPE, 2011.*

4.2 Future Energy Matrix: Expansion of Electricity Supply

According to EPE projection, installed capacity in the Brazilian National Interconnected System (SIN) will grow at 4.6% per annum on an average, roughly 3200 MW each year.[20] At this rate, installed capacity in SIN will reach 171 GW in 2020 with regional growth projections shown in Figure 15.10. The growing role of the North zone is noteworthy with several major hydropower projects expected to be constructed in Brazil in the next decade likely to be located in this area. Installed capacity in the North will increase about 28 GW reaching 23% of total Brazilian generating capacity by 2020.

Hydropower will remain predominant in generation capacity expansion reaching 115 GW in 2020. It is noteworthy that the large hydropower projects Belo Monte (Xingu River), Santo Antônio and Jirau (Madeira River) will together account for about 10% of the installed capacity of SIN in 2020.

Thermal power capacity will increase considerably in early planning years, reaching 25 GW in 2013 and remaining at this level until 2020. The first new energy auctions were characterized by a large presence of fuel oil fired plants[21] accounting for more than 6 GW of capacity in early auctions. It resulted from the restraints that more competitive sources faced at that

---

20. The chapters on other BRICs countries show projections of growth in the other major developing economies.
21. In the first four new energy auctions, fuel oil plants represented 63% of total trade.

time. Hydropower plants faced major difficulties to get licensing from the environmental agency[22] and natural gas supply became uncertain as Bolivia changed its policy on energy exports.[23]

The conclusion of Angra 3 in 2016 will add 1405 MW of nuclear power to the system. The share of renewables, wind, biomass, and small hydropower, is expected to increase significantly, growing at 12% per annum on average, mainly from wind (Table 15.9).

As shown in Table 15.10, despite the continued expansion of hydropower capacity, its share in the Brazilian generating mix will decline from 76% today to 67% in 2020, replaced by other renewables and thermal plants.

4.3 Environmental Impacts

In recent years, there are growing concerns about the environmental impacts of energy use and the effect of greenhouse gas (GHG) emissions on climate. As a number of chapters in this book describe, many governments are trying to alter the future energy mix in favor of low-carbon resources.[24] Increasing the share of renewable energy sources in the global energy matrix is considered as a necessary condition in this regard.

Brazil is fortunate in this context with an electricity generation matrix, which is mostly renewable. The Brazilian challenge is to keep this record, meeting growing energy demand while minimizing environmental and social impacts. Further development of the vast hydroelectric potential that remains in the Amazon region comes with many associated social and environmental impacts. An alternative option may be to increase power integration with neighboring countries notably Argentina, Peru, and Uruguay. This may result in postponing the construction of large hydroelectric plants in Brazilian Amazon by taking advantage of resources available in neighboring countries. Latin America energy integration is the subject of Chapter 14 by Raineri et al where the potential benefits of integration are discussed.

---

22. By the second reform rules, hydropower plant must get the environmental license before taking part in the auctions.

23. The authorities believe that thermal power expansions were excessive in the early auctions, which explains why there are no new projects in the Ten Year Plan after 2013.

24. One issue rarely addressed in Brazil refers to the possibility of distributed generation to assume the position of inducing social inclusion, in an optimal and decentralized way of leveraging the diverse resources distributed to the extent of the country, providing access to electricity to a large contingent of the population, lined also with issues of social and environmental sustainability. At the global level, it should be noted that in 2011, one quarter of the world's population, about 1.3 billion people, lacked access to electricity. According to Pereira et al. (2010), the availability of access to electricity is a key element for economic development and rural poverty reduction, and access to electric power shall be understood as an absolute right.

TABLE 15.9 Evolution of Installed Capacity by Generation Source (MW) – 2010–2020

Source	2010	2011	2012	2013	2014	2015	2016	2017	2018	2019	2020
Hydro[a]	82,939	84,736	86,741	88,966	89,856	94,053	98,946	104,415	109,412	111,624	115,123
Uranium	2007	2007	2007	2007	2007	3412	3412	3412	3412	3412	3412
Natural gas	9180	9384	10,184	11,309	11,309	11,659	11,659	11,659	11,659	11,659	11,659
Coal	1765	2485	3205	3205	3205	3205	3205	3205	3205	3205	3205
Fuel oil	2371	3744	5172	8790	8790	8790	8790	8790	8790	8790	8790
Diesel oil	1497	1497	1471	1471	1471	1121	1121	1121	1121	1121	1121
Process gas	686	686	686	686	686	686	686	686	686	686	686
PCH	3306	4201	4230	4376	4633	4957	5187	5457	5737	6047	6447
Biomass	4496	5444	6272	6631	7053	7353	7653	8003	8333	8703	9163
Wind	831	1283	3224	5272	6172	7022	7782	8682	9532	10,532	11,532
Total[b]	109,578	115,467	123,792	132,763	135,182	140,853	148,441	155,430	161,887	165,779	171,138

[a]Includes the estimated energy import of Itaipu generation that is not consumed by Paraguay.
[b]Excludes self-production which, for energy studies, is represented as load alleviation.
Notes: The values in the table indicate the insta led power in December of each year.
Source: EPE (2011)

TABLE 15.10 Evolution of Installed Capacity by Generation Source (%)

Source	2010	2011	2012	2013	2014	2015	2016	2017	2018	2019	2020
Hydro (a) (%)	75.7	73.4	70.4	67.0	66.5	66.8	66.7	67.2	67.6	67.3	67.3
Uranium (%)	1.8	1.7	1.6	1.5	1.5	1.4	2.3	2.2	2.1	2.1	2.0
Natural gas (%)	8.4	8.1	8.3	8.5	8.4	8.3	7.9	7.5	7.2	7.0	6.8
Coal (%)	1.6	2.2	2.6	2.4	2.4	2.3	2.2	2.1	2.0	15	1.9
Fuel oil (%)	2.2	3.2	4.2	6.6	6.5	6.2	5.9	5.7	5.4	5.3	5.1
Diesel oil (%)	1.4	1.3	1.2	1.1	1.1	0.8	0.8	0.7	0.7	0.7	0.7
Process gas (%)	0.6	0.6	0.6	O.5	0 5	0.5	0.5	0.4	0.4	0.4	0.4
PCH (%)	3.5	3.6	3.4	3.3	3.4	3.3	3.5	3.5	3.5	3.6	3.8
Biomass wind (%)	4.1	4.7	5.1	5.0	5.2	5.2	5.2	5.1	5.1	5.3	5.4
Wind (%)	0.8	1.1	2.6	4.0	4.6	5.0	5.2	5.6	5.9	6.4	6.7
Total (b) (%)	100	100	100	100	100	100	100	100	100	100	100

Source: EPE (2011)

The relationship between energy and environment is one of the most complex issues in developing countries. On the one hand it is undeniable that energy is fundamental to the Brazilian economic development. On the other hand, increased energy generation and consumption results in complex impacts on the environment, from local to global problems. Regarding hydropower, the main impact is on flooding of forests and displacement of local populations. It is worth pointing out that there are no electricity generation options without some adverse environmental side effects. The central question concerns the options available and the choices undertaken by the society.

Although there are still more than 150 GW of hydropower resources to be exploited, most of this potential is located in the northern region of the country, which is geographically flat. This physical characteristic will limit the construction of large reservoirs. This and other factors will make it necessary to diversify the Brazilian energy matrix with the inclusion of other resources, especially those suitable for operating the system in the dry season or during extended droughts (Losekann, 2008). One of the crucial issues for planning the Brazilian power sector is determining an energy policy, which complements the hydro resource with other energy resources while reducing GHG emissions.

As stated before, the Ten Year Plan projects an increase of fossil fuels share in the generating mix. The results of new energy auctions, where the thermopower sources were very relevant, are incompatible with the rationality and the expansion potential of Brazilian electric sector, considering the hydropower potential, wind and bioelectricity.

It is worth mentioning that the Brazilian energy sector accounts for 16.5% of the country's total GHG emissions and, due to the high share of renewable sources in its energy matrix, Brazil has one of the smallest shares of anthropogenic GHG emissions: 254 $kgCO_2$-eq/10^3 US$ and 12 tCO_2-eq/hab.

5 CONCLUSION

Brazilian electricity system has several peculiarities: ample hydro resources, a large and growing demand, and a continental-sized interconnected transmission network.

The country's historical focus on developing its vast hydro resources have been highly successful resulting in significant gains in scale, cost reduction, increased productivity, and technological advancements leading to a low-carbon power generation mix, allowing Brazil to assume a global leadership role in efforts to reduce GHG emissions.

Brazil's growing economy, increasing per capita electricity consumption, extending universal service, and the income distribution implies an annual growth of 5% requiring continuous investments and system expansion for the next 10 years and beyond.

During the last 20 years, Brazilian power sector has gone through two phases of reform. The first, introduced in the 1990s sought to liberalize the power market and increase efficiency, following the *text book* recipe, which included privatization, creation of an independent regulatory agency, an independent system operator, splitting and ring-fencing natural monopoly functions from competitive activities, with the former subject to price-cap regulation. But while these measures were gradually being implemented and before a truly competitive market could emerge, the country experienced a severe power shortage due to insufficient investment in generation, which required mandatory power rationing with significant impact on economic growth. The power rationing led to a new government, new energy policies, and significant changes in the Brazilian electricity sector.

The presidential election in 2002 marked a turning point in the regulatory framework while extending electricity access to 14 million Brazilians. The second phase of reforms, which were subsequently introduced, returned the planning function to the government and put in place new requirements to encourage additional investments in the sector, to avoid future power shortages and to feed Brazil's rapidly growing economy with affordable electricity.

Current discussion in the country is to find ways to reduce electricity tariffs to end-consumers including adjustments in concessions already made as well as concessions that come for extension in the future. Many existing generation, transmission, and distribution concessions expire in 2015, and that may be the time to make the necessary adjustments. While this debate has not been resolved, past experience in Brazil and elsewhere suggests the need to reach a reasonable balance between low tariffs, sufficient investment to expand supply while maintaining energy security and protecting the environment. In this respect, Brazil has a lot in common with other BRICs nations, all aspiring to maintain their growth prospects while pursuing a sustainable energy policy.

REFERENCES

Araujo, J.L.R.H., 2006. The *case of Brazil: reform by trial and error?*. In: Sioshansi, F.P., Pfaffenberger, W. (Eds.), Electricity Market Reform: An International Perspective. Elsevier, Amsterdam, pp. 565–594.

Comissão de Análise do Sistema Hidrotérmico de Energia Elétricar, 2001. Relatório da Comissão de Análise do Sistema Hidrotérmico de Energia Elétrica. Comissão de Análise do Sistema Hidrotérmico de Energia Elétrica, Brasília.

Dias Leite, A., 2007. A Energia do Brasil, second ed. Elsevier, Rio de Janeiro.

EPE, 2011. *Plano Decenal de Expansão de Energia* 2020. Ministério de Minas e Energia. MME/EPE, Brasília.

EPE/MME, 2012. *Anuário Estatístico de Energia Elétrica*—2012. Ministério de Minas e Energia. MME/EPE, Brasília. Available from: <www.epe.gov.br/> (accessed 15.11.12.).

Firjan, 2011. Quanto custa a energia elétrica para a indústria no Brasil? vol. 8. Estudos para o Desenvolvimento do Estado do Rio de Janeiro. Available from: <www.firjan.org.br/> (accessed 15.11.12.).

Losekann, L., 2008. The second reform of the Brazilian electric sector. Int. J. Global Energy Issues 29 (1/2), 75–87.

Losekann, L.D., Oliveira, A., Borges, G., 2009. Security of supply in large hydropower systems: the Brazilian case. In: Hunt, L., Evans, J. (Eds.), International Handbook on the Economics of Energy, first ed. Edward Elgar Publishing, Cheltenham, pp. 650–662.

MME, 2012. Gestão da Comercialização de Energia—Leilões de Energia. Apresentação da Secretaria de Energia Elétrica—SEE, Departamento de Gestão do Setor Elétrico—DGSE.

ONS, 2012. Dados relevantes 2011. Available from: <www.ons.org.br/> (accessed 15.11.12.).

Pereira, M.G., Freitas, M.A.V., Silva, N.F., 2010. Rural electrification and energy poverty: empirical evidences from Brazil. Renewable Sustainable Energy Rev. 14 (4), 1229–1240.

Sauer, I., (Ed.), 2003. Reconstrução do Setor Elétrico Brasileiro. Paz e Terra, Rio de Janeiro.

Silva, Neilton Fidelis, 2006. Fontes de energia renováveis complementares na expansão do setor elétrico brasileiro: o caso da energia eólica. Tese de Doutoramento em Planejamento Energético (PPE)–Rio de Janeiro.

Chapter 16

The Russian Electricity Market Reform: Toward the Reregulation of the Liberalized Market?

Anatole Boute

Lecturer in Law at the University of Aberdeen, Legal Adviser to the IFC Russia Renewable Energy Program (The World Bank Group)

1 INTRODUCTION

With a total installed capacity of approximately 219 GW,[1] the Unified Power System of the Russian Federation is the fourth largest electricity system in the world, covering a vast region that, as illustrated in Figure 16.1, extends from Russia's borders with the European Union to China and the Pacific Ocean.

With a population of 141.9 million and GDP of 1.858 trillion US dollars,[2] total electricity consumption in 2011 amounted to more than 1000 TWh (Minenergo, 2012). As shown in Figure 16.2, the fuel mix of the Russian electricity sector is characterized by a large share of gas-fired power plants—47.2%. Coal represents 20.5% of the installed capacity, hydropower plants 21% and nuclear plants 11%.

Electricity demand is growing strongly and, according to official forecasts, a total of 324 GW installed capacity is expected to be needed by 2030 (Energy Forecasting Agency APBE, 2011a). Due to insufficient investments in the modernization of energy supply following the collapse of the Soviet Union, the Russian electricity sector is also one of the most energy and thus carbon intensive in the world. According to the IEA (2011), the Russian economy is more energy intensive and efficiency improvements are slower than in the other main emerging economies (BRICS, without Russia). Russia accordingly represents a

---

1. Total installed capacity including power plants that are not connected to the Unified Power System amounts to around 225 GW.
2. See http://data.worldbank.org/country/russian-federation.

Evolution of Global Electricity Markets.

461

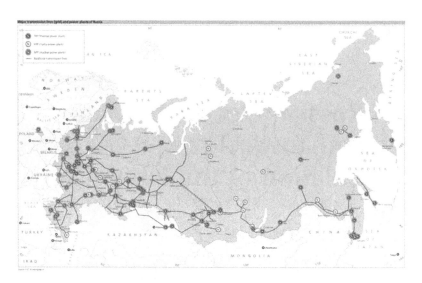

FIGURE 16.1 The Unified Power System of the Russian Federation: major transmission lines and power plants.[3] *Source: Kotikov and Trufanov (2012).*

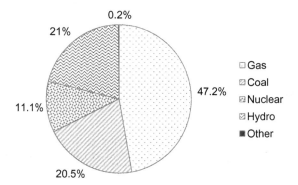

FIGURE 16.2 Fuel mix of the Russian electricity sector. *Source: Energy Forecasting Agency APBE (2011a) and Kotikov and Trufanov (2012).*

potentially huge market opportunity for power and energy service companies (Bernotat, 2007).

A huge amount of investment is needed to tackle the growth in electricity consumption and to replace obsolete capacity. The electricity and heating system is the biggest domestic gas consumer in Russia. Modernizing this sector with energy efficiency and low carbon investments is therefore not only essential from an international climate change mitigation perspective, it is also of

---

3. Available from: http://gmi.troika.ru/pdf/E/Sectors/Electricity/atlas_utilities.pdf.

TABLE 16.1 Major Milestones in the Russian Electricity Market Reform

Event	Year	Comment
President Edict on the Management of the Russian Electricity Sector in the Context of Privatization	1992	Creation of RAO UES
President Edict on the Main Principles underlying the Structural Reform of Natural Monopolies	1997	First liberalization reform initiative—postponed due to 1998 economic crisis
Government Decree on the Restructuring of the Electric Power Industry	2001	Reform principles adopted
Federal Electricity Law	2003	Regulatory architecture of the reform
Federal Law on the Regulation of the Electric Power Industry during the Transition Period	2003	Transitional arrangements
Government Decree on the Wholesale Market Rules during the Transition Period	2003	Gradual transition toward free market pricing
Blackouts hit Moscow, Tula, Kaluga, and Ryazan	2005	Highlights fragility of the system
Government Resolution on the General Scheme for the Location of Electricity Installations until 2020	2008	Strategy of development of Russian electricity ("second GOELRO")
Liquidation of RAO UES	2008	Privatization of the quasi-monopolist
Destruction of the Sayano–Shushenskaya hydropower plant	2009	Highlights necessity to modernize the system
Government Decree on the Procedure for the Implementation of State Regulation in the Electricity Sector	2009	Formalization of new phase of state regulation in the electricity sector
Government Decree on the Wholesale Market Rules	2010	Long-term wholesale market model adopted
End of the transition period	2011	free market pricing, in theory

particular geopolitical importance given the positive impact that energy savings can have on the availability of Russian gas for export (IEA, 2006, 2011).

To stimulate the modernization of electricity production and attract investment for additional capacity, Russia has reformed its electricity market on the basis of the Anglo-Saxon liberalization and privatization model (Engoian, 2006). As summarized in Table 16.1, the reform process started in

2003 with the adoption of the Federal Electricity Law FZ-35, following early reform attempts in the 1990s. In 2008, the former quasi-monopolist RAO Unified Energy Systems of Russia or RAO UES was liquidated, following the divestiture and privatization of its production assets. Moreover, in January 2011, wholesale market prices were fully liberalized, at least in theory. In practice, however, the implementation of the Russian electricity market reform appears to have considerably diverged from the fundamental principles behind the "textbook" for the restructuring of electricity markets. The Russian Government has reconsolidated its ownership of production assets through different state-controlled companies, mainly INTER RAO UES and Gazprom. It has also, to a large extent, reregulated the liberalized segment of the market.

The transition of the Russian electricity market from central command and control to liberalization and privatization generates conceptual questions that are of great relevance for the study of the evolution of global electricity markets. The strategic industrial and social importance of the electricity sector for the socialist project in the Soviet Union[4] continues to influence decision making in this field. Low energy prices are often considered as a "fundamental" right (von Hirschhausen and Opitz, 2001) and price increases—even if they are necessary to finance security of supply—are therefore an issue of particular political sensitivity. In this context, the Russian reform experience highlights the difficulty of balancing consumers' short-term concerns of affordability of electricity supply with investors' demand for long-term price certainty. In Russia, this challenge is exacerbated by the instability and unpredictability of the general investment climate, a tradition of state interference with the economy and the incompleteness of "first-generation" reforms (e.g., independence of the judicial system, protection of property rights, enforcement of contractual rights) (Ahrend and Tompson, 2005; World Economic Forum, 2009).

A study of Russia—as an energy superpower—also raises important conceptual issues regarding the transition of global electricity markets toward more sustainable patterns. The Russian case provides an interesting example of the challenges and opportunities that the decarbonization of electricity supply faces in energy-producing countries. Due to its high energy inefficiency, the Russian electricity sector presents considerable potential for energy savings and thus a reduction in greenhouse gas (GHG) emissions (IFC, 2008). Russia is also characterized by a large renewable energy resource base (IFC, 2011). The development of this potential depends on the creation of adequate incentives to level the playing field with Russia's relatively cheap and abundant fossil fuels. In an energy producing country, the price increases that are necessary in the short term to

---

4. Lenin famously stated that "*Communism is equal to the Soviet power plus the electrification of the whole country*" (Bushuev, 2006).

finance this low carbon transition cannot be so easily justified on the basis of energy security arguments, which may hold more weight in other parts of the world.

The focus of this chapter is on the evolution of the Russian electricity market following the implementation of the liberalization and privatization reforms.[5] The chapter begins, in Section 2, with an outline of the physical structure of the Unified Power System of Russia and an overview of the investments that are expected to be needed to modernize the system and ensure security and reliability of electricity supply. Section 3 introduces the restructuring process after outlining the preliberalization organization of the Russian electricity sector and describing the drivers of the reform. Section 4 examines the privatization restructuring process and focuses on the reconsolidation of state ownership. Section 5 analyzes the liberalization reform and argues that the Government is moving toward the reregulation of liberalized prices. Section 6 covers the transition of the Russian electricity toward sustainability by focusing on the promotion of renewable energy sources, energy efficiency, combined heat and power (CHP) generation and GHG emission reduction projects under the Joint Implementation (JI) mechanism of the Kyoto Protocol. Section 7 concludes. Where relevant, this chapter compares the organization of the Russian electricity market with other electricity systems covered in this book, in particular China and India.

2 THE UNIFIED POWER SYSTEM OF THE RUSSIAN FEDERATION

2.1 Key Characteristics

The current structure of the Russian electricity system is largely the legacy of Soviet planning and, in particular, the so-called GOELRO plan developed by the State Commission for the Electrification of Russia under the direction of Lenin in 1920. One of the core ideas of the GOELRO plan was the creation of a centralized electricity system that would connect the main industrial regions with power plants in optimal locations. The underlying reasoning was that such a centralized system would decrease the required reserve capacity by combining the maximum day and seasonal loads of regions located in different time zones. Based on the principles outlined in the GOELRO plan, the Unified Power System was further developed on the basis of the consolidation of local grids at a regional level and interconnection of these regional electricity systems through high-voltage transmission grids. The regional electricity systems were developed so as to guarantee the full exploitation of their respective potential and specific natural resources.

5. Network investments are largely excluded from the scope of the analysis.

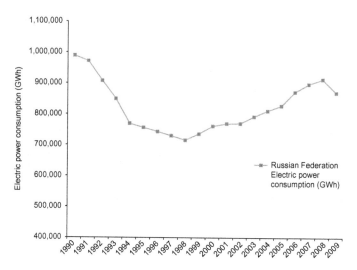

FIGURE 16.3 Electricity consumption in Russia for the period 1990–2009.[6] *Source: World Bank (2012).*

As shown in Figure 16.1, the Siberian electricity system was developed on the basis of huge coal and hydropower plants. Hydropower was also developed in the Volga region. Gas power plants were constructed in the center of Russia. Nuclear power was developed in the central and northwestern regions. Given the strategic importance of the electricity sector for the Soviet project, huge investments were made in this sector. However, toward the end of the 1970s, investment decreased substantially. This started the aging process and increased energy inefficiency in the Russian electricity production and transmission sector.

As illustrated in Figure 16.3, the fall in industrial output following the collapse of the Soviet Union generated a decrease in electricity consumption from approximately 1000 TWh in 1990 to 715 TWh in 1998.

As a consequence of this dramatic reduction in electricity demand, a considerable margin of reserve capacity appeared in the Unified Power System of Russia. This reserve margin ensured the reliability of electricity supply despite the lack of investment in new installations and in the modernization of existing plants (IEA, 2006). From 1998, domestic electricity consumption started to grow again following the strong revival of the Russian economy

---

6. See http://data.worldbank.org/country/russian-federation. According to the Ministry of Energy (2012), electricity consumption in 2010 and 2011 amounted to 988.96 and 1000.07 TWh, respectively. Data from the Ministry of Energy differs from World Bank data. The latter is only available until 2009. To ensure the consistency and reliability of the data, this graph is limited to electricity consumption until 2009 based on World Bank data. On statistical difficulties for energy research in Russia, see Pirani (2011).

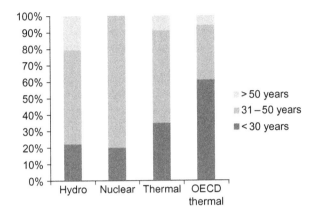

FIGURE 16.4   Age distribution of power plants in Russia, with OECD comparison. *Source: Energy Forecasting Agency APBE (2011b), Kotikov and Trufanov (2012), and IEA (2011).*

(see Figure 16.3). Investments, however, remained insufficient and, consequently, during certain periods of the year (notably winter peak demand) and in certain regions (e.g., the Moscow region), the supply–demand balance has been tightening. On May 25, 2005, serious blackouts hit Moscow, Tula, Kaluga, and Ryazan.

In 2008–2009, the global financial and economic crisis affected the Russian economy particularly severely. Following the decrease in industrial output, electricity consumption dropped significantly (System Operator, 2010). In 2011, electricity consumption returned to precrisis levels thereby again raising concerns regarding future supply–demand adequacy in certain parts of the electricity system.

Moreover, given inadequate investments in the modernization of electricity production, the Weighted Average Age of power plants in Russia and thus the efficiency of Russian electricity production fall far short of European averages (see Figure 16.4).[7] The electricity sector is, according to former President and current Prime Minister Medvedev (2009), in the "stone age." Fifty percent of the installed hydropower capacity and 40% of thermal capacity is considered to be "out of date" (Russian Government, 2008). Sixty to sixty-five percent of electricity equipment is depreciated. The accidental destruction of the 6400 MW Sayano–Shushenskaya hydropower plant in August 2009—and its social, environmental, and economic impacts—highlighted the need to modernize the Russian electricity infrastructure.

---

7. See IFC (2008) and McKinsey (2009) for a comparison of Russia's energy intensity in relation to OECD and other BRICS countries.

Thermal plants—of which about half is CHP—represent approximately 68% of the total installed capacity. As introduced above, the sector is characterized with a relatively large share of gas-fired power plants. Importantly, the fuel mix of the thermal production capacity differs amongst the regions of the Russian Federation. As illustrated in Figure 16.5, more than 60% of the total installed capacity in the European part, including Urals, is fuelled by gas. In contrast, as shown in Figure 16.6, in the East of the country, most thermal plants are fuelled by coal.

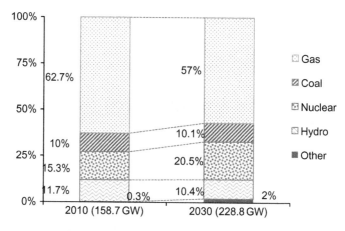

FIGURE 16.5 Fuel mix of the Europe-Ural zone. *Source: Energy Forecasting Agency APBE (2011a) and Kotikov and Trufanov (2012).*

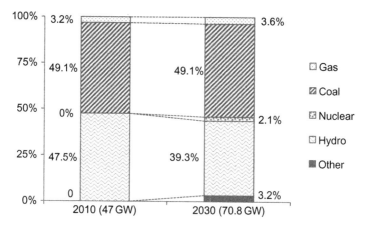

FIGURE 16.6 Fuel mix of the Siberia zone. *Source: Energy Forecasting Agency APBE (2011a) and Kotikov and Trufanov (2012).*

2.2 Perspectives of Development

The Russian Energy Forecasting Agency APBE (2011a) estimates electricity consumption in 2020 to be 1288 TWh in its base scenario. In 2030, electricity consumption is expected to grow to 1533 TWh. To maintain supply–demand adequacy, APBE considers that a total amount of 281 GW installed production capacity would be needed to meet the consumption requirements in 2020 and 324 GW in 2030. Under this scenario, 173 GW of capacity should be constructed between 2009 and 2030, taking into account the decommissioning of 67.7 GW of existing capacity. In contrast, the IEA (2011), in its New Policies Scenario, estimates that installed capacity will increase to 280 GW only in 2035, that is, 15 years later than the forecasts by the APBE.

As regards the fuel mix, the absolute priority of the Russian Government (2008, 2009) for the period until 2020 is to diversify from natural gas in order to free up more gas for exports. Therefore, the government plans to make maximum use of nuclear and hydropower plants and to develop thermal capacity primarily on the basis of coal. By 2030, the total installed nuclear capacity should amount to 50.5 GW and the total installed hydropower capacity should amount to 58.6 GW. Total installed thermal capacity should amount to 208 GW. In addition, as will be seen below, the production of electricity from renewable energy sources is increasingly considered to be a priority of the Russian energy strategy.

3 RESTRUCTURING OF RUSSIA'S ELECTRICITY INDUSTRY

3.1 Background and Drivers of the Reform

During the Soviet period, the electricity sector was organized as a state monopoly (Kuzovkin, 2006; Opitz, 2000; Yi-Chong, 2004). Production, supply, and network activities were vertically integrated and centrally managed by the Ministry of Energy and Electrification, in a comparable way to the Chinese electricity structure before 1997 (see Chapter 18 by Andrews-Speed). The Ministry made the investment decisions and financed the capital necessary to facilitate these investments with the state budget. Electricity tariffs were centrally determined. To support the energy intensive and inefficient industry and households, the authorities generally kept tariffs at low levels, below international averages (IEA, 1994, 2005). As illustrated in Figures 16.7 and 16.8, low energy prices and cross-subsidies are a trend that continues to influence electricity prices in Russia.

Following the implosion of the Soviet Union and the gradual reorganization of the economy toward more market-oriented principles, Russia initiated the corporate restructuring process of its electricity sector by establishing the RAO UES Russia joint stock company. The RAO UES group controlled the main assets of the electricity sector, including the high voltage (220 kV and

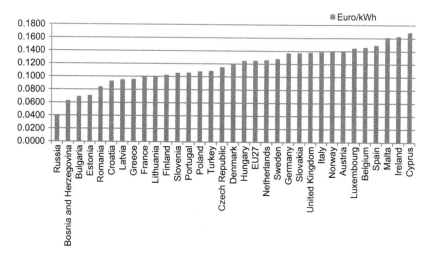

FIGURE 16.7 Electricity prices for household consumers in Russia in 2010 in comparison to Eurostat data.[8,9] *Source: Eurostat; Energy Forecasting Agency APBE (2011b).*

above) transmission grid, the Dispatching System, and the large federal thermal (1000 MW and above) and hydropower (300 MW and above) plants. The state-owned company Rosenergoatom controlled the nuclear power plants. Besides these assets of federal importance, RAO UES held controlling shares in most of the regional electricity companies.[10] All told, the RAO UES structure controlled almost three quarters of Russia's installed generation capacity. It was majority owned by the Russian state by means of a 52% controlling stake.

During the post-Soviet transition period, Russia also initiated the reform of the electricity market structure by creating the federal wholesale electricity market or FOREM. FOREM was designed as market platform where large power plants could sell electricity to large industrial consumers and regional energy systems that had a deficit of production capacity. FOREM was administered by RAO UES, creating conflicts of interest due to the dominance of RAO UES for the production and supply of electricity on the wholesale market. Electricity prices remained centrally regulated, mainly on the basis of the "cost-plus" tariff methodology.

The Russian authorities recognized that "cost-plus" tariffs, cross-subsidization, together with the dominant role of RAO UES on the market, constituted important obstacles to investments in the necessary modernization of

8. Not taking purchasing power parity into account.
9. See http://epp.eurostat.ec.europa.eu/cache/ITY_OFFPUB/KS-QA-10-046/EN/KS-QA-10-046-EN.PDF.
10. The so-called "AO-Energos."

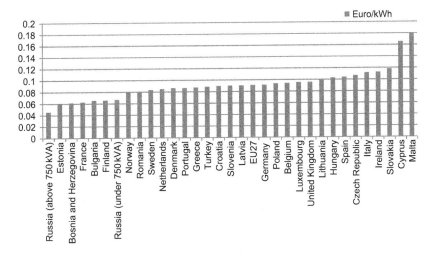

FIGURE 16.8 Electricity prices for industrial consumers in Russia in 2010 in comparison to Eurostat data.[11] *Source: Eurostat; Energy Forecasting Agency APBE (2011b).*

electricity supply. RAO UES discriminated in favor of its own thermal power plants, thus excluding cheaper and more efficient capacity operated by other entities. The "cost-plus" tariff structure and cross-subsidies did not provide an incentive to reduce costs, but led to increased energy intensity and inefficient investment policy, an issue also identified in China—see Chapter 18 by Andrews-Speed—and India—see Chapter 17 by Sen and Jamasb.

3.2 The Reform Plans and the Result

To address these obstacles to electricity investments, on July 11, 2001, the Russian Government adopted Decree No. 526 on the Restructuring of the Electric Power Industry. This strategic roadmap pursued the liberalization of the electricity sector, in parallel with the privatization of the former quasi-monopolist RAO UES. The objective of this restructuring process was to create a competitive electricity market that would attract private capital and technology in the electricity sector. With the liberalization and privatization of the electricity market, Russia aimed to create the necessary investment conditions to improve the efficiency of energy production and to ensure the security and reliability of the system. Comparable objectives underlie reform initiatives in many emerging and transition economies, for example, China

---

11. See http://epp.eurostat.ec.europa.eu/cache/ITY_OFFPUB/KS-QA-10-046/EN/KS-QA-10-046-EN.PDF.

(see Chapter 18 by Andrews-Speed) and India (see Chapter 17 by Sen and Jamasb).

Based on these principles, the Federal Electricity Law FZ-35 of March 26, 2003 created the main regulatory architecture governing the Russian electricity market reform. At least on paper, the Law reproduces most parts of the "textbook architecture" for electricity reform—as defined by Joskow (2006). Indeed, as will be analyzed in Section 4, the Law creates a regulatory structure that pursues the:

— privatization and restructuring of the generation segment into a number of competing producers;
— unbundling of competitive production/supply activities from regulated network activities and creation of an independent system operator; and
— nondiscriminatory connection and access to the network and market infrastructure.

Moreover, as will be examined in Section 5, the Federal Electricity Law aims to create a liberalized market by establishing:

— transitional mechanisms in the reform process;
— a wholesale spot market;
— a balancing market and a market for the provision of ancillary services;
— a system of capacity payments to address the risk that wholesale prices would not send sufficient price signals to investors;
— separated wholesale and retail markets, with the right for suppliers on the regulated segment of the retail market to purchase electricity on the competitive wholesale market; and
— an independent ("self-regulating") regulatory authority for the wholesale market (the Market Council), in addition to the Federal Service for Tariffs (FST) and the Federal Anti-monopoly Service (FAS).

However, in the implementation of the Federal Electricity Law, the Russian Government has moved away from the principles underlying the liberalization textbook. It has reconsolidated its control over electricity production, reregulated the liberalized electricity prices, and reinforced its control over the regulatory authorities.

4 THE PRIVATIZATION PROCESS: TOWARD THE RECONSOLIDATION OF STATE OWNERSHIP?

4.1 Privatization and Restructuring of Electricity Production

According to the privatization strategy of the Russian electricity industry, the production assets of RAO UES were organized around 7 Wholesale Generation Companies—WGC or OGK to use the Russian acronym—and 14 Territorial Generation Companies, TGC or TGKs. The OGKs are large companies[12] based on the assets of federal importance of RAO UES. To limit their dominant position on the market, the assets of these companies are located in different regional energy systems. Along with the state-controlled nuclear operator Rosenergoatom, the OGKs are the largest producers in the wholesale market. Six of these OGKs produce electricity from thermal power plants. One—RusHydro—is purely based on hydropower capacity and controlled by the Russian state. The TGKs have been formed on the basis of the interregional integration of the generation assets of the regional electricity companies controlled by RAO UES and, in particular, CHP plants. TGKs are thus also active in the heating market and generally control district heating distribution networks.

This corporate restructuring process ended on July 1, 2008 with the liquidation of RAO UES. E.ON, ENEL and Fortum have acquired electricity production assets from RAO UES. Russian companies that have entered the electricity production sector include Gazprom, the Russian Railways, the metallurgical company Norilsk Nickel, EN + (related to the aluminum producer RUSAL) the oil company Lukoil, the coal concern SUEK and the energy concern Integrated Energy Systems (PriceWaterhouseCoopers, 2007; 2008). At the moment of purchasing the production assets of RAO UES, investors committed to implementing the investment programs of these generating companies. The Russian Government has been very strict in ensuring adherence to these investment commitments. It is very difficult for OGK and TGK investors to persuade the authorities about the necessity to build power plants at locations other than those they committed to when purchasing the production assets from RAO UES.

An important development following the liquidation of RAO UES in 2008 is the increasing role in the electricity production sector played by INTER RAO UES—a state-owned company that has historically held a *de facto* monopoly over electricity export and import. The thermal production assets that did not find acquirers during the privatization of RAO UES were transferred to INTER RAO UES[13] thereby reinforcing state ownership of electricity production assets (Kornilov and Kulieva, 2009). Through Gazprom, Rosenergoatom, RusHydro and INTER RAO UES, the state has reconsolidated its position on the production market. The role of state-controlled companies (e.g., Gazprom) has been central for the sale of strategically sensitive assets, for example, power plants supplying the major cities of Moscow and Saint Petersburg, as well as industrial strongholds (Standard and Poor's, 2006, 2008).

Besides the consequences that this reconsolidation process has on the privatization of the Russian electricity sector, it is important to highlight the impact that reinforced direct state control overproduction assets can

---

12. With an installed capacity between 8.5 and 22 GW.
13. In a preliminary stage, these assets were transferred to the Federal Grid Company.

have on the functioning of the liberalized market. Standard and Poor's (2008) warns that "Government control in some generation companies presents risks of economically inappropriate, but politically motivated decisions." Gazprom's entry into the power production market also raises a potential competition issue due to the company's dominant role for gas supply and the risk that this position could be used to penalize competitors on the electricity production market (Seliverstov, 2010). This concern became particularly acute in the context of the eventually aborted merger plans between Gazprom and SUEK—the coal producer (Fitch, 2008; Ketting, 2008). Moreover, the IEA (2005) considered that, since nuclear and hydropower plants remain largely under the control of the state, this capacity could be used to influence wholesale market prices by "dumping large volumes of cheap electricity onto the market" during periods of peak demand (see also Tompson, 2004).[14]

4.2 Unbundling of Network Activities

On the network side, the Federal Grid Company—charged with the development and maintenance of the transmission network infrastructure—and the System Operator have been spun off from the production and supply companies. Russia thereby introduced ownership unbundling in addition to the functional unbundling provisions of the Federal Electricity Law that forbid network companies to be active in the electricity production and supply sector. Importantly, state ownership in both the production and network segments of the electricity sector limits the claim of ownership unbundling in Russia (Pittman, 2007).

In accordance with the Federal Electricity Law, the Russian state shall own at least 75% of the Federal Grid Company and 100% of the System Operator shares. At regional level, the distribution network assets were integrated into the Interregional Distribution Grid Company or MRSK to use the Russian acronym. In May 2012, plans to integrate MRSK under the Federal Grid Company started to be implemented. Simultaneously, the Federal Grid Company was classified as a company of strategic importance that could not be open to privatization thereby stalling previous plans to open its capital—up to 25% of it—to private investors (Figure 16.9).

4.3 Nondiscriminatory Treatment by Natural Monopolies

The Federal Electricity Law requires the providers of natural monopoly services on the electricity market to respect the fundamental principle of nondiscrimination. This principle, in particular, applies to connection and access

---

14. However, as Pittman (2007) notes, "in many situations it may be appropriate to consider hydro plants as base load rather than flexible generation plants."

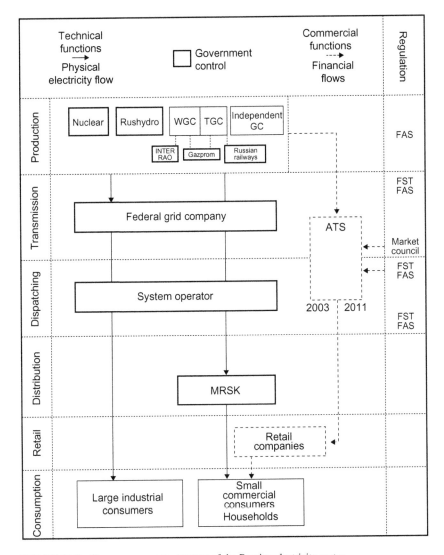

FIGURE 16.9   The new corporate structure of the Russian electricity sector.

to the network, as well as to access to the commercial infrastructure of the wholesale market. Nondiscrimination is not an absolute principle. It does not prevent differential treatment if this is objectively justified. The Federal Electricity Law, for instance, recognizes the need to take into account the specific production characteristics of nuclear, hydropower, and CHP plants in the regulation of the wholesale market.

The nondiscriminatory treatment of wholesale market participants is aimed at restoring the confidence of investors affected by the conflicts of

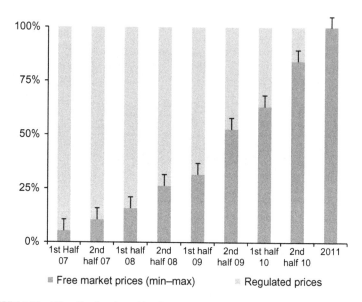

FIGURE 16.10 Liberalization timetable. *Source: Transition Wholesale Market Rules (2003).*

interest which characterized the functioning of the post-Soviet FOREM system. In practice, however, the relevance of the principle of nondiscrimination for the creation of a functioning electricity market depends on its application and enforcement by the public authorities. The FAS is charged with control over natural monopolies, including the assessment of allegations of discriminatory treatment. Importantly though, FAS is not independent from the Government. It is formally part of the executive branch and therefore must follow the orders issued by the Government. It could, therefore, be biased toward the state-controlled production assets.

5 THE LIBERALIZATION REFORM: TOWARD THE REREGULATION OF LIBERALIZED PRICES?

5.1 Transition Period

As shown in Figure 16.10, the Federal Electricity Law requires the gradual replacement of regulated prices with free market prices during a transition process that ended at the end of 2010.

From January 1, 2011, the Russian wholesale market has been supposed to function on an almost fully liberalized basis, at least in theory. According to the philosophy of the reform, as outlined in Decree No. 526 on the Restructuring of the Electric Power Industry, free market price formation would create an incentive for investments in new state-of-the-art production capacity and so stimulate the efficiency of the sector. Despite some initial

delays (Kurronen, 2006) and heavy lobbying from energy-intensive industries, the Russian authorities eventually respected the liberalization
timetable adopted. Lobbying in favor of postponing the liberalization agenda
was particularly intense in the context of the 2009 global economic and
financial crisis. According to analysts, resisting this pressure reinforced the
credibility of the Russian liberalization policy by sending a positive signal to
the investment community that Russia would respect its engagements (Fitch,
2010). In practice, however, mechanisms of price monitoring and price control seriously limit the freedom of market parties to determine electricity
prices on the wholesale and retail markets.

5.2 The Wholesale Day-Ahead Market

The Federal Electricity Law organizes the electricity market into a wholesale
and a retail segment. The wholesale market is open to production installations larger than 5 MW and to consumers with a total capacity of at least
20 MW. It organizes day-ahead and balancing trading, and establishes a system of capacity payment mechanisms. The Administrator of the Trading
System or ATS is the commercial operator of the wholesale market. It organizes the day-ahead market in close cooperation with the System Operator.
ATS operates under the regulatory supervision of FAS and the Market
Council—an independent—"self-regulating"—authority composed of the
wholesale market participants.

The wholesale market is divided into the European–Ural price zone and
the Siberian price zone due to the different structure and costs of electricity
production in these zones—coal and hydropower in Siberia versus more
expensive gas and nuclear capacity in the European part of the Russian electricity system. These price zones must be distinguished from "nonprice
zones" and isolated areas, referring to the electricity systems not connected
to the Unified Power System and that largely remain subject to regulated
prices.

Within each price zone, ATS determines equilibrium prices per node of
the calculation model following the Locational Marginal Pricing approach
(Kuleshov et al., 2012). In order to ensure that commercial transactions on
the day-ahead market match the technical reality of the system, ATS works
closely together with the System Operator. As illustrated in Figure 16.11, the
main steps of this procedure are as follows:

1. Not later than 24 h before the beginning of the trading day, electricity
   producers notify the System Operator of their minimum and maximum
   expected production capacity and communicate the maximum prices that
   they intend to submit in their bids. Electricity consumers communicate
   their maximum hourly electricity consumption.

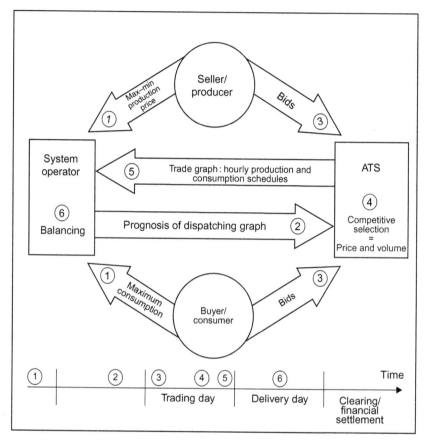

FIGURE 16.11 The organization of the day-ahead market (numbers represent time-ordering). *Source: Based on Russian Government (2010).*

2. Based on this information, the System Operator chooses the "structure of active and reserve production installations" in order to cover the minimum and maximum consumption loads, taking into account sufficient reserve capacity, electricity import—export flows, network congestion and losses.

   The production installations included in the "structure of active and reserve production installations" are thus the installations to be committed to electricity generation during the period concerned. In determining this structure for production installations, the System Operator not only aims to guarantee the reliability of electricity supply but also strives to minimize the cost of electricity generation. The System Operator integrates the installations to be assigned to electricity generation in the "prognosis of dispatching graph." This graph is a provisional representation of the hourly production schedules required to meet the estimated electricity demand. For every electricity production installation, it mentions the minimum and maximum available production capacity during the period concerned. The System Operator communicates this graph to the ATS.

3. During the trading day, the wholesale market participants submit their bids to the ATS. Participation in the day-ahead market is mandatory. Electricity producers must submit bids for the entire production capacity of their installations in order to prevent abuses of market power by withdrawing capacity from the market (Abdurafikov, 2009). Electricity sold under bilateral contracts is "passed through"[15] the day-ahead market in order to compensate congestion and network losses related to these contractual transactions.[16]

4. ATS selects the bids and includes them in the hourly production schedule following a certain order of priority: firstly, the electricity produced by power plants that provide ancillary services and by nuclear power plants to meet technical and safety requirements; secondly, the electricity produced by CHP installations in heat-extraction mode, by power plants that burn the "associated gas" produced in the course of oil extraction activities, and by hydropower plants for engineering or environmental reasons; thirdly, the electricity produced in accordance with bilateral contracts; finally, all other bids. For the first three orders of priority, the operators of power plants submit price-taking bids. In price bids, producers are in theory free to ask the prices they want.[17]

5. The ATS integrates the hourly production and consumption schedules in a trade graph. It communicates this trade graph, along with the nodal equilibrium prices, to the System Operator.

---

15. Producers and consumers submit "price-taking" bids to ATS.

16. For every bilateral contract, the System Operator confirms the presence or absence of substantial network constraints. In cases where the System Operator confirms the absence of network congestions, the supplier is only required to purchase an amount of electricity corresponding to the network losses calculated by ATS. If the System Operator did not confirm the absence of network constraints, the parties to bilateral contracts are obliged to pay the difference between the nodal prices related to the point of delivery of the buyer and to the point of delivery of the seller. Payment for the costs of congestion and network losses will be subjected to the volatility of the day-ahead market prices. The point of delivery chosen for the execution of the supply—purchase obligations will determine the repartition of this volatility risk between the parties. Indeed, if the contractually agreed point of delivery coincides with the buyer's point of delivery, the buyer will be hedged from any price variation. All the risk will be transferred to the seller. The parties can contractually agree on an "arbitrary" point of delivery for the execution of their contractual obligations. The choice of an "arbitrary" point of delivery for the execution of the contractual obligations permits the sharing of the volatility risk.

17. However, as will be discussed below, the FAS and the Market Council, in cooperation with the Ministry of Energy and the FST, closely monitor the evolution of prices and assess whether they are "reasonable," for example, in relation to benchmarks representing standard production costs.

6. The System Operator uses this data to organize the selection of bids in the context of the balancing market.

With the end of the transition period and abolition of regulated prices having taken place in December 2010, ATS should now, in theory, determine the electricity wholesale market prices based solely on the forces of supply and demand, taking into account technical network constraints. The Russian authorities have, however, been reluctant to fully transfer electricity price formation to the market. Even now the transition period has finished, the introduction of price limits continues to be authorized in a large range of situations, namely in case of spikes in electricity prices, shortage of capacity and absence of competition, as well as in case of price manipulation.

The monitoring and control of electricity prices became a particularly acute issue following the accidental destruction of the Sayano–Shushenskaya hydropower plant in 2009. With Decree No. 929 of November 14, 2009 on the Procedure for the Implementation of State Regulation in the Electricity Sector, the Russian Government adopted strict rules on the formation of electricity prices that marked, in the words of the Government (2010), a new "phase of state regulation in the electricity sector." Decree No. 929 provides that in the event of capacity shortages or emergencies, the Ministry of Energy, in tandem with the FST, will propose to the Government that they reregulate electricity prices for a limited period of time. Moreover, Decree No. 929 introduces a system of price monitoring and requires the reintroduction of regulated prices in case the average market price during a 3-day period exceeds the reference electricity price calculated for a standard 3-day period. In this case, electricity sellers may not submit price bids that exceed the regulated price level that is determined by the FST.

Based on the analysis of the evolution of electricity markets in this book (e.g., Chapter 11 by Kristov and Keehn, Chapter 17 by Sen and Jamasb, Chapter 18 by Andrews-Speed, and Chapter 22 by Kim et al.), it can be argued that this reregulation of prices does not contribute to sending the right investment signals to market players. Moreover, it contradicts the paradigm of liberalized electricity markets according to which regulators must allow prices to rise in times of scarcity. However, as will be examined below, the presence of a capacity market limits the impact of such intervention on the financial viability of investment in peak electricity production capacity.

5.3 The Balancing Market

The System Operator is in charge of the organization of the intraday balancing market. The System Operator determines the dispatching volumes and balancing prices on the basis of the price bids submitted by the wholesale market participants for their participation in the day-ahead market. The price bids submitted for the sale and purchase of electricity in the day-ahead market are thus "reused" in the context of the balancing market. In addition, producers and consumers can submit specific price-taking bids to participate in the balancing market. With these "balancing-specific" price-taking bids, producers/consumers communicate their readiness to increase or decrease the amount of electricity mentioned in the hourly production/consumption schedules. Since dispatching volumes and balancing prices are calculated on the basis of the price bids submitted for the day-ahead market, the results of the competitive selection of bids for the balancing of the system actually correspond to an adjustment of the trade graph.

On the basis of the price bids for the day-ahead market, the System Operator determines hourly balancing cost indicators for every node of the calculation model. A different pricing system applies depending on the nature of the initiative for the deviation.[18] Based on the information transmitted by the System Operator on the operative dispatching commands issued, the ATS calculates the volumes of deviations at own initiative and following dispatching orders. On this basis, it establishes the financial rights and obligations of the wholesale market participants.

The System Operator selects, coordinates, and remunerates the providers of ancillary services. The selection of the services related to frequency regulation takes place on a competitive basis. The System Operator must announce that it will organize a selection process for providers of such ancillary services. This announcement must describe the type of services requested and the technical characteristics of the production installations needed, as well as the location and duration of these services. The operators of production installations that are willing to provide these services submit price bids to the System Operator. The latter selects the

---

18. For electricity producers, the prices for balancing upward at external initiative equal the maximum value of the nodal cost indicators. These balancing prices cannot be lower than the day-ahead market price and the price mentioned in the price bid submitted by the producer concerned for participation in the day-ahead market. For hydropower plants and pumping storage plants, balancing prices are calculated on the basis of regulated tariffs. The prices for balancing downward at external initiative correspond to the minimum value of the cost indicators, the day-ahead market price and the price bid submitted by the producer concerned for participation in the day-ahead market. Electricity producers that, at their own initiative, do not produce enough must pay the highest price for balancing upward. If they produce an excess, the cost will equal the lowest price for balancing downward. Electricity consumers that, at their own initiative, consume in excess of their hourly consumption schedule must pay the highest price for balancing upward. If, following commands from the System Operator, electricity consumers consume less than mentioned in their hourly consumption schedule, they will, for the amount of electricity not consumed, receive balancing prices corresponding to the maximum value of the balancing cost indicators. If the price mentioned in their price bids is higher than the maximum value of the balancing cost indicators, this higher price will apply.

bids that are adequate to provide the services requested and that minimize costs. The price for the provision of ancillary services is, however, not entirely subject to market forces. Indeed, the total amount of financial resources that the System Operator can use to remunerate the providers of these services is regulated.

The services related to voltage regulation and to the prevention of emergency situations are not selected following a competitive process. Given the specific technical characteristics of the power plants that are capable of providing these services and the local character of the provision of these services, the System Operator directly approaches the operators of the power plants that answer to these specific requirements. If the System Operator does not manage to select electricity producers that would provide the requested services and so ensure the reliability of electricity supply, the System Operator can request "system producers" provide these services. "System producers" are production installations that are characterized by a specific functioning regime and location in the electricity system without which it would be impossible to ensure the functioning of the Unified Power System of the Russian Federation in accordance with the applicable parameters. The services provided by system producers are remunerated at regulated tariffs. These tariffs must enable the operators of these installations to recover their investment costs.

5.4 The Wholesale Capacity Market

The Russian electricity market is not an "energy only" market. In the capacity market, investors get paid for the availability and readiness of their installed capacity to produce electricity. A revenue stream, additional to the price of the electricity commodity, is created by requiring electricity buyers to purchase an amount of production capacity proportional to their monthly peak load, increased by a reserve coefficient.[19] In exchange, electricity producers, as sellers of capacity, must maintain their production installations in a condition of readiness to produce an amount of electricity corresponding to the capacity they have sold.

---

19. This reserve coefficient is aimed at providing an additional margin of capacity to ensure the reliability of the system. It is calculated for every electricity buyer as a proportion of their respective peak consumption to the total amount of production capacity that the System Operator considers necessary to guarantee the reliability and security of the Unified Power System of the Russian Federation. Electricity buyers have the possibility of notifying to the System Operator of the maximum (or peak) amount of electricity that they expect to consume. The System Operator will in this case select the capacity by taking into account the information submitted by the consumers, instead of its own prognosis. If the electricity buyers misjudge their future peak consumption and consume more in peak hours than expected, they will be penalized by the application of a more stringent reserve coefficient.

The System Operator organizes the competitive selection of capacity 4 years before the actual capacity supply.[20] The amount of capacity that is traded depends on how much production capacity the System Operator estimates will be needed to ensure the reliability and security of electricity supply to the Unified Power System of the Russian Federation.[21] The System Operator calculates this amount of capacity per "free capacity flow zone," that is, the parts of the Unified Power System within which the electricity produced by certain plants, and thus their capacity, may be replaced with similar plants without congestion constraints affecting this replacement. It then selects the production installations that are needed to supply this capacity by organizing a competitive selection of price bids. The installations that successfully pass this preliminary competitive selection procedure are entitled to sell their installed capacity.

To participate in this competitive selection, the operators of production installations must, not later than 10 working days before the beginning of this process, submit price bids to the System Operator.[22] Electricity consumers do not participate in this selection process. For every production installation, producers' bids specify the amount of capacity that they expect to be available for electricity production during the trading year and specify the technical parameters of the production equipment concerned, as well as the type of fuel used, the location of the power plant and the expected date of commissioning for new or modernized installations. The bids also mention the price per megawatt of capacity per month that the investors require to recover their investment costs and a certain profit. The System Operator selects the bids by comparing their technical parameters and their price. A capacity price is determined for every free capacity flow zone. The capacity that the System Operator selects during the competitive selection process is remunerated at this free capacity flow zone price.[23] Operators of the electricity production installations that are selected during

---

20. Every year preceding the actual supply of capacity, the System Operator organizes a corrective selection of capacity to provide for additional capacity where necessary.

21. This amount of capacity corresponds to the maximum amount of electricity consumed during the year, increased by a reserve coefficient that aims to guarantee the functioning of the electricity system within certain reliability parameters (the "N-1" criterion).

22. To avoid the abuse of any dominant positions, investors that control more than 15% of the installed capacity in a given free capacity flow zone can only submit "price-taking" bids. This percentage applies to the free capacity flow zones located in the first wholesale market price zone (Europe-Ural). A percentage of 10% applies to the second wholesale market price zone (Siberia).

23. In certain situations, the System Operator may select the production capacity of investors that required a higher price than the "free capacity flow zone" price. With certain exceptions, this selected capacity can be remunerated at the higher price required by the investor. It can thus be said that the "free capacity flow zone" price is not a genuine marginal price.

the competitive selection of bids may also sell their capacity on the basis of bilateral contracts.[24]

The electricity producers that have been selected by the System Operator must fulfill their capacity supply obligations by first guaranteeing that their production installations are designed to produce the amount of capacity they agreed to supply. Moreover, they must demonstrate that these installations really comply with the technical parameters that they claim to possess. In order to confirm the fulfillment of this obligation, the System Operator, before the beginning of each trading period, certifies whether the installed capacity and the technical parameters of the production installations concerned comply with the capacity and parameters specified in the price bids. If the amount of capacity offered in the price bids and selected by the System Operator during the competitive selection of capacity exceeds the real amount of installed capacity certified by the System Operator, the electricity producers concerned must pay a "fine" for the "noncertified" amount of capacity that they proposed for sale. This fine is levied in proportion to the capacity that they failed to supply. This can be regarded as an incentive for producers to submit proper price bids during the competitive selection of capacity.

Secondly, electricity producers must maintain their production installations in such a state that allows them to be called upon for the production of electricity in accordance with the applicable reliability requirements. The System Operator confirms the fulfillment of this obligation by, amongst others, examining the bids that the operators of the production installations concerned have submitted to sell electricity in the day-ahead or balancing market.[25] If electricity producers fail to guarantee the availability of their production installations, the remuneration of the capacity to which these producers are entitled is decreased by a coefficient calculated by the ATS. The values of these coefficients depend on the reason for not complying with their obligations, as well as on the type of production installation concerned.

Although the Federal Electricity Law organizes capacity trading on a free market basis, the Government has reintroduced a system of capacity price regulation. Firstly, in capacity flow zones that are characterized by economic concentration, price caps apply. The FAS is charged with the evaluation of the dominant position of producers and has qualified 24 out of 27 capacity

---

24. Contracts may only be concluded between an electricity buyer and an electricity producer that are located in the same free capacity flow zones.

25. The System Operator verifies the readiness of production installations to produce electricity by confirming that the installations concerned are able to supply active and reactive energy, and that the installations can respect the generation graphs developed by the System Operator (maneuverability). The supply of capacity by the operators of electricity production installations and its receipt by the consumers (buyers of capacity) is acknowledged in an "act of receipt and transfer of capacity."

flow zones[26] as zones with economic concentration (Gore, 2011). In practice, the capacity market is thus largely regulated.

Moreover, the Russian Government has introduced a mechanism of regulated capacity payments for the power plants that investors build to implement the investment programs to which they committed when they purchased OGKs and TGKs during the RAO UES privatization process. This regulated capacity payment mechanism is based on 10-year[27] contracts that investors conclude with the System Operator and ATS and that provide for the monthly remuneration at regulated tariffs of the installed capacity of new power plants. The Government determines the location and type of power plants that are eligible for the conclusion of these long-term regulated capacity contracts. Most new investments in the Russian electricity production sector are covered by these long-term tariff guarantees. These guarantees considerably reduce the scope of application of the competitive segment of the wholesale capacity market. However, given the risk of price interference outlined above and the instability that characterizes Russian electricity regulation, long-term regulated capacity prices appear to be necessary to attract investments in electricity production.

To finance the regulated capacity prices of the long-term capacity agreements, electricity buyers must purchase a certain amount of the installed capacity of the power plants built in accordance with the regulated capacity contracts. Electricity buyers purchase an amount of this capacity that is proportional to their peak consumption. All electricity buyers thus contribute to the financing of the investment programs created by the former RAO UES in proportion to their peak consumption.

Furthermore, the Russian authorities have created a technical capacity reserve scheme in order to generate specific investments that are necessary to cover forecasted shortages of capacity in certain regions and ensure the reliability and security of the Unified Energy System.[28] This scheme is based on the concern that competitive electricity prices—even when supplemented by an additional revenue stream for capacity—could fail to attract investors in certain projects that are needed to ensure the supply–demand balance in certain regions. In the context of a competitive selection of bids, the Ministry of Energy and the System Operator aim to attract investors on the basis of regulated capacity and electricity price guarantees, as well as a yearly fee for the "services provided to the formation of a technical reserve." The state guarantees the recovery of the invested capital, and a "necessary"

---

26. There are no price caps in the Siberia, Urals, and Centre free capacity flow zones.

27. New nuclear and hydropower plants are remunerated at the regulated price for a period of 20 years. In certain situations, investors that have concluded regulated capacity contracts can renounce to the regulated price and opt to sell capacity at "free" market prices on the competitive segment of the market.

28. Decree No. 738 of December 7, 2005 on the Formation of a Technical Capacity Reserve.

return on investment. The technical capacity reserve scheme thus reintroduces elements of preliberalization central planning and intervention of the executive branch in electricity generation. However, at the time of writing, this scheme largely remained "law in the books" as no power plant had yet been built on this basis.

5.5 The Retail Market

The Federal Electricity Law opens the retail market to competition. In theory, free market pricing is the general rule. However, in practice, a large segment of retail consumers retain the right to purchase electricity at regulated prices through the mechanism of the "guaranteeing supplier"—or supplier of last resort. Guaranteeing suppliers are companies that are obliged to conclude electricity contracts with any requesting household consumer. Regional Energy Commissions—corresponding to the different regions of the Russian Federation[29]—regulate the price at which these companies supply electricity to their consumers (wholesale market price + a regulated "supply premium"). Regional retail prices thus vary within the Russian Federation. However, these prices must remain within limits that the FST determines per region. Guaranteeing suppliers purchase at regulated prices the electricity corresponding to their supplies to household consumers—including the electricity they purchase on the wholesale market. This considerably limits the scope of liberalized prices in the Russian electricity market. Despite the announced policy to reduce cross-subsidies, retail prices for households remain subject to cross-subsidization and do not fully reflect the costs of production, transmission, distribution, and supply (Kuleshov et al., 2012) (Figure 16.12).

6 TRANSITION TOWARD LOW CARBON ELECTRICITY PRODUCTION

In the last few years, low carbon electricity production—in particular renewable energy—has received significant attention on the Russian national energy policy agenda. Russian policymakers recognize the environmental, social, economic and energy security benefits that the development of alternative modes of electricity production could generate. Russia's Energy Strategy to 2030 outlines the strategic objectives in this field, including: reducing the anthropogenic impact on climate change while meeting growing energy demand; improving the rational use of fossil fuels; safeguarding the health and quality of life of the population and reducing government health expenditure; reducing network costs by decentralizing electricity production and using local energy sources; diversifying the energy mix; and enhancing

---

29. For an overview of all the Regional Energy Commissions of the Russian Federation, see http://www.fstrf.ru/regions/region/showlist.

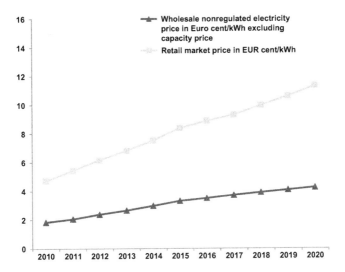

FIGURE 16.12   Wholesale and retail price forecasts until 2020. *Source: Energy Forecasting Agency APBE (2011b).*

energy security (Russian Government, 2009). In practice, however, the Russian low carbon policy in the electricity sector (in particular the renewable energy agenda) has suffered from a lack of implementation by the federal authorities.

6.1 Renewable Energy

The modernization program of the Russian electricity sector and the necessity to add capacity to answer future electricity demand provides an opportunity to meet part of the total investment requirements with renewable energy installations (IFC, 2011). The Russian authorities recognize this opportunity and have integrated the promotion of renewable energy sources in the development strategy of the electricity sector. The Government has adopted a strategic renewable energy target of 4.5% of the total electricity produced and consumed in 2020 (from less than 1% in 2010). Reaching this target would, according to official forecasts, require up to 25 GW of newly installed renewable energy capacity—excluding hydropower installations above 25 MW. This relatively modest target in relation to OECD and other BRICS countries can be explained by the availability of relatively cheap fossil fuels in Russia and by the cost of these investments in connection to the high political sensitivity of price increases in the energy sector.

The Federal Electricity Law establishes different mechanisms to stimulate the deployment of renewable energy sources in the Russian electricity sector.

Firstly, since November 2007, the Law requires the creation of an electricity "premium" scheme (Boute, 2011). In accordance with this scheme, the price of electricity produced from renewable energy sources shall be determined by adding a regulated premium to the electricity wholesale market price. Electricity buyers are supposed to finance the premium scheme in proportion to the amount of electricity they purchase on the wholesale market. To prove that a certain amount of electricity has been produced from renewable energy sources, the Law creates a system of "qualification" of renewable energy installations and "certificates" per megawatt hour of electricity produced from renewable energy sources. Despite these legislative provisions being introduced in 2007, Russia has so far failed to implement the premium scheme in practice. Concerned with the potential impact that the premium scheme might have on electricity prices (IEA, 2011), the Russian Government did not adopt the regulatory documents necessary for the determination of the level and duration of the premium. Moreover, so far, only one installation—a small hydropower plant in Karelia—has been "qualified" by the Market Council—the authority in charge of this procedure.

In December 2011, the Federal Electricity Law was amended to introduce a new support mechanism in addition to the premium scheme. The Federal Electricity Law now provides for the possibility of long-term contracts being concluded with investors for the remuneration of capacity at regulated prices, thus encouraging investment in renewable energy infrastructure. This new support mechanism is based on a similar contractual approach to the agreements that OGK and TGK investors signed to finance the construction of new power plants following the privatization of RAO UES. Under the new capacity-based support scheme, investors will, in return for long-term regulated tariffs, commit to build certain types of renewable energy installations at specific locations determined by the Russian Government (Boute, 2012a; IFC, 2011). Moreover, investors commit to guarantee the availability of their installations for electricity production. Under current capacity supply rules—based on the dispatchability of power plants, this requirement represents an important obstacle for variable renewable energy sources, such as wind and solar photovoltaic (PV) (IFC, 2011). Because wind and solar PV installations are characterized by less flexible production patterns and thus challenging dispatchability,[30] existing reliability requirements and coefficients could penalize these installations. As argued by Keay et al. in Chapter 2 of this book, the primary objective of a capacity market is to ensure resource adequacy, not to decarbonize the electricity system (see also Chapter 7 by Bauknecht et al.). If Russia remains committed to its policy to support

---

30. Importantly, Riesz et al. in Chapter 25 of this book note that, in addition to dispatchability challenges, variable generation brings new benefits to the system thanks to the capability of modern wind and solar PV plants to rapidly ramp downward if required.

renewable energy sources through the capacity market, it must amend the existing availability requirements in order to accommodate the more challenging dispatchability of variable renewable energy installations. On the other hand, the regulation of capacity supply needs to maintain sufficient incentives for electricity generation. Past experiences with capacity-based support mechanisms (e.g., in India—see Chapter 17 by Sen and Jamasb) have led to reduced efficiency of the constructed turbines, with investors focusing on installed capacity ("steel-in-the-ground") rather than energy production.

The Federal Electricity Law also creates a legal basis for the support of renewable energy through the retail market (IFC, 2012). Firstly, the Law requires the subsidization of network connection costs for renewable energy installations with an installed capacity of up to 25 MW. Secondly, the Law introduces an obligation on network companies to compensate electricity losses on their network by purchasing, in the first instance, electricity produced from renewable energy sources. Some regions of the Russian Federation (e.g., Vologda, Belgorod, and Karelia) have started to implement this scheme only with very limited success so far. Indeed, regulatory intervention by the federal Government is necessary, particularly to determine the methodologies to be used in determining renewable energy tariffs. In the absence of this federal basis, regional tariff schemes risk being nullified. This risk is not purely hypothetical: renewable energy tariffs in Vologda were nullified in the absence of federal approval (qualification) by the Market Council (IFC, 2012).

6.2 Combined Heat and Power Generation

The Unified Power System of the Russian Federation is characterized by a high penetration rate of CHP installations (IEA, 2009, 2011).[31] These installations operate at a level of energy efficiency well below international averages (McKinsey, 2009; Pirani, 2011). Moreover, in contrast to trends in other electricity systems (e.g., in the European Union) where CHP and district heating are seen as a solution to the decarbonization of energy supply, the share of centralized heat supply has been continuously decreasing since the collapse of the Soviet system (APBE, 2011a). This continuing decrease can be explained by the fact that reliability of supply concerns and tariff issues led to consumers increasingly switching from centralized heat supply to individual boilers (Boute, 2012b; IEA, 2009). According to the IEA (2009), heat supply from CHP installations dropped by more than 30% between 1990 and 2007. The Russian authorities consider this "boilerization" trend (i.e., the replacement of central heat supply by individual boilers) as a

---

31. The large share of CHP, together with district heating, in Russia is a legacy of the Soviet organization of the economy, which aimed at ensuring centralized energy supply to the major cities and industrial centers (Zeigarnik, 2006).

chaotic development that threatens the central heating system. This trend has serious implications from an energy efficiency perspective. The drop in demand for centralized heat supply affects the profitability of CHP installations and large boilers. It prevents investments in the modernization of existing plants.

Acknowledging the need to stimulate the efficiency of the heating sector and to safeguard the centralized heating system, Russia adopted the Federal Heat Law on July 27, 2010. The Law, firstly, introduces a system of heat supply schemes. Local authorities must outline how they intend to develop their district heating infrastructure. These schemes must prioritize the deployment of CHP installations and improve the energy efficiency of the system. The Federal Heat Law also establishes new tariff methodologies for heat supply. In contrast to the electricity sector, heat prices remain subject to state regulation. Traditionally, heat tariffs were determined on a "cost-plus" basis and were heavily cross-subsidized and therefore did not incentivize energy savings (IFC, 2008; IEA, 2004). In addition, tariffs were adopted on a yearly basis, which did not provide investors with sufficient financial predictability. The Federal Heat Law aims to improve the investment climate in the heating sector by phasing out "cost-plus" tariffs and replacing them with a "return on investment" methodology, thereby providing stronger guarantees to investors that they will recover the capital invested, together with a certain profit.

However, heat prices must, as a general rule, remain within federal tariff limits. The FST adopts minimum and maximum heat prices for every individual CHP installation. Regional tariff authorities can only exceed these limits under exceptional circumstances (Boute, 2012b). Another challenge for the development of CHP in Russia is the cross-subsidization between the electricity commodity—sold in a liberalized market environment—and the heat commodity—sold at regulated prices. The allocation of operating and investment costs to heat tariffs and electricity prices influences the competitiveness of CHP installations on the liberalized electricity market. The allocation of costs also influences the attractiveness of heat produced from CHP in relation to individual boilers.

Currently, Russian allocation rules aim to reduce heat tariffs so as to address the problem of "boilerization" that affects the Russian district heating sector. However, the low rate of recovery of costs in heat tariffs penalizes CHP installations on the electricity market. According to the Russian wholesale market regulation, the operators of CHP installations submit price-taking bids for the electricity they produce in heat extraction mode which complicates the recovery of production costs through electricity prices. Cross-subsidizing heat prices by allocating a larger share of the costs to the electricity output of CHP installations impacts on the profitability of these investments. To stimulate the modernization of CHP plants and maintain the integrity of the district heating system, the Government will have to guarantee an appropriate balance between the competitiveness of CHP on the heat market and on the electricity market.

6.3 Demand-Side Management and End-User Energy Efficiency

Energy efficiency has become a high priority on the Russian political agenda, particularly in the context of the modernization agenda promoted by former President and current Prime Minister Medvedev. Russia's 2030 Energy Strategy estimates the potential for the improvement of the energy efficiency of the general Russian economy by 2020 to be at least 40% in comparison to Russia's energy consumption in the year 2007 (i.e., according to the Russian strategy, 40% of consumption can be cost-effectively saved through energy efficiency improvements). Realizing this potential is recognized as being the central challenge to Russia's energy policy. It is also essential in the context of Russia's new "Climate Doctrine" that provides for a 25% reduction in GHG emissions by 2020 in relation to 1990 levels. The major issue is to tackle low energy prices and the cross-subsidization in the energy sector while at the same time protecting consumers from unreasonably high prices taking into account the relatively low purchasing power of the Russian population, in particular in the regions of the Russian Federation.

On November 23, 2009, Russia adopted the Federal Law No. 261-FZ on Energy Efficiency. This Law creates a system of labelling and information for energy-consuming products and forbids the sale of incandescent lamps. It also establishes a system of energy efficiency requirements for buildings and introduces metering obligations. The Law also provides for a system of mandatory energy audits and creates a system of contracts for energy services. In addition, the Federal Energy Efficiency Law requires public entities and companies that provide regulated services to implement energy efficiency improvement measures. Tariff authorities must take the investment costs of energy efficiency projects into account when determining the tariffs that apply to the regulated activities concerned. The Law also requires tariff authorities to adopt long-term tariff methodologies so as to improve the financial predictability of energy-saving investments. These tariff provisions provide important guarantees to stimulate the modernization of the Russian energy infrastructure (Boute, 2012a; IFC, 2012). However, the scope of these guarantees is limited to the regulated segment of the industry.

In addition, the Federal Electricity Law regulates the provision of demand-side management services. It seeks to identify "consumers of electricity with controlled load," that is, consumers who, due to their electricity consumption patterns, affect the quality and reliability of electricity supply within the Unified Power System of the Russian Federation. These consumers can provide ancillary services—for example, services related to the emergency recovery of the Unified Power System—on a remunerated and contractual basis to the System Operator.

In practice, the effectiveness of the Federal Energy Efficiency Law in generating energy savings will depend on the specific measures that the Government and the regional authorities are taking to implement the Law (Ignat'eva, 2011). In any event, the adoption of this Law has stimulated action at the federal and regional levels and has reinforced the sense of urgency that measures need to be taken to tackle the energy intensity of the Russian economy.

6.4 Carbon Policies: Joint Implementation Under the Kyoto Protocol

Russia adhered to the United Nations Framework Convention on Climate Change (UNFCCC) and ratified the Kyoto Protocol. It falls under the category of so-called Annex I Parties to the UNFCCC and is thus bound by limitations on its GHG emissions. Russia committed to stabilize emissions to 1990 levels. Due to the collapse of the Russian economy following the end of the Soviet Union, Russia's stabilization commitment to Soviet emission levels generated a surplus of emission allowances—"hot air."

As an Annex I Party, Russia may host JI projects executed together with other Annex I Parties. Russia has taken the necessary steps to fulfill the requirements for the implementation of JI projects under the "Track 1 procedure," that is, national JI approval in contrast to international approval (Gutbrod et al., 2010), and is the second largest generator of Emission Reduction Units. Firstly, it has ratified the Kyoto Protocol on November 4, 2004, after a long and controversial process. Secondly, it calculated its assigned amount in 2007. During the same year, it also put in place a GHG inventory and a national registry. Russia also submits annual reports on its inventory and on the calculation of its assigned amount to the UNFCCC Secretariat (UNFCCC, 2010). Moreover, Russia has, after considerable delay (Firsova and Taplin, 2008), adopted a national regulation for the implementation of JI projects—regulation that the Government has repeatedly amended.

Sberbank—the entity in charge with the JI qualification procedure—has approved different projects in the electricity and heating sector, including the construction of new combined cycle gas turbine (CCGT) installations, the modernization of hydropower plants and coal/diesel-to-biomass biomass fuel switch for heat production (Sberbank, 2012). However, the relevance of these financing options for the modernization of the Russian energy infrastructure is limited in time. The first commitment period of the Kyoto Protocol expires at the end of 2012 and Russia has expressed its opposition to new commitments during the international climate negotiations in Durban (Table 16.2).

TABLE 16.2 Major Policy Milestones for the Decarbonization of Electricity Supply

Event	Year	Comment
Federal Law on the Ratification of the UNFCCC	1994	Russia becomes UNFCCC Annex I country
Federal Law on the Ratification of the Kyoto Protocol	2004	Russia commits to stabilize emissions to 1990 level
Government Decree on the Procedure for the Approval and Verification of the Realization of JI Projects	2007	First JI-qualification procedure
Federal Law No. 250-FZ Amending the Federal Electricity Law	2007	Legal basis for the renewable energy premium scheme
Government Decree on Measures for the Implementation of Article 6 of the Kyoto Protocol	2009	Second JI procedure, repealing the first
Federal Law No. 261-FZ on Energy Efficiency	2009	Legal basis for energy saving measures
Resolution of the President on the Climate Doctrine	2009	25% GHG reductions in 2020 in relation to 1990
Federal Heat Law No. 190-FZ	2010	Legal basis for priority development of CHP and new tariff methodology
Government Decree on Measures for the Implementation of Article 6 of the Kyoto Protocol	2011	Third JI procedure, repealing the second
Federal Law No. 394-FZ Amending the Federal Electricity Law	2011	Introduction of a capacity-based scheme for renewable energy

7 CONCLUSION

Russia has developed a sophisticated electricity market architecture that, to a large extent, reflects the main components of the liberalization "textbook." However, in the implementation of this model, the Russian Government has been reluctant to transfer price formation to the market without controls over price levels and retention of the possibility to interfere with "unreasonable" price increases. The temptation of governments to interfere with the price formation mechanism and expose investors with the risk of price caps below investment and operating costs is not unique to Russia. As highlighted in this book (e.g., Chapter 11 by Kristov and Keehn, Chapter 17 by Sen and Jamasb, Chapter 18 by Andrews-Speed, and

Chapter 22 by Kim et al.), the political sensitivity of increases of end-user prices has affected electricity market reform in other main emerging and developed countries.

To stabilize the unpredictable investment climate that characterized the transition toward liberalization and privatization, Russia provided long-term regulated capacity guarantees to the investors that implemented the investment program of the former RAO UES. This measure was initially designed as a transitional and thus temporary arrangement. However, given the uncertainty that governs public interference with liberalized prices in Russia, it appears that investors will continue to require similar long-term regulated price guarantees to finance investments in electricity production.

In recent years, energy efficiency has become a priority of the Russian energy strategy. However, in a similar way to the electricity reform, the political sensitivity of short-term price increases hinders the implementation of this policy. Due to regulatory changes and the unpredictability of investment conditions, risk premiums increase the cost of Russia's sustainability agenda. Long-term guarantees of regulated prices for low carbon investments, for example, under capacity contracts, power purchase agreements with network companies or heat tariffs, could provide a solution to regulatory instability in Russia—solution that would be tailored to the specific institutional characteristics of Russian electricity regulation.

ACKNOWLEDGMENTS

The author is grateful to Alexey Zhikharev, Anton Chernyshev, and Russell Pittman for very useful comments on a previous version of this draft. Many thanks to Steffen von Buneau and Neale Tosh for outstanding research assistance.

REFERENCES

Abdurafikov, R., 2009. Russian Electricity Market—Current State and Perspectives. 121 VTT Working Paper.

Ahrend, R., Tompson, W., 2005. Fifteen Years of Economic Reform in Russia: What Has Been Achieved? What Remains to be Done? OECD Economics Department Working Paper No. 430.

Bernotat, W., 2007. Transcript of the Telephone Conference on the Acquisition of Shares in OGK-4 by E.ON, September 17, 2007.

Boute, A., 2011. A comparative analysis of the European and Russian support schemes for renewable energy: return on European experience for Russia. J. World Energy Law Bus. 4, 157–180.

Boute, A., 2012a. Promoting renewable energy through capacity markets: an analysis of the Russian support scheme. Energ. Policy 46, 68–77.

Boute, A., 2012b. Modernizing the Russian district heating sector: financing energy efficiency and renewable energy investments under the new Federal Heat Law. Pace Environ. Law Rev. 29, 746–810.

Bushuev, V. (Ed.), 2006. Energetika Rossii 1920–2020—Tom 1 Plan GOELRO. Energiia, Moscow.

Energy Forecasting Agency (APBE), 2011a. Stsenarnye usloviia razvitia elektroenergetiki na period do 2030 goda. Available from: <http://www.e-apbe.ru/5years/detail.php?ID = 40223/> (accessed 20.06.12.).

Energy Forecasting Agency (APBE), 2011b. Funktsionirovanie i razvitie elektroenergetiki Rossisskoi Federatsii v 2010 godu. Available from: <http://www.e-apbe.ru/analytical/detail. php?ID = 174784/> (accessed 20.06.12.).

Engoian, A., 2006. Industrial and institutional restructuring of the Russian electricity sector: status and issues. Energ. Policy 34, 3233–3244.

Firsova, A., Taplin, R., 2008. A review of Kyoto protocol adoption in Russia: joint implementation in focus. Transit. Stud. Rev. 15, 480–498.

FitchRatings, 2008. The Russian Power Generation Landscape 2008. FitchRatings Corporates. Available from: <www.fitchratings.com>.

FitchRatings, 2010. Prognoz po Rossiiskomu sektoru energetiki na 2010 god.

Gore, O., Viljainen, S., Makkonen, M., Kuleshov, D., 2011. Russian electricity market reform: deregulation or reregulation? Energ Policy 41, 676–685.

Government of the Russian Federation, 2003. Decree on the Transition Wholesale Market Rules (accessed through < www.consultant.ru >).

Gutbrod, M., Sitnikov, S., Pike-Biegunska, E., 2010. Trading in Air: Mitigating Climate Change Through the Carbon Markets. Infotropic Media, Moscow.

Ignat'eva, I., 2011. Pravovoe regulirovanie v oblasti energosberezheniia i povysheniia energeticheskoi effektivnosti: osobennosti i problemy, Energeticheskoe pravo. Available from: <http://www.consultant.ru/> (accessed 20.06.12.).

International Energy Agency (IEA), 1994. Russian Energy Prices, Taxes and Costs 1993. IEA, Paris.

IEA, 2004. Coming in from the Cold—Improving District Heating Policy in Transition Economies. IEA, Paris.

IEA, 2005. Russian Electricity Reform: Emerging Challenges and Opportunities. IEA, Paris.

IEA, 2006. Optimising Russian Natural Gas—Reform and Climate Policy. IEA, Paris.

IEA, 2009. The International CHP/DH Collaborative—CHP/DH Country Profile Russia. IEA, Paris.

IEA, 2010. World Energy Outlook 2010. IEA, Paris.

IEA, 2011. World Energy Outlook 2011. IEA, Paris.

International Finance Corporation (IFC), The World Bank, 2008. Energy Efficiency in Russia: Untapped Reserves. The World Bank Group, Washington, DC.

IFC, 2011. Renewable Energy in Russia: Waking the Green Giant. The World Bank Group, Washington, DC.

IFC, 2012. Financing Renewable Energy Investments in Russia: Legal Challenges and Opportunities. The World Bank Group, Washington, DC.

Joskow, P., 2006. Introduction to electricity sector liberalization: lessons learned from cross-country studies. In: Sioshansi, F., Pfaffenberger, W. (Eds.), Electricity Market Reform—An International Perspective. Elsevier, Amsterdam, pp. 1–33.

Ketting, J., 2008. Liberalization turns into renationalization. Eur. Energy Rev., 69–71.

Kornilov, A., Kulieva, E., 2009. INTER RAO: Aggressive Expansion Play. Alfa Bank, Moscow.

Kotikov, A., Trufanov, A., 2012. Russia Utilities Atlas. Troika Dialog.

Kuleshov, D., Viljainen, S., Annala, S., Gore, O., 2012. Russian electricity sector reform: challenges to retail competition. Util. Policy. Available from: <http://www.sciencedirect.com/science/article/pii/S0957178712000288> (accessed 20.06.12).

Kurronen, S., 2006. Russian Electricity Sector—Reform and Prospects. Bank of Finland Institute for Economies in Transition Online No. 6.

Kuzovkin, A., 2006. Reformirovanie elektroenergetiki i energeticheskaia Bezopasnost'. Institut Mikroekonomiki, Moscow.

McKinsey & Company, 2009. Pathways to an Energy and Carbon Efficient Russia—Opportunities to Increase Energy Efficiency and Reduce Greenhouse Gas Emissions.

Minenergo (Ministerstvo Energetiki RF), 2012. Order No. 387 of August 13, 2012 Adopting the Scheme and Programme for the Development of the Unified Energy System of Russia for the Period 2012–2018.

Medvedev, D., 2009. Vystuplenie v Universitete Khel'sinki i Otvety na Voprosy Auditorii, Helsinki, April 20, 2009.

Opitz, P., 2000. The (pseudo-)liberalization of Russia's power sector: the hidden rationality of transformation. Energ. Policy 28, 147–155.

Pirani, S., 2011. Elusive Potential: Natural Gas Consumption in the CIS and the Quest for Efficiency. Oxford Institute for Energy Studies. Available from: <http://www.oxfordenergy.org/wpcms/wp-content/uploads/2011/08/NG-531.pdf>.

Pittman, R., 2007. Restructuring the Russian electricity sector: re-creating California? Energ. Policy 35, 1872–1883.

PriceWaterhouseCoopers, Power Deals, 2007. Annual Review.

PriceWaterhouseCoopers, Power Deals, 2008. Annual Review.

Russian Government, 2008. Resolution No. 215-r of February 22, 2008 on the General Scheme for the Location of Electricity Installations until 2020.

Russian Government, 2009. Resolution No. 1715-r 13 November 2009 on Russia's Energy Strategy until 2030.

Russian Government, 2010. Decree No. 1172 of December 27, 2010 Approving the Wholesale Market Rules.

Sberbank, 2012. Svedeniia o proektakh i zaiiavkakh na proekty (zaiavkakh), osushchestvliaemykh v sootvetstvii so stat'ei 6 Kiotskogo protokola k Ramochnoi konventsii OON ob izmenenii klimata. Available from: <http://www.sbrf.ru/moscow/ru/legal/cfinans/sozip/>.

Seliverstov, S., 2010. The electricity sector reforms in Russia: European legal concepts and Russian reality. In: Delvaux, B., Hunt, M., Talus, K. (Eds.), EU Energy Law and Policy Issues. Euroconfidential, Rixensart, pp. 37–50.

Standard and Poor's, 2008. Why Russia's Power Sector Restructuring Presents Governance and Minority Shareholder Risks. RatingsDirect.

System Operator, 2010. Potreblenie elektroenergii v Rossii v dekabre 2009 goda na 4,6% prevysilo uroven' 2008 goda, a v tselom za 2009 god snizilos' na 4,6%, January 12, 2010.

Tompson, W., 2004. Restructuring Russia's Electricity Sector—Towards Effective Competition or Faux Liberalization? 403 OECD Economics Department Working Papers.

UNFCCC, 2010. Report of the Individual review of the Annual Submission of the Russian Federation Submitted in 2009 (FCCC/ARR/2009/RUS), January 28, 2010.

von Hirschhausen, C., Opitz, P., 2001. Power Utility Reregulation in East European and CIS Transformation Countries (1990–1999): An Institutional Interpretation. 246 Discussion Papers of DIW Berlin German Institute for Economic Research.

World Bank, 2012. Russian Federation. Available from: <http://data.worldbank.org/country/russian-federation>.

World Economic Forum, 2009. Global Competitiveness Report 2009–2010.

Yi-Chong, X., 2004. Electricity Reform in China. India and Russia—The World Bank Template and the Politics of Power. Edward Elgar Publishing, Cheltenham, UK.

Zeigarnik, Y., 2006. Some problems with the development of combined generation of electricity and heat in Russia. Energy 31, 2387–2394.

Chapter 17

Not Seeing the Wood for the Trees? Electricity Market Reform in India

Anupama Sen[a] and Tooraj Jamasb[b]

[a]*Senior Research Fellow, Oxford Institute for Energy Studies;* [b]*Professor of Energy Economics, Durham University Business School*

1 INTRODUCTION

In July 2012, the simultaneous failure of three of India's five regional transmission grids left roughly 600 million people across 20 states without access to electricity, over a period of approximately 2 days. Although this led to widespread disruption, including, most significantly, the mass failure of public transport systems, what should probably have been considered even more disconcerting is that a substantial proportion of urban and industrial consumers immediately switched over to India's unofficial backup power system—comprising an estimated 40,000 MW[1] of power from fuel oil and diesel-operated generator sets, and 20,000 MW of captive generation, both temporary solutions which are very inefficient and polluting, and in most cases, very expensive. This ability to "make do" has characterized much of the experience with electricity reforms in India over the last two decades, where there has been a tendency for policy to lose its way in the detail.[2] The main policy failure has been in the effective implementation of institutional reforms, which in turn has constrained the technical and economic benefits of power sector reform.

India is one of the BRIC countries[3]—the group of economies that experienced the fastest rates of economic growth in the decade of the 2000s; this is evidenced by the fact that in 2009, when the Organization for Economic Cooperation and Development (OECD) economies collectively entered recession, the Indian economy grew at 9%. While India is distinctive from the other

---

1. Daniel (2012).
2. Baker Institute (2012) discusses how this approach has characterized India's economic growth, arguing that growth has occurred despite the State and not because of it.
3. Refers to Brazil, Russia, India, and China, covered in Chapters 15-18 in this volume.

Evolution of Global Electricity Markets.

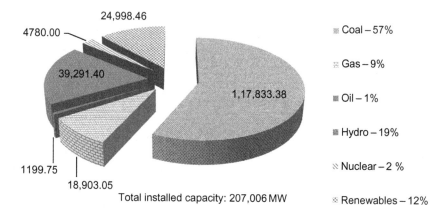

24,998.46

4780.00

39,291.40

1,17,833.38

1199.75

18,903.05

Total installed capacity: 207,006 MW

Coal – 57%

Gas – 9%

Oil – 1%

Hydro – 19%

Nuclear – 2 %

Renewables – 12%

FIGURE 17.1   Breakup of total installed capacity (in megawatts), 2012 (MoP, 2012).

economies in the BRIC group, it combines some of the characteristics of other member nations. With its division of political power between the federal government and the 29 different states that make up the Indian Union, it bears similarities to the federal governance structure of Brazil, while with its burgeoning population of 1.2 billion, it is demographically more similar to China.

On the one hand, solutions to energy problems in India arguably represent "universal" solutions to socioeconomic, political, and governance issues; on the other hand, the challenges faced are equally daunting as they represent the collective problems of the other BRIC economies. India also represents one of the world's largest consumer markets—over the last two decades significant proportions of its population have been lifted out of poverty and into the middle classes, expected to reach 500 million people by the year 2025; yet in 2012, over half of the population did not have access to any form of modern commercial energy (Sen and Jamasb, 2012). This suggests that the evolution of electricity market reforms in India have implications that go far beyond its national boundaries.

Coal forms the main fuel for electricity generation in India—Figure 17.1 shows that in 2012, coal made up the major proportion of total installed capacity of 207,000 MW, followed by hydro, gas, nuclear, and oil. Renewables made up a little over 10% of installed capacity (MoP, 2012).[4]

Electricity market reforms pertaining mainly to conventional energy have been in force for two decades in India, yet, the results have been disappointing—the federal and state sectors, which own and operate over 70% of the electricity sector (Figure 17.2), have failed to make the required investments or attract the private capital to extend the benefits of electricity to the vast sections of the population that still lack it.

---

4. These included small hydro, biomass, and wind. The contribution of these sources in terms of energy output is small, as not all installed capacity has been brought onto the grid.

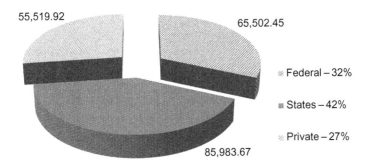

FIGURE 17.2   Share of federal, state, and private sectors in installed capacity (in megawatts), 2012 (MoP, 2012).

Political intervention has, in addition, acted as a constant obstacle to any progress in reforms. However, in recent years, there has been a nascent and an emerging market for trade in renewables on national platforms that cut through the constraints of the federal system of governance as well as circumvent political institutions. This new development presents new opportunities and ways of thinking about electricity market reform (Figure 17.3).

This chapter reviews past experience and explores some of these emerging paradigms in electricity market reform in India. Section 2 sets out the background and evolution of India's electricity market and reviews the three major phases of reforms. Section 3 discusses key legislation that laid the groundwork for a transition to a competitive market in the electricity sector, focusing on the Electricity Act 2003 (EA2003). Section 4 reviews the impact of these measures, and the argument that a midway reorientation of reforms toward capacity addition vis-à-vis improvements in commercial efficiency has led to a dichotomy in electricity market reform (Joseph, 2010). It also shows that despite this dichotomy a small but active market in electricity trading has been established. Section 5 describes this market in further detail—particularly in relation to trading in Renewable Energy Certificates (RECs)—and argues that the future expansion of this market could present a new opportunity to reclaim the process of electricity market reform. Section 6 concludes the chapter.

2 BACKGROUND AND EVOLUTION OF INDIA'S ELECTRICITY MARKET

The evolution of India's electricity market closely mirrors that of its socio-economic policy and institutions over the decades since the country gained its independence. Electricity provision in India originally began in 1899. As in some other countries, it initially consisted of a system of profit-oriented private sector firms which focused exclusively on supplying cities. Supply

FIGURE 17.3 Power outages across Indian states, July 2012. *Source: Authors; map based on Survey of India, 2011.*

was metered, and electricity prices were set at par with those in London; there was also regulatory oversight on these urban systems, carried out by municipal authorities (Tongia, 2003).

   This structure persisted until independence in 1947—after which the new government began a program of building infrastructure and expanding access to basic goods to the vast majority of the population that lived below the poverty line. For the electricity sector, this was done through two avenues: the first was through laying down a constitutional division of power between the federal and state governments.[5] This made electricity provision a "concurrent" subject, which gave each of India's 29 states autonomy in administering their electricity sectors, subject to national legislation and federal directives. The states' jurisdictions included generation, intrastate

---

5. Article 246 of the Indian Constitution.

transmission, distribution, and intrastate trading, whereas the federal government's jurisdiction included generation plants serving more than one state, interstate transmission, and interstate trading of electricity (Singh, 2010).

The second avenue was through the enactment of the Electricity Supply Act, 1948, modeled on the UK's Electricity Supply Act, 1926, establishing large vertically integrated public sector enterprises—called State Electricity Boards (SEBs)—to take on the functions of generation, transmission, and distribution of electricity in every state. SEBs were permitted to set their own tariffs and were required to generate a 3% annual return on net capital. The neglect of governments to regularly review and revise this target was one of the root causes, as this section later argues, for the crisis that later occurred in the electricity sector in India.[6]

SEBs took over the operations of most private electricity companies in India between 1948 and 1956. An Industrial Policy Resolution in 1956 marked a major reversal of ownership in the economy, with large-scale nationalizations —no new private licenses to supply electricity were granted after this until the 1990s, and only four original preindependence era licensees have survived to 2012, whose licenses are subject to 10-year renewals.[7] SEBs functioned as extensions of the government, operating on "soft budgets" including grants or transfers from the federal and state governments—their failure to operate on commercial principles and their capture by political factions seeking the agricultural vote are well documented in the literature as leading to their eventual insolvency in the early 1990s (Ruet, 2005; Tongia, 2003). However, the origins of this notorious electricity–politics nexus can be traced to the mid 1960s and a very high period of agricultural sector growth —also known as the Green Revolution.[8]

SEBs were instrumental in enabling the Green Revolution as they controlled electricity supply—as a result, the agricultural sector enjoyed high growth rates. This success translated into "vote banks" as the majority of the population were, at the time, employed in agriculture. The correlations between politics and the power sector originally developed as a result of this and eventually led to political decisions to provide free or subsidized power to agricultural consumers (GoI, 2011). The first explicit case of the promise

---

6. This section outlines the history and evolution of the structure and organization of the Indian electricity sector; the reader can refer to Figure 17.9 in Section 4.2 for a visual of the structure and organization as of 2012.

7. These are in the cities of Mumbai (Tata Power Company), Ahmedabad (Ahmedabad Electricity Company), Surat (Surat Electricity Supply Company), and Kolkata (Calcutta Electricity Supply Company).

8. The Green Revolution was a global phenomenon between the 1940s and the 1970s that transformed agricultural practices and increased crop production worldwide, using techniques and technologies which originated in Mexico. It was launched in India after severe famine in the 1950s as an intensive agricultural cultivation program base on heavy irrigation, aimed at attaining self-sufficiency in food production.

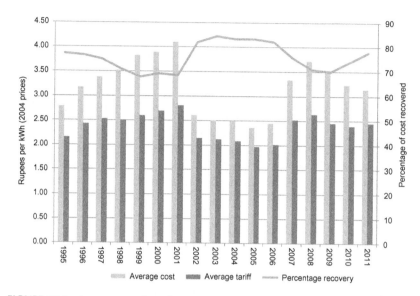

FIGURE 17.4 Average cost of supply, average tariff, and percentage of cost recovered from tariff[9] (GoI, 2002, 2011; PFC, 2005).

of free electricity for farmers in a political campaign has been traced to a state election in Andhra Pradesh in 1977 (Tongia, 2003).

Figure 17.4 illustrates the distortions that resulted from the provision of free electricity using data for 19 states—an inverse relationship developed between the average cost of supply and the average tariff between 1995 and 2011, as tariffs were insufficient to recover the costs of supply.[10]

These distortions had two negative effects:

- First, SEBs inevitably began making losses and were unable to invest in the infrastructure necessary to expand the electricity provision, defeating their original purpose. Evidence shows that these investments failed to occur despite consistent increases, between 1956 and 1985, in the expenditure allocated to the power sector in the Five Year Plans, indicating inefficiencies in the delivery process (GoI, 2011). This failure led to an increase in illicit methods of obtaining electricity, namely, power theft through the tapping of transmission lines.

- Second, the practice of subsidized or free electricity quickly became institutionalized; typically, as agricultural consumption remained unmetered due to the lack of capital for the upgrading of infrastructure, tariffs

---

9. Approximate exchange rate in 2012; 1 US\$ = ₹48; 1 PPP\$ = ₹17 (approximately).

10. Technically, the State Electricity Boards were responsible for unrecovered revenues; however, the burden of losses ultimately went onto the state budgets and formed part of the national budget deficit, under India's system of fiscal federalism.

continued to be charged at flat rates on the basis of the horsepower of agricultural pumpsets. This led to what has been described as "uneconomical practices" adopted by the SEBs although in reality they signaled endemic problems; industrial consumers were charged very high tariffs to cross subsidize the agricultural sector, and agricultural consumption was estimated rather than accurately metered—the true levels of agricultural consumption were therefore unknown. As the SEBs accumulated huge transmission and distribution losses and lacked the ability to identify and quantify them accurately, these were often clubbed with "agricultural consumption" in an attempt to disguise them. It is estimated that in some states, agricultural consumers accounted for as much as 30% of consumption but only 5% of revenue (Tongia, 2003).

One of the effects of this mismanagement of SEBs was the entry into captive generation of large process industries, from the 1980s onward, which could not function on the poor quality of service provided by SEBs (GoI, 2011). Obtaining permission to set up captives was a lengthy bureaucratic process, involving technical, economic, and environmental clearances from several agencies.[11]

In addition to financial difficulties, SEBs were plagued with an ineffective institutional environment that inevitably engendered wastefulness and poor productivity. For instance, in 1991—prior to the first attempt at reforms—the number of employees per million units of electricity sold was 5, compared to just 0.2 in Chile, Norway, and the United States (GoI, 2011).

It must be noted that the state operation of the electricity sector did not result in entirely negative outcomes. Regional Electricity Boards were established in 1964 to facilitate interstate transfers of electricity, and states were given interest-free loans by the federal government to build and expand networks of transmission lines. A Rural Electrification Corporation was set up in 1969 as a response to the famines of the previous decade to extend electrification to rural and semi urban areas. A number of federally owned and operated generation companies were also set up in the 1960s and 1970s to boost electricity supply.[12] As a result of these measures, installed generation capacity grew from 1362 MW in 1957 to 173,626 MW by 2011 (GoI, 2011). The federal government also set up the Power Finance Corporation in 1986 to fund transmission infrastructure and the Power Grid Corporation in 1989 to encourage the development of interstate electricity transfers.

---

11. Captive capacity in 1980 was estimated at 2859 MW or 10% of installed generation capacity (GoI, 2011).
12. The National Hydroelectric Power Corporation (NHPC) played an important role in laying the groundwork for the development of hydroelectric generation capacity, and the National Thermal Power Corporation (NTPC) was set up with a World Bank loan to augment thermal generation capacity. NTPC went on to become one of India's most efficient public sector companies.

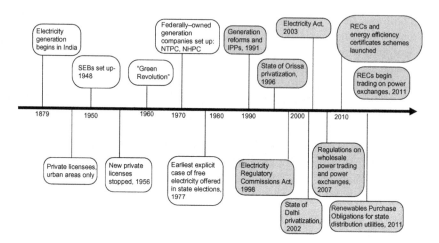

FIGURE 17.5 Timeline of electricity sector evolution and reforms in India. Note: Shaded boxes represent the reform and post-reform period. *Source: Authors.*

By 1991, SEBs were either already bankrupt or on the verge of bankruptcy, as losses had reached unsustainable levels (GoI, 2011; Tongia, 2003). The beginnings of a chronic supply deficit had also begun to take shape. However, the catalyst for electricity reforms came not from the crisis in the power sector but from a balance-of-payments crisis faced by the Indian economy in 1991. Similar to other developing countries, such as in Latin America, electricity reforms were therefore initiated as a result of conditions imposed by multilateral agencies that had agreed to provide assistance to the federal government during the macroeconomic crisis. Electricity reforms were therefore undertaken as part of a Structural Adjustment Program, which involved the liberalization of the Indian economy, in 1991 (Figure 17.5).[13]

---

13. Two wider consequences of the power crisis relate to first, the impact on demand, and second, the impact on economic growth. Although the power deficit at peak times has been estimated at between 9% and 12%, and the overall energy shortage at between 8% and 10%, it has been argued that the true level of demand remains unknown, largely due to the fact that very low priced or free electricity is likely to have created excess demand, and thus an accurate assessment of demand is not possible (Jain and Sen, 2011). A revision in the pricing system and the elimination of power subsidies are the only measures that are likely to reveal true levels of demand in the long term. To some extent, reforms have attempted to address this by rebalancing the tariffs.

The second consequence relates to the economic impact of power outages. The World Bank has estimated that South Asian countries experience an average of 1219 outages per year, with an average duration of 2.4 h per outage (World Bank, 2009). In India, it has been estimated that power outages and the failure to invest in the upgrading and new construction of electricity network infrastructure have led to a 1.2% reduction in GDP growth (Kumar, 2012).

2.1 Third Time Lucky? Three Phases of Electricity Market Reforms in India

The first phase of electricity reform was launched through a change in legislation—the Electricity Laws (Amendment) Act, 1991, to allow private participation in generation. Arguably, the generation sector was selected for reform as it provided the quickest route to bringing in foreign investment, through Independent Power Producers (IPPs) and a single-buyer market model, and this was in line with the wider program of structural adjustment in the economy.

Projects were based on long-term Power Purchase Agreements (PPAs) between IPPs and SEBs. The federal government aggressively pursued overseas investment, offering a 16% after tax return on equity at a Plant Load Factor (PLF) of 68.5% (considerably higher than the 12% return applicable to NTPC projects), a 5-year tax holiday, a debt equity ratio of 4:1, and the full repatriation of profits in US dollars. Provisions were also made for escrow cover, and the state and federal government guarantees to reassure foreign investors despite the dire financial situation of SEBs. The government received 189 proposals for 75,000 MW of capacity, but eventually, financial closure was achieved for just 15,000 MW (GoI, 2011). "Fast track" status was given to eight projects—most of which eventually failed to achieve completion.

Inevitably, the fundamental problem with SEBs' financial insolvency and the poor health of states' own finances remained unresolved, and in some cases, power from IPPs turned out to be more expensive than power generated from federal and state utilities—some of this was partially attributable to the high cost inputs that were used in these projects, such as naphtha, and to the fact that the prices of equipment were bid up to high levels. Littlechild (2000) and Newbery (2000) argue that pricing remains at the heart of electricity market reforms—in countries where electricity was subsidized and prices were not cost-reflective to begin with, reforms are likely to lead to rising prices. Opposition to these price rises creates an impediment *ex ante* in the implementation of reforms.

Eventually, a combination of economic factors and political opposition, including regime change, led to the failure of the IPP reform effort to produce any significant impacts—an infamous example that is often used to illustrate the problems with the IPP policy is the Dabhol power plant, built by Enron, which eventually exited the project and its investment in India over disputes relating to the price of electricity and the inability of the Maharashtra SEB to pay for it (Parikh, 1997; Tongia, 2003).

A second phase of reform took place in the mid-1990s, when some states—Orissa in 1996, Haryana in 1997, and Andhra Pradesh in 1998, decided to undertake the restructuring of their SEBs and set up independent regulation, in an effort to replicate some elements of the generic model of electricity reform being pursued around the world. Essentially, this involved the unbundling of the SEBs into generation, transmission, and distribution companies and the setting up of independent State Electricity Regulatory Commissions (SERCs).

These reforms began in Orissa as part of conditions attached to a World Bank loan for a hydropower project; it was also significant that Orissa had a very small agricultural consumer base—less than 5%.[14] Orissa effectively went the furthest out of the three states on restructuring—it unbundled the SEB into two generation companies, one transmission company and four distribution companies. Controlling stakes were sold to private investors in the generation and distribution companies, which effectively a mounted to privatization. At the time of privatization, the PLF in Orissa was estimated at 36%, transmission and distribution losses at 39.5%, and bill collection efficiency at 17% (Tongia, 2003). Orissa's attempts at privatization were largely unsuccessful as the unbundled companies were not restructured prior to privatization—thus, investors discovered that the baseline data upon which they had based their investment plans were grossly underestimated. Actual transmission and distribution losses turned out to be 49.5%, for instance.

Unlike Orissa, Andhra Pradesh unbundled but did not privatize—rather, the government adopted measures to improve efficiency through more rational tariffs and incentives for the unbundled utilities to improve their performance (Tongia, 2003). It also "corporatized" its unbundled utilities and promoted a commercially oriented form of operation, as opposed to one based on "soft budget" constraints. Andhra Pradesh also passed a stringent antitheft legislation under which it was able to reduce thefts by 3% in the years following reforms (Tongia, 2003); it also managed to dramatically bring down its transmission and distribution losses and in general was long upheld as the model for successful power sector reform in India, before it succumbed yet again to political intervention.

The federal government decided to encourage states to establish independent regulation by passing the Electricity Regulatory Commissions (ERC) Act, 1998, which also established the Central Electricity Regulatory Commission (CERC). While SERCs were responsible for intrastate matters, the CERC regulated the federal utilities as well as matters relating to interstate electricity transfers and the electricity grid. Among the eight states that adopted the 1998 act was Delhi, which again had very few agricultural consumers—Delhi unbundled and privatized its generation and distribution utilities in 2002. In contrast with Orissa, the Delhi privatizations were initially considered a relative success due to the manner in which they were carried out.

---

14. The World Bank viewed the case of Orissa as a window of opportunity to implement reforms in India, believing that such reforms, once implemented, would be difficult to reverse.

An audit was carried out prior to the privatizations, and the state government determined targets for loss reduction over a period of 5 years—similar to a notional reserve price. Bids were then invited from companies to match or better the reserve price, with the proviso that companies which achieved targets would earn a 16% return on equity if they met their bids and share the gains with consumers if they beat them. Targets were to be met over a 5-year period during which a government subsidy was to make up the shortfall. Investors were thus incentivized to operate efficiently in order to earn a return on their investments. This method was effective—aggregate technical and commercial losses were nearly halved from their original levels of 50%.[15] The sustainability of loss reductions and continued improvements in efficiency were questioned later in the process, however, as utilities in Delhi argued that continued investments could only be made at higher tariffs, and the increases in tariffs failed to come through. Littlechild (2000) points out this fundamental problem of market reform particularly in developing countries, where it may lead to increases in prices, if prices were subsidized to begin with, making reform unpopular often before it is even implemented. Indeed, the problem of achieving cost-reflective pricing lies at the heart of electricity reform in developing countries (Newbery, 2000) and is something the reformer states such as Delhi have continued to struggle with.

Several other states adopted the ERC Act but stopped short of privatization. The ERC Act thus set the basis for independent regulation in the sector, by detailing the structure, composition, and responsibilities of independent regulators.

Therefore, the second phase of reform had mixed success. Essentially, all the efforts at reform thus far had concentrated on the Single-Buyer Model—there was no competition in retail supply, as unbundled distribution companies were each assigned supply zones and consumers effectively had no choice of supplier. In the case of privatizations, although these resulted in efficiency improvements, there was also a change of market structure from public monopoly to private monopoly.

A parallel federal reform initiative was introduced during the second phase of reforms, in the year 2000, which consisted of an incentive-based investment program—the Accelerated Power Development Program, involving federal grants made to states that signed up to it, against a prespecified amount of reduction in the losses of state utilities, on an annual basis. The program aimed to incentivize states to invest in efficiency improvements and infrastructure expansion. The federal government also established the Power Trading Corporation (PTC) of India in 1999 to organize and encourage interstate power trading.

---

15. Defined as the sum total of technical losses, commercial losses, and shortage due to the non-realization of the total billed amount of energy.

The third and most recent phase of reform was the introduction of the Electricity Act 2003 (EA2003), a legislation which consolidated and replaced all previous federal laws governing the electricity sector, and was a fairly momentous step forward in India's hitherto unsteady progress in reforming the sector. The provisions of the Act aimed to transform the electricity market from a noncompetitive, Single-Buyer model to a Multiple-Buyer model with several competing participants in the generation, transmission, and distribution segments. The main contents of the Act are outlined in the next section.

3 LAYING THE FOUNDATIONS FOR A COMPETITIVE MARKET

The Electricity Act 2003 (EA2003) laid down provisions to enable the development of a competitive market in electricity. These provisions can be outlined in terms of how they relate to generation, transmission, and distribution.

The most significant provision in generation was that it was made a non-licensed activity; essentially, any entity, public or private, could set up a generation plant subject to environmental clearances, with the exception of hydro electric plants above a certain amount of capital investment.[16] Further, generators were allowed to sell electricity to any distribution licensee, and where permitted by SERCs, even directly to consumers.

The setting up of captive power plants[17] was also de-licensed and made easier—and the definition of captive generation was extended to include not just industrial users, but also "plants set up by a cooperative society or an association of persons primarily for the members of that society or association" (Bhattacharya, 2003). This provision was mainly directed at meeting a 2012 target of an additional 100,000 MW of capacity addition (Bhattacharya, 2003). It was also an attempt to bring the large amount of captive capacity—thought to be a third of total installed capacity—onto the grid. Captive producers were also permitted to feed or sell excess power to the grid. The SERCs were tasked with determining the tariffs for the sale of power. Essentially, this measure allowed for greater competition in generation.

In transmission, which probably contained the most radical changes, the Act marked a departure from the situation in which the State Transmission Utility (STU) performed both the functions of transmission company and bulk supplier. Transmission was made a regulated activity requiring a license, and a stand-alone function, distinct from bulk supply or trading (Bhattacharya, 2003). The system operation function was to be carried out by a separate government company—the Act empowered the federal government to designate a government company as a Central Transmission Utility

---

16. To be specified by the federal government on a case-by-case basis.
17. Captive power plants or captive generation refers to an individual or a group constructing a power plant, obtaining fuel, and generating electricity for the exclusive use of the individual or group in question.

(CTU),[18] and each state to designate one government company as the STU.[19] Additionally, the Act entitled the federal government to establish a National Load Dispatch Centre (NLDC) in addition to the five existing Regional Load Dispatch Centres (RLDCs) to improve the scheduling and dispatch of electricity.

Transmission utilities and Load Dispatch Companies were prohibited from engaging in the business of generation, to avoid conflicts of interest. Nondiscriminatory Open Access or Third Party Access to transmission networks was also made mandatory in the Act, and provisions were made that allowed for private companies to obtain power transmission licenses. In a sense, existing organizations and functions were formalized and new ones introduced, such that the business of transmission was clearly separated from other segments. The EA2003 also recognized power trading as a separate activity—the purchase of electricity for resale. The CERC issued regulations for interstate electricity trading, and the SERCs began the process of doing so shortly after the Act was introduced.

The EA2003 did not make any distinction between distribution and retail supply, making it different from the generic reform model. Thus, distribution companies could engage in distribution, trading, and retail supply through one license (Bhattacharya, 2003). There are likely benefits in the unbundling of distribution from retail supply. However, the benefits of introducing competition in the latter are inconclusive (Joskow, 1998; Littlechild, 2000; Newbery, 2000). Distribution companies were, in addition, permitted to carry out their business through agents—or the "franchisee" model. The Act allowed for multiple licensees or competing companies in distribution and permitted the development of parallel networks—encouraging the duplication of infrastructure, the benefits of which have been questioned in economic literature as it comes with a clear opportunity cost.

Distribution companies were allowed to enter the business of generation and vice versa— signaling implicit support to the reintegration of these functions. Open Access or third party access to intrastate network infrastructure was permitted in distribution, and states were required to mandatorily introduce it in phases varying with consumer size. However, recognizing the fact that this could lead to consumer flight from state utilities, particularly within the industrial consumer segment, and a consequent loss in the revenues used to cross subsidize agriculture, states were permitted to impose a "cross-subsidy surcharge" on consumers who opted for Open Access. Subsidy amounts for agriculture were required to be paid to state utilities in advance, and the Act encouraged the delivery of subsidies to be made in a direct manner to low-income consumer groups, in the long term.

---

18. This was the Power Grid Corporation of India Limited.
19. The CTU and STU provisions existed in previous legislation, but the EA2003 mandated them.

The Act also made the metering of electricity compulsory, to be carried out within timelines approved by the SERCs. It removed the licensing requirements for generation and distribution companies that operated or wished to operate in rural areas, prompting questions on the wisdom of permitting unregulated markets to develop in these areas (Bhattacharya, 2003).

Among other measures, the EA2003 promised "electricity to all" by 2012—a somewhat grandiose commitment which failed to materialize. The Act, however, outlined detailed rules on the constitution and functioning of independent SERCs—which have arguably been one of the most important institutional reforms in the Indian economy—empowering them with most functions, including tariff determination. This marked a separation of the electricity sector from the political establishment, at least on paper.

Provisions were also made for the constitution of consumer appellate tribunals and for the holding of public hearings by the CERC and the SERCs to invite comments from members of the public prior to the approval of important regulations—evidence suggests that these hearings have functioned well in improving the interface between policymakers and citizens.[20] The EA2003 also required the federal government to introduce a National Tariff Policy, a National Electricity Policy, and a National Electricity Plan, within a short time of its introduction. Of these, the National Electricity Policy and National Tariff Policy detailed the objectives and provisions of the EA2003. These are outlined briefly below; the next section discusses the outcomes of these measures in further detail.

3.1 National Electricity Policy

The National Electricity Policy was released in February 2005 shortly after the EA2003, as a statement of long-term objectives in electricity policy. They included the following:

- "Electricity for all" by 2012
- The development of a national, fully integrated grid
- The adoption of indices of service quality to replicate international best practices[21]
- The progressive and gradual reduction of cross subsidies
- The provision of a direct subsidy to households consuming 30 units a month
- Investments in upgrading network infrastructure
- Investments in the completion of rural electrification

---

20. Some SERCs have regularly published details of the public representations made at these hearings, and their responses, on their web sites.
21. These included the number and duration of interruptions in supply.

- The adoption of a two-part tariff structure similar to that used in interstate trading—Availability-Based Tariff (ABT), at the state level[22]
- Development of a spinning reserve of at least 5% by 2012 with an availability of installed capacity of 85%
- The promotion of trading in electricity—with a recommended 15% of new capacities to be traded outside PPAs.

3.2 National Tariff Policy

The National Tariff Policy was announced in January 2006. It detailed the general approach to tariff reform prescribed in the EA2003, serving as a guideline to the SERCs who were responsible for implementing it at the state level.

An important element of this approach relates to the computation of the cross-subsidy surcharge applicable to Open Access for states—it suggests the subsidy be based upon the difference between consumer-wise tariff and the marginal cost of supply that is avoided by the consumer moving off the state utility (Crisinfac, 2006).

A second element is the requirement that all new distribution companies after 2008 procure power from generators through competitive bidding based on tariffs. This exempted expansions of existing projects, and all public sector projects for a period of 5 years or until the SERC saw fit to withdraw the exemption. For private sector projects, expansions in relation to this tariff structure could only be carried out for no more than 50% of existing capacity. This measure therefore capped the amount of private installed capacity that could be built in relation to procurement based on competitive bidding.

The Tariff Policy also set out the requirement for SERCs to implement Multi-Year Tariff Orders rather than Annual Orders, in order to provide long-term signals to potential investors in electricity and in the main user industries. It also proposed a new procurement policy for peak load versus base load generation, suggesting differential tariffs for the two, reflecting their different costs of operation. Similarly, the Policy proposed the use of Time-of-Day tariffs by SERCs, to influence consumption behavior, particularly by industry.

The Electricity Act 2003 (EA2003) and the plethora of legislation and federal directives that followed it established and initiated much of the groundwork for the transition to a more competitive form of market operation in the electricity sector in India. The model included the unbundling of vertically integrated SEBs into several generation and distribution companies and one transmission company. The Act allows for "private participation" in generation and distribution but does not explicitly mention privatization. In a sense, the model is similar to the generic reform models adopted by developed countries (Jamasb, 2006) but lacks the separate function of retail supply.

---

22. This is discussed in Section 4.

Table 17.1 illustrates the status of implementation of the EA2003 in India's 29 states in mid-2011. The implementation of reform at the state level has been less than promising—the impacts of reform are discussed further in the next section.

4 IMPACT OF REFORMS AND THE DEVELOPMENT OF ELECTRICITY MARKETS IN INDIA

Despite the market structure envisaged in the EA2003 and the measures laid out to achieve it, the outcomes have been less than satisfactory; one of the reasons for this is that there have been changes in the main drivers of market reform over the decade following the announcement of the Act, which has altered the objectives of electricity market reform midway.

4.1 *Ex Post* Changes in Direction—Back to Square One?

Two main factors have influenced the direction of electricity market reform after the Act was passed in Indian states, which has arguably been diverted from its original aim of the creation of a competitive electricity market.

First, the fiscal and policy federalism that characterizes India has not always played to the advantage of electricity market reform, and states have been inconsistent in the implementation of competitive reform measures. Table 17.1 illustrates the status of reforms in Indian states in mid-2011. While all 29 states had established SERCs, just 19 had unbundled and/or corporatized. In most cases, the unbundled generation and distribution companies were assigned "supply zones"—this included the two states that privatized distribution, Delhi and Orissa. As of 2012, only the state of Maharashtra had permitted multiple distribution licenses where domestic retail consumers could choose their suppliers in designated distribution areas.[23] Regular reviews of tariffs were being carried out by SERCs in 23 states, and 11 of these were using Multi-Year Tariff Orders.

Although regulations on Open Access or Third Party Access had been announced by 25 states, in practice, only 15 states had actively used these regulations.[24] It has been argued that states that implemented market reform earlier on began to see benefits after an initial period of deterioration in performance indicators (Sen and Jamasb, 2012). However, even among states that have been progressive reformers, there have been reversals—for instance, Andhra Pradesh, a state that successfully cut its transmission and distribution

23. Maharashtra also implemented the "franchisee" model in distribution—it contracted out an entire distribution zone to a private company, which was tasked with managing administrative and some technical roles, including bill collection.
24. These included Andhra Pradesh, Assam, Bihar, Chhattisgarh, Delhi, Gujarat, Haryana, Himachal Pradesh, Jharkhand, Karnataka, Kerala, Madhya Pradesh, Maharashtra, Orissa, Punjab, Rajasthan, Tamil Nadu, Uttar Pradesh, Uttarakhand, and West Bengal.

TABLE 17.1 Status of Reforms in Indian States, 2011 (GoI, 2011)

	Arunachal Pradesh	Andhra Pradesh	Assam	Bihar	Chhattisgarh	Delhi	Gujarat	Goa	Haryana	Himachal Pradesh	Jammu \amp Kashmir	Jharkhand	Karnataka	Kerala	Meghalaya	Manipur	Mizoram	Maharashtra	Madhya Pradesh	Nagaland	Orissa	Punjab	Rajasthan	Sikkim	Tamil Nadu	Tripura	Uttar Pradesh	Uttarakhand	West Bengal	Total
1. SERC																														
Constituted	✓	✓	✓	✓	✓	✓	✓	✓	✓	✓	✓	✓	✓	✓	✓	✓	✓	✓	✓	✓	✓	✓	✓	✓	✓	✓	✓	✓	✓	29
Operationalized	✓	✓	✓	✓	✓	✓	✓	✓	✓	✓	✓	✓	✓	✓	✓	✓	✓	✓	✓	✓	✓	✓	✓	✓	✓	✓	✓	✓	✓	29
Issuing tariff orders	✓	✓	✓	✓	✓	✓	✓	✓	✓	✓	✓	✓	✓	✓			✓	✓			✓	✓	✓	✓	✓	✓	✓	✓	✓	23
2. Unbundling/ corporatization																														
Implemented	✓	✓	✓	✓	✓		✓	✓	✓	✓	✓	✓	✓	✓	✓		✓	✓	✓	✓	✓	✓	✓	✓	✓	✓	✓	✓	✓	21
Privatization of distribution						✓															✓									2
3. Distribution reform																														
Multi-year tariff order issued	✓			✓	✓	✓	✓			✓			✓					✓	✓				✓				✓		✓	12
Open access regulations	✓		✓	✓	✓	✓	✓		✓	✓	✓	✓	✓	✓	✓	✓	✓	✓	✓		✓	✓	✓	✓	✓	✓	✓	✓	✓	25

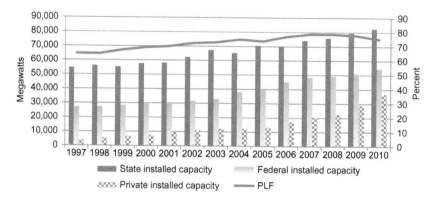

FIGURE 17.6 Capacity addition in the Indian electricity sector, 1997–2010 (GoI, 2002, 2011; CEA, 2006).

losses from a staggering 38% in 2000 to 18% in 2006 and also successfully raised agricultural tariffs, was ranked the best performing state on electricity market reforms in a 2006 government study on the impacts of reform. However, free electricity for farmers was announced prior to a state election shortly afterward, essentially negating the benefits of reform up to that point.

The implementation of reforms in states has therefore been somewhat fragmented and haphazard. More generally, regional politics has often presented obstacles to the process of reform in states, shifting it away from its intended objectives.

The second factor which has influenced the direction of electricity market reform *ex post* is a chronic supply deficit that developed in the decade of the 2000s. The deficit developed primarily as a result of the failure to achieve targets in capacity addition by the state sector. Although historically, the planned capacity additions in India's Five Year Plans have never been achieved, the Eleventh Plan—2002 to 2007—saw the largest underachievement by far; 34,462 MW of capacity was added, which was 56% less than the planned target (GoI, 2011). Additionally, private sector-installed capacity remained very low and barely grew from 2002 to 2005. Figure 17.6 shows capacity addition by the states, federal government, and private sector from 1997 to 2010.

Figure 17.6 shows that private sector-installed capacity began to gradually increase from 2006 onward—possibly reflecting the effects of de-licensing in generation. Other evidence of the lack of investments in capacity lies in the extent of deficits in the power sector.[25] Table 17.2 shows the peak and total deficits in each of India's five regions for the last 9 months of

25. Fuel shortages have also contributed to these deficits, particularly in gas-fired generation, which makes up 10% of installed capacity.

TABLE 17.2 Peak and Total Energy Deficit in India's Regions, 2010 (CEA, 2011)

	Peak Deficit (%)	Total Deficit (%)
Northern region	8.9	8.0
Western region	14.7	13.3
Southern region	6.4	5.2
Eastern region	5.0	4.5
Northeastern region	18.5	8.8
All India	9.8	8.5

2010.[26] The western region has faced the highest deficits over the decade following EA2003; in particular, Maharashtra, one of India's most industrialized states, faced a shortfall of capacity equivalent to 5000 MW (Planning Commission, 2007).

The crisis in electricity supply has arguably led to a change in the relative importance of the objectives of electricity reforms—more emphasis has begun to be placed on increasing installed capacity and generation, rather than on the restructuring of existing utilities. This essentially marks a return to the first phase of electricity reform, where the focus on IPPs grew out of the pressure to attract investment in capacity. Essentially, the issue at the core of the electricity crisis remains unaddressed—the financial positions of state-owned utilities, which continue to supply the bulk of electricity. Figure 17.7 shows the commercial losses of state-owned utilities from 1995 to 2011, along with transmission and distribution losses for the same period.

While transmission and distribution losses have declined since 2003—although they still remain extremely high in comparison with international benchmarks—commercial losses have increased.[27] The issue of electricity theft and its role in losses has been a persistent problem in Indian states, and several attempts have been made to quantify it. Although even efficient power systems are said to experience theft of between 1 and 2% of electricity generated (Golden and Min, 2012; Smith, 2004), in India this has been estimated at a staggering 30% for some states (TOI, 2012). Reforms to address theft have included the separation of feeders for agricultural and industrial consumers to

26. Data from a later year was unavailable.
27. Transmission losses in the OECD average 3% and distribution losses 7%; average transmission and distribution losses in the OECD therefore add up to roughly 10%. Very efficient power systems have transmission and distribution losses of roughly 6% (Smith, 2004).Transmission and distribution losses in the United States were 6% in 2009 (WDI, 2009).

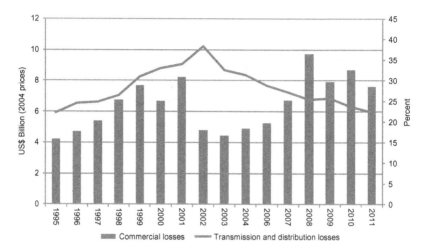

FIGURE 17.7   Combined commercial losses of state utilities, 1995–2011[28] (GoI, 2002, 2011; PFC, 2005).

enable greater accuracy in metering, attempts to extend metering to the agricultural sector, and antitheft legislation, with mixed results.[29]

Although the high level of commercial losses can be partly attributed to the continuance of subsidies and artificially low electricity prices, it is also because of increasing, expensive imports of steam coal and liquefied natural gas (LNG), due to failures to meet capacity addition targets in the coal and gas sectors. It has been argued that the latter have largely been due to the neglect of the Indian government to regularly review and revise the prices of coal and gas (Jain and Sen, 2011).

Coal has been indigenously mined for decades, and the prices kept at very low levels—at roughly 30% of the prices of international benchmarks such as US Appalachian coal and the Northwest European Marker in 2010, to sustain its use as a low cost input for several large industrial sectors, where the State has a dominant presence. In addition to similar institutional rigidities in the operation of the coal sector, Indian coal, in terms of heat content, is also very inefficient. This has led to an increase in imports of Australian and Indonesian coal in the last few years.[30]

---

28. Data for 2005 was unavailable.

29. Antitheft legislation, for instance, proved effective in the southern Indian state of Andhra Pradesh, which was able to dramatically reduce its levels of transmission and distribution losses, as discussed earlier in this chapter. However, reforms to curb electricity theft have failed in other states, such as Uttar Pradesh in the north, where studies have shown that theft is related to deeper socioeconomic and cultural factors, as well as to severe governance problems (Golden and Min, 2012).

30. India imported 11% of its coal requirements (83 million tonnes) in 2011.

Similarly, the prices of natural gas were unrevised for several years, and it was only in 2010 that gas prices were more than doubled from US$ 1.80 to 4.20/MMBtu after a period of 5 years. It has therefore been argued that artificially low prices have, in addition to inflating demand, been insufficient to incentivize the extraction and production of indigenous gas deposits, forcing government utilities to increase imports of spot LNG, with a consequent negative impact on their financial positions as the price increases have not been fully passed through but met with increases in subsidies instead (Sen, 2012). A policy for shale gas was yet to be finalized as of 2012, although pricing is again likely to be an important issue in the fiscal regime for shale exploration.

In reaction to the crisis in capacity with private investment, an "Ultra Mega Power Policy" was launched in 2006, as a stand-alone measure, which envisaged the construction of nine large coal-fired power plants of 4000 MW each. Bids were invited from private companies for the construction of these plants on a build, own, and operate basis; the rationale was that generation tariffs—one of the bid criteria—could be kept low due to the economies of scale that would result from these projects. Companies were also required to secure supplies for the projects; some of these were eventually obtained through the acquisition of captive mines overseas.

As of 2012, bids had been awarded for four plants—Mundra in the state of Gujarat (Tata Power Company), Sasan in Madhya Pradesh (Reliance Power Limited), Krishnapatnam in Andhra Pradesh (Reliance Power Limited), and Tilaiya in Jharkhand (Reliance Power Limited). However, in 2012 these began to run into obstacles with regards to pricing—the contracted tariff for the Mundra power plant was based on coal imports from Indonesian mines; Indonesia however announced an export tax on coal produced from its captive mines, essentially doubling the cost of fuel for the Mundra plant, putting its economic viability into question (Goyal et al. 2011; SC, 2012). The prices of coal imports may well continue to rise over the next decade, and it is unclear how the concurrent increase in capital costs will affect the completion of these Ultra Mega Power Plants (SC, 2012).

As a consequence of the power deficit and focus on additions to installed capacity, there has been a renewed emphasis on a previously little regarded section of electricity users—namely, captive power plants. It has been argued that the encouragement of captive power plants combined with relevant regulations on third party access to transmission and distribution networks could potentially bring a large amount of capacity onto the grid (Hansen, 2008). The EA2003, as discussed, lays out these regulations, and Indian policymakers have actively begun to encourage their adoption, in an attempt to meet capacity addition targets set out in the Five Year Plans (Nag, 2010). Figure 17.8 shows the increases in tracked captive capacity from 1997 to 2011; this comprised approximately 20% of total installed capacity. There is also a large amount of captive capacity that is not tracked, and existing

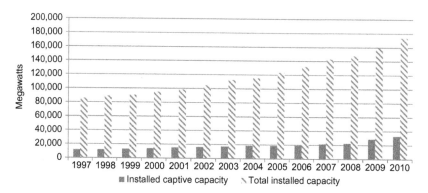

FIGURE 17.8   Growth of captive generation capacity in India, 1997–2010 (TERI, 2009; GoI, 2011).

captive plants have typically operated on low PLFs—these averaged 40% in 2007 (Nag, 2010).

The focus on captive generation has given rise to small-scale models on the utilization of captive power to address power shortages at a localized level. As discussed in Section 3, the EA2003 recognized the use of captive capacity not just by large industrial users but also by residential associations and cooperatives.

A model involving the coordination of surpluses in industrial captive capacity was implemented in the city of Pune in Maharashtra in 2006. At the time, Maharashtra faced a deficit equivalent to 3700 MW, and Pune, a small urban agglomeration, faced a deficit equivalent to 90 MW or between 2 and 4 h of power cuts a day.[31] Industries located in the region, on the other hand, had an underutilized captive capacity in excess of 100 MW (Nag, 2010). Captive power plants, under the umbrella of the Pune Confederation of Indian Industry, entered into an agreement through the SERC with residential and commercial consumers in Pune to operate and generate power for the city for the duration of scheduled power cuts. Consumers bore the incremental cost of generation—equivalent to the difference between the variable costs of generation and the average industrial tariff of electricity (Nag, 2010). The model succeeded in reducing and eventually eliminating power cuts in Pune.

As discussed above, the chronic power deficit has led to an increased focus on capacity additions and power generation, as opposed to pricing reform and the commercial losses of state utilities. The active encouragement of captive generation is a result of this shift in direction. On the one hand, Open Access regulations that encourage the sale of surplus power, in combination with relaxed norms on captive generation, have added to the pool of

---

31. This is referred to as "load shedding" by utilities in India and occurs across most states.

available supply and created a fledgling trade in electricity on power exchanges.

On the other hand, the focus on capacity addition as a priority objective, and away from the original aim of dealing with the finances of state utilities, has arguably created a worrying dichotomy of electricity market reform in India, particularly as the roots of capacity shortages lie in the system of electricity pricing by state utilities, which as Section 2 discussed is related to the broader issue of state-level political intervention in the provision of electricity to consumer segments that form the majority of the electorate. It has in fact been argued that this dichotomy in electricity markets—where a market-based electricity system is allowed to exist alongside a state run system—has an underlying political basis, as it allows the separation of the problem of electricity supply and shortages from the problem of the deterioration in the finances of state utilities and also the electricity–politics nexus to continue (Joseph, 2010). In a sense, this dichotomy takes the process of electricity market reform back to square one.

4.2 Nascent Market for Electricity Trading

Despite the unintended consequences of electricity market reform that have been outlined above, the EA2003 and the National Electricity Policy that followed it have encouraged the development of a relatively small but active market for electricity trading in wholesale markets.[32] The two most significant measures that have helped the opening up of avenues for electricity trading are the setting up of power exchanges and the implementation of Open Access or Third Party Access to transmission and distribution networks.

The foundations of the mechanism for electricity trading were originally established by the CERC in 2000, prior to the implementation of the EA2003, through the introduction of an ABT for interstate transfers of electricity. This tariff was introduced primarily to bring about "grid discipline," as states frequently drew electricity in excess of their stated requirements and generators failed to back down generation during off peak periods. ABT consists of a three-part tariff structure:

ABT = capacity charge + energy charge + unscheduled interchange (UI) charge

Typically, states submit in advance their expected demand schedules based on 96 slots consisting of 15 min each, to the State Load Dispatch Centres (SLDCs) which forward these to the RLDCs—these in turn receive supply schedules from federal generation companies located outside states'

---

32. The arrangements allow utilities to trade with each other, as well as large user industries (firms) to purchase power from captive power plants through trading companies and power exchanges.

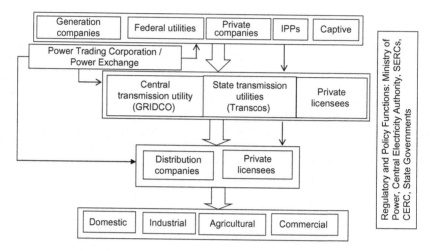

FIGURE 17.9 Organization of the electricity market in India. Single arrows represent contracts; wide arrows represent organizational flow. Based on Singh (2006, p. 2483).

boundaries for the same period, which are matched with the demand schedules and returned to the SLDCs. The capacity charge is a fixed charge based on the entitlements of each state in installed generation capacity. The energy charge is a variable charge that calculated according to states' demand schedules, for each 15 min slot. States are permitted to revise their schedules up to a minimum time limit on the day, and the adjustments are carried out by SLDCs and RLDCs, respectively.

If states draw electricity equal to their demand schedules, they pay the fixed charge plus the variable charge. If states draw excess capacity, they pay the UI charge as a penalty—this is determined based on deviations from the system frequency of the grid at the time of excess demand; the greater the deviation from the stipulated operating frequency, the higher is the UI charge. Similarly, if states draw less than their demand schedules—mainly in cases where generation companies fail to deliver on the supply side—they are credited an amount equal to the UI charge. The CERC regulates the UI charge within a broad range, and it is expected that consumers will respond to changes, adjusting their schedules accordingly.

Shortly before the grid collapse of July 2012, states on the northern regional grid continued to draw in excess of their demand schedules, despite the imposition of high UI charges. The failure of the ABT mechanism to impose grid discipline was therefore identified as one of the main reasons for the blackout. The ABT has nevertheless provided the conceptual basis for power trading, and the UI charge was later used as a benchmark in setting charges in power exchanges. Figure 17.9 shows the structure of the post-2003 electricity market, including the role of the power exchanges.

As discussed in Section 3, the EA2003 made trading a separate and regulated activity—regulations to enable this were passed in 2004, and by 2011, there were 44 trading licensees or companies. The National Electricity Policy, in addition, recommended that 15% of capacity be traded outside PPAs, and in relation to this, regulations enabling the establishment of power exchanges were passed in 2007.

The Indian Electricity Exchange was set up in 2007 and the Power Exchange India Limited in 2008.[33] Both exchanges operate with day-ahead and week-ahead products, based on prices arrived at through double-sided auctions. The share of short-term transactions (bilateral contracts, UI, and power exchanges) in total generation, from August 2008 to September 2009 ranged between 6.55 and 9.09%, with the remainder comprising of long-term PPAs (Shukla and Thampy, 2011).[34] Of these, trading on power exchanges comprised less than 1%.

The second important measure in enabling electricity trading has been the implementation of open access regulations—allowing the direct sale of electricity from nonstate generators to bulk consumers, and the wheeling of electricity from captive power plants to private facilities. As discussed in Section 3, state utilities are allowed to add on a cross-subsidy surcharge to compensate for a potential loss of revenues, as well as an electricity tax, which varies across different states.[35] In terms of transmission capacity, existing long-term contracts are given preference, followed by long-term open access transactions, and short-term open access transactions, in that order (Singh, 2010). Figure 17.10 shows details of applications for open access that were approved by states for the year 2009, along with their tariffs.

The figure compares the tariffs for open access consumers with the tariffs charged by the state utility, and as expected, open access tariffs are higher, on average. Despite this, the demand for open access is very high in few states, as is evident by the volume of applications as a percentage of installed capacity—notably, the resource-rich state of Chhattisgarh and the industrialized states of Gujarat and Maharashtra. The former signals opportunities for captive generation, and the latter, a market for captive power. It is therefore likely that this market will continue to expand.

It is evident that the fledgling electricity market for short-term transactions is still relatively small[36]—yet, its establishment and functioning have brought about significant benefits, in terms of reducing information asymmetry and providing insights into consumer demand and behavior. It also

---

33. Permission had been granted for a third exchange in 2012.
34. These were the latest published numbers at the time of writing.
35. This is a result of the system of fiscal federalism in India.
36. Between 6% and 9%, as discussed earlier.

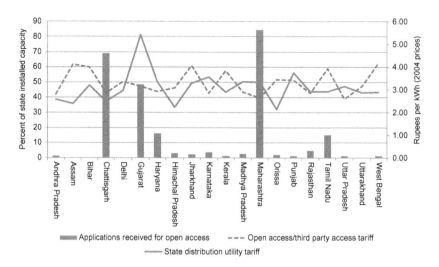

FIGURE 17.10 Approved applications for open access and corresponding tariffs, 2009 (Forum of Regulators, 2009a,b).

arguably presents a significant future opportunity, particularly in trading for unconventional or renewable energy, as the next section discusses.

5 OPPORTUNITIES FOR MARKETS IN RENEWABLE ENERGY—A NEW BEGINNING?

Initiatives to encourage the development of renewable energy in India were first introduced as early as the 1980s. These mainly consisted of incentives for private investments in installed capacity and remained, for a long time, largely ineffective due to the lack of grid connectivity and incentives for generation—they essentially ended up being "balance sheet" investments, similar with early experiences in other developing economies such as China.

The renewables sector came to the fore when concerns over Greenhouse Gas (GHG) emissions began to play a greater role in energy policy. India's GHG emissions have been estimated at 1727 million tonnes (MoF, 2010), although per capita emissions, at 1.4 million tonnes, are much lower than industrialized economies such as the United States. Figure 17.11 shows the composition of GHG emissions by sector, and it is evident that the energy sector contributes the most to total emissions. Within the energy sector, electricity generation makes up 65.4% of CO_2 equivalent emissions (MoF, 2010).

Renewables made up less than 10% of electricity generation in 2011. At GDP growth rates averaging 8% in the decade of the 2000s and an economy that is still driven mainly by the extraction of coal, there has been an urgent

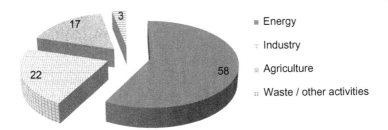

- Energy
- Industry
- Agriculture
- Waste / other activities

FIGURE 17.11   GHG emissions by sector in India (%), 2007 (MoF, 2010).

TABLE 17.3 Installed Capacity for Unconventional Energy, 2010 (MNRE, 2010)

Grid-Interactive Installed Capacity (MW)		Off Grid/Distributed Power (MW/MW Equivalent)	
Wind	12,010	Biomass	238
Small hydro	2767	Biomass gasifier	125
Solar	12	Waste-to-energy	53
Biomass	901	Solar photovoltaic plants	3
Bagasse	1412	Aero generators/hybrid systems	1
Waste-to-energy	72		
Total	17,174		420

need to address concerns over climate change in energy policy. Table 17.3 shows the installed capacity for unconventional energy in India in 2010. The potential installed capacity has been estimated at close to 100,000 MW (MNRE, 2010).

Given these developments, in 2008 the Indian government announced a National Action Plan on Climate Change (NAPCC) in an unprecedented attempt to address environmental concerns. The NAPCC consisted of eight "national missions"—setting out objectives with measureable targets to be carried out over the following decade, each addressing a distinct area of environmental and energy policy. These missions included a National Solar Mission, National Mission for a "Green India," and a National Mission on Energy Efficiency.

These, along with the conceptual basis for trading in renewable energy that had already been set out in the EA2003 and the National Electricity Policy, led to the emergence of markets for renewable and energy efficiency certificates. A distinctive characteristic of these markets is that unlike most developed economies, which target the supply side—requiring generators to engage in trading certificates—in India these programs have targeted electricity distribution companies and consumers and focus more on creating a demand for renewables.[37]

Renewables are not generally considered to be central to government strategy on electricity market reform, relative to other fuels such as natural gas—particularly after recently discovered indigenous resources that have made the latter more cost-competitive with coal. Table 17.4 shows some comparative tariffs for power generated for renewables for a sample of Indian states.[38]

The initiatives made to encourage the demand for renewables, however, appear somewhat promising in relation to progress on reforms.

5.1 Renewable Energy Certificates

The market for renewable electricity originated in a Renewable Purchase Obligation (RPO) established shortly after the National Electricity Policy, which required SERCs to specify a minimum annual percentage of power generated through renewables that would have to be procured by each distribution utility in the state—three states extended the RPO to include captive power plants and open access consumers as well. The intention was to then increase this minimum each year—although not all states have complied with these increases—in line with the long-term targets set out in the NAPCC, which aimed at raising the share of renewables in electricity generation to 15% by 2020 (MNRE, 2009).

The RPO stipulated by different SERCs in 2012 ranged between 2% and 10%.[39] RPOs could be met in one of two ways by distribution utilities; first, the utility could enter into a PPA with generators within the state to purchase some of or the entire stipulated amount. However, interstate variations in the availability of renewable electricity posed an obvious constraint to this method—therefore, a second option was to instead purchase RECs to fulfill the RPO. Utilities that failed to meet their RPO targets were charged a penalty set by the SERC, the revenues from which went into a special fund used to finance renewable energy projects and infrastructure (MNRE, 2009).

For distribution utilities, RECs therefore embodied the environmental benefit that came from meeting RPOs when it was impossible to procure

---

37. Denmark is one of the few developed economies that targets trading in certificates at consumers rather than producers.

38. Figure 17.4 earlier in this chapter contained the average tariffs for thermal (coal fired) power, which are relatively lower. Relatively expensive renewables, particularly solar energy, are largely financed through subsidies on the government budget.

39. Although the 10% targets were rarely met—Madhya Pradesh, for example, achieved just 1% of its stipulated 10% RPO target.

TABLE 17.4 Comparative Tariffs for Power from Renewables, 2009 (MNRE, 2009)[40]

States	Tariff (₹/kWh)[a]					
	Wind	Small Hydro	Biomass	Bagasse	Solar Photovoltaic	Solar Thermal
Andhra Pradesh	2.47	1.91	2.11	2.02	5.14	5.14
Gujarat	2.47		2.27	2.20		9.83
Karnataka	2.49	2.05	2.09	2.05	11.30	
Kerala	2.30	1.79		2.05	11.14	
Madhya Pradesh	2.91		2.49			9.54
Maharashtra	2.57	2.20	2.23	2.24	11.01	
Rajasthan	2.68		3.29		11.52	
Tamil Nadu	2.13		2.31	2.31	2.31	2.31
West Bengal	2.93	2.93	2.93	1.91	8.07	8.07
Haryana	2.99	2.69	2.93	2.74	11.71	

[a]2004 prices

physical supply from within the state (MNRE, 2009). RECs began trading on the two power exchanges in early 2011; Table 17.5 shows a summary of trading activity or auctions over the first 6 months of 2012, on the Indian Electricity Exchange Limited (IEXL).

RECs are matched with physical volumes of electricity that are fed into state grids by generators of renewable electricity who choose to enter the market, in different states, with one REC set equivalent to 1 MWh. The power exchanges thus provide a digital marketplace uniting sellers of renewable electricity with buyers of RECs. Two types of RECs are traded on the power exchanges—solar and nonsolar. Floor and ceiling prices are set differently for each type—in 2012, floor prices were ₹1500 and ₹12,000/MWh for nonsolar and solar RECs, respectively, and ceiling prices were ₹3900 and ₹7,000/MWh, again for nonsolar and solar RECs, respectively. RECs have a maximum validity of 1 year, in keeping with the annual nature of the RPO.

40. 1US$ = ₹48.64.

TABLE 17.5 Trade in RECs on IEXL, January–July 2012 (IEXL, 2012)

Summary		No. of Buy Bids (REC)	No. of Sell Bids (REC)	Cleared Volume (REC)	Cleared Price (Rupees/REC)
Average	Solar	4138	369	144	12,850
	Nonsolar	298,238	249,105	162,033	2574
Maximum	Solar	9489	541	336	13,000
	Nonsolar	414,387	435,348	223,164	3066
Minimum	Solar	289	149	5	12,750
	Nonsolar	149,628	105,844	62,277	2000
Total	Solar	28,968	1109	434	–
	Nonsolar	2,087,666	1,743,738	1,134,231	–

Table 17.5 shows that the response to the introduction of RECs has been good, in general. There have occasionally been concerns over the sustainability of demand for nonsolar RECS as there have been dips in demand, although demand for solar RECs remains strong. As this is a very new and emerging area in the Indian electricity market, further research is required into the understanding of the drivers of demand for the two products.

5.2 Energy Efficiency

The market for RECs is still very new; as discussed in Section 4.2, it represents less than 1% of the wider electricity market. However, it also represents an extremely significant move away from the status quo in electricity market reform over the past two decades. At the time of writing, there was discussion over the fungibility of RECs with parallel markets in energy efficiency certificates—or "ECerts"—which have emerged as a result of the National Mission on Energy Efficiency. The ECert scheme is part of a new demand-based initiative called "Perform Achieve Trade," which sets periodic efficiency targets for firms in the nine largest energy-intensive industries in the Indian economy. ECerts are released into the market by the Bureau of Energy Efficiency—they are purchased by firms that fail to achieve their efficiency targets and traded by firms that overachieve, thus incentivizing firms to use energy more efficiently, with the aim of achieving substantial economy-wide energy savings.

The renewables market despite its small volumes has facilitated in the space of a few years, what reforms in conventional electricity have been trying to achieve for over two decades, a competitive market for the sale and purchase of electricity, with prices set not on the basis of political agendas but on demand signals. The operation of the power exchanges and the trading of RECs occur on national platforms requiring participants to simply register as members, cutting through many of the constraints posed by India's federal system of governance and antiquated transmission infrastructure. In a sense, this emerging market for renewables could present an opportunity for a fresh start in seeking innovative market-based solutions.

6 CONCLUSION

The experience of electricity market reforms in India has mirrored that of the other BRIC member nations discussed in this book, particularly in the common problems they face with regard to the electricity—politics nexus, distorted pricing systems, and institutional rigidities. Electricity market reform in India also mirrors a wider transition in the Indian economy—the struggle to move from a centrally planned and administered form of operation to one that is based on the functioning of market principles.[41] This is, ironically, opposite to the direction of the current debate in developed-economy electricity markets, where market solutions are arguably being replaced with greater state intervention—see, for instance, Chapter 2.

Given that the reform model being adopted in India is based upon the same one that was adopted by the United Kingdom in the late 1990s, this poses a very interesting question: what are the implications of the fact that approaches to electricity market reform in developed versus developing economies appear to be going in opposite directions, despite their eventual shared goal of increasing the emphasis on renewable energy solutions? This requires serious thought going forward, particularly by economies such as India, where the amount of effort invested is often disproportionately larger than the size of the outcomes that it produces. Despite this question, given the evidence, there is little doubt that the mismanagement of the electricity sector in India by successive governments led to the failures of SEBs and the supply crisis, and that markets present a more transparent delivery mechanism.

Electricity reforms in India have been a difficult process, often resulting in counterintuitive outcomes; nevertheless, the July 2012 blackout underscored the need to persist with the implementation of reform. For the BRIC economies in particular, reform is also likely to be closely tied in with economic development—and experience in India (for instance, in Delhi and Orissa) shows that as parts of a developing country gradually urbanize, the implementation of reforms becomes marginally easier. However, the

---

41. Nepal and Jamasb (2012) showed that power sector reforms were more effective when part of a wider institutional reform in the economy.

experience of the last few years and the emergence of environmental concerns point to a marked change in the way reform is beginning to be viewed—going forward, policymakers in India (and other developing economies) should expect reform to be a dynamic and constantly evolving process rather than a one-off process involving the implementation of a specific set of measures. This is true even in developed countries such as the United Kingdom, which is currently reinventing electricity market reform to incorporate its targets on renewable energy. For policymakers in India and the developing world, the difficulties encountered in the process of electricity market reform should therefore not be construed as failure or as a need to return to the *status quo* that existed prior to its implementation; but rather, treated as a situation where *reforms* may need to be reformed. This chapter has argued that, at least for the case of India, the incorporation of targets on renewable energy reflects an opportunity for electricity market reform to move beyond the prescriptive solutions of the past.

REFERENCES

Baker Institute, 2012. Limits of the Jugaad Growth Model: No Workaround to Good Governance in India, Baker Institute Policy Report Number 52. James A. Baker III Institute for Public Policy, Rice University, Houston, TX. Available from: <http://www.bakerinstitute.org/policyreport52/>.

Bhattacharya, S., 2003. The Electricity Act 2003: will it transform the Indian power sector? Util. Policy 13 (3), 260–272.

CEA, 2006. All India Electricity Statistics—General Review 2006. Central Electricity Authority, Ministry of Power, Government of India, New Delhi.

CEA, 2011. Load Generation Balance Report 2011–12. Central Electricity Authority, Ministry of Power, Government of India, New Delhi.

Crisinfac, 2006. Impact Analysis: National Tariff Policy. Report—CRISIL Infrastructure Advisory Limited, India.

Daniel, F., 2012. Generators Whirring, India's Factories Shrug off Blackouts, Reuters, August 8, 2012.

Forum of Regulators, 2009a. State-wise Open Access. Available from: <http://www.forumofregulators.gov.in/state_wise_OA.aspx/>.

Forum of Regulators, 2009b. Comparative Open Access Charges at State Level. Available from: <http://www.forumofregulators.gov.in/Data/OpenAccess/Comparative_Open_Access_Charges_at_State_Level.pdf>.

GoI, 2002. Annual Report 2002 on the Working of State Electricity Boards and Electricity Departments, Planning Commission, Government of India.

GoI, 2011. Annual Report 2011–2012 on the Working of State Power Utilities and Electricity Departments, Planning Commission, Government of India.

Golden, M. and Min, B., 2012. Theft and Loss of Electricity in an Indian State, Working Paper, International Growth Center. Available from: <http://www.theigc.org/publications/working-paper/theft-and-loss-electricity-indian-state>.

Goyal, M., Dujari, H., Misra, S., 2011. Indian UMPP dream turned sour: a case study based discussion. Energ. Policy 46, 427–433.

Hansen, C.J., 2008. Bottom-Up Electricity Reform Using Industrial Captive Generation: A Case Study of Gujarat, India; Oxford Institute for Energy Studies Working Paper EL07. Oxford, UK.

IEXL, 2012. REC Data at Indian Energy Exchange; Indian Energy Exchange Limited. Available from: <http://www.iexindia.com/>.

Jain, A.K., Sen, A. 2011. Natural Gas in India: An Analysis of Policy. Oxford Institute for Energy Studies Working Paper No. 50; Oxford, UK.

Jamasb, T., 2006. Between the state and market: Electricity sector reform in developing countries. Util. Policy 14, 14−30.

Joseph, K.L., 2010. The politics of power: Electricity reform in India. Energ. Policy 38 (1), 503−511.

Joskow, P.L., 1998. Electricity sectors in transition. Energy J. 19 (2), 25−52.

Kumar, D., 2012. Blackouts challenge Indian economy's clout; Eurasia Review News and Analysis; 16 August 2012; Available from: <.http://www.eurasiareview.com/16082012-blackouts-challenge-indian-economys-clout-analysis/>.

Littlechild, S., 2000. Privatization, Competition and Regulation in the British Electricity Industry, with Implications for Developing Countries. ESMAP, the World Bank, Washington, DC.

MNRE, 2009. Report on Development of Conceptual Framework for Renewable Energy Certificate Mechanism for India. Ministry of New and Renewable Energy, Government of India. Available from: <http://mnre.gov.in/file-manager/UserFiles/MNRE_REC_Report.pdf>.

MNRE, 2010. Analysis of current trends in renewable energy sector—impacts on employment opportunities. Human Resource Development Strategies for Indian Renewable Energy Sector. Confederation of Indian Industry and Ministry of New and Renewable Energy, Government of India, New Delhi.

MoF, 2010. India: Greenhouse Gas Emissions 2007, Ministry of Environment and Forests, Government of India. Available from: <http://www.moef.nic.in/downloads/public-information/Report_INCCA.pdf>.

MoP, 2012. Power Sector at a Glance—All India: As on July 31, 2012. Available from: <http://www.powermin.nic.in/index.htm>.

Nag, T., 2010. Captive generation in India: The dilemma of dualism. In: Basu, S., Sarkar, R., Pandey, A. (Eds.), India Infrastructure Report 2010: Infrastructure Development in a Sustainable Low Carbon Economy. Oxford University Press (OUP) India, New Delhi.

Nepal, R., Jamasb, T., 2012. Reforming the power sector in transition: Do institutions matter? Energy Econ. 34 (5), 1675−1682.

Newbery, D., 2000. Privatization, Restructuring and Regulation of Network Utilities. MIT Press, Cambridge, MA.

Parikh, K., 1997. The Enron story and its lessons. J. Int. Trade Econ. Dev. 6 (2), 209−230.

PFC, 2005. Report on Performance of State Power Utilities for the Years 2003−04 to 2005−06; Power Finance Corporation, Government of India.

Planning Commission, 2007. Maharashtra Annual Plan 2007−08. Planning Commission State Plans Division, Government of India, New Delhi.

Ruet, J., 2005. Privatising Power Cuts: Ownership and Reform of State Electricity Boards in India. Academic Foundation, New Delhi.

SC, 2012. Locked in: The Financial Risks of New Coal-fired Power Plants in Today's Volatile International Coal Market. The Sierra Club, Washington, DC.

Sen, A., 2012. Gas pricing in India. In: Stern, J. (Ed.), The Pricing of Internationally Traded Gas. Oxford University Press, Oxford, UK.

Sen, A., Jamasb, T., 2012. Diversity in unity: An empirical analysis of electricity sector reform in Indian States. Energy J. 33 (1), 83–130.

Singh, A., 2006. Power sector reform in India: current issues and prospects. Energ. Policy 34 (16), 2480–2490.

Singh, A., 2010. Towards a competitive market for electricity and consumer choice in the Indian power sector. Energ. Policy 38 (8), 4196–4208.

Smith, T.B., 2004. Electricity theft: A comparative analysis. Energ. Policy 32 (18), 2067–2076.

Teri, 2009. Tata Energy Data Directory and Yearbook. Teri Press, New Delhi.

TOI, 2012. Power Grid Failure: 30% Power Lost to Theft, Politics. *Times of India*, August 1, 2012. Available from: <http://articles.timesofindia.indiatimes.com/2012-08-01/india/32979728_1_power-sector-transmission-technical-loss/>.

Tongia, R., 2003. The Political Economy of Indian Power Sector Reforms. Working Paper No. 4, Program on Energy and Sustainable Development, Stanford Institute for International Studies, University of Stanford.

WDI, 2009. Electric power transmission and distribution losses. World Development Indicators. *The World Bank*. Available from: <http://data.worldbank.org/indicator/EG.ELC.LOSS.ZS/>.

World Bank, 2009. Costs of Power Outages. Energy Strategy Approach Paper—Sustainable Development Network.

Chapter 18

Reform Postponed: The Evolution of China's Electricity Markets

Philip Andrews-Speed

Principal Fellow, Energy Studies Institute, National University of Singapore, Singapore 119620, Singapore

1 INTRODUCTION

China now has the world's largest electrical power sector in terms of total capacity (more than 1000 GW) and annual output (nearly 5000 TWh). This arises from the size of the economy and of the population (1.4 billion) and from the rapid electrification, which has accompanied modernization (Figure 18.1). Coal has consistently provided 70–80% of the feedstock for power generation (Tables 18.1 and 18.2), and most of the coal production takes place in the north of the country, to the north and west of the capital, Beijing (Figure 18.2). This lies far from the main centers of economic growth, in the east and south. On account of the high proportion of coal in the fuel mix (Tables 18.1 and 18.2), the way in which this sector is governed is of global importance on account of China's influence on international coal markets and on carbon emissions. The sector also provides insights into how plans for electrical power liberalization can be derailed by changing domestic circumstances and by actors with vested interests in the *status quo*.

The governance of China's power sector has been undergoing incremental change since the 1980s, but the most substantial reforms took place between 1997 and 2003 when the industry was corporatized, commercialized, and restructured. The original aims of these reforms were to enhance the economic and technical efficiency of the electricity industry and to reduce the need for subsidies by introducing competition.

In 1998, the government corporatized and commercialized the sector, but left it dominated by a wholly state-owned, vertically integrated power company. Four years later this company was broken up into a number of generating and transmission entities. But these reforms were not followed by the introduction of competitive markets for generation. Instead, reform was halted

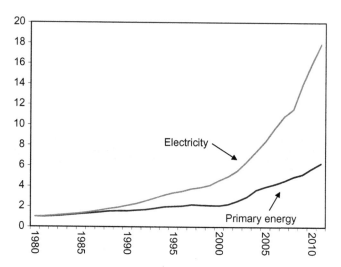

FIGURE 18.1 Primary energy consumption and electricity consumption in China, 1980–2011, normalized to 1980 levels. *Source: Energy Information Administration (2012), BP (2012).*

TABLE 18.1 Fuel Mix for Electricity Generation in China, 1990–2011 (in Percentage)

	1990	2000	2002	2004	2006	2008	2009	2010	2011
Coal	72.5	77.9	77.2	77.7	80.2	78.9	78.7	76.4	80.2
Oil	7.5	3.3	3.0	3.2	1.8	0.7	0.5	0.4	–
Gas	0.5	1.4	1.0	0.8	0.9	1.2	1.7	1.7	2.3
Nuclear	0.0	1.2	1.5	2.2	1.9	1.9	1.9	1.8	1.9
Hydro	19.5	16.0	17.2	15.8	15.0	16.7	16.5	15.6	14.4
Other renewables	0.0	0.1	0.1	0.2	0.2	0.4	0.8	1.1	1.6

Source: International Energy Agency, *World Energy Outlook* (various years), China Electricity Council (2011a, 2012a).

and all subsequent initiatives have taken the form of either *ad hoc* adjustments introduced to address short-term policy challenges or experiments in competition which have then been abandoned. Meanwhile, political debate on whether and how to continue reforms continues, mainly behind closed doors. In many respects, the stalling of power sector reforms in China resembles what has happened in South Korea (see Chapter 22).

In documenting the evolution of the electricity market in China, the principal aim of this chapter is to show how and why the urge to liberalize the electrical power sector rose up the agenda of China's central government in

TABLE 18.2 Fuel Mix of Electricity Generation Capacity in China, 2000–2011 (in Percentage)

	2002	2004	2005	2006	2007	2008	2009	2010	2011
Coal	68.6	69.5	71.2	72.2	71.1	72.2	68.5	67.6	66.9
Oil	4.7	3.6	2.3	2.6	2.8	2.6	1.7	1.4	–
Gas	2.2	1.8	1.9	2.3	3.4	3.7	3.1	2.8	3.3
Nuclear	1.1	1.4	1.4	1.1	1.1	1.2	1.0	1.1	1.2
Hydro	22.8	23.8	22.6	21.2	20.5	18.6	22.5	23.8	21.8
Other renewables	0.6	0.2	0.6	0.8	1.0	1.7	3.1	3.3	4.3

Source: International Energy Agency, *World Energy Outlook* (various years), China Electricity Council (2011a, 2012a).

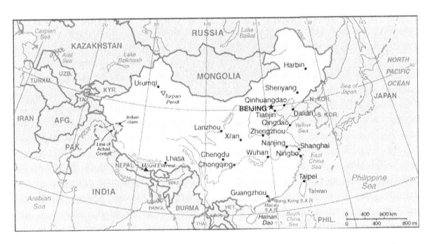

FIGURE 18.2   Map of China. *Source: U.S. Energy Information Administration, China Country Analysis.*

the 1990s and then slipped down the agenda after 2003. It also examines how the absence of a competitive power market has allowed the government to address a number of urgent policy priorities after 2004 but has probably made implementing further reforms progressively more difficult.

The chapter is arranged chronologically. Section 2 describes the reforms carried out between 1997 and 2002, whereas Section 3 is devoted to reforms between 2002 and 2005. Section 4 examines the wide range of measures introduced by the government since 2005 to address its new priorities. Section 5 provides conclusions and the outlook for future reform.

2 REFORMS BETWEEN 1997 AND 2002

2.1 Pressures for Reform

The reform of China's power sector in the 1990s was directly affected by the evolving understanding of the liberalization process around the world, especially in the international financial organizations, such as the World Bank. However, the desire to reorganize the power sector was part of a much deeper plan to reform the entire economy and to restructure all the state-owned enterprises, which in earlier decades had dominated the national economy.

The key elements of industrial reform in China included the diversification of enterprise ownership, increasing autonomy and commercialization of enterprise management, and the gradual alignment of prices with market forces. The government progressively removed itself from both the operational management of the industries and from the financing of their investments. These and other reforms were implemented incrementally, often with local experiments. Though the reform process started in the early 1980s, the most radical steps were taken during the 1990s and were accompanied by reforms to the banking sector, the launch of domestic stock markets, and the establishment of new accounting rules, as well as by the growing foreign involvement in China's economy both through direct investment and through local and international stock markets (Chiu and Lewis, 2006).

The structural reforms were particularly pronounced in 1998. That year saw the abolition of a number of industrial ministries, the creation of new state companies, and the restructuring and commercialization of existing state-owned enterprises. The structure of the energy sector was completely transformed by these changes (Andrews-Speed, 2004).

During the 1990s, the primary objectives of China's government in reforming the power sector were to increase the quantity and quality of power supply in order to support economic growth, to raise the technical and commercial performance and thus to constrain the costs of the industry, and to pass the benefits of these cost reductions to the consumer (Shao et al., 1997). As was the case with other industrial sectors, these reforms were directed at industry structure and at pricing (Andrews-Speed, 2004; Xu, 2002). The main ideas were outlined in the Electric Power Law which came into effect in 1995. The late 1990s was an auspicious time to embark on power sector reform because the rate of growth of demand for electricity in China had declined as a consequence of the Asian financial crisis (Figure 18.1), and a surplus of generating capacity was available.

2.2 Industry and Market Structure Until 1997

Before 1997, much of the electrical power industry lay within the Ministry of Electrical Power, which acted as policy maker, regulator, and operational manager. Under the Ministry, the provincial power bureaus held monopoly power over transmission, distribution, and supply within their respective domains. However, the government allowed significant diversification of the participants in and sources of funding for power generation in order to boost investment in new capacity. Provincial and lower levels of government and state-owned enterprises were permitted to invest in new capacity, subject to the approval of the State Planning Commission for projects over 50 MW. Finance was made available from state-owned banks, the capital markets, and local financial sources (Zhang and Heller, 2007).

In the early 1990s, foreign participation was seen as vital to ensure that the level of investment in generation reached a sufficiently high level. Until this time most foreign funds flowing to the power sector had come from international financial organizations, such as the World Bank and the Asian Development Bank. Between 1994 and 1997, the government issued a number of regulations which were intended to encourage foreign direct investment by private sector (Andrews-Speed, 2004).

Electricity tariffs had already been undergoing reform for several years. Since 1986, the tariff paid to power generators had been based on a "new price for new power" policy, which provided significantly higher tariffs for new plants in order to provide these plants with the revenue to pay off their debts. These new and higher prices applied to plants constructed between 1986 and 1992 that did not use central government funds, and to all plants built after 1992. This scheme was successful in encouraging investment but provided no incentive for investors to reduce their costs or to seek more favorable financing terms (Ma, 2011).

Despite these progressive changes to wholesale tariffs, the system for setting consumer prices changed little during the 1990s and continued to be based on an approach started in the 1960s called the "Catalogue." The Catalogue system is a method of setting consumer tariffs according to different categories of end user and allows the government giving preferential treatment to heavy industry, chemical plants, agriculture and irrigation both in terms of allocation of power, and the price of power. It evolved to comprise eight main categories of consumer with three voltage classifications, making 24 basic categories (Table 18.3). The Catalogue formed the basis of end-user tariffs throughout China. Each of the categories was assigned a Catalogue Price, which formed the starting point for calculation of the final price. To this price were added a range of charges and fees to reach the final end-user price (Table 18.4). Each province established its own Catalogue of end-user prices, subject to the approval of the Pricing Department of the State Planning Commission.

The success of these measures can be seen in a number of improvements from the late 1980s to the late 1990s. First, the generating capacity grew significantly from 100 GW in 1987 to 200 GW in 1994, and 300 GW in 1999 (Figure 18.3). Second, the proportion of central government investment in the power sector declined as the role of local governments and enterprises

TABLE 18.3 The End-User Electricity Tariff Catalogue for Beijing in 1997, in yuan/kWh

	<1 kV	1–10 kV	35–110 kV	>110 kV
Residential	0.310	0.300	0.300	–
Nonresidential lighting	0.514	0.502	0.501	–
General industrial	0.471	0.463	0.452	–
Heavy industrial	–	0.349	0.337	0.327
Chemical industry	–	0.339	0.327	0.317
Commercial	0.514	0.502	0.501	–
Agriculture	0.396	0.387	0.374	–
Irrigation	0.236	0.234	0.231	–

Note: In 1997, the exchange rate was approximately US$1 = yuan 8.3.

TABLE 18.4 Example Calculations of Final Tariffs Paid by Selected Categories of End User in Beijing in 1997 in yuan/kWh, Illustrating the Importance of the Additional Fees

	Residential "existing user"	Industrial "existing user"	Commercial "new user"
Catalogue price	0.300	0.463	0.502
Local government surcharge	0.024	0.018	0.040
Construction fee: central government	0.020	0.020	0.020
Construction fee: local government	–	0.020	0.020
Construction fee: Three Gorges Dam	0.007	0.007	0.007
"New power" fee for new users	–	–	0.120
Connection fee	–	–	0.300
Final tariff	0.351	0.528	1.009

Note: All examples are in the 1–10 kV band.
Source: Andrews-Speed et al. (1999).

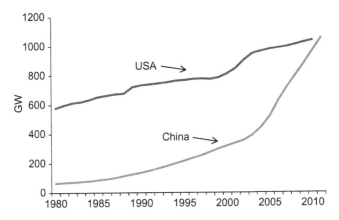

FIGURE 18.3   Total electricity generating capacity in China, 1980–2011, and in the United States, 1980–2010, in Gigawatts. *Source: Energy Information Administration (2012), China Electricity Council (2012a).*

grew, and progressively more of the central government funds came from banks rather than directly from the government itself (Xu, 2002). Finally, great progress was made in providing access to electricity to rural communities. In 1998, only an estimated 9 million rural households lacked access to electricity (Pan et al., 2006).

2.3 First Steps in Reform

In 1997, the State Power Corporation of China (SPCC) was created to take over the enterprise management functions from the Ministry of Electrical Power, and the provincial and lower level power bureaus were renamed as companies within the SPCC. The wholly state-owned SPCC retained most of the transmission and distribution infrastructure and about 50% of the nation's generation capacity. The rest of the assets was owned by a wide variety of state-owned enterprises, linked to different levels of government (Andrews-Speed, 2004; Xu, 2002). The year 1998 saw the abolition of the Ministry of Electric Power and the transfer of its government functions to the State Economic and Trade Commission (SETC), which was charged with the operational and regulatory oversight of these corporatized enterprises. Meanwhile, national planning and strategy development for the energy sector continued under the renamed State Development and Planning Commission (SDPC), as did responsibility for power sector reform.

From 1998 to 2002, a number of measures were taken to reorganize the SPCC, to corporatize the subsidiary provincial power companies, to implement a limited separation of generating assets from transmission and distribution, and to embark on experimental "market" trials in a number of provinces. Beginning in 1999, bidding by power generators was carried out on an experimental basis in four regions of China: Shanghai, Shandong Province, Zhejiang Province, and in the Northeast (Jilin, Heilongjiang, and Liaoning Provinces). Though the detailed rules varied from case to case, a number of common features ran across all the experiments. Only a small percentage of total available power was bid into the "pool" and tariffs were capped (Zhang and Heller, 2007).

Aside from these trial pools, the pricing system for power generators was changed nationwide. The government introduced a new policy in 1998 known as the "operating period tariff." This approach sought to base the wholesale tariff on the expected lifetime of the plant, rather than on the debt repayment period. The objective of this approach was to control and lower the capital cost of new plants and to place the responsibility for negotiating suitable financing terms on the project sponsors.

Few changes were made to the system of pricing for end users except to add some additional fees, such as for the construction of the Three Gorges Dam (Xu, 2002). The lack of a change to the way in which consumer prices were set did not prevent the government from raising these prices in order to allow the power industry to recoup its costs and to encourage energy efficiency. Prices in 1997 were set at levels 40–50% higher than those for 1995, at a time when inflation was running at an annual rate of 10% p.a. This reflected a real increase of 15–25% over a 2-year period, except for household consumers who were protected with an increase equivalent to inflation (Andrews-Speed, 2004). As a consequence, most end-user tariffs did reflect long-run marginal costs (Zhang and Heller, 2007).

2.4 Significance of the Original Reforms

The reorganization of the electrical power industry in the late 1990s was much more limited in extent than that of the petroleum and coal industries, and involved no unbundling of assets as would be required by most reform models (Andrews-Speed, 2004; Zhang and Heller, 2007). Although the government had partially removed itself from the operational management, the power industry was still dominated by a state-owned monopoly, the SPCC, albeit corporatized and commercialized. On the one hand, the government retained control over investment capital, project approval, the siting of plants, fuel mix, and the appointment of senior management. On the other hand, the SPCC sought to improve technical and financial efficiencies and allowed some decentralization of decision-making. The limited scope of these reforms arose in part from the intense resistance from conservative elements within the SPCC and SETC to the more ambitious agenda pursued by the SDPC (Chen, 2010).

With its new structure and new allocation of functions, the central government was unable to fulfill the new roles required of it. This failure resulted from a combination of bureaucratic competition between the SDPC and the SETC, ambiguity concerning the role of government, a lack of authority over the industry, and a shortage of staff. With respect to the last two points, the President of the SPCC retained ministerial status, many staff in the electrical power department of the SETC were drawn from the SPCC itself, and the restructuring of government agencies in 1998 was accompanied by a dramatic reduction in the numbers of civil servants across all departments. These factors further exacerbated the information asymmetry between the government and the SPCC (Chen, 2010).

Local government agencies and officials were able to use their authority over approvals for the construction of power stations to secure bribes or to gain an equity share in the enterprises. They also gave preference to local state-owned enterprises, thus creating barriers to entry to privately owned companies and to those from other parts of the country. The most notable example involved the inability of the newly completed Ertan dam to dispatch much of its potential power output as a result of protectionism in Sichuan Province (Zhang and Heller, 2007). The large state-owned energy companies were not immune to corruption. Indeed the power sector has been one of the most corrupt of all China's state-owned industries, as exemplified by the President of the SPCC in 2002 (Chen, 2010). It can also be argued that local governments colluded with both local and centrally owned power enterprises to capture rent and to undermine the regulatory influence of the central government (Cheng and Tsai, 2009).

As a consequence, the reforms enacted in the late 1990s led to a progressive decline of central government control over the national electricity sector. Local governments and state-owned energy enterprises were able to take advantage of this trend through local protectionism and uncontrolled capacity expansions, which undermined central government strategies relating to economic efficiency and fuel mix (Downs, 2006; Meidan et al., 2009; Xu, 2002).

3 FURTHER REFORMS BETWEEN 2002 AND 2005

In 2002, the government was ready to embark on the next stage of reform of the power sector. In order to circumvent the delays caused by bickering between the different actors, the highest government organ, the State Council, took direct control of the reform process (State Council, 2002). Its plan followed most of the ideas which had been proposed by the World Bank and other external advisers, and comprised three main elements:

1. The restructuring of the State Power Corporation into five generating companies, two grid companies, and a number of service companies.
2. The immediate establishment of a State Electricity Regulatory Commission (SERC) under the State Council to formulate market rules and to regulate the developing markets.

3. A new approach to power pricing and the development of competitive markets for power generation across five to six separate regions of China, with participation of most major power plants in this competition by the end of 2005.

In addition to these major reform measures, the government also introduced other changes to government institutions.

3.1 Industry Restructuring

Just 5 years after it was created, the SPCC was itself dismantled in 2002 to further separate generation from transmission and distribution and to reduce the concentration of ownership of power generating capacity. The generating assets of the State Power Corporation were unbundled from the grid and, together with those of the preexisting Huaneng Group, were assigned to five companies whose sole business was to be power generation:

1. The China Huaneng Power Group
2. The China Datang Corporation
3. The China Huadian Corporation
4. The China Guodian Corporation
5. The China Power Investment Corporation.

The redistribution of generating assets to the five new companies was carried out in such a way that no single company held more than 20% of the generating capacity in one of the planned regional power markets. Immediately after this restructuring, each company owned about 20 GW of generating capacity, though through their majority share in consortia the amount of capacity each company controlled was higher, ranging between 30 and 38 GW (Table 18.5). These and other power generating companies have since been listed on domestic stock exchanges and a small number of them on international exchanges (Andrews-Speed and Cao, 2005). Although these five companies nominally lay at arm's length from the government, their connection with the conservative element of the political elite was exemplified by the appointment of the son and daughter of the previous Premier, Li Peng, to senior management positions in Huaneng and China Power Investment Corporation, respectively (Yeh and Lewis, 2004).

The transmission and distribution assets of the State Power Corporation were divided between two new companies. The State Grid Corporation was to own and control the majority of the regional grids in the country, as well as the interregional transmission lines. The Southern China Power Grid Company took over the assets in the far south of the country. The two new grid companies were required to progressively sell off most of the generating capacity that had been previously assigned to the transmission and distribution subsidiaries of the State Power Corporation.

TABLE 18.5 Structure of Generating Capacity of China's Five Major Power Generating Companies, 2002 and 2010, and Their Share of Total Generating Capacity

	2002				2010				
	Total capacity	Thermal	Hydro	Nuclear	Total capacity	Thermal	Hydro	Wind	Nuclear
	GW	%	%	%	GW	%	%	%	%
Huadian	38.0	82	18	0	113.4	86.2	9.5	4.2	0
Datang	32.5	79	21	0	105.9	80.8	14.5	4.8	0
Huadian	31.3	81	19	0	88.2	80.2	17.4	2.3	0
Guodian	30.8	85	15	0	95.3	80.0	10.6	9.4	0
China Power Investment	30.2	70	26	3.8	70.7	72.6	25.1	2.3	0
Share of five companies	41.4%				49.0%				

Source: Ni (2006); China Electricity Council (2012b).

Although the five new generating companies were created from the pre-existing State Power Corporation, they, together with the two new grid companies, only controlled about 40% of the generation capacity across the country. A wide range of industrial and financial enterprises owned the remaining generating capacity. These players formed consortia to own and operate individual plants, with or without the involvement of one of the new large five generating companies. Some of these players were state-owned at national level, such as the Three Gorges Dam Corporation, the Shenhua Group, the China Nuclear Power Corporation, and the State Investment and Development Company, but most participants in these consortia were owned at local rather than at national levels.

Despite the radical nature of the restructuring, it did not include two steps which form a part of most programs of sector reform. Distribution was not separated from transmission, and the function of dispatch was not separated from grid ownership. The state dispatching center within the State Grid Company remained responsible for dispatching the interregional transmission lines and facilities, and regional dispatching centers within each regional grid subsidiary continued to be responsible for dispatch within the region.

3.2 Restructuring of Regulatory Agencies

The period 2002–2005 was marked by a series of reforms to the structure and function of government agencies charged with oversight of the electrical power industry, again pushed through by the State Council. The result was an increase in the number of agencies responsible for regulating the electricity industry, a redistribution of existing functions and the creation of some new functions.

The most important of these measures was the creation of the SERC in November 2002. SERC reported directly to the State Council and was charged with wide ranging responsibilities relating to both strategy and regulation. It was to become the major source of proposals for the development of power markets and for further reforms to the power sector. At the same time, it was responsible for the routine technical regulation of the operations of the power industry, including both technical and environmental standards, as well as collecting data. With respect to economic regulation, its powers were deliberately constrained. SERC could investigate "irregular" or anti-competitive behavior in the power markets and could help to resolve disputes, but was empowered only to make proposals relating to tariffs and then to supervise the implementation of the agreed tariffs. Ultimate authority for all electricity tariffs remained with the Pricing Department of the National Development and Reform Commission (NDRC), the successor to the previous SDPC (Table 18.6).

Two further agencies were created in March 2003: the Energy Bureau and the State-owned Asset Supervision and Administration Commission

TABLE 18.6 Summary of the Allocation of Government Functions Relating to the Power Sector Between 2003 and March 2008

Function	Responsible Agency	Participating Agency
Power sector reform strategy	State Council	
Energy policy formulation	NDRC (Energy Bureau)	Energy Leading Group, State Energy Office
Power sector policy formulation	NDRC (Energy Bureau)	SERC
Power sector planning	NDRC (Energy Bureau)	SERC
Price regulation	NDRC (Price Dept)	SERC
Investment approval	NDRC (Energy Bureau)	
Market entry approval	SERC	
Service obligations and quality	SERC	
Law-enforcement and administration	SERC or Local Economic and Trade Commission	
Demarcation of geographic area of power supply	SERC or Local Economic and Trade Commission	
Approval of new technologies	NDRC	
Approval of CDM projects	NDRC	
Technical and quality standards	NDRC	
Regulation of financial system of enterprise	Ministry of Finance	SERC
Regulation and management of national assets	SASAC	
Environmental regulation and management	SEPA	
Approved scope of enterprise operation	State Administration of Industry and Commerce	
Electrical power standards	Ministry of Science and Technology	
Safety regulation	SERC	
Public service	SERC	

Source: State Electricity Regulatory Commission (2007a, 2008a).

(SASAC). The Energy Bureau was created within the NDRC. This brought together many, but not all, of the energy functions, which had been scattered across the previous SDPC and SETC. The functions of the Energy Bureau included formulating policy, drawing up plans for sector reform and development, and managing the strategic oil stocks as well as routine oversight of the country's energy sector including the approval of major investments (Downs, 2006). The Energy Bureau continued the NDRC's traditional role of approving major construction projects, including power stations and transmission lines. Despite the importance of pricing to the energy sector, it was the Pricing Department, not the Energy Bureau, which retained control of energy prices.

It soon became clear that the Energy Bureau, with a staff of less than 30, could not possibly fulfill its mandate. Two years later, in 2005, the government set up an Energy Leading Group within the State Council, supported by a State Energy Office. The role of this Leading Group was to set strategic directions and to improve policy coordination (Downs, 2006; Rosen and Houser, 2007).

SASAC was established with the role of executing the functions of government as a shareholder in state corporations, and it executes this function at central, provincial, and municipal levels. It has authority to approve a wide range of actions by the relevant corporations including the appointment and removal of directors and senior managers, plans for restructuring or public listing, mergers and acquisitions, and asset disposals.

In addition to these changes, the status and resources of the agency charged with environmental regulation, the State Environmental Protection Agency (SEPA), were enhanced in 2003. This expansion gave the agency greater administrative capacity to monitor and investigate the environmental consequences of large construction projects. SEPA thus became more capable of evaluating proposed power construction projects and the environmental behavior of power plants, in order to seek compliance with the relevant laws and regulations.

3.3 Price Reform and Market Development

Proposals for price reform over the period 2003−2005 took two forms: strategic proposals for substantial reform to the approach to electricity pricing and for the introduction of competitive markets in generation, and short-term measures to address specific concerns relating to coal.

3.3.1 Strategic Proposals for Price Reform

In 2003, the State Council issued the "Scheme for power price reform" (State Council, 2003) which provided the outline of a strategy to overhaul the current tariff system for the electrical power sector and to develop competitive markets for generation and retail. This was followed by a further notice issued by the NDRC in March 2005 which described these plans in some detail (National Development and Reform Commission, 2005). The strategy foresaw the creation of three separate sets of tariffs: for generation, for transmission and distribution, and for retail, with the eventual separation of transmission and distribution tariffs.

The wholesale generation tariff would have two parts: a capacity payment and an energy fee. The capacity payment would be determined by government, while the energy fee would be set by market competition in regional pools. A formula was provided for the calculation of capacity payments, which included depreciation and financing costs. The nature of the market and the bidding rules were not specified, but were to be determined separately for each regional market. Bilateral sales from generators to large consumers were to be permitted.

Coal, oil, natural gas, nuclear, and hydroelectric power stations would participate in the market competition. Wind, geothermal, and other new and renewable forms of energy would be subject to separate rules. Foreign-invested power plants approved and constructed before 1994 and which had signed power purchase agreements or which had received other forms of government undertaking would be obliged to renegotiate these arrangements.

A tariff for transmission and distribution would be set on the basis of cost recovery, reasonable profit, and tax liability. Initially the "postage stamp" approach would be used, by which the tariffs in a region are shared according to the capacity of the user or producer. A specific service tariff would be set separately and would include a connection fee. Formulas were provided for the calculation of permitted profit and capital cost.

The Catalogues were to be retained for end-user pricing, but the number of categories would be reduced to three: residential, agricultural, and all industrial and commercial users. The first two categories would be subject to a single tariff, and the third category to a two-part tariff for users with a transformer capacity of 100 kVA or greater, or a capacity of 100 kW or more. A range of new tariffs would be introduced where appropriate, including peak and off-peak, dry and wet seasons, high reliability, and interruptibility.

The Pricing Department of the NDRC was to retain responsibility for setting or regulating end-user prices as well as wholesale prices until competitive bidding is introduced. This agency would also retain responsibility for transmission tariffs until such time as distribution was separated from transmission. From that time provincial pricing departments would be responsible for distribution tariffs.

3.3.2 Market Development

SERC set out its vision for the establishment of regional power markets in 2003. A document entitled "Guidelines for Establishing Regional Power Markets" (State Electricity Regulatory Commission, 2003) described the objectives, the main models, and the main trading types in the planned regional markets. By the end of 2005 or early in 2006, six regional power markets would be established with regulatory systems and institutions in place. The capacity charge would be determined by government and the energy charge would be set through competition. The majority of generation companies would bid to be dispatched and qualified large end consumers (including independent supply companies) could directly purchase electricity from generators.

The first trials of the new markets were held in Northeast China and East China. The Northeast China power market was put into monthly bidding simulation in January 2004. It initially adopted a one-part price model with 15% of total electricity bid into the market. Following the recommendation of NDRC, the market changed to a two-part price model with all electricity bid into the regional power market. At the beginning, only those generators with capacity of 100 MW or above were allowed to participate in the pool. During the simulation period, only the bidding system was put into operation and there was no actual settlement. The East China power market was put into monthly bidding simulation in May 2004, again without actual dispatch and settlement.

Both these pilot markets took the form of a mandatory pool with a single buyer. Bidding to the pool was compulsory for qualified generators, which, in the case of East China, covered coal-fired plants with capacities of 100 MW or greater. The grid company was the single buyer. Trading arrangements were dominated by contract trade, and supplemented with trading in the spot market. The trading types included yearly contracts, monthly bidding contracts, day-ahead bidding, and real-time balancing. Monthly bidding and day-ahead bidding were operated in the regional trading centre with all the coal-fired units of capacity of 100 MW or above participating. The provincial dispatching centre was responsible for scheduling the implementation of the annual contracts and for real-time balancing to control the provincial power system.

Further trials were launched in South China in 2005. Unlike the pilot programs in Northeast and East China, this program had the intention to stimulate a greater degree of competition. Two characteristics distinguished it from the earlier pilot programs. Firstly, it engaged not only multiple sellers, but also multiple buyers in the market. The program required grid companies from four provinces (Guangdong, Guangxi, Yunnan, and Guizhou) to participate into the market and these grid companies competed with each other for power purchase. Secondly, the program separated the dispatch function from the market operator.

The development of these pilot regional markets faced a number of challenges. The varying levels of economic development in different provinces in same region made it difficult to implement a unified pricing system because the poorer provinces were not able to afford a higher price.

Allegations emerged that grid companies were favoring their own generators. The weakness of interprovincial transmission capacity led to grid congestion. Finally, the growing shortages of power rendered these pilot markets irrelevant and the trials were soon abandoned (Wang, 2007; Zhang et al., 2005).

In anticipation of actual implementation of power markets, the government sought to bring a greater degree of order to prices offered to generators and at the same time improve incentives for efficiency. The new approach which involved setting benchmark tariffs was described in a document issued by the NDRC in April 2004 (Ma, 2011; National Development and Reform Commission, 2004). New plants in the same region and using the same fuel were to receive the same price, and the prices paid to existing plants should gradually be brought into line with these regional average levels. The aim was to provide power generators with a guaranteed rate of return (Zhang, 2012). The method of setting wholesale prices was based on a cost-plus method, which involved bargaining between the power generators and the government and which did not provide incentives for improving efficiency (Ma, 2011; Zhang, 2012). Further, coal-fired plants, which installed and operated desulfurization equipment, were to receive a higher price, which would be set at a national level.

3.3.3 Measures Concerning Coal

Coal has long been the primary source of energy, in recent years accounting for about 80% of electricity generated in China (Table 18.1), and therefore the pricing of coal has a direct bearing on the financial health of the electricity industry. Since 1994, a large proportion of the nation's coal output has been sold through wholesale markets, and prices in coastal provinces have been close to international levels. Despite this "liberalization," coal continued to be sold to large power stations at subsidized prices. The SDPC (later NDRC) ran an annual meeting at the end of each year at which the principal producers, transporters, and consumers of coal reached agreement, under SDPC guidance, on coal prices for the following year (Thomson, 2003).

The rapid rise of coal prices during 2003 and 2004 put a great strain on power generating companies and on their relationship with coal producers. To solve this problem, the NDRC agreed to allow the price of coal for power stations to be set by market forces and announced, in December 2004, a new scheme to link wholesale power prices to coal prices. The link was defined by a formula which included coal digestion ratio, standard coal consumption, and the calorific value of the coal. The scheme provided for approximately 70% of a rise in coal price to be passed through to the grid. A change in coal price of 5% or more would trigger an immediate adjustment of wholesale prices. Lesser changes of coal price would be addressed in six monthly reviews.

3.4 Significance of These Reforms

The measures drawn up over the period 2002−2004 marked a new determination on the part of the government to push ahead with the reform and liberalization of the electricity sector, with the aim of introducing a competitive power market. The State Power Corporation was unbundled, generation was separated from transmission, and an entirely new regulatory agency, SERC, was created. Pilot markets for power generation were run. Yet, much remained unchanged. The NDRC retained authority over both pricing and project approval, and the proposals for power pricing and markets for power generation were shelved in 2005 on account of the growing shortages of electrical power across the country.

As a result, the industry saw a change of structure but little modification in the way that electricity was bought and sold nor in the way that the industry was regulated. In some ways, China's power industry resembled the single buyer model of Hunt and Shuttleworth (1996), with a purchasing agency (the grid companies) buying power from the newly unbundled generating companies. The key difference was that, in the case of China, the processes for purchasing this power were neither transparent nor predictable, nor were they underpinned by binding contracts. In other words, industry restructuring had not been accompanied by the introduction of competitive markets.

Despite the elevation of the reform agenda from the NDRC to the State Council, true competition in power generation had been successfully resisted by a combination of the State Grid Corporation, the large generating companies, and local governments (Chen, 2010; Zhang and Heller, 2007).

4 CHANGING PRIORITIES BETWEEN 2005 AND 2012

The 5-year period from 2004 to 2008 was characterized by a dramatic increase in demand for all forms of energy across China, including for electrical power. The national energy intensity had started to rise after two decades of steady decline (Figure 18.4) and remained high relative to other large economies (Figure 18.5). The resulting shortages of energy caused both the government and the power industry to switch their attention from sector reform to security of supply and, in particular, to investing in new generation and transmission capacity. At the same time, in order to address the energy shortages, the government introduced a number of policies to enhance energy efficiency in all sectors of the economy and to reduce the energy level of intensity. The surge in demand for electricity was tempered but not ended by the financial crisis of mid-2008.

In the international arena, two further trends were affecting the government's approach to power sector management. Firstly, the years 2000−2005 saw severe blackouts and politically unacceptable price volatility in a number of liberalized power markets in some OECD countries; for example, the

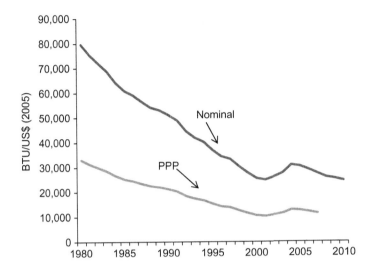

FIGURE 18.4   Energy intensity in China, 1980–2010, both nominal and purchasing power parity (PPP). *Source: Energy Information Administration (2011).*

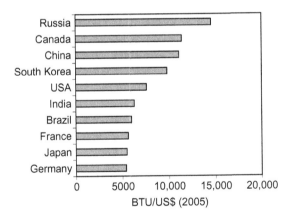

FIGURE 18.5   Energy intensity in 2008 for the main energy consuming nations, at purchasing power parity, in British Thermal Units per US$ (year 2005). *Source: Energy Information Administration (2011).*

United States (California), Canada (Ontario), the United Kingdom, Scandinavia, and Italy. China's government took from these experiences the lesson that electricity sector liberalization was fraught with risks, and that such risks would be exacerbated in an environment with weak regulatory and legal systems (Yeh and Lewis, 2004). Given the weak state of key legal and regulatory institutions in China, this hesitation to push ahead with power

TABLE 18.7 Technical Performance and Sulfur Dioxide Emissions from the Power Sector, 2002–2011

Year	2002	2003	2004	2005	2006	2007	2008	2009	2010	2011
Average utilization hours	4860	5245	5455	5425	5221	5020	4648	4527	4660	4731
Net coal consumption (g/kWh)	383	380	376	370	367	356	345	340	333	330
Line losses, %	7.52	7.60	7.57	7.21	7.04	6.97	6.79	6.72	6.53	6.31
Total SO_2 emissions (Mt)	8.20	10.0	12.0	13.0	13.5	12.3	10.5	9.4	9.2	–
SO_2 emissions (g/kWh)	6.1	6.3	6.6	6.4	5.7	–	3.8	3.1	2.7	–

Source: State Electricity Regulatory Commission (2008b); China Electricity Council (2004, 2006, 2010a, 2012b); State Electricity Regulatory Commission (2008c, 2009, 2010); Liu (2012).

sector reform was justifiable (Andrews-Speed, 2012). Secondly, the growing consumption of energy within China was taking the country to the top of the league table of emitters of greenhouse gases (GHGs) as well as increasing the amount of sulfur dioxide emissions (Table 18.7). As a consequence, both international and domestic pressures on China's government were mounting to take steps to constrain these emissions.

For these reasons, the policy context had changed significantly since the late 1990s and early 2000s when the reform strategy was first drawn up. As a consequence, the period 2005–2011 was marked by stagnation in the reform process. Whilst the power companies focused their attention on increasing the capacity of the industry to satisfy the rapidly rising demand (Figures 18.1 and 18.2), the government sought to drive though policies to enhance energy efficiency and the use of renewable energy. At the same time, the government used its power to control retail prices to address social equity concerns by tightly constraining the rise of electricity prices for households.

4.1 Energy Security and Energy Efficiency

The growing shortages of energy drove the government to undertake a thorough review of its energy strategies in 2004. At the same time, the NDRC issued their "Medium and Long-Term Energy Conservation Plan," which not only demonstrated that energy efficiency and energy conservation did indeed lie at the heart of China's new energy policy but also laid out specific targets and objectives and identified the key steps to be taken (National Development and Reform Commission, 2004). A revised version of the 1997 Energy Conservation Law was approved in October 2007.

The overriding goal of the new strategy to reduce energy intensity by 20% between 2005 and 2010, to combat the high and rising levels of energy intensity. This Energy Conservation Plan and subsequent documents set targets for individual energy intensive industries, such as electrical power generation, as well providing proposals for technological process or management improvements needed to achieve these targets. These priorities were further elaborated in the Five-Year Plan for the period 2006–2010 (National Development and Reform Commission, 2007a).

In addition to setting energy-saving targets for the most energy intensive industries, the government also began to apply higher retail tariffs to selected metallurgical, chemical, and cement industries. The differential tariff was gradually increased until, by 2010, plants in the "restricted" category were paying a surcharge of at least 10% and those in the "eliminated" category at least 30% (Price et al., 2010). In 2010, an additional category was introduced in which punitive pricing was applied. While these differential tariffs do appear to have been applied across much of the country causing plants to close or upgrade their processes, this policy encountered resistance from many local governments. In order to reduce the level of obstruction, the central government allowed the funds raised from the differential tariffs to flow to Provincial governments so that they could provide support to enterprises trying to upgrade their plants (Chen, 2011; Price et al, 2010).

A number of detailed regulations were issued which related to the power sector with the objective of encouraging high-specification generation technologies with large capacities, high efficiency, low water usage, and effective environmental controls (International Energy Agency, 2009). It is these regulations that encouraged the construction of the supercritical and ultra-supercritical plants. At the same time, small, old, and inefficient plans were closed (Table 18.8).

Although the experimental competitive power pools had been abandoned, the government introduced new types of mechanisms to promote energy efficiency in power generation. Between 2005 and 2007, 23 provinces ran experiments in the trading of generation rights, a system whereby a more efficient thermal plant which had reached its annual quota of generating hours could buy more hours from the quota of a less efficient plant. This led to significant reductions in coal use and SO_2 emissions (Gao and Li, 2010).

The government also took steps to adjust the system for the dispatch of power plants. In August 2007, a new trial method for dispatch was

TABLE 18.8 Coal-fired Power Generation Technologies

	Units	2005	2006	2007	2008	2009	2010	2011
Coal-fired capacity	GW	368	454	524	574	599	650	730
Ultra-supercritical capacity	GW	0	3	10	11	21	34	39
Installed FGD capacity	GW	53	162	270	363	470	560	–
Coal-fired capacity with FGD	%	14.4	35.7	51.5	63.2	66.9	86.1	–
Small thermal units shut down	GW		3.1	14.1	16.5	26.1	16.9	16.0
Transmission capacity, 220 kV and above	1000 km	254	286	328	359	401	445	480

Source: International Energy Agency, 2009; China Electricity Council (2012a, b); State Electricity Regulatory Commission (2009a, 2010a, 2011a); Liu (2012).

announced (National Development and Reform Commission, 2007b), which set out the following order for dispatch:

1. Renewable energy
2. Nuclear power
3. Coal-fired co-generation units and those using waste heat
4. Natural gas and gasified coal units
5. Conventional coal-fired units
6. Oil-fired plants.

For thermal plants within the same category, the order of dispatch was on the basis, first, of energy consumption and, second, of pollution levels. Trials were started in late 2007 in five provinces (Henan, Jiangsu, Guangdong, Sichuan, and Guizhou) and ran for 2 years with significant reductions in coal use per kilowatt hour generated and in SO_2 emissions. Problems with pricing for peak-load plants, network constraints, and rising wholesale tariffs led to the abandonment of these experiments (Gao and Li, 2010).

4.2 Climate Change and Environmental Protection

In response to the growing awareness of China's contribution to current (not historic) GHG emissions, the State Council approved a national plan to address the challenges posed by climate change at the end of May 2007 (National Development and Reform Commission, 2007c). Ambitious though some of these targets were, most of those relating to energy were consistent with the newly developed energy strategies (Lewis, 2007). Three components of the climate strategy that are of relevance to electrical power, are renewable energy, the clean development mechanism (CDM), and carbon trading.

The Renewable Energy Law passed in 2005 marked a new determination by the government to substantially enhance the role of renewables in the national energy supply. This law created, for the first time, a relatively coherent framework for promoting investment in renewable energy. It provided an obligation for grid companies to connect all renewable plants and to purchase all electrical power generated by these plants. Incentives for research and development were also provided in order to encourage the domestic manufacturing of the required technologies.

Despite these positive components, the law did not provide for a fixed and predetermined feed-in tariff. Rather the tariff was set by competitive bidding. This resulted in the state-owned power companies driving prices down to levels below what most would estimate to be commercially viable for wind power or other renewables, and private sector investors, both domestic and foreign, failed to gain significant opportunities (Lema and Ruby, 2007; Li and Ma, 2007).

An added potential incentive for the construction of renewable energy capacity is the CDM, the instrument established by the Kyoto Protocol to encourage financial support from developed economies for investment in clean energy in developing economies. Wind power has been the prime beneficiary within the power generation sector of the CDM mechanism in China (International Energy Agency, 2007). Administrative obstacles and policy ambiguity have prevented rapid implementation for renewable energy within China to date (Resnier et al., 2007; Zhang, 2006).

The years 2007 and 2008 saw the launch of two major initiatives relating to carbon trading. In collaboration with the UNDP, the government established exchanges in Beijing and Shanghai to provide platforms for carbon trading, as well as to collect and publicize relevant information and undertake advisory and consultancy services. The second initiative called MGD Carbon (Carbon Finance for Achieving Millennium Development Goals) was intended to establish service centers in poorer parts of the country to enable them to take part in the carbon trading schemes. No progress was made in implementing carbon trading in the power sector, but the idea was resurrected in 2011 (Stanway, 2011).

In 2006, the government renewed its efforts to reduce sulfur dioxide emissions by decreeing that all new coal-fired plants be fitted with flue-gas desulfurization equipment (FGD) and that all plants built after 1997 should be retrofitted. This policy has been supported by agreements with seven provincial governments and six large power generating enterprises. Specific economic incentives for fitting and operating FGD include a higher wholesale tariff, priority for connection, more annual operating hours and, in some localities, priority for despatch. By 2010, the proportion of coal-fired capacity with FGD installed had risen to above 80%, up from 14% in 2005, and over a period when at a time when total coal-fired capacity had almost doubled (Table 18.8; Li et al., 2011; Zhang, 2011). This program contributed to the 32% fall in total SO_2 emissions from the power sector between 2006 and 2010 (Table 18.7).

4.3 Capacity Growth

To satisfy the rising demand for electricity, power companies of all types across the country embarked on a massive campaign to invest in new generation capacity, assisted by easy access to low cost capital (Zhang and Heller, 2007). The quantity of additional capacity becoming available grew steadily each year until 2006 when a total of 104 GW of new capacity was commissioned. The annual rate of commissioning then declined slightly to 80–90 GW resulting in an aggregate capacity of 962 GW by the end of 2010 (Figure 18.2). As a result, the generating capacity of China's powers sector doubled between the end of 2002 and the end of 2007.

This growth of generation capacity was characterized by two trends, one unfavorable and the other favorable. The unfavorable trend was the rise in the proportion of coal in the fuel mix (Tables 18.1 and 18.2). This arose from two factors. First, coal has long been the major feedstock of the country's power stations and domestic reserves of coal are plentiful. Second, the time and cost involved in building a coal-fired plant is significantly less than for the other preferred fuel in China which is hydropower. The alternative fuels were not suitable for such a large and rapid expansion of capacity for a variety of reasons: natural gas was not available in sufficient quantities; oil was becoming increasingly expensive and, though its use in power generation did surge in 2003 and 2004, the government was seeking to reduce this application of oil; and the renewable energy industry in China lacked the capacity to deliver such a vast capacity in such a short time.

A trend favorable from the point of view of energy efficiency and emissions was the substantial improvement in the nature of the coal-fired stations being constructed with respect to both scale and technology. A majority of new plants were 600 MW or larger, and between 2002 and 2006 the proportion of plants with a size of 300 MW and above rose from 41% to 51%. Many of the new plants incorporated advanced technologies which greatly enhance thermal efficiency and reduce pollution, for example ultra-supercritical technology (Table 18.8). A small number of plants using circulating fluidized bed combustion were also coming into operation (International Energy Agency, 2009; State Electricity Regulatory Commission, 2008b). These policies led to improvements after 2006 in average utilization hours for the power sector and in unit net coal consumption (Table 18.7).

This expansion of generating capacity allowed each of the five main generating companies to increase their sales revenue over the period 2003–2006. Each company also succeeded in raising their profits

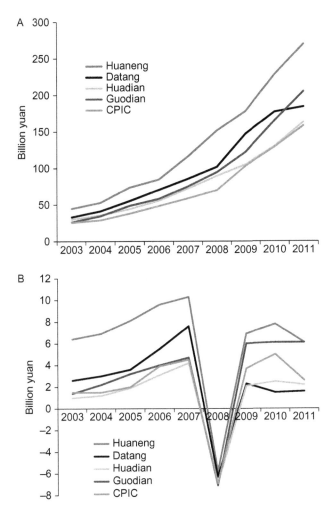

FIGURE 18.6 (A) Annual revenues for the five major power generating companies, 2003–2011, in billions of yuan. (B) Annual profits for the five major power generating companies, 2003–2011, in billions of yuan. *Source: State Electricity Regulatory Commission (2007a, 2008d, 2012); China Electricity Council (2010b, 2012b).*

(Figure 18.6; State Electricity Regulatory Commission, 2007b). Recent analysis suggests that this improvement in profitability was driven, at least in part, by significant improvements in the efficiency of the use of inputs, such as labor and nonfuel materials, but not in the use of fuel (Du et al., 2009). The profits of these generating companies declined in 2007 and 2008 as coal prices continued to rise faster than the wholesale electricity price.

The period 2002–2006 saw substantial investment in the transmission and distribution network creating a total of nearly 100,000 km of additional line and marking a 50% increase in line length. During the 5 years from 2006 to 2010 the length of transmission line at 220 kV or above grew at an average annual rate of nearly 12% (Table 18.6). Most of the expansion was at 500 and 220 kV, with some at higher voltages of up to 1000 kV (Electricity, 2011; State Electricity Regulatory Commission, 2008b).

4.4 Efforts to Diversify Fuel Mix

Both efficiency and environmental agendas have required China's government to diversify the fuel mix of the power sector away from its long-term reliance on coal. Natural gas is seen as an important clean fuel for the future, but policy decisions and pricing structures have led to a relatively low level of supply of gas to power stations. In particular, natural gas has been directed to cities for commercial and household use rather than power stations (National Development and Reform Commission, 2007d). As a result, natural gas accounted for only 2.3% of power generation in 2011 (Tables 18.1 and 18.2). New plans to boost gas-fired generation capacity are likely to face the same problems of high price and low availability (Wang, 2011).

Hydroelectricity has been the main source of primary electricity supply in China for many years, yet its share of electricity supply has declined from nearly 20% in 1990 to 14−17% in between 2006 and 2011 (Tables 18.1 and 18.2), the recent fluctuations being linked to variable rainfall. The government continues to boast of the country's vast potential for hydroelectricity, but a range of obstacles may result in a continuing decline in the relative contribution of hydroelectricity unless the government expends great efforts. Not only do these large dams require huge investment, but they are also attracting growing levels of opposition and criticism from within China and from downstream states in Southeast Asia.

Meanwhile the government is pushing forward with plans to rapidly expand the capacity of nuclear power. As of July 2012, China had 15 operating nuclear power plants with a total capacity of about 12 GW. Nearly 30 GW of nuclear power capacity was under construction at this time, and most of this capacity is due to be commissioned by 2015. This would bring the total to about 40 GW. Estimates of nuclear power generation capacity available by 2020 range lie in the range 60−70 GW (World Nuclear Organisation, 2012). Whether these targets will be met will depend on the ability of the government and the power companies to mobilize the resources in a timely manner and on the continuation of the apparently high level of acceptance of nuclear power within China.

Renewable energy, aside from hydroelectricity, continues to account for a small, though growing, proportion of China's power generation capacity and output (Tables 18.1 and 18.2). The Renewable Energy Law passed in 2005 marked a new determination by the government to substantially enhance the role of renewables in the national energy supply. Of particular importance is the role of wind power in electricity generation (Cherni and Kentish, 2007).

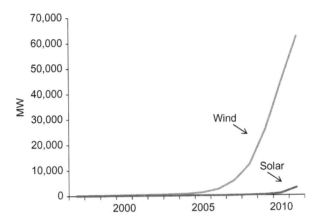

FIGURE 18.7 Installed capacity of wind and solar energy in China in megawatts, 1997–2011. *Source: BP (2012).*

At the end of 2005, total installed wind power capacity was about 1 GW. Since then the rate of growth has been dramatic. Total capacity reached 12 GW in 2008, 20 GW in 2009 and 62 GW by the end of 2011 (Figure 18.6). During 2010, China's wind power capacity took it to the top of the world rankings for installed wind power, marginally ahead of the United States (McNulty, 2011). This expansion has been assisted in part by the CDM (Zhang, 2011). The growth of solar power capacity has been rather slower, though the years 2010 and 2011 saw a substantial growth in installed capacity from 300 to 3000 MW (Figure 18.7). Feed-in-tariffs are now paid for both these forms of renewable energy, which should ensure that the construction of new capacity continues apace (Bai and Walet, 2011; Interfax, 2009). Despite the rapid growth of installed capacity, the amount of wind power actually dispatched has remained well below expectations, mainly as a result of poor planning and coordination (Cheung, 2011; Zhang, 2011). The targets for wind and solar power capacity continue to be increased. As of August 2012, they stand at 200 and 50 GW, respectively (Hua and Chen, 2012). It is likely that these targets for capacity will be met, but whether this capacity will be efficiently used remains to be seen.

4.5 Rising Prices and Social Equity Concerns

The pricing policies introduced in 1998 have led to the power industry being vulnerable to international markets and dependent on government policy. The price of the main primary energy feedstock, coal, is set by international markets, while the end-user prices for electricity are set by government at levels intended to address social equity concerns and with little regard for actual costs.

As international prices for coal and crude oil rose over the period 2002–2008, so did domestic prices for the producers of coal and oil. The government allowed coal prices to react to supply and demand, and so border prices for steam coal increased from about US$40 per tonne in 2004 to US $110 in July 2008 in line with international prices. Inland, near the areas of production, coal prices were at lower levels but rose by a similar proportion. The government has sought to constrain the price of coal sold to power stations but otherwise have not directly capped coal prices.

The electricity industry continues to apply the Catalogue system for end-user tariffs (Table 18.9), though the proportion of the final charge made up of additional has declined substantially. In its concern to protect private citizens and, to a lesser extent, industrial and commercial enterprises, the government has proven very reluctant to raise end-user prices for electricity, as well as for oil products and for natural gas. Though consumer prices for most forms of energy were raised at irregular intervals, end-user electricity prices continued to lag behind wholesale electricity prices and they in turn lagged behind the rise in coal prices. A reluctance to raise energy prices further was enhanced during 2007 by the government's fear of the social impacts of rising inflation.

By March 2008, power shortages were appearing across the country, despite the massive and ongoing investment in new power generation capacity over the last previous 5 years. In part these shortages were caused by the severe winter weather in the southern part of the country. But a further cause was the unwillingness of power generators to operate at a time of rapidly rising coal prices and static electricity prices. Wholesale tariffs for coal-fired

TABLE 18.9 Average Catalogue End-User Prices of Electricity, 2007–2010, in yuan per 1000 kWh

	2000	2004	2007	2008	2009	2010
Commercial	670	643	851	847	842	812
Large industrial	372	457	514	535	555	617
General industrial	430	614	692	718	747	771
Nonresidential lighting	500	776	724	729	736	743
Residential	379	447	470	469	467	475
Agriculture	329	365	401	399	398	436
Irrigation in poor rural areas	146	166	177	160	165	194

Note: From the beginning of 2007 to the end of 2010 the exchange rate changed from US $1 = yuan 7.84 to US$1 = yuan 6.67.
Source: Zhang (2012); State Electricity Regulatory Commission (2008e, 2009b, 2010b, 2011b).

power plants had risen by only 10—20% since 2005 in a period when coal prices almost doubled. As a consequence, coal-fired power generating companies were making significant financial losses by 2008 (Figure 18.6B; Ma, 2011). The risks associated with the discontinuity between coal and electricity prices have encouraged vertical integration between coal mining and power generation: coal mining companies acquiring or building power stations and power companies acquiring coal mines. This has allowed the construction of mine-mouth power plants and a reduction of transaction costs and price risks. Although these developments have brought financial advantages to the enterprises, they have also introduced new risks of anticompetitive behavior by these integrated companies (Zhang and Chen, 2011; Zhao et al., 2012a).

In June 2008, China's government could no longer resist the pressure for further substantial tariff adjustments and it announced a round of price rises for energy products. From July 1, 2008, wholesale electricity tariffs were allowed to rise by 5%. This increase provided some compensation to the power generators, but the industry argued at the time that a further rise of 50% would be required to match the amount that coal prices had risen over the previous 12 months. The burden of these tariff increases was borne mainly by the industrial and commercial sectors and by the grid companies, as rural and urban households remained protected (Table 18.9).

A three-tiered pricing system for households was formally proposed in 2010 with the aim of providing stronger incentives to save energy for the larger users (Table 18.10). This suffered from a number of weaknesses. On the one hand, price differentials between the three tiers were too small, as was the proportion of the population paying the highest tariffs. On the other hand, it was deeply unpopular among the middle classes (Andrews-Speed, 2010; Wang et al., 2012). The reformulated plan issued in the summer of

TABLE 18.10 Key Features of the Tiered Tariff Schemes for Residential Electricity Users in 2010 and 2012

	2010 Proposal			2012 Proposal	
	Tariff	Monthly use, kWh		Tariff	Monthly use
	yuan/kWh	Option 1	Option 2	yuan/kWh	kWh
Tier 3	+ 0.10	> 210	> 270	+ 0.08	> 400
Tier 2	+ 0.05	110—210	140—270	+ 0.05	240—400
Tier 1	+ 0.0	< 110	< 140	+ 0.0	< 240

The tariff differentials are relative to the Tier 1 tariff.
Source: Andrews-Speed (2010), Lelyveld (2012).

2012 took into account these latter complaints and, as a consequence, some 80% of the households are likely to pay the lowest tariff (Lelyveld, 2012). Thus, although this tiered scheme reflects a new approach, it is unlikely to achieve much in the short term.

4.6 Evolving Institutions for Sector Regulation

The sudden rise in importance of energy as a national priority saw the top leadership and the State Council becoming more involved in issues relating to energy than before, as was shown by the creation of the Energy Leading Group and its supporting State Energy Office. But the NDRC retained control over most of the key aspects of policy making and regulation in the electricity sector. Specifically, the NDRC continued to be responsible for formulating energy policy and policy for the power sector, including sector reform. At the same time, it held approval authority over pricing, investment, new technologies, and CDM projects.

Though these roles were concentrated in the NDRC, other tasks were dispersed among a number of other government departments, such as the SASAC, the Ministries of Finance and of Science and Technology, the State Administration of Industry and Commerce, and the SEPA.

SERC itself was left with few clear responsibilities except for drawing up proposals for the NDRC, drafting rules of minor importance, and undertaking certain minor regulator functions. SERC appears to have carried out the former two tasks with great enthusiasm to judge by the large number of documents it has produced since January 2006, but its capacity to undertake the minor regulatory tasks has been restricted by the shortage of staff and of offices at local level. As a result, these functions continue to be carried out by local Economic and Trade Commissions wherever no local office of SERC has been established.

Thus, despite a rearrangement of roles and responsibilities, the long-standing systems of regulation of China's power sector changed little over the period 2003–2008. Authority remained concentrated in the NDRC for the most important regulatory functions, while other functions were highly dispersed. The result was excessive government interference where it was not needed, and inadequate regulation where it was needed (State Electricity Regulatory Commission, 2008b). After the fanfare that accompanied the creation of SERC as an "independent regulator," this new agency has been treated as a peripheral advisory body rather than a regulator of critical importance to the development of the sector.

The new government installed in March 2008 undertook a radical reorganization of some ministries and agencies, but the energy sector only saw minor changes. The Energy Leading Group was transformed into the National Energy Commission and the Energy Bureau was upgraded and enlarged to become the National Energy Administration (NEA). This new

NEA incorporates the previous Energy Bureau and State Energy Office, as well as the nuclear power administration. At its launch in July 2008, the NEA had a staff of 112 in 9 departments: energy policy, project planning, project approval, electricity, coal, oil, nuclear power, alternative resources, and international cooperation. Though its rank has been raised to Vice-Ministerial status, the NEA is likely to continue to lack the capacity and authority to fulfill its mandate, not least because its remains within the NDRC and because the Price Department of the NDRC retains control over energy pricing (Miller, 2008).

4.7 Significance of the Recent Policy Adjustments

A combination of domestic policy pressures and a decline of international confidence in the efficacy of liberalized electricity markets have been the main reasons for China's government halting the process of liberalization of the energy sector. But this change of direction was almost certainly also influenced by the very energy companies that had been facing further reform. Rather than a wholesale switch in policy thinking, this retreat from a reform agenda probably reflects a modest shift in the balance of influence between those parties in favor of reform and those parties against it.

The electrical power industry continues to be a notable location for struggles for influence that date back to the 1990s and before (Chen, 2010). Since the mid-1990s, the debate over power sector reform has seen the SDPC and its successor, the NDRC, actively promoting the liberalizing agenda, with the State Power Corporation and its successor, the State Grid Corporation, resisting substantial liberalization while putting forward their own more modest reform proposals. The consequence has been significant restructuring of the industry and some experimental power pools, but no true competition between power generators.

As a result, the reforms have not made the power companies sufficiently responsive to changing prices or demand, and efficiency gains have not been optimized (Zhao et al., 2012b). On the other hand, the current framework allows the continuation of massive investment in new infrastructure at relatively low costs of capital (Zhang and Heller, 2007). In the meantime, the State Grid Corporation has retained its monopoly power over transmission and distribution across most of the country, and the role of the newly created "regulator," SERC, has been restricted to that of an advisory agency (Chen, 2010).

The large generating and grid companies appear to prefer the current, partially reformed state of the power sector, despite its idiosyncrasies, to the uncertainty of further reform which might lead to a truly competitive market. Their commercialized status relieves them of many aspects of direct government control and noncommercial obligations, and allows them to seek profits and rents. At the same time, their national economic importance and political connections provide them with the power to negotiate new terms if the economic conditions turn against them.

The last 15 years of power sector reform in China closely resembles that in South Korea (Chapter 22). Reforms to the power sector there also started in the late 1990s as part of a wider economic reform process aimed at introducing competition. As in China, the vertically integrated, state-owned electricity monopoly was broken up into a small number of generating companies and one transmission and distribution company (not two, as in China). The reforms were halted in 2004 before competition could be introduced on account of growing opposition both within government and from power stakeholders.

5 CONCLUSIONS AND OUTLOOK

China's power sector has undergone intermittent reforms since the 1980s involving changes to the structure of the industry, to its relationship with government and to the nature of economic and other incentives. The most important changes took place between 1997 and 2003 when the productive assets of the Ministry of Electrical Power were progressively corporatized, commercialized, and restructured, and some of the new companies were partially listed on domestic and international stock exchanges. The original aim of these reforms was to pave the way for competitive electricity markets in order to improve economic and technical efficiency and to constrain. But no steps have been taken since 2003 to introduce competition in generation despite the existence of a large number of different generating enterprises, including the major five large generating companies.

This unwillingness to pursue the reforms to their logical conclusion after 2003 can be attributed to a major change in the domestic context in which the need to encourage massive investment in new generation and transmission infrastructure outweighed the desire promote efficiency through competition. Such a change in objectives was also seen at this time in India (Chapter 17). In addition, the various crises experienced in liberalized power markets around the world in the early years of the century played into the hands of the conservative elements in China's government and in the power sector which had been resisting the reforms for many years.

In many respects, the path of power sector reform in China closely resembles that in South Korea. In both countries, the sector was restructured but not liberalized and entrenched interests have obstructed further reform. In Russia (Chapter 14) and India, the reform process was pursued further with the privatization of generation companies and some degree of competition introduced. Despite the greater vigor with which these reforms were implemented in the early stages, the process has also stalled: in Russia, as the state claws back controls over pricing on account of social equity concerns; and in India on account of fragmented implementation and the need to promote and manage capacity growth more effectively.

At the time of writing (October 2012), the future for power sector reform in China is uncertain. By the spring of 2013 a new government will be in place with new policy priorities. These priorities are likely to include addressing disparities in income and wealth, promoting energy and resource efficiency, and environmental protection, in addition to the perennial objectives of sustaining economic growth and restructuring the economy. In such a context, it is hard to see the new government possessing the political will and enthusiasm to push through reforms to the power sector that creates genuine competition between power generators, which break the monopolistic position of the State Grid Corporation, for example, by separating transmission from distribution, and which allow end-user prices to match long-run marginal costs.

Instead, it is highly likely that China's new government will continue to make only minor adjustments to structures and systems. Although this might be seen as weakness in the face of active resistance from power companies, local governments, and citizens, such an approach avoids the risks associated with the introduction of competitive electricity markets in a country in which institutions, such as the legal system and capital markets, remain immature (Andrews-Speed, 2004; Victor and Heller, 2007). It also allows the government to retain the ability to pursue policy objectives beyond pure economic efficiency, such as social equity, security of supply, and environmental protection. In this way, the electrical power industry, the wider energy sector, and parts of the rest of the economy may remain in a "trapped transition," stranded between the plan and the market (Andrews-Speed, 2012; Pei, 2006). China is not alone in possessing a power sector in which the sector reform process has stalled or been subverted. Korea, Russia, and India all illustrate the same phenomenon, albeit with different characteristics.

Critics might claim that a power industry with such a hybrid mix of state and market characteristics is fundamentally unstable and requires further reform. However, Victor and Heller (2007) have argued that these features are typical of electrical power in many developing and transition economies and that such systems are relatively stable as they are favorable for the power companies themselves and provide environments in which investment in new capacity can be made. Their stability is provided by the high entry barriers to new players and by ability of the incumbent companies to resist further reform.

Further, it can be argued that China's preference for incremental reform interrupted by long periods of stability during which the government and state-owned enterprises have retained a high degree of control has allowed it to manage its power sector more effectively than its large neighbors, as described in chapters on Russia and India in this volume. In both these countries, liberalization was taken further than in China but a combination of government interference and weak institutions undermined the ability of these power sectors to attract sufficient investment in new capacity and, in the case of Russia, to start to decarbonize the sector. In contrast, investment in new generation and transmission capacity has broadly kept pace with demand in China, significant steps have been taken to promote nonfossil fuel generation, and technical efficiency continues to improve.

REFERENCES

Andrews-Speed, P., 2004. Energy Policy and Regulation in the People's Republic of China. Kluwer Law, London.

Andrews-Speed, P., 2010. China's Latest Electricity Tariff Reform: Lifeline Rates at Last. Available from: <http://www.andrewsspeed.com/index.php/permalink/3035.html/> (accessed 18.06.12).

Andrews-Speed, P., 2012. The Governance of Energy in China. Transition to a Low-Carbon Economy. Palgrave MacMillan, Basingstoke.

Andrews-Speed, P., Cao, Z., 2005. Prospects for privatisation in China's energy sector. In: Green, S., Liu, G.S. (Eds.), Exit the Dragon? Privatization and State Ownership in China. Royal Institute for International Affairs, London, pp. 196–213.

Andrews-Speed, P., Dow, S., Wang, A., Mao, J., Wei, B., 1999. Do the power sector reforms in China reflect the interests of consumers? China Q. 158, 430–446.

Bai, J., Walet, L., 2011. Unified power tariffs mean clearer guidance for solar projects. Chin. Daily August (2), 13.

BP, 2012. BP Statistical Review of World Energy 2012. BP, London.

Chen, J., 2011. China's experiment on the differential electricity pricing policy and the struggle for energy conservation. Energ. Policy 39, 5076–5085.

Chen, L., 2010. Playing the market reform card: the changing patterns of political struggle in China's electric power sector. China J. 64, 69–95.

Cheng, T.J., Tsai, C.M., 2009. Powering rent seeking in the electricity industry. In: Ngo, T.K., Wu, Y. (Eds.), Rent Seeking in China. Routledge, London.

Cherni, J.A., Kentish, J., 2007. Renewable energy policy and electricity market reforms in China. Energ. Policy 35, 3616–3629.

Cheung, K., 2011. Integration of Renewables. Status and Challenges in China, Working Paper, OECD/IEA, Paris.

China Electricity Council, 2004. Main Statistical Index of National Electricity Production 2003. China Electricity Council, Beijing.

China Electricity Council, 2006. Statistical Bulletin of National Power Industry 2005. China Electricity Council, Beijing.

China Electricity Council, 2010a. Statistical Bulletin of National Power Industry 2009. China Electricity Council, Beijing.

China Electricity Council, 2010b. Annual Development Report on China's Power Industry 2009. China Electricity Council, Beijing.

China Electricity Council, 2011a. Statistical Bulletin of National Power Industry 2010. China Electricity Council, Beijing.

China Electricity Council, 2012a. Statistical Bulletin of National Power Industry 2011. China Electricity Council, Beijing.

China Electricity Council, 2012b. Annual Development Report of China's Power Sector 2011. China Electricity Council, Beijing.

Chiu, B., Lewis, M.K., 2006. Reforming China's State-owned Enterprises and Banks. Edward Elgar, Cheltenham.

Downs, E., 2006. The Energy Security Series: China. The Brookings Institution, Washington, DC.

Du, L, Mao, J., Shi, J., 2009. Assessing the impact of regulatory reforms on China's electricity generating industry. Energ. Policy 37, 712−720.

Electricity, 2011. National electric power statistics bulletin 2010. Electr. J. Electr. Power Chin. 22 (1), 54.

Energy Information Administration, 2011. International Energy Statistics 2011. US Energy Information Administration, Washington, DC.

Energy Information Administration, 2012. International Energy Statistics 2012. US Energy Information Administration, Washington, DC.

Gao, C., Li, Y., 2010. Evolution of China's power dispatch principle and the new energy saving power dispatch policy. Energ. Policy 38, 7346−7357.

Hua, J., Chen, A., 2012. China's renewable targets for 2015 and 2020. Reuters Beijing August (14).

Hunt, S., Shuttleworth, G., 1996. Competition and Choice in Electricity. John Wiley, Chichester.

Interfax, 2009. China to give new wind farms power tariffs between $0.075 and $0.089 per kWh. Interfax Chin. Energ. Weekly VIII (29), 23−29 (July 5−6, 2009).

International Energy Agency, 2007. World Energy Outlook 2007. OECD/IEA, Paris.

International Energy Agency, 2009. China Cleaner Coal Study. OECD/IEA, Paris.

Lelyveld, M., 2012. China Pushes Power Plan. Radio Free Asia. Available from: <http://www.rfa.org/english/energy_watch/power-06252012112025.html/> (accessed 25.06.12.).

Lema, A., Ruby, K., 2007. Between fragmented authoritarianism and policy coordination: creating a Chinese market for wind energy. Energ. Policy 35, 3879−3890.

Lewis, J.I., 2007. China's climate change strategy. Chin. Brief VII (13), 9−13.

Li, J., Ma, L., 2007. In focus: China's wind industry. Renew. Energ. Focus 8, 46−48, September/October.

Li, L., Tan., Z., Wang, J., Xu, J., Cai, C., Hou, Y., 2011. Energy conservation and emission reduction policies for the electric power industry in China. Energ. Policy 39, 3669−3679.

Liu, Z., 2012. Clean Coal Power Generation in China. APEC Expert Working Group of Clean Fossil Energy. Available from: <http://www.egcfe.ewg.apec.org/publications/proceedings/CFE/Austrailia_2012/2C-1/_zhan.pdf/> (accessed 10.06.12).

Ma, J., 2011. On-grid tariffs in China: development, reform and prospects. Energ. Policy 39, 2633−2645.

McNulty, S., 2011. China beats the US in wind energy. Fin. Times January (25), Available from: <http://www.ft.com/intl/cms/s/0/1976c12c-2819-11e0-8abc-00144feab49a.html#axzz2OqL47Zjw> (accessed 29.01.11).

Meidan, M., Andrews-Speed, P., Ma, X., 2009. Shaping China's energy policy: actors and processes. J. Contemp. Chin. 18, 591−616.

Miller, L., 2008. Demystifying China's energy wars. Far Eastern Econ. Rev. 171 (7), 44−48.

National Development and Reform Commission, 2004. Notice Regarding Accelerating a Healthy and Orderly Development of Restructuring of the Power Industry. National Development and Reform Commission, Beijing (in Chinese).

National Development and Reform Commission, 2005. Notice on the Enforcement Method for Power Price Reform and on Standardization of Power Price. National Development and Reform Commission, Beijing (in Chinese).

National Development and Reform Commission, 2007a. Eleventh Five-Year Plan for Energy Development. National Development and Reform Commission, Beijing.

National Development and Reform Commission, 2007b. Method of Electricity Dispatch for Energy Saving in Electricity Generation (Trial). National Development and Reform Commission, Beijing (in Chinese).

National Development and Reform Commission, 2007c. China's National Climate Change Programme. National Development and Reform Commission, Beijing.

National Development and Reform Commission, 2007d. Natural Gas Utilisation Policy. National Development and Reform Commission, Beijing (in Chinese).

Ni, C.C., 2006. China's Electric Power Industry and Its Trends. Institute of Energy Economics Japan, Tokyo. Available from: <http://enchen.ieej.or.jp/en/data/326.pdf/> (accessed 19.05.12).

Pan, J., W. Peng, Others, 2006. Rural Electrification in China, 1950–2004. Program on Energy and Sustainable Development. Working Paper No. 60, Stanford University, Stanford.

Pei, M., 2006. China's Trapped Transition. The Limits of Developmental Autocracy. Harvard University Press, Cambridge, MA.

Price, L., Wang, X., Yun, J., 2010. The challenge of reducing the energy consumption of the top-1000 industrial enterprises in China. Energ. Policy 38, 6485–6498.

Resnier, M., Wang, C., Du, P., Chen, J., 2007. The promotion of sustainable development in China through the optimization of a tax/subsidy plan among HFC and power generation CDM projects. Energ. Policy 35, 4529–4544.

Rosen, D.H., Houser, T., 2007. China Energy: A Guide for the Perplexed. Peterson Institute for International Economics, Washington, DC.

Shao, S., Lu, Z., Berrah, N., Tenenbaum, B., Zhao, J. (Eds.), 1997. China: Power Sector Regulation in a Socialist Market Economy. World Bank, Washington, DC.

Stanway, D., 2011. China set to cap energy use in national low-carbon plan. Reuters Beijing August (4).

State Council, 2002. Notice of the State Council on the Electric Power Structural Reform Scheme. State Council, Beijing (in Chinese).

State Council, 2003. Scheme for Power Price Reform. State Council, Beijing (in Chinese).

State Electricity Regulatory Commission, 2003. Guidelines for Establishing Regional Power Markets. Beijing (in Chinese).

State Electricity Regulatory Commission, 2007a. Study of the Capacity Building of the Electricity Regulatory Agency (SERC) P.R. China. State Electricity Regulatory Commission, Beijing.

State Electricity Regulatory Commission, 2007b. Production and business operation of enterprises are improving. Electricity 4, 35–37.

State Electricity Regulatory Commission, 2008a. Policy Recommendations on the Comprehensive Reform of the National Electricity System. State Electricity Regulatory Commission, Beijing.

State Electricity Regulatory Commission, 2008b. Review on development of Chinese power industry in 2002–2006. Electricity 1, 13–32.

State Electricity Regulatory Commission, 2008c. Circular on Electricity Saving and Emission Reduction 2007. State Electricity Regulatory Commission, Beijing.

State Electricity Regulatory Commission, 2008d. Annual Report on Electricity Regulation 2007. State Electricity Regulatory Commission, Beijing.

State Electricity Regulatory Commission, 2008e. Circular on Annual Tariff Implementation and Electricity Billing 2007. State Electricity Regulatory Commission, Beijing.

State Electricity Regulatory Commission, 2009a. Circular on Electricity Saving and Emission Reduction 2008. State Electricity Regulatory Commission, Beijing.

State Electricity Regulatory Commission, 2009b. Circular on Annual Tariff Implementation and Electricity Billing 2008. State Electricity Regulatory Commission, Beijing.

State Electricity Regulatory Commission, 2010a. Circular on Electricity Saving and Emission Reduction 2009. State Electricity Regulatory Commission, Beijing.

State Electricity Regulatory Commission, 2010b. Circular on Annual Tariff Implementation and Electricity Billing 2009. State Electricity Regulatory Commission, Beijing.

State Electricity Regulatory Commission, 2011a. Circular on Electricity Saving and Emission Reduction 2010. State Electricity Regulatory Commission, Beijing.

State Electricity Regulatory Commission, 2011b. Circular on Annual Tariff Implementation and Electricity Billing 2010. State Electricity Regulatory Commission, Beijing.

State Electricity Regulatory Commission, 2012. Annual Report on Electricity Regulation 2011. State Electricity Regulatory Commission, Beijing.

Thomson, E., 2003. The Chinese Coal Industry: An Economic History. Routledge Curzon, London.

Victor, D.G., Heller, T.C., 2007. Major conclusions: the political economy of power sector reform in five developing countries. In: Victor, D.G., Heller, T.C. (Eds.), The Political Economy of Power Sector Reform. Experiences of Five Major Developing Countries. Cambridge University Press, Cambridge, pp. 254–306.

Wang, B., 2007. An imbalanced development of coal and electricity industries in China. Energ. Policy 35, 4959–4968.

Wang, V., 2011. China's gas-fueled power plans could backfire. Interfax Chin. Energ. Weekly X (35), 12 (September 28, 2011).

Wang, Z., Zhang, B., Zhang, Y., 2012. Determinants of public acceptance of tiered electricity price reform in China: evidence from four urban cities. Appl. Energ. 91, 235–244.

World Nuclear Organisation, 2012. Nuclear Power in China. Available from: <http://www.world-nuclear.org/info/inf63.html/> (accessed 10.06.12).

Xu, Y.C., 2002. Powering China. Reforming the Electrical Power Sector in China. Ashgate/Dartmouth, Aldershot.

Yeh, E.T., Lewis, J.I., 2004. State power and the logic of reform in China's electricity sector. Pac. Affairs 77, 437–465.

Zhang, C., Heller, T.C., 2007. Reform of the Chinese electrical power market: economics and institutions. In: Victor, D.G., Heller, T.C. (Eds.), The Political Economy of Power Sector Reform. Experiences of Five Major Developing Countries. Cambridge University Press, Cambridge, pp. 76–108.

Zhang, C., Heller, T.C., May, M.M., 2005. Carbon intensity of electricity generation and CDM baseline study: case studies of three Chinese provinces. Energ. Policy 33, 451–465.

Zhang, L., 2012. Electricity pricing in a partially reformed plan system: the case of China. Energ. Policy 43, 214–225.

Zhang, V.Y., Chen, Y., 2011. Vertical relationships in China's electricity industry: the quest for competition? Util. Policy 19, 142–151.

Zhang, Z.X., 2006. Towards an effective implementation of clean development mechanism projects in China. Energ. Policy 34, 3691–3701.

Zhang, Z.X., 2011. Energy and Environmental Policy in China. Towards a Low-Carbon Economy. Edward Elgar, Cheltenham.

Zhao, X., Lyon, T.P., Wang, F., Song, C., 2012a. Why do electricity utilities cooperate with coal suppliers? A theoretical and empirical analysis from China. Energ. Policy 46, 520–529.

Zhao, X., Lyon, T.P., Song, C., 2012b. Lurching towards markets for power: China's electricity policy 1985–2007. Appl. Energ. 94, 148–155.

Part IV

The Evolution of Electricity Markets in Australasia

Evolution of Australia's National Electricity Market

Alan Moran[a] and Rajat Sood[b]

[a]*Institute of Public Affairs, Melbourne, Australia,* [b]*Frontier Economics, Melbourne, Australia*

1 INTRODUCTION

The Australian electricity supply industry has undergone substantial pro-competitive reform over the last two decades. In the early 1990s, virtually all electricity was supplied through vertically integrated state monopolies. Within a decade, the integrated monopolies had been disaggregated into different businesses with the competitive aspects of supply (generation and retailing) reconstituted into dozens of independent firms, many of them privately owned and the rest "corporatized" and operating at arm's length from their government owners. Monopoly aspects of supply, transmission and distribution networks were also in some cases privatized with their prices set by independent economic regulators.

Nevertheless, the process of reform remains incomplete in key areas and certain aspects of market design remain unsettled. The incomplete status of reforms largely reflects political reluctance to engage in further privatization and retail tariff liberalization, while ongoing debates over aspects of market design, such as nodal pricing, are attributable to policy makers' self-acknowledged doubts about the benefits of greater sophistication.

Against this background, the Australian electricity sector faces fresh challenges as a result of policies aimed at reducing greenhouse gas emissions. The Federal Government introduced the world's most broadly based tax on emissions in July 2012. At the same time, consumer costs are increased by a 20% Renewable Energy Target (RET). Renewables also benefit from more direct tax-payer- or customer-financed subsidies.

This chapter commences by describing the most important elements of the Australian electricity supply industry (Section 2), before outlining the key structural, institutional, and regulatory reforms undertaken to date (Section 3). The chapter then explains the design and performance of the National Electricity Market (NEM, Section 4), followed by a brief discussion of outstanding areas of reform (Section 5). The chapter goes on to explain

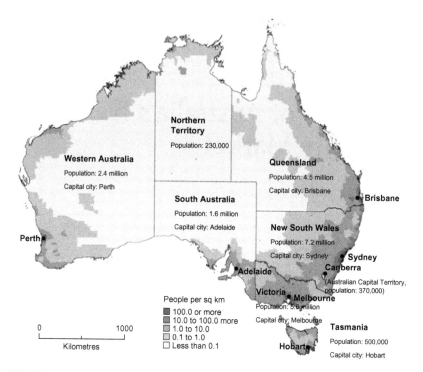

FIGURE 19.1 Map of Australia—Population Density. *Source: Australian Bureau of Statistics, 2012. Year Book Australia. Figure 7.14, p. 247.*

the nature and effects of Australian climate change policies on the power industry (Section 6) and concludes with some final observations (Section 7).

2 THE AUSTRALIAN ELECTRICITY SUPPLY INDUSTRY

2.1 Industry Outline

Electricity production across Australia's six states and two territories was approximately 228 TWh in 2010–11.[1] Australian grid-connected generation capacity as at June 2011 was approximately 54 GW, with the bulk of capacity being in the form of coal-fired steam turbines.[2]

Australia's population is concentrated within a narrow strip of land near the eastern, southeastern, and southwestern coastlines (Figure 19.1).

A high-voltage network, rated at 220 kV or higher, connects most consumers and producers of electricity in Queensland, New South Wales (NSW),

---

1. Energy Supply Association of Australia, 2012. Electricity Gas Australia. Table 2.5, pp. 24–25.

2. Energy Supply Association of Australia, 2012. Electricity Gas Australia. Table 2.1, pp. 20–21.

FIGURE 19.2   Map of the NEM. *Source: AER, 2012. State of the Energy Market. Figure 1.3, p. 31.*

the Australian Capital Territory (ACT), Victoria, South Australia, and (via the undersea Basslink cable) Tasmania (Figure 19.2). This power system operates within the NEM and it incorporates the longest interconnected AC network in the world, stretching 4500 km.[3]

3. Australian Energy Regulator, 2012. State of the Energy Market. p. 28. Available from: http://www.aer.gov.au/node/18959.

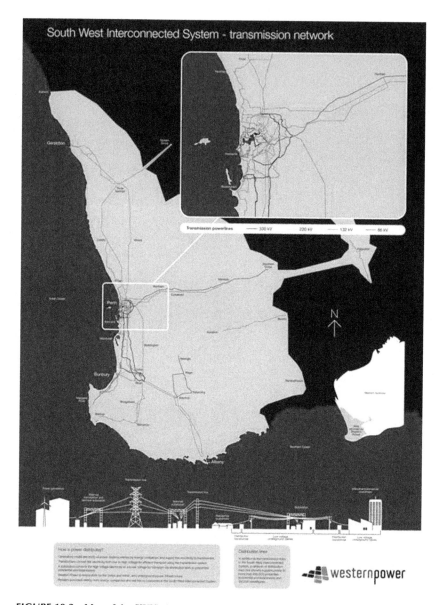

FIGURE 19.3　Map of the SWIS. *Source: Western Power.*

The main power system in Western Australia is known as the South West Interconnected System (SWIS), which operates within the Wholesale Electricity Market (WEM) (Figure 19.3). Various smaller systems supply consumers in northern parts of Western Australia and the Northern Territory as well as in other isolated areas.

The NEM commenced in December 1998 following the earlier creation of state-based wholesale markets in Victoria, NSW, and Queensland. Queensland was interconnected to the southern states in 2000 with the commissioning of the Terranora interconnector (formerly Directlink), and later the Queensland–NSW Interconnector (QNI). Tasmania entered the NEM in May 2005, followed by the commissioning of Basslink.

The NEM supplies approximately 9 million end-use customers. Residential households represent approximately 90% of total end-use customers but account for less than 30% of annual electricity consumption, with business, commercial, and large industrial customers making up the remainder.

While almost all electricity assets were originally owned and operated by state government instrumentalities, there are now multiple private and government-owned businesses operating as generators and/or retailers in the NEM. Distribution and transmission network businesses in the NEM also reflect a mix of government and private ownership. Table 19.1 in Section 2.2.1 outlines the ownership structure of key participants in each NEM jurisdiction.

All customers in the mainland NEM are now contestable. In Tasmania, only customers with annual consumption levels above 150 MWh are contestable.[4] However, retail tariff controls continue to apply to smaller customers outside Victoria and South Australia.

In Western Australia, the WEM is dominated by the state-owned generation and retailing businesses, Verve Energy (Verve) and Synergy, respectively. Verve and Synergy were formally the generation and retailing arms of the vertically integrated state utility Western Power, which was disaggregated in 2006. The SWIS transmission and distribution networks are operated by Western Power, which also remains government owned. Most energy in the WEM is traded under contracts, but the Independent Market Operator operates a real-time balancing mechanism and a market for capacity credits. Customers in the SWIS with annual consumption levels of 50 MWh or higher are contestable. However, customers consuming between 50 and 160 MWh per annum can choose to be supplied by Synergy at a regulated tariff. The state government approves changes in regulated retail tariffs for both residential and business customers.

2.2 Key Supply and Demand Characteristics

2.2.1 National Electricity Market

Most of Australia's electricity generation capacity is located in the NEM. Registered NEM generation capacity is approximately 48 GW across 308 registered generators and total electricity generated over 2011/2012 was 199 TWh.[5]

---

4. Energy Supply Association of Australia, 2011. Electricity Gas Australia. Table 4.2, p. 60.
5. AER, 2012. State of the Energy Market. p. 28.

TABLE 19.1 Key Participants in the NEM and Their Ownership

Generators	Ownership	Retailers	Ownership	Transmission	Ownership	Distribution	Ownership
Queensland							
Stanwell	Queensland Government	AGL	Private	Powerlink	Queensland Government	Energex	Queensland Government
CS Energy	Queensland Government	Origin Energy	Private			Ergon Energy	Queensland Government
Origin	Private	Ergon Energy	Queensland Government				
InterGen	Private						
Arrow	Private						
Alinta	Private						
New South Wales							
Macquarie Generation	NSW Government	AGL	Private	TransGrid	NSW Government	AusGrid	NSW Government
Origin Energy	Private	Origin Energy	Private			Endeavour Energy	NSW Government
EnergyAustralia	Private	EnergyAustralia	Private			Essential Energy	NSW Government
Delta Electricity	NSW Government						
Snowy Hydro	Split between the NSW, Victorian and Federal Governments						

Victoria

EnergyAustralia	Private	AGL	Private	SP AusNet	Private	Powercor	Private
GEAC	Private	Origin	Private			SP AusNet	Private
International Power	Private	EnergyAustralia	Private			United Energy	Private
Snowy Hydro	Split between the NSW, Victorian and Federal Governments	Momentum Energy	Tasmanian Government			CitiPower	Private
AGL	Private					Jemena	Private

South Australia

AGL	Private	AGL	Private	ElectraNet	Majority Owned by Powerlink (Queensland Government)	ETSA Utilities	Private
International Power	Private	Origin	Private				
Alinta	Private	EnergyAustralia	Private				
Origin	Private	Aurora Energy	Tasmanian Government				
EnergyAustralia	Private	Momentum Energy	Tasmanian Government				

(Continued)

TABLE 19.1 (Continued)

Generators	Ownership	Retailers	Ownership	Transmission	Ownership	Distribution	Ownership
Infigen	Private						
Infratil	Private						
Tasmania							
Hydro Tasmania	Tasmanian Government	Aurora Energy	Tasmanian Government	Transend	Tasmanian Government	Aurora Energy	Tasmanian Government
Aurora Energy (Tamar Valley)	Tasmanian Government						

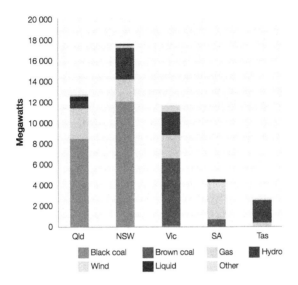

FIGURE 19.4   Registered capacity by fuel source 2012. Note: New south wales and victoria include snowy Hydro capacity allocated to those regions. *Source: AEMO; AER. 2012. State of the Energy Market. Figure 1.5, p. 32.*

Most generation in the NEM is coal-fired steam turbine plant, although significant gas-fired capacity has been developed over the past decade in response to rising peak demand. Black and brown coal plants are the primary sources of baseload power, with 59% of registered capacity and 78% of output. Gas-fired generation mainly provides intermediate and peaking capacity, with 22% of registered capacity but only 12% of output. Hydroelectric generation capacity is relatively small despite the large land area and is nearly fully developed, accounting for 16% of capacity and 8% of output. Scheduled wind now accounts for 4% of capacity and 2.5% of output.

Government policies to promote renewable generation have driven the development of wind, particularly since 2001, further described in Section 6.2. Liquid and other sources of power are minimal and nuclear power is forbidden by legislation.

As shown in Figure 19.4, the States of Queensland and NSW are dominated by black coal-fired plant, whereas Victoria is mainly supplied by brown coal plant. South Australia has limited coal reserves and has traditionally relied on gas-fired plant. Finally, Tasmania is chiefly supplied by hydroelectric plant. Most wind plant in the NEM has been developed in South Australia and Victoria, where wind capacity factors tend to be highest.

Even though the coal used in Australian power stations is mainly below "export quality," rising international prices for black coal in recent years as a result of the worldwide commodity boom have increased the fuel costs of a

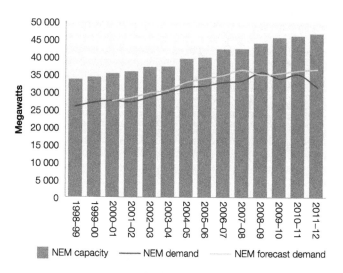

FIGURE 19.5   NEM peak demand and capacity. Notes: Demand forecasts are two years in advance,based on a 50 per cent probability that the forecast will be exceeded and an average diversity factor. NEM capacity excludes wind generation and power stations not managed through central dispatch. *Source: AEMO, Electricity statement of opportunities for the National Electricity Market, various years. AER, 2012. State of the Energy Market. Figure 1.23, p. 56.*

typical NSW black coal plant from \$A15 to over \$A25/MWh.[6] Gas prices on the Australian east coast are expected to rise toward export parity in response to the development of liquefied natural gas (LNG) terminals in Queensland. However, wholesale electricity prices have so far not risen to match these cost increases. This is partly due to the expansion of subsidized renewable generation, which is typically bid into the market at a low or negative marginal cost, in recent years and partly due to demand-side factors.

Peak demand in the NEM typically occurs during summer, driven by air-conditioning load during heat waves. In recent years, peak summer demand has reached 35.6 GW (Figure 19.5), but was only 30.3 GW in the summer of 2012.[7]

Total electricity production reached 212 TWh in 2008/2009 (Figure 19.6) but has subsequently tapered off to 199 TWh.[8] The flattening of consumption and, to a lesser extent, peak demand has been in part attributed to cooler summer weather in recent years as well as to the installation of domestic solar photovoltaic (PV) panels encouraged by generous federal and state

---

6. At the time of writing, 1 Australian dollar was similar in value to 1 US dollar. For the remainder of this chapter, the notation "\$" will be used to refer to Australian dollars.
7. AER, 2012. State of the Energy Market. p. 28.
8. AER, 2012. State of the Energy Market. p. 28.

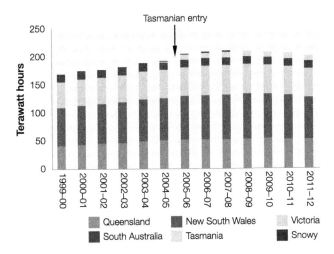

FIGURE 19.6   NEM electricity consumption. Note: The snowy region was abolished on 1 july 2008. Its energy demand was redistributed between the victoria and new south wales regions from that date. *Source: AEMO; AER. 2012. State of the Energy Market. Figure 1.1, p. 28.*

government subsidies. Total consumption is now forecast to reach only 222 TWh by 2021/2022.[9]

Another key factor in slowing electricity consumption has been steadily rising retail tariffs. The principal driver of these increases has been sharply higher network charges. In particular, expenditure on low voltage distribution networks has increased by over 50% in real terms in NSW and Queensland over the last decade as investment in the grid has ballooned to meet summer demand peaks. Higher retail tariffs also reflect the costs of renewable subsidies, such as the RET and generous fed-in tariffs (FiTs) for solar PV installations. These schemes and the new carbon pricing mechanism are discussed in Section 6.

2.2.2 Wholesale Electricity Market

The WEM incorporates over 6 GW generation capacity.[10] This is dominated by coal, gas, and dual-fueled (gas and liquids) generation. Western Australia's heavy reliance on gas for electricity generation was exposed in the winter of 2008, when an explosion at Varanus Island reduced gas supplies to the southwest of the state by 30%, resulting in many forced outages and higher wholesale prices. Significant wind capacity has been developed in recent years, largely due to incentives under the WEM Capacity Mechanism and the Federal Government's

9. Australian Energy Market Operator, 2012. National Electricity Forecasting Report. Table 3.1, pp. 3–4. Available from: http://www.aemo.com.au/en/Electricity/Forecasting/2012-National-Electricity-Forecasting-Report.
10. See Independent Market Operator web site at: http://www.imowa.com.au/rc-capacity-in-the-swis.

RET policy. As a result of new entry since market commencement, incumbent Verve Energy's share of overall capacity has fallen from nearly 90% in 2005/2006 to just over 50% in 2012/2013.[11]

Like the NEM, peak demand in the WEM occurs during summer. Presently, peak demand is about 4 GW and total consumption in 2011/2012 was approximately 18 TWh.[12] The Independent Market Operator expects peak demand to increase by 1.5 GW over the next decade, an increase of nearly 40%.[13] This compares with NEM peak demand, which is expected to grow by 16% over the same period.[14]

As noted above, retail supply to household customers in Western Australia is not contestable. Regulated retail tariffs have been rising steeply in Western Australia, but from a low base. Residential tariffs did not change for over a decade, from 1997/1998 to 2008/2009 and business tariffs did not increase between 1991/1992 and 2007/2008. To help bring tariffs back into line with the costs of supply, the state government approved increases to residential tariffs totaling 25% in 2009, a further 17.5% in 2010, and 5% in 2011. Businesses have seen similar aggregate increases.[15]

3 STRUCTURAL, INSTITUTIONAL, AND REGULATORY REFORMS

3.1 Initial State-Based Reforms

3.1.1 Disaggregation and Privatization[16]

Prior to the 1990s, electricity supply in each state was provided by vertically integrated, government-owned monopolies. The initial impetus for reform came from:

- the recognition that other countries were achieving considerably greater efficiencies than Australia in electricity supply[17,18];

---

12. Independent Market Operator, January 2013. Wholesale Electricity Market: Request for Expressions of Interest for the 2013 Reserve Capacity Cycle, p. 5.

13. Independent Market Operator, January 2013. Wholesale Electricity Market: Request for Expressions of Interest for the 2013 Reserve Capacity Cycle. Figure 4, p. 8.

14. Australian Energy Market Operator, 2012. National Electricity Forecasting Report, Table 3.1, pp. 3–4.

15. See Western Australian Department of Finance web site at: http://www.finance.wa.gov.au/cms/content.aspx?id=15096.

16. For more detail on the history of early reforms, see Moran, A., 2006. The Electricity Industry in Australia: Problems Along the Way to a National Electricity Market (Chapter 6). In: Sioshansi, F.P., Pfaffenberger, W. (Eds.), Electricity Market Reform: An International Perspective. Elsevier.

17. Project Victoria: A Rebuilding Strategy for Electricity in Victoria, Tasman Institute/Institute of Public Affairs, 1991.

18. Industry Commission, Energy Generation and Distribution Report, Report No. 11, May 17, 1991. Available from: http://www.pc.gov.au/ic/inquiry/11energy.

- the implementation of National Competition Policy, which required a general review of the operations of "essential facilities" and a requirement that they be opened to nonaffiliates on reasonable terms; and
- the consequences of poor financial circumstances in the States of Victoria and South Australia, which resulted in new governments that sold energy assets partly in pursuit of a privatization agenda and in part to reduce debt.

Following initial moves to reduce overmanning, Victoria was the first state to undertake vertical and horizontal disaggregation of its electricity business. The Kennett Liberal (conservative) Government, elected in 1992, broke up the State Electricity Commission (SECV) into separate generation, transmission, and distribution/retail businesses before proceeding to sell those businesses over the mid- to late-1990s. While the government had a strong philosophical belief in the benefits of privatization, it also wanted to avoid the mistakes made in Britain, where the reforms had resulted in a highly concentrated structure of generation ownership. Great care was taken to establish multiple generation, distribution, and retailing firms, with this being given a higher priority than maximizing sale proceeds. Notwithstanding these steps to minimize concentration, the sale process earned $23 billion, far more than the $9−10 billion that had been widely expected. The Kennett Government also set up an independent regulator to set prices for the monopoly network businesses as well as maximum retail tariffs. The regulator gathered performance data that would prove invaluable in defusing claims that privatization had led to higher prices and lower reliability. The setting up of independent regulators was a step followed by all other states as they introduced their reforms.

The Victorian disaggregation and privatization took place as a result of financial pressures and pro-competition/privatization views that were especially prevalent in that state. In the case of other states, the reforms followed on a program of National Competition Policy, agreed to by the state and federal governments in 1996. The competition policy reforms called for natural monopolies to be opened to all users on terms that were fair and reasonable. Regulatory arrangements were put in place to determine where such access was required and, in the event of disputes, the prices at which access was to be made available.

In the late 1990s into 2000, South Australia disaggregated its monopoly ETSA Corporation and privatized its component businesses. However, the extent of disaggregation and commitment to competition was not as strong as in Victoria. The government also encouraged the development of a private unregulated interconnector from Victoria (Murraylink) instead of a regulated open access link from NSW.

In Queensland, the Electricity Commission was disaggregated in several steps. By the start of the NEM, the industry comprised two main distributor/retailers, a transmission business and four generation businesses, all in government ownership. In 2007, the retail businesses were separated from their distribution networks and one was sold. New generators were built but most generation capacity remains in government ownership. In 2010, the government reaggregated its three generation businesses into two in response to falling wholesale prices.

The New South Wales Government's formerly integrated Electricity Commission has undergone several iterations of disaggregation. At NEM start, there were three major generation businesses (excluding the separate Snowy Hydro business, which remains co-owned with the Victorian and Federal Governments), a single transmission company, and five distributor/retailers. The three smallest distributor/retailers were merged in 2001. In late 2010 and early 2011, the retail businesses and trading rights to several of the generators were sold under the "Gentrader" model further described in Section 4.4. However, the largest generator, the transmission network and all distribution networks presently remain in government ownership.

Tasmania entered the NEM in 2005, 7 years after vertically separating the monopoly Hydro-Electric Corporation into a single generation business, transmission business, and combined distribution/retail businesses. All components remain in government ownership at the time of writing, but the government has raised the prospect of a sale of the Aurora retail business. There is no plan to disaggregate or sell the Hydro Tasmania generation business or the networks.

Outside the NEM, the Western Australian Government disaggregated its State Energy Commission to form Western Power in 1996. Western Power was vertically disaggregated in 2006 to form Western Power Networks, Verve (generation) and Synergy (retailing), plus Horizon Energy, which operates outside the SWIS. At the time of writing, the government had proposed remerging Verve and Synergy.

Table 19.1 sets out the current ownership status of key participants in the NEM by state.

3.1.2 Early Productivity Gains

The initial state-based reforms yielded substantial productivity gains. Electricity output grew due to better plant management while capital inputs remained flat and employment fell steeply as overmanning was reduced. As a result, output per worker increased from 2 GWh in 1989/1990 to 5.5 GWh per employee in 1997/1998. Generator productivity continued to increase across all Australian states in the period after 1997. By 2004, after which comparable employment figures are unavailable, all state generation systems were operating at 30–40 GWh per employee with productivity growth highest in the fully privatized systems of Victoria and South Australia[19] (Figure 19.7).

---

19. For more details see Moran, A., Skinner B., 2008. Resource adequacy and efficient infrastructure investment (Chapter 11). In: Sioshansi, F.P. (Ed.), Competitive Electricity Markets: Design, Implementation, Performance. Elsevier.

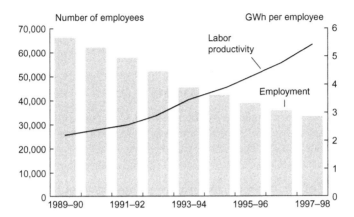

FIGURE 19.7 Electricity industry employment and labor productivity. *Source: Treasury Economic Roundup Spring 1999. Developments in Electricity. Chart 4, p. 57.*

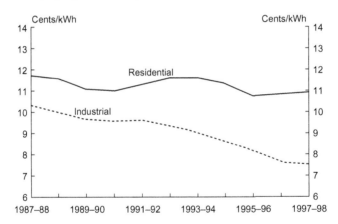

FIGURE 19.8 Australian electricity prices (real terms). *Source: Treasury Economic Roundup Spring 1999. Developments in Electricity. Chart 5, p. 58.*

Reflecting the lower wholesale prices, retail prices also broadly fell in real terms over this period, especially for commercial and industrial users (Figure 19.8).

3.1.3 Creation of Spot Markets

Alongside structural reform, the states were also involved in experimenting with the creation of their own spot markets. The Victorian system operator, VPX, began operating the VicPool in early 1995 and the NSW ELEX market commenced in 1996. Both markets were compulsory spot markets, based on the design of the England and Wales pool. VPX and the NSW system operator, TransGrid, jointly introduced systems to form the NEM1 market in

March 1997. The NEM1 market allowed for managed interstate trade of electricity, but VPX and TransGrid retained responsibility and control over power system security and reliability in their own states.

3.2 Creation of the National Market

The NEM was devised jointly by the state and Commonwealth governments and implemented through cooperative mirror legislation known as the National Electricity Law (NEL). The NEL established the NEM and provided for the National Electricity Code (Code), which set out the arrangements for:

- operation of the wholesale spot market and management of the interconnected power system and
- terms of access to the transmission and distribution networks across the NEM jurisdictions.

The NEL also established the NEM governing bodies, the National Electricity Market Management Company (NEMMCO), and the National Electricity Code Administrator (NECA). NEMMCO's role was to operate the market, manage the power system, and approve new interconnectors. NECA's role was to administer changes to and enforce the Code. The national regulatory body, the Australian Competition and Consumer Commission (ACCC), was also required to "authorize" the original Code and any Code changes because of the competitive implications of the provisions dealing with price-setting and terms of network access. The ACCC was also given the job of regulating transmission network revenues using a methodology that incorporates an inflation adjustment and a required productivity increase. This is referred to as CPI-X; CPI is the Consumer Price Index.

3.3 Subsequent Institutional and Regulatory Reforms

The autonomous and overlapping style of decision-making practiced by NECA and the ACCC led to pressures from a number of sources for an overhaul of their powers. In 2005, the NEM jurisdictions agreed to change the NEM institutional arrangements. The Ministerial Council on Energy (MCE) was set up to provide high-level policy direction to the market institutions. The Code was replaced by the National Electricity Rules (the Rules). NECA was dissolved and replaced by the Australian Energy Market Commission (AEMC), which was given responsibility for making changes to the Rules without ACCC approval and for conducting broader reviews requested by the MCE. At the same time, the Australian Energy Regulator (AER) was set up as a division within the ACCC to monitor and enforce compliance with the Rules as well as to perform the ACCC's transmission revenue regulatory functions. In 2008, the economic regulation of distribution networks was transferred from the jurisdictional

TABLE 19.2 NEM Institutional Responsibilities

Policy Planning and Regulatory Institutions in the NEM

Institution	Role/Function
AEMO	Wholesale market operation and settlement
	Power system operation
	Transmission planning in Victoria (only)
AEMC	Making changes to the National Electricity Rules
	Conducting reviews mandated by the SCER
AER	Enforcement of the National Electricity Rules
	Economic regulation of transmission and distribution networks
SCER	Overall policy and reform direction

regulators to the AER. However, the ACCC maintains its role in approving mergers and acquisitions in the energy industry and elsewhere. In 2011, the MCE was replaced by the Standing Council on Energy and Resources (SCER).

NEMMCO's functions as market and system operator for the NEM were largely untouched by the 2005 reforms. In 2009, a number of state planning bodies and gas market operators were combined with NEMMCO to form the Australian Energy Market Operator (AEMO). In addition to market and system operation across the NEM, AEMO is responsible for transmission network planning in Victoria. This differs from the arrangements in place in other states, where the primary transmission network owner and operator also has responsibility for network planning. In Victoria, planning functions have long been separated from network ownership and operation, which is left to other privately owned parties.

Table 19.2 summarizes the current institutional roles in the NEM.

Despite the implementation of full retail contestability across the mainland NEM in the early 2000s, maximum retail tariffs for small customers remain the subject of jurisdictional regulation outside Victoria and South Australia. Nonpricing regulations apply throughout the NEM. Further discussion of retail reforms is contained in Section 4.3.

4 DESIGN AND PERFORMANCE OF THE NEM

4.1 Wholesale Market

4.1.1 Market Design and Operation

The NEM is an energy-only gross pool limited nodal market with security-constrained dispatch. Generators and scheduled loads submit bids on a day-ahead basis in up to 10 price bands. Participants can change their bids or "rebid" up until real time. AEMO, as market and system operator, schedules plant in accordance with their positions in the bid-based merit order to minimize the overall costs of dispatch. The final generator bid accepted by AEMO to meet demand sets the spot price. The spot market is cleared on a 5 min ("dispatch interval") basis and the simple average of the six dispatch interval prices within a half-hour ("trading interval") is used to settle wholesale transactions. Typical spot prices are $30-60/MWh.

The energy-only feature of the NEM means that there is no separate capacity market; generators must earn their entire revenues either through spot market transactions or through derivative contracts settled against spot market outcomes. Because of this, spot prices are permitted to reach very high levels—and often do reach such levels at times of peak demand and supply scarcity—to enable generators to earn prices in excess of their operating costs and hence recover a contribution toward their fixed costs. To minimize cash flow risks, generators hedge most of their output with derivative contracts—further explained below.

The AEMC Reliability Panel is responsible for setting the market price cap at a level designed to encourage sufficient generation capacity to maintain the NEM reliability standard, which is that unserved energy does not exceed 0.002% of total energy in a year. The market price cap in the NEM is now $12,900/MWh, having risen from $5000/MWh at NEM start. This is well in excess of the offer caps applying or being contemplated in the energy-only Texas market (Chapter 10). If the sum of wholesale prices over the previous 336 half-hours (1 week) exceeds the Cumulative Pricing Threshold (CPT)—$187,500 indexed from 2012—an Administered Pricing Cap of $300/MWh applies. This threshold has rarely been reached, as noted in Section 4.1.2. AEMO also has the ability to intervene in the market to procure reserves if this standard does not appear likely to be maintained up to 9 months in advance. This Reliability and Emergency Reserve Trader function is scheduled to expire on June 30, 2013.

There are no provisions in the Rules or elsewhere preventing generators bidding some or all of their capacity up to the market price cap if they so choose. Spot prices are also permitted to fall to as low as-$1000/MWh to facilitate generators running at minimum levels at low-load times (such as overnight) or when transmission constraints bind.

The gross pool nature of the NEM means that most sales and purchases of electricity by significant grid-connected participants must be settled at the relevant NEM spot price. Both generator and retailer market participants usually hedge the bulk of their spot market exposures with derivative contracts, such as swaps and caps. Over-the-counter contracts have historically been the most common form of hedging tool. However, in recent years, the trading volume of exchange-traded contracts has increased dramatically. In both cases, swap strike prices traditionally incorporate a positive premium over expected spot prices due to the upward skew of spot price outcomes.

There is considerable dispute on the question of whether it is necessary to have a market that rewards suppliers with capacity payments as well as for the energy they actually have scheduled.[20]

Electricity retailers have to be able to constantly meet a variable demand at a final consumer price that is normally inflexible. The case for capacity payments is based on concerns that an energy-only market design might not be able to achieve this or that it would require very high maximum prices to ensure that supplies are continuously available (or to provide a sufficient incentive for demand to back off where this is possible).

A capacity market seeks to ensure continuous availability by providing a direct reward for suppliers simply for being available to be scheduled. Its rationale is that this is achieved at a lower cost and with greater certainty than is the case in an energy-only market. Adib et al.[21] express doubt about this taking place when they refer to the "bipolar nature" of capacity markets. They note that capacity prices tend toward zero where capacity is ample and infinity where it is short.

In general, it might be argued that if additional payments are made for supplying energy for one set of reasons, compensating reductions will occur in real-time energy prices as firms jockey for revenues that cover their costs.

Other deficiencies of a capacity market include:

- It is a blunt instrument requiring, if it is to work well, bureaucratic judgments on the reliability and dispatchability values of particular types of plant.
- Typically it has to be administered on a short forward period (sometimes day-ahead, rarely, as with the Western Australian capacity market, more than year ahead) while the requirement is to obtain assurances of capacity some time further into the future.

The Australian NEM has operated very satisfactorily with an energy-only market. This entails retailers constantly searching for lower cost supplies of energy and for a pattern of supply availability that meets their forecast demand. For their part, generators also look at the forward demand in its different time-of-day and seasonal patterns, and in the context of the competitive environment they face. Generally, this means the parties arrange for contracts of varying lengths—generators need these to offer assurances to

20. It was addressed in detail in Moran, A., Skinner, B., 2008. Resource adequacy and efficient infrastructure investment (Chapter 11). In: Sioshansi, F.P. (Ed.), Competitive Electricity Markets. Elsevier.

21. Adib, P et al., 2008. Resource adequacy, alternate perspectives and divergent paths (Chapter 9). In: Sioshansi, F.P. (Ed.), Competitive Electricity Markets. Elsevier.

their lenders, especially when looking to construct new capacity, while retailers need them to ensure their ongoing capabilities to supply.

Another alternative is for a public agency to contract for additional supply where it considers existing supplies to be insufficient. To do so it must either:

- move into the market and contract supplies at a higher price than the supplier was able to get from actual customers or
- build its own capacity.

If the government agency contracts additional supplies this will encourage firms or demand-side suppliers to hold back offering contracts to the market, hoping that the government will offer them a better price; such an outcome was reported by Joskow (2006)[22]: "In New England, the amount of generating capacity operating subject to special reliability contracts with the ISO has increased from about 500 MW in 2002 to over 7000 MW projected for 2005 (ISO-New England, 2005), amounting to over 20% of peak demand." Such responses will undermine the commercial market as a whole.

If the reserve power agency was to hold its own capacity to be used only in special circumstances, for example, when the price equals the spot market cap for an agreed period of time, this would simply create an added insurance cost. If the reserve capacity was to be used more liberally than this, it would distort investment incentives bringing future supply shortages.

The NEM is a limited nodal market in that all wholesale transactions are settled at one of five regional reference prices (RRPs), each within a geographically defined region. Generators and retailers within each region receive and pay, respectively, the relevant RRP for that region adjusted by an annually set static marginal loss factor based on their precise location within the region. The RRP in each region for a given trading interval is determined by the marginal cost of supply at a point in the region known as the regional reference node (RRN). The marginal cost of power at the RRN is determined on the basis of participants' bids and applicable transmission constraints. RRPs may vary from one another due to transmission constraints that affect flows between regions and dynamically adjusted marginal loss factors. Participants can hedge interregional exposures by acquiring interregional settlement residue (IRSR) units, which are offered through quarterly auctions run by AEMO. IRSR units are available for each flow direction between adjacent regions and are funded by the rentals accruing from interregional power transfers at times of RRP divergences.

To the extent that transmission constraints limit flows within a region, they are taken into account in the dispatch process and may affect the RRP. However, such intraregional constraints are not reflected in more localized

22. Joskow, P., 2006. Competitive Electricity Markets and Investment in New Generating Capacity, April 6–9, http://dspace.mit.edu/bitstream/handle/1721.1/45055/2006-009.pdf?sequence=1.

settlement prices. This means that when constraints bind, a disjunction can arise—along the lines well documented in the Enron-inspired collapse of the Californian market—between the marginal value of power at a participant's location (the nodal shadow price) and the RRP upon which participants are settled. When this occurs, generators can have incentives to bid in a "disorderly" manner. For example, generators in a load-rich part of a region, with a higher nodal shadow price than the RRP may have incentives to bid up to the market price cap to avoid being dispatched at a price they find inadequate. Conversely, generators in generation-rich parts of a region may have incentives to bid down to the market floor price in order to be dispatched at an attractively high RRP.

Originally, the NEM had five regions, largely based on state boundaries. This made sense because the historical state-based development of transmission networks meant that the primary points of transmission congestion emerged at state boundaries. NECA proposed expanding the number of regions in the early 2000s to reflect additional points of network congestion. However, state governments' concerns about intrastate price differences, generator market power, inadequate hedging instruments, and the lack of clearly demonstrable benefits from more regions stopped NECA from proceeding. The number of regions increased to 6 when Tasmania joined in the NEM in 2005, but reverted to 5 in 2008 when the Snowy region was abolished and its area absorbed into the NSW and Victorian regions. The AEMC is presently reviewing the appropriateness of more localized pricing of energy as part of its Transmission Frameworks Review.

In addition to the spot market for energy, the NEM incorporates eight ancillary services markets for frequency control. As generators can trade-off their provision of frequency control against their provision of energy, AEMO's dispatch engine co-optimizes the provision and pricing of frequency control ancillary services (FCAS) with outcomes in the energy market. Network control and system restart ancillary services are procured by AEMO through contracts with participants.

4.1.2 Pricing and Investment Outcomes

Spot prices in the NEM have generally remained under $100/MWh for most of the time in most regions (Figure 19.9).

Following market start, prices in South Australia and Queensland were higher than elsewhere due to relatively tight supply conditions and limited interconnection. This led to new generation investment in both states. Hot summer conditions in early 2001 led to higher prices across the NEM, particularly in South Australia and Victoria, stimulating investment in gas peaking plant in those states. Further interconnector investment was also developed over 2002/2003, both between Victoria and South Australia, and between Victoria and the Snowy region. These investments combined with milder summer conditions over the next few years led to fairly flat average wholesale prices from 2002 to early 2007.

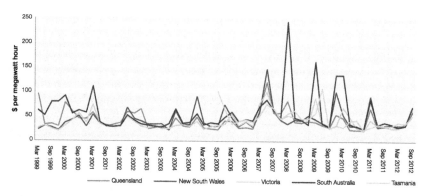

FIGURE 19.9   Quarterly spot prices since NEM start. Note: Volume weighted average prices. *Source: AEMO; AER. 2012. State of the Energy Market. Figure 1.15, p. 46.*

In 2007, an emerging drought over southeastern Australia constrained the hydroelectric generation capacity in the Snowy, Victorian, and Tasmanian regions. The drought also limited the availability of water for cooling in some coal-fired generators, especially in Queensland. These conditions were exacerbated in June 2007 by a number of generator and network outages and generator limitations.[23] As a result of all of these factors, spot prices rose considerably.

South Australian prices spiked in March 2008 due to a record heatwave combined with reduced interconnector capacity from Victoria. The AER also observed that the new owner of the 1280 MW Torrens Island gas-fired power station, AGL, bid a significant proportion of its capacity near the market price cap. At the time, Torrens Island represented 39% of South Australian electricity capacity. The AER continued to highlight the role of AGL's bidding of Torrens Island when prices spiked again in 2009 and 2010.[24]

The high South Australian spot prices in March 2008 led to the first ever triggering of the CPT and the implementation of administered pricing. At that time, the CPT was $150,000, being 15 times the then-prevailing value of the market price cap.

Following the 2003–2008 period during which relatively little generation was developed in the NEM, the period since 2008 has seen substantial amounts of new generation gradually become commissioned, particularly in NSW and Queensland. In the 3 years to June 2011, 4700 MW of capacity was commissioned, with more than half in 2008/2009 alone. Most of the new capacity has been in gas-fired plant (Figure 19.10). The bulk of the remaining generation investment in recent years has been wind plant in Victoria and South Australia in response to the RET.

More recently, wholesale prices have been more subdued. The summers of 2010/2011 and 2011/2012 were much milder than in previous years,

23. AER, 2007. State of the Energy Market, pp. 7–10.
24. AER, 2009. State of the Energy Market, pp. 3, 9, 81–82.

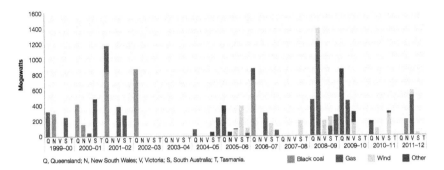

FIGURE 19.10 Annual investment in registered generation capacity. Note: Data are gross investment estimates that do not account for decommisioned plant. *Source: AEMO; AER. 2012. State of the Energy Market. Figure 1.20, p. 51.*

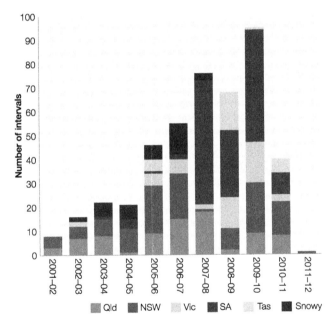

FIGURE 19.11 Number of half-hourly trading intervals above $5000/MWh. Note: Each trading interval is a half hour. *Source: AEMO; AER. 2012. State of the Energy Market. Figure 1.16, p. 46.*

leading to fewer price spikes (Figure 19.11). Another important factor in pushing prices down has been the rapid growth of nonscheduled renewable energy, particularly wind and domestic solar PV units in response to government policies further described in Section 6.

The result of recent subdued prices is that generation businesses have been earning less than their total levelized costs. Analysis commissioned by the AEMC (Figure 19.12) shows that prices across the NEM have fallen

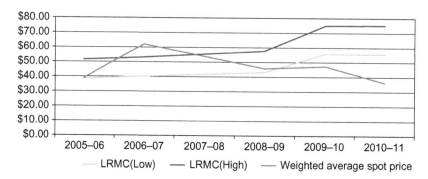

FIGURE 19.12 NEM weighted average prices compared with Long-Run Marginal Cost. *Source: AEMC, 2012. Draft Rule Determination, Potential Generator Market Power in the NEM. Figure 5.1, p. 25 (June 7).*

below estimates of the long-run marginal cost (LRMC) of supply. While some participants have suggested that these outcomes demonstrate the need for capacity mechanisms to support the viability of existing operators, there is as yet little evidence of a generalized shortfall in supply.

Looking forward, exchange-traded futures prices provide some indication of the market's view of the direction for wholesale prices. In general, futures prices indicate that having risen due to the recent imposition of the carbon tax of \$23/t CO_2-e, spot prices will gradually fall over the next 3 years. This is likely a reflection of expected continuing weak demand growth and futures markets pricing in an increasing likelihood of the carbon pricing scheme being repealed should a change of federal government occur in late 2013 as described in Section 6.1.

4.1.3 Market Power Concerns

The NEM design reflects perhaps the most *laissez-faire* attitude to generator bidding behavior of any electricity market in the world. As noted above, despite having a high price cap (essential in an energy-only market design), there is effectively nothing that prevents generators bidding their capacity at high prices as and when they choose. The absence of such constraints combined with the occurrence of occasional price spikes has led to an ongoing debate about the extent of generator market power in the NEM and whether the market design ought to be changed to prevent certain types of bidding behavior.

Under the Rules, the AER is required to publish a report whenever the spot price exceeds \$5000/MWh in any region. The AER has often used these reports, as well as its annual *State of the Energy Market* report, to express concerns about the exercise of market power by generators. Typically, the AER focuses on some form of generator withholding behavior at times of high demand and/or curtailed supply, when transmission constraints limit imports into the region experiencing the high price.[25]

In accordance with its Rules enforcement role, the AER has previously launched investigations into bidding behavior by certain generators where that behavior is alleged to breach Rule obligations to bid in "good faith." Good faith in this context simply refers to a requirement for generators to submit bids that reflect their honest intention at the time those bids are made. A generator can only be found to breach this provision where it makes an initial bid in an attempt to mislead or deceive other parties and follows up with a rebid that exploits the situation created by its initial bid. Although the AER has found *prima facie* grounds for prosecuting a generator for violating the "good faith" obligation, it has not yet brought a successful prosecution.

Other than the good faith rule and other provisions that similarly require generators to behave honestly, the Rules do not prevent generators from withholding or repricing their output at high prices as opportunities arise. In an attempt to address this perceived shortcoming in the market design, the Major Energy Users Inc. (MEU) submitted a Rule change proposal seeking to limit the bids of "dominant generators" in a similar manner to the pivotal generator mitigation measures in place in many northeastern US markets. The AEMC published a draft determination in June 2012 provisionally rejecting the change.

The historical lack of constraints on generator bidding behavior in the NEM reflects a preference for structural solutions and a degree of skepticism toward intrusive regulation by the responsible NEM institutions.

Importantly, as evidenced by Figure 19.12, the lack of constraints on generating bidding has not prevented average spot prices falling below the LRMC of supply in recent years, suggesting that market power has not been exercised in a sustained fashion.

Concerns about market power have also arisen in the context of the ACCC's mergers approval processes, further discussed in Section 4.4.

4.2 Network Pricing

4.2.1 Form of Regulation

Under the original Code, the ACCC was obliged to adopt a CPI-X building block approach to setting revenues for transmission businesses. The ACCC was also obliged to accept the transmission networks' asset valuations prepared by each jurisdictional regulator when determining the appropriate return on capital and depreciation under the building block methodology.

However, the Code otherwise allowed the regulator considerable discretion in how it implemented economic regulation. Originally, the ACCC

---

25. For example, see the AER's report into the events of November 9, 2011 in NSW. Available from: http://www.aer.gov.au/node/11018.

engaged in *ex post* reviews of transmission investments to determine whether all capital expenditure incurred was efficient and ought to be recoverable through regulated charges. In 2003, the ACCC moved away from *ex post* reviews in favor of a "lock in and roll forward" approach, which shifted regulatory emphasis to the setting of appropriate expenditure forecasts. While this shift was broadly preferred by transmission companies, they and some jurisdictions were concerned that the ACCC faced so few obstacles to changing its approach to regulation. This was one of the drivers for the 2005 changes to the NEM institutional arrangements and the "hardwiring" of the network regulatory regime in the Rules in 2006.

Under the Rules, both the methodology and the procedure for setting regulated network revenues are heavily prescribed. Businesses submit a revenue proposal document setting out their calculation of building block revenues for the next 5-year control period. The AER, being now the regulator of both transmission and distribution networks, must accept the proposal unless it can demonstrate that a relevant parameter does not "reasonably reflect" the service provider's efficient costs. This approach effectively reverses the onus for demonstrating that a particular parameter is reasonable from the network business to the AER.

The network regulatory approach in the Rules has arguably led to higher regulated revenues and prices than would have otherwise been the case (see below). In response, the AEMC recently made some changes to the Rules to address concerns about overinvestment and excessive regulated returns. The SCER is also considering changes to the scope of merit reviews of regulatory decisions to tilt the balance of regulatory outcomes toward consumers.

4.2.2 Network Expenditure and Pricing Outcomes

Recent regulatory determinations have allowed substantial increases in regulated revenues for transmission and distribution network businesses across the NEM (Figure 19.13). However, what is unclear is the extent to which higher allowable expenditures are the result of necessary investment to meet rising peak demand and the extent to which they reflect inefficient "gold-plating" by the businesses under the post-2006 regulatory regime.

The data reveal a significant divergence between the expenditures and allowed revenues of privately owned network businesses (in Victoria and South Australia) and network businesses that remain under government ownership (in NSW and Queensland). Part of this may reflect the incentives on government-owned businesses to maximize revenues via expansion of their asset bases rather than establish savings and take the profits. Part may reflect the cost of satisfying increases in network planning standards that governments in NSW and Queensland have imposed on their businesses in light of high-profile outages. Even if the latter explanation is correct, it is unclear whether the imposition of higher planning standards has brought consumer benefits commensurate with

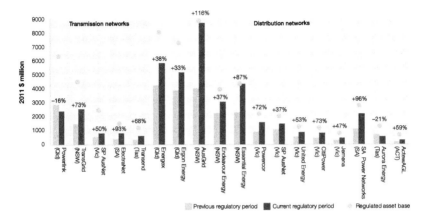

FIGURE 19.13 Electricity network investment. Notes: Regulated asset bases are as atthe beginning of the current regulatory periods. Investment data reflect forecast capital expenditure for the current regulatory period (typically,five years), amended for merits review decisions by the australian competition Tribunal. see tables 2.1 and 2.2 for the timing or current regulatory periods. The data include capital contributions and exclude adjustments for disposals. AusGrid's distribution network includes 962 kilometers of transmission assets. *Source: Regulatory determinations by the AER. 2012. State of the Energy Market. Figure 2.4, p. 70.*

their much higher costs. The AEMC is also currently reviewing distribution reliability standards in the NEM to address this question.[26]

4.3 Retail Market

4.3.1 Retail Price Outcomes

As noted in Section 3.1.2, Australian retail customers enjoyed falling or flat real tariffs in the 1990s, as the initial reforms promoted large productivity gains. Business customers benefited in particular, due to the unwinding of decades-long cross-subsidies to households. However, as wholesale prices rose in the early 2000s, retail prices began to follow (Figure 19.14). Overall, between 1990/1991 and 2005/2006, real business tariffs fell by 23% whereas real household tariffs rose by 4%.[27]

Retail prices rose more dramatically from the middle of the last decade as network investment and regulated network charges ballooned. Household electricity bills are now dominated by distribution and transmission network charges (Table 19.3).

---

26. There is strong evidence that productivity in the privatized networks is considerably higher than in the government owned systems. See Mountain, B., Littlechild, S., 2010. Comparing electricity distribution network revenues and costs in New South Wales, Great Britain and Victoria. Energy Policy, doi: 10.1016/j.enpol.2010.05.027.

27. AER, 2007. State of the Energy Market. Figure 6.15, p. 191.

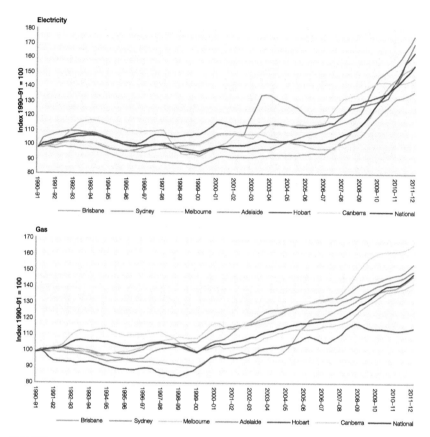

FIGURE 19.14   Retail electricity price indexes (inflation adjusted), Australian capital cities. Note: Consumer price index electricity and gas series, deflated by the consumer index for a groups. *Source: ABS, consumer price index, cat.no.6401-0, various years 2012. State of the Energy Market. Figure 5.4, p. 131.*

TABLE 19.3 Retail Bill Proportions

Retail Bill Proportions				
Wholesale Energy Costs	Network Costs	Retail Operating Costs	Retail Margin	Green Costs (Excluding Carbon)
32–42%	46–51%	4–7%	4–5%	4–8%

Source: AER, 2011. State of the Energy Market. Table 4.2, p.110

4.3.2 Retail Competition

Competition in the retail supply of electricity was gradually introduced in the late 1990s and early 2000s for progressively smaller customers. Contestability timetables varied by jurisdiction, with NSW and Victoria the first to introduce full retail competition ("FRC") in January 2002. By mid-2007, all customers in the mainland NEM were free to choose their electricity retailer. In Tasmania, small customers (those consuming less than 50 MWh per annum) remain noncontestable, with the state government recently announcing an administrative reduction in the rate of increase of regulated retail tariffs.

Victoria and South Australia are the only states where maximum retail tariffs are not regulated, and even in South Australia tariffs are fixed by agreement with the state government for two years. In this regard, only Victoria has a fully liberalized retail market along the lines of the Texas retail market further described in Chapter 10 by Adib et al. All other NEM jurisdictions regulate the default tariffs offered by incumbent ("first tier") retailers to small customers. Regulated tariffs are generally based on the jurisdictional regulator's estimate of the long-run marginal cost of energy of serving small customers—based on their load shape—plus transmission and distribution network charges and other pass-through items.

The retail tariff-setting methodologies applied in most jurisdictions initially failed to provide sufficient "headroom" to enable nonincumbent ("second tier") retailers to undercut regulated rates. This curbed customers' incentives to switch retailer. Lack of headroom was primarily due to jurisdictional governments' concern that FRC would become associated with higher prices, coupled with their reluctance to rely on competition to discipline first tier retailers. This has gradually changed and regulated tariffs now allow reasonable scope for second tier retailers to offer attractive deals to customers, with discounts of 10–20% now available to direct debit customers. Customer churn rates have increased in recent years and now exceed 60% in every mainland NEM jurisdiction (Figure 19.15).

Regulation of nontariff conditions for retail supply has traditionally been left to individual jurisdictions. In July 2012, the National Energy Retail Law and Rules commenced under the auspices of the National Energy Customer Framework (NECF). This established a single regulatory framework for retail supply contracts, including customer protections, to reduce barriers and costs to retailers operating across multiple jurisdictions. Different states are introducing the NECF at different times.

4.3.3 Tariff Innovation

Although originally the poor cousin to wholesale market reform, retailing has received greater attention in recent years due to rising tariffs and growing interest in demand-side response.

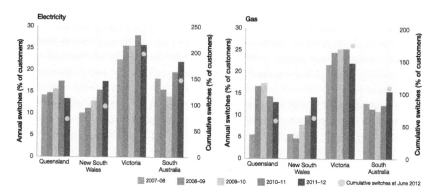

FIGURE 19.15 Cumulative customer switching rates. *Source: Customer switches: AEMO. MSATS transfer data to july 2012 and gas market reports, transfer history to july 2012; customer numbers: estimated from retail performance reports by IPART(New south wales), the ESC (victoria), ESCOSA (South Australia) and the QCA. AER, 2012. State of the Energy Market. Figure 5.3, p. 126.*

Policy makers are aware that a substantial share of new investment in distribution networks and peaking generation is undertaken to meet system peak demands for just a few hours per year. Yet most end-use consumers do not face the costs that their peak demand calls forward. This is due to the lack of time-sensitive pricing for most small customers, who tend to face flat tariffs based on deemed load profiles. The lack of time-sensitive pricing is largely attributable to the accumulation metering technology at most small customers' premises, which is not capable of recording when during a reading period electricity consumption occurs.

Originally, state governments considered that retail competition would drive the voluntary take-up of interval meters, which are capable of recording the timing of consumption and facilitating more cost-reflective tariffs. However, in the wake of consecutive hot summers and record peak demand in the early 2000s, the Victorian government decided to mandate a rollout of smart meters—interval meters capable of two-way communications including being remotely read—to all customers. The rollout, which is due to be completed by 2014, has been plagued by substantial cost overruns and a cost−benefit analysis commissioned by the incoming state government in 2011 found that the rollout was not likely to be net beneficial.[28] Other states have observed developments in Victoria and appear unlikely to mandate their own rollouts, although individual businesses often install interval meters on a "new and replacement" basis. The AEMC's "Power of Choice" review

28. See Deloitte, 2011. Advanced Metering Infrastructure Cost Benefits Analysis, Final Report, 2 August. Available at the Department of Primary Industries web site: http://www.dpi.vic.gov. au/smart-meters/publications/reports-and-consultations/advanced-metering-infrastructure-cost-benefit-analysis.

recommended the voluntary take-up of smart metering and supported the introduction of more dynamic tariffs.

Apart from funding the smart meter rollout in Victoria through distribution network charges, retail customers are also bearing the cost of a variety of renewable subsidies and other green policies. These policies are discussed in more detail in Section 6.

4.4 Structural Developments

At the time of the initial structural reforms and later the authorization of the NEM Code, the market arrangements contained no specific long-term measures to prevent reaggregation in the NEM. This was because there were no firm views as to the most productive structure of the industry, only that the previous state-owned integrated monopolies were not optimal. Regulatory oversight of structural changes was left to the ACCC, to assess in accordance with the general competition protections set out in the *Trade Practices Act*, now the *Competition and Consumer Act* (CCA). Section 50 of the CCA contains the key prohibition against mergers and acquisitions that are likely to lead to a "substantial lessening of competition" in a relevant market. As discussed below, to date the ACCC has not successfully opposed any electricity merger in the NEM under Section 50.

4.4.1 Current Market Shares

Following from the structural reforms and privatizations discussed in Section 3.1, a wide range of investors have entered and exited the NEM. The high prices originally paid for the Victorian electricity generators could not be justified in light of subsequent low wholesale prices and several rounds of restructurings and sales followed. South Australian generators have also seen ownership changes. The NSW generation sector operates in a hybrid space, with trading rights to several plants held by private participants but the physical assets and the largest generation portfolio (Macquarie Generation) still within government hands at the time of writing. The majority of Queensland and Tasmanian generation assets also remain in state ownership. However, all major electricity retailers outside Tasmania are privately owned.

Current market shares for each NEM jurisdiction are as set out in Figure 19.16 (generation) and Figure 19.17 (retail) below. Leaving aside assets still within government ownership, both figures show that three private firms now dominate generation and retailing activities in the mainland NEM—AGL, Origin, and EnergyAustralia.

Since NEM commencement, horizontal aggregation of distribution networks has occurred across NSW, Victoria, and South Australia. One consortium now owns two Victorian distribution networks and the single South Australian distribution business, ETSA Utilities. The state-owned

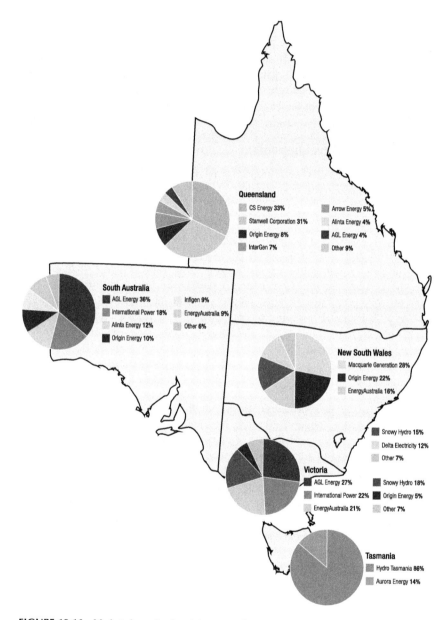

FIGURE 19.16   Market shares in electricity generation capacity by region. Note: Capacity that is subject to power purchase agreements is attributed to the pary with control overoutput. Excludes powerstations not managed through central dispatch. *Source: AER, 2012. State of the Energy Market. Figure 1.10, p. 37.*

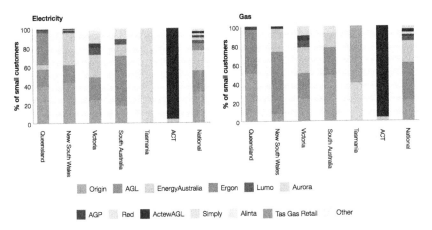

FIGURE 19.17 Market shares in electricity retailing to small customers by jurisdiction. *Source: AER estimates 2012. State of the Energy Market. Figure 5.1, p. 121.*

Queensland transmission network, Powerlink, is a part-owner of the South Australian transmission network, ElectraNet. All network businesses in Queensland, NSW, and Tasmania remain within government ownership.

4.4.2 Retail-Distribution Separation

One key structural development since the original reforms has been the unbundling of retail and distribution activities. First tier retailers and distribution networks within a particular geographic area were originally sold as combined entities in Victoria and long operated as combined entities under state government ownership in NSW and Queensland. However, to promote retail entry and competition, the combined entities were "ring-fenced" to prevent the distribution business favoring its retail affiliate, though the different nature of the businesses, most of which are now totally separate, probably made the statutory requirement unnecessary. Each distribution area has seen considerable second tier entry.

While ring-fencing undoubtedly promoted retail competition, it may have inhibited the rollout of smart meters and dynamic pricing due to the dispersal of the benefits from demand response across more parties. However, the more significant reasons for the lack of take-up of smart meters and dynamic pricing are more likely to have been:

- the high cost of the technology relative to the benefits (Section 4.3) and
- the lack of appropriate tariff unbundling to enable retailers interested in offering smart meters to customers to avoid paying regulated network metering charges.

Over the last decade, corporate transactions and further privatizations have led to different owners of retail and distribution businesses throughout the mainland NEM.

This reflects two factors. The first factor is the very different characteristics and requirements of each type of business: successful retailing is primarily about strong marketing capability, customer handling skills, and efficient wholesale market risk management. By contrast, running a distribution business is much more concerned with the planning, procurement, maintenance, and operation of physical network assets. The nature of regulation affecting each type of business is also substantially different. Retailers are subject to extensive customer protection regulations regarding marketing, billing, privacy, and disconnection. Outside Victoria, a key regulatory issue is the setting of an appropriate level of wholesale energy purchase costs for regulated retail tariffs. Retailers are also the conduit through which the costs of green policies are passed through to customers. All of this means that retailers need a good understanding of the wholesale market and all schemes and subsidies affecting end-use consumers. On the other hand, distribution networks are subject to periodic regulatory reviews of revenues and prices. These reviews are long and extremely complex processes, typically stretching for well over 1 year, with enormous implications for the financial returns of the business. Distributors are also subject to reliability, safety, and other technical standards. For these reasons, ownership of distribution networks typically appeals to investors seeking a relatively steady and secure return, whereas retailing inevitably involves managing wholesale spot market risks.

The second issue is that some state governments have been more reluctant to privatize distribution businesses than retail businesses. Distribution businesses own visible physical assets and employ large numbers of unionized workers. By contrast, retail businesses have few physical assets and provide less visible "back office" services, such as billing.

4.4.3 Vertical Integration

A more contentious structural development over the last decade has been vertical integration between retail and generation activities. When the NEM commenced, it was expected that retailers and generators would manage their complementary wholesale risk exposures by entering derivative contracts with one another. However, investors have preferred acquiring physical hedges rather than financial hedges.[29] As noted above, the largest three private energy businesses in the NEM (AGL, Origin and EnergyAustralia) are all considerably vertically integrated in their retailing and generation assets.

While reduced transaction costs from contracting represents a legitimate rationale for vertical integration, the ACCC has been wary of vertical integration on competition grounds. In 2003, it opposed energy retailer AGL's

29. For a discussion of some of the advantages of vertical integration, see Simshauser, P. Vertical integration, credit ratings and retail price settings in energy-only markets: navigating the Resource Adequacy problem. Available from: http://www98.griffith.edu.au/dspace/bitstream/handle/10072/34985/65004_1.pdf?sequence=1.

acquisition of a minority share in the Loy Yang A power station in Victoria. The ACCC was concerned that the transaction would lead to a "thinning" of the wholesale contract market, which would have the following two detrimental effects:

- encourage generators to withhold output to spike the spot price and
- make it harder for stand-alone retailers to obtain hedges, thereby increasing barriers to stand-alone new entry.

The matter became the subject of litigation and the Federal Court found comprehensively in AGL's favor, noting that:

- a generator's occasional ability to spike to the spot price under particular favorable conditions was not evidence of market power as defined under the precursor to the CCA;
- barriers to entry in retail and generation were relatively low; and
- residual demand analysis was not a suitable tool for analyzing how real-world generators compete in a market such as the NEM.

A number of other moves toward vertical integration followed the AGL decision. In 2004, Singapore Power, the owner of the Victorian transmission network, sought to acquire TXU's Australian energy assets. These assets included merchant retail and generation interests. The ACCC was concerned about Singapore Power's transmission business favoring its affiliated generators and imposed a number of enforceable undertakings as a condition of its approval for the transaction. These undertakings included an obligation on Singapore Power to divest the former TXU generation assets within 2 years. Those assets were subsequently sold to China Light & Power in 2005, forming the TRUenergy business (now called EnergyAustralia).

Perhaps the most controversial private transaction since the NEM commenced was the so-called "asset swap" between AGL and TRUenergy in 2007. This involved TRUenergy selling the 1280 MW mid-merit Torrens Island power station to AGL in exchange for cash plus AGL's 200 MW Hallett peaking plant. Both plants are located in South Australia. As AGL was and remains the largest retailer in South Australia, the transaction resulted in both AGL and TRUenergy having more vertically balanced portfolios. The ACCC was concerned by the potential for AGL to exploit the market power of the Torrens Island plant, which is the largest generator in South Australia. However, the ACCC were unable to show that AGL would have stronger incentives than TRUenergy to engage in strategic bidding at Torrens Island. Accordingly, the ACCC did not oppose the transaction. Subsequently, a series of hot summers and opportunistic bidding by AGL led to very high prices in South Australia in 2008 and 2009. This resulted in a flurry of investigations by the ACCC to check whether AGL misrepresented its position at the time of the transaction, but ultimately no action was taken against AGL.

By and large, early concerns that retailer and generator amalgamations would lead to monopoly power and market inefficiencies have abated as a result of evidence that competition is bringing lower prices and retail churn. Like other firms that assemble supplies sourced from affiliated and nonaffiliated businesses, electricity retailers are forced by market circumstances to ensure that the affiliates are not favored. The risk of high price occurrences are too great for electricity retailers to gamble on fulfilling most of their needs in-house and the costs of alienating potential counterparties by discriminatory behavior far outweigh any short-term benefits possible through collusion with in-house suppliers. Indeed, when AGL acquired the entire Loy Yang A power station in 2012, the ACCC focussed only on the horizontal aggregation aspects of the acquisition and cleared the transaction. As such, vertical integration has given way as a regulatory concern in favor of more general concerns about generator market power (Section 4.1.3).

4.4.4 NSW Gentrader Model

In 2009, after having failed on several occasions to gain the support of trade unions and its political opponents for privatization, the NSW Labor Government embarked on a sale of the trading rights to the state's generation assets. The "Gentrader model" involved separating the physical power station assets from the rights and obligations associated with the energy trading activity. The former were retained in public ownership, whereas the latter were sold along with the three incumbent NSW retail businesses through a combinatorial auction process. Similar approaches have been used in Alberta, France, and many other European and North American jurisdictions.[30]

The sales process led to Origin Energy acquiring the Eraring power station and TRUenergy (now EnergyAustralia) acquiring the Mt Piper and Wallerawang power stations. Several generators, including the largest in the state, Macquarie Generation, did not achieve their reserve prices and remained in government ownership. These transactions went ahead following the ACCC's approval, which was somewhat controversial given the horizontal aggregation occurring at the retail level. However, the ACCC's approach was consistent with the reasoning embraced in the Federal Court's original 2003 AGL decision.

Total sales proceeds for the Gentrader rights were just shy of $1.5 billion. While clearly a second-best outcome from a reform perspective, the Gentrader model did not preclude the sale of the remainder of generation or network assets in NSW. In May 2012, the government passed legislation to sell the generation assets. These are expected to fetch another $3 billion,

---

30. See New South Wales Government, New South Wales Energy Reform Strategy, Defining an Industry Framework, March 2009, pp. 12–14. Available from: http://www.nsw.gov.au/sites/default/files/NSW-Energy-Reform-Strategy.pdf.

resulting in proceeds of $4.5 billion for the NSW generation system. If achieved, this would be a fraction of the $9 billion achieved for the much smaller Victorian system more than a decade earlier.

5 UNFINISHED AND UNRESOLVED AREAS OF REFORM

While the Australian electricity supply industry has undergone substantial pro-competitive reform over the last two decades, reforms remain incomplete in key areas and certain aspects of market design remain unsettled. These outstanding areas are:

- ongoing government ownership of electricity assets;
- incomplete retail tariff liberalization;
- uncertainty over transmission pricing and access; and
- greenhouse gas mitigation policies.

Section 6 is devoted to a discussion of greenhouse gas mitigation policies. The other outstanding areas are addressed below.

5.1 Government Ownership

As noted above, many or all generation and network assets in NSW, Queensland, Tasmania, and Western Australia remain within government ownership. Apart from the sale of remaining NSW generation assets, no other generation or network assets appear likely candidates for privatization in the near term. In particular, the new Queensland and NSW state governments have ruled out privatizing their distribution assets in their respective first terms of office.

Government ownership has been shown to have a multitude of drawbacks:

- First, overstaffing and rigid work practices are harder to address than under private ownership, slowing efficiency gains.
- Second, government shareholders tend not to exercise as tight a level of corporate governance as private investors, which can result in wasteful gold-plating and empire-building by the managers of government-owned businesses.
- Third, a government presence in generation deters private investment, as investors face the risk that governments will order their businesses to undertake uneconomic investment to "keep the lights on."
- Fourth, government ownership keeps taxpayers exposed to the risks of owning electricity assets, such as the introduction of carbon pricing and higher fuel costs—both of which were likely responsible for cutting billions of dollars from the sales value of the NSW generation assets.

Many of these issues are evident from a comparison of NSW and Victorian energy businesses, especially distribution network businesses described in Section 4.2.2.

These drawbacks combined with state governments' fiscal constraints may eventually bring about further privatization later this decade.

5.2 Retail Tariff Liberalization

Outside Victoria and more recently South Australia, default retail tariffs for small customers are regulated throughout Australia. While administrative approaches to setting regulated tariffs have become more generous in some states in recent years—encouraging competition—the reverse has occurred in other jurisdictions. For example, the Queensland Government recently announced that default residential tariffs that cover around half of household customers, the others having accepted a lower cost option, would be frozen for 12 months except for the effect of the carbon price. The Tasmanian Government also announced a reduction in the rate of increase of residential tariffs prior to the introduction of FRC in 2014. Residential and business tariffs in Western Australia did not change for over a decade and have only recently started to approach the real costs of supply.

For so long as governments regulate retail tariffs, retail competition and innovation is likely to be handicapped. Retailers will remain exposed to risks that governments intervene to reduce default tariffs for political ends, squeezing retailers' margins. An unsupportive environment for innovation and investment by private retailers perpetuates the need for regulation. The result can be an unhealthy cycle of tighter price regulation, less competition, and innovation, followed by even tighter regulation.

5.3 Transmission Pricing and Access

The pricing of transmission access and network congestion has been controversial since the start of the NEM. Presently, loads pay transmission charges that are partly postage-stamped and partly location-specific based on load-flow modeling. Generators only pay "shallow" connection charges and seldom contribute to the costs of upstream augmentation, which are borne by consumers. Network congestion is reflected in wholesale price divergences between regions, but prices within regions are kept uniform (subject to the application of static marginal loss factors). The result is that the cost implications of generation and load locational decisions within a region are under-signaled relative to the theoretical ideal.

To date, the jurisdictions and the AEMC have been reluctant to shift transmission costs to generators or to increase the locational granularity of wholesale pricing. The argument against making generators pay higher transmission charges is that generators in the NEM do not "cause" transmission investment—transmission investment is undertaken by regulated entities when justified through cost-benefit or cost-effectiveness analysis. This means that the test for transmission investment itself sends a signal to generators to locate close to loads and existing network infrastructure. The case against more granular pricing of congestion is partly driven by the need to devise suitable financial risk management instruments and partly driven by political concerns about intrastate wholesale price differentials. The AEMC is currently reexamining these issues in its Transmission Frameworks Review. Whatever this review finds, transmission pricing and access will likely remain unsettled for some time.

6 GREENHOUSE GAS MITIGATION POLICIES

Greenhouse gas mitigation policies can be neatly split between carbon pricing on the one hand and policies to promote renewable energy on the other. This section discusses both types of policy in turn. Australia is not alone in this respect as reflected in a number of chapters in this volume, such as Great Britain (Chapter 1), Germany (Chapter 4) and Korea (Chapter 22).

6.1 Carbon Pricing

The Gillard Labor Government introduced a carbon pricing mechanism on July 1, 2012 as a means of fulfilling a commitment to reduce greenhouse gas emissions by 5% of 2000 levels by 2020.[31] The mechanism commenced with a fixed price of \$23/t CO_2-e imposed on electricity generators and other large emitters. The price is scheduled to rise by 2.5% per annum (real) over the following 2 years and convert to an emissions trading scheme (ETS) with international linkage on July 1, 2015. The ceiling price for permits under the ETS will be \$20/t CO_2-e above the expected international price.[32] A floor price was originally set at \$15/t CO_2-e in real terms, but abandoned in August 2012, less than 2 months after the scheme commenced.[33] Given the scope for international trading under the ETS and the current state of the European ETS, Australian permit prices could drop well below \$15/t CO_2-e in 2015. By comparison, a (credible and sustained) permit price of at least \$30/t CO_2-e would be needed to switch investment in new plant from coal to gas, and about \$50/t CO_2-e would be required before existing coal-fired plant began to be

---

31. See the web site of the Department of Climate Change and Energy Efficiency at: http://www.climatechange.gov.au/government/reduce/national-targets.aspx.

32. See Securing a Clean Energy Future, The Australian Government's Climate Change Plan (2011), p. 27. Available from: http://www.cleanenergyfuture.gov.au/wp-content/uploads/2011/07/Consolidated-Final.pdf.

33. See the web site of the Department of Climate Change and Energy Efficiency at: http://www.climatechange.gov.au/government/submissions/closed-consultations/price-floor-carbon-pricing.aspx.

significantly displaced by new entrant gas. A carbon price of at least \$80/t CO_2-e would be needed to make renewables cheaper than gas-fired plant.

Since its introduction, the carbon price has been close to fully passed through to wholesale electricity prices. That is, wholesale prices have been (to date) about \$23/MWh higher in 2012/2013 than they would otherwise have been. Given the current low wholesale prices, this represents an increase of over 50%. Meanwhile, regulated retail prices have risen by 8–10% more than otherwise, with further increases expected. Over time, increases in the carbon price—to the extent they arise—are likely to be less than fully passed through to wholesale prices due to falling carbon intensity of the marginal plant in the market.

To help offset the costs of the "carbon tax" (as it is popularly known), the government provided compensation to lower- and middle-income households, energy-intensive trade-exposed industries and private generators. Consequently, the burden of the carbon tax is principally being incurred by higher-income households, less trade-exposed industries, and government-owned generators.

Prior to implementation, the Federal Treasury undertook computable general equilibrium modeling of the carbon price and compensation package. Based on extremely optimistic assumptions about the extent of international action to curb emissions, the analysis found that mining, construction, and electricity supply would be the hardest hit sectors and real wages across the economy would be 1.1% lower than otherwise by 2020. Modeling based on more realistic assumptions regarding the extent of international action would likely yield negative impacts many times those tabled by Treasury.

The introduction of carbon pricing in Australia has been highly controversial and the federal opposition has promised to repeal the carbon tax should it win office at the next election. The Opposition's present policy for meeting the bipartisan emissions reduction target is called "Direct Action," and involves government-run and funded tenders for reduced emissions.

6.2 Renewable Generation Subsidies

Like many other countries, Australia has a slew of policies favoring the development of renewable generation, most of which have been modified over the years as investors have taken advantage of the generous subsidies on offer. The most significant scheme was originally the Mandatory Renewable Energy Target (MRET), which was designed to promote the development of 9500 GWh of new renewable generation.

In 2009, the Federal Government legislated the Expanded RET, which was designed to ensure that 20% of Australia's electricity supply will come from renewable sources by 2020 (set at 45,000 GWh). The RET obliged "liable entities" (mainly retailers) to acquire a certain volume of Renewable Energy Certificates (RECs) based on the size of their energy purchases. RECs could be produced by renewable forms of generation. However, the mass installation of domestic solar PV units induced by state schemes led to a collapse in the REC price, discouraging investment in wind and other larger-scale renewable plant. This led the government to replace the RET with a dual set of policies: the Large-Scale Renewable Energy Target (LRET) and the Small-Scale Renewable Energy Scheme (SRES).[34]

The LRET, which covers large-scale renewable energy projects, such as wind farms, commercial solar, and geothermal, will deliver the majority of the 2020 target. The SRES provides households, small business and community groups $40 for each REC created by small-scale technologies such as solar panels and solar water heaters. In light of concerns about rising electricity tariffs and falling demand, some participants are now calling for the winding back of these policies.

In addition to the Federal Government's LRET and SRES, each state also has an FiT for energy produced by domestic solar PV units. FiTs were initially extremely generous, causing a 100-fold increase in the installed capacity of domestic solar PV from 10 MW in 2007 to more than 1000 MW in 2011. The most generous was the NSW FiT, which was originally set at 60 c/kWh of energy produced by the unit, including energy consumed by the resident owner. This was reduced to 20 c/kWh in late 2010, and further reduced in June 2012 to a range of 8−13 c/kWh on net exported energy only.[35] In other states, FiTs were not quite as generous to begin with, but all had reduced rates or tightened eligibility by mid-2012.

Table 19.4 sets out the key renewable energy policies in Australia.

TABLE 19.4 Key Renewable Energy Policies

Key Renewables Policies

Policy	State/Federal	Objective
LRET	Federal	Promote large-scale renewable projects (e.g., wind)
SRES	Federal	Promote small-scale renewable project (e.g., solar PV and solar hot water)
Solar FiTs	State	Promote domestic PV units

---

34. See the web site of the Department of Climate Change and Energy Efficiency at: http://www.climatechange.gov.au/government/initiatives/renewable-target.aspx.

35. See IPART, 2012. Media Release, A Fair and Reasonable Solar Feed-In Tariff for NSW, 27 June. Available from: http://www.ipart.nsw.gov.au/Home/Industries/Electricity/Reviews/Retail_Pricing/Solar_feed-in_tariffs_-_2012-2013.

Numerous other renewable subsidies are in place or are being set up. For example, the Federal Government is setting up the Clean Energy Finance Corporation (CEFC) to invest up to $10 billion in renewable energy ventures. While the CEFC is meant to be "commercially orientated," it is likely to become a vehicle of industry policy and as such will probably fail to achieve its modest 4% nominal return target. The Federal Government also has a Solar Flagships Program to support the construction and demonstration of large-scale, grid-connected solar power stations.

All of these policies represent extremely expensive means of curbing greenhouse gas emissions. In general, the narrower a scheme in terms of complying technologies, the higher the cost of abatement in $/t CO_2-e. For example, the LRET provides the cheapest abatement of the existing policies at about $90/t CO_2-e, whereas the cost of abatement under the SRES is about $300/t CO_2-e. The abatement cost under most FiTs is over $350/t CO_2-e.

Further, the costs of renewable energy plant are not limited to the high cost of energy or greenhouse gas abatement from such plant. Large-scale renewable plant are often developed far from the existing transmission network, requiring lengthy and expensive network extensions to enable them to inject power into the grid. Another hidden cost of renewable plant arises from their intermittent supply of energy. The average capacity factor of wind plant in the NEM is about 30%, but it may be less than 10% at peak demand times. This means that provision of a firm 100 MW of capacity requires 90 MW of flexible thermal plant, such as open-cycle gas turbines (OCGT) for every 100 MW of wind plant. The growing penetration of wind, especially in Western Australia and South Australia, is also creating problems for system operators. In Western Australia, the issue is the unscheduled nature of wind generation, which is leading to coal-fired plant being shut down overnight when demand is low. This is inefficient and may lead to reliability problems the following day when supply is needed quickly. In South Australia, the issue is more about maintaining voltage control in a region with a large and growing stock of wind plant.

Perhaps the best that can be said about Australian policies subsidizing renewable energy is that they are less generous than in many other western countries. This has limited the magnitude of the still-substantial costs of these subsidies that are being borne by consumers and conventional generators.

7 CONCLUSION

The Australian electricity industry has undergone substantial reforms over the last two decades aimed at increasing productive efficiency, consumer choice, and decentralized investment decision-making. While different states have moved at different rates, the reforms have for the most part been highly successful. The NEM now represents a not-quite-textbook example of an energy-only gross pool market in operation. In particular, the NEM has shown that if the market price cap is set appropriately and bidding rules are not prescriptive, market forces can provide the right type, quantity and location of new generation in a timely manner. Nevertheless, the true test of the market's robustness and the full benefits of reform will only come when the remaining steps of full privatization and retail tariff liberalization are completed.

The two most unsatisfactory elements of the NEM lie outside the wholesale market design. These elements are the regime for the economic regulation of networks and the impacts of governments' climate change policies.

Economic regulation of networks is often a thorny issue, and regulatory regimes commonly oscillate between emphasizing cost efficiency and the provision of adequate reliability. The NEM arrangements now appear to be moving back toward cost efficiency.

Climate change policies, especially in view of Australia's generous endowment of fossil fuels, represent considerable additional costs to power supplies.

Regulatory and tax matters have led to a major increase in electricity prices since 2007. Network price increases have been considerable, caused in part by political insistence on very high reliability standards and possibly also in part by some catch-up of previous under-remuneration of networks and some gold-plating, especially in the government-owned networks in NSW and Queensland. In addition, carbon-related policies have had a progressively greater impact.

The combined effects are illustrated for a typical household electricity bill in Table 19.5.

The carbon and other green schemes account for 22% of the increase according to these estimates, though the proportion may be slightly

TABLE 19.5 Increased Costs in NSW 2007/2008−2012/2013

Component of Increased Cost	Increased Cost in Nominal $	Percent of the Increase
Energy	140	12
Carbon and other green schemes	248	22
Retail margin	94	8
Network	654	58
Total	1130	100

Source: IPART

understated since a portion of the increased energy costs and the increased retail margin would be indirectly caused by these measures.

ACKNOWLEDGMENTS

The authors would like to thank Liam Blanckenberg of Frontier Economics for helpful comments and feedback on this chapter. However, all errors remain the responsibility of the authors.

Chapter 20

Is Electricity Industry Reform the Right Answer to the Wrong Question? Lessons from Australian Restructuring and Climate Policy

Iain MacGill[a,c] and Stephen Healy[b,c]

[a]*School of Electrical Engineering and Telecommunications;* [b]*School of Humanities,* [c]*Centre for Energy and Environmental Markets, University of New South Wales*

1 INTRODUCTION

Electricity industries around the world are undergoing rapid and widespread change from drivers including growing global energy security and climate change pressures. In response, many countries have implemented renewable policy support measures that both reduce emissions and fossil fuel dependence (REN21, 2012), and some have adopted emission reduction targets for their electricity industries.

At the same time, many jurisdictions have also undergone a process of restructuring[1] away from traditional integrated monopoly arrangements toward more market-oriented approaches over recent decades, as documented in other chapters of this book. The stated intention of these changes has typically been to increase economic efficiency and customer choice by introducing greater competition into the industry.

---

1. A range of terms are used for this process including reform, privatization, liberalization, and deregulation. Our preference is *restructuring* as *reform* is generally defined to mean improvement which may or may not be the case, *privatization* is not a necessary condition for restructuring as competing entities may remain state owned, *liberalization* can get caught up in differing views of liberalization more generally (particularly in the context of the United States) while *deregulation* is generally a poor description of the process which often involves moving from state-owned monopolies under limited direct regulation to far more complex regulatory arrangements required to direct competitive behavior and privately owned monopoly components of the industry.

Evolution of Global Electricity Markets.

Australia presents an interesting example in both these regards. Its electricity industry has an emissions intensity double the OECD average and per capita emissions three times the OECD average (Garnaut, 2008), due primarily to its high dependence on coal generation. However, Australia was also one of the first countries to ratify the United Nations Framework Convention for Climate Change (UNFCCC), in late 1992, with the aim of stabilizing *"greenhouse gas concentrations in the atmosphere at a level that would prevent dangerous anthropogenic interference with the climate system"* (Australian Government, 2012).

At around the same time, Australia embarked on a major process of electricity industry restructuring. Two decades later, Australia has a National Electricity Market (NEM) that is widely regarded as a successful international example of restructuring. However, its electricity industry has also seen the highest emissions growth of any OECD country, over 55% from 1990 to 2010 (Australian Government, 2012b).

This chapter explores how a restructuring process with what the Australian government initially claimed were shared economic efficiency and environmental objectives, somehow ended up with such economic "success" yet environmental "failure." In particular, given growing international alarm on climate change, was Australian restructuring the right answer to the wrong question?

A key challenge in addressing this question was an early decision by Australian policy makers, seen also in many other jurisdictions, to focus electricity industry restructuring policy on improving economic efficiency through competitive pressures, while pursuing environmental objectives through "external" policy efforts. As such, restructuring itself might have played little role in these poor environmental outcomes that would then merely reflect a lack of political will on environmental aims.

This chapter therefore focuses on the currently contested question of whether Australian electricity industry restructuring has:

1. inherently assisted in reducing environmental harms such as climate change from what they otherwise would have been (in which case restructuring was arguably the right answer to both questions);
2. played a neutral role, effectively facilitating environmental improvements where the political will to implement appropriate external policies exists (suggesting restructuring was a reasonable answer to the right question); or
3. actually worked against improved environmental outcomes (in which case restructuring was arguably the wrong answer to the right question).

A definitive answer would of course require counter-factual knowledge of what would have happened in the absence of a restructuring process running over two decades. Instead, the intent of this chapter is more exploratory, seeking to draw out insights and possible lessons from the Australian experience for policy makers here, and elsewhere, contemplating future steps for both restructuring and climate policy. Such lessons are of particular relevance given the Australian government's current target of 80% emission reductions from 2000 levels by 2050.

Section 2 of this chapter provides a general framework for considering electricity industry objectives and the potential role that restructuring might play in delivering these. Section 3 describes two parallel policy processes that unfolded over the 1990s in Australia—microeconomic reform and ecologically sustainable development (ESD)—and their impacts on the design of the NEM. Section 4 discusses the NEM design whereas Section 5 assesses the direct environmental outcomes of restructuring. Section 6 assesses how well the NEM has facilitated a range of external environmental policies whereas Section 7 considers whether, and how, restructuring might have worked against environmental objectives. Section 8 explores how market arrangements might better manage the challenge of achieving an industry transition adequate to the challenge of climate change followed by the chapter's conclusions.

2 ELECTRICITY INDUSTRY OBJECTIVES AND RESTRUCTURING

Energy security, economic efficiency, and environmental drivers for the electricity industry need to be placed in a broader context of societal objectives and priorities. The World Energy Council has presented a pyramid-based hierarchy of policy priorities that begins with accessibility to commercial energy followed by energy security, economic efficiency, environmental viability, and finally social acceptability (Frei, 2004). Access and security represent deficit needs about which debate and compromise are particularly difficult. Once satisfied, however, there is the possibility of debating potential trade-offs between efficiency, environmental, and social objectives.

Experience to date generally seems to support this proposed prioritization. Societal objectives of access and energy security were important drivers for the near universal emergence of government-owned or highly regulated integrated monopoly electricity industry arrangements over the early course of industry development. Recent efforts by different countries to address the environmental challenge of climate change within the Kyoto Protocol also established differentiated responsibilities for emissions reductions between developed countries and developing countries which are in many cases still struggling to deliver access and security needs (Frei, 2004). And with some surprising, primarily South American, exceptions, electricity industry restructuring efforts received most early attention in well-established electricity industries whose main failure was argued to be unnecessary economic inefficiencies.

Given growing climate change and broader sustainability concerns, a key first question is how policy makers should prioritize the objectives of economic efficiency and environmental performance for their electricity industries. It can, of course, be argued that these objectives are actually one and the same—that is, environmental harm represents an unpriced externality and hence economic inefficiency (Stern, 2006). However, in practice there has been little consensus on how these objectives can be merged (IEA, 2005, 2012).

Instead, narrow economic efficiency and environmental objectives typically remain in tension (e.g., see Chapter 2 for the UK experience). And while there has been widespread electricity industry restructuring around the world over the past two decades, there has been only modest progress on reducing global electricity industry emissions. Certainly, there is little evidence of widespread industry transition towards the low-carbon future required to effectively address climate change (IEA, 2012).

As this book highlights, the question of whether these varied restructuring efforts have actually delivered improved economic efficiency outcomes is still debated. There are some clear examples of failure, such as California, yet other apparent "successes." This chapter does not focus on this question with respect to Australian restructuring, which is covered by Moran and Sood in Chapter 19 of this volume, but, rather, on how economic and environmental objectives have conflicted in what is generally held to have been a reasonably "successful" case of electricity industry restructuring (MacGill, 2010).

Some advocates of restructuring have certainly argued that there are inherent environmental benefits from the process itself. For example, the Australian government argued early in their electricity industry "reform" process that increased competition would assist in delivering climate change objectives as detailed later in this chapter.

Others have noted, however, that there is no reason to expect that competitive electricity markets would voluntarily consider environmental impacts falling outside existing commercial arrangements. Indeed, to do so might result in competitive disadvantage against those competitors who don't (IEA, 2005). In this case, the impact of restructuring on environmental objectives depends on two factors. The first is how well or badly environmental impacts align with underlying competitive advantage. This is often highly problematic. For example, low cost but high emission coal-fired generation is extremely competitive in many electricity industries including Australia. The second factor is what policies are introduced to induce a competitive electricity market to deliver environmental objectives, and how effective they are. A key challenge here, as noted by the IEA (2001, p. 13), is that:

In a competitive electricity supply industry, policies have to be implemented with competitively neutral instruments that do not discriminate among market players and minimise market distortions. Implementing suitable instruments has proved to be a difficult task.

A difficult task indeed—for example, it can and has been argued that pricing carbon emissions unfairly discriminates against those participants with high emissions intensity generating plants. However, there are a growing range of market-oriented regulatory approaches, such as emissions trading and green certificates, that are argued to be particularly appropriate for restructured industries (MacGill et al., 2006).

Finally, yet others have argued that restructuring does more than just ignore environmental objectives unless they are appropriately priced but can actually work against achieving them. One reason might be when restructuring processes inherently ignore or trivialize environmental objectives (WRI, 2002). Another reason might be the limitations of competitive market arrangements to deliver appropriate investment. For example, Chapter 2 highlights the limitations of current UK market arrangements to deliver high upfront cost renewable energy investments elsewhere. Another reason might be that the environmental policy measures suitable for competitive electricity markets are inherently less effective than conventional regulatory approaches. Finally, more fundamental concerns have been raised regarding the concept of restructuring itself including the inherent limits of markets in reflecting broader societal concerns regarding energy services (e.g., Bozeman, 2002; Byrne and Mun, 2003; Kay, 2007).

This chapter will investigate the Australian case for and against each of these three propositions regarding restructuring and environmental outcomes—inherently supportive, neutral, or in opposition to improved environmental performance.

3 THE AUSTRALIAN ELECTRICITY INDUSTRY: MICROECONOMIC REFORM VERSUS ECOLOGICAL SUSTAINABLE DEVELOPMENT

The early development of the electricity industry in Australia had high government involvement focused, primarily, upon accessibility and energy security objectives.[2] The limitations of the resulting state-owned monopoly utility arrangements with regard to economic efficiency were highlighted in an Inquiry of the NSW Electricity Commission in the mid-1980s. The inquiry concluded that: generation planning suffered from procedural problems as well as a lack of consultation with the public; it was proving difficult to exercise adequate political control, and that the commission had inefficiently allocated billions of taxpayers' dollars through excessive construction of coal-fired generation. More generally, the inquiry noted that "without clearly

---

2. CIGRE Australia (1996) notes that "In those [early] decades that the industry was run as a series of state-owned enterprises, governments used it as an instrument of policy to promote rural electrification, the loads and revenues of the metropolitan customers being used to support financially the construction and operation of vast tenuous electrical systems."

laid down guidelines and external scrutiny, large technological organizations often develop monopolistic practices and an internal culture remote from broader social and political concerns" (McDonnell, 1986; see also Chapter 19 regarding some of these issues, and consequent electricity industry restructuring).

The early 1990s saw two developments that seemed destined to greatly change the electricity industry. One was an ambitious microeconomic "reform" agenda that included the highly influential Hilmer Review and embraced significant electricity industry restructuring. The other was a major ESD process that commenced around the same time under Federal Government direction.

3.1 Ecologically Sustainable Development

The Australian ESD process emerged from growing concerns about the environmental impacts of prevailing patterns of economic growth and development in the 1970s and 1980s. In 1990, the Federal Labour Government established nine "ESD Working Groups" to focus on the implementation of ESD including one on "Energy Production" and another on "Energy Use." These groups had wide stakeholder participation from government, industry, and environmental stakeholders. The final reports of the ESD Working Groups, containing over 500 recommendations, were released in November 1991 (Harding et al., 2009) and handed over to 37 interdepartmental committees of federal and state governments where the recommendations largely languished.

The National Strategy for ESD (NSESD), endorsed by all Australian governments in 1992, had stated objectives of economic development that safeguarded the welfare of future generations, provided equity within and between generations, and protected biodiversity and essential life-support systems (CoA, 1992). It also had guiding principles encompassing requirements to integrate long- and short-term economic, environmental, and social considerations, and that a lack of full scientific certainty should not be used as a reason for postponing action. However, many working group recommendations were discarded or weakened. The ESD process also informed the National Greenhouse Response Strategy (NGRS) although this "was predominantly influenced Industry Commission findings rather than the ESD process," with the Industry Commission having conducted a study, contemporaneous to the ESD process, that had found that meeting the then interim target for reducing greenhouse emissions would reduce national product by 1.5% and that some sectors of the economy, including the coal industry, would suffer (Reidy 2005, p. 206). As a result the NGRS did not support the "no regrets' measures," (those with conventional, economic, and environmental benefits) originally proposed for ESD but stated, instead, that "first phase measures will meet equity objectives by causing minimal disruption to the wider community, any single industry sector, or any geographical region." This acted to entrench protection of the coal industry over other, including environmental, objectives (Diesendorf, 2000).

The failure to address environmental concerns was widely remarked upon at the time and during subsequent years. For example, Wilkenfield et al. (1995) noted that:

[a]fter two years of its operation, there is no evidence that even one tonne of carbon emissions has been saved as a result of the NGRS ... the failure of the NGRS derives from a failure of governments to show leadership, to reconcile conflicting policy objectives and to distinguish the public interest from narrow commercial interests.

In summary, the ESD process which began in the early 1990s with a great deal of energy and enthusiasm is widely regarded to have, in the end, fizzled out and to have involved more rhetoric than substance. The primary reason for this was the primacy accorded the other reform agenda running parallel to the ESD process explored below.

3.2 Microeconomic Reform

Two key early steps in the restructuring of the Australian electricity industry were an Industry Commission (now Productivity Commission) report on "Energy Generation and Distribution" (Industry Commission, 1991), complemented and significantly reinforced by the later report of the Independent Committee of Inquiry into a National Competition Policy for Australia (Hilmer, 1993).

One early interaction with the ESD process was the 1991 Industry Commission report on the "Costs and Benefits of Reducing Greenhouse Gas Emissions" for the Australian Federal Government. This was generally unsupportive of action to address climate change, something the final sentence of the report underscores:

While microeconomic reform will not necessarily lead to reductions in greenhouse gas emissions, it is expected to produce higher income levels. These should make the costs of any emissions reductions more affordable

(Industry Commission, 1991b, p. 242).

This "microeconomic reform" was secured and given significant impetus by the 1993 Hilmer report that was also commissioned by the Federal Government. Prior to this there had been some constitutional limits on trading between some state-owned enterprises and a major focus of these competition reforms was structural reform of public monopolies. The report did note that:

Competition policy is not about the pursuit of competition per se. *Rather, it seeks to facilitate effective competition to promote efficiency and economic growth while accommodating situations where competition does not achieve efficiency or conflicts with other social objectives.*

However, it made no mention of "accommodating" environmental challenges or parallel environmental processes. The Hilmer recommendations were endorsed by all Federal and State governments and drove major restructuring in a range of sectors including electricity through an agreed National Competition Policy. By comparison with the largely side-lined ESD process, the impacts of this microeconomic reform agenda have been significant.

3.3 Implications for Australian Electricity Industry Restructuring

The National Grid Management Council (the NGMC) was established in 1991 to facilitate a "co-ordinated electricity market spanning the eastern States, South Australia and the Australian Capital Territory." Importantly, the draft national grid protocol (NGP) included some explicit environmental objectives (NGMC, 1992, p. 3):

 — to encourage the most efficient, economical and environmentally sound development of the electricity industry consistent with key national and state policies and objectives and

 — to maintain and develop the technical, economic and environmental performance, and/or utilization of the power system.

The National Electricity Law that came into force prior to the commencement of the NEM in late 1998, however, had legislated objectives that represented a marked departure from these environmental objectives of the NGP. In particular, there was no longer any explicit reference to environmental objectives at all.

The Australian Energy Market Agreement of 2004 represented the next stage of electricity industry restructuring and established new national governance, regulatory, and legislative frameworks for the NEM. Given growing concerns about climate change this presented a valuable opportunity to return environmental objectives to the NEM. However, it's new National Electricity Objective (NEO) stated (TEC, 2006, p. 7):

The national electricity market objective is to promote efficient investment in, and efficient use of, electricity services for the long term interests of consumers of electricity with respect to price, quality, reliability and security of supply of electricity and the reliability, safety and security of the national electricity system.

Although again there is no explicit mention of environmental objectives, the NEO's focus on the "long-term interests of consumers" might seem amenable to consideration of environmental objectives. In practice, however, key stakeholders have typically not taken this view. As one earlier example, the National Electricity Market Management Company Limited (NEMMCO) noted (TEC, 2006, p. 9):

Under the Rules, NEMMCO's charter focuses specifically on efficiency, security and reliability of power supply, and excludes favouring one fuel source over any other.

Consequently, NEMMCO has neither the power nor the authority to make decisions based on considerations of sustainability and balance in resource management.

The Australian government itself noted that the:

NEL and Rules do not deal with social equity or environmental policy issues. Such issues (for example, community service obligations or measures for the reduction of greenhouse emissions), are more appropriately addressed under other policy instruments

(MCE, 2004, p. 8).

In summary, Australian policy makers initiated two policy development processes in the early 1990s of specific relevance to the restructuring of the Australian electricity industry. One focused on ESD, the other on microeconomic reform with a major focus on electricity industry restructuring. The primacy of the economic agenda became apparent early in the process, resulting in an Australian electricity industry objective that explicitly excluded environmental considerations and instead, made these subject to external policies. Other jurisdictions have, in many cases, made similar choices as highlighted in other chapters of this volume. However, others have explicitly included these wider objectives in the restructuring process, and it would certainly seem to be a high risk approach to leave key environmental industry objectives outside the formal direct governance arrangements for the electricity industry. The outcomes of this choice will become evident in the following sections.

4 THE AUSTRALIAN NEM

The NEM includes all states and territories other than Western Australia and the Northern Territory, with almost 50 GW of generation serving over 90% of Australian electricity demand. The generation mix is dominated by coal-fired plant (around 80% of electricity provision) with contributions from gas-fired generation and hydropower supplying most of the remainder (AER, 2011). There is a growing contribution from wind energy and other renewables driven in large part by external renewable energy policy support measures.

The NEM largely operates under a uniform set of National Electricity Rules and has, as its centerpiece, a set of regional gross-pool spot energy and ancillary service markets. All plant of greater than 30 MW capacity are required to participate as scheduled (or in the case of intermittent technologies, semi-scheduled) generators in the dispatch process. The NEM is an energy-only market and participants are required to manage their own unit commitment and other inter-temporal scheduling challenges (within a range of technical dispatch constraints). There are no formal day-ahead markets or capacity markets. The long-term commercial regime in the NEM is, instead, implemented via derivatives for electricity and ancillary service spot markets. There has also been some movement towards large generators and retailers joining together to form "gentailers" who then have a partial physical hedge against spot price variability and volatility.

The NEM's rather unusual design by comparison with other restructured industries appears to have achieved reasonable success to date in efficiently operating the electricity industry, within an effective security regime (AER, 2011; see also Chapter 19). Wholesale prices have been generally low by international standards although a key factor here is the low costs of coal-fired generation in Australia. Reliability and security outcomes seem reasonable by international standards despite challenges including growing peak demand and, until recently, continuing energy consumption growth.

Key concerns with the efficiency of the NEM include the emergence of large gentailers with significant potential market power (AER, 2011), the current lack of demand-side participation in market dispatch (AEMC, 2012), and growing expenditure by the monopoly "economically regulated" network service providers (NSPs) within a regulatory framework that encourages them to overinvest in network assets which then earn them a guaranteed rate of return (Garnaut, 2011; Productivity Commission, 2012). All of these, particularly network expenditure, have contributed to very significant retail price increases over the past 5 years.

Of greatest relevance to this chapter, however, are concerns that various climate policies targeting the Australian electricity industry are also reducing the economic efficiency of the NEM (Productivity Commission, 2011; Chapter 19). As noted earlier, it was the stated government intention that environmental objectives were better served by external policies than direct inclusion in NEM governance. Certainly, a number of external policies have been implemented including a Renewable Energy Target (RET) that has seen considerable new generation, largely wind, in the NEM over the past decade. Other relevant environmentally oriented policies include a number of state-based energy efficiency schemes and renewable feed-in tariffs for photovoltaics (PV) and a gas-generation scheme in Queensland. Furthermore, a carbon price was implemented on energy sector emissions including electricity in July 2012.

The following three sections directly address the question posed in the Introduction—has Australian restructuring assisted in improving environmental outcomes from the electricity industry compared to what would have happened otherwise, played a neutral role with these outcomes determined by the effectiveness of external policies, or actually worked against environmental outcomes.

5 HAS AUSTRALIAN ELECTRICITY INDUSTRY RESTRUCTURING DIRECTLY ASSISTED ENVIRONMENTAL OBJECTIVES?

There was, at least initially, a stated expectation from the Australian Federal Government that the electricity industry restructuring process would

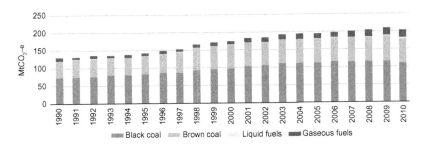

FIGURE 20.1   Australian electricity industry emissions over the past two decades (Australian Government, 2012b, p. 37).

contribute to climate change objectives by promoting changes including efficient competition in supply by: embedded cogeneration and renewable energy sources; more appropriate patterns of energy use through incentives for investment in energy efficiency, and through the penetration of natural gas into the energy sector that would lower the average greenhouse emissions intensity of energy (MacGill et al., 2006).

Unfortunately, it was soon clear that these outcomes had not, and were unlikely to, eventuate. Instead, it is widely accepted that restructuring actually increased emissions from expected "business-as-usual" outcomes (COAG, 2002). Reasons for this increase, some which might seem relatively obvious in retrospect, included: the significant overcapacity and low operating costs of high emission brown and black coal generation; less emphasis on energy efficiency given lower industrial and commercial electricity prices, and immature and inflexible gas industry arrangements that hampered a growth in its use (MacGill et al., 2006).[3] As shown in Figure 20.1, electricity industry emissions have grown over the past two decades, particularly in the early phases of restructuring.

A review by the Productivity Commission concluded that NCP had been highly successful in driving economic development. However, while:

the NCP's procedural framework explicitly provided for consideration of social, environmental, equity, regional and adjustment objectives in assessments of particular reform options ... public interest test requirements have not always been rigorously applied. Consequently, the NCP may have had some unanticipated adverse effects on the environment—in particular, higher greenhouse gas emissions associated with reform-related increases in demand for electricity

(Productivity Commission, 2005, p. 85).

---

3. International experience with restructuring has also highlighted the impact of preexisting circumstances including fuel mix, national endowment of resources, and existing infrastructure on environmental outcomes (IPCC, 2001).

Other assessments are less charitable. For example, energy consultant Alan Pears, who represented environment groups in the Energy Use Working Group of the ESD process summarized the results of the first decade of restructuring as follows:

Basically we've thrown away the last ten years, and if anything we've gone backwards. [f]or example, we've now acknowledged that introduction of electricity markets has increased Greenhouse gas emissions by 6 million tonnes of CO_2 equivalents per year above the predictions, yet the restructuring of the electricity industry was the centrepiece of the 1992 National Greenhouse Response Strategy! At that time supporters of electricity reform claimed it was the most powerful Greenhouse strategy we could apply!

(Findlay, 2000).

6 HAS THE NEM EFFECTIVELY FACILITATED EXTERNAL CLIMATE POLICIES?

While the Australian experience doesn't support the proposition that restructuring will inherently assist in emissions reductions, it is notable that emissions growth began to slow after 2003. While a number of factors are likely to have contributed, this period did coincide with a growing number of external environmentally oriented policies. A key question is, thus, how well NEM arrangements have facilitated these external measures in changing electricity industry outcomes towards environmental ends.

Such external policies can change participant behaviors in a number of ways. Informational, sometimes termed "suasive," measures generally seek to better inform decision-makers regarding their options and opportunities. Examples include energy efficiency star ratings for appliances or buildings. Regulatory measures, by comparison, effectively restrict available decisions. Examples include minimum energy performance standards (MEPS) for new equipment, and maximum permitted operational emissions for electricity generators.

Another policy approach, and one that found particular favor in Australia, is the use of economic instruments that change the effective "price" seen by decision-makers for different choices (MacGill et al., 2006). Examples include renewable energy certificate (REC) schemes and feed-in tariffs that provide an additional revenue stream to renewable energy generators. Another example is making electricity generators pay a price on their operational carbon emissions. Such measures are argued to be particularly appropriate for restructured electricity industries with market competition. Note, however, that it is not only inevitable but actually an objective of such measures to change the competitiveness of market participants according to environmental performance. As highlighted next, however, some key market participants have strongly opposed such policies on the basis that they represent inefficient market distortions that also, as it happens, adversely impact on their profitability.

Some key current Australian stationary (nontransport) energy sector policies are shown in Table 20.1 and highlight that two measures—regulatory energy efficiency and the RET—have contributed almost two-thirds of estimated stationary energy emission reductions over 2008−2012. The electricity industry is responsible for around two-thirds of total stationary energy sector emissions and has been the sole or primary focus of these measures. The different policies, however, have had different implications for NEM arrangements and imposed different obligations on market participants as outlined next.

TABLE 20.1 Estimated Emission Reductions Now and in 2020 Arising from a Range of Key Climate Policies Focused on Stationary Energy Sector Emissions

Stationary Energy Sector-Oriented Policy Measure	Estimated Kyoto Period Annual Emission Reductions from BAU (Mt CO_2-e)	Projected 2020 Annual Emission Reductions from BAU (Mt CO_2-e)
RET	8.8	29.9
Regulatory energy efficiency efforts		
Equipment Energy Efficiency (E3) Program	6.3	20.3
Energy efficiency requirements: building codes	4.2	11.8
Phaseout of inefficient incandescent lighting	1	1.9
Energy efficiency Certificate Schemes		
NSW Energy Saving Scheme	0.1	1.2
Victorian Energy Efficiency Target and Energy Saver	0.2	1.6
Queensland Gas Scheme	2.2	4.3
Total estimated stationary energy sector abatement of these measures from BAU	20.6	71
Percentage of total estimated emission reductions from all measures	71%	85%

Note: BAU, Business as Usual.
Source: Adapted from Australian Government, 2012c.

It is also notable that the scale of emission reductions to date is very modest—as noted earlier, these estimated reductions have still seen electricity industry emissions grow significantly over the last two decades. These estimates do not include the impact of the carbon price that was introduced in mid-2012. This price, intended to transition to an emissions trading scheme in 2015, has complex implications for the emissions reductions of these other measures (Twomey et al., 2012) and will be considered following assessment of the other measures.

6.1 Regulatory Energy Efficiency Measures

These measures include MEPS on a range of household and commercial equipment, and mandatory building requirements for energy efficiency. They have been the most significant greenhouse abatement policy efforts to date within the stationary energy sector. However, direct NEM participants have had only a very limited role in the design, implementation, and operation of these measures. Instead, the key industry stakeholders have been equipment manufacturers and providers. End users have themselves faced few decision-making challenges—MEPS and building codes merely act to remove the least efficient equipment or building design options from the market.

One notable exception to the minor facilitating role that the NEM has played with these policies has been the need to introduce MEPS for distribution transformers. Before electricity industry restructuring that separated distribution (investment and operation of the distribution wires) from retail (the wholesale purchase of electricity for sale to consumers), transformer selection by the publicly owned and monopoly electricity utilities was generally optimized with regard to losses as well as capital and maintenance costs. Within the restructured industry arrangements, however, distributors select the transformers while retailers (and hence end users) pay for the losses. Distributors were motivated to choose transformers on the basis of cost and reliability, rather than efficiency, and there was "some anecdotal evidence that commercial pressures in the corporatized and privatized distribution businesses [were] starting to drive efficiency levels down." This led to the decision to implement an MEPS (EEEP, 2011). This provides empirical evidence of the way widely accepted competitive principles can act to undermine environmental and even efficiency outcomes without specific regulatory redress.

Interestingly, the decline in energy consumption within the NEM over the past 3 years has been attributed in part to the impact of these policy measures, and this decline has had some significant impacts on market outcomes including reduced wholesale prices (AEMO, 2012). At least some market participants have argued that such outcomes are contrary to restructuring objectives. For example, International Power Australia which owns significant fossil fuel generation assets noted, in a submission to the Prime Minister's Energy Efficiency Task Group that it:

... rejects any proposal to introduce climate change policy, under the guise of energy efficiency measures, which has the potential to destroy the value of existing investments in the generator sector

(IPRA, 2010).

Another set of energy efficiency policies has, by contrast, required significant facilitation by NEM market participants and achieved far less. Three Australian States—NSW, Victoria, and South Australia—have recently established energy efficiency schemes that impose energy savings targets on retailers. The NSW and Victorian schemes also permit trading of these "energy savings" between the liable parties and accredited certificate providers. Of key relevance to this chapter has been the fairly unimpressive performance of the schemes to date in delivering significant genuine energy efficiency improvements, and the generally limited efforts that retailers have made to more closely engage in improving the energy efficiency of their customers (MacGill et al., 2012). The issue of retailer engagement in energy efficiency, and energy services more generally, is considered further in the next section.

6.2 Renewable Energy Target

The Federal Mandatory Renewable Energy Target (MRET) scheme that commenced in 2001 requires all Australian electricity retailers and wholesale electricity customers to source an increasing amount of their electricity from new renewable generation sources. The design of this "baseline and credit" scheme is based on tradable RECs that represent 1 MWh of "new renewable electricity generation." These can come from either new renewable generators, or by increasing the output from existing renewable generators. The original target was intended to be 2% of electricity sales in 2010 which was translated into a fixed target of 9500 GWh of additional renewable generation that year, to be then maintained until 2020. In 2011, this target was extended to 45,000 GWh in 2020 to be maintained until 2030, and separated into large project and small-scale system targets.

The Australian MRET was one of the world's first national Tradable Green Certificate (TGC) schemes, and therefore a highly innovative policy measure. Such schemes have been adopted by a number of other jurisdictions including some European countries and US states. They have considerable theoretical advantages over other approaches including feed-in tariffs and capital subsidies by offering technology-neutral support to a wide range of potential renewable energy sources and creating competitive pressures to reduce costs. They may also be more suitable for facilitating the integration of high renewable energy penetrations within restructured industries than

some other approaches which shield renewable energy projects from economic market "signals" (MacGill et al., 2006). This general approach was
also taken with the Queensland 13% Gas Scheme.[4]

Indeed, MRET appears to have performed reasonably up to date in effectively achieving its target at low public costs by international standards
(MacGill and Passey, 2009). Project developers have included some existing
generators and retailers as well as new project developers. In general, the
NEM would appear to have performed reasonably well in facilitating the
integration of reasonable amounts of new renewable generation, particularly
given that a considerable proportion of that generation was located in the
South Australian region of the NEM (MacGill, 2010; see also Chapter 25 on
electricity market design to facilitate renewable energy integration). Wind
integration has certainly posed some challenges and policy and regulatory
responses have included establishing new NEM technical connection standards, a new category of "semi-scheduled" generation for technologies, such
as wind and a centralized wind forecasting system.

However, MRET only had to achieve a modest target and did demonstrate some significant, largely governance related, failings. Of particular
note was the inclusion of solar hot water in a "renewable electricity" target
and the "confidential" baselines established for existing hydro generation
which allowed them to earn significant numbers of RECs from existing
operations.

More generally, Quota or TGC schemes have been argued to have failed
in Europe in comparison with feed-in tariffs with respect to both effectiveness in driving deployment, and efficiency due to high public costs.
Suggested reasons for this poor performance include not only the novelty of
the schemes, but also developer demands for a higher internal rate of return
given the greater investor insecurity than seen with other approaches, such as
feed-in tariffs, a "single" price for different technologies and particular projects that leads to windfall profits, and the susceptibility of the scheme
design process to be captured by incumbents who lobby for regulations that
they know they can satisfy but that smaller less-established competitors will
not be able to manage (MacGill and Passey, 2009).

With regard to this, the retailers within the NEM have had a key role in
which technologies are chosen and where they been deployed, as they hold
both an obligation for the RECs as well as having customers for the

---

4. The Queensland 13% Gas Scheme uses a rather similar "designer" market to that of the RET
but in this case a market based around "new gas-fired electricity generation" contributing to
Queensland demand. Again, the target is imposed on the retailers who then source their required
Gas Electricity Certificates (GECs) from accredited parties able to provide new renewable generation. By many measures the scheme is seen as having been relatively effective in delivering
lower-emission gas-fired generation investment (Grattan Institute, 2011) although assessment is
complicated by the significant Queensland government roles in ownership of both the large retailers and generators in Queensland at the time the scheme was implemented.

renewable energy generated. The key to successful project development has generally been obtaining a Power Purchase Agreement from a retailer. The expanded national RET for Australia now includes a far more significant target to be implemented within an increasingly stressed electricity industry infrastructure, including transmission, and a rapidly evolving industry structure with less government ownership and growingly powerful vertically integrated "gentailers" that own both significant generation and retail businesses. The emergence of three large gentailers within the NEM has raised particular competition concerns for renewable project development (AER, 2011). Indeed, a number of renewable energy project developers are now obtaining retail licenses in order to facilitate the Power Purchase Agreement (PPA) process given the apparent challenges of negotiating with the major gentailers.

The MRET has had significant impacts on NEM investment, generation mix and commercial outcomes including prices—a result due in part to the merit order effect seen when renewable generation with very low operating costs enters a wholesale market and displaces higher cost generation, hence reducing other generator dispatch and spot prices (Forrest, 2010). A number of incumbents have argued against expansion of RETs in the NEM on the basis of its impact on their profitability (Victorian Auditor General, 2011).

6.3 Carbon Pricing

A national carbon pricing scheme commenced in Australia in July 2012, directly covering around 500 large emitters in all energy-related sectors of the economy. The electricity sector represents some half of included emissions. The scheme faces an uncertain future given the current political divide in Federal Parliament over carbon pricing, but is currently intended to have a 3-year period with a fixed price, starting at AU\$23/tCO_2e, before transition to market-set prices and linkage to both the EU Emissions Trading Scheme and international carbon markets, such as the Clean Development Mechanism. This carbon price is part of a broader policy package that includes a range of complementary policies and so-called compensation for some key stakeholders.

It was some 14 years between the publication of four internationally regarded government-commissioned reports on carbon pricing options for Australia in the late 1990s and its eventual introduction. Debate regarding carbon pricing ranged from whether action on climate change was required at all through to the effectiveness and efficiency of emissions trading versus other options, then finally to specific design questions including targets, price caps, linkages, and the vexed issue of compensation to large emitters.

For the electricity industry, most electricity generators are required to pay for their emissions rather than having permits administratively allocated—a welcome contrast to the first two phases of the European Union (EU) Emissions Trading Scheme. The exception is so-called transitional assistance of over $5 billion in the form of free permits and cash to coal generators who face what has been termed significant asset impairment.

The NEM has now been operating under a carbon price for less than 6 months so it is too early to provide definitive assessments of how well or badly the NEM has managed its introduction. Certainly wholesale prices have risen significantly and there is some early evidence of reduced coal generation. It would certainly seem to have an important role to play in meeting Australia's current 2020 emissions targets (a national reduction of 5–25% from 2000 levels depending on international developments) yet even more importantly, the current government's target of 80% emission reductions by 2050.

However, perhaps the most important lessons of the carbon price to date have been in the process that was involved in implementing them. Not only electricity industry participants—fossil fuel generators but also, particularly, large electricity consumers—played very vocal roles in delaying the introduction of a carbon price, and demanding compensation for any potential loss of competitiveness associated with it—a point to which this chapter will return.

To conclude, it is not possible to draw definitive conclusions on how well the NEM has facilitated these various environmental policies focused on electricity industry outcomes that were, by conscious design, implemented externally to NEM governance. The measures to date have had relatively modest environmental objectives that haven't been a major test of NEM arrangements. The NEM does appear to have facilitated the RET to date, although there are some concerns regarding its future performance in this regard (AEMC, 2009; MacGill and Passey, 2009). The most successful emission reduction policies in the NEM to date have been energy efficiency regulations that haven't required the active involvement of major electricity industry participants. By contrast, those measures directly involving the retailers having achieved little to date. While this clearly reflects the modest objectives set to date for energy efficiency certificate trading, it is less why the objectives have been so modest.

The question of whether the market-oriented policies favored for restructured industries are more or less effective than approaches better suited for monopoly industry arrangements is still hotly debated. For example, Pollitt (2012) has argued that restructuring has significantly improved the quality of environmental policy instruments through the emergence of trading mechanisms. Others including Lohmann (2006) and Spash (2010) have, by contrast, highlighted key potential failings in the carbon emissions trading. The authors of this chapter have also explored the limitations of energy efficiency certificate trading (MacGill et al., 2012).

What is clear is that these types of environmental trading mechanisms are still relatively novel by comparison with more conventional regulatory, taxation, and government expenditure mechanisms. Furthermore, key market-oriented schemes have demonstrated mixed success to date (Pollitt, 2012) and key challenges for successful policy implementation have been rent seeking and capture by large industry incumbents (Helm, 2010; Helm and Hepburn, 2012). These challenges are certainly relevant in the Australian context as highlighted above.

7 HAS AUSTRALIAN ELECTRICITY RESTRUCTURING INHERENTLY WORKED AGAINST IMPROVED ENVIRONMENTAL OUTCOMES?

Given mixed findings for the first two of the three propositions of the relationship between restructuring and environmental objectives posed at the beginning of this chapter, the third can now be considered—has restructuring in Australia actually worked directly against improved environmental outcomes.

Possible reasons for this that are considered here are the creation of large powerful industry incumbents with an interest in maintaining the status quo, the failure to establish retail markets that sell energy services rather than just electricity (kilowatt hour), and poor governance in external policies. Beyond this are more fundamental concerns about the limitations of market arrangements, and the process by which they are implemented, toward environmental objectives.

7.1 The Impact of Commercially Focused and Politically Powerful Incumbents

The breakup of the existing state-owned monopoly generation and retail entities in Australia varied by state. However, all involved the creation of a relatively small number of major players. There has also been a continuing trend of vertical integration between electricity generators and energy retailers into "gentailers." Three companies—Origin Energy, AGL Energy and TRUenergy—together now supply some 80% of small customers and control over 30% of generation capacity. Furthermore, they have undertaken the majority of new generation capacity since 2007, and there has been almost no investment by players that don't also have a retail arm (AER, 2011).

There are economic efficiency concerns regarding these developments which work against new market entry by nonincumbents. Alongside these adverse competitive impacts are also potentially adverse environmental outcomes. Any transformation of the electricity industry towards a low-carbon future hinges on investment, and it seems likely that such decisions are increasingly available only to a small number of major players who are, by definition, current beneficiaries of the status quo.

The creation of these large incumbents has also had other adverse outcomes on the policy making process itself. For example, and as noted earlier, some led calls for compensation to be paid to high emission generators that would be adversely impacted by the carbon price scheme. The eventual Clean Energy Future package included some A$5b of assistance to the most pollution intensive generators in the NEM—some of which have been amongst the most profitable under the original NEM arrangements.

7.2 Failure to Establish Competition in the Desired Objective—Energy Services Not Delivered Energy

Even within the terms of the NEO, the role of competition is to "facilitate the efficient investment in, and efficient use of, electricity *services*"—that is, the services that energy consumers actual desire. Electricity consumption is merely a means to that end. Unfortunately, the emphasis on retail competition has been based largely on price of electricity (c/kWh) rather than a more holistic view of energy service delivery which entails energy efficiency opportunities, demand management, and distributed generation options, as well as per unit energy costs. Such a framework does require far greater end-user engagement and this is something that the current NEM retail design doesn't facilitate (AEMC, 2012; Healy and MacGill, 2011; MacGill et al., 2013).

Many retailers offer "energy services" consulting to larger customers, however, current market arrangements mean that it is not generally in a retailers' financial interests to promote energy savings (Victorian Parliament, 2005). Furthermore, assessments of the effectiveness of retail competition by the Australian Energy Market Commission have explicitly argued that their

> ...*finding that competition is effective is supported by evidence of strong rivalry between retailers. Because the provision of energy is viewed as a homogenous, low engagement service ... retailers have a strong incentive to be pro-active in seeking and retaining customers in competition with their rivals*

(AEMC, 2008).

Energy services are, by definition, not homogenous and effective and efficient delivery of these therefore requires high engagement.

Recently, even the AEMC (2012) has acknowledged the importance of greater end-user engagement in the NEM albeit in terms of improved efficiency.[5] However, there is continuing disagreement on the need and potential nature of support for end-user decision-making. For example, the AEMC Review of Demand-Side Participation claimed that:

> *A key assumption behind this review is that consumers will always make the best decision from their viewpoint, based on the prices they face, the technology and equipment they have access to, the information they have and their individual*

---

5. The AEMC (2012) notes that "Efficient markets are characterized by effective participation of both the supply and demand side .... While there is some evidence of uptake of demand side participation (DSP) in the NEM over recent years, the efficiency of the electricity market can be improved by effective use of the demand side."

transaction costs. ...This will also allow third parties to assist consumers make
optimal decisions under innovative business models.

(AEMC, 2011).

This narrow economically focused framework misses the reality that the NEM is a highly complex "designer" market with major shared network infrastructure managed by regulated monopolies and major asymmetries between supply and demand in terms of information, knowledge, and resources. There is an evident need for Energy Service Companies (ESCOs) to provide support—advice and resources—to allow end users to effectively participate in the NEM (Outhred and MacGill, 2006). As the Energy Efficiency Council (EEC, 2011) noted in their submission to the review,

In this context, expecting energy consumers to optimize their level of DSP without
any support from third-parties and specific DSP schemes is preposterous.

Furthermore, the interface between the retail market and the economic regulation of the monopoly NSPs is also working against efficient delivery of energy services. In particular, the NSPs have little interest in engaging with energy users to reduce peak demand because this requires network expenditure, which then earns them a guaranteed rate of return (AEMC, 2012; Productivity Commission, 2012). This works against both efficiency and environmental outcomes, and has underpinned recent significant retail market price increases.

To conclude, the IEA (2012) amongst many has highlighted the key role that improved end-use energy efficiency, demand management, and distributed generation (from sources including renewables such as PV and highly efficient fossil fuel options such as cogeneration), can play in delivering the same end-use energy services while reducing adverse environmental impacts. Market arrangements that work against such options therefore also work against improved environmental outcomes from the electricity industry.

7.3 Poor Governance of External Policies

One immediate challenge for external policies is that of legitimacy. Key participants in Australian electricity industry restructuring have argued that external policies, such as carbon pricing and RETs, are inefficient distortions of the electricity market that should be avoided, or require compensation for those whose investments are adversely impacted. Within the legal framework of the NEO, such policy efforts are more accurately seen as an obligation upon policy makers and participants if the NEM, with its major environmental and social externalities, is to contribute to broader societal objectives. Nevertheless, this question of legitimacy continues to be raised.

The next challenge is that of policy design and implementation. Market-based approaches such as TGCs are sometimes argued to be simpler than regulatory or direct fiscal approaches because governments just have to set the target, and then let the markets work out how best to achieve these objectives. The reality is very different. These markets are exemplary "designer" markets—they arise from policy, and design choices can markedly affect their effectiveness, efficiency, and equity impacts. Of perhaps greatest concern with these designer markets is the potential for influential stakeholders to capture the process and manipulate initial design choices to their own advantage, and then delay or defer changes to the rules that work against their interests.

In the Australian context, the policy process for designing the expanded RET scheme not only failed to correct identified problems with the original scheme but actually made the scheme design worse in some key regards (MacGill and Passey, 2009). Governance appears to be going backwards. With regard to the torturous policy process for establishing a carbon price in Australia, Professor Garnaut who was an advisor to the Federal Government on climate policy noted "*I think this whole process of policy making over the emissions trading scheme has been one of the worst examples of policy making we have seen on major issues in Australia*" (ABC, 2009). The key reason was the failure of the governance process to resist lobbying by special interests including large electricity industry incumbents.

By many measures the governance arrangements and market design of the NEM are far more robust than those implemented for these external policies. Maintaining NEM security has priority over commercial arrangements—unlike the environmental markets that have been designed in Australia over the last decade, widespread market failure is not an option for the NEM. Furthermore the rule change process reflects the need to be able to change market rules if and as required through a clear and transparent process. Importantly, and in marked contrast to some environmental market processes, investor and market certainty concerns do not prevent rule changes where they are warranted—surely a requirement for "serious" governance. While the NEM processes are not without failings—they can certainly be slow—they do represent serious governance and there might have been considerable value in using this framework for environmental policy as well through mechanisms including the direct incorporation of environmental objectives in the NEM NEO.

7.4 Foundational Concerns

The problems with governance flagged above may well signal more foundational concerns with restructuring. One has been the impact of a narrowly focused restructuring process on the legitimacy, coherence, and comprehensiveness of climate policy. Given the evident lack of political will on environmental objectives for much of the past two decades in Australia, one envisages that we would have seen less progress on restructuring, or more progress on climate policy if the two processes had been more formally integrated. The potential for poorly focused restructuring processes to work against environmental outcomes has also been observed internationally.[6]

Furthermore, competition policy tends to assume some sort of idealized market and, when it comes to specific instances, such as electricity industry restructuring, assume that this can be effectively approximated via certain specific prescriptions (such as, in this case the vertical and horizontal desegregation of the industry and the facilitation of competition between certain of the resultant entities). However, electricity is so far removed from a conventional commodity, and the electricity industry from any other form of "market" (as witnessed by the complexity of the real-time trading arrangements characteristic of them) that this assumption is questionable. In reality, electricity markets are "designer markets" *par excellence* in which the most fundamental aspects of the market, including the content and character of both the commodities traded and of the arrangements for trading them, have to not only be defined but then institutionally, and in this case also technologically, instantiated.[7]

In other words restructured industries involve all manner of complex decisions and trade-offs regarding what to include, ignore, and/or marginalize, necessarily benefitting some stakeholders at the expense of others. The Australian experience, through, for example, the way the Industry Commission acted to privilege the interests of the coal industry above precautionary action detailed in Section 3.1 above, suggests that these choices can actively privilege a narrow economic perspective over environmental objectives.

8 POSSIBLE WAYS FORWARD FOR AUSTRALIA

The chapter is not an argument for a return to the Australian electricity industry arrangements prior to restructuring. The limitations of these monopolistic state-led arrangements with regard to both economic efficiency and

6. For example, a report exploring the experience of six developing and transition countries with so-called electricity industry reform and sustainable development objectives found that financial concerns and donor conditions were driving the process and that "*Managed by closed political processes and dominated by technocrats and donor consultants, environmental considerations play almost no role in a re-envisioned electricity sector.*" The report concludes that to facilitate sustainable development, sustainability concerns needed to be factored into the reform process and backed up with political commitment (WRI, 2002).
7. For example, Byrne and Mun (2003) question restructuring based on the underlying principle of trading electricity as a commodity: "*A commodity policy relies for its claim of being a distinctive source of public benefits on two premises—cornucopianism and individualism [that is, the logic that more is better and that individual choice is paramount]. Experience with liberalization has underscored the existence of vital public values that are neither cornucopian nor individualistic. These include the value of reducing energy use in the interest of sustainability...*"

environmental outcomes have been widely recognized.[8] Furthermore, the failure of Australia to effectively address climate change is hardly unique as evident in a number of other chapters in this volume.

There are some examples of electricity industries elsewhere in the world that have achieved significant, if as yet still far from complete, low-carbon transition over the past two decades of growing climate change concerns (Helm and Hepburn, 2012). Some countries have certainly achieved significant deployments of renewable energy over this time within so-called liberalized markets including Denmark, Germany, Spain, and Italy. As a particularly pertinent example, emissions from the Danish electricity industry have fallen almost 40% from 1990 to 2010 (DEA, 2011) and they have extremely ambitious decarbonization plans for coming decades (DMCEB, 2012). Over that time, the Danish electricity industry has restructured from around 100 vertically integrated utilities, predominantly cooperative or municipal government-owned and largely run on not-for-profit basis to more competitive arrangements. Another example of electricity industry transformation that is often cited is that of California, which has major energy efficiency and renewable energy goals. By comparison with the examples above, California is a leading jurisdictional proponent of Integrated Resource Planning of its regulated utilities—an approach to which it returned following a largely failed restructuring process (Byrne and Mun, 2003).

What can be said about the Australian experience of restructuring is that it didn't inherently facilitate improved environmental outcomes. Furthermore, while the NEM has facilitated a range of external policies intended to drive change in the electricity industry, there have been challenges and the scale of change to date has been modest—largely due to a lack of political will to achieve environmental aims. However, restructuring to date would also seem to have actively worked against environmental outcomes in key ways.

8.1 Opportunities for Addressing Current NEM Inadequacies

NEM arrangements continue to evolve under both internal market drivers as well as external policies. Some promising processes underway include the demand-side participation review that appears to now better appreciate the important role of ESCOs (AEMC, 2012). Furthermore, proposed changes to NSP regulation would improve the transparency and stakeholder engagement

---

8. It should, however, be noted that some valuable progress was also made by the State Utilities. For example, the State Electricity Commission of Victoria undertook Australia's largest ever Demand Management program in the early 1990s with an action plan developed in the late 1980s that acknowledged "... *demand management investment measures commencing in the mid 1990s is likely to be an integral part of Victoria's least cost strategy for balancing future electricity supply and demand. There is a distinct possibility however that a more aggressive demand management effort will be needed as a key plank of the SECV's effort to combat the Greenhouse Effect.*" (SECV, 1989). These efforts, however, were largely ended by the Victorian restructuring process that followed.

of their decision-making—a first step towards a formal Integrated Resource Planning approach to network expenditure which represents around half of total financial flows through the industry.

Beyond these developments, there are a number of relatively modest changes to the NEM which would assist in assessing and enhancing the NEM's support for low-carbon transition. One would be better public performance reporting against the broader objectives that are implicit, although not yet explicit, in the "longer-term interests" of consumers.

The current policy framework of seeking to achieve social and environmental outcomes within the NEM through only external policies has not worked well. We have shown how a lack of political intent and poor external policy processes can be understood to be major reasons for this. Nevertheless, explicit incorporation of social and environmental policies into the NEO could assist in focusing NEM governance on the importance of facilitating transition. As noted earlier, NEM governance appears more robust than that used for external policies to date. There are useful international precedents from other jurisdictions including the United Kingdom where the regulator, the Office of Gas and Electricity Markets (OFGEM) has as its principle objective:

... to protect the interests of existing and future consumers in relation to gas conveyed through pipes and electricity conveyed by distribution or transmission systems. The interests of such consumers are their interests taken as a whole, including their interests in the reduction of greenhouse gases and in the security of the supply of gas and electricity to them

(ISF, 2011).

The AEMC undertook an inquiry into the impact of the RET and carbon pricing on the NEM back in 2008 (AEMC, 2008). There would be value in undertaking another inquiry regarding whether the NEM can currently facilitate fundamental low-carbon transformation, and how its capabilities in this regard might be improved. As highlighted in this chapter, such an inquiry should encompass matters of fundamental market structure and design.

One important area of market design and structure is that of the governance process itself, and the ability of this to deliver a rapid response to emerging challenges and opportunities (Passey et al., 2008). Another challenge for governance is the need for decision-making processes that are open, transparent, and engage all stakeholders in a sustainable energy future—that is, all of us. Commercial arrangements that limit such engagement with end users to that of a customer are almost certainly not sufficient and rather act to bolster the inherent tendency of such arrangements to privilege economic over other interests.

8.2 Reconsidering the Role of Markets

A longer-term priority for restructuring should be reconsideration of the role of markets in a future low-carbon Australian electricity industry. As detailed previously, "markets," particularly designer ones, have integral shortcomings that are rarely if ever acknowledged by market adherents. For example, the "economically rational" framing of humanity as no more than an ensemble of utility optimizing autonomous individuals; the collective aggregation of the preferences of such abstracted individuals; the unrealistic conditions these necessitate (including complete information for all market participants, perfect competition, and the notion of market "equilibrium") are all conceptual abstractions that render opaque much of the complexity of real-world contexts (Colander et al., 2009; Jackson, 2009; Stiglitz, 2001; Veblen, 1898). One of the results of this being that abstractions such as "economic efficiency" can act to cloud a host of deeply contingent judgments regarding what is traded, how this is defined, and how trading is configured and structured.

So while markets can, evidently, be a very powerful means of allocating scarce goods and services they also have a number of limitations and draw-backs that need to be accounted for in practice. This is particularly evident when economic principles stand in tension with others, such as environmental considerations as shown here. In the Australian case the evident attachment of policy makers to narrow economic principles appears to have ensured environmental considerations were marginalized such that a narrow economism became institutionally entrenched and resulted in the "capture" of public policy by dominant players, notably those attached to the coal industry and utility interests (e.g., Hamilton, 2007; Pearce, 2009).

While there may have been recent progress in these regards it is clear that such tensions still exist. Facilitating an industry transition of the scale and speed required to avoid dangerous climate change necessitates not only the transcendence of these tensions but also far more care and attention to the weaknesses and limitations of market arrangements.

9 CONCLUSION

As a number of other chapters in this volume highlight, it is unclear whether restructuring is the right answer to the question of economic efficiency. Australia's restructuring process over the past two decades appears to be one of the more successful efforts worldwide towards achieving narrow economic efficiency objectives, although there are some evident failings. However, climate change should be a higher order policy priority given our growing climate challenges, and the fundamental threat they pose to economic welfare and progress. Hence this chapter has focused on a different question—that of whether Australian electricity industry restructuring has:

1. inherently assisted in reducing environmental harms such as climate change;
2. proved neutral, facilitating environmental improvements where the political will to implement appropriate policies exists; or
3. inherently worked against improved environmental outcomes?

The Australian experience has been that without political will and effective external policies, restructuring certainly did not assist in reducing environmental harms such as climate change. On the contrary, the narrow economic framing of restructuring in Australia appears to have actively worked against improved environmental outcomes. At least in part, this outcome was also the result of failure to implement effective external policies. However, in turn, this environmental policy failure was at least in part an outcome of a restructuring process that prioritized narrow economic efficiency perspectives over environmental concerns. Given growing climate change concerns, a key objective of further restructuring should be to support large and rapid emission reductions from the electricity industry. While there are a range of straightforward improvements to the NEM that would facilitate this new role, a fundamental revisiting of the role of markets is almost certainly required.

REFERENCES

Australian Broadcasting Corporation, 2009. Ross Garnaut interviewed on The 7:30 Report, Broadcast: 12/10/2009. Transcript Available from: <www.abc.net.au/>.

Australian Energy Market Commission, AEMC, 2008. Review of the Effectiveness of Competition in Electricity and Gas Retail Markets in Victoria, Second Final Report, February, Sydney. Available from: <www.aemc.gov.au/>.

Australian Energy Market Commission, AEMC, 2009. Review of Energy Market Frameworks in light of Climate Change Policies, September. Available from: <www.aemc.gov.au/>.

Australian Energy Market Commission, AEMC, 2011. Power of choice—giving consumers options in the way they use electricity—Issues Paper, July, Sydney. Available from: <www.aemc.gov.au/>.

Australian Energy Market Commission, AEMC, 2012. Power of Choice Draft Report, September, Sydney. Available from: <www.aemc.gov.au/>.

Australian Energy Market Operator, AEMO, 2012. Electricity Statement of Opportunities, August. Available from: <www.aemo.com.au/>.

Australian Energy Regulator, AER, 2011. State of the Energy Market, December. Available from: <www.aer.gov.au/>.

Australian Government, 2012. United Nations Framework Convention on Climate Change. Available from: <www.climatechange.gov.au/government/initiatives/unfccc.aspx/>.

Australian Government, 2012b. Inventory Report 2010—Volume 1. Australian Government Submission to the United Nations Framework Convention on Climate Change, April. Available from: <www.dccee.gov.au/>.

Australian Government, 2012c. Australian Stationary Energy Emission Projections. Available from: <http://climatechange.gov.au/publications/projections/australias-emissions-projections/stationary-energy.aspx#elec-generation/>.

Bozeman, B., 2002. Public-value failure: when efficient markets may not do. Publ. Adm. Rev. 62, 145–161.

Byrne, J., Mun, Y.M., 2003. Rethinking reform in the electricity sector: power liberalization or energy transformation. In: Wamukonya, J. (Ed.), Electricity Reform: Social and Environmental Challenges. United Nations Environment Program, Denmark.

CIGRE Australia, 1996. Contribution on Australia: A Dictionary on Electricity. Brady, F. (Ed.), For CIGRE and The Association for the History of Electricity in France.

CoA, (Commonwealth of Australia), 1992. National Strategy for Ecological Sustainable Development. Prepared by the Ecologically Sustainable Development Steering Committee. Endorsed by the Council of Australian Governments. Available from: <http://www.environment.gov.au/about/esd/publications/strategy/intro.html/>.

Colander, D., Follmer, H., Haas, A., Goldberg, M.D., Juselius, K., Kirman, A., et al., 2009. The financial crisis and the systemic failure of academic economics. (March 9, 2009). University of Copenhagen Department of Economics Discussion Paper No. 09-03. Available from: <http://papers.ssrn.com/sol3/papers.cfm?abstract_id = 1355882/>.

Council of Australian Governments, COAG, 2002. Towards a truly national and efficient energy market, December. Available from: <www.ret.gov.au/>.

Danish Ministry of Climate, Energy and Building, DMCEB, 2012. Denmark Energy Agreement. March 22.

Danish Energy Agency, DEA, 2011. Energy Statistics 2010, November. Available from: <www.ens.dk/>.

Diesendorf, M., 2000. A critique of the Australian government's greenhouse policies. In: Gillespie., A., Burns, W.C.G. (Eds.), Climate Change in the South Pacific: Impacts and Responses in Australia, New Zealand, and Small Islands States. Kluwer, Dordrecht, pp. 79–93.

Equipment Energy Efficiency Program, EEEP, 2011. Review of Minimum Energy Performance Standards for Distribution Transformers. Equipment Energy Efficiency Program consultation Regulatory impact statement, May. Available from: <www.finance.gov.au/>.

Energy Efficiency Council, EEC, 2011. Submission to the AEMC Power of Choice Issues Paper. Available from: <www.aemc.gov.au/>.

Findlay, M., 2000. ESD—The last 10 years. Nat. Parks J. 44 (5). Available from: <http://dazed.org/npa/npj/200010/Octfeatures-ESD1.htm/>.

Forrest, S.R., 2010. Quantifying the Impact of Intermittent Wind Generation in the Australian National Electricity Market. Engineering Honours Thesis, UNSW. Available from: <www.library.unsw.edu.au/>.

Frei, C.W., 2004. The Kyoto protocol—a victim of supply security? or if Maslow were in energy politics. Energ. Policy 32 (11), 1253–1256.

Garnaut, R., 2008. The Garnaut Review. September. Available from: <www.garnautreview.org.au/>.

Garnaut, R., 2011. The Garnaut Review 2011—Australia in the Global Response to Climate Change. Cambridge University Press, Australia, June.

Grattan Institute, 2011. Learning the Hard Way: Australia's Policies to Reduce Emissions, April. Available from: <www.grattan.edu.au/>.

Hamilton, C., 2007. Scorcher: The Dirty Politics of Climate Change. Black Inc. Agenda, Melbourne.

Harding, R., Hendriks, C.M., Faruqi, M., 2009. Environmental Decision-Making—Exploring Complexity and Context. The Federation Press, Sydney.

Healy, S., MacGill, I., 2011. From smart grid to smart energy use. In: Sioshansi, F.P. (Ed.), Smart Grid: Integrating Renewable, Distributed, and Efficient Energy. Academic Press, Oxford.

Helm, D., 2010. Government failure, rent-seeking, and capture: the design of climate change policy. Oxford Rev. Econ. Policy 26 (2), 182–196.

Helm, D., Hepburn, C. (Eds.), 2012. The Economics and Politics of Climate Change. Oxford University Press, Oxford.

Hilmer, F., 1993. National Competition Policy, Report by the Independent Committee of Inquiry. Available from: <http://www.australiancompetitionlaw.org/>.

Industry Commission, CoA, 1991. Energy Generation and Distribution. Report of the findings of the Industry Commission's inquiry into Energy Generation and Distribution. Available from: .

Industry Commission, CoA, 1991b. Costs and Benefits of Reducing Greenhouse Gas Emission. Report of the findings of the Industry Commission's inquiry into Costs and Benefits of Reducing Greenhouse Gas Emissions. Available from: .

Institute for Sustainable Futures, ISF, 2011. The NEM Report Card: How Well Does the National Electricity Market Serve Australia? UTS and Monash University, Sydney. Available from: .

International Energy Agency, IEA, 2001. Competition in Electricity Markets, Paris. Available from: .

International Energy Agency, IEA, 2005. Lessons from Liberalised Electricity Markets, Paris. Available from: .

International Energy Agency, IEA, 2012. Energy Technology Perspectives—Pathways to a Clean Energy System. Paris. Available from: <http://www.iea.org/>.

Intergovernmental Panel on Climate Change, IPCC, 2001. Climate Change 2001—IPCC Third Assessment Report Working Group 3: Mitigation. Available from: <http://www.ipcc.ch/ipccreports/tar/>.

International Power Australia, IPRA, 2010. Comment on the Prime Minister's Energy Efficiency Task Group's Issues Paper, May. Available from: <www.cedaily.com.au/>.

Jackson, T., 2009. Prosperity Without Growth? Economics for a Finite Planet. Earthscan, London.

Kay, J., 2007. The failure of market failure. Prospect Mag. August (137).

Lohmann, L., 2006. Carbon Trading: A Critical Conversation on Climate Change, Privatisation and Power. Available from: <www.thecornerhouse.org.uk/>.

MacGill, I.F., Outhred, H.R., Nolles, K., 2006. Some design lessons from market-based greenhouse regulation in the restructured Australian electricity industry. Energ. Policy 34 (1), 11−25.

MacGill, I.F., 2010. Electricity market design for facilitating the integration of wind energy: experience and prospects with the Australian national electricity market. Energ. Policy 38 (7), 3180−3191.

MacGill I.F., R. Passey, 2009. CEEM Submission to the Senate Economics Committee Inquiry into the Renewable Energy Bill, August. Available from: <www.ceem.unsw.edu.au/>.

MacGill, I.F., Healy, S., Passey, R., 2013. Trading in energy efficiency—a market based solution, or just another market failure? In: Sioshansi, F.P. (Ed.), Energy Efficiency: Towards the End of Electricity Demand Growth. Elsevier Academic Press, Oxford.

McDonnell, G., 1986. Commission of Inquiry into Electricity Generation Planning in New South Wales. NSW Government Publishing Service, Sydney.

Ministerial Council on Energy, MCE, 2004. National Electricity Law and National Electricity Rules—Information Paper, December. Available from: <www.ret.gov.au/>.

National Grid Management Council, NGMC, 1992. National Grid Protocol—First Issue, November. Available from: <www.efa.com.au/>.

Outhred, H.R., MacGill, I.F., 2006. Electricity industry restructuring for efficiency and sustainability—lessons from the Australian experience. In: Proceedings of the ACEEE'06 Summer Study, California, August.

Passey, R., MacGill, I., Outhred, H., 2008. The governance challenge for implementing effective market-based climate policies: a case study of The New South Wales Greenhouse Gas Reduction Scheme. Energ. Policy 36 (8), 3009−3018.

Pearce, G., 2009. Quarry vision—coal, climate change and the end of the resources boom, Quarterly Essay, vol. 33. Black Inc. Agenda, Melbourne.

Pollitt, M.G., 2012. The role of policy in energy transitions: lessons from the energy liberalisation era. Energ. Policy 50, 128–137.

Productivity Commission, 2005. Review of National Competition Policy Reforms, February. Available from: <www.pc.gov.au/>.

Productivity Commission, 2011. Carbon Emission Policies in Key Economies, June. Available from: <www.pc.gov.au/>.

Productivity Commission, 2012. Electricity Network Regulatory Frameworks Draft Report, October. Available from: <www.pc.gov.au/>.

REN21, 2012. Renewables Global Status Report, Paris, June. Available from: <www.ren21.net/>.

Riedy, C., 2005. The Eye of the Storm: An Integral Perspective on Sustainable Development and Climate Change Response. PhD Thesis, Institute for Sustainable Futures, University of Technology, Sydney.

SECV, 1989. Demand Management Development Project, 3 Year Demand Management Action Plan, Information Paper No. 5, December. Available from: <http://www.efa.com.au/>.

Spash, C.L., 2010. The Brave New World of carbon trading. New Pol. Econ. 15, 2.

Stern, N., 2006. The Stern Review on the Economics of Climate Change. H.M. Treasury, London.

Stiglitz, J.E., 2001. Nobel Prize Lecture: Information and Change in the Paradigm in Economics. Available from: <http://www.nobelprize.org/nobel_prizes/economics/laureates/2001/stiglitz-lecture.html/>.

Total Environment Centre, TEC, 2006. How Should Environmental and Social Policies be Catered for as the Regulatory Framework for Electricity Becomes Increasingly National?, November. Available from: <www.tec.org.au/>.

Twomey, P., Betz, R., MacGill, I.F., Passey, R., 2012. Additional action reserve: a proposed mechanism to facilitate additional voluntary and policy emission reductions efforts in emissions trading schemes. Clim. Policy 12 (4), 424–439.

Veblen, T., 1898. Why is economics not an evolutionary science? Q J. Econ. 12 (4), 373–397.

Victorian Auditor General, 2011. Facilitating Renewable Energy Development, April. Available from: <www.audit.vic.gov.au/>.

Victorian Parliament, 2005. Report of the Environmental and Natural Resources Committee on the Inquiry into Sustainable Communities. Parliamentary Paper No. 140.

Wilkenfeld, G., Hamilton, C. and Saddler, H., 1995. Australia's Greenhouse Strategy: Can the Future be Rescued?. Report for The Australia Institute. Available from: <www.tai.org.au/>.

World Resources Institute, WRI, 2002. Power Politics: Equity and Environment in Electricity Reform. Dubash, N.K. (Ed.). Available from: <www.wri.org/>.

Chapter 21

Weak Regulation, Rising Margins, and Asset Revaluations: New Zealand's Failing Experiment in Electricity Reform

Geoff Bertram
Institute for Governance and Policy Studies, Victoria University of Wellington

1 INTRODUCTION

Reform of the New Zealand electricity sector began in 1986, as part of a wave of neoliberal policy changes pushed through by the Fourth Labour Government (1984–2000). In that year, two major pieces of legislation—the State-Owned Enterprises Act 1986 (SOE Act) and the Commerce Act 1986—transformed the institutional and policy environment within which electricity, along with other publicly owned essential services, was produced and delivered.

The SOE Act changed the primary objective of these enterprises from social service to profit: state-owned enterprises (SOEs) were to be "as profitable and efficient as comparable businesses that are not owned by the Crown." SOEs were transformed from government departments to corporations, with management systems modeled on the private sector.

Meantime, the Commerce Act 1986 brought fundamental changes in regulatory philosophy. The stated purpose of the Act was "to promote competition in markets for the long-term benefit of consumers within New Zealand," but this left open the issue of what was to be done in markets where competition was weak or absent (Stevens, 2003, pp. 93–98). There were no provisions to outlaw price gouging or profiteering at consumers' expense. The presumption was that promotion of competition would generally suffice to restrain the use of market power by large firms to transfer

wealth from consumers to themselves. If abuse of market power was sus-
pected, Part 4 of the Act empowered the Minister of Commerce to instruct
the Commerce Commission to inquire into particular industries and, if
appropriate, recommend price regulation. If the Commission judged regulation to be warranted, a ministerial decision was still required before the
Commission could proceed. Decisions on whether and how to regulate
prices and profits were thus removed from the judicial to the political
sphere. The common-law right of consumers, faced with high prices for
essential services, to seek redress through the courts was extinguished
(Taggart, 2008).

Electricity policy-making since 1986 has involved ongoing tension
between those who saw privatization under generic competition law as the
logical destination for policy, and those who retained the idea of electricity as
an essential public service for the price and quality of which the state
remained responsible, implying ongoing state participation and/or regulation.
Successive changes to the structure and regulation of the electricity sector
over two and a half decades have not resolved these tensions.

Several key features of the New Zealand experiment made it unusually
neoliberal in its approach, compared with most other countries' approach to
electricity restructuring:

- First, New Zealand shared with Germany the absence of a specialist
  regulator to oversee the early stages of reform. When eventually forced
  to establish an Electricity Commission in 2003, New Zealand policy
  makers gave it a minimal mandate, with its price-regulating power narrowly confined to transmission grid pricing and a governance-focused
  agenda. As described later, the Commission was abolished after only 6
  years, to be replaced by a more generator-friendly Electricity
  Authority.

- Second, an antiregulatory mindset applied broadly to all facets of the
  electricity industry reform process. "Light-handed regulation" was interpreted as a reduction not only in regulatory intervention, but in regulatory
  capacity as well,[1] so that the rare government interventions in the face of
  anticompetitive market outcomes have been poorly designed, and have
  come too late to prevent large wealth transfers from being banked by key
  private-sector players.

- Third, there was no official enthusiasm in New Zealand for the institutional innovations that many other countries adopted to widen the range
  of electricity market participation. No feed-in tariff has yet been tried, nor

[1]. A 2009 consultants' review of electricity regulation noted that "New Zealand has no consistent or coherent measure of electricity sector performance against which to measure the results
of policy changes. None of the many regulatory agencies which routinely comment on the
electricity sector has developed, or report[s] in a structured way, indicators of electricity sector
performance." (Murray et al., 2009, p. 1.).

any serious restriction on the power of incumbent large vertically integrated generator-retailers (gentailers) to foreclose wholesale market entry by independents, using tactics which in other jurisdictions would arguably be in breach of antitrust law. Smart metering is yet to arrive as of 2012 (Concept Consulting Group Ltd, 2008; Parliamentary Commissioner for the Environment, 2009). The demand side of the market remains undeveloped, and the ripple-control peak-shaving technology that was universal in 1986 has been largely eliminated under commercial incentives, moving the market away from real-time pricing.
- Fourth, the balance of representation and policy influence in relation to electricity matters has been loaded against small consumers and in favor of large industry and the big electricity suppliers. Over time the official stance has hardened against any protection for small consumers exposed to the exercise of market power by the generator-retailers. Terms like "fair" that at one stage were included in the government's objectives for the industry have disappeared from the policy lexicon, and small-consumer representation on governance bodies has been token or nonexistent.

Overall, therefore, the most interesting lessons from the New Zealand experiment have to do not with the technical engineering detail of the pool, the wholesale market, the system operator, or the grid, but rather with the outcomes in terms of prices, profits, asset valuations, investment timing, industry coordination in meeting demand forecasts, and entry by independent entrepreneurs at any level of the supply chain or on the demand side. These areas are accordingly the focus of this chapter.

A brief overview of the industry provides the background in Section 2. Section 3 summarizes the history of structural and regulatory change since 1986. Section 4 traces industry performance in terms of pricing, profitability, and investment outcomes. Section 5 concludes.

2 BACKGROUND: SYSTEM DESCRIPTION AND HISTORY

New Zealand is an island nation of 4 million people with no interconnection to any other national electricity system. Matching generation, transmission, and distribution capacity to local requirements, both at the national aggregate level and locally, presents major engineering challenges in a long, thin country with rough topography, low population density, and population clustered in a few major cities the largest of which, Auckland, is at the opposite end of the country from the largest hydroelectric resources in the south. Those challenges were successfully overcome in the course of the twentieth century by construction of a publicly owned integrated system with large generating stations located to take advantage of the country's natural resources—hydro, geothermal, coal and natural gas—and with the main load centers supplied over a

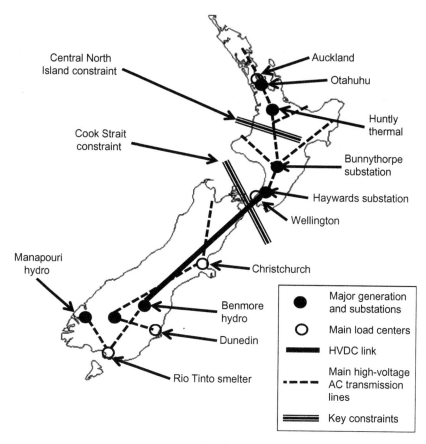

FIGURE 21.1   New Zealand electricity system.

nationwide transmission grid and local distribution networks.[2] Figure 21.1 shows the grid layout.[3]

Electricity demand is currently nearly 40,000 GWh per year, of which roughly two-thirds are used as an input by productive sectors (43% in industry, agriculture, forestry, and fishing; 24% in commerce and transport) and one-third is residential final consumption.[4] Installed generating capacity at December 2011 (Table 21.1) was 9751 MW, implying overall capacity

2. The history of the building of the national system is set out in Martin (1998).

3. Detailed maps of the transmission system are available online at: http://www.transpower.co. nz/maps-diagrams#cs-91581. An online map of distribution networks and their associated grid exit points is at: http://www.electricity.org.nz/Site/Map/default.aspx. Both sites accessed July 2012.

4. Ministry of Economic Development, *Energy Data File* June (2012, pp. 105–106) http://www. med.govt.nz/sectors-industries/energy/energy-modelling/publications/energy-data-file/new-zeal-and -energy-data-file-2012 (accessed 07.12).

TABLE 21.1 Trends in the New Zealand Electricity Sector 1980–2011

Total Installed Generating Capacity	Peak Load, MW	Generation, GWh	Consumption, GWh	Total Sales Revenue $m	Average Price, c/kWh	Real Average Price, c/kWh at March 2011 Prices
1975 4784	3391		17,306	215.2	1.2	11.5
1980 6576	3677	22,700	19,415	744.2	3.8	17.0
1985 8038	4642	27,689	24,205	1393.8	5.8	14.3
1990 8001	5122	31,459	27,745	2267.6	8.2	13.8
1995 8061	5240	35,250	30,370	2748.7	9.1	13.8
2000 8323	5766	38,069	34,011	3189.4	9.4	13.4
2005 8851	6084	41,514	37,626	4981.1	13.2	16.7
2011 9751	6654	43,138	39,005	6357.7	16.3	17.2

Source: Capacity, generation, consumption, and average prices from Ministry of Economic Development *Energy Data File* (2012, Tables G.3a, G.2a, and 1.1a)http://www.med.govt.nz/sectors-industries/energy/energy-modelling/publications/energy-data-file/new-zealand-energy-data-file-2012 (accessed 07.12), respectively. Total sales revenue calculated from sectoral data in the same sources; real average price derived using CPI for residential sales and PPI Inputs for commercial and industrial sales. Peak load 1975–1995 from Bertram (2006, Table 7.1) and 2000–2011 from http://www.ea.govt.nz/industry/monitoring/cds/centralised-dataset-web-interface/peak-electricity-demand-nationally/ (accessed 08.12).

utilization of less than 50%, which is to be expected in a system heavily reliant on renewable resources, such as wind and rain. Peak demand, also shown in Table 21.1, reached 6654 MW in 2011, equal to 68% of installed capacity (but far closer than this to actually available capacity at the time of the peak). Softer demand growth since the global financial crisis has cut previous projections of peak load by about 1000 MW for the next couple of decades (Transpower New Zealand Ltd, 2012, p. 33), implying that the system is currently carrying excess capacity relative to that needed to meet normal and projected demand.

Figure 21.2, charting the growth of generation capacity since 1945, shows that the pioneering construction and expansion phase of the industry's history had been completed when the Government embarked on increasingly radical restructuring experiments from 1986 onward. By then the system was engineered and operated to a very high standard, having overcome the recurrent supply shortfalls that had been the central challenge from the 1940s to the 1970s. By the late 1980s, there was ample generation capacity in hand, an expansion of the interisland HVDC link was underway, and sophisticated software programs had been developed for the merit-order scheduling of generation and frequency control. Thus when, during the 1990s, economic policy makers disrupted the established ownership, governance, and objectives of the industry, often displaying scant regard for engineering subtleties, they were fortunate to be operating on a patient in robust physical health, with a strong core of technically skilled employees who were able to maintain supply quality in the face of the often-perverse incentives thrown up by market institutions in a deregulated policy environment.

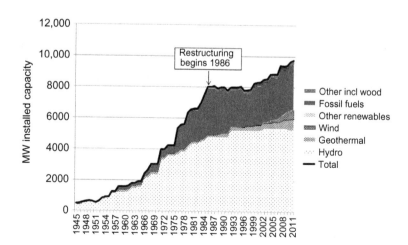

FIGURE 21.2 Installed generating capacity in the New Zealand electricity system, 1945–2011. Source: 1945–1975 data assembled from annual reports of the New Zealand Electricity Department (NZED); 1975–2011 from Ministry of Economic Development Energy Data File (2012, p. 112, Table G.3a).

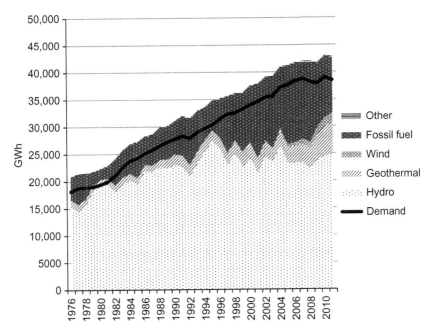

FIGURE 21.3 Generation by fuel type, 1976–2011. *Source: Ministry of Economic Development* Energy Data File *(2012, p. 110, Table G.2c).*

Figure 21.3 shows generation by fuel type, and total electricity demand, which is less than total generation due to line losses in transmission and distribution. The New Zealand electricity system has always been primarily renewables based, but a large tranche of fossil-fueled capacity was added in the 1970s.

The availability of low-priced gas from the Maui field raised nonrenewables to over a third of generation by the late 1990s, part of a 30-year trend seen in Figure 21.4. The trend then reversed, bringing the renewables share back up to 77% in 2011. Depletion of the Maui gasfield raised the operating cost of thermal plant after 2005 and this, combined with the beginning of carbon pricing policies and generators' preemptive occupation of the best windfarm sites, has contributed to a rapid decline in the share of nonrenewables in the past 5 years. The system is now on track back towards the 90% renewables share last seen in the 1970s (Bertram and Clover, 2009), even in the absence of Government policy support.[5]

Comparison of Figure 21.3 with Figure 21.2 shows a mismatch between trends in consumption and trends in capacity over the reform period.

---

5. The Labour administration which was defeated in 2008 had adopted an explicit 90% target (http://www.beehive.govt.nz/release/90-renewable-energy-target-achievable). The new National administration initially backed away from the target, but later included it as a general aspiration in its 2011 Energy Strategy (http://www.eeca.govt.nz/sites/all/files/nz-energy-strategy-2011.pdf, p. 6). There are no explicit policies directed to achieving the aspiration, however.

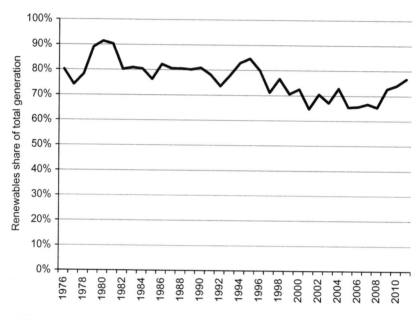

FIGURE 21.4 Share of renewables in total electricity generation, 1976–2011. *Source: Calculated from Ministry of Economic Development* Energy Data File *2012, p. 110, Table G.2c).*

Consumption rose steadily throughout the 1980s and 1990s while capacity stagnated; then capacity grew rapidly from 2000 to 2011 while consumption leveled off in response to rising electricity prices, the global financial crisis, and some gains in energy efficiency. The failure of capacity to track consumption is significant because one major argument advanced in the 1980s and 1990s in favor of corporatization, deregulation, and privatization was that commercial management guided by market signals would deliver investment in a more timely and orderly way than had been accomplished by the old government monopoly NZED. The reality has turned out to be, if anything, the reverse, an issue discussed in more detail later in this chapter.

Turning to the institutional setup prior to the onset of reform, at the time when New Zealand embarked on restructuring of its electricity sector in the mid-1980s the industry had a simple structure that had been built up pragmatically over the preceding 80 years to fit the geographical, physical, and social realities of the country.[6] Supply was entirely in public hands, and operated according to politically determined rules for pricing and investment. Large generating stations and the national high-voltage transmission grid were owned and operated by a government department, NZED. Local-area distribution networks, electricity retailing, and appliance sales and servicing were controlled by a special class of local Electricity Supply Authorities

6. The history of the industry is described in detail in Martin (1998).

(ESAs), called Electric Power Boards in rural and small town areas, and Municipal Electricity Authorities in the main centers.

At both wholesale and retail levels of the industry, prices were set by administrative procedures designed to recover costs on a nonprofit basis, with prices to residential consumers held below those paid by industrial and commercial users. Electricity was, in short, an essential service made available at cost. At all levels the industry's operations were transparently accounted for by the annual publication of detailed statistics relating to its operations, revenues, costs, and sales volumes.[7] Downward pressure on prices at wholesale level was provided by the accountability of the responsible Minister to Parliament, and at retail level by the requirement for Electric Power Board members and city councils to face the local electorate in local-body elections.

These arrangements displayed the familiar advantages and disadvantages of public ownership. Prices were set in advance for long periods by administrative decision—the Bulk Supply Tariff (BST) at wholesale level and retail prices at local level as a markup on the BST to cover distribution costs. Investment decisions were taken on the basis of engineering considerations subject to political constraints, and hence were somewhat isolated from commercial market disciplines. The planning agency responsible for forecasting demand and advising on the commissioning of new capacity investments— the Committee to Review Power Requirements—was subject to a degree of capture by engineers and the heavy construction sector, and was risk-averse in the sense of trying to establish and maintain a substantial margin of excess generating capacity so as to minimize the prospect of power blackouts. Management of lake levels by NZED was similarly risk-averse, aimed to keep the system secure against the risk of dry years when meeting winter demand would stretch the system to its limits. ESAs similarly gave high priority to supply security, investing in excess capacity and employing enough staff to ensure in-house capability to handle emergencies.

The industry in that old form was exposed to three lines of criticism, which in the 1980s developed into a political platform for change.

- First, administrative decision-making was only as good as the people holding responsible posts, and could potentially be subject both to political influence from above and to "capture." During the 1970s, the ambitions of engineers in NZED and the Ministry of Works to sustain a rapid hydroelectric construction program, combined with the wish of government energy planners to use the newly discovered Maui natural gas reserves for electricity generation, made the growth of future electricity demand a political issue. Increasingly strong criticisms of the escalating

---

7. Statistics appeared annually under the cumbersome title *Statistics in relation to the electric power industry in New Zealand*. The minister's annual reports to Parliament can be found in the *Appendices to the Journal of the House of Representatives*.

cost of new hydro plants, and of the Government's promotion of electricity-intensive large industries to ensure a market for expanded generation, meant that NZED became politically exposed going into the 1980s as the legitimacy of its demand forecasts was eroded.

- Second, the allocation of supply costs across different categories of consumer at retail level had arbitrary elements that were subject to political challenge. ESAs purchased power in bulk at a uniform wholesale price, the BST, and then distributed it at different retail prices to commercial, small-industrial, and residential consumers. Residential customers dominated the local-body electorate and probably had higher demand elasticity than commercial users; these two factors combined to produce a pricing structure under which residential pricing was significantly cheaper, leading commercial and small-industrial sector organizations to lobby for tariff rebalancing in their favor. Because such rebalancing was politically difficult so long as ESAs were elected bodies, a constituency developed in favor of abolishing electoral accountability in favor of a governance structure more susceptible to commercial pressures.

- Third, and crucial, was the rise of neoliberal ideology in New Zealand policy circles during the 1980s, inspired by overseas discussion of deregulation and the privatization policies of the Thatcher government in the United Kingdom. Led by the New Zealand Treasury, neoliberal thinking supported corporatization and if possible privatization of state-owned operations, minimal regulation, and heavy reliance upon market forces to guide investment, pricing, and industrial structure. The average cost-pricing model of the BST was rejected in favor of supposedly more "efficient" pricing procedures focusing on the margin of supply in both short and long runs, and competitive market forces were promoted as the alternative to central planning as the means to direct investment in new capacity. The promise was that a market-driven alternative to the NZED/ESA organizational structure would deliver electricity as reliably, but at lower cost to the economy (and to consumers).

The next section covers the institutional detail of the industry restructuring program.

3 INDUSTRY RESTRUCTURING SINCE 1986

3.1 Supply-Side Structure

The history of the New Zealand electricity reforms is extremely complex and only a quick overview of major trends will be given here.[8] The evolving

---

8. More detailed descriptions are provided in Bertram (2006), Evans and Meade (2005), Nillesen and Pollitt (2011), and on the Electricity Authority web site at: http://www.ea.govt.nz/about-us/structure/background-to-governance-and-regulation/. A detailed and accessible description of the generation system and the operation of the New Zealand pool and wholesale market is in Wolak (2009, Section 2 and Appendix 1).

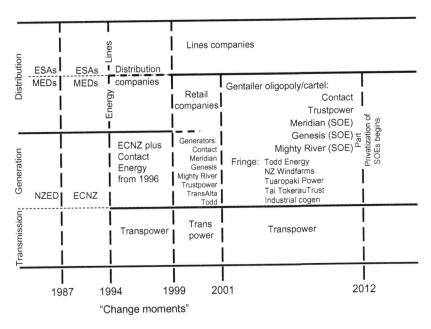

FIGURE 21.5 Evolving industry structure.

structure of the industry is set out in Figure 21.5. Initially generation and transmission were combined in the state monopoly, NZED, while distribution lines
and retail were combined in the hands of local electric supply authorities.
NZED was corporatized in 1987, becoming the Electricity Corporation of New
Zealand, ECNZ, then split into separate generation and grid companies in 1994
after a period of uncertainty during which privatization of the generation assets
and club ownership of the grid operator Transpower were unsuccessfully pursued. ECNZ was split into two supposedly competing generators in 1996 by the
creation of Contact Energy. Contact was privatized in 1999 at the same time as
the remaining ECNZ generation assets were broken up among three new SOEs
and a small group of independent private companies. In 2012, legislation was
passed allowing part-privatization of the three generation SOEs by selling up to
49% stakes to private investors; this sell down of the government stake is
intended to be complete by 2014.

Meanwhile, the ESAs and Municipal Electricity Departments that operated retail distribution were converted into commercial companies in
1993–1994, lost their local-monopoly franchises, and were required to ring-
fence their lines businesses which were made subject to "light-handed regulation" through information disclosure. The Electricity Industry Reform Act
1998 forced breakup of these companies, requiring ownership separation of
distribution networks from energy (generation and retailing). An intensive
period of takeovers and mergers followed as industry players jockeyed for

position. By 2000, there were six large vertically integrated players in generation and energy retailing—three private and three state-owned—plus a tiny fringe. The elimination of OnEnergy in 2001, a victim of imbalance between its large retail customer base and its tiny generation portfolio (Bertram, 2006, pp. 218–219), and a shuffling of retail portfolios among the remaining gentailers, produced a cartel of five large operators controlling around 95% of generation and a similar share of the retail market, with a very limited set of small independents occupying the fringe. The dominance of the five-gentailer group over the past decade has remained unchallenged.

3.2 Regulatory Arrangements: Generation and Retail

The successive restructurings summarized above have been motivated by the desire to bring market forces to play at all levels to the greatest extent possible, starting with abandonment at the outset of the previous social contract under which electricity was publicly supplied at cost and treated as an essential service. Once profit-maximization replaced social service as the industry objective, policy makers aimed to create competitive markets in generation and retail— the former by separating generation from involvement with lines networks and breaking-up ECNZ, the latter by abolishing the franchise areas which had given distributors a local monopoly. Since the commercialization of generation in 1987, of retail in 1993–1994, and ownership separation in 1999, these two "energy" levels of the industry have been left free of so-called "heavy-handed regulation" of their prices and profitability.

The only regulatory intervention the gentailers have faced over the past decade has related to industry governance. The engineering imperatives of coordinating dispatch and delivery of generated electricity have had to be matched by institutional arrangements for the operation of a pool which sets a spot price, the provision of ancillary services such as voltage support for frequency control, the maintenance of quality standards of supply, and some measures to reduce the uncertainty surrounding investment decision-making and to improve coordination among nominally competing parties in a market effectively foreclosed to large new independent entry.[9]

Until 1999 ECNZ's dominance, and state ownership of all major generation, enabled these issues to be handled within firm by management, but thereafter some more formal arrangements were required. After a ministerial inquiry in 2000 an Electricity Governance Establishment Committee was set

---

9. Foreclosure is a consequence of vertical integration of generation with retail, which forces any potential entrant to begin operations in two industries simultaneously if it is to escape the sort of fatal imbalance that destroyed TransAlta/OnEnergy in 2001. That company's failure to secure sufficient generation capacity out of the breakup of the old ECNZ portfolio left it overweight in retail and dependent on the spot market for wholesale supply during the 2001 dry winter; bankruptcy was the outcome.

up, within which the industry was invited to develop its own set of rules and institutions for self-governance, with Government standing aside but available to become involved if agreement could not be reached. After several years of fruitless negotiations, the Government was forced to establish, for the first time, a specialist electricity-sector regulator, the Electricity Commission, in 2003.[10]

The Electricity Commission was tasked with implementing and overseeing the rules, already largely developed by industry players themselves, covering the New Zealand Electricity Market (NZEM), the Metering and Reconciliation Information Agreement (MARIA), and the Multilateral Agreement on Common Quality Standards (MACQS). It was responsible for managing the supply into the market of reserve energy from a 155 MW generation plant at Whirinaki installed by the Government in the wake of the 2001 dry year crisis.[11] The Commission had no brief to oversee pricing[12]; this area was covered—if at all—by New Zealand's generic competition law, the Commerce Act 1986, under which the Commerce Commission (a general-purpose regulator under the Act) would be responsible for undertaking any pricing inquiries and possible direct regulation that the Minister of Commerce might consider desirable. No such political decision to consider or introduce price regulation of generation or retail has been forthcoming to date.[13]

Most importantly, the Commission was mandated to facilitate efficient new investment. Investment in new generation was to be coordinated by indicative planning, comprising economic modeling and publication of demand projections and of a "statement of opportunities" to indicate when and where new capacity would be required, leaving it up to the gentailers to decide where and when to actually undertake the investments. Investment by the monopoly grid operator was, however, to be directly regulated, with all significant Transpower investment projects requiring approval by the Commission, subject to a rigorously specified cost−benefit procedure, the "Grid Investment Test."

The Commission's first 3 years were dominated by an acrimonious dispute over Transpower's desire to build a new 400 MW transmission line

---

10. A comprehensive overview of the Commission's activities, including a concise summary of the industry's officially perceived failings as of 2009, is in Office the Auditor-General, *Electricity Commission: Review of the First Five Years*, June 2009, online at: http://www.oag. govt.nz/2009/electricity-commission/docs/electricity-commission.pdf.
11. The new plant was commissioned in 2004, on the same site as had previously been occupied by a similar dry-year reserve plant decommissioned by Contact Energy in 2001. After the Electricity Commission was abolished the plant was sold in 2011 to Contact Energy for $33 million, less than one quarter of its original cost of $150 million.
12. Apart from approving the grid pricing methodology for Transpower.
13. "Threshold" price regulation of lines networks by the Commerce Commission came into effect in 2003 and is discussed further below.

across the Waikato region from Whakamaru to Auckland. Transpower's plans were judged incompatible with the legally specified Grid Investment Test. After repeated unsuccessful attempts to persuade the Commission, Transpower called on political support from its owner, the Government, which dismissed the Commission Chair, Roy Hemmingway.[14] A chastened Commission, headed by a political appointee installed to replace Mr. Hemmingway, proceeded to approve the project.[15]

After this defeat the Commission survived only another 2 years, facing growing opposition within the industry. The price at which Whirinaki reserve power was to be dispatched was a major bone of contention with the large gentailers, and a 2009 ministerial inquiry concluded that the existence of a reserve generator was deterring new investment and thereby reducing rather than improving system security (Electricity Technical Advisory Group, 2009, vol. 1, pp. 16−22; Office of the Minister of Energy and Resources, 2009, paragraph 88). The Commission's analyses and modeling occasionally ruffled feathers as well, uncovering preliminary evidence of excessive retail margins being charged by the gentailers. Following an election win by the conservative National Party at the end of 2008, a new ministerial inquiry recommended abolition of the Commission and a return to more business-friendly approach (Office of the Minister of Energy and Resources, 2009). The Electricity Act 2010 provided for the termination of the Whirinaki contract, some shuffling of generation assets between two SOEs (South Island-based Meridian Energy and North-Island-based Genesis Energy), and the replacement of the Electricity Commission by a new Electricity Authority with the general objective of "promot[ing] competition in, reliable supply by, and the efficient operation of, the electricity industry for the long-term benefit of consumers."[16]

Included in the functions of the new Authority alongside monitoring the market, maintaining registers, and implementing an industry code of conduct, was "to promote to consumers the benefits of comparing and switching retailers."[17] Together with asset switching between two of the gentailers, this was intended to revive some appearance at least of competition in the retail market by raising the rate at which consumers were switching from one retailer to another. In association with a consumer watchdog organization, the

---

14. A detailed insider account of these events was provided by Mr. Hemmingway in a sworn affidavit dated December 2007 for a High Court case in which local interests sought to overturn the Commission's eventual decision in favor of the Waikato line. The judgment in the case, which refused judicial review, is *New Era Energy Inc. v Electricity Commission* (2010) NZRMA 63 [HC].

15. de Lacy, H., 2010. The Waikato Pylon War. *Energy New Zealand* 4(5) September−October, http://www.contrafedpublishing.co.nz/Energy + NZ/Vol.4+No.5+September-October+2010/ The+Waikato+pylon+war.html.

16. Electricity Industry Act 2010, Section 15.

17. Electricity Industry Act 2010, Section 16(1)(i).

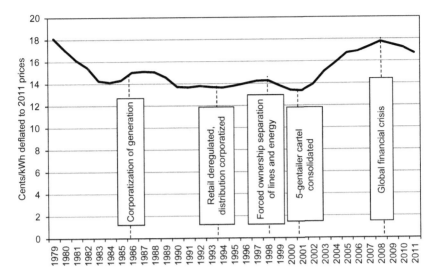

FIGURE 21.6   Average real electricity price 1979–2011, 2011 cents per kWh. *Source: As for Table 21.1; author's calculations.*

Consumer Institute, the Authority set up a "Powerswitch" web site[18] on which consumers could compare retail prices in their locality. The gentailers contributed to the impression of competitive activity by diversifying the brand names under which they sold electricity at retail. These activities, plus an intensive advertising campaign entitled "What's my Number?" (Electricity Authority, 2012), succeeded in raising the rate of customer churn amongst retailers, at the cost of a very large deadweight burden of information-gathering, calculation, and anxiety borne by individual consumers and voluntary budget advisory services; the Authority nevertheless judged its efforts a success. It is not yet clear whether increased customer switching has had any effect on the rate of increase of retail prices over the long run, but it may have had at least a temporary effect similar to the initial impact of the Electricity Industry Reform Act in 1998–2000 (Figure 21.6).[19]

3.3 Lines Regulation

The ongoing nonregulation of generation and retail pricing and profitability (in the name of promoting, and relying on, market competition and efficiency) was from the outset separate from the regulatory treatment of

---

18. http://www.powerswitch.org.nz/powerswitch.
19. In both 1998–2000 and 2009–2012, there were downward pressures on price from surplus generation capacity in periods of economic recession, which makes it difficult to identify any separate effect from customer churn.

transmission and distribution lines, whose natural-monopoly character made some sort of regulation inescapable, but for the first decade of corporatization, 1994–2003, policy makers relied on information disclosure alone as a means of disciplining market behavior.

Lines businesses were required to disclose financial, and some physical, information for publication in the official *New Zealand Gazette*; but the government department responsible for overseeing the information disclosure process (the Ministry of Commerce) did not maintain a public registry of disclosed information, nor undertake analysis of financial disclosures. As lines companies became increasingly bold in their asset revaluations, price-cost markups, and creative accounting, government's ability and will to regulate declined steadily through the 1990s. The official electricity sector statistics shrank rapidly in coverage and accuracy after 1989, and were phased out entirely in 1994.

By 2000, the lines companies were carrying on their books some $2 billion of asset revaluations that their auditors had signed off without challenge from government—effectively a wealth transfer from electricity consumers, underpinned by increased prices and margins—and had become the object of widespread public disquiet about their apparent profiteering at consumers' expense. A ministerial inquiry largely exonerated the companies, even going so far as to deny that capital gains captured by natural monopolists should be treated as income for regulatory purposes (Caygill, 2000). Nevertheless, some regulatory intervention was judged politically unavoidable, and amendments to the Commerce Act were passed requiring the Commerce Commission to inquire into the need for regulation. In due course, after a 3-year process of deliberation and ministerial approval, the Commission introduced a version of CPI-X price regulation (prices allowed to rise by X% less than the Consumer Price Index), based on thresholds that accepted as a fait accompli the asset revaluations and high-priced takeovers of the preceding decade, with the accumulated revaluations reclassified as historic cost (Bertram and Twaddle, 2005, p. 298).

Subsequently, two detailed empirical studies utilizing disclosed information (Bertram and Twaddle, 2005; Nillesen and Pollitt, 2011) have traced large increases in price-cost margins, and in asset values, in the late 1990s, as the new company managements drove costs down and (to a lesser extent) prices up. Figure 21.7 charts the revaluation boom which doubled the book value of distribution lines assets 1994–1999 before the Government referred lines pricing to the Commerce Commission. After regulation began in 2003, steady annual revaluations were still allowed to reflect the increasing replacement cost of assets. Of the 2011 regulatory asset base of around $8 billion, nearly half consisted of revaluations (basically bare wealth transfers) as distinct from net actual investment. Around $2 billion of those revaluations represent the initial wealth transfer from consumers—the rate shock attributable to the reform process—as valuations rose from historic cost to

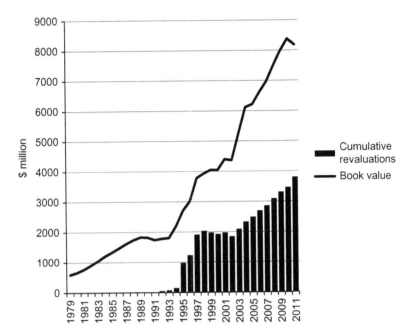

FIGURE 21.7   Distribution networks book value of fixed assets. *Source: 1979–1993 from annual electricity statistics; 1994–2002 assembled from individual company disclosure statements; 2003–2011 from Commerce Commission disclosure data at http://www.comcom.govt.nz/information-disclosure/.*

Optimized Deprival Value (ODV) or more during the unregulated 1994–2003 period.

The next section turns to the outcomes of industry restructuring and the evolving regulatory regime since 1986, asking whether there is clear evidence of improved performance emerging from two decades of neoliberal "reform."

4 OUTCOMES OVER THE REFORM PERIOD

This section summarizes the trends in prices, costs, margins, profits, asset values, and investment performance over the 25-year period 1987–2012. At aggregate level, Figure 21.6 showed that the real price of delivered electricity was falling prior to reform, stabilized over the first decade of restructuring, fell briefly during a competitive interlude 1998–2001, and then climbed steadily from 2001 to 2009. In the last 2 years of the chart, it is evident that the global financial crisis, emerging excess generation capacity, and some pro-competitive regulatory activism have curbed the industry's rate of price inflation to below the economy-wide inflation rate.

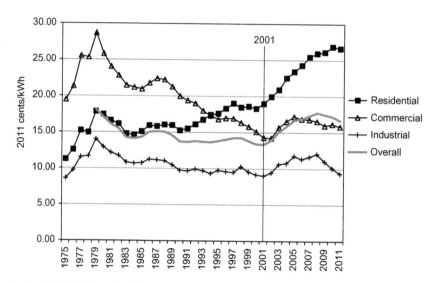

FIGURE 21.8 Real electricity price by end-use sector, 1979–2011. *Source: Prices and volumes from* Energy Data File *(2012, Tables I.1a and G.5a). Deflated by the author using CPI for residential and Producer Price Index (PPI) (Inputs) for commercial and industrial, but using CPI for years before the PPI series begins.*

Figure 21.8 shows real price trends by end-use sector, with the overall average from Figure 21.6 overlaid. It is apparent that the flat trend of prices overall through the 1990s (Figure 21.6) did not hold for all end users. Residential prices began to rise in 1990 and over the following two decades they increased 3% per year ahead of the inflation rate. The only time when competitive market pressures seriously checked the trend was 1998–2001, when a scramble for retail-market shares was underway following the owner-ship separation of lines and energy businesses. Once the *de facto* cartel of five vertically integrated gentailers consolidated in 2001, price rises for resi-dential consumers accelerated to a steady 5% per year above the general inflation rate until the global financial crisis struck.

Commercial prices fell steeply in real terms through the 1990s, offsetting the residential price hikes. However, after the gentailer cartel consolidated in 2001, they too turned up sharply until 2008. Industrial prices stayed much more constant in real terms throughout, but with a period of real price hikes 2001–2008. The average industrial price is held down by the heavy weight of the Rio Tinto aluminum smelter at Bluff (Figure 21.1) which uses around 17% of New Zealand's total electricity at a very low contract price.

The key to the divergent trends pitting households against business has been the lobbying and bargaining power exercised by the industrial and com-mercial sectors to keep their costs down, and to the almost complete loss by

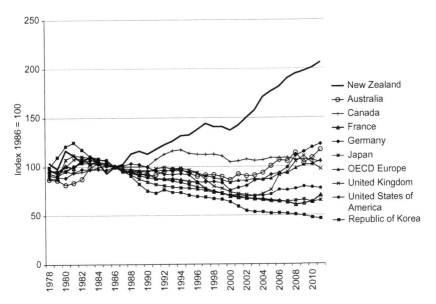

FIGURE 21.9   Real electricity price to residential consumers: New Zealand compared with some other countries covered in this book. *Source: International Energy Agency,* Energy Prices and Taxes *online database (accessed July 2012).*

residential consumers of any effective voice in policy or regulatory arrangements. In government, both main political parties have allowed the large electricity suppliers to set retail prices without regulatory intervention.

How did the New Zealand price trends in Figure 21.8 compare with those elsewhere in the world? Using International Energy Agency data, Figure 21.9 shows the path of residential prices in New Zealand compared with a number of the other countries covered in this book. Figure 21.10 carries out the same exercise for industrial electricity. Both charts are constructed to show trends before and since the restructuring of the New Zealand industry began in 1986. New Zealand is clearly an extreme outlier in terms of the extent to which residential prices have been driven up ahead of the inflation rate in the reform era, whereas other countries in this book have held residential prices stable or reduced them over the same period. In contrast, industrial electricity prices in New Zealand have roughly tracked those in neighboring Australia.

A profitability analysis of the three SOE gentailers (Meridian, Genesis, and Mighty River) conducted for the Treasury by Ernst and Young in 2011 estimated that "economic profit" (the companies' return over and above their cost of capital) had totaled some $3.8 billion over the 10 years (Ernst and Young, 2011; Treasury, 2011, p. 39, Table 8) on total revenues of $42 billion. Their "invested capital" was stated to have risen from $4 billion in

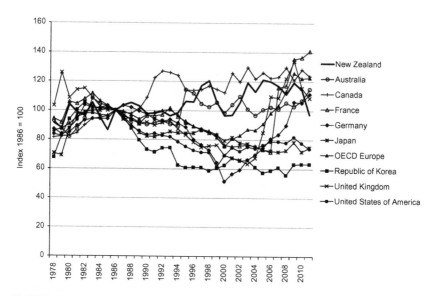

FIGURE 21.10 Real electricity price to industrial consumers: New Zealand compared with some other countries covered in this book. *Source: International Energy Agency,* Energy Prices and Taxes *online database (accessed July 2012).*

2002 to nearly $12 billion in 2011 (Treasury, 2011, p. 38, Table 8); but the great bulk of this increase—$6.2 billion, according to the companies' annual reports—was asset revaluations, with less than $2 billion representing the historic cost of net actual investment. The revaluations were so-called "fair value" exercises which wrote up the book value of assets to reflect the present value of expected cashflows, and in this sense they provide a gauge of the actual and anticipated profitability of the vertically integrated state-owned businesses.

Analysis of the annual financial reports of all five large gentailers, and of their predecessor ECNZ, enables construction of Figure 21.11, which compares the recorded book value of fixed assets with the cumulative revaluations taken to book over the years. It can be seen that by 2011 roughly half of the total book value was revaluations. The revaluation process got under way in 1999, when generation assets transferred from ECNZ to their new owners were marked up by $1.5 billion; but the really big surge came after 2001 as the cartel consolidated and was able to exercise its market power to generate increasing cashflows that could be capitalized into "fair value" entries on their balance sheets.

In interpreting Figure 21.11, two peculiar features of the New Zealand regulatory environment need to be borne in mind. First, there is no capital gains tax in New Zealand, so that revaluations are in effect untaxed income,

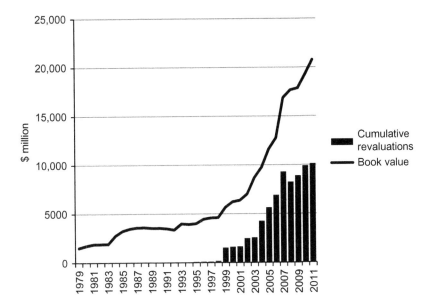

FIGURE 21.11   Gentailers' book value of fixed assets decomposed between net capital expenditure and revaluations. *Source: 1979–1993 from annual electricity statistics; 1994–2011 collated by the author from company annual reports.*

which under New Zealand's generally accepted accounting practice (GAAP) does not have to be entered in the profit and loss account. Second, as noted earlier, there is no law prohibiting the taking of excess profits. As a Commerce Commission report noted in 2009:

> *The exercise of market power to earn market power rents is not by itself a contravention of the Commerce Act, but is a lawful, rational exploitation of the ability and incentives available to the generators*
>
> (Commerce Commission, 2009, p. 6, paragraph v.)

Under new financial reporting standards introduced about 2007, and used by all the gentailers from 2008 on, company accounts now show, in addition to the book value of fixed assets including revaluations, a figure for the value that would have appeared under historic-cost accounting conventions of the sort that apply in, for example, the United States of America. Table 21.2 assembles, from the five large gentailers' accounts, comparative data on the two measures for the years 2008–2012. If one imagines a hypothetical regulator controlling retail prices on the basis of a historic-cost regulatory asset base, it is immediately apparent that such a regulator would have disallowed a substantial proportion of the price increases charged by these companies over the past decade.

TABLE 21.2 "Fair Value" Book Value of Generation Fixed Assets Compared with Historic Cost

	2008	2009	2010	2011	2012
"Fair value" at which generation fixed assets are carried					
Contact	4.1	4.1	3.7*	4.1*	4.2
Genesis	1.5	1.5	1.4	2.5	2.5
Meridian	6.0	5.9	7.7	7.3	7.3
Mighty River	3.0	3.5	4.1	4.4	4.5
Trustpower	2.0	2.3	2.3	2.4	2.5
Total	16.5	17.2	19.2	20.8	21.0
Book value if a historic cost basis were used					
Contact	1.6	1.6	1.6*	1.6*	1.6*
Genesis	1.2	1.2	0.9	1.7	1.6
Meridian	2.3	2.2	2.9	2.9	2.7
Mighty River	1.2	1.5	1.7	1.7	1.7
Trustpower	1.1	1.2	1.2	1.2	1.2
Total	7.4	7.7	8.3	9.1	8.9

Contact Energy in its 2010 *Annual Report*, p. 60 announced a voluntary switch to historic cost and an end to "fair value" adjustments. In constructing this table, the 2009 revaluation reserve has been carried forward to make up for the removal of any corresponding item in Contact's financial statements.
Source: Company Annual Reports, collated by the author.

To capture as cash the capital gains embodied in "fair value" revaluations, either the companies must make return-of-capital distributions to their owners or the assets must be on-sold to a third party at a price that includes the revaluations. The New Zealand Treasury, hungry for revenue from SOEs, has encouraged and benefited from the sequence of ownership changes forced on the industry by restructuring policies. When in 1996 the first split of ECNZ took place with the creation of Contact Energy, the new company raised debt finance to pay out ECNZ in cash, enabling ECNZ to pay its owners cash dividends totaling $3 billion over 2 years (Figure 21.12). The divestment of the remaining ECNZ assets to new state-owned companies in 1999 produced another $2.5 billion. The Treasury in 2012 hoped that the pending sale of 49% of the three generator SOEs could produce another $5−6 billion in cash.

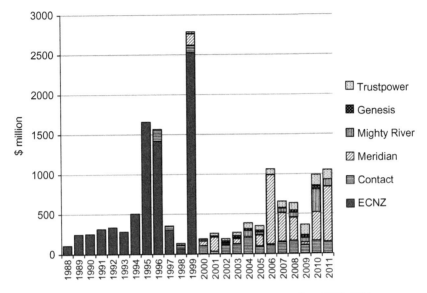

FIGURE 21.12 Distributions to owners by the large generator-retailers, 1988–2011.
Source: Collated by the author from annual reports.

The circular process of revaluing assets, declaring higher capital charges to service the resulting higher book values, then raising prices to recover those charges before proceeding to the next "fair value" exercise, slowed down after 2008—more because of macroeconomic conditions affecting the market than due to any regulatory restraint.

To look more closely at cost/price trends in the four levels of the industry (generation, transmission, distribution lines, and retail), a first step is to decompose total electricity sales revenue between lines charges and energy charges. Figure 21.13 carries out this exercise by assembling scattered data published in official statistics, company annual financial statements, and information disclosure documents. The solid bars at the bottom of this chart represent lines charges for transmission (black) and distribution lines (gray), while the hatched bars at the top represent energy charges. For the first half of the 1990s, official statistics permit a further breakdown between whole-sale and retail charges for both lines and energy, with the retail margin derived as a residual from total sales after subtracting transmission, distribution lines charges, and ECNZ's wholesale energy sales revenue.

It is not possible to disaggregate the "energy" charges in Figure 21.13 for years since 1997 because the vertically integrated gentailers combine their generation and retail businesses into a single set of annual financial statements, with only limited segmental reporting of revenues and costs.

It is obvious, however, from inspection of Figure 21.13 that trends in total sales revenue have been driven more by energy charges than by lines charges.

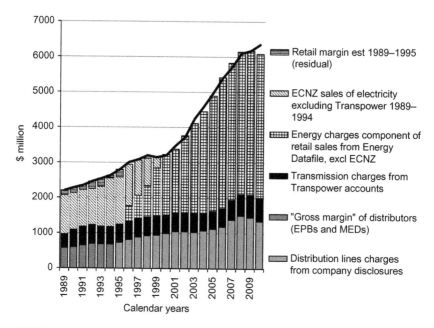

FIGURE 21.13 Decomposition of total sales revenue, 1989–2010. *Source: Compiled by the author from electricity statistics, company annual reports, information disclosures, and* Energy Data File *various issues*

The 1998–2000 dip in total revenue and its very rapid rise 2001–2009 were due primarily to developments in the pricing of energy. In 2009–2010, when average overall prices softened (Figure 21.7 above) it was distribution lines, rather than energy revenues, that took the hit. Transmission charges, apart from their sharp increase 1989–1991 (when ECNZ restructured to load costs onto its transmission subsidiary and thereby improve the privatization prospects of the generation operation) have remained constant in nominal terms, and declined in real terms, over the two decades.

Using the Producer Price Index (Inputs) to deflate the figures, and dividing by annual final sales volume, Figure 21.14 shows the breakdown of the average real price using the same data as Figure 21.13. The relative constancy of distribution lines charges and the decline in transmission charges are clear, as is the sharp increase in energy charges after 2000.

Recently, data have become available for the segmental breakdown of the gentailers' aggregated sales revenues between the cost of wholesale electricity generated and purchased, and the retail margin (covering charges for various market services, metering, consultants, industry governance levies, advertising, plus the gross profit margin secured). This new data makes it possible to compare, in Figure 21.15, the breakdown of final price paid by all consumers on average in 1990 and 2010.

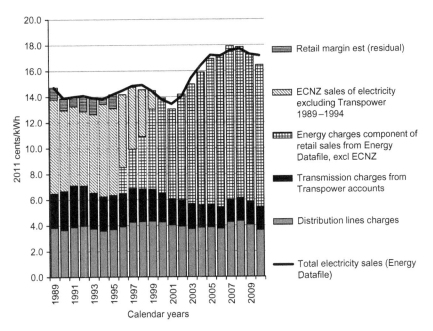

FIGURE 21.14 Decomposition of the real average price of electricity, 2011 cents/kWh. *Source: As for Figure 21.13, deflated using the PPI (Inputs) published by Statistics New Zealand.*

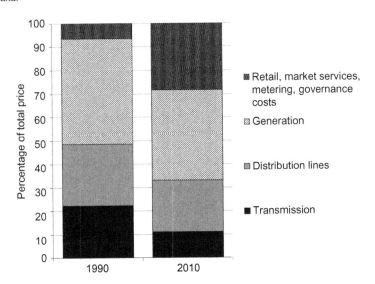

FIGURE 21.15 Proportional breakdown of final electricity price amongst the four components, 1990 and 2010. *Source: 1990 from the data used to construct Figure 21.12. 2010 based on a chart in Electricity Authority (n.d, p. 1), with the four sectors aggregated using sectoral consumption weights from* Ministry of Economic Development *(2011, p. 118, Table G.6a).*

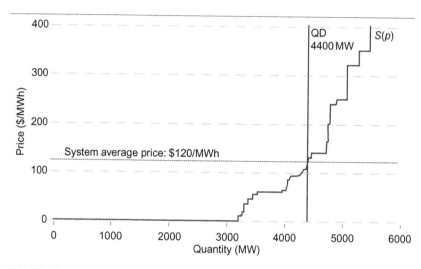

FIGURE 21.16 Aggregate offer curve for all generators in a half-hour period, with total market demand at 4400 MW. *Source: McRae and Wolak (2009, p. 36, Figure 3.4).*

The striking feature in Figure 21.15 is the enormous increase in the proportion of the final price that is absorbed by the retail segment of the industry. Under the old industry structure, with retailing undertaken by ESA distributors on a nonprofit basis, retailing absorbed 7% or less of the consumer dollar. By 2010, after two decades of restructuring supposedly for the benefit of consumers, these charges had risen to 28% of the bill. In real 2011 dollar terms, retail and other minor charges cost the consumer 0.9 cents/kWh in 1990; by 2010 this had risen to 4.2 cents. For residential consumers, the increase was far greater than this, taking retail charges from under a cent to around 8 cents/kWh.

An important component of the price of electrical energy is the wholesale price emerging from the operation of New Zealand's pool.[20] The wholesale market is an energy-only market which clears every half hour at the price offered by the marginal tranche of generation. A typical diagrammatic cross section of the cost and demand structure is Figure 21.16,[21] which shows clearly the large component of generation capacity available at very low short-run cost, and the steeply rising segment at the right-hand end of the supply curve as successive tranches of

20. It is unclear how the wholesale price enters into the operating costs reported by gentailers in their annual financial statements and shown in Figure 21.15. The transfer of electricity from wholesale to retail divisions of each company is an internal transaction, while electricity acquired from other generators is subject to complex contracting and hedging arrangements.
21. The basic shape of the short-run marginal supply curve has remained unchanged since the 1980s; see Bertram (1988) for a very similar diagram.

thermal plant (and of hydro generation with high opportunity cost of water) are called upon.

It is obvious from the diagram that a large part of the total revenue secured in any period is rent on low-operating-cost fixed assets. Since the market is energy-only, incumbent generators are expected to utilize this large producer surplus to fund new capacity, and the availability of such large rents has allowed the system to function without separate arrangements to incentivize investment like capacity rights. There is no reason, however, to believe that the total rents bear anything other than an arbitrary relation to the actual financing needs of generation. There is, on the contrary, much reason to see them as pure wealth transfers from consumers, positively related to the spot price.

Acknowledgment that the spot price might be subject to the exercise of market power led the Commerce Commission, in 2005, to initiate an inquiry into the issue of whether market power was being exercised in the wholesale market and, if so, whether this involved anticompetitive behavior as defined under New Zealand competition law. The resulting quantitative study of the wholesale market (McRae and Wolak, 2009; Wolak, 2009) found strong evidence that at times of relative shortage of supply, the spot price had been driven well above a hypothetical competitive level (the "counterfactual price" in Figure 21.17).

Based on this modeling work, Wolak estimated market power rents, shown in dark shading in Figure 21.18, at $4.3 billion over the period—18%

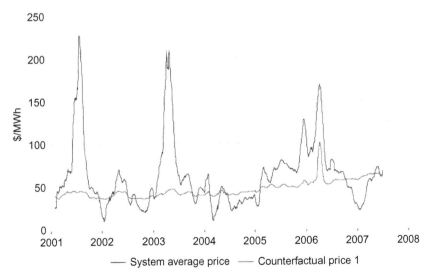

FIGURE 21.17 Deviation of actual wholesale spot price from hypothetical competitive price, 2001–2007. *Source: Wolak (2009, p. 201, Figure 5.13).*

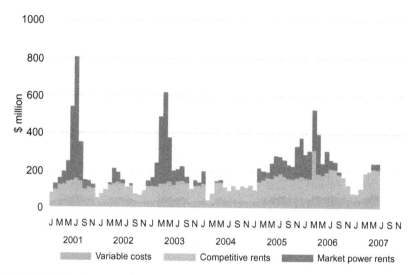

FIGURE 21.18 Wolak estimates of "market-power rents". *Source: Wolak (2009, p. 200, Figure 5.11).*

of the total wholesale market revenues received by all generators over the entire period (Commerce Commission, 2009, p. 6, paragraph ii). A later, independent, study using a different methodology (agent-based modeling) found monopolistic rents on the same scale as Wolak, but with a different distribution of rents across seasons and years (Browne et al., 2012).

Because it found no evidence that the market power identified by Wolak had been "exercised for any anticompetitive purpose," the Commission ended its investigations with no further action being taken, but it did comment that

there are serious systemic issues arising out of the current market structure, market design, and market rules that provide the generators with the ability and incentive to exercise market power under certain periodic and recurring conditions

(Commerce Commission, 2009, p. 6, paragraph v.)

The Commission pointed also to the inadequacy of information available, even a decade and a half after the New Zealand Government had adopted a light-handed regulatory stance built around the idea of information disclosure as the key restraint on monopolistic conduct:

[U]nlike the situation in many other jurisdictions, the regulatory bodies monitoring the electricity industry in New Zealand have not historically collected, and still do not collect, all of the information typically required by a competition authority to fully assess competition in the wholesale and retail markets

(Commerce Commission, 2009, p. 8, paragraph xvii).

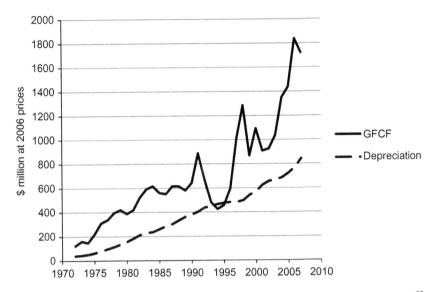

FIGURE 21.19 Investment performance of "electricity, gas, and water" sector, 1972–2007.[22] *Source: National accounts published by Statistics New Zealand. The published sectoral data ends at 2007.*

Turning to investment, Figure 21.19 shows the trends in depreciation and gross fixed capital formation for the sector "electricity, gas, and water" in the New Zealand national accounts. Within that sector, electricity accounts for 85% or more of total sales and value added,[23] which means it is the dominant influence on the aggregate figures.

The uncertainty and disruption caused by restructuring of the electricity sector had a major impact on maintenance spending and investment in new capacity, both in the generation fleet and in the transmission grid. As was seen in Figure 21.2, over the first 15 years of "reform" from 1986 to 2000 generation capacity remained flat with no net gain from investment, as commercial decisions eliminated over 600 MW of reserve high-operating-cost thermal plant whose disappearance fully offset what new investment was undertaken in lower cost thermal baseload units (Bertram, 2006, p. 225, Table 7.5). This profit-motivated scrapping of inherited reserve plant had the effect of narrowing the safety margins in a system that was heavily dependent upon hydroflows. In due course a dry winter in 2001 brought a supply crisis, high prices, and blackouts, spurring a resumption of new construction.

---

22. Official national accounts data at sector level are currently not available for years after 2007.
23. This figure comes from the 1996 input–output matrices published by Statistics New Zealand.

The 1990s swing to negative net investment in Figure 21.19 amounted to an "investment strike" not only by the newly corporatized generation sector—resulting in a 15%-point increase in capacity utilization as the margin of excess generation capacity was stripped back—but also by the state-owned grid operator Transpower, as it struggled to meet stringent financial performance requirements—high capital charges on revalued grid assets—imposed by Government as its shareholder. The grid was allowed to run down for a decade and a half while large cash dividends were extracted by the New Zealand Treasury. The grid assets inevitably deteriorated and eventually key components began to fail, a process dramatically illustrated by a blackout of much of Auckland in June 2006, and by the 2007 reduction in the capacity of the interisland high-voltage DC link which is a key part of the grid backbone.[24]

The HVDC link had been installed in the 1960s, and upgraded in 1987–1992 by addition of a new thyristor converter alongside the original mercury valves, plus the laying of three new undersea cables. Pole 1 of the system continued to operate with the original 1960-vintage equipment until 2007, leaving the link increasingly vulnerable to failure as Pole 2 also began to age. The coldest day of the 2006 winter, June 19, brought a grid emergency and blackouts in the North Island due to a sudden outage on the HVDC; and in September 2007, Pole 1 finally had to be taken out of service. The next year it was reactivated for a while using old equipment salvaged from Denmark, but the link effectively was reduced to a single pole, meaning that normal industry standards of grid security were not met, and wear and tear on the operational pole increased sharply. In 2008, planning finally commenced on a new pole for the HVDC; work began in 2010, and the new pole is finally due for commissioning in 2013, restoring an $(n-1)$ level of security.

Figure 21.20 shows the grid's book value since Transpower was established in 1987. The steep increase from 1990 to 1993 reflects a combination of asset revaluation (to ODV) and heavy expenditure on the interisland link. Thereafter until 2006, the book value drifted steadily down as depreciation outran new investment spending. A major downward revaluation was taken to book in 1998. Only after 2006 did the value of assets begin to increase as investment in replacement and extension of grid assets was resumed. Short-term thinking and Treasury's hunger for revenue eventually had their consequences in the 2006 Auckland blackout and 2007 HVDC downgrade.

The proposition that market mechanisms would coordinate investment more efficiently than central planning had done prior to 1987 receives no support from Figures 21.19 and 21.20.

---

24. A useful overview of the HVDC is at http://en.wikipedia.org/wiki/HVDC_Inter-Island (accessed 07.12).

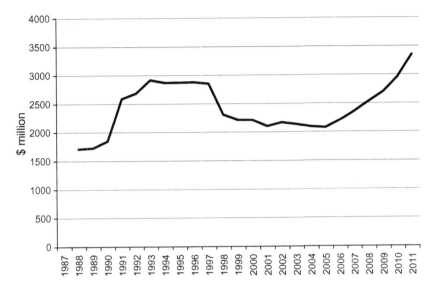

FIGURE 21.20   Transpower fixed assets book value 1987–2010. *Source: Transpower NZ Ltd Annual Reports.*

5 CONCLUSION

At the beginning of the restructuring process, the 1989 Electricity Industry Task Force report recommended privatization of ECNZ with some limited spin-off of its assets, removal of retail franchises, separation of lines and energy, corporatization and privatization of distribution, light-handed regulation of lines, and no regulation of retail prices (Bertram, 2006, p. 209, Box 7.1). While not all details have matched the Taskforce blueprint, the general outline of the New Zealand restructuring program has remained generally consistent with its general strategic thrust.

The absence of a fully resourced specialist regulator has proved a significant drawback. The regulatory arrangements covering prices, costs, and profitability in the electricity sector have undergone repeated adjustments without as yet settling into a sustainable long-term shape.

Unresolved issues lie in wait as the state-owned generators are put through part-privatization over the coming years. The tax treatment of windfall profits on sunk-cost renewables as the carbon price rises is one of these. Another is the long-term exemption of hydrogenerators from any charge on the water they use. A third is the rise in energy poverty among low-income households faced with rising retail prices. A fourth is the reemergence of surplus generation capacity, on a scale potentially greater than that of the 1980s which was used to justify industry restructuring in the first place.

In terms of the international reform agenda reviewed in this book, New Zealand can point to success in maintaining supply through a period of drastic upheaval in the industry structure; a mixed record on operating costs (which were reduced in the lines sector but have risen sharply in vertically integrated generation and retail); progress towards a 90% renewables share (based on resource endowments rather than policy); but little progress to date on smart metering, demand-side participation in the market, freedom of entry for independent generation, or planning for major future market development like the pending arrival of electric vehicles. Much remains to be done.

REFERENCES

Bertram, G., 1988. Rents in the New Zealand energy sector, in Royal Commission on Social Policy. The April Report, vol. IV. Government Printer, Wellington, pp. 293–325.

Bertram, G., 2006. Restructuring of the New Zealand electricity sector, 1984–2005. In: Sioshansi, F.P., Pfaffenberger, W. (Eds.), International Experience in Restructured Electricity Markets: What Works, What Does Not, and Why? Elsevier, Amsterdam.

Bertram, G., Clover, D., 2009. Kicking the fossil fuel habit: New Zealand's case. In: Sioshansi, F.P. (Ed.), Electricity Generation in a Carbon Constrained World. Elsevier, Amsterdam (Chapter 14).

Bertram, G., Twaddle, D., 2005. Price-cost margins and profit rates in New Zealand electricity distribution networks since 1994: the cost of light handed regulation. J. Regul. Econ. 27 (3), 281–307.

Browne, O., Poletti, S., Young, D., 2012. Simulating market power in the New Zealand electricity market. New Zealand Econ. Pap. 46 (1), 35–50 (53–55, April 2012).

Caygill, D., 2000. Report of the Ministerial Inquiry into the Electricity Industry.

Commerce Commission, 2009. Investigation Report: Commerce Act 1986 s.27, s.30 and s.36 Electricity Investigation. <http://www.comcom.govt.nz/investigation-reports/> (accessed 03.13).

Concept Consulting Group Ltd, 2008. Smart Metering in New Zealand: a Report Prepared for the Parliamentary Commissioner for the Environment. <http://www.pce.parliament.nz/assets/Uploads/Reports/pdf/Concept_Smart_Metering.pdf/> (accessed 03.13)

Contact Energy Ltd, 2010. Annual Report 2010. <http://www.contactenergy.co.nz/web/shared/annualreports?vert = in#a2010/> (accessed 03.13).

Crown Ownership Monitoring Unit, 2011. 2011 Annual Portfolio Report. New Zealand Treasury, Wellington. <http://www.comu.govt.nz/resources/pdfs/apr-11.pdf/> (accessed 08.11.).

Electricity Authority, n.d. Comparing Prices Paid by Other Consumers: Fact Sheet 3. <http://www.ea.govt.nz/document/16851/download/consumer/factsheets/ (accessed 08.12.).

Electricity Authority, 2012. What's my Number? A Changing Landscape for New Zealand Electricity Consumers. <http://www.ea.govt.nz/document/16344/download/about-us/documents-publications/> (both accessed 03.13).

Electricity Technical Advisory Group, 2009. Improving Electricity Market Performance: A Preliminary Report to the Ministerial Review of Electricity Market Performance, 2 vols. <http://www.med.govt.nz/sectors-industries/energy/pdf-docs-library/electricity-market/implementing-the-electricity-market-review-recommendations/background-papers-on-2009-review/improving-electricity-market-performance-volume-one-discussion-document-1152-

kb-pdf.pdf/> and <http://www.med.govt.nz/sectors-industries/energy/pdf-docs-library/
electricity-market/implementing-the-electricity-market-review-recommendations/background-
papers-on-2009-review/improving-electricity-market-performance-volume-two-appendices-
2252-kb-pdf.pdf/> (both accessed 08.12.).

Ernst, Young, 2011. SOE Profit Analysis. Treasury, Wellington. Available from: <http://www.
comu.govt.nz/resources/pdfs/ey-soe-epa-nov11.pdf/> (accessed 03.13).

Evans, L.T., Meade, R.B., 2005. Alternating Currents or Counter Revolution? Contemporary
Electricity Reform in New Zealand. Victoria University Press, Wellington.

Martin, J., 1998. People, Politics and Power Stations: Electric Power Generation in New Zealand
1880–1998. Electricity Corporation and Bridget Williams Books, Wellington.

McRae, S.D., Wolak, F.A., 2009. How Do Firms Exercise Unilateral Market Power? Evidence
From a Grid-based Wholesale Electricity Market, EUI Working Paper RSCAS 2009/36.
<http://cadmus.eui.eu/bitstream/handle/1814/12098/RSCAS_2009_36.pdf?sequence=1/>
(accessed 03.13).

Ministry of Economic Development, 2011. Energy Data File 2011. Wellington.

Ministry of Economic Development, 2012. Energy Data File 2012. Wellington.

Murray, K., Scott, G., Stevenson, T., 2009. Determining Outcomes or Facilitating Effective
Market Processes: A Review of Regulation and Governance of the Electricity Sector.
LECG, Wellington.

Nillesen, P.H.L., Pollitt, M.G., 2011. Ownership unbundling in electricity distribution: empirical
evidence from New Zealand. Rev. Ind. Org. 38 (1), 64–66, January 2011.

Office of the Minister of Energy and Resources, 2009. Ministerial Review of the Electricity
Market, Cabinet Paper. <http://www.med.govt.nz/sectors-industries/energy/pdf-docs-library/
electricity-market/implementing-the-electricity-market-review-recommendations/back-
ground-papers-on-2009-review/Elec.0025%20-%20Electricity%20Market%20Review%20-%
20Cabinet%20Paper.pdf/> (accessed 08.12.).

Parliamentary Commissioner for the Environment, 2009. Smart electricity meters: how households and the environment can benefit. <http://www.pce.parliament.nz/publications/all-pub-
lications/smart-electricity-meters-how-households-and-the-environment-can-benefit/>.

Stevens, L., 2003. The goals of the commerce act. In: Berry, M.N., Evans, L.T. (Eds.),
Competition Law at the Turn of the Century: A New Zealand Perspective. Victoria
University Press, Wellington.

Taggart, M., 2008. Common law price control, state-owned enterprises and the level playing
field. In: Pearson, L., Harlow, C., Taggart, M. (Eds.), Administrative Law in a Changing
State: Essays in Honour of Mark Aronson. Hart Publishing, Portland OR.

Transpower New Zealand Ltd, 2012. Annual Planning Report, Incorporating the Grid Reliability
Report and the Grid Economic Investment Report. <http://www.transpower.co.nz/
f5620,69352750/APR2012CompleteFINAL.pdf/> (accessed 07.12.).

Wolak, F., 2009. An assessment of the performance of the New Zealand wholesale electricity
market. Report for Commerce Commission. Wellington. <http://www.comcom.govt.nz/
assets/Imported-from-old-site/BusinessCompetition/Publications/Electricityreport/ContentFiles/
Documents/comcom-wolakreportassessmentofthenzelectricitymarketperformancept1.pdf/> and
<http://www.comcom.govt.nz/assets/Imported-from-old-site/BusinessCompetition/Publications/
Electricityreport/ContentFiles/Documents/comcom-wolakreportassessmentofthenzelectricitymar-
ketperformancept2.pdf/> (accessed 03.13).

Chapter 22

The Korean Electricity Market: Stuck in Transition

Suduk Kim[a], Yungsan Kim[b] and Jeong Shik Shin[c]

[a]*Department of Energy Studies, Ajou University;* [b]*Department of Economics and Finance, Hanyang University;* [c]*School of Economics, Chung-Ang University*

1 INTRODUCTION

The restructuring of electricity industry in Korea officially started in 2001 with the enactment of the Electricity Industry Restructuring Act. But it came after years of preparatory works of planning and lawmaking dating back to early 1990s. Before the restructuring, Korea's electricity industry used to consist of the state-owned vertically integrated monopoly of Korea Electric Power Corporation (KEPCO). The restructuring was initiated by the government as part of a general economic reform of deregulation. However, a top-down reform could succeed only as much as the strength of the political will to push it through. Though the first phase of the restructuring was successfully implemented, the general public was not convinced of the need for the reform, while some stakeholders strongly opposed it. By the time the restructuring was to shift gears to its second phase in 2004, it became a political burden and lost the driving force. Further restructuring was put on indefinite hold. As a result, the electricity industry still maintains a mixed system of the market and regulation, the balance between which was determined by the unanticipated interruption of the restructuring process. Only the generation sector was opened to market competition. In 2001, six wholly owned generation subsidiaries were spun-off from KEPCO, and independent generation companies were also allowed to enter the market. Together they compete to sell electricity wholesale to KEPCO, which retains its monopoly status in the retail sector.

Is the current state of unfinished reform viable? Eleven years after the restructuring began and 8 years after it stopped, it's high time to assess its achievements and future perspectives. The current system has revealed limitations. The wholesale market depends heavily on administrative interventions, and the retail rates have failed to reflect the rising costs. KEPCO began incurring huge losses, while the demand growth outpaced the capacity increase, resulting in power shortage during peak seasons. This chapter discusses these issues and also the challenges and opportunities facing Korea's electricity industry for the future.

The remainder of the chapter is organized as follows. Section 2 provides an overview of Korea's electricity industry. Section 3 explains the restructuring history and its current state. Section 4 assesses the achievements so far and discusses the problems including the challenges for the future. Section 5 concludes the chapter.

2 OVERVIEW OF KOREA'S ELECTRICITY INDUSTRY

2.1 The History

Korea has achieved exceptional economic growth during the past 50 years, transforming itself from one of the poorest countries into the 15th largest economy of the world. Its per capita GDP increased from less than US$100 in 1960 to US$20,757 in 2011. Average annual primary energy consumption growth rate since 1998 is 3.53% with renewable energy growing at 10.61%. Figure 22.1 shows the major economic indicators of Korea during the past 30 years.

Korea's electricity sector has successfully supported the fast economic growth. Its capacity has kept pace with the economic growth in spite of the lack of domestic energy resources and the tumultuous world energy markets. It has also succeeded in developing electric engineering capability to

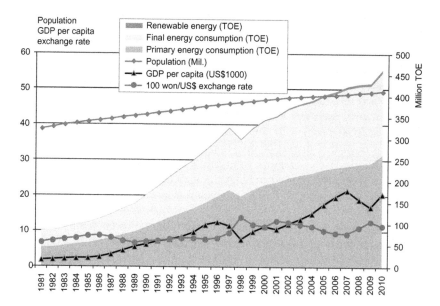

FIGURE 22.1   Major economic indicators of the Korean economy.

construct and maintain power plants and networks. In 2010, Korea became the sixth country to export nuclear power plants. The government took the leading role in the rapid growth of the electricity industry as well as the economy in general.[1] It established KEPCO in 1961 by combining three existing electric companies—one generation and transmission company and two distribution and sales companies—and put it in charge of the entire electricity sector. This system of state-owned, vertically integrated monopoly had been maintained until 2001, when restructuring of the electricity industry first took effect.

Figure 22.2 shows the growth of power generation, fuel mix, and the average domestic retail electricity price since 1978. It also shows the crude oil price for reference. Korea's electricity generation depended heavily on oil until early 1980s, after which the dependency has rapidly decreased with diversification into other fuels. Oil-fired generation accounted for 72.5% of total electricity generation when the oil price hike of 173.9% from 13.03 $/bbl to 35.69$/bbl hit the world energy market during the Second Oil Shock of 1978−1980. Due to this price shock, the average retail electricity price

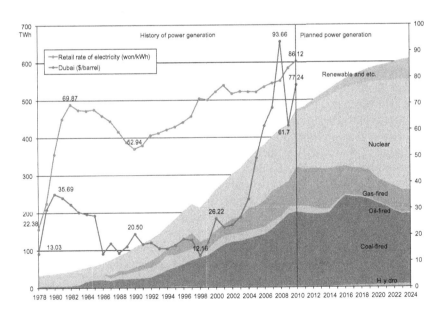

FIGURE 22.2 Power mix change since late 1970s. Note: Figures after 2010 are projections taken from MKE (2010). *Source: Data compiled from KESIS.*

1. The Korean government made extensive use of public enterprises during the process of economic development in energy and material industries. See Lee (2011).

jumped from 22.33 won/kWh in 1978 to 69.87 won/kWh in 1985, which is a 212.89% increase in a short period of time.[2]

Subsequently, the Korean government drastically changed the electricity sector fuel mix. By 2010, oil-fired generation accounted for only 3.57% of total electricity generation, while nuclear power generation accounted for 31.36%, LNG 21.69%, and coal-fired generation 41.85%, as shown in Table 22.1. Thanks to this change in the fuel mix, the average price of electricity has not changed much, even though the crude oil price continuously increased from 12$/bbl in 1998 to over 93$/bbl in 2008.

Nuclear activities were initiated when Korea became a member of the International Atomic Energy Agency in 1957. In 1958, the domestic Atomic Energy Law was passed, and the Korean government established the Office of Atomic Energy in 1959. The first Korean nuclear reactor to achieve criticality was a small research unit built in 1962. Ten years later, the construction of the first nuclear power plant, Kori-1, began with the capacity of 573 MW. This was a Westinghouse unit built on a turnkey contract. Its commercial operation began in 1978. As of 2010, 32 years after the first nuclear power plant was put into commercial operation, there are a total of 21 operating nuclear power plants. The total installed capacity of nuclear power plants is 18,715 MW.

Figure 22.3 shows the geographical distribution of transmission lines and generation capacities. The load centers are located in large cities where population and economic activities are concentrated. The Capital Area in and around Seoul (the upper left area of the map) accounts for about 40% of the load.[3] However, many of the large generation plants are located in the sparsely populated southern parts of the country. As a result, the transmission network is burdened with northward currents, sometimes straining the system operation.

Figure 22.4 summarizes the significance of electricity in Korea's energy sector in the 2010 Energy Balance Flow. The electricity sector is shown in the lower right corner. Korea depends on imports for 96.5% of the total primary energy supply. Out of 262.6 mil. TOE of total primary energy supply, around 40% is comprised of oil. After going through the transformation process, electricity accounts for 19.3% of the total final energy of 193.8 mil. TOE. As depicted in the figure, the industrial sector consumes 51.4% of the power supply, while commercial and residential sectors consume 27.8% and 14.1%, respectively.

In July 2012, Korea's electricity sector has installed generation capacity of 79.5 GW. Coal, LNG, and nuclear power capacity are 31.6%, 28.6%, and 22.3% of total capacity, respectively. While coal and nuclear power generation take 53.9% of total capacity, their electricity generation explains more

---

2. All the price figures here are nominal.
3. 38.4% of the national peak load in July 2012 was from the Capital Area. Korea Power Exchange (2012).

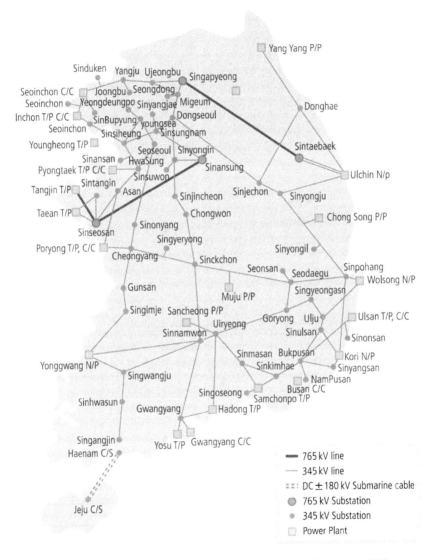

Yang Yang P/P

Sinduken   Yangju Ujeongbu   Singapyeong

Seoinchon C/C   Joongbu Seongdong
Seoinchon   Yeongdeungpo Sinyangjae Migeum   Donghae
Inchon T/P C/C   SinBupyung youngsea Dongseoul
Seoinchon   Sinsiheung   Sinsungham
Younghyeong T/P   Sinsungham
               Sinansan   Seoseoul Sinyongiri   Sintaebaek
Pyongtaek T/P C/C   HwaSung
               Sintangin   Sinsuwon   Sinansung   Ulchin N/p
Tangjin T/P   Asan   Sinjechon Sinyongju
                                  Sinjincheon
Taean T/P   Chongwon   Chong Song P/P
                   Sinonyang
               Sinseosan
                          Singyeryong
Poryong T/P, C/C   Sinyongil
               Cheongyang   Sinckchon
                                Seonsan   Seodaegu   Sinpohang
                Gunsan   Wolsong N/P
                          Muju P/P   Singyeongasn
           Singimje Sancheong P/P
                          Uiryeong   Goryong Ulju   Ulsan T/P, C/C
                    Sinnamwon   Sinulsan
                                                     Sinonsan
                            Sinmasan Bukpusan   Kori N/P
Yonggwang N/P   Sinkimhae   Sinyangsan
                Singwangju   NamPusan
                          Singoseong   Busan C/C
           Sinhwasun   Samchonpo T/P
                     Gwangyang   Hadong T/P

       Singangjin
       Haenam C/S   Yosu T/P Gwangyang C/C
                                         ━━ 765 kV line
                                         ── 345 kV line
                                         ∷∷ DC±180 kV Submarine cable
                                         ◉   765 kV Substation
        Jeju C/S   •   345 kV Substation
                                         ▢   Power Plant

FIGURE 22.3   National transmission map. *Source: KPX, http://www.kpx.or.kr/english/.*

than 73% of the total as indicated in Table 22.1. Table 22.2 shows the recent trends of the capacity and the load in Korea. Though the capacity has increased by 38% from 2004 to 2012, demand has grown faster. It grew 44% during the same period. As a result, the supply reserve ratio steadily declined to reach the precarious level of 3.8% in 2012. Figure 22.5 shows these trends in monthly frequency. The tick marks on the horizontal axis represent the summer month of July. After the summer of 2008, annual peak loads are

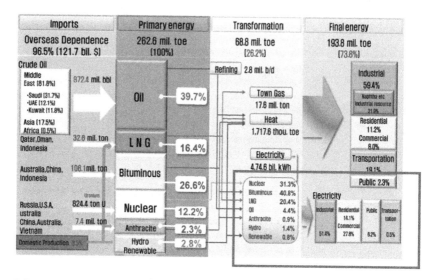

FIGURE 22.4   2010 Energy balance flow. *Source: KEEI (2011).*

TABLE 22.1 Energy Mix in Electricity Sector (by generation)

	Hydro (%)	Coal (%)	Oil (%)	LNG (%)	Nuclear (%)	New and Renewable (%)
1978	0.00	12.76	75.08	4.79	7.38	0.00
1980	0.17	11.69	77.81	0.99	9.34	0.00
1990	1.56	20.90	16.75	11.54	49.25	0.00
2000	0.61	38.00	8.86	10.89	41.64	0.00
2010	0.59	41.85	3.57	21.69	31.36	0.95
2020	1.06	36.93	0.52	10.54	44.05	6.90
2024	1.35	30.96	0.48	9.73	48.54	8.95

Note: Figures for 2020 and 2024 are projections taken from MKE (2010).
Source: Data compiled from KESIS

observed during the winter unlike in the previous years. This is due to sharp increases in electricity use for heating. The reserve ratio used to hit the annual low during the summer, but after 2008 it tends to happen during the winter.

2.2 The Outlook

Table 22.1 and Figure 22.2 show the projection of future fuel mix by generation up to 2024, based on the Fifth Electric Power Demand–Supply Plan

TABLE 22.2 Recent Trends of Capacity and Load (Unit: MW)

	Total Installed Capacity	Total Supply Capacity	Peak Load (P)	Reserve (R)	R/P (%)	Peak Day and Hour
2004	59,129	57,528	51,264	6264	12.2	July 29 (15:00)
2005	61,737	60,818	54,631	6187	11.3	August 17 (12:00)
2006	64,778	65,183	58,994	6189	10.5	August 16 (12:00)
2007	67,196	66,778	62,285	4493	7.2	August 21 (15:00)
2008	70,353	68,519	62,794	5725	9.1	July 15 (15:00)
2009	73,310	72,071	66,797	5274	7.9	December 18 (18:00)
2010	76,078	75,747	71,308	4439	6.2	December 15 (18:00)
2011	76,131	77,179	73,137	4042	5.5	January 17 (12:00)
2012	81,552	77,082	74,291	2791	3.8	August 6 (15:00)[a]

[a]Up to 14th October. It may be revised by the end of 2012.
Source: EPSIS

(2010–2024).[4] According to the projection, nuclear and renewable are supposed to grow the most, while coal and especially gas are to decline substantially in relative terms (see Table 22.1). Out of the total expected power capacity of 112.593 MW by 2024, nuclear power is planned to be 35.916 MW or 31.9% of the total. Table 22.3 shows the details of the nuclear power plant from WNA (2011a), which is somewhat different from MKE (2010).

Renewables are another area to see the most growth. In Copenhagen in 2008, the Korean government announced a goal of reducing CO_2 emissions by 30% based on the expected emissions of 2020. As the world's 11th largest emitter, Korea is obligated to join the effort to reduce CO_2 emissions. In addition, it will help enhance Korea's energy security by reducing its heavy

---

4. MKE (2010).

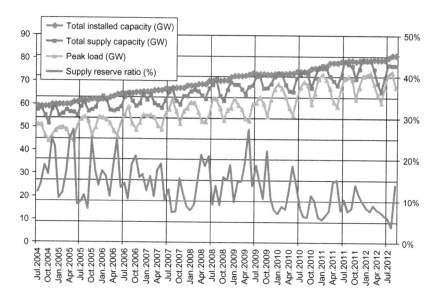

FIGURE 22.5   Monthly trends of capacity, load and reserve ratio. Note: Supply reserve is the difference between total supply capacity and peak load.

TABLE 22.3 Nuclear Capacity of Korea (Unit: MW)

	Number of Reactors	Capacity (MW)	Percentage
Operating	21	18,785	56.95
Under construction	5	5700	17.28
Planned	6	8500	25.77
Total	32	32,985	100.00

Note: According to MKE (2010), total capacity of nuclear power by 2024 is expected to be 35,916 MW.
Source: WNA (2011a)

reliance on foreign energy sources.[5] Table 22.4 shows the targeted plan for renewable energy promotion and its details by renewable power sources. The promotion of nuclear power is also an important factor in the perspective of CO_2 reduction.

Among the various measures the government has implemented or trying to implement in the future to reduce CO_2 emissions are Renewable Portfolio

5. IEA (2010).

TABLE 22.4 Renewable Energy Promotion Targets

	2008	2010	2015	2020	2030
Primary Energy (Mil. TOE): A	247	253	270	287	300
Renewable Energy (Thou.TOE): B	6360	7566	11,731	17,520	33,027
B/A (%)	2.58	2.98	4.33	6.08	11.00
					(Unit: MWh)
Solar Heat	–	–	15,330	391,890	2,046,139
PV	202,443	476,709	961,773	1,424,471	1,971,513
Wind	425,297	880,641	4,336,243	8,138,081	16,619,638
Small Hydro	289,949	393,836	653,552	913,269	1,926,163
Wood Related	–	62,306	166,396	1,146,446	2,628,920
Biogas	294	3,449	31,372	64,222	161,129
Geothermal	–	70,080	744,600	1,401,600	2,803,200
LFG	597,460	684,853	903,336	1,121,819	1,340,302
Marine	876	278,102	1,571,488	3,629,361	6,159,599
Hydro	3,494,833	3,494,833	3,631,991	3,746,289	3,860,587
Renewable Power Total: C	5,011,152	6,344,809	13,016,081	21,977,446	39,517,190
C/Total Power Gen. (%)	1.20	1.50	2.90	4.70	7.70

Source: MKE (2008)

Standard (RPS), a mandatory energy target management program for energy-intensive companies, and the plan to start carbon emission trading by 2015. The implementation of RPS started in 2012. Before 2012, feed-in tariff has been the main tool for promoting renewable energy sources. The RPS duty is currently imposed on 13 electricity generation companies including both KEPCO affiliates and private sector companies. The Korean government announced an energy target management policy in 2009 and launched a pilot program with 48 energy-intensive companies in 2010. The policy came into effect in March 2011. Selected companies under the energy target management program are required to submit their target management plan and to report on the result each year. Table 22.5 summarizes the schedule and the criteria for selecting target companies or plants of this program. Both RPS

TABLE 22.5 Schedule of Energy Target Management

Criteria	By the end of 2011		From February 2012		From January 2014	
	Company	Plant	Company	Plant	Company	Plant
By CO_2 (ton)	125,000	25,000	87,500	20,000	50,000	15,000
By energy (terajoules) (TOE)	500	100	350	90	200	80
	12,000	2400	8400	2100	4800	1900
Expected number of target companies or plants	170	240	220	250	300	280

Sources: BIS, http://www.bsigroup.co.kr/, KEEI, http://www.keei.re.kr/keei/download/seminar/100831/S2-1.pdf (in Korean).

and the target management program are said to be continued after the start of carbon emission trading, but it remains to be seen.

The Korean National Assembly has approved the start of a carbon emission trading scheme from January 2015. Despite fears that it would hurt the economy, the expectation of long-term benefits to the country's development of green and energy-efficient technology must have driven such decision-making. The details of the trading scheme are not yet determined, but it would be a great challenge for the electricity sector to accommodate this upcoming measure considering that around 40% of total primary energy was used for electricity generation in 2010.[6]

The Korean government is also promoting smart grid to improve energy efficiency and as one of the next strategic growth engines of Korea. A pilot program was initiated in Jeju island in December 2009. This project has five subprojects: Power Grid, Smart Place, Smart Transportation, Smart Renewables, and Total Operation Center. The national road map of smart grid was announced in January 2010 followed by the enactment of Smart Grid Construction and Promotion Act in May 2011. Recently, the First Master Plan for Smart Grid was announced in June 2012 by the Korean government.

3 RESTRUCTURING OF KOREA'S ELECTRICITY INDUSTRY

3.1 Background and History

The system of state-owned vertically integrated monopoly, i.e., the KEPCO, worked fairly well until early 1990s, when the discussion of electricity sector

6. KEEI (2011).

restructuring started. The electricity quality was good, and the rates were relatively low thanks to the high proportion of nuclear power and the low oil price during late 1980s and early 1990s. KEPCO employees were proud and enjoyed the privileges of the public sector. There might have been inefficiencies associated with state monopoly and the big size, but at least they didn't show. What prompted the discussion of restructuring was not KEPCO's poor performance but concern about its future.

To keep up with the fast growing economy in the future, the electricity sector had to keep expanding its capacity at a pace surpassing KEPCO's own financing capability. Depending only on debt would have increased its leverage level too much. The government had to find the money to inject into the electricity sector to continue the existing system. This task was becoming more difficult. However, the news of electricity sector deregulation and reform in such countries as England and the United States suggested another solution: open the electricity sector to competition and invite investments from the private sector. Once this alternative became a possibility, the debate about the pros and cons of the two alternative approaches began in earnest. Some experts raised the issue of potential problems and inefficiencies of a big state monopoly.[7] Others argued that there was no need to change the current system which had served the economy so well.

In 1993, the Korean politics reached a milestone with the election of President Young Sam Kim, who was the first president without military backing in more than 30 years. He called his government the "Civilian Government" and vowed to lead sweeping reforms to remove the legacies of past authoritarian regimes. Among his agendas was the reform of state-owned enterprises (SOEs). In 1994, the government commissioned leading consulting companies and think tanks managerial evaluation of major SOEs, including Korea Telecom, POSCO[8] and KEPCO.[9] The evaluation process took almost 2 years from July 1994 to June 1996. The final report recommended structural changes to meet the rapidly increasing electricity demand and address the problem of concentration of economic power. It recommended liberalization of the electricity industry to introduce market competition and privatization.[10]

---

7. Young Jun Kim, who led the government's task force for the restructuring during the early years, mentioned three rationales for the restructuring in his memoir (Kim, 2002). They are (i) relatively small economies of scale in the generation sector, (ii) inefficiencies of large state-owned monopoly, and (iii) the need to finance future growth of the electricity sector. The scale elasticities in the generation sector and the power industry in Korea were estimated and shown to be decreasing by Sonn and Chung (1993).
8. POSCO is the biggest still company in Korea
9. KEPCO was assigned to Anjin Accounting Company, Samil Accounting Company, and Korea Industrial Economy Research Institute.
10. Korea Telecom and POSCO were privatized in 2002 and 2000, respectively.

In 1998, the government finally decided to go ahead with the electricity industry restructuring and began the preparations. It consulted the Rothschild Group and other international consulting companies, which recommended a mandatory electricity pool model like the early English model.[11] The government also held public hearings to solicit opinions. It is noteworthy that the electricity industry restructuring was initiated by the government as a part of a general economic reform program. It was regarded as a necessary step to break off the heavy-handed regulatory approaches of the past and enter a mature market economy.

The restructuring plan faced strong opposition from the labor union, denouncing it as a sellout of national wealth to the few conglomerates or the foreign capital. Some experts argued that the restructuring plan had been hastily drawn, simply copying foreign models without regard to the national differences.[12] However, after experiencing the 1997/1998 foreign currency crisis, in which Korea was bailed out by the IMF, a sense of urgency to reform every aspect of the country prevailed. After 3 years of legal, institutional, and organizational groundworks, the first phase of the plan was launched in 2001.

Despite the opposition from some quarters of the society, the National Assembly passed the Electricity Industry Restructuring Act and the Amendment of the Electricity Business Act unanimously at the end of 2000. On April 1, 2001, six generation companies were spun-off from KEPCO as wholly owned subsidiaries. On the same day, Korea Power Exchange (KPX) and Korea Electricity Regulatory Commission (KOERC) started operation.

According to the original plan stipulated in the Restructuring Act, the electricity industry restructuring was to proceed in three phases over a decade as shown in Figure 22.6.

3.1.1 Phase I: Generation Competition (2001–2003)

In this phase, the generation sector of KEPCO is separated and divided into six generation companies, which remain as KEPCO's wholly owned subsidiaries. Independent generation companies are also allowed to enter. KEPCO maintains monopoly in all other sectors: transmission, distribution, and retail operations. The generation companies compete with each other to sell electricity in the wholesale market, in which KEPCO is the sole buyer. KPX is set up to operate the wholesale market. It also takes over the function of system operation. KOREC is established as the regulatory body.[13]

---

11. Rothschild was consulted for the basic plan, Anderson Consulting for the financial aspect, KEMA for the technical aspect, and Freehills for the legal aspect.
12. These criticisms were raised in a hearing arranged by the National Assembly on November 23, 2000.
13. The details of these institutions will be explained in Section 3.2.

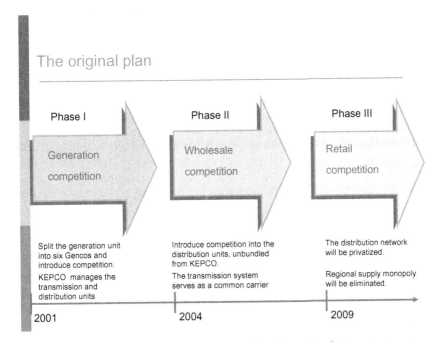

The original plan

Phase I

Generation

competition

Phase II

Wholesale

competition

Phase III

Retail

competition

Split the generation unit
into six Gencos and
introduce competition.

KEPCO manages the
transmission and
distribution units

Introduce competition into the
distribution units, unbundled
from KEPCO.

The transmission system
serves as a common carrier

The distribution network
will be privatized.

Regional supply monopoly
will be eliminated.

2001   2004   2009

FIGURE 22.6   Original plan of power sector restructuring. *Source: Korea Power Exchange.*

3.1.2 Phase II: Wholesale Competition (2004–2008)

Regional distribution companies are to be unbundled from KEPCO to take charge of distribution and retail operations. The distribution companies are to be monopolies of retail service in their respective regions. They are to participate in the wholesale electricity market to purchase electricity.

3.1.3 Phase III: Retail Competition (2009–)

The retail market is to be opened to competition. Third party access to the transmission/distribution networks is to be established, so that the regional distribution companies can service customers outside of their own franchises. Independent retail companies can also enter the market.

Privatization of the unbundled companies was also included in the plan. The generation subsidiaries were to be sold in several stages starting from 2002.[14] Between 2003 and 2004, the government actually put up one of the generation subsidiaries for sale, but few parties showed interest, and none made a final offer. Maybe this was because of the uncertainty about the

---

14. The nuclear generation company was not included in the privatization plan.

future of the Korean electricity market. The privatization plan was shelved. Most of the Phase I plan was successfully implemented except the privatization part. The wholesale market started trading, and the transition of system operation from KEPCO to KPX was completed successfully.

In 2004, however, the restructuring came to a halt before the start of Phase II. The restructuring, initiated by the government, did not command popular support. While most electricity consumers were indifferent to the restructuring, other stakeholders were divided based on their own interests. Furthermore, the California power crisis of 2000/2001 showed that ill-prepared restructuring could cripple the electricity sector instead of improving it. The news of rolling blackouts from California was quite disturbing when Korea was about to plunge into a full-fledged reform of its own. Voices of opposition became louder and many called for caution.[15] Some pointed to other countries adopting more guarded approaches, such as France and Japan. The KEPCO labor union strongly opposed further restructuring and privatization. It went into a strike in February 2002 demanding cancelation of the privatization plan. The strike lasted for more than a month, and was supported by nationwide labor organizations.[16] In 2003, the new Korean president who was elected on a more progressive platform ordered reevaluation of the restructuring plans of the electricity and the railroad industries.[17] Amid heated debate involving both political and economic circles, the president empowered the Tripartite Commission of Labor, Management, and the Government to decide whether to continue with the Phase II restructuring. In May 2004, after months of deliberation, the Commission decided against the continuation.[18]

The end result was that restructuring stopped in 2004, and no substantial changes have been made since. Further, no decisions have been made as to when and how to resume the process. In 2009, opponents of restructuring urged the government to repeal the restructuring all together and restore the pre-restructuring industry structure. The government sought recommendations for policy direction from a top government think-tank, the Korea Development Institute (KDI). KDI assembled a team of experts and conducted a comprehensive study. The resulting report, submitted in 2010, recommended staying the course of restructuring.[19] This muted the arguments to roll back the restructuring for a while, but it did not mean that the restructuring was put back on track. The government adopted KDI's recommendation to guarantee the generation subsidiaries more managerial

---

15. For example, Lim (2004) suggested considering alternative models depending less on the spot market and more on bilateral contracts and vertical integration.
16. Since then, the union threatened to take collective action several times against further restructuring or privatization.
17. See Lee (2011) for how the election of the new president affected the restructuring efforts.
18. The restructuring of the railroad industry was also stopped before it started.
19. KDI (2010).

independence from KEPCO so that they could compete with each other like unaffiliated companies. Although they are wholly owned subsidiaries of KEPCO, they now have to report directly to the government instead of to the KEPCO management.

As for the restructuring of downstream retail sectors, however, the government just announced that it would consider it as a long-term project. In the meantime, the Electricity Industry Restructuring Act expired on December 31, 2009.[20] Without a legal mandate or strong advocates, the electricity industry restructuring lost its driving force and became stranded.

3.2 Current Market Structure and Design

3.2.1 The Participants

Table 22.6 shows the major generation companies and their capacities. KEPCO's six generation subsidiaries account for more than 83% of total capacity. One of them (KHNP) has all the nuclear power plants and thus the biggest capacity. The other five are mostly thermal with similar capacities and fuel mixes. They have coal plants for the base load and natural gas plants for the peak load. Unaffiliated generation companies account for less than 17% of total capacity, although they exceed 400 in number.[21] Most of them are small, regional companies such as Community Energy Service (CES) companies, but a few of them, such as POSCO Power and GS, are sizable and operate as merchant generation companies selling to the wholesale market.[22] Entry into the generation business is regulated, requiring government approval.

KEPCO, the parent, owns and operates the transmission and the distribution networks. It is also the only supplier in the retail sector,[23] which in turn makes it the only buyer in the wholesale market. To prevent KEPCO from exercising too much market power, regulations restrict KEPCO's roles in the wholesale market. It is not allowed to bid the purchasing price but takes the price determined by the market operator. It does not participate in the system operation though it owns and maintains all the networks. The retail rates are strictly regulated. KEPCO is listed both on the Korea Exchange and the New York Stock Exchange. The government controls 51% of KEPCO's voting shares[24] as of 2011. Foreigners own 23.4%, and the rest is owned by the public.

20. Electricity Industry Restructuring Act, Addendum.
21. Their total generation is even smaller because most of their plants are peaking units.
22. Four of them have current capacities over 900 MW.
23. There are Community Energy Service companies which coproduce heat and power to sell to final consumers in certain areas, but their share is negligible.
24. The government directly owns 21.2% and controls 29.9% through the shares owned by Korea Finance Corporation.

TABLE 22.6 Generation Companies and Their Capacity (Unit: MW)

	Nuclear	Coal	Hydro	LNG	Oil	New and Renewable	Others	Total
KEPCO and affiliates (% to total)	100.0	96.4	81.6	65.1	75.7	19.3	0.0	83.2
KEPCO	0	0	0	0	177	1	0	178
KHNP	17,716	0	529	0	0	14	0	18,259
KOSEP	0	6905	600	922	529	8	0	8964
KOMIPO	0	4400	1000	4350	295	9	0	10,054
KOWPO	0	4000	1200	2998	1400	2	0	9601
KOSPO	0	4000	400	4448	745	0	0	9593
KOEWPO	0	4900	700	2100	1800	6	0	9506
Independents (% to total)	0.0	3.6	18.4	34.9	24.3	80.7	100.0	16.8
POSCO Power	0	0	0	1800	0	0	0	1800
GS	0	0	0	1934	0	0	0	1934
Others	0	916	1001	4215	1590	168	1721	9611
Total	17,716	25,121	5430	22,768	6536	209	1721	79,500
Fuel source/total (%)	22.3	31.6	6.8	28.6	8.2	0.3	2.2	100.0

KOSEP: Korea South East Power Co.
KOMIPO: Korea Midland Power Co.
KOWPO: Korea Western Power Co.
KOSPO: Korea South Power Co.
KOEWPO: Korea East West Power Co.
KHNP: Korea Hydro and Nuclear Power Co.
Note: Coal power includes bituminous, anthracite, dual fuel including coal. Hydro includes pumped storage. Petro products include B–C, kerosene, byproduct B–C and kerosene, diesel, heavy oil, orimulsion, and low sulphur waxy residue (LSWR). Others include (high pressured) steam, byproduct gas, etc. New and Renewable includes photovolatic (PV), small hydro, biogas, land fill gas, wood chip, and combustible wastes.
Source: Data Compiled from EPSIS based on July 2012

KPX, established in April 2001, is the market operator and the system operator for the whole power system. It is an independent nonprofit organization with over 400 member companies. Its operating expenses are funded through electricity trading fees. KPX also houses four committees in charge of rule revision, cost estimation, information release, and dispute resolution, respectively. These committees comprise the representatives of the member companies and the government, and independent experts.

KOREC is the regulatory body established within the Ministry of Knowledge Economy[25] with the mandate to create an environment of fair competition in the electricity business. It is responsible for the implementation of the restructuring plan of the electricity industry, and is in charge of supervising the operation of the electricity market and of consumer protection. It makes the final decisions regarding most of the important rules and regulations of the electricity sector.

3.2.2 Wholesale Market

Since Korea's current electricity market system was designed for use during a short transitional period, its design is relatively simple, leaving many decisions to regulation. The key features of the wholesale market are as follows:

- It consists of a single day-ahead market for the whole system.
- It is a mandatory pool, and does not allow physical bilateral contracts.
- Only the generation companies make bids, and the buyer (KEPCO) purchases all the electricity it needs at prices set by the market rule.
- The generation companies bid available capacities and not the prices.
- The bidding price for each generation plant is predetermined by regulation based on the variable cost.
- The market clearing price, System Marginal Price (SMP), is the bidding price of the last plant in the merit order to operate to meet the demand forecast.
- The SMP applies to the entire system except the province of Jeju island.[26]
- Capacity payments are made to all capacity bids at uniform regulated prices.

---

25. It was succeeded by Ministry of Trade, Industry and Energy (MOTIE) in 2013.
26. Jeju island is located about 100 km south from the mainland and home to about 0.55 million people. Its power system is connected to the mainland system by a HVDC cable. Jeju is considered a separate system and thus has its own SMP.

These features are best summarized by the term "Cost-Based Pool (CBP)," whose details are discussed below.[27]

3.2.2.1 Market Participation

CBP is a mandatory pool. Anybody who wants to generate and sell electricity should sell into the pool.[28] The only exception is CES companies that are allowed to sell electricity directly to the final consumers in the designated regions. However, CES license is limited to small-scale generation mostly from cogeneration facilities, and its total market share is less than 1%. CBP is also the only place to purchase electricity wholesale. Since only KEPCO has the retail license, it is naturally the only buyer in the wholesale market except CES. Large customers are allowed to buy directly from CBP, but the direct purchase rates are higher than the retail rates. No direct purchase has been made so far. Financial contracts—contracts for difference (CfD)—between two parties are allowed to emulate long-term physical contracts. Again, due to lack of incentives to hedge against future uncertainty, no CfD contract has ever been made. As a result, almost all the wholesale electricity in Korea is traded at prices set by the day-ahead market.

3.2.2.2 Market Equilibrium and System Operation

In the day-ahead market, generation companies submit their bids for each day by 10:00 AM of the previous day. They bid available capacities at predetermined prices for each of the 30 min. intervals of the following day. The bidding prices are determined by the Cost Estimation Committee of KPX every month for each unit reflecting its variable costs. The variable costs consist of three components: incremental fuel cost, no-load cost, and start-up cost. Fixed costs are compensated separately through capacity payments. The bids also include the technical details of each generation unit for consideration in economic dispatch. Stacking all the capacity bids by merit order renders a supply curve of the entire power system.

By 15:00 PM each day, KPX determines a daily demand forecast for each of the 30 min. intervals of the following day. This forecast is based on a computer program reflecting past records and the weather forecast. However, since it does not consider the effect of the price, the demand forecast does not render a downward sloping demand curve. For each interval,

---

27. In PJM's day-ahead market, the generators make both cost-based offer and market (price)-based offer. In general, market-based offers are used to determine the bid production cost in the operation of the PJM Energy Market. Cost-based offers of a generating unit are used only in circumstances where the transmission constraints limit available market supply in the subregion of the market and the owner of the generation unit is deemed to possess structural market power. See PJM (2009).

28. Generation companies should participate in bidding except for small units of less than 20 MW capacity or renewable energy.

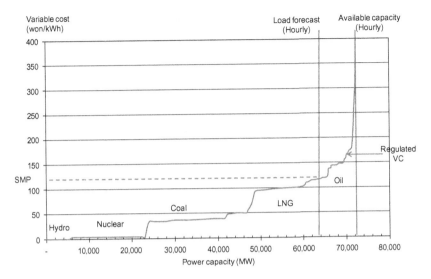

FIGURE 22.7   Illustration of SMP, based on the merit order of March 2010. *Source: KPX, not in public domain.*

the forecasted demand curve is vertical at the forecast level. Intersecting the supply and the demand curve produces the equilibrium price: the bidding price of the last unit to be dispatched to meet the demand forecast. It is the SMP. All the units dispatched according to their bidding get paid the SMP in principle.[29] Figure 22.7 shows how SMP is determined based on the information of March 2010.[30]

The actual operation of the system, however, differs from the plan based on the market equilibrium. Firstly, the market equilibrium plan does not consider the technical constraints, such as transmission capacity and ramping rates. So KPX, as the system operator, finalizes the operation plan for the following day, incorporating the technical constraints by 18:00 PM. Secondly, the demand forecast is different from the actual load, and the real-time operation following the actual load differs from the operation plan of the previous day. Since CBP does not have a real-time balancing market or an ancillary service market, there is no way to fill the gap through market interaction. CBP addresses this problem by an administrative formula. Any unit ordered to generate more or less than is prescribed by the market equilibrium plan through no fault of its own gets compensation of the difference between SMP and its own bidding price. For instance, if a unit is ordered to

---

29. Some adjustment is made in the final payment, which will be explained later.
30. Note that hydroelectric power is treated differently in the sense that it is not fully dispatchable and is being utilized as peak load during the summer time when the most of the annual precipitation is observed in Korea.

TABLE 22.7 Adjustment Factors for Capacity Payment in 2012						
Time of day	Peak (5 h/day)		Middle (9 h/day)		Base (10 h/day)	
Season	Normal	Peak	Normal	Peak	Normal	Peak
Adjustment factor	1.187605	2.378612	1.013809	2.012950	0.350934	0.798301

Source: Korea Power Exchange (2011)

produce more, it will receive the higher of SMP and its bidding price for the portion of overproduction. If it is ordered to produce less, it will receive the difference between SMP and its variable cost. However, if the discrepancy occurs owing to the generator's own fault, it will receive the lower of SMP and its bidding price for overproduction, and nothing for underproduction.

3.2.2.3 Capacity Payment

CBP does not allow price spikes or operate a separate capacity market to compensate the generation companies for the fixed costs. Instead, it offers capacity payment to all capacity bids. The payment level is administratively decided every year to cover the construction costs and the operation and maintenance costs of a designated peaking generation unit. Currently, the average capacity payment is 7.46 won/kWh. Then it is adjusted according to a predetermined schedule across different seasons and times of the day reflecting capacity scarcity by multiplying an adjustment factor (Table 22.7). The capacity payment is uniformly applied to all kinds of generation capacity.

3.2.2.4 Wholesale Purchasing Prices

The current wholesale market is designed with a single buyer to consider. So the purchasing price from CBP is calculated simply to cover all the expenses made out to generation companies in each trading interval. Since KEPCO is the only buyer, there is no issue of how the total expenses should be distributed among different buyers. As mentioned before, consumers with more than 30 MVA are qualified to purchase directly from CBP. However, the direct purchase rates are higher than the retail rates, granting no incentive to make a direct purchase.

3.2.2.5 Wholesale Price Controls

The restructuring plan for Phase I did not include vesting contracts. As a result, no binding contractual arrangements existed to prevent excessive profits for generation companies which had benefited from their past status of state monopoly or from any other advantages endowed to the incumbents. Instead, CBP imposed price controls on the wholesale market as a temporary

TABLE 22.8 Correction Factors Applied to Different Fuel Types

Fuel Type Year	Nuclear		Coal	LNG	Anthracite
	LWR	HWR			
2008	0.2184	0.2184	0.0894	0.0894	0.75
2009	0.2798	0.4820	0.1865	0.3270	0.75
2010	0.1963	0.1510	0.1315	0.3200	0.50
2011	0.2457	0.3516	0.1881	0.3033	0.50

Note: Figures for 2011 are for the first half of the year.

safety measure against excessive SMPs. The specific forms of the price controls changed over time. First, CBP imposed a price cap for the base load set by the highest variable cost among the base load generators. This price was called the Base Load Marginal Price (BLMP). CBP also paid different capacity payments to the base load capacity and the peak load capacity. The BLMP could keep the average price low while allowing the marginal price to move freely reflecting the market condition.

In 2008, CBP revised the market rules to remove BLMP and make the capacity payment uniform. This time, however, CBP introduced a different and more elaborate kind of price control. It reduces the margin between the SMP and the bidding price (i.e., the variable cost) by a certain proportion called the "correction factor," which has a value between 0 (full reduction) and 1 (no reduction). This proportion differs across fuel types and is decided annually by the Cost Estimation Committee. Table 22.8 shows the correction factors applied to different fuel types since its introduction. Since CBP can impose different correction factors by the fuel type, it can now fine-tune market outcomes even more elaborately than with BLMP and the differential capacity payments.

3.2.3 Retail Sector

The retail sector is still a monopoly under tight regulation. The retail market is not open to competition, and the government virtually sets the retail rates. Formally, when KEPCO experiences significant increase in costs, its board of directors can submit a rate adjustment request to the Ministry of Knowledge Economy. Then the Ministry consults with the Ministry of Strategy and Finance,[31] and lets the KOREC review the request and make the final decision of accepting or rejecting the proposal (Figure 22.8).

---

31. The Ministry of Strategy and Finance is in charge of managing the government budget and the macro aspect of the economy.

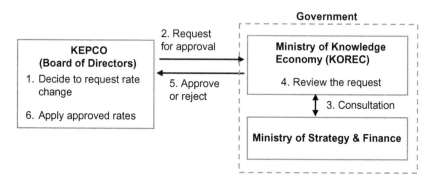

FIGURE 22.8   Procedure of changing electricity retail rates.

There are six categories of retail rates by use: industrial, commercial, res-idential, agricultural, educational, and street lights. The residential rate is an inverted block rate. The rate increases in six steps to reach a punishingly high level of maximum 670.6 won/kWh. It is more than 11 times the starting rate of 57.3 won/kWh.[32] It is as if electricity is rationed at low rates up to a certain amount, beyond which the rate jumps up to discourage further con-sumption. As a result, residential electricity consumption per household is still very low compared with other countries, though the overall electricity consumption per capita is among the highest in OECD.

The commercial and industrial rates apply time-of-use (TOU) rates, except for very small customers. The rate can change across seasons and, for large customers, across times of the day. The agricultural rate is flat at a very low level far below the cost. Even though agricultural consumption is only a small portion of the total electricity use at about 2.5% in 2010, it accounts for a significant portion of KEPCO's loss. Figure 22.9 compares the industrial electricity prices available from IEA Energy Prices and Taxes (2012). This figure presented with PPP adjustment clearly shows that tariff has been very stable compared to those of other countries. The price level for 2009 is 24th out of 33 countries.

4 ACHIEVEMENTS AND ISSUES

4.1 Generation Unbundling

Since the restructuring of Korea's electricity industry stopped after creating six generation subsidiaries of KEPCO, it is difficult to evaluate whether it has brought any benefit or harm to the industry. When a reform is only half finished, its positive or negative effects, if any, can be ascribed either to the finished half or the unfinished half of the reform, depending on the point of

---

32. Low voltage residential rates as of July 2012.

(Unit: US $/MWh)

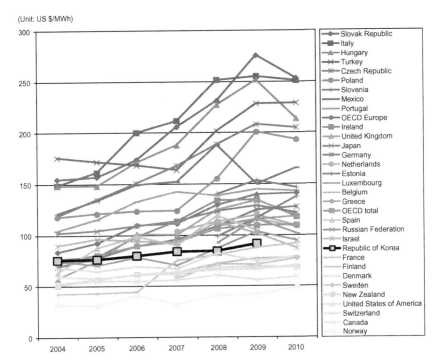

FIGURE 22.9 International comparison of industrial electricity price (using PPP). Note: Figure for Korea 2010 is not reported at IEA. *Source: IEA Energy Prices and Taxes Statistics, 2012.*

view. Other countries examine the retail price changes for a sign of the impact of a reform. In Korea, regulation has kept the retail rates very low, but obviously it has nothing to do with the restructuring. Instead, most debates have revolved around the issue of whether the economies of scale or the economies of scope have been sacrificed by separating the generation sector and the system operation from KEPCO.

The 2010 KDI report, previously mentioned, includes assessment of the outcome of the restructuring so far. It addresses some of the criticisms raised by the opponents of the restructuring. They argued that KEPCO lost its bargaining power in the international coal market by breaking up its generation sector into smaller generation companies.[33] Though this argument was plausible and intuitive, no convincing evidence of actual bargaining power reduction was found to support it. The opponents also warned that the power system reliability and long-term resource adequacy would be compromised when the system operation was separated from KEPCO. Again, 11 years

---

33. Generation companies purchase most of the LNG from KOGAS which is a state-owned monopoly in procuring LNG from abroad.

after the start of the reform, no systematic evidence or event has emerged to substantiate the argument. There is no sign of deterioration of electricity quality, either in terms of frequency, voltage, or outage after the separation of the system operation.[34] However, the KDI report agrees to the concern that the separation of system operation and transmission/distribution operation may cause delays and inefficiencies in responding to network-related emergencies. Several other studies provided results suggesting that the efficiency of the generation subsidiaries had improved after the restructuring.[35]

Overall, Phase I restructuring accomplished most of its objectives without major problems. There were no serious technical problems or cost disadvantages to justify aborting the entire restructuring plan. On the other hand, the operation of the CBP market and the lack of retail competition have revealed many limitations of the temporary market design. Also the current system seems unfit to handle the important new issues of climate change and the smart grid. The next sections will discuss these problems and the need to upgrade the market design.

4.2 Wholesale Market Issues

The CBP market is a mixture of the market and regulation. There is some room for the market, in which the generation companies bid their available capacities at predetermined bid prices. The bids and the forecasted load determine the SMP, and regulation determines everything else. However, since the scope of the regulation is not clearly specified, there could be lack of regulation in some areas and too much regulation in others. The boundary between the market and regulation is often blurred. Lack of clarity and predictability of regulation creates uncertainty for the market participants and causes disputes among them.

For instance, there is the problem of determining the variable cost of each generation plant. Not only is it a big administrative burden, it also suffers from conceptual ambiguities such as the distinction between the variable cost and the fixed cost. Some costs fit neither category perfectly. For example, no-load cost and start-up cost are quasi-fixed costs but are regarded as variable costs. On the other hand, operation and maintenance costs are all regarded as fixed costs, when some parts of them are of variable nature.[36] Cogeneration and pumped storage hydroelectricity also pose difficult questions about their cost evaluation.

---

34. The rolling blackout on September 15, 2011 sparked a heated debate on the viability of separate system operation.
35. Kim et al. (2006) and Kim and Kim (2008).
36. Since the variable cost defined in CBP market design consists of three components related to fuel cost as was discussed in Section 3, there is no room for O&M to be included as variable cost.

More controversial is the wholesale price control. The price cap for the base load or the correction factor has been used as substitute for vesting contracts. But unlike vesting contracts which specifies all terms and conditions, the rationale and scope of the price control are not explicitly stated. The price cap for the base load was criticized for the lack of rationale for discrimination by the fuel type.[37] The correction factor that replaced the base load price cap was subject to the same criticism for discriminating between different fuel types but caused even more controversy. It is applied only to the KEPCO affiliates and not to independent generation companies, introducing another layer of discrimination.[38] The "corrected" wholesale prices are not applied to large customers bypassing KEPCO to purchase directly from CBP, either. This is one of the reasons why the direct purchase rates are higher than KEPCO's retail rates.[39] Being applied only between KEPCO and its affiliates, the correction factor has become a means of implementing internal transfer prices between them. The idea that CBP allows exclusive internal transfer pricing among some of its members contradicts the notion of an open pool in which all participants are treated equally. It is also against the principle that KEPCO should be restrained from exercising its market power as the single retail supplier.

Another problem with the current system of price control is that it deprives the market of its ability to achieve long-term equilibrium. When there are not enough generators, the market induces more investments by promising high profits to new entrants. But the price control and the accompanying uncertainty dampen the incentive to invest in new capacity. There is no knowing when the price control will end or when a new kind of price control will be imposed to whom. The correction factors are determined by a committee and, as seen in Table 22.8, are largely unpredictable, sometimes causing acrimonious disputes among market participants.

4.3 Retail Market Issues

The lack of the market mechanism on the retail side makes it difficult to incorporate demand responses in the market. The retail sector provides the linkage between the consumers and the producers. An ideal retail sector conveys the information about consumer preferences to the producers to help develop the products that the consumers want. It also passes the costs of various products through to the consumers responsible for their incurrence.

37. Kim et al. (2008) suggested that the differential capacity payments had been used as a means to balance out the interests of the market participants under the base load price cap, especially among the KEPCO participants.
38. The government is currently considering whether to apply the correction factors to independent generation companies, too.
39. Another reason is that KEPCO's retail prices are even below its own cost. See the following section for details.

Barring a market failure, competition is supposed to perform these functions the best.

Some argue that, since electricity is a homogeneous good with transparent cost structures, retail competition may not bring enough benefit to warrant the cost. The product differentiation is minimal, and the price can be easily regulated to pass through the cost.[40] Korea's retail sector, with a competitive generation sector and a fully regulated retail sector, can be a showcase for such a theory, but the result so far does not seem promising. The CBP does not allow KEPCO's active bidding on the demand side of the wholesale market. This virtually blocks the channel of demand response to the wholesale market. The vertical demand curve makes the SMP highly volatile as the demand fluctuates. It also increases the social cost of maintaining system reliability, as the system operation relies on nonprice measures to secure minimum reserves. The government is willing to enforce mandatory load cuts across the board in case of acute shortage.[41] The lack of competition in the retail sector also hinders the development of various demand response programs, because there is little room for entrepreneurial activities such as curtailment service or demand aggregation.

Korea's experience also shows that the retail rate tightly controlled by the government can, ironically, run out of control. As explained in Section 3.2.3, any rate change should go through many steps and entities including two government ministries. The slow and opaque decision-making process did not cause too big a problem when the market environment changed slowly. When fuel prices change once in a while, the retail rates can be adjusted slowly over several years to smooth out the shocks. It is different, however, when fuel prices keep increasing by leaps and bounds as in the 2000s. Owing to the government's tendency to procrastinate public rate hikes, the electricity rate has failed to keep up with the fast increasing fuel cost in the 2000s. According to KEPCO's own estimations, the electricity rate of 2011 is lower than the cost by more than 10%. Further, since other energy prices have kept up with the increasing costs, electricity has become relatively cheaper than its substitutes.

Figure 22.10 shows the trends of electricity rates and SMP over the past 10 years. Comparing the fast growing SMP with the stable trends of the nominal retail rates, it can be easily understood that electricity rates declined in real terms. This gave the consumers strong incentives to substitute electricity for other energies. Air conditioners have become ubiquitous, and people have begun using electricity for heating. Though

---

40. See Joskow (2000) and Littlechild (2000) for the original debate, and Defeuilley (2009) for a recent summary.
41. The government actually implemented this in the winter of 2011/2012, for which a serious supply shortage was anticipated.

(Unit: won/kWh)

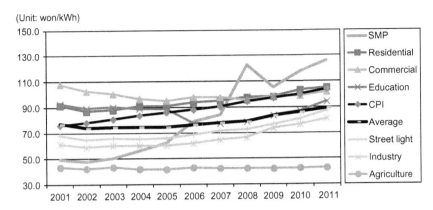

FIGURE 22.10   Retail electricity rates by use compared with SMP. *Source: EPSIS.*

(Unit: Million US$)

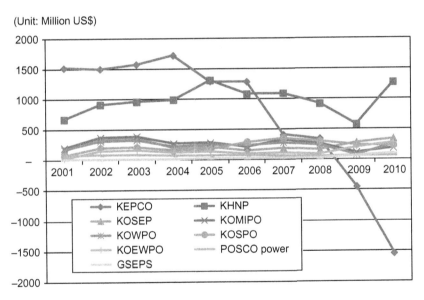

FIGURE 22.11   Trends of operating profits for KEPCO and generation companies.
Source: EPSIS.

residential demand increase has been moderated by the inverted block rate, commercial and industrial demands have increased at a pace far exceeding the forecast.

The persistent low retail rate has begun wreaking havoc in the industry's health. The supply shortage has reached a crisis level as will be discussed next. KEPCO's loss has accumulated to an unsustainable level as seen in Figure 22.11. In 2008, the government provided a one-time subsidy of 835 billion won[42] to KEPCO to make up for some of the losses, but it brought only temporary relief. In 2011, several of KEPCO's shareholders sued its chairman for not doing his best to increase the retail rates which, they argue, constituted dereliction of managerial duty. The below-cost retail rate, which is the outcome of a regulated monopoly, would become an entry barrier to the market even if entries were allowed. With the rate so low, nobody could enter the retail market and make money. Not even the option of direct purchase from CBP can beat the current retail price.

4.4 Resource Adequacy

As of 2012, Korea is experiencing serious power supply shortage. The growth of load outpaced the long-term load forecast that had underestimated the effect of the low electricity prices and overestimated the efficacy of the demand-side management. As a result, the total capacity, especially the base load capacity, has fallen short of the optimal level. As shown in Table 22.2, the reserve ratio has declined steadily since 2003 and fell below 10% in 2007. The frequency of peaking LNG generators setting the SMP has increased from 58% of the time in 2003 to 87% in 2011, as shown in Figure 22.12. This drove the SMP level much higher than the average variable cost for most of the generators. For the base load generators, the margins between the SMP and the variable cost became even greater. The uncertainty and difficulty in the construction of new power plants or transmission lines are another cause of the current problem. It is becoming more and more difficult to find sites for new power plants or transmission lines as the public awareness of environmental issues increases.

An investigation of the demand pattern reveals an important factor behind the current shortage. Figure 22.13 shows the monthly peak load and average hourly electricity consumption since 1998. Until 2001, the monthly peak and monthly average electricity consumption of the summer surpassed those of the previous winter. However, since 2002, the monthly average of the winter exceeded that of the summer. Since 2009, even the annual peak occurred in the winter. This clearly indicates that the electricity demand for heating has increased much faster than the demand for cooling. Considering that there are more substitutes for heating fuels than for cooling, this changing pattern is consistent with the explanation that consumers are switching their heating fuel from gas or oil to electricity.

The supply shortage problem has emerged to the surface as the reserve ratio fell below 10% in 2007, but it took an acute power shortage to drive home its seriousness to most of the people. Korea experienced a rolling blackout for the first time in history on September 15, 2011. It was just after the end of the official summer peak season. KPX had lowered its alert level

---

42. This figure is equivalent to US$757.3 million.

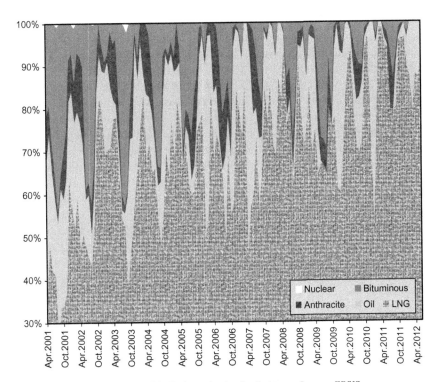

FIGURE 22.12 Frequencies of SMP determination by fuel type. *Source: EPSIS.*

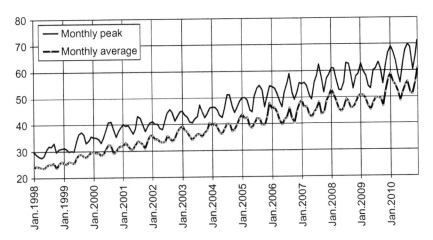

FIGURE 22.13 Trends of monthly peak and monthly average hourly consumption. Note: Each tick mark denotes January. Summer time is in between the tick marks.

and begun overhauling generators according to schedule. Then a sudden heat wave struck the country on September 14, 2011. The load surged and the reserve plummeted. It got worse the next day. Sensing a crisis, the system operator invoked the emergency plan to implement a rolling blackout over the entire country. It created chaos. Most of the people in the affected areas were not warned in advance. The consumers were outraged and lashed out at the government for mishandling the crisis. Most of the criticism centered on the decision to cut the power supply. Some blamed the restructuring, reiterating their call to roll it back. Not many blamed the dysfunctional retail rates.

The recent development of supply shortage is ominous for the prospect of resuming the restructuring. In view of the high SMP levels being obtained in the CBP, it is obvious that liberalizing the wholesale market now will lead to windfall profits to the generation companies and huge losses to the retail companies and the consumers. That is, unless the liberalization is accompanied by strong measures of preventing windfall profits and mitigating market power. It must be a challenge to design and implement a mechanism that can obtain the right equilibrium price, give enough incentive for new capacity investments and, at the same time, prevent excessive profits of the incumbent generation companies. The supply shortage is expected to last until 2013, after which the total capacity is set to increase by almost 10 GW.

4.5 Other Issues

4.5.1 Future of Nuclear Power

After the nuclear accident in Fukushima caused by the tsunami's knocking down the auxiliary cooling facility, serious concerns and discussions became apparent in Korea's future energy mix. Some NGOs started to claim that the energy mix in Korea should be managed without the operation of nuclear power. A wider use of new and renewable energy sources should be an alternative to nuclear power. However, this would mean a huge increase in electricity price. Accommodating over 30% of renewable power source into the existing grid would also be a great challenge considering that Korea is currently an isolated island in terms of international power interconnection.

Recent research about the nuclear sectors of three northeastern Asian countries, Korea, China, and Japan, reveals that there is an urgent need to have a legal framework for the future development of nuclear power in this region. In December 2011, the National Energy Administration (NEA) of China said that China will make nuclear energy the foundation of its electricity generation system in the next 10−20 years, adding as much as 300 GW of nuclear capacity over that period.[43] With more than half of such plan based on inland nuclear power plants, and the prevailing westerly wind from

---

43. According to Min et al. (2012), currently available figure for China's nuclear plan is 284.461 MW which is a much higher figure than reported by WNA (2011b).

the eastern part of China toward Korea and Japan, the safety issue of nuclear power is not a local issue any more (Table 22.9).[44]

4.5.2 Climate Change and Smart Grid

The issue of climate change poses a challenge to the current electricity market system. As mentioned earlier, the link between the consumers and the producers does not exist in the current CBP market. So the regulators should adjust the retail prices to incorporate the environmental costs associated with the climate change. But can the regulators perform this task properly? Can they design and implement environmental surcharges that accurately reflect the environmental costs incurred by various groups of consumers?

The RPS duty is imposed on the electricity generation companies instead of on the retailers in Korea. Maybe it's because of the concern that the regulated monopsony of KEPCO in the generation market would not promote a competitive and liquid market for renewable generation. But since the generation market is also highly concentrated on a few incumbent companies, putting the incumbents in charge of the renewable generation has the risk that they leverage the existing market power to dominate the renewable generation market.

Smart grid offers both opportunities and challenges for the electricity industry reform. Smart grid can transform electricity into a highly differentiated product by enabling to measure and control individual customers' electricity consumption in real time. This will surely strengthen demand response in all stages of the electricity market. Smart grid also plays a key role in promoting renewable electricity generation as it enables the power system to accommodate distributed generation. As explained in Section 2.2, the government announced many plans to promote smart grid in Korea. However, it is not certain how these plans will proceed in coordination with the electricity sector restructuring. Although the First Master Plan for Smart Grid mentioned the need to open the retail market to realize the full potential of smart grid, no concrete plan for the future of restructuring has been announced yet.[45]

4.6 Some Proposed Improvements

The government and KPX are seeking to improve the current market design and practices in order to address the above-mentioned problems. To prevent further expansion of the retail rate deficit, the government announced a plan

---

44. The cooling tower of inland nuclear power plants is designed to be 200 m high × 160 m wide.

45. The private companies participating in the Jeju pilot program wanted the retail license at least for this target region so that they could develop new business models around electricity retail. But the request was denied.

TABLE 22.9 China, Japan an Korea in Ranking of World Nuclear Power Capacity (Unit: MW)

	Operable		Operable + Under Construction		Operable + Under Construction + Planned		Operable + Under Construction + Planned + Proposed	
1	USA	101,421	USA	102,639	USA	109,839	China	222,971
2	France	63,130	France	64,850	China	99,971	USA	148,439
3	Japan	44,642	Japan	47,398	France	66,570	Russia	76,044
4	Russia	23,084	China	39,981	Japan	61,170	India	72,985
5	Germany	20,339	Russia	32,044	Russia	48,044	Japan	67,930
6	Korea	18,785	Korea	24,585	Korea	32,985	France	67,670
7	Ukraine	13,168	Germany	20,339	India	23,985	Ukraine	37,868
8	Canada	12,679	Canada	14,179	Germany	20,339	Korea	32,985
9	China	11,271	Ukraine	13,168	Canada	17,479	UK	29,425
10	UK	10,745	UK	10,745	UK	17,425	Canada	21,279
Total	World	371,584	World	432,918	World	605,013	World	993,318

Source: WNA (2011b)

to introduce a scheme to tie the retail rate change to the fuel cost change. KPX is considering changing the bidding rule to allow the generation companies limited price bidding within a certain band in the Pool. This will eliminate the need to determine the bidding prices of each power plant administratively. KPX is also contemplating ways to replace the wholesale price control with some kind of vesting contracts. Well-designed vesting contracts should be able to prevent unfair transfer of wealth from consumers to generation companies without distorting the market incentives. They should also facilitate future retail competition by establishing nondiscriminatory treatment among retail suppliers. These changes, if successfully implemented, may be able to improve the efficiency of the CBP market. But unless they are followed by more fundamental structural changes, they will be just stopgap measures.

5 CONCLUSION

The restructuring of the electricity industry in Korea began more than 10 years ago but could not finish its planned course. As a result, Korea now has a unique system in which generation companies sell electricity to a state-owned retail monopoly, which is the parent company of most of the generation companies. The generation competition takes place in a strictly controlled mandatory pool, and the retail operation is fully regulated.

The biggest problem with the current market structure and design is that they were not intended as the final destination but for use during a short transitional period. As such, they are not comprehensive and self-contained. They depend on administrative decisions almost as much as before the restructuring. Yet, the regulatory system is not prepared to take on such tasks. This often results in confusion, indecision, and uncertainty about the future, which distort the market and impede investments. Lately, the retail rates have declined below the cost, and the retail company (KEPCO) is accumulating huge losses. At the same time, the country is experiencing power shortage during peak hours. It is imperative to put the market and the regulatory system in order as soon as possible.

Many things have happened in the electricity industries around the world during the past decade. Smart grid has emerged to revolutionize the electricity industry, and climate change has become an imminent threat. Restructuring and liberalization of the electricity markets made progress in many parts of the world, while problems and setbacks were encountered in some others. Vast amount of valuable lessons from these experiences have accumulated as compiled in this book. If we could take best advantage of these lessons to upgrade our electricity industry, the delay of restructuring with all the debates and controversies could turn out to be something positive.

REFERENCES

ABN AMRO Rothschild, 1998. Final Report to KEPCO on the Restructuring of the Electric Supply Industry in Korea, 12.

Defeuilley, C., 2009. Retail competition in electricity markets. Energ. Policy 37, 277–386.

Electric Power Statistics Information System (EPSIS). Available from: <http://epsis.kpx.or.kr/> (accessed 14.10.12).

Government of the Republic of Korea, 2012. The 1st Master Plan for Smart Grid, June. (in Korean).

IEA, 2010. CO_2 Emissions from Fuel Combustion, IEA Statistics. OECD/IEA.

Joskow, P., 2000. Why do we need electricity retailers? Or can we get it cheaper wholesale? Discussion Draft, MIT, Center for Energy and Environment Policy Research, Massachusetts.

KEMA Consulting, 2000. KEPCO Restructuring Program.

Kim, D., Cho, C., Cho, H., Knittel, C., 2006. The characteristics of Korean electricity industry and the efficiency of its restructuring. Research Paper, Korea Institute for Economics and Trade, December 2006 (in Korean).

Kim, J., Kim, K., 2008. The Electricity Industry Reform in Korea: Lessons for Further Liberalization. Mimeograph.

Kim, Suduk, Sonn, Yang-Hoon, 2008. An empirical analysis on the determination process of wholesale power price: with its focus on capacity payment and settlement price. Korean Energy Econ. Rev. 7 (2), 27–52 (in Korean).

Kim, Y.J., 2002. Opening the 21st Century for the Electricity Industry, December 2002 (in Korean).

Korea Development Institute (KDI), 2010. A Recommendation on Electricity Industry Restructuring Policy (in Korean).

Korea Energy Economics Institute (KEEI), 2010. Energy Balance Table.

Korea Energy Economics Institute (KEEI), 2011. Energy Info. Korea.

Korea Energy Statistics Information System (KESIS). Available from: <http://www.kesis.net/>.

Korea Power Exchange, Electric System Operation Report for July 2012, August 2012 (in Korean).

Korea Power Exchange, Electricity Market Operation Rules, 2011, December (in Korean).

Lee, S.H., 2011. Electricity in Korea, Presented in Symposium on APEC's New Strategy for Structural Reform, May 2011.

Lim, W., 2004. Electricity industry restructuring: key issues and alternatives. Report, Korea Development Institute, December 2004 (in Korean).

Littlechild, S., 2000. Why we need electricity retailers: a reply to Joskow on wholesale spot pass-through. Working Paper, Department of Applied of Economics, University of Cambridge, Cambridge.

Min, Eunju, Yanping, Zhang, Hyun-Goo, Kim, Suduk, Kim, 2012. Can China, Korea and Japan avoid the controversy over nuclear energy. J. Energy Develop. 37 (2), 143–178.

Ministry of Knowledge Economy (MKE),2010. The 5th Electric Power Demand–Supply plan (2010–2024). MKE publ. No. 2010–490 (in Korean).

Ministry of Knowledge Economy (MKE), 2008. The 3rd Master Plan for the Technology Development, Utilization and Promotion of New and Renewable Energy (2009–2020), December (in Korean).

PJM, 2009. A Review of Generation Compensation and Cost Elements in the PJM Markets. PJM, Pennsylvania.

Sonn, Y.H., Chung, T., 1993. A Study on the Economies of Scale in the Electricity Industry, Policy Report 93–07, Korea Energy Economics Institute (in Korean).

World Nuclear Association (WNA), 2011a. Nuclear Power in Korea, Updated 1st August. Available from: <www.world-nuclear.org>.

World Nuclear Association (WNA), 2011b. Nuclear Power in China, Japan, South Korea, Updated August 1, 2011. Available from: <http://www.world-nuclear.org/>.

Chapter 23

After Fukushima: The Evolution of Japanese Electricity Market

Hiroshi Asano[a] and Mika Goto[b]

[a]Central Research Institute of Electric Power Industry (CRIEPI), and the University of Tokyo;
[b]Central Research Institute of Electric Power Industry (CRIEPI)

1 INTRODUCTION

Japan has few energy resources and is dependent on imports for 96% of its primary energy supply. Even with nuclear energy included as domestic energy, energy import dependency is still at 82% (Energy White Paper, 2010). Thus, Japan's energy supply structure is extremely vulnerable. Following the two oil crises in the 1970s, Japan diversified its energy sources through increased use of nuclear energy, natural gas and coal, as well as the promotion of energy efficiency and conservation.

Despite these improvements, oil still accounts for about 40% of Japan's primary energy supply, and nearly 90% of imported oil comes from the politically unstable Middle East (Energy White Paper, 2010). Moreover, prospects for importing electricity from neighboring countries are not realistic because Japan is an island nation. In addition, since the Kyoto Protocol came into effect in 2005, there has been an urgent need to implement global warming countermeasures, such as reducing carbon dioxide emissions from the use of fossil fuels. To ensure Japan's stable electricity supply under these environmental constraints, it is crucial to establish an optimal combination of power sources that can concurrently deliver energy security, economic efficiency, and environmental conservation. After experiencing the Great East Japan Earthquake and nuclear accident at Fukushima Daiichi nuclear power plant in March 2011, Japan has been challenged to drastically change its energy policy toward more renewable resources and less dependency on nuclear energy.

To establish an "Innovative Strategy for Energy and the Environment," hereafter referred to as ISEE, the government has been discussing what would be the best mix of energy sources, robust power supply systems, and nuclear power policy suited for Japan. In particular, detailed studies on the cost of the nuclear fuel cycle and the amount of renewable energy that can be introduced across the country are needed, while thoroughly implementing safety measures for nuclear power generation.

This chapter is organized as follows: Section 2 describes the historical development of the Japanese electric power industry and market liberalization since 1995 when the Electricity Business Act was amended and competition was introduced in generation market. Section 3 explains the government's ISEE including the current debate on Japan's future energy options. Section 4 covers current and future issues related to the realization of a sustainable society under the new strategy including reform proposals for the electric sector followed by the chapter's conclusions.

2 HISTORICAL DEVELOPMENT OF JAPANESE ELECTRICITY SECTOR

2.1 Before World War II

In Japan, Tokyo Electric Lighting, as the nation's first electric power company (EPCo), commenced operations in 1886 as a private company, and began supplying electricity to the public in the following year. By 1896, 33 electric utilities were established throughout the nation.

The early twentieth century marked the establishment of long-distance transmission technology. As larger thermal and hydropower plants were introduced, generation costs fell and electricity came into wider use throughout the country. Consequently, electricity became an indispensable energy source for peoples' lives and the industry.

In the years that followed, the electricity utility business grew in tandem with the modernization of Japan and its industry. At the same time, the electric utility industry experienced a major restructuring that led to the dissolution of 700 electric utilities, which merged to create five large electric utilities after the First World War. During the Second World War, the electric utility industry was completely state-controlled and utilities were integrated into Nihon Hatsusoden Company, a nationwide generating company, and nine distribution companies.

2.2 After World War II

After the end of World War II in 1945, electricity supplies were tight in Japan leading to the 1950 Electricity Utility Industry Reorganization Order, which divided the country into nine service areas in May 1951, each served by a vertically integrated EPCo that generated power, transmitted and distributed to end users. This resulted in the nine regional monopoly, investor-owned General Electricity Utilities (GEUs)—Hokkaido, Tohoku, Tokyo, Chubu, Hokuriku, Kansai, Chugoku, Shikoku, and Kyushu EPCos, which remain the dominant players to this day.

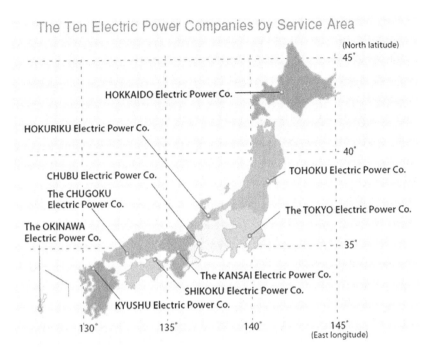

FIGURE 23.1 Ten regional power companies in Japan. *Source: FEPC (2013) (The Federation of Electric Power Companies of Japan).*

This arrangement was reaffirmed by the Electricity Business Act in 1964 and remains in place to this day. With the return of Okinawa to Japan in 1972, Okinawa Electric Power Company became the 10th GEU in Japan. The 10 EPCos are referred to as GEUs when appropriate. Figure 23.1 and Table 23.1 show their supply areas, generation capacity, and fuel mix.

Figures 23.2 and 23.3 show the evolution of installed capacity and generation in Japan for the period 1952−2009 showing the growth of nuclear, coal, and gas generation from LNG imports while the share of oil has declined over time.

Until the amendment of the Electricity Business Act in 1995, Japan's electric power industry was composed of the GEUs and the following wholesale electric utilities: Electric Power Development Company, Ltd., fully privatized in October 2004, and referred to below as J-POWER, Japan Atomic Power Company (JAPC), and various wholesale electric utilities such as joint thermal power generation companies, all of which have received investment from GEUs and publicly owned hydroelectric power generators.

This arrangement contributed to a reliable power supply in Japan. In the 1990s, prompted by restructuring developments in other parts of the world, the Japanese government decided to introduce competition in the electricity sector as described by Goto and Yajima (2006). The intent of the 1995

TABLE 23.1 Generation fuel mix in 10 GEUs in 2010

Year	Company Fuel	Hokkaido	Tohoku	Tokyo	Chubu	Hokuriku	Kansai	Chugoku	Shikoku	Kyushu	Okinawa	Total
						Licensed Maximum Capacity (MW)						
2010	Hydro	1234	2423	8981	5219	1904	8196	2906	1141	3279	0	35,282
		16.63%	14.08%	13.82%	15.90%	23.64%	23.50%	24.24%	16.39%	16.13%	0.00%	17.08%
	Thermal	50	224	3	0	0	0	0	0	210	0	487
		0.67%	1.30%	0.01%	0.00%	0.00%	0.00%	0.00%	0.00%	1.03%	0.00%	0.24%
	Fossil fuel	4065	11,286	38,696	23,969	4400	16,907	7801	3797	11577	1919	124,417
		54.80%	65.59%	59.54%	73.01%	54.61%	48.48%	65.08%	54.53%	56.95%	99.97%	60.23%
	Nuclear	2070	3274	17,308	3617	1746	9768	1280	2022	5258	0	46,343
		27.90%	19.03%	26.63%	11.02%	21.67%	28.01%	10.68%	29.04%	25.86%	0.00%	22.43%
	Wind	0	0	1	22	5	0	0	0	3	0	32
		0.00%	0.00%	0.00%	0.07%	0.07%	0.00%	0.00%	0.00%	0.02%	0.03%	0.02%
	Photovoltaic	0	0	0	1	1	6	0	2	3	0	13
		0.00%	0.00%	0.00%	0.00%	0.01%	0.02%	0.00%	0.03%	0.01%	0.00%	0.01%
	Total	7419	17,206	64,988	32,828	8057	34,877	11,986	6963	20,330	1919	206,575

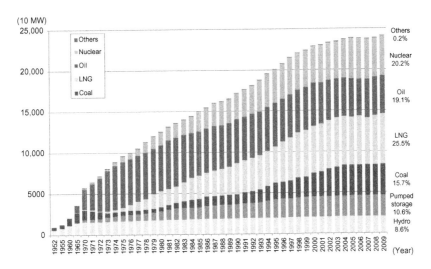

FIGURE 23.2   Installed capacity of 10 Japanese GEUs. *Source: Energy White Paper (2010).*

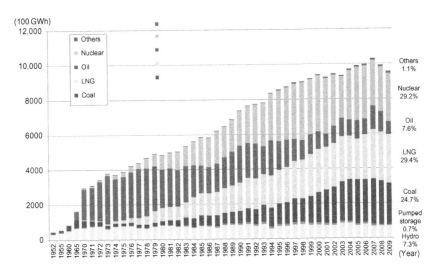

FIGURE 23.3   Generated electricity of 10 Japanese GEUs. *Source: Energy White Paper (2010).*

Electricity Business Act, which was the first comprehensive amendment in 30 years, was to reduce electricity prices to internationally comparable levels through competition among the stakeholders:

- The Act introduced partial competition in the generation sector by allowing independent power producers (IPPs) to participate in the wholesale

TABLE 23.2 Historical Development of Regulatory Reform in Japan

First Institutional Reform: 1995

- Introduction of competition into generation sector; auction of new power plant procurement
- Foundation of new system that allows new entrants to supply electricity to customers in specified area

Second Institutional Reform: 1999

- Partial liberalization of retail electricity market for extra-high-voltage users with maximum demand of 2000 kW
- Shift from approval system to notification system of rate changes when the rate decreases

Third Institutional Reform: 2003

- Gradual extension of retail liberalization to high-voltage users with maximum demand of 50 kW
- Establishment of neutral organization for monitoring of transmission (ESCJ)
- Establishment of wholesale power market (JEPX)

Fourth Institutional Reform: 2008

- Wholesale power market reform for activating power trading
- Improvement of competition conditions on transmission usage for new entrants
- No extension of retail market liberalization; reexamination after 5 years

Note: ESCJ, Electric Power System Council of Japan; JEPX, Japan Electric Power Exchange.

market. This was, at least on paper, tantamount to liberalization of wholesale supply;

- The Act authorized "special electric utilities" to engage in electricity retailing in designated service areas; and
- Encouraged GEUs to improve operational efficiencies through the introduction of a competitive yardstick for assessing their performance when they applied to raise their tariffs.

Since 1995, the electric power market in Japan has been progressively liberalized in stages in 1999, 2003, and 2008 as shown in Table 23.2. This has resulted in the gradual growth of IPPs supplying vertically integrated utilities through competitive bidding and long-term bilateral contracts. IPP generating capacity has grown to 6.6 GW, roughly 5% of total thermal-installed capacity.

The second institutional reform introduced in March 2000, partially liberalized the retail market by allowing Power Producer and Suppliers (PPSs) to sell electricity to extra-high-voltage users with contracted demand of over 2 MW. At the same time, access to the transmission/distribution network owned by the EPCos was liberalized to ensure network neutrality.

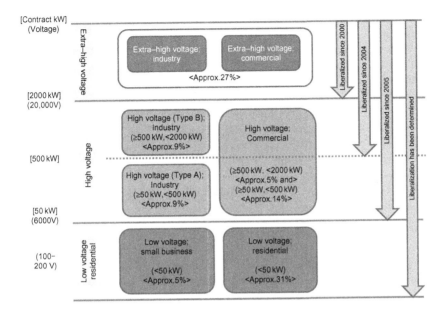

FIGURE 23.4   Progress of retail liberalization. Note: Numbers in < > indicate shares in electricity demand.

The scope of retail liberalization was subsequently expanded in April 2004 to users with contracted demand of 500 kW or more and subsequently extended to customers with 50 kW or more in April 2005 as a result of the third institutional reform.

As part of the fourth institutional reform in 2008, the pros and cons of further expanding the scope of retail liberalization to the residential sector was considered and it was decided that the subject should be reexamined after around 5 years.[1] Currently, approximately 60% of total electricity demand in Japan has been liberalized since the last expansion of eligibility in 2005 (Figure 23.4).

Ministry of Economy, Trade, and Industry (METI), the ministry responsible for market reform, released two reports in March and April 2008 with the following proposals:

- Creation of an hours-ahead market as a means of procuring power sources in the event of unforeseen imbalances between supply and demand;
- Encouragement of forward transactions[2];
- Reexamination of methods of calculating imbalance charges;

---

1. The reason for the decision was that effects of retail competition in the residential sector were unclear and metering costs were expensive compared with expected benefits.

2. These two items concern the development of a competitive environment for power generation and operation of the wholesale electricity market.

- Reporting of receipts and payments related to imbalance charges by GEUs[3];
- Review of the criteria governing the issuance of orders to change the rules for wheeling service charges and clarification of the disposition of excess profits that are generated in the transmission and distribution sectors[4];
- Development of two coordination processes for the construction of interconnected transmission lines to support supply stability through wide area interchange by the Electric Power System Council of Japan (ESCJ);
- Improvement of the institutional infrastructure to ensure supply stability in liberalized markets[5]; and
- Experimental trial of "CO_2-free electricity" and Kyoto Mechanism credit transactions.[6]

ESCJ was established with the third institutional reform in 2003 as the sole private organization to make rules and supervise operations from a neutral position and to maintain fair and transparent use of the electric transmission and distribution system, which the GEUs own and operate. ESCJ started full-scale operation in April 2005 and supervises GEU's network operation and congestion management of intertie lines. ESCJ has also proposed expansion plans of intertie lines and frequency converter capacity.[7]

While maintaining the vertically integrated GEUs, the Act introduced a "code of conduct" for the stakeholders, including the prohibition of the use of information acquired by the transmission and distribution sectors through wheeling services for purposes other than the original intent; prohibition of discriminatory treatment concerning wheeling services; and prohibition of internal cross-subsidies by GEUs between the transmission/distribution sector and other power supply sectors, that is, accounting separation. This left the GEU monopolies intact while promoting the PPSs.

As a guide to future direction of reforms in Japan, the Basic Energy Plan (BEP) adopted by the cabinet in March 2007 states the following objectives:

---

3. These two items concern the implementation of a system for immediate supply balancing and payment adjustment in the event of imbalances.

4. This item concerns the appropriate tariff structure of wheeling services.

5. These two items concern supply stability.

6. This item concerns the environmental compliance of the electric power industry.

7. The ESCJ's role is not limited to the supervision of GEUs' operations. Their role includes formulating various regulations governing electric power systems, monitoring market activities, and mediating and arbitrating disputes concerning matters related to electric power systems including transmission service. In June 2004, the METI designated ESCJ as the "organization responsible for supporting power transmission, distribution, and related services" under Article 93 of the Electricity Business Act. ESCJ's membership includes neutral members, such as academics, GEUs, PPSs, wholesale electric utilities, wholesale suppliers, and self-generators.

FIGURE 23.5   Basic structure of JEPX.

First, a stable supply of electricity should be promoted with the GEUs as the key
players, acting as responsible suppliers providing a reliable supply of electric power
in a vertically integrated system extending from generation to transmission and
distribution. Second, regulations should be applied properly and a code of conduct
(such as prohibition of the use of information for purposes other than the original
intent) should be implemented in an appropriate manner to ensure fair, transparent
access to transmission and distribution networks.

In addition, Japan Electric Power Exchange (JEPX) was set up as a private, voluntary wholesale exchange to stimulate trade and establishing market prices to help stakeholders in assessing investment risks. Participants in JEPX include the GUEs (except Okinawa, which is not connected to the main network), J-POWER, PPSs, and self-generators.

JEPX offers "spot market" products for trading electricity to be delivered the next day in 30 min intervals, "forward market (fixed-form)" products for trading electricity at 1-month and 1-week intervals, and "forward market (bulletin board)" products for free trading of electricity for future delivery. In line with the recommendations described above, JEPX also began trading "CO_2-free electricity" and Kyoto Mechanism credit transactions in November 2008, and an hours-ahead (intraday) market in September 2009. Figure 23.5 depicts basic structure of JEPX as of October 2012.

The volume of trade on JEPX has gradually increased. In fiscal 2010, the average daily volume of spot market transactions was 15,070 MWh, up from 9710 MWh in fiscal 2009, and the average daily trading price was 8.4 yen/kWh, up from 6.5 yen/kWh in fiscal 2009.[8] The total volume of transactions conducted on the spot market in 2010 was 5,501,210 MWh, up from 3,545,120 MWh in 2009. Despite the growth, this accounted for approximately 0.5% of the total volume of retail electricity sales in Japan primarily because electricity transactions among GEUs, wholesale electricity utilities, and wholesale suppliers, which account for a large fraction of power trading, are primarily conducted through bilateral contracts.

---

8. It is approximately 8.1 cent/kWh under the exchange rate of 80 yen/US$.

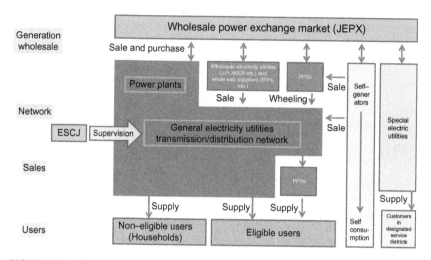

FIGURE 23.6   Current electricity supply system in Japan as of August 2012 (effective April 1, 2006).

As shown in Figure 23.6, there are currently 10 investor-owned vertically integrated utilities responsible for supplying electricity to consumers in their respective service areas. GEUs must obtain approval from the Japanese government by providing supply conditions such as electricity rates as "general supply provisions" to those consumers who are excluded from the retail liberalization. The regulated utilities are also responsible for supplying electricity to consumers subject to retail liberalization, based on the "provisions for last resort service" if they cannot secure contracts with PPSs.

The entry of PPSs has introduced an element of competition in the generation market but the monopolies have managed to maintain their market share for the most part. Currently, 53 PPSs are registered and 27 supply electricity to retail customers. Total generation capacity of PPSs increased from 940 MW in 2006 to 2010 MW in 2010, representing 0.71% market share. There are a number of reasons for the limited success of PPSs to date including the fact that most of them rely on gas-fired generation and recent increases in gas prices have hit their profitability. Although the share of electricity supplied by PPSs has remained limited, the mere potential threat of competition has resulted in gradual decreases in retail tariffs to both residential and industrial consumers as illustrated in Figure 23.7.

The competitive pressures stemming from the market reforms have forced the incumbent utilities to increase their operational efficiency while lowering electricity rates and offering a variety of pricing plans. These developments have resulted in a 15% drop in the system average prices from 1990 to 2011 as shown in Figure 23.7. Part of the savings may be attributed to drastic cuts in equipment investment as illustrated in Figure 23.8 since 1990.

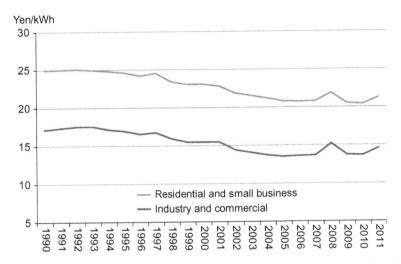

FIGURE 23.7 Trends of system average prices of electricity. *Source: Author's calculation based on data from FEPC Electricity Statistics.*

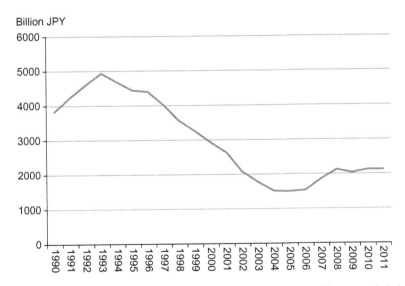

FIGURE 23.8 Trend of equipment investment in total of 10 GEUs. Note: The amount includes extension work, improvement work, and others. *Source: Author's calculation based on data from FEPC Electricity Statistics.*

3 AFTER FUKUSHIMA: INNOVATIVE STRATEGY FOR ENERGY AND THE ENVIRONMENT

After the Great East Japan Earthquake in March 2011, eastern Japan faced severe electricity shortages because power plants were damaged by the

FIGURE 23.9   National transmission grid in Japan. *Source: JEPIC (2011).*

earthquake and tsunami. Meanwhile, the western half of Japan, which did not suffer from the earthquake and tsunami, could not help eastern Japan because of limited capacity of frequency converters between the 50 Hz eastern grid and the 60 Hz western grid as shown in Figure 23.9.[9]

Converting power from one system to the other is a complex and expensive task. Japan has only a few frequency converter facilities with limited capacities. This has not been a major issue in the past because there were enough power plants in each region with limited need to shift power around during spikes in demand or plant outages. The massive unplanned shortage of power following the accident at the Daiichi nuclear power plant has raised interest to expand the capacity of frequency converters between the 50 and 60 Hz systems.[10]

Aside from the limited frequency converters, the Japanese government started discussions on the country's future energy with essentially a clean slate. Before the earthquake, the BEP, which was revised in June 2010, envisioned increasing the share of nuclear power generation up to 50% by 2030 to ensure energy security in Japan and to contribute to combating climate

9. This uncommon situation is the legacy of history: eastern Japan built its grid based on the German 50 Hz system and western Japan followed the American 60 Hz system.
10. For the time being, the government has proposed to enhance the capacity of existing facilities and improve their operation.

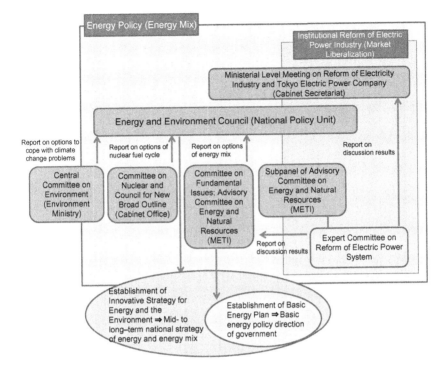

FIGURE 23.10   Discussion process toward ISEE.

change. That was before the accident. As of October 2012, only two nuclear units with capacity of 2360 MW are operating in Japan[11] with the other 48 units totaling 43,788 MW have stopped operations.

In this context, The Energy and Environment Council was established in June 2011 under the National Policy Unit, which promotes cross-ministerial planning and coordination, to draft Japan's future energy policy as schematically shown in Figure 23.10.

The main focus of the debate is the future share of nuclear power generation in the energy mix with three scenarios under consideration: 0%, 15%, and 20–25% by 2030 as shown in Table 23.3. In September 2012, the Japanese government announced a proposal, in which the nuclear power is to be phased out by 2030s, followed by a second announcement 1 week later stating that the government has decided to postpone a formal decision on the nuclear phase out policy, proposing a nonbinding target. The lack of clear

---

11. The two operating units, Ooi Nos. 3 and 4, are owned by Kansai EPCo, whose demand and supply conditions in summer 2012 were extremely precarious causing the Japanese government to allow the units to resume operation to avoid a risk of power shortage.

TABLE 23.3 Three Scenarios of the Mix of Electric Energy in 2030

	2010	2030				
		0% scenario				
		Before Additional Measures	After additional Measures	15% Scenario	20–25% Scenario	Current Strategic Energy Plan of Japan
Share of nuclear energy	26% Note 1	0% (− 25%)	0% (− 25%)	15% (− 10%)	20–25% (− 5 to −1%)	45%
Share of renewable energy	10%	30% (+ 20%)	35% (+ 25%)	30% (+20%)	30–25% (+ 20 to +15%)	20%
Share of fossil fuels	63%	70% (+ 5%)	65% (current level)	55% (− 10%)	50% (− 15%)	35%
Share of nonfossil energy resources	37%	30% (− 5%)	35% (current level)	45% (+ 10%)	50% (+ 15%)	65%
Electric energy generated	1100 TWh	Approx. 1000 TWh (− 10%)	Approx. 1000 TWh (− 10%)	Approx. 1000 TWh (− 10%)	Approx. 1000 TWh (− 10%)	Approx. 1200 TWh
Final energy consumption	390 million kl	310 million kl (− 72 million kl)	300 million kl (− 85 million kl)	310 million kl (− 72 million kl)	310 million kl (− 72 million kl)	340 million kl
Greenhouse gas emissions Note 2 (compared to 1990)	−0.3%	− 16%	−23% (− 21%)	−23% (− 22%)	−25% (− 25%)	(Around −30%)

Note 1: The share of nuclear energy under the current Strategic Energy Plan of Japan (53%) is the share of large-scale power sources (excluding cogeneration and self-generation).
Note 2: Figures in parentheses indicate only energy-related CO_2 emissions.
Source: National Policy Unit (2012)

signals increases the uncertainties about the future energy policy and nuclear energy's role in Japan.

4 CONTINUED REFORM OF THE ELECTRIC POWER SYSTEM

4.1 Current Reform Plan

Despite the ambiguity of government policy on nuclear power, there is interest to proceed with market reform process primarily focused on unbundling of transmission and introduction of full retail competition. This has been among the goals of a committee established in January 2012, which published an interim report in July 2012 with proposals on demand-side reform, supply-side reform, and transmission/distribution reform. The committee is expected to continue its deliberations including detailed institutional designs and operations of the new system.

Among its findings, the interim report concluded that full retail liberalization by expanding the current boundary for customer eligibility to small commercial and residential users was necessary. Unbundling of transmission/distribution from the other functions and reform of the wholesale power market are among other important reform issues that are expected from the committee before the end of 2012. This section provides a brief summary of the important topics under consideration.

4.1.1 Wholesale Power Market

The new electric power supply system under consideration would consist of:

- Full liberalization of the generation sector—Currently, long-term wholesale transactions are regulated based on cost-plus tariff regulation and supply obligations. In parallel with full retail liberalization, these regulations are expected to be abolished resulting in increased business autonomy in wholesale transactions leading to more vigorous competition in generation.[12] The scheme will be implemented in stages over time.
- Enhancement of wholesale power transactions—Under the new system, it is expected that GEUs will participate in the wholesale power market leading to increased trading volume at the JEPX more efficient use of generation capacity at the national level. For example, GEU generation capacity beyond the level needed to meet the reserve requirements would be traded in the wholesale power market. In addition, new policies are needed by which wholesale electricity utilities and wholesale suppliers can sell electricity to markets and new entrants.

---

12. Tokyo Electric Power Company (TEPCO) is expected to conduct a competitive bidding of IPPs to supplement the lost capacity of Fukushima Daiichi nuclear power plant in 2013. These IPPs will start operation between 2019 and 2022 under long-term bilateral contract with TEPCO.

- Energy savings as an alternative generation capacity—Under the new scheme, demand-side resources would be treated on par with supply-side options allowing "negative load" or demand bidding by consumers to effectively participate in the wholesale market.
- Procurement of supply and reserve capacity—In the current system, GEUs and wholesale electric utilities who have invested in generation capacity are essentially ensured to recover construction and maintenance costs based on supply obligation and cost-plus tariff regulation.

In this context, a real-time market has been proposed whereby power trading would be conducted until 1 h before dispatch.[13]

To further promote competition, several additional provisions have been proposed including "partial supply rules" by which GEUs and new entrants share responsibility to supply power to an entity by the type of load and time of supply. In addition, the payment of new entrants for receiving constant backup service from GEUs is reduced to the level of base load generation cost.

Under the proposed reform, the generation sector will become subject to competition and this raises concerns about adequate investment in generation capacity. The committee has considered the necessity of introducing capacity remuneration mechanisms. Reliability pricing model in Pennsylvania, New Jersey, Maryland Interconnection (PJM) and other capacity market design are surveyed to assess their effectiveness in the Japanese context.[14] But even if a capacity remuneration mechanism is introduced, there are concerns about long-term capacity procurement and its cost recovery. In the new system, the proposed wide area transmission organization that covers the whole nation needs to evaluate future demand and necessary generation capacity. When lack of capacity is anticipated, ESCJ or a new system operator needs to be responsible to secure the construction of generation plants and ensure cost recovery. The cost recovery mechanism is proposed in a manner that the costs are embedded in the transmission surcharges.

4.1.2 Unbundling of Transmission

As described in Section 2, the Japanese electric power industry consists of 10 GEUs, which are vertically integrated from generation to retail sales. For the purpose of neutral transmission functions, the ESCJ was established and has been operating since 2005. However, the Expert Committee on Reform of the Electric Power System has proposed two types of unbundling options in its July 2012 report. In both options, the ESCJ will be abolished and replaced by a new wide area transmission organization. The purpose of the

---

13. Under the current JEPX system, market participants can trade electricity until 4 h before dispatch.
14. Chapter 9 describes PJM's capacity market.

unbundling is to promote competition in generation and retail sectors. To realize this goal, the committee has identified four issues for further discussion, briefly described below.

1. Wide area operation of the transmission sector—Under the current system, each GEU procures generation capacity for each supply area. The proposed reform will dissolve the ESCJ and establish a single nationwide transmission organization responsible for (a) power system planning, (b) operation of transmission lines, (c) supervision of regional grid operators, and (d) operation of the wholesale electricity market.

2. Neutralization of the transmission sector—In a liberalized market, various types of entities should have nondiscriminatory access to the transmission facilities. As described above, the proposed wide area transmission organization will operate the nationwide transmission service independently, neutrally, and transparently. Moreover, the neutrality of transmission/distribution operations will be required within each GEU region, where GEUs will have to choose between (a) an Independent System Operator (ISO)-type functional separation or (b) a legal separation of transmission/distribution.[15]

Under the ISO-type functional separation, the roles of regional transmission investment planning and operation are transferred to the wide area transmission organization. A disadvantage of this approach is the separation of an entity that conducts investment planning and operation from one that owns and maintains the network assets.

Legal separation requires establishing an independent company responsible for transmission investment planning, operation, ownership, and maintenance. Under this approach, all the roles from transmission investment planning to maintenance are controlled by a single entity enabling these functions to be better coordinated. However, there is the concern that such an entity may not treat all transmission users fairly and evenly, which requires rigorous monitoring. The regulatory costs of such supervision may exceed the benefits.

Regardless of which approach is adopted, a comprehensive code of conduct covering all important aspects of business particularly fair treatment of contracts among transmission/distribution, generation, and retail sectors needs to be in place. Full ownership unbundling is not discussed in the present reform process, possibly to be considered at a future date.

---

15. It is important to note that the legal separation discussed in Japan is not necessarily similar to that adopted in the third EU electricity directive or independent transmission operator (ITO) model, because the ITO model applies very strict rules to the operation of the ITO, particularly with respect to the business relationship between the generation and retail utility company and the ITO.

3. Strengthening of regional interconnection lines—It is necessary to strengthen regional interconnections to facilitate competition in markets. This is particularly important after the Great East Japan Earthquake because supply and demand conditions have become very tight under the unique constraints of eastern and western Japan using different frequencies, as well as the limited capacity of regional interconnection lines. In addition, a robust transmission system is necessary to integrate increasing shares of renewable energy. The capacity of frequency converters is expected to increase from 1200 to 2100 MW by 2020 and possibly 3000 MW thereafter.[16]

4. Revision of the balancing payment mechanism—Imbalance pricing is expected to be revised so that the pricing mechanism is more open and transparent.[17]

4.1.3 Retail Competition

The share of retail sales supplied by PPSs has increased to approximately 3.5% of total eligible users and approximately 4.2% for extra-high-voltage users as of 2010 (Figure 23.11). The share varies from region to region, for example, ranging from 0% of Hokuriku and Shikoku to approximately 6% of Tokyo—relatively low percentages compared to markets, such as Texas and Victoria, which have achieved more vigorous retail competition.[18] Several reasons may explain the small percentages:

- First, individual GEUs have optimized their operations to fit their own service areas under the long-term supply obligations, which means that they may not have strong incentive to compete beyond their own service areas.
- Second, as described in Section 2.2, PPSs mostly depend on expensive natural gas generation resulting in little or no competitive advantage.[19]

16. It is aimed to increase interconnection capacity from 600 to 900 MW between Hokkaido and Honshu as soon as possible to supply stable electricity. This expanded capacity will help additional operation of wind farm in Hokkaido which is good for wind generation. The cost would be recovered from wheeling service fees.

17. Under the current system, supply deviation is measured from the real demand within a time unit of 30 min. In the new system, this measurement method is retained for GEUs but options are provided to new entrants who can choose either the old method or a new method that calculates supply deviation from the planned generation amount within 30 min.

18. Numbers in Figure 23.11 are accumulated sales volume of customers who are supplied by new entrants compared to total sales, with no duplicated counts due to multiple changes of suppliers of same customers.

19. It is difficult for PPSs to construct plants with low variable costs, such as nuclear and hydro, because these require substantial initial investment. Construction of coal-fired power plants requires meeting stringent environmental requirements. Consequently, PPSs depend on self-generators and GEUs for more than 70% of electricity procurement, that is, 62% on self-generators and 9% on GEUs in 2010, whereas electricity generation by their own plants and procurement from JEPX are less than 30% of total.

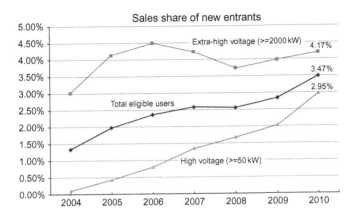

FIGURE 23.11   Trend of accumulated sales share of new entrants (PPSs). *Source: METI(a) (2012).*

- Third, since there are constraints in capacities of interconnection lines and frequency converters, it is costly for PPSs to supply customers beyond their immediate area.

Other concerns that require further study before introducing full retail competition include the issue of provider of last resort and supply for remote areas or isolated islands.[20] There are a number of options for the provider of last resort including (1) dominant supplier of the region, (2) retail supplier with a certain level of scale, and (3) regional transmission/distribution company. Discussion of these issues is underway and will be finalized by the end of 2012.

4.2 Future Direction of Japanese Energy System

4.2.1 Smart Meters and Demand Response

The introduction of smart grid and smart meters had been discussed before the Fukushima disaster to integrate renewable energies, encourage demand participation in markets, improve customer service, and enhance operational efficiency. The Fukushima disaster enhanced the business case and the necessity of a smart grid.

The operational capacity of Tokyo Electric Power Company (TEPCO) dropped by approximately 21 GW after the massive March 11, 2011 earthquake with little warning. The resulting shortage of supply resulted in rotating outages for 10 days from March 14 through March 28. The supply capacity, excluding Fukushima No. 1 and No. 2 nuclear stations, had been

---

20. Supplier of last resort is needed to protect consumers in case there is no supplier willing to provide service or a supplier goes bankrupt or disengages from business.

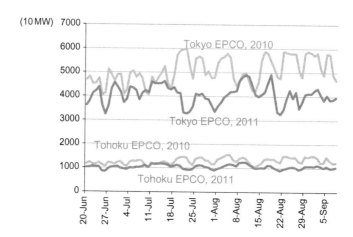

FIGURE 23.12   Daily peak demands in Tokyo EPCo and Tohoku EPCo in 2010 and 2011.

quickly restored, but it was predicted that the supply capacity would be short in the peak demand period in summer of 2011. Large-scale consumers were forced by law to save electricity by 15%.[21] Subsequently, the government and power companies changed the policy from the mandatory measures such as the national power-saving scheme to market-based demand response programs. Figure 23.12 shows daily peak demand in Tokyo and Tohoku service areas in 2010 and 2011. Figure 23.13 shows the adverse effects of electricity saving measures as surveyed by CRIEPI (2012).

The Energy and Environment Council proposed to promote the installation of smart meters, covering 80% of total electricity consumption within 5 years. The communication interface with the meters was standardized by the government in 2012 allowing home energy management systems to optimize energy usage within customers' premises.

The first case of negawatt trading or demand reduction by users was introduced in the Kansai area in the summer of 2012 involving a building energy management system and 16 aggregators to supply negawatts from commercial buildings. Total contracted demand of interruptible programs in the nine EPCo areas was 5110 MW, roughly 3.2% of summer peak demand in 2012.[22]

4.2.2 Supporting Renewable Generation

The Renewables Portfolio Standard (RPS) was enacted in April 2003 in Japan, where renewable energy is defined as electricity produced from

---

21. Customers' peak demand reduction during daytime compared with the previous year.
22. Total contracted demand for planned load adjustments in which large customers shift summer holidays was 6290 MW in summer of 2012 and actual peak demand reduction was 5290 MW, roughly 3.4% of peak demand of nine EPCos.

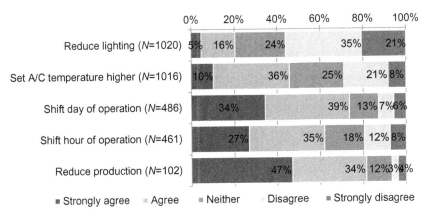

FIGURE 23.13   Had electricity saving actions adverse effect? *Source: CRIEPI Report Y12002 (2012).*

photovoltaics (PVs), wind, biomass, geothermal, and small hydropower. The initial RPS goal was 3.3 TWh, approximately 0.39% of total generation, raised to 11 TWh in 2011 or 1.3% of supply.

To further promote PVs, the government has set a goal of 28 GW PV deployment by 2020 and 53 GW by 2030 encouraged through feed-in-tariffs (FITs). A FIT scheme applicable to surplus energy from residential PV systems was introduced in November 2009. The support mechanism has three main features: first, the price is fixed at 48 yen/kWh, double the former level, and applies for 10 years. Second, the price is revised annually and gradually reduced. Third, the cost of the scheme is recovered from all consumers through a surcharge on electricity rates. The FIT surcharge in 2012 was around 0.22 yen/kWh.

An additional FIT Act was introduced in July 2012 to purchase all output of any renewable resource replacing the prior RPS scheme.

The renewable support scheme has resulted in a construction boom and job creation in solar installations in the devastated areas following the Fukushima accident by virtually guaranteeing a reasonable profit to investors. As of March 2012, there were about 80 solar parks in operation, under construction or planned with a capacity of 500 MW—approximately one-third owned by EPCos.

As further described in other chapters in this volume, coordination and expansion of transmission system as well as sufficient backup and storage will be required to integrate large amounts of wind and PV into the network. Another concern is the increased financial burden that the new FIT would place on consumers, who already pay high retail tariffs by international standards. Table 23.4 shows the current purchase prices and internal rate of return (IRR) for each technology defined for Fiscal Year 2012.

TABLE 23.4 Conditions on Purchase in FIT for FY2012

Generation Technology	Photovoltaics		Wind		Geothermal		Small Size Hydro		
Group	≤10 kW	<10 kW (surplus)	≤20 kW	<20 kW	≤15 MW	<15 MW	≤1 MW, <30 MW	≤0.2 MW,<1 MW	<0.2 MW
IRR (before tax)	6%	3.2%	8%	1.8%		13%	7%		7%
Purchase price (kWh) (JPY, include tax)	42.00	42.00	23.10	57.75	27.30	42.00	25.20	30.45	35.70
Purchase period (years)	20	10	20	20	15	15	20		

Generation Technology	Biomass						
Group	Gasification (Sewage Sludge)	Gasification (Livestock Excreta)	Combustion of Solid Fuel (Unused Timber)	Combustion of Solid Fuel (General Timber)	Combustion of Solid Fuel (General Waste)	Combustion of Solid Fuel (Sewage Sludge)	Combustion of Solid Fuel (Recycled Timber)
IRR (before tax)	1%		8%	4%		4%	4%
Purchase price/ kWh (JPY, include tax)	40.95		33.60	25.20		17.85	13.65
Purchase period (years)	20						

IRR, internal rate of return.
Source: METI(b) (2012)

As shown in Table 23.4, the purchase prices and IRR of PVs are enough to attract new investment. The prices will be reviewed annually and may be periodically adjusted by the Ministry of Economy, Trade and Industry. For the first 3 years, however, the government has decided to ensure the profitability of renewable investors to encourage additional investments renewable generation.

5 CONCLUSION

The Fukushima accident in 2011 has been a major event with long-term implications not just for the future of nuclear but overall energy policy in Japan. Japanese government started policy discussions including establishment of objective and clear safety standards for existing nuclear power plants to resume operations. Currently, only 2 of the 50 nuclear units are in operation and if the situation is not rectified soon, it will have a significant negative impact on the Japanese economy.

To date, the government has been ambiguous on how it intends to proceed.[23] Regardless of who is in power, Japan needs to shape its future energy policy to sustain economic growth while protecting the environment.

The electricity market reform in Japan has had limited success to date, but there are hopeful signs that the process may resume and may lead to better outcome in the future. It is important to introduce energy policies that are suited for Japan's energy situation after Fukushima, perhaps dissimilar to those of the other countries. To this end, more matured policy discussions are necessary for Japan.

REFERENCES

CRIEPI Report Y12002, 2012. Questionnaire survey on firms' activities to save electricity in the summer of 2011—Focusing on results from eastern Japan. <http://criepi.denken.or.jp/jp/kenkikaku/report/detail/Y12002.htm> (in Japanese).

Energy White Paper, 2010. <http://www.enecho.meti.go.jp/topics/hakusho/2010energyhtml/index.html/> (in Japanese).

Expert Committee on Reform of the Electric Power System, 2012. Basic Principles of Electric Power Systems Reform: Towards Electric Power Systems Opened to the Public <http://www.meti.go.jp/committee/sougouenergy/sougou/denryoku_system_kaikaku/pdf/report_001_00.pdf> (in Japanese).

FEPC (The Federation of Electric Power Companies of Japan), 2013. Electricity Review Japan. <http://www.fepc.or.jp/english/library/electricity_eview_japan/__icsFiles/afieldfile/2013/03/28/2013ERJ_full_1.pdf>.

FEPC Electricity Statistics. <http://www.fepc.or.jp/library/data/tokei/> (in Japanese).

---

23. Prime Minister Noda announced the dissolution of the House of Representatives on November 14, 2012.

JEPIC (Japan Electric Power Information Center), 2011. The Electric Power Industry in Japan 2011.

JEPX (Japan Electric Power Exchange). <http://www.jepx.org/English/index_e.html/>.

METI(a), 2012. On Liberalization of Electric Power Retail Sales. <http://www.enecho.meti.go. jp/denkihp/genjo/seido.pdf/> (in Japanese).

METI(b), 2012. On FIT of Renewable Energy. <http://www.enecho.meti.go.jp/saiene/kaitori/dl/ 120522setsumei.pdf/> (in Japanese).

National Policy Unit, 2012. Options for Energy and the Environment.

The Energy and Environment Council, 2011. Interim Compilation of Discussion Points for the Formulation of Innovative Strategy for Energy and the Environment.

Chapter 24

The Singapore Electricity Market: From Partial to Full Competition

Youngho Chang[a] and Yanfei Li[b]

[a]*Division of Economics, School of Humanity and Social Sciences, Nanyang Technological University, Singapore;* [b]*Energy Research Institute @ NTU, Nanyang Technological University, Singapore*

1 INTRODUCTION

Over the last two decades, Singapore has gradually deregulated its electricity market from a vertically integrated monopoly to a fully divested generation with competition in the wholesale and retail electricity sectors and a monopoly in the transmission and distribution (T&D) sector.

Singapore is a city-state in Southeast Asia, with 714 km^2 of land area, a population of 5.2 million and Gross Domestic Product (GDP) per capita of US $50,123 in 2011.[1] Being a rich and industrialized economy, Singapore's per capita annual electricity consumption is around 7866 kWh in 2011, matching those of some Organisation for Economic Co-operation and Development (OECD) countries.

Singapore's annual electricity consumption in 2011 was 41,787 GWh as shown in Figure 24.1. Over 80% of the electricity is consumed by manufacturing and other industrial activities. As shown in Figure 24.2, from 1986 to 2010, domestic and industry sector's electricity consumption has increased by about 4.7 and 4.3 times, respectively. Overall, Singapore's total electricity consumption continues to increase steadily, growing 4.3 times over the same period. Electricity demand in 2009 dropped below the trend line due to the economic downturn but rebounded in 2010.

With projected economic and population growth, electricity demand is forecast to grow steadily reaching more than 60,000 GWh in 2020 as shown in Figure 24.3.

---

1. Singapore department of statistics. http://www.singstat.gov.sg/stats/keyind.html#econind (accessed 01.11.12).

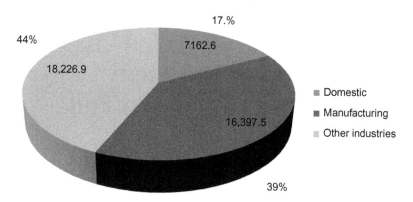

FIGURE 24.1 Electricity consumption of Singapore in 2011. *Source: Authors' estimation based on Yearbook of Statistics Singapore (2012).*

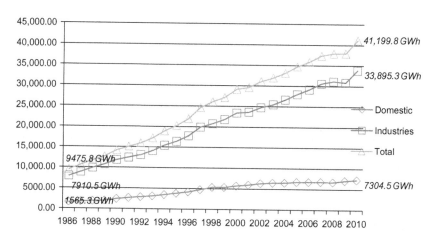

FIGURE 24.2 Yearly electricity consumption (in GWh) in different sectors from 1986 to 2010. *Source: Historical Yearly Electricity Consumption from 1986 to 2010, Energy Market Authority, Singapore.*

On the supply side, the total installed generation capacity in the National Electricity Market of Singapore (NEMS) is 10,216 MW in 2012 with peak demand of 6639 MW and off-peak demand of 4220 MW recorded in May 2012 (Figure 24.4).

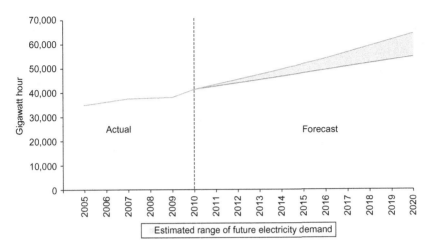

FIGURE 24.3   Total electricity demand trend. *Source: Statement of Opportunities (2011); Energy Market Authority, Singapore.*

In terms of fuel mix, almost 80% of electricity is generated from natural gas, with the rest from fuel oil and others.[2] Liquefied Natural Gas (LNG) is expected to provide some diversification in the fuel mix starting in 2013.

The NEMS has a few unique features that need to be considered when the electricity market is to be further liberalized:

- First, electricity consumption is expected to grow at a fast rate in the next decade. Importing electricity from neighboring countries, such as Malaysia and Indonesia, could be one of the solutions to meet the fast growing demand. The regulator and market participants should prepare for cross-border power trade and settlement;
- Second, the supply side of the NEMS is highly concentrated at the moment. The market power issue has been controlled through vesting contracts but these are expected to be eventually eliminated; and
- Third, natural gas currently dominates the fuel mix, which suggests the need for diversification while maintaining reliable services at minimum costs.

This chapter reviews the history, structure, and performance of NEMS and analyzes the above challenges. It is organized as follows: Section 2 reviews key reforms introduced since 2000 and the current market structure. The pricing issues are examined in detail in Section 3. Section 4 assesses the performance of the NEMS, followed by chapter's conclusions.

---

2. A small portion of the installed capacity consists of an incinerator (National Environment Agency and Senoko) or waste-to-power generator (Keppel Seghers).

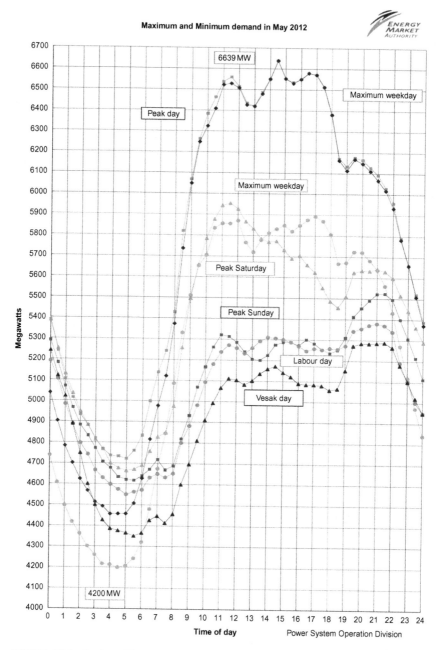

FIGURE 24.4 Load profile of the NEMS in May 2012. *Source: http://www.ema.gov.sg/reports/id:72/.*

2 THE NATIONAL ELECTRICITY MARKET OF SINGAPORE

2.1 The Chronicles of Market Reforms

Before 1995, Singapore's electricity market was vertically integrated, operated, and regulated by the Public Utilities Board (PUB). The Singapore Electricity Market Reform Milestones since 1995 are as follows (Energy Market Authority of Singapore, 2010, 2011a):

- Corporatization of the PUB's gas and electricity undertakings, October 1995
- Singapore Electricity Pool (SEP) commenced operation, April 1998
- Government decision on further deregulation, March 2000
- Energy Market Authority (EMA) was formed, April 2001
- Energy Market Company (EMC), a joint venture between EMA and Marketplace Company of New Zealand (M-Co), was established, April 2001[3]
- Commencement of the NEMS, January 2003
- Commencement of the Phase I Retail Market Liberalization, June 2003
- Commencement of the Phase II Retail Market Liberalization, December 2003
- Vesting Contracts and Interruptible Load scheme introduced, January 2004
- Completion of Temasek's divestment of the three gencos, December 2008
- Contestability to all consumers is being currently studied.

Over the last two decades, it has undergone a gradual and steady spate of market deregulation. As a precursor to the vertical separation of the supply chain, Singapore Power (SP), owned by the Temasek Holdings, took over the electricity and piped gas functions from the PUB in October 1995. In addition, SP owned the electricity generation companies.

Starting in April 1998, initial steps toward deregulation were introduced in the generation sector, with the setup of a day-ahead market in the SEP. SP PowerGrid (SPPG) became the grid owner and system operator, while SP Services is the sole purchaser in the electricity pool. Both companies are subsidiaries of SP.

Further changes were introduced in April 2001 including the divestment of PowerSeraya and Power Senoko, companies owning electricity generation assets, to her parent company Temasek Holdings. In hindsight, this was a significant step towards the eventual full privatization of the generation component of the electricity market. As for the T&D ownership, the assets remain in the hands of SPPG. Up to this point, Singapore's electricity market has introduced competition in the generation sector but retained the

3. EMA owns 51% of share and M-Co has 49%. The joint venture status has ended on August 6, 2012 when Singapore Exchange (SGX) has acquired 49% stake of EMC from M-Co.

integrated state of the T&D assets. In the same year, the EMA was established under the Energy Market Authority of Singapore Act.

Liberalization of the retail electricity market took place in July 2001, with 250 of the largest (>2 MW demand) nondomestic consumers deemed to be contestable. The NEMS replaced SEP in January 2003, allowing generation companies to bid every half hour to sell electricity into the new wholesale market. Between June 2003 and February 2006, the retail market was further liberalized in two phases. For the Phase 1 liberalization of the electricity retail market, nondomestic consumers with average monthly consumption above 20,000 kWh became contestable among the electricity retailing companies. In the Phase 2 (starting on December 2003 and completed by February 2006), consumers with average monthly consumption above 10,000 kWh became contestable. The process rendered 10,000 large consumers of electricity to be contestable. In sum, they represent about 75% of Singapore's total electricity demand.

Developments in the electricity sector picked up speed again by late 2007, perhaps hastened by concern over a sustained global rise in energy prices and growing international awareness of climate change. Importing LNG was viewed as a strategy to ameliorate Singapore's dependence on (piped) natural gas as the fuel source for electricity generation. In September 2007, PowerGas Ltd. was designated as the LNG terminal operator to build and operate Singapore's first LNG terminal. This was followed up with the appointment of BG Asia Pacific in April 2009 as the LNG aggregator for Singapore, with an exclusive license to import and sell up to 3 million tons per annum of LNG locally. Upon facing the financial crisis and a sharp fall in the price of natural gas, the Singapore government was impelled to take over the development and ownership of the LNG terminal in June 2009, with the EMA setting up the SLNG Company to own and oversee the development.

More changes were afoot in the generation segment with foreign suitors courted for the purchase of power generation companies in Singapore. Tuas Power was sold by the Temasek Holdings to SinoSing Power, a wholly owned subsidiary of the Chinese Huaneng Group in March 2008; Senoko Power was divested to the Lion Power Consortium (consisting of Marubeni—30%, GDF Suez—30%, Kansai, Kyushu, Japan Bank for International Cooperation [JBIC]) in September 2008. In March 2009, PowerSeraya Group was divested from the Temasek Holdings to the YTL Power International Berhad. The divestment of the three largest electricity generation companies in Singapore, namely Tuas Power, Senoko Power, and PowerSeraya, into private hands (and foreign ones at that) marked the conclusion of the state's ownership in the generation segment of the electricity industry.

Leveraging on new technology became a key aspect of new policies introduced by the regulator. In November 2008, EMA conducted trial runs involving 1000 households over 6 months for the development of the Electricity Vending System (EVS), a precursor to the possibility of extending retail contestability to domestic consumers. In May 2009, a multiagency taskforce was set up to spearhead the test-bedding of electric vehicles (EVs) in Singapore to study the impact of e-mobility[4] on the electricity system. In November 2009, a pilot project for Intelligent Energy Systems (IESs) was launched to evaluate new applications and technologies that assimilate advances in info-communication technologies with the power system, with a particular focus on the integration of intermittent power sources into a "smart grid." Accenture was appointed to deploy this large-scale project in Q4 2010.

2.2 The Structure of the NEMS

Today, the NEMS is almost fully liberalized and horizontally unbundled in generation, T&D, with wholesale and retail competitions. The key stakeholders include:

- The Industry Regulator (EMA);
- Market Operator (EMC);
- The Power System Operator (PSO), a division of EMA in charge of dispatch of facilities in the wholesale market;
- The Grid Operator/Owner (SPPG);
- Market Support Services Licensee (MSSL), which provides market support services (MSSs) such as metering and billing to various parties. SP Services is the only company playing this role;
- Competing generators (12 generators);
- Competing retailers (seven retailers); and
- Consumers (contestable and noncontestable).

Figure 24.5 presents a schematic of the current electricity industry structure in Singapore.

The contestable consumers can purchase electricity from an electricity retailer, indirectly from NEMS through the MSSL or directly from the NEMS provided they are allowed and registered to trade in the NEMS. The noncontestable consumers are mainly domestic consumers and small industrial and commercial consumers whose monthly electricity demand is less than 10,000 kWh. They are still serviced by the MSSL under regulated tariffs.

The industry players in the contestable sector are generation licensees, electricity retail licensees, wholesale (generation) licensees,[5] and wholesale

---

4. E-mobility is an urban transportation solution based on EV and plug-in hybrid electric vehicle (PHEV) technologies and the relevant battery charging technologies to substitute the current transportation system that is based on internal combustion engine (ICE) technologies.

5. Generators with facilities smaller than 10 MW are required to hold a wholesale (generation) license and register with EMC to supply electricity to the grid.

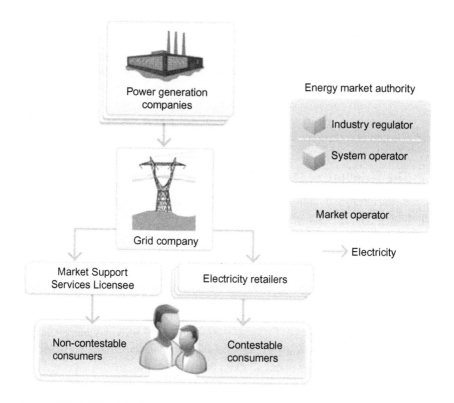

FIGURE 24.5   Electricity industry structure in Singapore. *Source: Energy Market Authority, Singapore.*

(interruptible load service) licensees.[6] The players in the noncontestable sector are the EMC, a transmission licensee (SP PowerAssets: SPPA), a transmission agent licensee (SPPG), and the MSSL that provides MSSs, such as metering and billing, to various parties including noncontestable consumers. Companies in the noncontestable sector are natural monopolies.

As for generation companies, there are currently 12 licensees,[7] of which YTL PowerSeraya, Tuas Power Generation, and Senoko Energy make up

---

6. A wholesaler (interruptible load) licensee is issued to companies who can offer their own load to be interrupted or provide services to other consumers interested in offering their load to be interrupted. Interruptible Load Service Providers are Diamond Energy, Air Products Singapore, and Chesterfield Manufacturing.

7. Generation Licensees are YTL PowerSeraya, SembCorp Cogen, GMR Energy (formerly known as Island Power Company), Keppel Merlimau Cogen, National Environment Agency, Tuas Power Generation, Shell Eastern Petroleum, Senoko Waste-To-Energy, Senoko Energy, ExxonMobil Asia Pacific, Keppel Seghers Tuas Waste-to-Energy Plant and Tuasspring.

more than 80% of the total installed capacity. The nameplate rating of each generating unit should be 10 MW or above to be a generation licensee.

Table 24.1 shows current composition of generation in Singapore. More than 60% of the total installed capacity is natural gas-fired. The rest are fuel oil-fired or diesel-based plants commissioned in the 1980s and 1990s.

Due to concerns about potential abuse of market power in the generation sector, the NEMS introduced vesting contracts in January 2004. Vesting contracts are bilateral electricity agreements between generation companies and MSSL (SP Services), which require generation companies to sell a set amount of electricity at a specified price. The initial vesting level was set at 65% and it has been adjusted downward to 60% and 55% for 2011 and 2012, respectively. This policy limits the ability of the larger generation companies to exercise market power via withholding their capacity during periods of capacity scarcity to push up spot prices in the wholesale electricity market.

The market regulator reviews both the vesting contract level and the parameters used to set the vesting price regularly. The vesting price is determined by calculating long-run marginal cost (LRMC) of the most efficient generation technology such as combined cycle gas turbine. The vesting contract level is set to effectively curb the exercise of market power based on projected electricity supply and demand.

There are seven retailers catering to the contestable consumers who must register with the EMC to purchase electricity from the NEMS. The retailers are Keppel Electric, SembCorp Power, Tuas Power Supply, Senoko Energy Supply, Seraya Energy, GMR Supply (formerly known as Island Power Supply), and Diamond Energy Supply. In addition, there are two different classes of wholesalers, split into the interruptible load service providers[8] and generation providers.[9] The capacity and market shares of these wholesalers are a small part of the overall NEMS. The total capacity of these generators is about 106 MW.

Projections for electricity demand and generation capacity by 2020 are shown in Figure 24.6. The total generation capacity is projected about 14,000 MW by 2018. The peak demand is projected to increase to about 8000 MW by 2020. As the size of the market is projected to increase significantly, opportunities exist for more gencos to enter the market. On the one hand, the concern of market power of the current three largest gencos could be relieved. On the other hand, this raises the issue of whether the NEMS is

---

8. See Footnote 3.
9. Generation Providers wholesaler (generation) licensees as mentioned earlier. They are Biofuels Industries, Pfizer Asia Pacific, Banyan Utilities, ISK Singapore, Singapore Oxygen Air Liquide, MSD International, Green Power Asia, CGNPC Solar-Biofuel Power (Singapore), Eco Special Waste Management and Singapore LNG Corporation. (A firm whose aggregate generating capacity exceeds 10 MW has to bid to secure dispatch.)

TABLE 24.1 Generation Capacity by Type, Fuel, Age, and Ownership

Generator	Unit Capacity	Quantity	Capacities (MW)	Fuel	Age	Ownership
Incineration Plant	47.8	1	47.8	MSW	1987	NEA
Incineration Plant	132	1	132	MSW	2000	NEA
Jurong Diesel	90	2	180	Diesel	1986	Seraya
Keppel Merlimau Cogen	250	2	500	NG	2007	Keppel
Keppel Seghers Waste	22	1	22	Waste	2009	Keppel
Pasir Diesel	105	1	105	Diesel	1983	Senoko
SempCorp Cogen	392.5	2	785	NG	2001	SempCorp
Senoko CCGT	425	2	850	NG	1996	Senoko
Senoko CCGT	365	1	365	NG	2002	Senoko
Senoko CCGT	365	2	730	NG	2004	Senoko
Senoko Steam 3	250	2	500	Fuel oil	1979	Senoko
Senoko Incineration Plant	55	1	55	Waste	1983	Senoko
Seraya CCGT	366	2	732	NG	2002	Seraya
Seraya CCGT	370	2	740	NG	2010	Seraya
Seraya Steam 1	250	2	500	Fuel oil	1987	Seraya
Seraya Steam 1	250	1	250	Fuel oil	1988	Seraya
Seraya Steam 2	233	1	233	Fuel oil	1991	Seraya
Seraya Steam 2	233	2	466	Fuel oil	1992	Seraya
Tuas CCGT 1	360	1	360	NG	2001	Tuas
Tuas CCGT 1	360	1	360	NG	2002	Tuas
Tuas CCGT 2	360	2	720	NG	2005	Tuas
Tuas Steam	600	2	1200	Fuel oil	1999	Tuas
Total			9832.8			

Source: Compiled from http://www.ema.gov.sg/page/115/id:129/#generation

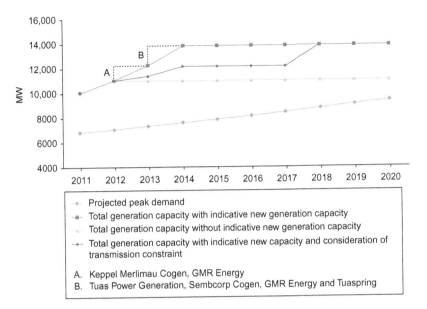

Projected peak demand
Total generation capacity with indicative new generation capacity
Total generation capacity without indicative new generation capacity
Total generation capacity with indicative new capacity and consideration of transmission constraint

A. Keppel Merlimau Cogen, GMR Energy
B. Tuas Power Generation, Sembcorp Cogen, GMR Energy and Tuaspring

FIGURE 24.6 Projected generation capacity. *Source: Energy Market Authority of Singapore (2011b).*

able to deliver sufficient incentives for the development of new capacity and new entries.

SPPA is the transmission licensee and SPPG is the transmission operator. The "open access" principle applies to the transmission network. This means SPPA is responsible for planning transmission network up to connecting sub-station from each genco's switchhouse. Beyond the connecting substation, gencos are competing with other gencos for transmission network if there is a constraint (EMA, 2011).

Tariffs are applied to noncontestable consumers, while the Uniform Singapore Energy Price (USEP) is applied to contestable consumers in the wholesale market. Tariffs are fixed unit rates which change every quarter.[10] The USEP is a weighted average of the nodal prices determined in the wholesale market for energy only.[11]

The tariffs for domestic consumers in 2011 ranged from S¢24.10/kWh to S¢27.28/kWh reflecting the fluctuations of oil prices.[12] Unlike the unit price for domestic consumers, there are various components for the tariffs of

---

10. The quarterly vesting price, which is a regulation mechanism against market power of large gencos and will be discussed in detail in Sections 3 and 4, is used to determine the electricity tariff each quarter.
11. The details of the USEP are presented in Section 3.
12. US$1=S$1.225 as of November 11, 2012.

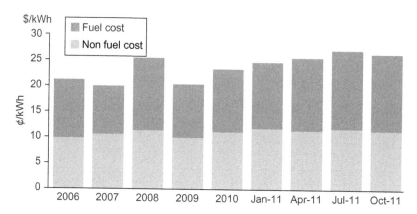

FIGURE 24.7   Fuel and nonfuel cost components in fuel tariff. *Source: SP Services.*

nondomestic consumers, such as capacity charges, energy charges for peak and off-peak periods, and reactive power charges.

Tariffs can be split into two components—fuel (top) and nonfuel (bottom) components as shown in Figure 24.7. The fuel component is based on the average fuel price in the previous 3 months. The nonfuel cost component reflects the cost of generating and delivering electricity to homes and has remained relatively constant over time.

Fuel cost in the tariff calculation is mainly determined by the price of natural gas, which is pegged to that of fuel oil. The fuel component of the vesting price of every quarter is determined by the quarterly forward fuel oil price in all the months of the preceding quarter.[13] The forward fuel oil price is taken from the Intercontinental Exchange (ICE).

There are two parts to the nonfuel costs. The first part consists of the network cost, MSS fee, and the market administration and PSO fee which remain largely unchanged over the past few years. The second part of nonfuel costs consists of the costs of operating the generating plants, manpower costs, capital costs, and an adjustment component to account for any variation between forecast and actual electricity consumption whose cost will be considered over a longer period of time.

In sum, the components of the tariffs are energy cost, network costs, MSS fee, and market administration and PSO fee. Energy costs go to generation companies, network costs, and MSS fee go to SPPA and SP Services, respectively, and administration and PSO fee go to PSO and energy market. Energy costs usually take up about 70% of the tariff.

---

13. For example, to set the vesting price (and the tariff) for the second quarter of a year, the fuel oil price data for January, February, and March of the year is used.

3 THE PRICING STRUCTURE UNDER DEREGULATION

The Singapore electricity market has gone through three different types of market regime since 1998 when a precursor of a deregulated electricity market started as a day-ahead market. The three types are a day-ahead market, real-time pricing with a cap, and real-time pricing with vesting contracts.

The current market is a competitive with large number of sellers and buyers, a homogenous product, sufficient knowledge, and low barriers to entry. EMA is implementing tight regulations such as vesting contracts on 55–60% of demand, drastically reducing firms' ability to mark up the price above LRMCs. Anticollusion laws are already in place to prevent collusion between firms and ensure competition. Thus an individual firm's demand curve takes a near vertical shape with a low degree of elasticity.

Figure 24.8 shows a sample supply and demand curve in the wholesale market. The supply curve is the stepped offer curve (a merit order) from generators and the demand curve is the forecast load. The market-clearing price and dispatch schedule quantities are set by the intersection of the two curves.

Gencos offer electricity blocks at costs pegged to their LRMC and earn producer surplus equal to the difference between their offer and the market-clearing price. Supply from gencos that is above the market-clearing price is not dispatched.

The above process is repeated every half hour, providing both consumers and suppliers with the "best prices" and "best demand," respectively. The existence of negative offer prices would be due to the fact that companies may find it cheaper to produce at these rates as opposed to shutting off the generator completely before ramping it up again during peak loading.

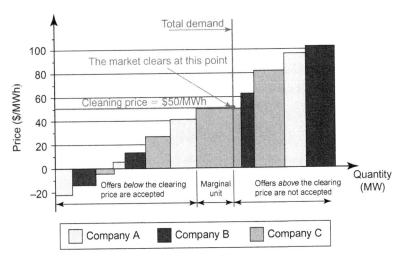

FIGURE 24.8 A schematic diagram of competitive wholesale electricity market.
Source: Energy Market Authority, Singapore.

The market-clearing price determination process depicted in Figure 24.10 is repeated across all the nodes which result in different prices—there are about 720 nodes across the island—in the market, and the weighted average of the pricing of all nodes is called the USEP. Although generating companies are paid the nodal prices at each node, retailers buy the energy at a uniform price such that no consumers are locationally disadvantaged. It is to be noted that the USEP is the price paid by the retailers only for the energy. The final amount paid by contestable consumers also includes distribution, administration, and other costs. The price retailers pay in the market is the Wholesale Electricity Price (WEP), which is the net purchase price that includes all administrative costs incurred in the wholesale market. This WEP consists of the following components:

- Uniform Singapore Electricity Price (USEP),
- Allocated Regulated Price (AFP),
- Hourly Electricity Uplift Charge (HEUC),
- Monthly Electricity Uplift Charge (MEUC),
- EMC fees, and
- PSO fees.

The AFP is the cost of purchasing regulation products from the market that is co-shared by retailers and generators. The AFP is determined for retailers based on their consumption (metered withdrawal quantities), and for generators based on their injected energy quantity, up to 5 MWh.

The HEUC is the charge that captures any differences between total amounts received from retailers and total amounts paid to generators for energy, reserve, and regulation products. While the HEUC is called a charge, it is typically a return to retailers of the revenue arising from the sale of energy to cover transmission losses.

At the end of each calendar month, EMC calculates the MEUC for the following calendar month and recovers from the load on the basis of withdrawal quantity in megawatt hour. This charge, levied on retailers, covers the following potential payments and refunds each month:

- the cost of procuring contracted ancillary services (e.g., black-start services) and related costs,
- compensation claims and refunds,
- financial penalties and refunds, and
- the estimated monthly energy uplift shortfall to be recovered and/or deducted in the following calendar month.

EMC and PSO fees are the approved administrative costs for EMC and PSO to operate the NEMS in each fiscal year. These fees are recovered from both generators and retailers based on per megawatt hour generated or consumed.

4 MARKET PERFORMANCE AND THE WAY FORWARD

4.1 Contestable Consumers

The performance of the NEMS can be considered satisfactory by many measures. On reliability, the average interruption time per customer is less than 1 min/year.[14] On cost and price stability, the system also meets the expectations of the reforms.

In the market for contestable customers, the USEP has exhibited a general upward trend, as shown in Figure 24.9. Fuel, being the primary raw material for energy, affects energy prices to a large extent. An increase in fuel prices would naturally lead to an increase in energy prices, especially with Singapore's reliance on natural gas. Figure 24.9 also shows that fuel prices have generally risen over the years, and the rise and fall of the USEP follows the same trend as that of fuel price. Clearly, there is a strong correlation between the two while the pricing of vesting contracts reduces price volatility (Chang and Park, 2007).

A study by Chang (2007) examined the regulatory framework, market power, and competition in the NEMS using various measures of market power, such as concentration ratios, Herfindahl–Hirschman Index (HHI), supply margin assessment (SMA), residual supply index (RSI), and Lerner Index. The study found that the market has high concentration ratios but there is no pivotal player who can exercise market power. Looking at the Lerner Index, the study suggests that the NEMS is fairly competitive although the three largest generation companies have more than 80% of generation capacity.

Moreover, the Lerner Index suggests that the electricity prices are close to the marginal cost. The Lerner Index in the NEMS in 2003, 2004, and 2005 is 0.255, 0.199, and 0.188, respectively.[15] The downward trend in the Lerner Index suggests that the introduction of vesting contracts has put downward pressure on prices.

Another study examined the technical efficiency of the generation companies in the NEMS using the parametric and nonparametric approaches. Based on the parametric approach, the study found high levels of efficiency among generation companies. The nonparametric approach also concluded that there is little inefficiency or wastes in input (Chang and Toh, 2007).

Another study examined whether the NEMS can supply electricity adequately and reliably at a reasonable price. Using a 3×3 matrix consisting of

---

14. 2006 is an exception. The system average interruption duration index (SADI) in 2006 was 3.95 due to an island-wide blackout as the supply of natural gas was cut but a smooth transition to diesel-fired generators did not occur. The average number of interruption per customer is less than 0.05 except 2006. The system average interruption frequency index (SAIFI) in 2006 was 0.114 due to the cut of fuel supply.

15. These figures are derived from the short-run marginal cost of the long-run marginal cost that is used as a base price of vesting contracts. The long-run marginal cost is calculated based on the most efficiently configured combined gas cycle turbine and hence these figures should be considered the upper bound of the true Lerner Index.

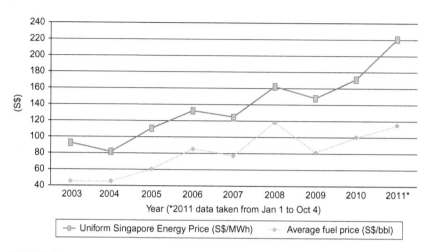

FIGURE 24.9 Yearly average USEP and fuel price 2003–2011. *Source: Energy Market Company, Singapore.*

energy adequacy, reliability, and reasonable price and under three scenarios, the study concluded that NEMS performs adequately under most scenarios. The only area of concern may be lack of investment to ensure grid reliability in the future (Chang and Lee, 2008).

The transparency of the market operations in terms of information disclosure appears satisfactory with the EMC, the market operator, posting all relevant market price information on its web site. All concerned parties can access market data. The EMA, the regulator, also appears open and transparent while encouraging input on proposed or newly introduced regulations or changes in the existing regulations before implementing them. This allows key players to make informed decisions, allocate risk, and make necessary investments at minimal costs.

4.2 Noncontestable Consumers

The tariffs paid by noncontestable consumers have generally increased over the years as shown in Figure 24.10. This increase can largely be attributed to the increase in fuel prices,[16] which exhibited a similar trend over the same time period. The smaller magnitude of tariff percentage change can be attributed to the nonfuel cost component, which remains relatively constant irrespective of fuel prices.

The regulated tariffs behave differently from market-determined USEP. As the nonfuel cost component of the tariff remains relatively unchanged and

---

16. The exceptionally high tariffs around Q3 of 2008 were caused by the record high oil prices at the time.

$/kWh

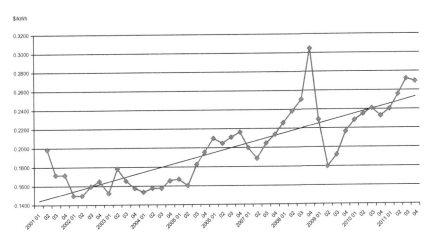

FIGURE 24.10   Noncompetitive electricity tariff 2001–2011. *Source: SP Services.*

is not affected by fuel prices, this would mean that as fuel prices increase, the relative increase of tariff would be lower, shielding noncontestable consumers from the full brunt of the price increase. The same can be said when the opposite happens. Consumers will pay relatively more when fuel prices fall.

Although tariffs for noncontestable consumers are revised every quarter based on fuel costs, this falls short of full retail contestability.[17]

4.3 From Partial to Full Competition

The gradual implementation of market reforms has resulted in a market that appears to be working. However, the NEMS remains subject to vesting contracts and the noncontestability of domestic and small industrial consumers renders it a partially competitive electricity market. Whether bringing full competition into the NEMS will enhance the efficiency and reliability of the market needs to be studied as suggested below:

- First, it is a question of how and when, not whether, to introduce full retail contestability to the market. Introducing futures market and demand response can also improve the market's performance.
- Second, the effectiveness and desirability of vesting contracts, which have been in place for nearly a decade, need to be evaluated including options to further reduce or eliminate them. Clearly, there is a need for fair returns to investment in generation in a small and isolated market such as Singapore,

---

17. The regulator has conducted a pilot study for EVS for implementing the full retail contestability. When the deregulation was originally announced, full retail competition was supposed to be completed by 2009. With the implementation of EVS and the establishment of IES, full retail competition could be introduced.

not only to replace aging capacity but also to build new capacity to meet the growing demand.
- Third, the diversification of fuel sources, generation technology, and interconnections to neighboring countries appear prudent.

5 CONCLUSION

Since the introduction of market reform in 1995, the Singapore electricity market has performed well and has achieved most of what was intended. The success of the NEMS may be attributed to the gradual implementation of reforms to different segments of market while managing the potential for exercise of market power by dominant generators in small and isolated markets. This means that the NEMS remains a partially competitive electricity market. Whether and how to move to a full competitive market needs to be evaluated.

The Singapore government is studying the feasibility of importing electricity from neighboring countries and to introduce electricity futures market and demand response bringing the NEMS closer to full competition. Another key milestone will be reached when noncontestable consumers are charged by the time of use and the quantity consumed. These advancements promise to transition Singapore beyond the legacy systems of the past.

REFERENCES

Chang, Y., 2007. The new electricity market of Singapore: regulatory framework, market power and competition. Energ. Policy 35, 403–412.
Chang, Y., Lee, J., 2008. Electricity market deregulation and energy security: a study of the UK and Singapore electricity markets. Int. J. Global Energy Issues 29 (1/2), 109–132.
Chang, Y., Park, C., 2007. Electricity market structure, electricity price and its volatility. Econ. Lett. 95, 192–197.
Chang, Y., Toh, W.L., 2007. Efficiency of generation companies in the deregulated electricity market of Singapore: parametric and non-parametric approaches. Int. J. Electron. Bus. Manage. 5 (3), 225–238.
Energy Market Authority of Singapore, 2010. Introduction to the National Electricity Market of Singapore, Version 6. Updated as of October 2010. Available from: <www.ema.gov.sg/media/files/books/intro_to_nems/Introduction%20to%20the%20NEMS_11%20Oct%2010_2.pdf>.
Energy Market Authority of Singapore, 2011a. Developments of the Singapore Electricity Transmission Network, April 2011. Available from: <www.ema.gov.sg/images/files/Developments%20in%20the%20Singapore%20Electricity%20Transmission%20Network_5%20Apr%202011.pdf>.
Energy Market Authority of Singapore (EMA), 2011b. Available from: <www.ema.gov.sg/>.
Energy Market Company (EMC). Available from: <www.emcsg.com/>.
SP Services. Available from: <www.singaporepower.com.sg/irj/portal/spservices>.
Energy Market Authority of Singapore, 2011b. Statement of Opportunities for the Singapore Energy Industry 2011. Available from: <www.ema.gov.sg/ema_soo/index.html>.
Yearbook of Statistics Singapore, 2012. Department of Statistics, Singapore. Available from: <www.singstat.gov.sg/Publications/publications_and_papers/reference/yearbook_2012/yos2012.pdf>.

Chapter 25

Market Design for the Integration of Variable Generation

Jenny Riesz[a], Joel Gilmore[b] and Magnus Hindsberger[c]

[a]*Senior Consultant, Energy Strategic Advisory, AECOM Australia, Sydney, Australia;*
[b]*Principal, Renewable Energy and Climate Policy, ROAM Consulting, Brisbane, Australia;*
[c]*Specialist, Network Development, Australian Energy Market Operator (AEMO), Brisbane, Australia*

1 INTRODUCTION

Investment in electricity generation around the world is increasingly moving toward new sources of renewable energy. Nonhydro renewable generation accounted for 44% of new generation capacity installed in 2011, with total investment in renewable energy rising to a record US$257 billion (Bloomberg, 2012). Wind and solar photovoltaics accounted for almost 40% and 30% of new renewable capacity in 2011, respectively (REN21, 2012). By the end of 2011, total renewable power capacity worldwide exceeded 1360 GW, up 8% over 2010; renewables comprised more than 25% of total global power-generating capacity and supplied an estimated 20.3% of global electricity. Nonhydropower renewables exceeded 390 GW, a 24% increase in capacity from 2010 (REN21, 2012). These figures are summarized in Table 25.1.

The trend to renewables is driven by a wide range of factors, including rising fossil fuel costs, falling renewable energy capital costs, energy security, climate change mitigation, avoidance of the externalities of fossil fuel stationary energy, and other political pressures. Modeling by the International Energy Agency (IEA) projects that renewable technologies are expected to account for almost half of all new capacity added worldwide through to 2035, as illustrated in Figure 25.1 (IEA, 2011b).

A wide range of renewable technologies exist at various stages of maturity. All have unique properties that mean they integrate into electricity

TABLE 25.1 Selected Indicators of Global Renewable Energy Development

		2009	2010	2011
Investment in new renewable capacity (annual)	Billion US$	161	220	257
Total renewable power capacity (including hydro)		1170	1260	1360
Total renewable power capacity (not including hydro)		250	315	390
Wind	GW	159	198	238
Solar PV		23	40	70
Concentrating solar thermal		0.7	1.3	1.8

Source: REN21 (2012)

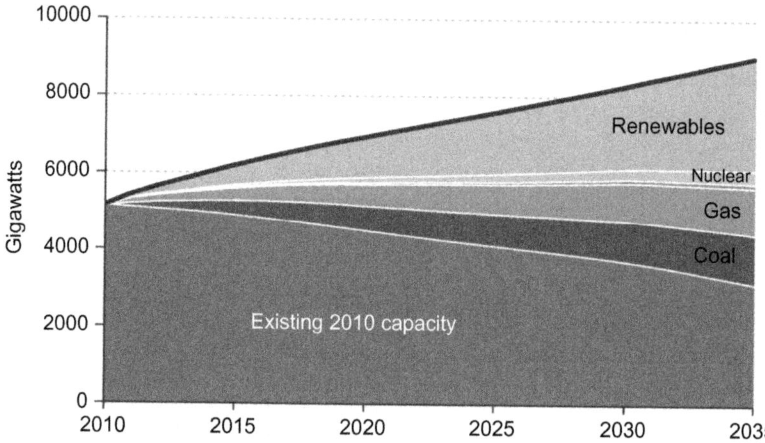

FIGURE 25.1 Global-installed power generation capacity and additions by technology (IEA "New Policies Scenario"). *Source: IEA (2011b).*

markets in different ways. However, many of the most mature and lowest cost forms of renewable generation are variable in operation.[1] For example, Figure 25.2 illustrates the present and projected levelized cost of energy from a selection of renewable technologies in the Australian market. Solar

---

1. The term "variable" in this context is defined in Section 1.

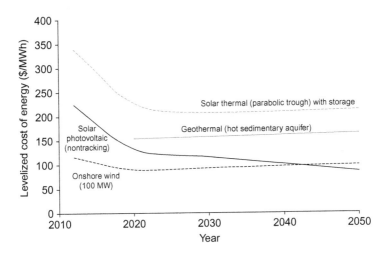

FIGURE 25.2   Levelized cost of energy for a selection of renewable technologies in Australia. *Source: Data from BREE (2012).*

photovoltaics and onshore wind are currently the least expensive forms of widely available renewable energy, and are projected to remain so for the foreseeable future. Both have variable generation profiles.

These factors are driving the necessity for electricity markets around the world to integrate significant quantities of variable generation. Germany is an example which is leading in this regard, integrating significant quantities of wind and photovoltaic generation, as described in Chapters 4, 5, 7, and 8. The issues explored in these chapters demonstrate the importance of renewable integration for the future design and operation of global electricity markets.

There is variability inherent in the loads of almost all systems, and certain types of industrial loads can be highly variable. For example, electric arc furnaces (used in steel making) repeatedly cycle from zero to full load virtually instantaneously (Milligan and Kirby, 2010). This imposes a large regulation burden on the power system; in some balancing areas in the United States a single arc furnace can account for over half of the regulation requirement (Kirby and Hirst, 2000). Power systems around the world currently deal with this variability, so there is no reason to suggest that variable generation cannot be similarly integrated into standard system operations.

A range of studies have demonstrated that very high penetrations of renewable generation are technically feasible (Connolly et al., 2011; Krajacic et al., 2011; Lund and Mathiesen, 2009; Mason et al., 2010; Mathiesen et al., 2011). A particularly comprehensive example was completed recently by the National Renewable Energy Laboratory (NREL), indicating that renewable electricity generation from technologies that are commercially available

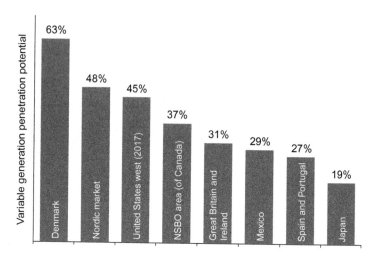

FIGURE 25.3 Variable generation deployment potentials, based upon technical flexible resource. *Source: Data from IEA (2011a).*

today is more than adequate to supply 80% of total US electricity generation in 2050 while meeting electricity demand on an hourly basis in every region of the country (NREL, 2012).

Furthermore, considerable technical flexible resources exist already in most markets. For example, a modeling study by the IEA indicates that from a purely "hardware" point of view, the existing system potential for integrating variable generation ranges from 19% in the least flexible area assessed (Japan) to 63% in the most flexible area (Denmark) (IEA, 2011a). Results for all areas studied are illustrated in Figure 25.3.

Significantly, although the IEA analysis indicated that significant technical flexibility exists, the study also found that none of these regions are currently able to access this full potential, primarily due to suboptimal grid strength and unfavorable market design (IEA, 2011a). Experience from markets around the world is now echoing this message that market design, rather than technical limitation, is often the main barrier to cost-effective integration of variable generation. Integration costs vary widely, ranging from negligible amounts, to as high as US$16/MWh of wind (ISO/RTO Council, 2007; Milligan and Kirby, 2010). A large proportion of the cost difference between markets is attributed simply to differences in market design. As stated by the European Wind Energy Association (EWEA) (Van Hulle, 2005):

The capacity of the European power system to absorb significant amounts of wind power is determined more by economics and regulatory rules than by technical or practical constraints.

This highlights the imperative for market design scrutiny to identify and remove inefficiencies that are simply economic or regulatory in nature. In many cases, market reform has the potential to achieve far more cost-effective variable generation integration.

It is noteworthy that most of the market characteristics that facilitate the integration of variable generation also promote more efficient operation *without* variable generation. However, higher levels of variable generation increase the value of these characteristics (Milligan and Kirby, 2010).

It should also be recognized that the unique properties of variable generation bring new benefits to the system, in addition to new challenges. For example, most modern wind and solar photovoltaic plants have fast and precise control capability, typically exceeding that of most thermal generators (Milligan and Kirby, 2010). They offer a high degree of flexibility, being able to rapidly ramp downward to any level, including zero, if required. By comparison large thermal units are typically far less flexible. In many cases they ramp slowly and are usually unable to move below a minimum level of generation without notice. These additional benefits of variable generation may not be accessible under existing market rules, but can be harnessed if market designs are suitably adjusted.

1.1 Characteristics of Variable Generation

Variable generation, as referred to in this chapter, is defined as technologies that generate electricity on an as-available basis, varying in output over time as the resources that drive them vary over time. Wind generation and solar photovoltaics are common examples, although a range of technologies including run-of-river hydro and wave power would also qualify.[2] The term "variable" is used in preference to "intermittent" since it captures more effectively the relatively slow (hour to hour) output changes that characterize these generation types. By contrast, the term "intermittent" implies sudden rapid changes, which are not typically observed at an aggregate level (Holttinen et al., 2009; Milligan and Kirby, 2010).

Variable generation is characterized by a range of features that differentiate it from thermal generation. Designing electricity markets to effectively manage two of these characteristics forms the key focus of this chapter. These are (Milligan and Kirby, 2010):

1. Uncertainty—Variable technologies exhibit relatively higher uncertainty over future production levels than other technologies. The resources that drive them may unexpectedly deviate from forecast levels, meaning that

---

2. Tidal power could also be considered variable, since its output will vary by time of day depending upon resource availability. However, tidal power will generally be far more predictable than the other technologies which form the key focus of this chapter.

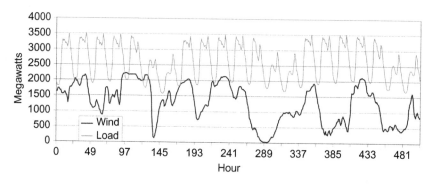

FIGURE 25.4   Wind and demand variability in West Denmark (January 3–23, 2005).
Source: Holttinen et al. (2009).

adjustments to the supply–demand balance may be required to correct
forecast errors.

2. Variability—As the name suggests, variable technologies exhibit relatively higher variability than other technologies. Even if perfect forecasts
were available, they would still exhibit a generation profile with relatively
higher variability than other technologies. An example of the variability
of wind generation in West Denmark is illustrated in Figure 25.4.

Uncertainty and variability are already present in power systems to some
degree. System load cannot be forecast with perfect certainty, and exhibits
variability (as illustrated in Figure 25.4). Similarly, thermal generators have
a level of uncertainty over future generation profiles related to the possibility
of unexpected forced outages. However, loads can usually be forecast more
accurately and exhibit less variability than variable generation (Milligan and
Kirby, 2010). Also, forced outages of large thermal units involve the sudden
removal of a large capacity, but usually occur rarely. This makes the nature
of the associated uncertainty quite different in nature from the uncertainty
related to variable generation.

The aggregate uncertainty of load, variable generation and generator-
forced outage components combined is less than the sum of the individual
uncertainties, as long as the errors in each component are not fully correlated
(which is typically the case) (Dany and Haubrich, 2000). Given that power
systems already have processes in place to manage load variability, this can
allow a certain amount of variable generation to be integrated into power
systems utilizing only the existing reserves for frequency control and balancing (Holttinen et al., 2011). However, as the penetration of variable generation grows beyond a certain level, the uncertainty in variable generation will
start to dominate.

Therefore, at high penetration levels, the uncertainty and variability of
variable generation provides new and different challenges for electricity

system operation, whether in scale or in type. This means that electricity market designs that have worked well in the past may not be effective for the generation mix of the future.

To integrate variable generation into electricity systems effectively, a three-pronged approach to optimization of electricity market design is apparent:

1. Minimize uncertainty in variable generation;
2. Minimize variability in variable generation; and
3. Maximize the flexibility of the remaining system.

Market design features that have the potential to achieve these goals are described in the remainder of this chapter. For the optimal outcome, each initiative should be pursued to the degree they are supported by the balance of costs against benefits.

There are several other specific qualities of renewable technologies that differ from thermal generation, and have not yet been mentioned. Firstly, variable generation is typically characterized by very low operating costs, with short-run marginal costs often near zero. This has implications for the operation of marginal pricing markets, particularly for energy-only markets. This is not addressed in this chapter, but is discussed in Chapters 5 and 7.

Secondly, variable generation is often locationally constrained, with the best sites being located far from load centers and the existing transmission network. This creates challenges for efficient network development. Again, this is not dealt with in detail in this chapter, but is discussed in Chapter 6.

Thirdly, variable renewable generation is "nonfirm" in nature, meaning that while it can be relied upon to produce substantial quantities of energy *on average* across a year, it cannot provide a high level of certainty that it will generate in any particular period of time. This creates challenges in the assessment of system adequacy, and the way in which investment in system capacity is incentivized. Again, this topic will not be dealt with in detail in this chapter, but instead is discussed in Chapter 7.

This chapter is structured as follows:

- Section 2 outlines the value of "fast markets," which feature short dispatch intervals and short delays from scheduling to dispatch.
- Section 3 outlines the value of large markets, illustrating how maximizing system size reduces uncertainty, reduces variability, and increases flexibility. Some potential barriers to the development of large markets are also discussed.
- Section 4 describes how accurate pricing signals can assist in signaling correct incentives to market participants, ensuring that the market services required are provided to the economically efficient level by each participant.

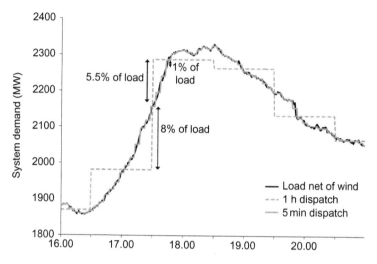

FIGURE 25.5 Shorter dispatch intervals decrease the requirement for secondary reserves. *Source: Data from ROAM (2010).*

- Section 5 addresses the important topic of frequency control ancillary services and outlines strategies to provide regulation services at lower cost.
- Section 6 describes the value of gross pool single platform markets in minimizing integration costs for variable generation.
- Section 7 outlines the conclusions of the chapter.

2 THE IMPORTANCE OF FAST MARKETS

The length of dispatch intervals is one of the most important determining factors in the cost of integrating variable generation. Short dispatch intervals allow more frequent redispatch of the whole system, enabling deviations to be dealt with by adjustment of every market participant in the system as appropriate. Long dispatch intervals mean that deviations in load and variable generation away from the central system set point for the interval will be significantly larger, requiring larger regulation reserves.[3] As eloquently stated by Milligan and Kirby (2010):

Scheduling rules that restrict generators to hourly movements artificially hobble the conventional generation fleet.

The value of shorter dispatch intervals is illustrated in Figure 25.5. This shows a randomly selected day in the South-West Interconnected System

---

3. Regulation reserves are capacity used to maintain the supply–demand balance minute to minute by managing short-term variability and correcting forecast errors and other unexpected events.

(SWIS) in Western Australia. Net load over this period is illustrated in black. The gray line indicates the dispatch of the aggregate system based upon a 5 min dispatch interval. This dispatch schedule closely follows the net load, with the largest deviation being only 1% of the load. By contrast, hourly scheduling and dispatch constrains most generators' output levels until the top of the next hour (Milligan and Kirby, 2010), as illustrated by the gray dotted line. Very large deviations of 5−8% of the load are now apparent, and would need to be met by regulation reserves as a part of routine operation. This will be complicated somewhat by the fact that generators do not instantaneously move to their new set point when the dispatch changes, but rather will ramp gradually to the new level (or may be instructed to maintain a ramp during a dispatch interval, particularly during rapid and reasonably predictable load ramps). However, it serves as a simple illustrative example of the concept.

There is a strong trend in the United States toward faster markets. Currently, over two-thirds of electricity demand in the United States is in regions with 5 min markets, with the Electric Reliability Council of Texas (ERCOT) and the Southwest Power Pool (SPP) having recently joined these ranks (Maggio, 2011; Milligan and Kirby, 2010; Monroe et al., 2011).

A range of studies have calculated the cost of integrating wind into electric systems. In general, the studies show lower integration costs in large market areas (US$0-5/MWh of wind) than in smaller, single utility service areas (US$8-16/MWh). This result is primarily attributed to the sub-hourly markets operating in the large balancing areas, while the smaller balancing areas studied tended to require generators to follow hourly schedules and obtain all sub-hourly balancing from regulating units (ISO/RTO Council, 2007; Milligan and Kirby, 2010). The large size itself is also a contributing factor, for the range of reasons described in Section 3.

Short dispatch intervals are particularly important for regions that are rich in variable generation but have little load, and primarily function to supply variable generation to other balancing areas. In this case, slow markets can impose an additional capacity requirement on the region exporting variable generation. This arises because of the necessity of internal balancing within the hour if the output of variable generation reduces. The exporting region must continue to supply this quantity of generation to meet their obligations until the beginning of the next dispatch interval, which can require additional system capacity. The additional capacity required is simply an artifact of the scheduling limitations, and constitutes an economic inefficiency. It can be eliminated if fast markets are implemented, allowing the importing region to provide the required response rapidly (rather than waiting until the beginning of the next hour) (Milligan and Kirby, 2009, 2010).

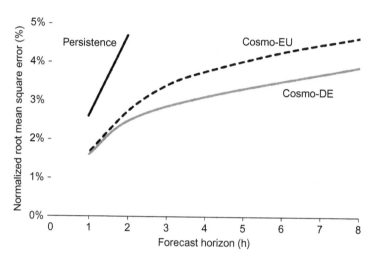

FIGURE 25.6 Forecast errors for wind power production forecasts for Germany using two (Cosmo-DE and Cosmo-EU) data as input, compared with a persistence forecast. *Source: Data from Holttinen et al. (2009), Wessel et al. (2008).*

2.1 Minimizing Delay from Scheduling to Dispatch

Uncertainty in variable generation manifests as forecast errors (differences between the forecast and the actual generation that eventuates). This means that sufficient reserves are required to meet the possible range of forecast errors that could occur. As might be expected, forecast errors for wind generation have been consistently demonstrated to decrease with shorter lead times (DENA, 2005, 2010; Focken et al., 2002; Hulle et al., 2010; Von Roon and Wagner, 2009). Forecasts 1–4 h ahead provide a significant improvement over day-ahead forecasts, and forecasts 5 min ahead provide an even higher level of accuracy (Hodge and Milligan, 2011).

The value of short forecast horizons in minimizing forecast errors is illustrated in Figure 25.6. Wind forecast errors from two models used in Germany are compared with persistence forecasting.[4] All models demonstrate reduced forecast errors when the forecast horizon is minimized.

Figure 25.7 illustrates this for the Australian Wind Energy Forecasting System (AWEFS) that is used by the Australian Energy Market Operator (AEMO) in the Australian National Electricity Market (NEM). In this system, minimizing the delay between scheduling and dispatch significantly reduces forecast errors. This allows the full system dispatch determined in the scheduling process to be closer to optimal, reducing the amount of

---

4. A persistence forecast assumes that variable generation remains at the present level for the future periods in question. This is reasonably accurate for short time periods. Forecasting models are often compared to persistence, to provide a common reference standard.

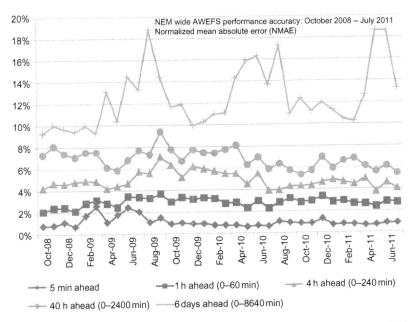

FIGURE 25.7   Forecast performance of AWEFS over a number of time frames. *Source: AEMO.*

balancing reserves required and therefore reducing costs (Borggrefe and Neuhoff, 2011; Muesgens and Neuhoff, 2002; TradeWind, 2009).

Analysis of the Pacific Northwest electricity system in the United States (Bonneville Power Administration, BPA) further supports this point, illustrating that delaying schedule setting until just before the operating hour can significantly reduce reserve capacity requirements. Data from 2007 indicates that if hourly schedules were set 10 min before the start of dispatch interval, a 10 min persistence model for wind provides a reduction in the reserve capacity requirement to 174 MW, compared with 451 MW if scheduling is performed 2 h prior to dispatch (Milligan and Kirby, 2010). This is illustrated in Figure 25.8.

3 THE IMPORTANCE OF LARGE MARKETS

A second key determining factor in the costs of integrating variable generation is the size of the system, whether (preferably) as a single market or through effective sharing of balancing with other markets. Larger electricity markets allow increased geographic diversity, and therefore take advantage of decreased variability in variable generation, and decreased uncertainty through reduced forecast errors. Larger markets also allow access to a larger fleet of conventional generators, increasing system flexibility.

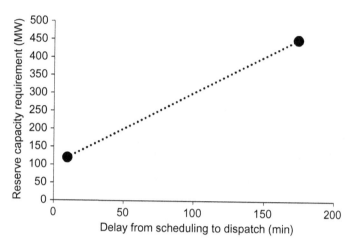

FIGURE 25.8　Reserve capacity requirements for BPA (USA), based upon a persistence model for wind forecasting. *Source: Data from Milligan and Kirby (2010).*

The first benefit of large markets is that they contribute to reducing uncertainty. Maximizing electricity system size allows aggregation of variable generation over a more diverse geographical area, thereby reducing errors in forecasting. By aggregating over a broader geographic region, wind forecast errors have been found to reduce by as much as 30–50% (Ahlstrom, 2008; Rohrig, 2005). For example, the integration of the German transmission system operators into one market in 2009 reduced the day-ahead forecast error from 6.6–7.8% to 5.9% (Borggrefe and Neuhoff, 2011). Figure 25.9 illustrates the reduction in forecast errors as region size increases, for the aggregated wind power production of 40 wind farms in Germany.

The second benefit of large markets is that they exhibit reduced variability. Maximizing the size of balancing areas accesses increased geographical diversity in the variable generation in the system. Wind generator variability has been shown to lose correlation as the distance between units increases (Ernst, 1999; Milligan et al., 2009). This means that due to statistical independence, the per-unit variability significantly decreases as wind generation is aggregated across larger areas. A similar characteristic is observed for solar generation (Milligan and Kirby, 2010), and presumably also for other variable generation types. Thus, if variable generation units are geographically dispersed, increasing the size of the balancing area can be an effective way of reducing the variability that needs to be managed by the system.

This concept can be phrased slightly differently in this way. Due to the principle of statistical independence, the net per-unit variability declines with aggregation, meaning that variability adds *less* than linearly as system size increases. However, system flexibility (measured as the combined ramping

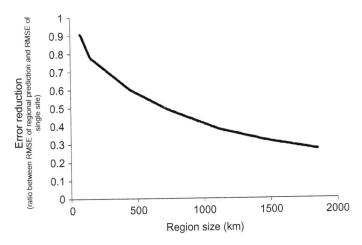

FIGURE 25.9   Decrease of forecast error of prediction for aggregated wind power production due to spatial smoothing effects. Based upon results of measured power production of 40 wind farms in Germany (RMSE: root mean square error). *Source: Holttinen et al. (2009).*

capability of the system) adds *linearly* with aggregation. This means that as system size increases, system flexibility increases faster than system variability, creating a net benefit in aggregation (Milligan and Kirby, 2010).

For example, consider the analysis of the combination of the 11 zones in the New York State power system (GE Energy, 2005). Aggregation of these zones was predicted to reduce wind variability by 33% on an hourly basis and 53% on a 5 min basis. System variability decreased even further when load and wind were considered together.

Modeling by the NREL shows the reduced variability benefit from aggregating smaller transmission areas into a single footprint, as illustrated in Figure 25.10. Variability for single small areas (such as Colorado-West or Wyoming) increases significantly as renewable penetrations increase. However, when the balancing areas across WestConnect are aggregated (Footprint), there is only a slight increase in variability with increased renewable penetration. If all the areas WECC (Western Electricity Coordinating Council)-wide are combined, variability exhibits a slight decrease (NREL, 2010). This modeling indicates that combining the provision of spinning reserves from the many smaller balancing areas in the Western United States into five large regions could save $1.7 billion (US$2009) (NREL, 2010).

Large markets also act to increase the flexibility of the system. Maximizing system size allows balancing errors between regions to cancel out as much as possible. Canceling of balancing errors means that system ramping is reduced when neighboring balancing areas are combined. For example, it has been estimated that if the four balancing areas in Minnesota

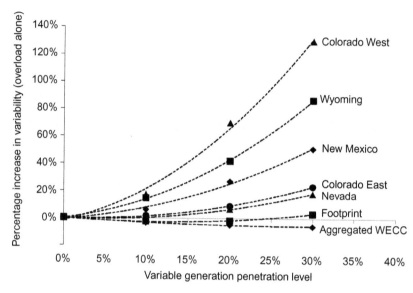

FIGURE 25.10 Percent increase in the standard deviation of the hourly changes of the net load in various areas of the Western United States, and aggregated (Footprint, WECC) as penetration of variable generation increases. WECC: Western Electricity Coordinating Council.
Source: Data from NREL (2010).

were to combine,[5] a 14% reduction in annual ramping requirements (both up and down) would result (Milligan and Kirby, 2007, 2008). This translates into lower system costs.

Larger markets also give access to a larger pool of generation assets to perform balancing (Green, 2008). Increasing the possible pool of generators that are able to respond allows the least cost assets to provide the services required.

Through a combination of these factors, it is estimated that the aggregation of the Nordic countries into the Nord Pool spot market has halved the quantity of reserve required, compared to if they operated independently (Holttinen et al., 2007). Remarkably, despite wind penetrations in the range 24−33%, Germany and Denmark have not needed to increase their amount of operating reserves because of wind (although they do use reserves more often) (Holttinen et al., 2007). Each of these areas has benefited significantly from robust interconnection and shared system balancing with surrounding areas, highlighting the benefits of large systems. Denmark benefits particularly from the fact that neighboring Norway and Sweden have large amounts of hydropower which offer significant flexibility (Green, 2012).

---

5. This aggregation did subsequently occur, as a part of the expansion of the Midwest Independent System Operator footprint.

It is not always economically viable to increase system size. For example, in Western Australia, the SWIS system is highly isolated. In this case, linking with neighboring systems would require extensive transmission network augmentation and is not likely to be a cost-effective alternative in the near term (an interconnector to the east coast grid would be around 2000 km long, and has been estimated at costing in the range AU$2 billion to AU$3.4 billion (BZE, 2010), and transmission losses could be high. This would be difficult to justify with the SWIS being only 4 GW in size).

However, in many parts of the United States there are closely neighboring small electricity systems that function in a relatively isolated way. The boundary between the systems is often purely regulatory in nature, and thereby reducing the ability of the system to integrate variable generation. While it may be politically challenging to combine the management of these systems entirely, many of the benefits could be accessed via a more limited sharing of certain functions, such as system balancing. In this way, balancing areas can retain their autonomy, and virtually aggregate via combination of their Area Control Errors (ACEs). When implemented with fast markets, this would allow trading of opposite ramping requirements which would reduce the costs for both parties (Milligan and Kirby, 2010). There is now significant interest in developing cooperative agreements between these small balancing areas that would provide some of the benefits of consolidation.

There is also interest in Europe for moving toward a competitive single European market for electricity. To this end, France, Belgium, and the Netherlands introduced market coupling of their day-ahead spot markets in 2006, and Germany and the countries in Nord Pool[6] joined in 2010 (Borggrefe and Neuhoff, 2011). Although further coupling is recognized as desirable, it has been stymied by the fact that many European markets significantly differ in their market setup, creating barriers to harmonization (Vandezande et al., 2010).

3.1 Incentives to Limit Network Congestion

In order to access the benefits of a large system, it is important that the transmission network is robust. Extensive transmission congestion will limit the value of large systems. This may mean that the value of robust transmission networks increases as variable generation penetration increases, and network investment studies should be revisited as penetration levels change.

Regulatory frameworks for networks are covered in Chapter 6. However, without requiring extensive new transmission augmentation, market incentives can be provided for generation to locate in areas that are relatively free of network congestion, helping to prevent significant congestion. For example, locational pricing provides a strong signal to accurately indicate the

---

6. Denmark, Finland, Norway, and Sweden.

value of geographical diversity. Prices at locations that are heavily oversupplied with renewable generation will be low, incentivizing development elsewhere. This has been observed in the Australian NEM, with renewable penetration levels in the South Australian pricing region exceeding 20%; low prices in that region are now incentivizing wind development in other regions despite lower quality wind resources (Cutler et al., 2011; ROAM, 2011a).

Nodal pricing systems are applied in the markets in New Zealand, Texas (ERCOT), Singapore, New York, New England, and the PJM Interconnection (PJM), among others (Frontier Economics, 2008; Synergies, 2009). The use of nodal pricing is also being considered in Germany to help address disconnects between the areas of renewable energy development and loads (Cochran et al., 2012). Other markets, such as Nord Pool (covering the Nordic countries) and the Australian NEM, apply a regional (zonal) pricing model. This provides a more general geographical indication to project developers of regions that are presently oversupplied with renewable generation (Cochran et al., 2012). Fixed feed-in tariffs (FiTs) may, however, hide or diminish these signals (Hiroux and Saguan, 2009) and additional revenue from access to higher quality energy resources may overshadow them (Newbery, 2011; ROAM, 2011a). Germany is investigating providing higher FiTs to regions of lower resource to incentivize geographically diverse generation (Cochran et al., 2012). However, a centrally determined process of adjusting FiTs is inevitably going to be slower, less granular, and more subject to vested influence than an automatic process such as nodal pricing. For further discussion on nodal pricing, refer to Chapter 6.

4 ACCURATE PRICE SIGNALS

Price signals that provide an accurate indication of the value of a particular action to the system are an effective way of incentivizing efficient behaviors from market participants. There are two distinct aspects to consider when seeking price signals ensuring sufficient system flexibility (Milligan and Kirby, 2010):

1. Short-run access to flexibility—Ensuring that the market rules allow access to the full degree of technical flexibility available in the installed generators in the system and incentivizes its operation appropriately. In many existing market designs, much of the flexibility available in the system cannot be accessed. Changes to market design (as outlined throughout this chapter) can ensure that the full potential for flexibility in the system is available and incentivized to operate appropriately.

2. Long-run investment in flexibility—Ensuring that there is sufficient technical flexibility installed in the system. At some level of variable generation penetration, it will become necessary to increase the technical

flexibility of the fleet. Thermal generators in the system will need the ability to ramp more quickly, achieve lower turndown ratios and cycle more efficiently (Milligan and Kirby, 2010).

If incentives are correctly designed, new thermal generation installed in the system will be able to undertake a market analysis to compare the anticipated market value of increased flexibility against its cost. Similarly, existing generators will be able to assess the costs and benefits of retrofitting to achieve a higher level of flexibility, once it becomes apparent to the market that it is required.

Fast markets (with short dispatch intervals) naturally provide incentives for flexibility. For example, if variable generation ramps suddenly upward, there will be a temporary oversupply in the system. The dispatch in the next interval will therefore be at a lower market price. For some generators, the market price will no longer exceed their short-run marginal costs. Although their bids may reflect this, such that the system operator will direct them to reduce their dispatch, if they do not possess sufficient flexibility to ramp down quickly they will remain exposed to periods of negative profitability.

Conversely, if variable generation suddenly ramps downward, there will be a temporary undersupply on the system. This will require additional units to be dispatched in the following dispatch interval, causing the market price to rise. If the marginal units already operating in the system cannot ramp upward quickly enough to balance the market, higher priced, more flexible bid bands will be dispatched. This can cause prices to temporarily spike to significant levels. In this situation, the generators that have the ability to ramp upward will capture the maximum benefit (Milligan and Kirby, 2010).

Operation in this model requires a dispatch algorithm that includes information about the ramp rate limitations of each unit. In many systems in the United States, these limits are provided with market participant bids and used in the co-optimized scheduling process (Borggrefe and Neuhoff, 2011). This model also requires the market to have a high market price cap, and a low market price floor to provide sufficient incentives. This may cause issues for markets that feature a capacity mechanism coupled with a low price ceiling in the energy market.

Analysis of 5 min price data from the New York (NYISO) market reveals a significant response incentive, with the high and low prices for each interval in an hour differing by US$17.41/MWh on average in 2006. This suggests that the 5 min market continuously sends a strong price signal to market participants to move up or down as required (Milligan and Kirby, 2010). Furthermore, since the NYISO 5 min market is observed to clear at a lower average price than the day-ahead hourly market, it is suggested that this 5 min flexibility is obtained at no additional cost to the system (Milligan and Kirby, 2010).

4.1 Exposure to Negative Prices

Variable generation can cause issues during low-load periods, particularly in systems that have a low-load factor and a significant quantity of inflexible "base load" generation installed. In these systems, during high generation, low demand periods, the sum of thermal unit minimum loads, and variable generation may exceed the load. This can be addressed by allowing significant negative prices to occur. Owners of thermal units and wind generation can then determine on an economic basis at what price they are prepared to curtail or shutdown.

For these reasons, many markets allow negative prices to occur. The Australian NEM, Germany, Ireland, MISO-USA, and Netherlands markets all allow negative prices to a specified market floor, and markets in Great Britain, Norway, PJM-USA, Slovenia, Sweden, and Switzerland operate markets that allow negative prices with no market floor being defined.

Since cycling costs can be significant, slightly negative prices will not be sufficient to incentivize the shutdown of many thermal units. For example, a study of the German market revealed that even when prices were as low as −500€/MWh, 46% of the available capacity remained online, including 83% of nuclear power plants and 71% of lignite power plants (Nicolosi, 2010).

The nature of renewable support schemes can also be a significant factor in determining the effectiveness of negative pricing. For example, FiTs that supply a renewable generator with a fixed price for each megawatt hour supplied to the system shield renewable generators from negative prices, removing incentives to curtail. Green certificate schemes can have a similar effect where renewable generators have entered into fixed price power purchase agreement (PPA) to manage market risk. For example, in the Australian NEM most wind farms are under PPAs, and despite a market price floor of − AU$1000/MWh, many wind farms do not regularly curtail output in response to negative pricing events, as illustrated in Figure 25.11. Historically, cumulative payments from wind farms due to generating during negative price periods have been AU$2000−10,000/MW per annum; these payments would be recovered from their PPA counterparties. As the penetration of variable generation increases, the nature of these renewable support schemes and the incentives that they provide for beneficial market behavior will need to be closely scrutinized.

5 EFFICIENT ANCILLARY SERVICES MARKETS

Most power systems manage frequency via a series of different kinds of reserves, as listed in Table 25.2 with the terminology adopted in this report. The different kinds of reserves are employed to manage variability in load and generation over different timescales.

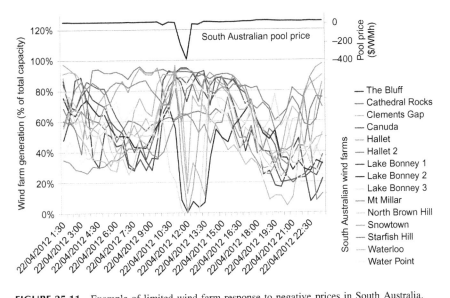

FIGURE 25.11   Example of limited wind farm response to negative prices in South Australia. *Source: Generation and price data from AEMO.*

Fast fluctuations (seconds to minutes) are evident from variable generation, however, a range of studies indicate that these are smoothed with system aggregation, and can be adequately handled by traditional methods used to manage fast load fluctuations already present in power systems (Frunt, 2011; Holttinen, 2003).[7] Similarly, variable generation typically does not ramp rapidly enough to constitute a contingency event, even during extreme weather events, due to geographical smoothing (Holttinen et al., 2007).

Longer timescale, gradual variations in wind generation are anticipated to cause a higher impact on power systems, particularly increasing the requirement for noncontingency secondary frequency control reserves (regulation). This service provides adjustments to the supply−demand balance over timescales of seconds to minutes to maintain system frequency. The amount of additional reserves that are required with an increase in variable generation will vary substantially from system to system, depending upon many of the market design characteristics outlined in this chapter. A review of a wide range of studies shows that with wind penetration of 20%, additional reserve

---

7. An increase in variable generation penetration could act to reduce the system inertia via replacement of thermal plant (Fox and Flynn, 2007). With decreased inertia, system frequency is likely to become less stable. At high penetration levels it may become necessary to provide direct incentives for the provision of system inertia, although this is not common practice at present.

TABLE 25.2 Types of Ancillary Services

Ancillary Service Type		Operation	Application	Time Frame
Primary frequency control		Not directed (automatic)	Instantaneous and continuous	<1 min
Secondary frequency control	Regulation reserves		Continuous	1−30 min
	Contingency reserves (spinning)	Directed	Event-driven	
	Contingency reserves (nonspinning)			
Tertiary network control	Replacement reserves		Event-driven	30−120 min
Voltage control		Directed	Event-driven	
Black start			Event-driven	

Source: Modified from CIGRE (2010)

requirements can range from 4% to 18% of installed wind capacity (Holttinen et al., 2009).

By comparison to the other ancillary services, regulation has relatively higher costs per megawatt of reserve provided (Milligan and Kirby, 2010). Therefore, as the penetration of variable generation increases and the requirement for regulation reserve increases, costs can increase considerably (Riesz et al., 2011; ROAM, 2010, 2011b). The increase in costs can be minimized by the implementation of fast, large markets, and an efficient competitive market for regulation services, allowing all market participants to contribute to its provision where technically feasible and economically justified. A competitive market for the procurement of regulation reserve is also important in incentivizing the installation of more flexible units. When prices in the regulation market become high (indicating a lack of units with sufficient capability), new units may be installed purely with the intention of providing this service to the market.

5.1 Co-Optimization of Regulation Reserve and Energy Scheduling

The provision of regulation service necessitates dispatch at a set point between minimum and maximum loads, to allow for responsive movement upward and downward (depending upon the quantity of raise and lower

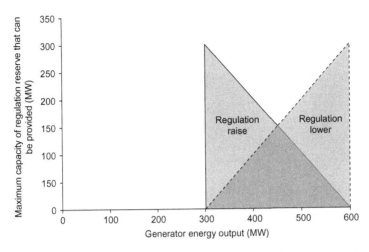

FIGURE 25.12   Example of the codependency of energy and regulation reserve provision for a hypothetical 600 MW thermal unit. *Source: Adapted from AEMO (2010).*

regulation to be provided). This means that the provision of regulation service affects the scheduling process. Figure 25.12 provides an illustrative example of the codependency of energy and regulation reserve provision. The example illustrated is for a hypothetical 600 MW unit with a 300 MW minimum load. When operating at minimum load, the unit can provide up to a maximum of 300 MW of "regulation raise" service but no "regulation lower" service. By contrast, when the unit is operating at its maximum load of 600 MW, it can provide up to a maximum of 300 MW of "regulation lower" service but no "regulation raise" service. The unit illustrated is assumed to be very flexible and capable of fast ramping; if it had ramp rate limitations, the maximum amount of regulation service available may be capped to a lower value (Thorncraft, 2007).

Ideally, energy and regulation are centrally co-optimized during the scheduling process, with generators submitting bids for each service reflecting their individual costs and economic drivers. Algorithms can determine the least cost dispatch that meets the energy and regulation requirements for each period. The Australian NEM and New Zealand markets both operate with co-optimization in real time. Other markets (particularly in the United States) co-optimize scheduling of reserves and energy on a day-ahead basis (not in real time).

5.2 Dynamic Setting of Regulation Reserves

The quantity of regulation reserve that is required to adequately maintain system frequency will change over time depending upon a range of factors. However, some systems set the regulation reserve requirement statically,

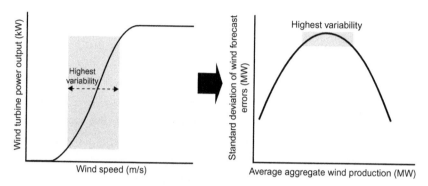

FIGURE 25.13   The shape of wind turbine power curves leads to a parabolic dependency of the standard deviation in wind forecast error upon aggregate wind production. *Source: Adapted from Ela et al. (2010).*

adjusted on a daily or even annual basis as the average statistical characteristics of the load and variable generation change gradually over time. This means that the regulation reserve in many periods will exceed the actual system requirement, creating unnecessary cost.

Regulating reserves can be set dynamically in each dispatch interval. For example, in the Australian NEM the regulation reserve is determined dynamically based upon the time error (aggregate frequency error), within a predetermined range. If the time error is within the ± 1.5 second band, the regulation reserve is set to 130/120 MW (raise/lower). If the time error is outside this band, an extra 60 MW of regulation per 1 s deviation outside the band is added, with an upper limit of 250 MW (AEMO, 2010). This fully dynamic, real-time approach to reserve setting ensures that only the reserves that are actually required in each dispatch interval are scheduled, thus minimizing costs.

An alternative approach to dynamic setting of the regulation reserve is to use a simple equation to capture the impacts of key system features that drive regulation requirements. For example, the level at which aggregate wind generation is operating correlates with the standard deviation of aggregate wind forecast errors, as illustrated in Figure 25.13. This is due to the underlying shape of the wind turbine power curve, in which the central (steepest) section features the highest variability. This trend can be determined based upon historical data for a particular power system and can be used to define a simple equation that determines the regulation reserve requirement based upon the level of wind generation in the dispatch interval of interest (combined with load characteristics) (Ela et al., 2010, 2011). This approach has been explored for application in the United States in the power system operated by the WestConnect group of utilities in Arizona, Colorado, Nevada, New Mexico, and Wyoming (NREL, 2010), and in the Eastern Interconnection (NREL, 2011). Although this approach is not as ideal as full optimization against the time error, it does allow reserves to be reduced at times when wind generation is low or high, and smaller reserves are known to be sufficient.

In determining the level of regulation reserves required, it could also be worth considering the other reserves already provided to the system. Variable generation ramps are typically relatively slow and moderate in scale, by comparison with forced outages at thermal units. However, rare "tail events" can occur occasionally. This large but infrequent response is likely to be more economically obtained from existing contingency reserves. Rapid response can be obtained from spinning reserves that are already required to manage thermal generator-forced outages, and would likely need to be employed only rarely for wind management. Slower large responses are likely to be more frequent, and can be provided at low cost via nonspinning reserves. In general, nonspinning reserve is an order of magnitude less expensive than spinning reserve (Milligan and Kirby, 2010). This means that nonspinning reserve could prove to be a valuable tool for managing variability over the 15−30 min time frame cost effectively (Walling et al., 2008). In systems where hydro generation is available, it may be possible to cost effectively provide nonspinning reserves with response times in the range 5−30 s.

In fast markets with a high price ceiling, it may not even be necessary to explicitly contract or schedule nonspinning reserve, instead allowing market signals to incentivize rapid market response in the event of sudden wind ramps. This approach is applied in the Australian NEM, where there is no defined nonspinning reserve ancillary service market. With detailed knowledge of the statistical behavior of variable generation on the system, it should be possible to optimize the allocation of rare and more frequent events to the different types of reserves, minimizing costs (Holttinen, 2003; Holttinen et al., 2009; Milligan and Kirby, 2010).

5.3 Real-Time, Causer-Pays Allocation of Regulation Reserve Costs

The uncertainty and variability of variable generation can be reduced to very low levels by curtailing their output to a level below their available generation (Milligan and Kirby, 2010). Most modern variable generation technologies have this capability. However, this is usually not economically attractive; given their relatively high capital cost and very low operating cost, it is usually most economically efficient to operate variable generators with minimal curtailment, except in the event of a threat to system security or during negative pricing events.

Curtailment of variable generation in a limited form may prove economic in some cases. For example, large sudden generation increases could be curtailed to a slower level, allowing the scheduling of smaller "regulation lower" reserves. The costs and benefits of this strategy would need to be carefully considered, ensuring that the value in reduced variability exceeds the value of the renewable energy curtailed. Any mandatory curtailment of variable generation is likely to be best done in aggregate (rather than by application of ramp rate limitations on individual plant), to fully utilize geographic diversity and limit curtailment to circumstances when it is actually economically justified.

It is possible to create appropriate incentives such that variable generation can conduct this cost–benefit analysis on an individual basis. For example, where variable generation is fully exposed to the real-time cost of providing regulation reserves associated with the operation of their individual plant, they may choose to limit ramp rates in order to reduce costs, where it is appropriate to do so. This would require:

- Regulation reserves to be charged on a real-time basis based upon the cost of provision in each dispatch interval;
- Regulation reserves to be charged to individual generators on an individual "causer-pays" basis; and
- Distinguishing between raise and lower regulation reserves, attributing separately the costs associated with providing each.

Causer-pays regulation pricing in this manner may also assist with the development of projects that minimize variability. For example, wind turbine power curves are nonlinear, suggesting that wind farm variability could be minimized by the careful selection of turbines to minimize variability associated with the site-specific wind resource. Decisions would need to be weighed against the alternative of selecting a turbine type that is optimized solely for maximum energy generation at the site. Accurate causer-pays pricing for regulation services allows developers to directly consider these factors in their decision-making process.

5.4 Implement Advanced Forecasting

More broadly, the uncertainty in wind production (and hence the reserve requirements) can be reduced through high-quality renewable generation forecasting systems. System scheduling (determination of the system dispatch) is performed at some time prior to the relevant dispatch interval occurring, meaning that a forecast of variable generation provides the best available indication of the quantity of aggregate variable generation that can be expected to be operating. The optimal system dispatch is therefore determined based upon that forecast.

High-quality forecasting systems are now available for wind generation (Von Roon and Wagner, 2009) and continuing improvement is expected (DENA, 2010). With increasing deployment of solar photovoltaics a similar focus is being placed on developing solar forecasting tools. When used in the scheduling process to determine unit commitment and reserve requirements, advanced forecasting tools can minimize the ramping requirements of thermal plants and reduce the need for reserves, potentially providing significant cost savings (NREL, 2010).

In order for maximum value to be extracted, forecasting tools are ideally integrated into centralized electricity system operation. To ensure that forecasts are accurate, variable generation operators need to provide continual information to system operators about turbine availability, power and wind speeds. A lack of turbine availability data, in particular, can be significantly detrimental to the accuracy and usefulness of forecasts, particularly during periods such as wind farm commissioning (Cochran et al., 2012).

6 GROSS POOL SINGLE PLATFORM MARKETS

Electricity markets can be fundamentally designed in a range of different ways, as illustrated in Figure 25.14. Some markets feature only a single platform via which all electricity is traded. The Australian NEM is an example, where all electricity is traded via a single gross pool market ("gross" in this context specifies that it is mandatory to trade all electricity via the pool. Contracts and other financial instruments can be utilized to manage spot price risk and increase certainty). Some electricity systems do not feature a market at all, and instead dispatch is managed centrally via system operator.

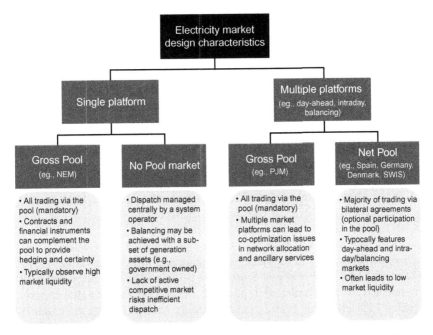

FIGURE 25.14 Fundamental characteristics of market design. *Source: Developed by authors.*

Many markets feature multiple trading platforms, such as day-ahead markets, intraday markets, and balancing markets. This can include bilateral trading outside of the pool (making it a "net pool" market) or trading via the pool may be mandatory (making it a gross pool market).

These fundamental design characteristics affect the ability of the market to efficiently integrate variable generation. For example, in net pool markets with a strong component of bilateral trading, a significant fraction of demand may be met by generators independently setting their output (e.g., based on supply contracts). This can restrict the options that the market operator has to respond to changes in the supply−demand balance. By contrast, mandatory gross pool markets where generator dispatch is determined by an independent system operator based on submitted price and generation offers can increase the flexibility of the system by allowing clear signaling between generators (Green, 2008).

Net pool markets, or bilateral trading markets, often feature intraday markets with low liquidity. For example, in 2009, exchanges in the Spanish intraday market represented 10−16% of exchanges in the spot market; in Germany they represented 4.2% of exchanges, and in Denmark they represented only 0.5% of exchanges (Barquin, 2011). Low market liquidity can lead to price volatility and uncertainty. In this situation, variable generators may prefer to avoid volatile intraday market prices, and simply pay imbalance payments in real time (Maupas, 2008). It is possible to increase market liquidity by making participation in the relevant market mandatory. For example, in 2009 it was mandated that German operators were required to use the intraday market for the procurement of balancing services. This resulted in significantly increased market liquidity (Borggrefe and Neuhoff, 2011). Spain has responded to the issue by introducing six consecutive intraday clearing auctions that allow for a full rescheduling. Since all interactions are focused on six intraday auctions, they exhibit more liquidity than observed in other European markets with continuous intraday trading (Borggrefe and Neuhoff, 2011).

Gross pool-type trading arrangements, such as those implemented in most markets in the United States generally do not suffer from low market liquidity. For example, consider the PJM market, one of the largest electricity markets in the United States of America comprising all or parts of 13 states and the District of Columbia. In this system, market participants provide firm schedules and flexible bids (including technical parameters, such as ramp rates and part load constraints) to the system operator, who calculates an optimal dispatch close to real time. Transmission allocation and reserves are optimized in the same process. Since most day-ahead and intraday activities are pursued on the centralized platform of the system operator, market liquidity is much higher than in most European markets (where the majority of trades are bilateral) (Borggrefe and Neuhoff, 2011).

Multiple pricing platforms can also be problematic for the optimization of reserve scheduling and transmission allocation. Many of the current market designs in the European Union with separate day-ahead, intraday, and balancing markets are based upon three different organizational schemes. This can mean that reserves and network allocation may be optimized in one platform (e.g., the day-ahead market), but may not be optimized across all three schemes. As stated by Smeers (2008):

these multiple arrangements violate the finance view that day-ahead, intra-day and real time are just different steps of a single trading process and hence require a single trading platform.

Particularly in European power market designs, transmission capacity is usually allocated for long-term and day-ahead energy sales. There are only limited instances where it is possible to reschedule power flows between countries during the day (Borggrefe and Neuhoff, 2011). As the quantity of variable generation in the system increases the original network allocation will become increasingly suboptimal, meaning that the network is being utilized less efficiently. Determination of network allocation in a single optimization process (on a single platform), co-optimized with energy and reserve scheduling in real time, alleviates this issue (refer to Chapter 8 for further details).

Single platform, mandatory gross pool-type trading arrangements can avoid these issues, particularly where a single market price applies for each interval. This is the case in the Australian NEM, for example. In this market, all energy is traded through a single real-time (5 min) market, and participants manage risk via separate contractual arrangements. A series of constraint equations identify dispatch combinations at the limit of network capability, and are used in real-time scheduling to ensure that the dispatch is within the necessary technical limits. This creates a highly transparent, highly liquid market that allows real-time optimization of network allocation and reserves.

6.1 Summary of Selected Markets

Table 25.3 provides a summary of the key features discussed in this chapter in different markets.

7 CONCLUSION

There are now significant drivers for the development of renewable generation in electricity markets around the world. This necessitates the evolution of global electricity markets toward designs that more efficiently integrate renewable technologies.

TABLE 25.3 Summary of Key Features in Various Markets

	Great Britain	Ireland	Nord Pool	PJM (USA)	Brazil	NEM (Australia)	SWIS (Australia)
Market speed							
Dispatch interval	30 min	30 min	60 min	5 min	Centralized dispatch by system operator	5 min	30 min
Delay from scheduling to dispatch	1 h	20 h	1 h	5 min		5 min	Day ahead
Market size							
Peak demand (GW)	59	6–7	72	158	75	30–35	4.5
Annual energy (TWh)	319	35	315	700–750	470	190–200	18
Locational pricing							
Area or nodal	Area	Area	Area	Nodal	Area	Area	Area
Negative pricing							
Market price floor	None	−€100/MWh	−€200/MWh	None	No negative prices	AU$ −1000/MWh	AU$ −314/MWh
Market structure							
Single/multiple platform	Multiple (bilateral + balancing)	Single, moving to multiple	Multiple	Multiple	Multiple	Single platform	Multiple platform
Gross/net pool	Net	Gross	Gross	Net	Gross	Gross	Net

Source: AEMO (2012), AER (2012), Barroso (2012), CIGRE (2012), EirGrid (2011), IMO WA (2012), National Grid (2011a, 2011b), Nord Pool (2009, 2012)

Many existing electricity markets have been designed with a portfolio of thermal generators in mind. Therefore, electricity market designs of the past have aimed to address the particular challenges that face these technologies. Many thermal units, particularly those intended for "base load" operation can be relatively inflexible, featuring slow ramp rates and high minimum loads. Unit cycling is considered an ordeal, performed rarely and associated with high risks and costs. Significant notice is preferred in order to coordinate a unit shutdown. For these reasons, many of today's electricity markets have been designed to allow market participants high certainty over generation levels a day ahead. Net pool bilateral trade arrangements have arisen in many markets, since they provide a simple transition from the contractual arrangements that may have existed previously, they remove the need for complex legal frameworks (addressing factors such as collusion), and can form an appropriate compromise in regions where the generation is predominantly government owned and operated.

These electricity market designs have worked adequately in the past, but are likely to create artificial barriers to the entry of variable generation and increase integration costs.

For effective integration of the variable generation technologies likely to predominate in future, global electricity markets must evolve toward designs that allow access to the full degree of technical flexibility in the system. The focus of this chapter has been to identify the features of electricity markets that achieve this. In many cases, market reform along the lines proposed in this chapter can significantly reduce the integration costs for variable technologies.

While the focus of this chapter has been in identifying market features that facilitate the integration of variable generation, the majority of the measures proposed will increase system efficiency even in its absence. Their value is greatly heightened if variable integration is included, but they are worthwhile pursuits for electricity market reform in any case. Thus, market evolution toward large, fast markets with accurate pricing signals applied to all technologies could be considered a more technology agnostic and therefore superior approach to electricity market design.

REFERENCES

AEMO, 2010. Guide to Ancillary Services in the National Electricity Market. Australian Energy Market Operator. Available from: <http://www.aemo.com.au>.

AEMO, 2012. National Electricity Forecasting Report for the National Electricity Market (NEM). Australian Energy Market Operator.

AER, 2012. Seasonal Peak Demand (NEM). Australian Energy Regulator. Available from: <http://www.aer.gov.au/node/9766/>.

Ahlstrom, M., 2008. Short-term Forecasting: Integration of Forecast Data into Utility Operations Planning Tools. Presented at the Utility Wind Integration Group/National Renewable Energy Laboratory Wind Forecasting Applications to Utility Planning and Operations, St. Paul.

Barquin, J., 2011. Current designs and expected evolutions of day-ahead, intra-day and balancing market/mechanisms in Europe. OPTIMATE—Model for Pan-European Electricity Markets, Seventh Framework Programme.

Barroso, L., 2012. The Brazilian Electricity Market—Ensuring Supply Adequacy Through Energy Contract Auctions. Available from: <http://www.florence-school.eu/portal/page/portal/FSR_HOME/ENERGY/Training/Summer_School/Presentations/2012.06.29%20Luiz%20Barroso.pdf/>.

Bloomberg, 2012. Global Trends in Renewable Energy Investment. Bloomberg New Energy Finance, Frankfurt School UNEP Collaborating Centre for Climate and Sustainability Energy Finance.

Borggrefe, F., Neuhoff, K., 2011. Balancing and Intra-day Market Design: Options for Wind Integration. European Smart Power Market Project, C.P. Initiative.

BREE, 2012. Australian Energy Technology Assessment. Bureau of Resources and Energy Economics.

BZE, 2010. Australian Sustainable Energy—Zero Carbon Australia Stationary Energy Plan. Beyond Zero Emissions, University of Melbourne.

CIGRE, 2010. Ancillary Services: An Overview of International Practices. Working Group C5.06.

CIGRE, 2012. Market Design for Large Scale Integration of Intermittent Renewable Energy Sources and Demand Side Management. CIGRE Working Group C5−11.

Cochran, J., Bird, L., Heeter, J., Arent, D.J., 2012. Integrating Variable Renewable Energy in Electric Power Markets: Best Practices from International Experience. National Renewable Energy Laboratory (NREL), Joint Institute for Strategic Energy Analysis (JISEA), Clean Energy Solutions Centre, Clean Energy Ministerial.

Connolly, D., Lund, H., Mathiesen, B., Leahy, M., 2011. The first step towards a 100% renewable energy-system for Ireland. Appl. Energy 88, 502−507.

Cutler, N.J., Boerema, N.D., MacGill, I.F., Outhred, H.R., 2011. High penetration wind generation impacts on spot prices in the Australian national electricity market. Energ. Policy 39, 5939−5949.

Dany, G., Haubrich, H.J., 2000. Anforderungen an die Kraftwerksreserve bie hoher Windenergieeinspeisung—Requirement for power reserve with high wind energy supply. Energiewirtschaftliche Tagesfragen 50 (12), 890−894.

DENA, 2005. Planning of the Grid Integration of Wind Energy in Germany Onshore and Offshore up to the Year 2020 (Dena Grid Study). German Energy Agency.

DENA, 2010. DENA Grid Study II—Study for the Integration of a Share of 30% Renewable Energy in the German Electricity Market. German Energy Agency.

EirGrid, 2011. EirGrid plc—Annual Report. Available from: <http://www.eirgrid.com/media/EirGrid%20Annual%20Report%202011.pdf>.

Ela, E., Milligan, M., Kirby, B., 2011. Operating Reserves and Variable Generation. National Renewable Energy Laboratory (NREL).

Ela, E., Milligan, M., Kirby, B., Lannoye, E., Flynn, D., O'Malley, M., et al., 2010. Evolution of Operating Reserve Determination in Wind Power Integration Studies. National Renewable Energy Laboratory (NREL).

Ernst, B., 1999. Analysis of Wind Power Ancillary Services Characteristics with German 250 MW Wind Data. National Renewable Energy Laboratory (NREL).

Focken, U., Lange, M., Monnich, K., Waldl, H.P., Beyer, H.G., Luig, A., 2002. Short term prediction of the aggregated power output of wind farms—a statistical analysis of the reduction of the prediction error by spatial smoothing effects. J. Wind Eng. Ind. Aerodyn. 90 (3), 231−246.

Fox, B., Flynn, D., 2007. Wind Power Integration. The Institution of Engineering and Technology, Stenvenage.

Frontier Economics, 2008. Generator Nodal Pricing—a review of theory and practical application. A report prepared for the Australian Energy Market Commission.

Frunt, J., 2011. Analysis of Balancing Requirements in Future Sustainable and Reliable Power Systems. TU Eindhoven.

GE Energy, 2005. The effects of integrating wind power on transmission system planning, reliability and operations: report on Phase 2. Prepared for The New York State Energy Research and Development Authority.

Green, R., 2012. How Denmark manages its wind power. Int. Assoc. Energy Econ. 9−11. Available from: <https://www.iaee.org/>.

Green, R.J., 2008. Electricity wholesale markets: designs now and in a low-carbon future. Q. J. IAEE's Energy Econ. Edu. Found. 29, 2.

Hiroux, C., Saguan, M., 2009. Large-Scale Wind Power in European Electricity Markets: Time for Revisiting Support Schemes and Market Designs? Larsen.

Hodge, B.-M., Milligan, M., 2011. Wind Power Forecasting Error Distributions over Multiple Timescales. National Renewable Energy Laboratory (NREL).

Holttinen, H., Lemstrom, B., Meibom, P., Bindner, H., Orths, A., van Hulle, F., et al., 2007. Design and operation of power systems with large amounts of wind power—State-of-the-art report. IEA Wind, Task 25.

Holttinen, H., 2003. Hourly Wind Power Variations and their Impact on the Nordic Power System Operation (Licenciate Thesis). Helsinki University of Technology, Department of Engineering, Physics and Mathematics.

Holttinen, H., Kiviluoma, J., Estanqueiro, A., Gomez-Lazaro, E., Rawn, B., Dobschinski, J., et al., 2011. Variability of load and net load in case of large scale distributed wind power. Proceedings of the 10th International Workshop on Large-scale Integration of Wind Power into Power Systems.

Holttinen, H., Meibom, P., Orths, A., Hulle, F.V.,Lange, B., 2009. Design and Operation of Power Systems with Large Amounts of Wind Power. Final report, Phase one 2006−08. IEA Wind Tas 25.

Hulle, F.V., Fichaux, N., Sinner, A.-F., Morthorst, P.E., Munksgaard, J., Ray, S., 2010. Powering Europe: Wind Energy and the Electricity Grid. European Wind Energy Association (EWEA).

IEA, 2011a. Harnessing Variable Renewables. International Energy Agency.

IEA, 2011b. World Energy Outlook. International Energy Agency.

IMO WA, 2012. Statement of Opportunities. Independent Market Operator of Western Australia.

ISO/RTO Council, 2007. Increasing Renewable Resources: How ISOs and RTOs Are Helping Meet This Public Policy Objective. Available from: <http://www.isorto.org/atf/cf/%7B5B4E85C6-7EAC-40A0-8DC3-003829518EBD%7D/IRC_Demand_Renewables_Glossy.pdf>.

Kirby, B., Hirst, E., 2000. Customer-Specific Metrics for the Regulation and Load-Following Ancillary Services. Oak Ridge National Laboratory.

Krajacic, G., Duic, N., Carvalho, M.D., 2011. How to achieve a 100% RES electricity supply for Portugal? Appl. Energy 88, 508–517.

Lehmann, P., 2012. Carbon lock-out: advancing renewable energy policy in Europe. Energies 5, 323–354.

Lund, H., Mathiesen, B.V., 2009. Energy system analysis of 100% renewable energy systems—the case of Denmark in years 2030 and 2050. Energy 34, 524–531.

Maggio, D., 2011. ERCOT Nodal—How's It Going. Presentation to 2011 Texas Renewables. Available from: <http://www.treia.org/assets/2011/Events/TR-2011/presentations/1.31c_mon_1030_maggio_ercot-nodal_hows-it-going.10282011.pdf>.

Mason, I.G., Page, S.C., Williamson, A.G., 2010. A 100% renewable electricity generation system for New Zealand utilising hydro, wind, geothermal and biomass resources. Energ. Policy 38, 3973–3984.

Mathiesen, B.V., Lund, H., Karlsson, K., 2011. 100% renewable energy systems, climate mitigation and economic growth. Appl. Energy 88, 488–501.

Maupas, F., 2008. Analyse de l'impact Economique de l'alea Eolien sur la Gestion de l'equilibre d'un Systeme Electrique. Paris-XI.

Milligan, M., Kirby, B., 2007. Impact of Balancing Areas Size, Obligation Sharing, and Ramping Capability on Wind Integration. WindPower. NREL. Available from: <http://www.nrel.gov/docs/fy07osti/41809.pdf>.

Milligan, M., Kirby, B., 2008. The impact of balancing area size and ramping requirements on Wind integration. Wind Eng. 32 (4), 399–414.

Milligan, M., Kirby, B., 2009. Capacity requirements to support inter-balancing area wind delivery. Presented at WindPower 2009, Chicago, IL.

Milligan, M., Kirby, B., 2010. Market Characteristics for Efficient Integration of Variable Generation in the Western Interconnection. National Renewable Energy Laboratory (NREL).

Milligan, M., Porter, K., DeMeo, E., Denholm, P., Holttinen, H., Kirby, B., et al., 2009. Wind power myths debunked. IEEE Power Energy Mag. 7 (6), 89–99.

Monroe, C., Cate, G., Brown, C.J., Moore, J., 2011. SPP EIS Overview BPA. Southwest Power Pool (SPP).

Muesgens, F., Neuhoff, K., 2002. Modelling Dynamic Constraints in Electricity Markets and the Costs of Uncertain Wind Output. EPRG Working Paper Series 0514.

National Grid, 2011a. 2011 NETS Seven Year Statement: Chapter 10—Market Overview.

National Grid, 2011b. National Electricity Transmission System Seven Year Statement.

Newbery, D., 2011. Reforming Competitive Electricity Markets to Meet Environmental Targets. University of Cambridge, Electricity Policy Research Group.

Nicolosi, M., 2010. Wind power integration and power system flexibility—An empirical analysis of extreme events in Germany under the new negative price regime. Energ. Policy 38 (11), 7257–7268.

Nord Pool, 2009. No. 16/2009 Nord Pool Spot implements negative price floor in Elspot from October 2009. Available from: <http://www.nordpoolspot.com/Message-center-container/Exchange-list/Exchange-information/No162009-Nord-Pool-Spot-implements-negative-price-floor-in-Elspot-from-October-2009-/>.

Nord Pool, 2012. Available from: <http://www.nordpoolspot.com/>.

NREL, 2010. Western Wind and Solar Integration Study. National Renewable Energy Laboratory, Work performed by GE Energy.

NREL, 2011. Eastern Wind Integration and Transmission Study. National Renewable Energy Laboratory, Work Conducted by EnerNex Corporation.

NREL, 2012. Renewable Electricity Futures Study. National Renewable Energy Laboratory.

REN21, 2012. Renewables 2012 Global Status Report. Renewable Energy Policy Network for the 21st Century. Available from: <http://www.ren21.net/REN21Activities/GlobalStatus Report.aspx>.

Riesz, J., Shiao, F.-S., Gilmore, J., Yeowart, D., Turley, A., Rose, I., 2011. Frequency control ancillary service requirements with wind generation—Australian projections. Proceedings of the Wind and Solar Integration Workshop, Aarhus, Denmark.

ROAM, 2010. Assessment of FCS and Technical Rules. ROAM Consulting Report to Independent Market Operator of Western Australia. Available from: <http://www.imowa. com.au/f3086,1258199/Report_Imo00016_to_IMO_2010-11-03a_FINAL.pdf>.

ROAM, 2011a. Assessing the Capacity of Commercially Profitable Wind Generation in South Australia. ROAM Consulting Report to ElectraNet. Available from: <http://www. electranet.com.au/assets/Uploads-2/Wind-Generation-in-SA.pdf>.

ROAM, 2011b. Impact of the LRET on the costs of FCAS, NCAS and Transmission. Report to the Australian Energy Market Commission (AEMC). ROAM Consulting. Available from: <http://www.aemc.gov.au/Media/docs/ROAM%20Report-566727a5-bfb8-4c7d-a88a-0b946 9e5d19c-0.pdf>.

Rohrig, K., 2005. Entwicklung eines Rechenmodells zur Windleistungsprognose. Abschlussbericht Forschungsvorhaben Nr. 0329915A, gefördert durch Bundesministeriums für Umwelt, Naturschutz und Reaktorsicherheit (BMU), Kassel, Germany.

Smeers, Y., 2008. Study on the General Design of Electricity Market Mechanisms Close to Real Time. Study for the Commission for Electricity and Gas Regulation (CREG). Available from: <http://www.creg.info/pdf/Etudes/F810UK.pdf>.

Synergies, 2009. Independent Expert Report for Directions in Queensland and Victoria. Draft Report under clause 3.15.17A of the National Electricity Rules. Synergies Economic Consulting, ROAM Consulting. Available from: <https://www.aemo.com.au/>.

Thorncraft, S.R., 2007. Experience with Market-Based Ancillary Services in the Australian National Electricity Market. Proceedings of the IEEE PES General Meeting, Tampa, FL, USA.

TradeWind, 2009. Integrating Wind. Project report for the trade-wind study coordinated by the European Wind Energy Association. Available from: <http://www.uwig.org/ TradeWind.pdf>.

Van Hulle, 2005. Large Scale Integration of Wind Energy in the European Power Supply: Analysis, Issues and Recommendations. Available from: <http://ebookbrowse.com/large-scale-integration-of-wind-energy-in-the-european-power-supply-analysis-issues-and-recom-mendations-pdf-d243742520>.

Vandezande, L., Meeus, L., Belmans, R., Saguan, M., Glanchant, J.M., 2010. Well-functioning balancing markets: a prerequisite for wind power integration. Energ. Policy 38 (7), 3146–3154.

Von Roon, S., Wagner, U., 2009. The interaction of Conventional Power Production and Renewable Power under the aspect of balancing Forecast Errors. Proceedings of the 10th IAEE European Conference on Energy, Policies and Technologies for Sustainable Economies, Vienna.

Walling, R., Banunarayanan, V., Miller, N., 2008. Analysis of Wind Generation Impact on ERCOT Ancillary Services Requirements. ERCOT.

Wessel, A.J., Jian, A., Conz, J., Dobschinski, B., Lange, H., 2008. Improving short-term forecast with online wind measurements. Proceedings of the German Wind Energy Conference, Bremen.

$$\hspace{5cm}\text{Epilogue}$$

ELECTRICITY REFORMS 20 YEARS LATER: THE FIVE BIG UNFORESEEN DEVELOPMENTS

When approached by the editor of this volume to contribute an Epilogue to the book, I could not resist: the topic of the book has been my passion for some time.

Twenty years ago a big surprise happened to me. In spite of the warnings from Joskow and Schmalensee's "*Markets for Power*" in 1983 and Schweppe et al's "*Spot Pricing*" in 1988, my own career has been devoted to a discovery of "market-based electricity reforms," and what a journey it has been!

At the beginning it took me 2 years to understand the UK Pool. I was convinced that no market could ever work in this industry. It took another 2 years to understand Nord Pool. Then 3 years to understand PJM, California, Texas, and of course, Enron. And again 2 years for the European Union's two first "Internal Market" Packages and ... Germany. France did not cause me much trouble or headache.

After so many years spent at understanding that I was always wrong and too late, I was ready to advise the European Commission on the coming Third Internal Market Package—by the way... why the two others? We now think in the EU that this European-wide power market should (might?) work in 2014/15 ... while we also know that it is going to implement the "old" goal of... 1996—the first EU Internal Market Directive! Is it the end of the journey or only a coffee break?

We also know that the power markets we will need tomorrow will be of a different nature than in 1996 in the EU or 1998 in California. Markets are only tools, they process data and signals to align investor behavior. They are not religion: we may change these tools when we expect different outcomes or foresee different environments. Only religions have to endure centuries as horizons of small significance.

I therefore do face today, 20 years later, five of my big unforeseen developments. They are not at the periphery of the system. They are at its core. They are generation mix, network neutrality, market design, market integrity, and dual fuel complementarity. Allow me to explain.

FIRST BIG UNFORESEEN: GENERATION MIX

Generation mix is the first big "unforeseen" shift from the initial power reform draft. I was pretty sure that most of the steam for "Markets for

791

Power" was there. At that time generation did not seem like any type of natural monopoly except in very small "pockets" where the size of the market is too small to duplicate the existing facility. Free entry in generation, free choice of the fuel or primary resource, of the technology and the plant size—if not the location—should act to break the old world of chartered territories for incumbent generation self-planning. This belief is heavily questioned today. The renewables push through feed-in tariffs largely locates the generation in the realm of a public authority. The renewable priority of dispatch both reduces the market size remaining for other generators and breaks the price trend at which they can run. The generation set is then deeply fractured into two opposite sets of generators. On the one hand, the "new" generators bearing no significant capacity, volume or price risks thanks to feed-in tariffs and priority of dispatch and, on the other hand, the "old" others bearing significantly increased risks as well as increased uncertainty. Actually this is much more important in Europe than in the United States where the growing generation revolution (shale gas) is mainly a market-based replicate of the "dash from coal" (and "dash for gas") seen in the 90s in the United Kingdom. Nuclear, which was looking for more market-based entry gates at the time of its so-called "renaissance" before Fukushima is now surviving only as an option in a low CO_2 generation future supported by "cap and floor" schemes shifting many lethal market risks to consumers or public authorities.

SECOND BIG UNFORESEEN: NETWORK NEUTRALITY

The idea that network neutrality is the second big unforeseen development may be controversial but it should not be because … it actually is a major departure.

I did see the network as a "Jurassic Park" for the remaining monopoly bones of the former power industry. Network monopolies should then be: First detached from market operation; Second neutral vis-à-vis fuel mix; Third cost based for hosting generation capacity. The outcome of this list of "don't do/don't act" is to reduce network costs to their bare bones, a la RPI-X formula. Let us see where we actually are today.

Transmission as well as distribution grids are now seen as the core of a significant shift of the whole industry to new business models. Firstly if significant new primary resources have to be exploited, either renewable or shale gas, transmission assets have to move or connect these new resources to the existing network. Furthermore the economic features of these new resources may presumably be different, that is, different timing of investment and construction, different load following, production ramping, and dispatch firmness profiles. The own development and operation of networks need to strongly adapt to this new users and new usages of transmission services. If so big a shift has to be implemented through grid investment and grid operation, the way we designed grid tariffs yesterday will not be sustainable for long. New grid tariffs have to be aligned with the new "system needs." We also need a system-wide coherence of grid pricing—not only inside a control zone but pretty much across all control zones. Big geographical and technical changes in the electrical system call for some continental-wide tariff design harmonization. Moreover, more interactive links are needed to coordinate demand response with the dispersed generation intermittency costs or ramping or with the greenhouse gas emission rate. Then both transmission and distribution are somehow supposed to become smarter platforms for deeper market interactions. Grids may be going to be remunerated for hosting more of the "socially preferred" generation mix or even to start innovating and running pilots or demonstration (e.g., off-shore grids). At the end of the day average grid operating costs will go up with increased investment costs. The low cost, postage stamp pricing, and energy mix neutral grid of the past may entirely fade away.

THIRD BIG UNFORESEEN: MARKET DESIGN

It is obvious that "market design" is one of the five big unforeseen developments: it could also be the very first one. I do remember Bill Hogan concluding, after a decade of reform experiments, that market players cannot design power markets by themselves because power markets are structurally incomplete. I did see that market players can easily trade energy prices and quantities until the "market gate closure;" but they cannot easily trade at exactly the same pace the corresponding transmission capacity and reserve availability needed to implement their "ex ante" energy trade. To alleviate these particular market difficulties power markets play the wholesale trade through a series of steps, which mimic the simple offer and demand arrangement of a *textbook* market. Power markets are actually "sequences of markets" from the precommitment of plants at Day(s) Ahead to the real-time balancing of actual injections and withdrawals; via the allocation of transmission capacity and the necessary management of unforeseen congestions.

Making electricity easily marketable among market players actually means to complete the textbook market with some central coordination, some third party intervention and market intermediation. If one wants to make electricity a more homogenous commodity, easier to contract and to trade, one has to somehow deal with the growing gap between the actual physical flows and the notional traded good. The very nature and the right amount of "third party" coordination in power markets are still under discussion after 20 years. We not only disagree on how to design complements or auxiliaries to the market but also disagree on what to keep free for trade: only this ... or also that? Fortunately enough of our disagreements of both types may be rationally aligned.

I did not really foresee how deeply market trade and market interactions will depend on the market arrangement decisions agreed by policy makers here and there. Even if I bypass more than a decade of wholesale story-board—UK Pool and New Trading Arrangement; Nord Pool; PJM; CA Pool; Texas flow gates; German dual competing PXs; etc.—today I am still discovering how to connect the existing market areas across the existing electrical control zones. Should we "couple" the existing markets within a harmonized nodal framework? Or only with an explicit... or implicit... transmission capacity auctioning? Only on the Day-Ahead time horizon? Should then it be "flow based" or with rigid predetermined "Net Transfer Capacity"? Should we also couple for intraday trade? With a few successive windows of price fixing? Or with continuous trading? Should we extend to pooling adjacent markets on their balancing horizon? Through a "loose" common pool of offers where several system operators may meet their needs? Or through a "tight" cross-border common management of all balancing options? And all this does not tell us if (and how) we might redesign all existing zones, inherited from the past, into new zones solely defined by a rational algorithm.

FOURTH BIG UNFORESEEN: MARKET INTEGRITY

How to protect market integrity is the fourth big unforeseen development. My initial wisdom was that a "good enough" generation structure is a necessary precondition to market competition: why bother to open markets that are structurally unable to be competitive? This was a key question in the United Kingdom in 1990s as it is in France today. The consensus was a magic number of five or more competitors none with more than 20% market share. It was also HHI and SCP, "Structures—Conduct—Performances," common sense. The Californian crisis with FERC blindly sticking to its HHI index prejudice opened many eyes to unacceptable deficiencies. F. Wolak et al. showed that more accurate definitions of market power and more sophisticated econometrics might be able to identify most of the "new industrial economics" in which market power is abused in power markets.

However many doors remained open between market and manipulation. First how to deal with thieves and criminals like Enron and many others seen operating in financial markets and how to deter them from destroying the market from inside? If we cannot guarantee ex ante transparency and integrity in power markets how can we rely on these markets to bridge the physical—for example, unit commitment, dispatch, transmission allocation, and congestion management—and the financial—for example, price arbitrage; portfolio and risk management; FTRs and other dimensions of trade? Today we still know more about the "fire alarm" strategy of monitoring—how to assess ex post the fairness of actual behaviors on existing markets—and less on the "police patrol" strategy—how to ex ante prevent manipulations or crimes. An obvious link between ex ante and ex post strategies is how we conceive the definition and the collection of data, the architecture and languages of databases and the screening tools, the market models and the software? Second we know more about a "single market" monitoring but did we achieve enough across borders and across markets? Third another key issue is how we manage the loop between market monitoring, market investigation and market fixing. Do we really know how to accurately repair the many details to be revealed as flawed?

FIFTH BIG UNFORESEEN: DUAL FUEL COMPLEMENTARITY

There are many good candidates for the fifth big unforeseen development. My favorite is the codesign of the gas and the electricity markets. We already use the concept of "dual fuel" for more than a decade to signal a particular market strategy where suppliers offer to consumers both gas and electricity in an energy bundle. I know well that "dual fuel competition" initiated in the United Kingdom in an era where the gas and the electricity incumbents started to move against each other before reaching a dual fuel market equilibrium.

I, however, largely used to think of gas and electricity as distinct fuels as in the case of gas and oil; there is of course a link through pricing but not much more. Today with the increased use of gas to fuel electricity generation, they are more and more complementary goods. Currently the decision of selling gas in this or that market or using the same gas to inject that electricity in these markets is taken on every day and every hour. It drives the relation and complementarity of gas and electricity down to their short-term markets operation and their cross-border arrangements. I did not pay much attention to the intricacies of the two correlated wholesale markets. If the CCGTs go out of the base load to sell their operational flexibility to the more intermittent electricity system, how would they find in the gas system the right characteristics to feed them? What about the corresponding increased fuel operational flexibility? With the typically rigid architecture of long-term contracts for gas transportation in the United States, CCGTs may rush to interruptible contracts where the gas transportation flexibility is actually "stored." But a growing fleet of CCGTs should someday inevitably reach the limits of this architecture of flexibility conceived for only a fringe of the US gas system. In the EU where CCGTs are increasingly used to balance daily as well as hourly—not to say: within a few minutes—the intermittency of renewables, the flexibility they offer to the electricity system becomes more and more "fast" and "peaky." In this case, how can the entire gas system adjust to provide this increased operational flexibility to CCGTs? How to get very flexible gas supply contracts, very flexible transportation contracts and an appropriate mechanism to manage the unforeseen (then the imbalances) intraday to real time? How would different market and transportation arrangements for crossing borders with gas or with electricity work when gas is also the main fuel, hence the main physical and economical constraint, for electricity balancing?

A SIXTH IN THE BIG 5?

Other unforeseen developments could have been substituted for my Big 5, including the evolution of retail markets. As Littlechild keeps reminding us "retail market" was not foreseen at all in the United Kingdom when the "Electricity White Paper" was being prepared in 1989 to introduce the Big Bang of the industry. For other people another good candidate might have been incentive regulation for its limited performance in the management of the loop between grid and generation investment. Many think today that with RPI-X or Performance-Based Regulation. "Incentive regulation" does not do enough to align the schedule, the amount, the technology, the location of new transmission grid capacity with the "real" needs of generation investment driven by wind farms and solar panels. These people argue that the two processes, investing in grids and investing in generation, are not sufficiently aligned with the existing "incentive regulation scheme." One may think that grid investment only has "to follow" decisions taken by generators and suppliers; what "Open Seasons" try to do for gas. However, it might not be as simple with electricity as the length of the investment period for CCGTs, wind farms, and solar panels is now several times shorter than for the grid investments. The chickens and the eggs do not easily align.

For some others, another potential candidate for the Big 5 could be the notion of a radically new and redefined scheme to provide reliable energy services with more flexibility, strong signals for investment, capacity and low-carbon fuels. They think that the typical pools and PXs are not anymore the key energy markets able to frame the development of the electricity system.

I have no doubt that many of you would even prefer other items to be included in my list. For me this will only confirm that using "markets as institutional structures" (R. Coase) of the electricity industry is feasible but very demanding. There are only a few ways for you and me and the markets to be perfect. While imperfect markets deal with their imperfections with much more creativity, tension and ... surprises. The main message conveyed with any list of big unforeseen developments is that imperfect markets cannot perfectly deal with everything. In practice we only choose a limited set of problems to solve and to find any workable way to deal with. Williamson calls this "the remediability criterion." We should then acknowledge at the start that our designs, as rational as they are in our minds, never deal with the entire set of problems and remedies. We cannot ignore our bounded rationality and the unlimited process of market errors and discovery.

At the beginning of the electricity market story, as well as at the end of this day, it is still impossible to design a perfect, waterproof market. We always have many unsolved problems. However less of these problems should stay unforeseen, if we carefully choose what to leave out.

This book draws on the experiences that lead us to approach this fundamental trade-off: doing better with what we want to achieve while not forgetting how wrong things can go in the areas that we did not choose as today's priorities.

ACKNOWLEDGMENTS

I do thank P. Sioshansi, S. Littlechild, and my Florence School—Loyola de Palacio "market design" research team (M. Hallack, A. Henriot and M. Vazquez) for their suggestions and comments. Of course all errors and misjudgments are still mine.

Jean-Michel Glachant

Loyola de Palacio Chair and Florence School of Regulation,
European University Institute in Florence, Italy

FURTHER READING

Joskow, P., Schmalensee, R., 1983. *Markets for Power*. MIT Press.

Schweppe, F.C., Caramanis, M.C., Tabors, R.E., Bohn, R.E., 1988. *Spot Pricing of Electricity*. Kluwer Academic Publishers.

Green, R.J., Newbery, D.M., 1992. Competition in the British electric spot market. J. Political Econ. 100, 929–953.

Hogan, W.W., 1995. A wholesale pool spot market must be administrated by the independent system operator: avoiding the separation fallacy. Electr. J. 8, 26–37.

Joskow, P., 1996. Introducing competition into regulated network industries: from hierarchies to markets in electricity. Ind. Corporate Change 5 (2), 341–383.

Littlechild, S., 2000. Why we need electricity retailers: a reply to Joskow on wholesale spot price pass-through. Judge Institute for Management Studies, University of Cambridge, Working Paper 21/2000.

Wolak, F.A., 2000. An empirical analysis of the impact of hedge contracts on bidding behavior in a competitive electricity market. Int. Econ. J. 14, 1–39.

Cramton, P., 2003. Electricity market design: the good, the bad, and the ugly. In: Proceedings of the IEEE Hawaii International Conference on Systems Science.

Smeers, Y., 2003. Market incompleteness in regional electricity transmission. Part II. The forward and real time markets. Network Spatial Econ. 3 (2), 175–196.

Stoft, S., 2002. Power System Economics: Designing Markets for Electricity. Wiley.

Wolak, F.A., 2008. Inefficiencies and market power in financial arbitrage: a study of California's electricity markets with S. Borenstein, J. Bushnell, and C. Knittel. J. Ind. Econ. 56 (June), 347–378.

Joskow, P., 2008. Introduction to new institutional economics: a report card. In: Brousseau, E., Glachant, J.-M. (Eds.), New Institutional Economics. A Guidebook. Cambridge University Press, pp. 2–19.

Joskow, P., 2009. US vs EU electricity reforms achievement. In: Glachant, J-M, Lévêque, F. (Eds.), *Electricity Reform in Europe. Towards a Single Electricity Market*. Edward Elgar, pp. xiii–xxix.

Glachant, J-M, Lévêque, F., 2009. The electricity internal market in the European Union: what to do next? In: Glachant, J.-M., Lévêque, F. (Eds.), *Electricity Reform in Europe. Towards a Single Electricity Market*. Edward Elgar, pp. 3–32.

Newbery, D., 2009. Refining market design. In: Glachant, J-M, Lévêque, F. (Eds.), *Electricity Reform in Europe. Towards a Single Electricity Market*. Edward Elgar, pp. 34–64.

Newbery, D., 2012. Reforming competitive electricity markets to meet environmental targets. *Econ. Energy Environ. Policy* I 1, 69–82.

Index

Note: Page numbers followed by "*f*" "*t*" and "*b*" refers to figures, tables and boxes, respectively.

A

ACCC. *see* Australian Competition and Consumer Commission (ACCC)
Accelerated Power Development Program, 507
Administrator of Trading System (ATS), 477
AEMC. *see* Australian Energy Market Commission (AEMC)
AEMO. *see* Australian Energy Market Operator (AEMO)
AER. *see* Australian Energy Regulator (AER)
Affiliated REPs (AREPs), 270, 336–337
Africa
  technically recoverable shale gas resources *vs.* existing reserves, production, and consumption, 423*t*
AGL, 604–606
Agreement on Internal Trade (AIT), 387
AIT. *see* Agreement on Internal Trade (AIT)
Alberta, 363–364, 372–380, 375*t*, 376*f*
Allocated Regulated Price (AFP), 752
American Electric Power (AEP), 269–270
Ancillary services, 46–47, 774–781
Andean Community (CAN), 408–409
  goals of, 408–409
  markets initiative, 409–410
Andean Electricity Integration Initiative, 409–410
ANEEL (Brazil), 435, 441–442
Annual DR product, 256
ARENH scheme. *see* Regulated Access to Incumbent Nuclear Electricity (ARENH) scheme
AREPs. *see* Affiliated REPs (AREPs)
Argentina, 397
  Chile and, energy integration between, 410–412
  gas pipelines, 404–405, 405*f*
  installed capacity, 419*t*
  main transmission lines length and conversion capacity, 421*t*
  peak load, 428*t*
  peak load growth rate in percent, 429*t*

Salto Grande, binational hydroelectric power plants of, 401
technically recoverable shale gas resources *vs.* existing reserves, production, and consumption, 423*t*
Asia
  technically recoverable shale gas resources *vs.* existing reserves, production, and consumption, 423*t*
Auction(s), 447
  default service models, 343–354, 343*f*
    Connecticut, 344–345, 345*f*, 346*f*
    Illinois, 346–349, 347*f*, 348*f*
    Pennsylvania, 349–354, 349*f*, 352*f*
  energy, results in Brazil, 448–450, 450*t*
  uniform price, 186
  wholesale, 446
Australia, electricity supply industry
  environmental objectives, 624–626
  environmental outcomes, 633–637
    commercially focused and politically powerful incumbents, 633–634
    failure to establish competition, 634–635
    foundational concerns, 636–637
    poor governance of external policies, 635–636
  greenhouse gas mitigation policies, 609–612
    carbon pricing, 609–610
    renewable generation subsidies, 610–612
  implications for restructuring, 622–623
  institutional and regulatory reforms, 586–587
  microeconomic reform *vs.* ESD, 619–623
  National Electricity Market (NEM), 575–581, 623–624
    addressing inadequacies, 638–639
    creation of, 586
    generation in, 579
    institutional responsibilities, 587*t*
    key participants in, 576*t*
    network pricing, 595–597
    peak demand in, 580
    retail market, 597–601
    structural developments, 601–607

Australia, electricity supply industry
(*Continued*)
wholesale market, 587–595
objectives and restructuring, 617–619
outline, 572–575
overview, 571–572
reforms, 607–609
government ownership, 607–608
retail tariff liberalization, 608
transmission pricing and access,
608–609
state-based reforms, 582–586
disaggregation and privatization,
582–584
early productivity gains, 584–585
spot markets, 585–586
technically recoverable shale gas resources
vs. existing reserves, production, and
consumption, 423*t*
Wholesale Electricity Market (WEM),
581–582
peak demand in, 582
Australian Competition and Consumer
Commission (ACCC), 586–587,
595–596, 601
vertical integration and, 604–606
Australian Energy Market Agreement of 2004,
622
Australian Energy Market Commission
(AEMC), 586–587, 596–597,
600–601
Australian Energy Market Operator (AEMO),
587
Australian Energy Regulator (AER),
586–587, 596

B

Balancing market
in EU countries, 42
Russia, 480–482
Base load generators, 186
Base Load Marginal Price (BLMP), 698–699
Base residual auctions (BRAs), 240
Basic Energy Plan (BEP), Japan, 722–723
BC Hydro, 371–372
BETTAs. *see* British electricity trading and
transmission arrangements (BETTAs)
BLMP. *see* Base Load Marginal Price
(BLMP)
Bolivia, 397, 408–410
gas pipelines, 404–405, 405*f*
installed capacity, 419*t*

main transmission lines length and
conversion capacity, 421*t*
peak load, 428*t*
peak load growth rate in percent, 429*t*
technically recoverable shale gas resources
vs. existing reserves, production, and
consumption, 423*t*
BRAs. *see* Base residual auctions (BRAs)
Brazil, 397, 433
antecedents, 439–440
bilateral initiatives, 401–406
distribution companies, 448, 450*t*
electricity consumption centers, 437*f*
first reform, 440–444
short-term price formation, 442*f*
timetable of, 441*t*
future trajectories, 451–457
consumption forecast, 451–452, 452*f*, 452*t*
energy matrix (expansion of electricity
supply), 453–454, 455*t*
environmental impacts, 454–457
gas pipelines, 404–405, 405*f*
generation companies, 447, 449*t*
hydropower plants, 438, 439*f*
installed capacity, 419*t*, 438, 439*f*, 456*t*
and electricity consumption evolution,
442–443, 443*f*
institutional framework, 444*f*
Itaipu power plant project, 401
largest installed capacity in, 396
"lost decade," 440
main transmission lines length and
conversion capacity, 421*t*
natural gas, 439
and neighboring countries, energy
integration in, 412
overview, 435–436
peak load, 428*t*
peak load growth rate in percent, 429*t*
power system, 436–439, 437*f*, 438*t*
price of supply contracts, 451*f*
reservoir levels, 442–443, 443*f*
second reform, 444–448, 444*f*
energy auctions results, 448–450, 450*t*
timetable of, 447, 448*t*
technically recoverable shale gas resources
vs. existing reserves, production, and
consumption, 423*t*
thermal power plants, 446–447
wholesale market, 445–446, 445*f*
agents registered on, 448*t*
Brazilian Developmental State policy, 439–440

Brazilian National Interconnected System (SIN)
  installed capacity projection, 453, 453f
BRIC countries. *see* Brazil; China; India; Russia
British Columbia (BC), 363, 368–372, 370f
British electricity market, 1. *see also* United Kingdom (UK)
  climate challenge, 10–15
    GHG emissions and, 11, 11f
    pressure for reform, 14–15
    renewable generation, 12–14
  EMR, criticisms of, 18–26
    bureaucracy *vs.* market, 24–26
    contract design, 18–22
    credit risk, 22–23
    funding risk, 23–24
    ROC rebanding and future wind CfDs, 20–22
  history (first two decades), 4–10
  low-carbon generation, role of policy for, 1
  market reform in Britain, 31
    alternative market-based approach, 48–53
    central-buyer approach, 53–55
    challenges, 32–41
    improved markets, 41–48
    overview, 31–32
  overview, 3–4
  reforming, 15–17
    carbon price floor, 16–17
British electricity trading and transmission arrangements (BETTAs), 10
BST. *see* Bulk Supply Tariff (BST)
Bulk Supply Tariff (BST), 653
Bureaucracy *vs.* market, EMR and, 24–26
Business-as-usual (BAU) scenario, 130–131, 309

C

CAISO. *see* California Independent System Operator (CAISO)
California, 297, 303–315. *see also* California Independent System Operator (CAISO)
  electricity crisis of 2000–2001, 299–300
  forces and developments, affects of, 315–319
  market redesign for efficiency and flexibility, 301–303
  new technologies, 313–315
    electric vehicles, 315

    smart grid, 313–314
    storage, 314–315
  overview, 297–299
  restructuring electricity sector, 299–301, 357–358
  retail choice models (hybrid/other), 357–358, 358f
  retail direct access, 304–307, 306f
  state environmental policy directives, 307–313
    cap-and-trade program, 309–310
    distributed generation, 311–312
    greenhouse gas regulation, 309–310, 310f
    loading order, 312–313
    once-through cooling plant retirements, 310–311
    thirty-three percent renewables portfolio standard, 307–309
California Independent System Operator (CAISO), 230. *see also* California
  challenges, 328
  design features, 301
  grid operations
    flexible ramping constraint and product, 320–321
    innovative study methodologies, 319–320, 320f
    state-of-the-art control room, 321
  infrastructure development, 325–328
    generator interconnection integration into transmission planning process, 326–328
    public policy-driven transmission planning, 325–326
  initiatives and innovations, 319–328
  market features, 301
  market redesign effort, 301–303
  "must-offer obligations," 323
  overview, 297–299, 298f
  Participating Intermittent Renewable Program (PIRP), 322
  Regulation Energy Management, 322
  Resource Adequacy Program, 323–324
  for restructuring of electricity sector, 299–301
  spot markets, 321–325
    design changes in progress, 323–324
    near-term market enhancements, 322
    potential real-time market design, 324–325
California Power Exchange, 299

California Public Utilities Commission (CPUC), 297, 305
  Self-Generation Incentive Program, 312
California Solar Initiative, 311–312
CAN. *see* Andean Community (CAN)
Canada, 363
  four energy areas, 367
  generation capacity, 366*t*
  geographic distribution of provinces, 364*f*
  global view of, 365–367, 366*f*
  hydropower-dominated provinces, 363, 368–372
    British Columbia (BC), 363, 368–372, 370*f*
    Manitoba, 363, 368–372
    market structure, 369*t*
    Newfoundland and Labrador, 363, 368–372
    overview, 368–372, 369*t*
    Quebec, 363, 368–372, 371*f*
    reform assessment, 372
    removing consumption subsidies in, 383–385
  market integration
    approaches to foster, 386–387
    consumption subsidies removal in hydro provinces, 383–385
    forces, 383–386, 384*t*
    hydro balancing maximization for wind power, 385
    less electrical heating and, 385
    obstacles, 386
    optimize renewable energy location and production, 385
    potential technical benefits, 384*t*
  overview, 363–365
  provincial per capita electricity consumption, 368*f*
  reform assessment, 382–383
  restructured provinces, 363–364, 372–380
    Alberta, 363–364, 372–380, 375*t*, 376*f*
    market structure, 373*t*
    New Brunswick, 363–364, 372–380
    Ontario, 363–364, 372–380, 377*t*, 378*f*
    reform assessment, 379–380
  traditional provinces, 364
    Nova Scotia, 364, 380–382, 381*t*
    Prince Edward Island, 364, 380–382, 381*t*
    Saskatchewan, 364, 380–382, 381*t*
Canadian Electricity Association (CEA), 385–386

Capacity credit market (CCM), 236, 243–244
Capacity growth, in China, 554–556
Capacity market prices, in PJM, 243–244, 244*f*
Capacity markets, 143–144, 227. *see also* PJM, capacity markets
  overview, 227
Capacity mechanisms
  renewable energy generation in Germany, 191–195
    capacity credits, 192
    capacity payments, 191–192
    reliability options, 192–193
    strategic reserves, 192
Capacity payments models, 42–45, 44*t*
Cap-and-trade program, California, 309–310
Carbon cost
  market-driven approach, 37–41
  policies, 40–41
  social cost/optimal quantity limit, 37–41
Carbon intensity, measures for European countries, 34*t*
Carbon-intensity reduction certificates (CIRCs), 50
Carbon markets, 47–48
Carbon policies, in Russia, 492
Carbon price floor (CPF), 15–17
Carbon pricing, in Australia, 609–610, 631–633
Catalogue system, 535, 536*t*
CBP. *see* Cost-Based Pool (CBP)
CCC. *see* Committee on Climate Change (CCC)
CCGT. *see* Combined cycle gas turbines (CCGT)
CCM. *see* Capacity credit market (CCM)
CDM, 553
CDS. *see* South American Defense Council (CDS)
CEAC. *see* Central-American Electrification Council (CEAC)
CEFC. *see* Clean Energy Finance Corporation (CEFC)
CEGB. *see* Central electricity generating board (CEGB)
Central America interconnection (SIEPAC project), 406–408
Central-American Electrification Council (CEAC), 406–407
Central-buyer approach, 53–55
  advantages, 54–55
  background and objectives, 53–54

Central electricity generating board (CEGB), UK, 6
Central Electricity Regulatory Commission (CERC), 506
CERC. *see* Central Electricity Regulatory Commission (CERC)
CfDs. *see* Contracts for differences (CfDs)
Chamber of Electric Energy Trade (CCEE), 444–445
Champsaur Commission, 76–77
Chile, 397, 409–410
  Argentina and, energy integration between, 410–412
  bilateral connection, 401
  gas pipelines, 404–405, 405*f*
  installed capacity, 419*t*
  main transmission lines length and conversion capacity, 421*t*
  peak load, 428*t*
  peak load growth rate in percent, 429*t*
  technically recoverable shale gas resources *vs.* existing reserves, production, and consumption, 423*t*
China's electrical power industry
  capacity growth, 554–556
  climate change and environmental protection, 552–554
  consumption, 532*f*
  fuel mix, 532*t*, 533*t*
  overview, 531–533
  policy adjustment, 561–562
  reforms (1997–2002), 534–539
    market structure, 534–537
    pressure for, 534
    pricing system, 538
    SETC, 538–539
    significance, 538–539
    SPCC, 537–539
  reforms (2002–2005), 539–548
    coal, 547
    industry restructuring, 540–542
    market development, 545–547
    price reforms, 544–547
    regulatory agencies, restructuring, 542–544
    significance, 548
  social equity concerns, 557–560
CHP generation. *see* Combined heat and power (CHP) generation
CIER. *see* Regional Energy Integration Commission (CIER)
CIRCs. *see* Carbon-intensity reduction certificates (CIRCs)

Clean Energy Finance Corporation (CEFC), 612
Climate change
  China's electricity industry, 552–554
  Korea's electricity industry, 709
Climate Change Act, 11, 32
Climate policies, Australia, 626–633
  crbon pricing, 631–633
  energy efficiency measures, 628–629
  MRET, 629–631
Coal
  in China, 547
  consumption of
    French electricity market and, 61
    India, 516
Colombia, 397, 408–410
  electricity interconnection, 409
  gas pipelines, 404–405, 405*f*
  installed capacity, 419*t*
  main transmission lines length and conversion capacity, 421*t*
  peak load, 428*t*
  peak load growth rate in percent, 429*t*
  technically recoverable shale gas resources *vs.* existing reserves, production, and consumption, 423*t*
Combined cycle gas turbines (CCGT), 4–6, 9, 138–139
Combined heat and power (CHP) generation, 465, 468
  Russia, 489–491
Commerce Act 1986, New Zealand, 645–646
Committee on Climate Change (CCC), 11
Competition and Consumer Act (CCA), 601
Competition transition charge (CTC), 300–301
Competitive market, defined, 334–335
Competitive REPs (CREPs), 270
Connecticut
  default service through auction or RFP model, 344–345, 345*f*, 346*f*
Contestable consumers, Singapore, 753–756
  USEP, 749, 753
Contract design, criticisms of EMR and, 18–22
Contracts for differences (CfDs), 16
Co-optimization of regulation reserve and energy scheduling, 776–777
Cost-Based Pool (CBP), 696. *see also* Wholesale market
  capacity payment, 698
  market participation, 696

Cost-Based Pool (CBP) (*Continued*)
  price controls, 698–699
  purchasing price, 698
"Cost-plus" principle, 79
CPF. *see* Carbon price floor (CPF)
CPUC. *see* California Public Utilities
  Commission (CPUC)
Credit risk, criticisms of EMR and, 22–23
CREPs. *see* Competitive REPs (CREPs)
Cross-border regulation system (2009 to
  present), in EU, 212–215
CTC. *see* Competition transition charge (CTC)
Current market shares, NEM, 601–603

D
DECC Department of Energy and Climate
  Change (DECC)
Delaware
  default service through auction or RFP
    model, 353
Demand curve shift
  PJM capacity markets and, 254–255
Demand resources (DRs), limited
  PJM capacity markets and, 255–257
Demand response (DR)
  ERCOT market, 281–283, 282*f*
Demand-side management, Russian market
  and, 491–492
Denmark, 413–414
Department of Energy and Climate Change
  (DECC), UK, 11
Differential rent, 72
Directive 2009/28/EC, 125
Disaggregated nodal pricing, 150–155
  disaggregated model framework, 153–155
  electricity systems (integrated *vs.*
    disaggregated perspective), 150–152
  elements, 156
  market power regulation and, 162–166
    fallacies of investment obligations,
      165–166
    fallacies of price structure regulation,
      164–165
    minimal regulatory basis, 162
    profit maximizing transmission network
      carrier, 163–164
  social welfare maximizing benchmark,
    156–162
    financial viability problem, 160–162
    generalized merit order rule and optimal
      consumption rule, 158–160

of optimal network investment, 156–158
Disaggregation and privatization, in Australia,
  582–584
Distribution system
  Germany, 100, 109–110
District of Columbia
  default service through auction or RFP
    model, 353
Draft Energy Bill, 4, 16, 19
Dual fuel complementarity, 795–796

E
Ecologically sustainable development (ESD),
  Australia, 620–621
  NSESD, 620
  vs. microeconomic reform, 621–622
Ecuador, 397, 408–410
  electricity interconnection, 409
  installed capacity, 419*t*
  main transmission lines length and
    conversion capacity, 421*t*
  peak load, 428*t*
  peak load growth rate in percent, 429*t*
EDF. *see* Electricité de France (EDF)
EEX. *see* European Electricity Exchange
  (EEX)
EIA. *see* Energy Information Agency (EIA)
EILS. *see* Emergency Interruptible Load
  Service (EILS)
Electricité de France (EDF), 59, 65–66
Electricité Réseau de Distribution France
  (ERDF), 59
Electricity Act 2003, 508–509, 517–519
Electricity Business Act in 1995, Japan,
  717–720
Electricity Business Act of 1964, Japan, 717
Electricity exchanges
  Latin America, 397–399, 402*t*, 403*f*
Electricity Industry Restructuring Act,
  Korea, 693
Electricity Laws (Amendment) Act, India, 505
Electricity Market Reform (EMR), 4, 16, 37, 40
  criticisms of, 18–26
    bureaucracy *vs.* market, 24–26
    contract design, 18–22
    credit risk, 22–23
    funding risk, 23–24
    ROC rebanding and future wind CfDs,
      20–22
Electricity Regulatory Commissions (ERC)
  Act, 1998, 506

Electricity Supply Act, 1948, 501
Electricity Utility Industry Reorganization
    Order, 716
Electricity Vending System (EVS), 744–745
Electric Power Boards, 652–653
Electric Power Development Company, Ltd.,
    717
Electric Power Law, China, 534
Electric Power System Council of Japan
    (ESCJ), 722
Electric Reliability Council of Texas
    (ERCOT), 265, 268–269. see also
    Texas
  all-in price for electricity in, 285f
  challenges, 281–293
    demand response, 281–283, 282f
    market design improvements (Nodal 2.0),
      290–291
    resource adequacy, 283–289, 285t
    retail market, 292–293
    significant increase in wind capacity,
      289–290
    wholesale market, 291–292
  competitive renewable energy zones
    transmission projects, 273–274, 274f
  efficiency and operational improvements,
    275–281
    gradual enhancement of retail operation,
      278–280, 279t
    interruptible loads utilization as resource,
      280–281
    transition to nodal market operation,
      276–278
  electricity generation fuel mix, 269f
  historical real-time zonal prices, 285t
  Independent Market Monitor (IMM),
    271–272
  legislative and regulatory activities,
    271–275
  nodal operation, 290–291
  overview, 265–267
  peaker net margin, 286f
  prices in restructured, 270
  wind capacity installed and future
    projection., 274, 275f
Electric Sector Monitoring Committee
    (CMSE), 444–445
Electric Security Plan (ESP), 354
Eletrobras, 440
EMC fee, Singapore, 752
Emergency Interruptible Load Service
    (EILS), 281

Emissions Performance Standard (EPS), 16
EnBW, 98, 100, 102
End-user energy efficiency
  Russian market and, 491–492
Energiewende 2011, the, 102–103
Energy and Environment Council, Japan, 727,
    734
Energy auctions (Brazil), 448–450, 450t
Energy Bureau, 542–544
Energy Conservation Plan, 550–551
Energy consumption
  forecast, in Brazil, 451–452, 452f, 452t
  in Latin America (Latam), 394–396
Energy efficiency certificates (ECerts), 526
Energy efficiency (EE)
  India market and, 526–527
  measures, Australia, 628–629
  resources, 255
  Russian market and, 491–492
"Energy Europe," 65
Energy Information Agency (EIA), 332
Energy Leading Group, 544
Energy-only markets
  renewable energy generation in Germany
    and, 185–188, 187f
Energy Research Company (EPE), 435
  Ten Year Energy Plan, 451
Energy security and efficiency, 550–552
Energy White Papers, 4
Enron, 268
Environmental objectives, Australian
    electricity industry restructuring and,
    624–626
Environmental protection
  China's electricity industry, 552–554
  impacts in Brazil, 454–457
Eon, 98, 100
EPS. see Emissions Performance Standard (EPS)
ERCOT, 230–231
ERDF. see Électricité Réseau de Distribution
    France (ERDF)
ESCJ. see Electric Power System Council of
    Japan (ESCJ)
ESP. see Electric Security Plan (ESP)
EU allowances (EUAs), 10–11
  price of, 10–11, 10f
EUAs. see EU allowances (EUAs)
EU Emissions trading system (ETS), 4,
    10–11
  improved markets, 41–48
    capacity payments, contracts, and
      markets, 42–45

EU Emissions trading system (ETS)
(*Continued*)
   carbon markets, 47–48
   incentives for low-carbon generation,
     45–46
   organized ancillary service markets, 46–47
   short-term energy markets, 41–42
Euler's theorem, 165–166
Europe
   disaggregated nodal pricing.
     see Disaggregated nodal pricing
   impact of renewable generation in, 125
     changes in price spreads, 141–143, 141*f*,
      142*f*
     critical issues, 127
     on fossil plants, 138–141, 139*f*, 140*f*
     future prospects, 128–132, 133*f*
     historical trends, 128–132, 129*f*, 130*f*,
      131*f*, 132*f*
     liberalization and, 133–136, 134*f*, 135*f*,
      136*f*, 137*f*
     on market price, 137–138, 138*f*
     new market structures, 143–145, 144*f*
     overview, 125–128, 126*f*, 127*f*
   liberalization stages of electricity in,
     66–72, 67*t*
     "electricity" directive, 67
     "Energy Package," 68
   technically recoverable shale gas resources
     vs. existing reserves, production, and
     consumption, 423*t*
European Commission (EC), 59, 413
   Directive 2009/28/EC, 125
   Roadmap for moving to a competitive low-
     carbon economy in 2050, 33
European Electricity Exchange (EEX), 93,
   98–99
European Legal Framework, 205*b*
European policy targets, 119–121
   GHGs emissions, reduction of, 119
   primary energy consumption, reduction of,
     120
   share of renewables, 120–121
European Regulators' Group for Electricity
   and Gas (ERGEG), 413
European Union (EU), 93
   central-buyer approach, 53–55
     advantages, 54–55
     background and objectives, 53–54
   Electricity Directives, 3
   energy integration, 413
   generation capacity in, 200*t*

improved markets, 41–48
   better balancing markets, 42
   capacity payments, contracts, and
     markets, 42–45, 43*f*
   carbon markets, 47–48
   incentives for low-carbon generation,
     45–46
   integration of, 42
   liquidity, 41
   nodal pricing, 42
   organized ancillary service markets,
     46–47
   short-term energy markets, 41–42
integrated markets in, 200–215
   cross-border regulation system (2009 to
     present), 212–215
   European Legal Framework, 205*b*
   initial setup (1999–2003), 201–204
   steps toward deepening regulation
     (2003–2009), 204–212
opening markets to demand participation,
   221–222
renewable energy integration, 215–221, 216*t*
   congestions within bidding zones,
     217–218
   curtailment of, 220
   marketing of, 220–221
   term shift in wholesale trading, 218–220
stages of liberalization, 201*f*
tradable carbon-intensity targets, 48–53, 49*t*
   benefits of scheme, 51–53
   business friendly, 52
   efficiency and cost-effectiveness, 52
   good for demand side, 52–53
   market neutral, 52
   methods, 50–51, 50*t*
   simplicity, credibility, and predictability
     for producers, 51–52
   technology neutral, 52
   transparency for investors, 52
   UK. *see* United Kingdom (UK)
Expert Committee on Reform, in Japan,
   730–731
Extended Summer DR product, 256
External market risks, 188

F

FAS. *see* Federal Anti-monopoly Service
   (FAS)
Fast markets, 764–767
Federal Anti-monopoly Service (FAS), 472,
   475–476

Federal Electricity Law, 463–464, 472, 474–476
Federal Energy Regulatory Commission (FERC), 301, 336, 364–365
Federal Grid Company, 474
Federal Service for Tariffs (FST), 472
Feed-in-tariffs (FiTs), 12, 33–34, 125, 735, 736t
FERC. see Federal Energy Regulatory Commission (FERC)
FGD, 553–554
Final Hourly Marginal Cost, 374
Final investment decision (FID), 24
Financial viability, of transmission network carrier, 160–162
Finland, 413–414
FIT Act, 735
FiTs. see Feed-in-tariffs (FiTs)
Flexible market mechanism, in PJM, 248–252
Flexible ramping product, by CAISO, 320–321
Forecasting, 780–781
FOREM, 469–470
Fossil plants, renewable generation impact on, 138–141, 139f, 140f
France, 59
    ARENH mechanism, 74–81
        breakdown of cost components in, 81t
        current cost of nuclear and, 82–84, 82t, 83t
        features, 78–81
        implementation of, 78–81
        subsidization and, 75–78
    carbon intensity measures for, 34t
    electricity costs, 87t
    energy industry
        French Energy Policy, stages of evolution, 61, 61t
        nuclear power stations, location of, 64f
        nuclear turnaround, 63
    history of, 60–66
        coal consumption, 61
        government role, 60–61, 63–65
        J.M. Jeanneney Plan, the, 62
        liberalization of network industries, 63
        Messmer Plan, the, 62–63
        nuclear turnaround, 63
        reconstruction of economy, 61
        Treaty of Rome, 62, 65
    market shares of electricity suppliers, 70t
    nuclear debate, 84–90
        electricity mix, 85–87, 87t
        peaking capacity, 89–90

subsidization of renewable electricity, 87–89, 88f
        nuclear differential rent vs. scarcity rent, 72–74
        overview, 59–60
        shale gas exploitation, 86
Free Trade of the Americas (FTAA), 406
French energy independence rate, 62
French Energy Policy
    stages of evolution, 61, 61t
FST. see Federal Service for Tariffs (FST)
FTAA. see Free Trade of the Americas (FTAA)
Fuel cost, in tariffs, 750
Fuel mix
    ERCOT, 269f
    of Europe-Ural zone, 468f
    evolution in UK, 4–6, 6f
    of Russia, 462f
    of Siberia zone, 468f
Fukushima disaster, 63–65, 67, 80–81, 83, 85, 93, 102

G
Galapagos Declaration, 409–410
Gazprom, 473
GB. see Great Britian (GB)
Gencos, 751
General Electricity Utilities (GEUs), 716–717, 718t, 719f
Generalized merit order rule, 158–160
Generation mix, 791–792
Gentrader model, NSW, 606–607
Germany, 93
    basic data, 95–96, 95t
    capacity in renewable generation, 98–99, 99f, 100t
    carbon intensity measures for, 34t
    CO_2 emission, 101, 102t
    distribution and retail, 100
    Energiewende 2011, the, 102–103
    European policy targets, 119–121
        GHGs emissions, reduction of, 119
        primary energy consumption, reduction of, 120
        share of renewables, 120–121
    legal framework, 97–98
    length of 220 and 380 kV network, 97t
    map of power stations, 96f
    overview, 93–94, 94f
    performance (prices, environment, and competition), 101–102

Germany (*Continued*)
  phasing out, effect of, 103, 104*f*
  prices development, 101–102, 101*f*
  renewable energy in. *see* Renewable energy
    generation, in Germany
  transmission, 100
  transmission grid operators, 97*f*
  turnaround policy, economic impacts of,
    112–118
    industry, trade, and commerce, 112–118,
      113*f*, 114*f*
    private households, 118
  turnaround policy, future implications of,
    104–112
    competition and markets, 110–111
    distribution, 109–110
    finance, 111–112, 112*f*
    generation, 104–106, 105*f*
    transmission, 106–109, 107*f*
  wind energy, 147–148
GOAL program, 8
GOELRO plan, 465
Government ownership, in Australia,
  607–608
Great Britian (GB), 3. *see also* British electricity
  market; United Kingdom (UK)
  high-tension transmission systems in, 4, 5*f*
Great East Japan Earthquake, 725–726
Greenhouse gas (GHG) emission
  Brazil, 454–457
  British electricity market and, 11, 11*f*
  India, 522, 523*f*
  reduction in, policy targets in UK, 32–35
  reduction of
    Germany, 119
Greenhouse gas mitigation policies, in
  Australia
  carbon pricing, 609–610
  renewable generation subsidies, 610–612
Green Revolution, 501–502
Grid operations, CAISO in
  flexible ramping constraint and product,
    320–321
  innovative study methodologies, 319–320,
    320*f*
  state-of-the-art control room, 321
Gross pool single platform markets, 781–783

H

HEUC. *see* Hourly Electricity Uplift Charge
  (HEUC)
HHI. *see* Hirschman–Herfindahl Index (HHI)

High-tension transmission systems, in GB, 4,
  5*f*
Hirschman–Herfindahl Index (HHI), 76,
  291–292
Hourly Electricity Uplift Charge (HEUC), 752
Houston Lighting & Power (HL&P), 269–270
HQ Energy Services (United States), 371
Hydroelectricit, in China, 556
Hydroelectric potential
  in South America, 398*t*
Hydro generation, 128
Hydropower-dominated provinces, Canada,
  363, 368–372
  British Columbia (BC), 363, 368–372, 370*f*
  Manitoba, 363, 368–372
  market structure, 369*t*
  Newfoundland and Labrador, 363,
    368–372
  overview, 368–372, 369*t*
  Quebec, 363, 368–372, 371*f*
  reform assessment, 372
  removing consumption subsidies in, 383–385
Hydropower plants
  Brazil, 438, 439*f*
Hydro-Quebec, 371

I

IEA. *see* International Energy Agency (IEA)
Illinois
  default service through auction or RFP
    model, 346–349, 347*f*, 348*f*
Illinois Power Agency (IPA), 346–347
IMM Independent Market Monitor (IMM)
Impact Assessment, 21
Implementation costs, defined, 194
Independent Market Monitor (IMM), ERCOT,
  271–272
Independent power producers (IPPs),
  368–370, 440–441, 505
India, 497
  background and evolution, 499–508, 500*f*,
    502*f*
    phases of reforms, 505–508
    timeline of, 504, 504*f*
  capacity addition in, 514*f*
  combined commercial losses of state
    utilities, 516*f*
  Electricity Act 2003, 508–509
  foundations for competitive market, 508–512
    National Electricity Policy, 510–511
    National Tariff Policy, 511–512
  GHG emissions, 523*f*

impact of reforms and development of
changes after, 512–519
nascent market for electricity trading,
519–522, 520f
status of, 513t
installed capacity
share of federal, state, and private sectors
in, 497–499, 499f
total, 497–499, 498f
overview, 497–499
renewable energy, opportunities in,
522–527
certificates, 524–526, 526t
comparative tariffs for power, 525t
energy efficiency, 526–527
transmission grids failure in, 497, 520
Indian Electricity Exchange Limited (IEXL),
163, 165
Industrial Policy Resolution (1956), 501
Infrastructure development, CAISO in, 325–328
generator interconnection integration into
transmission planning process, 326–328
public policy-driven transmission planning,
325–326
Innovative Strategy for Energy and the
Environment (ISEE), 715–716
Integrated Energy Systems, 473
Intelligent Energy Systems (IES), 744–745
Internal market risks, 188
International Atomic Energy Agency, 682
International Energy Agency (IEA), 85–86,
436
International Power Australia, 628–629
INTER RAO UES, 473
Interregional Distribution Grid Company, 474
interregional settlement residue (IRSR) units,
590
Investment Tax Credit (ITC), 289
IPA. see Illinois Power Agency (IPA)
IPPs. see Independent power producers (IPPs)
ISEE. see Innovative Strategy for Energy and
the Environment (ISEE)
ISO-NE, 230
Itaipu power plant project, as bilateral
initiative, 401
Italy, carbon intensity measures for, 34t
ITC. see Investment Tax Credit (ITC)

J

Japan Atomic Power Company (JAPC), 717
Japan Electric Power Exchange (JEPX), 723,
723f

JAPC. see Japan Atomic Power Company
(JAPC)
JEPX. see Japan Electric Power Exchange
(JEPX)
J.M. Jeanneney Plan, 62

K

KDI. see Korea Development Institute (KDI)
KEPCO. see Korea Electric Power
Corporation (KEPCO)
Korea Development Institute (KDI), 692–693
Korea Electricity Regulatory Commission
(KOERC), 690
Korea Electric Power Corporation (KEPCO),
679–681, 688–689
Korean National Assembly, 688
Korea Power Exchange (KPX), 690
proposed improvements, 690
Korea's electricity industry
achievements and issues, 700–711
climate change and smart grid, 709
history, 680
market structure and design, 693–700
participants, 693–695
retail structure, 699–700
wholesale market, 695–699
nuclear power, 708–709
overview, 679–680
proposed improvements, 709–711
resource adequacy, 706–708
restructuring, 688–700
background and history, 688–693
generation competition (2001–2003), 690
retail competition (2009), 691–693
retail market issues, 703–706
wholesale market issues, 702–703
Kyoto Protocol, 119, 492, 553, 715
joint implementation under (Russia), 492

L

Large Combustion Plant Directive (LCPD), 14
Large markets, 767–772
Large-Scale Renewable Energy Target
(LRET), 611–612
Latin America (Latam), 393. see also specific
countries
energy, integration, and geopolitics in,
413–418
integration advantages, 415–418
integration challenges, 414–415
energy consumption in, 394–396

Latin America (Latam) (*Continued*)
  final consumers' electricity price, 401*f*
  gas pipelines network, 426*t*
  installed capacity, 419*t*
  interconnection lines, 424*t*
  main transmission lines length and
    conversion capacity, 421*t*
  overview, 393−394
  per capita electricity consumption, 396*f*
  power plants, 426*t*
  regional energy integration, 400−412, 403*t*
    Andean electricity integration initiative,
      409−410
    Andean markets (CAN), 408−409
    bilateral initiatives, 401−406
    in Brazil and neighboring countries, 412
    Central America Interconnection
      (SIEPAC), 406−408, 407*f*
    Chile−Argentina energy integration,
      410−412
    South America, 404*f*, 405*f*
    Unasur, 406
  South America, 394−399, 395*t*
    electricity exchanges, 397−399, 402*t*,
      403*f*
    electricity interconnections in, 401−406,
      404*f*
    electrification levels 2009, 395*t*
    energy mix of countries, 396−397
    gas pipelines in, 404−405, 405*f*
    generation capacity by source percent, 400*f*
    hydroelectric potential in, 398*t*
    installed power generation capacity, 399*f*,
      399*t*
    Unasur, 406
  technically recoverable shale gas resources
    vs. existing reserves, production, and
    consumption, 423*t*
LCPD. *see* Large Combustion Plant Directive
  (LCPD)
LDAs. *see* Locational delivery areas (LDAs)
Legal framework, Germany, 97−98
Lerner Index, 753
Liberalization
  in Australia, retail tariff, 608
  Europe
    "electricity" directive, 67
    "Energy Package," 68
    impact of, 133−136, 134*f*, 135*f*, 136*f*,
      137*f*
    stages of electricity in, 66−72, 67*t*
  of network industries, in France, 63

reform in Russia, 476−486
  balancing market, 480−482
  retail market, 486
  transition period, 476−477
  wholesale capacity market, 482−486
  wholesale day-ahead market, 477−480,
    478*f*
Limited DR product, 255−257
Liquefied natural gas (LNG), 516, 741, 744
LMPs. *see* Locational marginal prices (LMPs)
LNG. *see* Liquefied natural gas (LNG)
Locational delivery areas (LDAs)
  PJM, 240−241, 241*f*
Locational marginal prices (LMPs), 229−230
Locational Marginal Pricing approach,
  477−480
LOLP. *see* Loss of load probability (LOLP)
Long-term contracts (LTC), 144
Long-term marginal costs (LTMC), 136, 140
Loss of load probability (LOLP), 8
Low-carbon generation
  British electricity market and role of policy
    for, 1
  incentives for, 45−46
  Russia, 486−492
    carbon policies (joint implementation
      under Kyoto Protocol), 492, 493*t*
    combined heat and power generation,
      489−491
    demand-side management and end-user
      energy efficiency, 491−492
    renewable energy, 487−489
  tradable carbon-intensity targets, 48−53,
    49*t*
    benefits of scheme, 51−53
    business friendly, 52
    efficiency and cost-effectiveness, 52
    good for demand side, 52−53
    market neutral, 52
    methods, 50−51, 50*t*
    simplicity, credibility, and predictability
      for producers, 51−52
    technology neutral, 52
    transparency for investors, 52
Low-carbon investment
  market signals for, in UK, 35−37, 36*t*
LRET. *see* Large-Scale Renewable Energy
  Target (LRET)
LTC. *see* Long-term contracts (LTC)
LTMC. *see* Long-term marginal costs
  (LTMC)
Lukoil, 473

M

Maine
  default service through auction or RFP model, 353
Major Energy Users Inc. (MEU), 595
Mandatory Renewable Energy Target (MRET), 610, 629–631
Manitoba, 363, 368–372
Manitoba Hydro, 371–372
MAPP. see Mid-Continent Area Power Pool (MAPP)
Market design, 793–794
Market integrity, 794–795
Market net revenue, PJM, 245–248, 246t, 247t
Market outcomes, in PJM, 243–253
  capacity market prices, 243–244, 244f
  flexibility, 248–252
  net revenue, 245–248, 246t, 247t
Market power regulation
  disaggregated nodal pricing and, 162–166
    fallacies of investment obligations, 165–166
    fallacies of price structure regulation, 164–165
    minimal regulatory basis, 162
    profit maximizing transmission network carrier, 163–164
Market price
  renewable generation impact on, 137–138, 138f
Market Rate Offer (MRO), 354
Market Supply Charge (MSC), 340
Market vs. bureaucracy, EMR and, 24–26
Maryland
  default service through auction or RFP model, 353
Massachusetts
  default service through auction or RFP model, 353
MCE. see Ministerial Council on Energy (MCE)
Merchant Function Charge (MFC), 340
Messmer Plan, the, 62–63
METI. see Ministry of Economy, Trade, and Industry (METI)
MEUC. see Monthly Electricity Uplift Charge (MEUC)
Mexico, 397
MFC. see Merchant Function Charge (MFC)
MGD Carbon, 553
Michigan

retail choice models (hybrid/other), 358–359, 359f
Microeconomic reform, in Australia, 621–622
Mid-Continent Area Power Pool (MAPP), 371
Minimum Offer Price Rule (MOPR), 262
Mining and Energy Ministry (MME), 444
Ministerial Council on Energy (MCE), 586–587
Ministry of Economy, Trade, and Industry (METI), 721–722
Ministry of Electrical Power, China, 534–535
MISO, 230
Missing-money problem, 72–73, 139–140
  renewable energy generation in Germany and, 185–188, 187f
Monopoly rent, 73
Monthly Electricity Uplift Charge (MEUC), 752
MOPR. see Minimum Offer Price Rule (MOPR)
MRET. see Mandatory Renewable Energy Target (MRET)
MRO. see Market Rate Offer (MRO)
MSC. see Market Supply Charge (MSC)
Mundra power plant, 517

N

NAPCC. see National Action Plan on Climate Change (NAPCC)
National Action Plan on Climate Change (NAPCC), 523
National Council for Energy Policy (CNPE), 444
National Development and Reform Commission (NDRC), 542–545, 561
National Electricity Code (Code), Australia, 586
National Electricity Law (NEL), Australia, 586, 622
National Electricity Market Management Company (NEMMCO), 586–587, 622–623
National Electricity Market (NEM), Australia, 575–581, 623–624
  climate policies, 626–633
    carbon pricing, 631–633
    energy efficiency measures, 628–629
    MRET, 629–631
  creation of, 586
  generation in, 579
  institutional responsibilities, 587t
  key participants in, 576t

National Electricity Market (NEM), Australia
(*Continued*)
network pricing, 595–597
peak demand in, 580
retail market, 597–601
structural developments, 601–607
wholesale market, 587–595
National Electricity Market of Singapore
(NEMS), 740
competition, 755–756
contestable consumers, 753–756
features, 741
generation capacity, 747–749, 748*t*
liberalization, 744
load profile of, 742*f*
market reforms, chronology, 743–745
noncontestable consumers, 754–755
structure of, 745–750, 746*f*
National Electricity Policy, India, 510–511,
519, 524
National Electrification Plan (NEP), 440
National Energy Administration (NEA) of
China, 560–561, 708–709
National Greenhouse Response Strategy
(NGRS), Australia, 620
National Grid Co. (NGC), 6
National Grid Electricity Transmission plc
(NGET), 4
National Grid Management Council (NGMC),
Australia, 622
National Policy Unit, Japan, 727
National Power (NP), 6
National Strategy for ESD (NSESD),
Australia, 620
National Tariff Policy, India, 511–512
Natural gas, 270
consumption of
Brazil, 439
India, 517
Natural monopoly services, nondiscriminatory
treatment by, 474–476
NBSO. *see* New Brunswick System Operator
(NBSO)
Near-term market design, CAISO and, 322
Negative prices, 774
NEMMCO. *see* National Electricity Market
Management Company (NEMMCO)
NEMS. *see* National Electricity Market of
Singapore (NEMS)
NEP. *see* National Electrification Plan (NEP)
NETAs. *see* New electricity trading
arrangements (NETAs)

Net pool markets, 782
Net revenue offset, lag in
PJM markets and, 257–258
Network model management system
(NMMS), 277
Network neutrality, 792–793
Network pricing, in Australia, 595–597
expenditure and pricing outcomes,
596–597
form of regulation, 595–596
New Brunswick, 363–364, 372–380
New Brunswick System Operator (NBSO),
378
New electricity trading arrangements
(NETAs), 9–10
Newfoundland and Labrador (NL), 363,
368–372
New Hampshire
default service through auction or RFP
model, 354
New Jersey
default service through auction or RFP
model, 354
New Organization of the Electricity Market
(NOME) law, 66, 77, 85
New York Independent System Operator
(NYISO), 230, 339
New York model (market based on cost pass-
through), 339–343, 340*f*, 342*f*
New Zealand, electricity sector
background, 647–654
Commerce Act 1986, 645–646
lines regulation, 659–661
reform outcomes, 661–674
regulatory arrangements, 656–659
SOE Act, 645
supply-side structure, 654–656
NFFO. *see* Nonfossil fuel obligation (NFFO)
NGET. *see* National Grid Electricity
Transmission plc (NGET)
Nihon Hatsusoden Company, 716
NMMS. *see* Network model management
system (NMMS)
Nodal pricing, 42, 150–151
disaggregated. *see* Disaggregated nodal
pricing
NOME law. *see* New Organization of the
Electricity Market (NOME) law
Noncontestable consumers, Singapore,
754–755
tariffs, 749, 754–755
Nonfossil fuel obligation (NFFO), 12–13

Nonfuel cost, in tariffs, 750
Nordic countries
    energy integration, 413–414
Norilsk Nickel, 473
North America, 414
    technically recoverable shale gas resources
        vs. existing reserves, production, and
        consumption, 423*t*
Norway, 413–414
Nova Scotia, 364, 380–382, 381*t*
Nuclear debate, France, 84–90
    electricity mix, 85–87, 87*t*
    peaking capacity, 89–90
    subsidization of renewable electricity,
        87–89, 88*f*
Nuclear generation
    ARENH and current cost of, 82–84
    average historical cost of, 82*t*
Nuclear power
    in Korea, 708–709
    world ranking, 710*t*
Nuclear rent, 65–66
    scarcity rent and, 72–74
    ARENH mechanism, 74–81
NYISO. *see* New York Independent System
    Operator (NYISO)

O

OATT. *see* Open Access Transmission Tariffs
    (OATT)
Obligation-based systems, 46
Offre de Marche, 68
Ofgem, 3–4, 9, 22, 41
    Project Discovery, 14–15
Ohio
    hybrid market, 354–356, 356*f*, 357*f*
Okinawa Electric Power Company, 717
Once-through cooling plant retirements, in
    California, 310–311
Ontario, 363–364, 372–380, 377*t*, 378*f*
Ontario Power Authority (OPA), 375–376
Open Access Transmission Tariffs (OATT),
    364–365
Optimal consumption rule, 158–160

P

Pacific Gas & Electric (PG&E), 299
Pan-European competition policy, 63
Paraguay, 397
    bilateral initiatives, 401–406
    gas pipelines, 404–405, 405*f*

installed capacity, 419*t*
Itaipu power plant project, 401
main transmission lines length and
    conversion capacity, 421*t*
peak load, 428*t*
peak load growth rate in percent, 429*t*
technically recoverable shale gas resources
    vs. existing reserves, production, and
    consumption, 423*t*
Peaking capacity, debate about, 89–90
Peaking units, 186
Pennsylvania
    default service through auction or RFP
        model, 349–354, 349*f*, 352*f*
Perform Achieve Trade, 526
Performance incentives, in PJM capacity
    market, 258–260
Peru, 397, 408–410
    bilateral connection, 401
    electricity interconnection, 409
    installed capacity, 419*t*
    main transmission lines length and
        conversion capacity, 421*t*
    peak load, 428*t*
    peak load growth rate in percent, 429*t*
PJM
    capacity market design, current issues with,
        229–230, 239–242, 253–263
        demand curve shift, 254–255
        lag in net revenue offset, 257–258
        limited DR, 255–257
        performance incentives, 258–260
        reliability issues for individual states,
            260–262
    capacity markets in, history of, 227–228,
        230, 232–242
        implementation of, 236–239, 237*t*
        net revenue performance, 239–242
        RPM capacity market design, 239–242
    footprint and zones, 233*f*
    goal of, 229
    installed capacity, 234*t*
    LDAs, 240–241, 241*f*
    market outcomes, 243–253
        capacity market prices, 243–244, 244*f*
        flexibility, 248–252
        net revenue, 245–248, 246*t*, 247*t*
    overview, 231–232
Plant Load Factor (PLF), 505
PLF. *see* Plant Load Factor (PLF)
Policy adjustments, in China, 561–562
Pool markets, 781–783

Potential real-time market design, CAISO
and, 324–325
Power Exchange India Limited, 163
Power Finance Corporation, India, 503
PowerGas Ltd., 744
PowerGen (PG), 6
Power Grid Corporation, India, 503
Power Producer and Suppliers (PPSs), 720,
724
Power Purchase Agreements (PPAs),
368–370, 505
Power Senoko, 743–744
PowerSeraya, 743–744
Power Trading Corporation (PTC), India, 507
PPAs. *see* Power Purchase Agreements
(PPAs)
PPSs. *see* Power Producer and Suppliers
(PPSs)
Price signals, 772–774
Price spread, 135
changes in, 141–143, 141*f*
Primary energy consumption
reduction, in Germany, 120
Prince Edward Island (PEI), 364, 380–382,
381*t*
Privatization
in Australia, 582–584
of Russian electricity industry
of electricity production, 472–474
nondiscriminatory treatment by natural
monopolies, 474–476
unbundling of network activities, 474,
475*f*
Production Tax Credit (PTC), 289
Profit maximizing transmission network
carrier, 163–164
Project Discovery (Ofgem), 14–15
PSO fee, Singapore, 752
PTC. *see* Production Tax Credit (PTC)
Public Utilities Commission of Ohio (PUCO),
354
Public Utility Commission of Texas (PUCT),
268–270
PUCO. *see* Public Utilities Commission of
Ohio (PUCO)
PUCT. *see* Public Utility Commission of
Texas (PUCT)

Q

Quantity-based models, 42–46, 44*t*
Quebec, 363, 368–372, 371*f*

R

RAO UES group (Russian company),
469–473, 485
liquidation of, 473
RECs. *see* Regional electricity companies
(RECs); Renewable Energy
Certificates (RECs)
Regional electricity companies (RECs), 6, 9
Regional Energy Integration Commission
(CIER), 417–418
Regional reference node (RRN), 590
Regional reference prices (RRPs), 590
Regulated Access to Incumbent Nuclear
Electricity (ARENH) scheme, 60,
74–81
breakdown of cost components in, 81*t*
current cost of nuclear and, 82–84, 82*t*, 83*t*
features, 78–81
implementation of, 78–81
subsidization and, 75–78
Regulated Rate Option (RRO), 374
Regulation reserves
causer-pays regulation pricing, 779–780
co-optimization of, 776–777
dynamic setting, 777–779
Reliability issues, for individual states
PJM capacity markets and, 260–262
Reliability pricing model (RPM), 236
Renewable Energy Act (EEG), 172
Renewable Energy Certificates (RECs),
610–611
Renewable energy generation
Britain, 12–14
in Germany. *see* Renewable energy
generation, in Germany
impact of, in Europe, 125
changes in price spreads, 141–143, 141*f*,
142*f*
critical issues, 127
on fossil plants, 138–141, 139*f*, 140*f*
future prospects, 128–132, 133*f*
historical trends, 128–132, 129*f*, 130*f*,
131*f*, 132*f*
liberalization and, 133–136, 134*f*, 135*f*,
136*f*, 137*f*
on market price, 137–138, 138*f*
new market structures, 143–145, 144*f*
overview, 125–128, 126*f*, 127*f*
integration, in EU, 215–221, 216*t*
congestions within bidding zones,
217–218
curtailment of, 220

marketing of, 220–221
  term shift in wholesale trading, 218–220
opportunities in India, 522–527
  certificates, 524–526, 526t
  comparative tariffs for power, 525t
  energy efficiency, 526–527
  GHG emissions and, 522, 523f
optimize location and production of, in
  Canada, 385
Russia, 487–489
share of, in Germany, 120–121
subsidization of, debate about, 87–89, 88f
Renewable energy generation, in Germany,
  125–126, 169
  capacity mechanisms, 191–195
    capacity credits, 192
    capacity payments, 191–192
    reliability options, 192–193
    strategic reserves, 192
  characteristics and flexibility potential, 179t
  energy-only markets and "missing-money
    problem," 185–188, 187f
  flexibility of renewables and market risk,
    177–180, 177f
  impact on market prices, 188–191, 189f,
    190f
  investment incentives
    concerns of insufficient, 187–188
    increasing renewables impact on,
      188–191
  market integration, 170–174
    guidelines for, 184–185
    and market design, 180–181
    operation of plants, 181–183
    rationale for, 174–177
  merit order of conventional generation in,
    186, 187f
  operation of plants, 181–183
  overview, 169–170
  support schemes, 174–185, 175t
  variability and uncertainty, 182t
Renewable Energy Law, China, 553
Renewable energy promotion targets, Korea,
  687t
Renewable energy supply (RES), 4
Renewable generation subsidies, Australia,
  610–612
  CEFC, 612
  key policies, 611t
  LRET, 611–612
  MRET, 610
  RECs, 610–611

RET, 610–611
  SRES, 611
Renewable Portfolio Standard (RPS),
  686–688
Renewable Purchase Obligation (RPO), 524
Renewables Directive, 13–14
Renewables Obligation (RO) Scheme, 12–13
Renewables Portfolio Standard (RPS),
  734–735
REPs. see Retail electric providers (REPs)
Request for proposal (RFP), default service
  models, 343–354
  Connecticut, 344–345, 345f, 346f
  Illinois, 346–349, 347f, 348f
  Pennsylvania, 349–354, 349f, 352f
RES. see Renewable energy supply (RES)
Re'seau de Transport de l'Electricite'
  (RTE), 59
Reserve margin, 229
Resource adequacy
  ERCOT market, 283–289, 285t
  Korea's electricity industry, 706–708
Resource Adequacy Program, 302
Responsive Reserve Service, 280–281
Restructured provinces, Canada, 363–364,
  372–380
  Alberta, 363–364, 372–380, 375t, 376f
  market structure, 373t
  New Brunswick, 363–364, 372–380
  Ontario, 363–364, 372–380, 377t, 378f
  reform assessment, 379–380
Restructuring Act, Korea, 690
Retail choice models, in United States, 332f
  auction or RFP, 343–354, 343f
    Connecticut, 344–345, 345f, 346f
    Illinois, 346–349, 347f, 348f
    Pennsylvania, 349–354, 349f, 352f
  hybrid/other, 354–359
    California, 357–358, 358f
    Michigan, 358–359, 359f
    Ohio, 354–356, 356f, 357f
  market based on cost pass-through (the
    New York model), 339–343, 340f,
    342f
  no default service (Texas model), 335–339,
    337f, 338f
  potential regulatory expansion, 359–360
Retail competition, in Japan, 732–733
Retail direct access
  in California, 304–307, 306f
Retail-distribution separation, in Australia,
  603–604

Retail electric providers (REPs), 269–270
Retail market
　in Australia, 597–601
　　bill proportions, 598*t*
　　competition in, 599
　　price indexes, 598*f*
　　retail price outcomes, 597–598
　　tariff innovation, 599–601
　challenges for ERCOT, 255–257
　Germany, 100
　in Korea, 699–700
　　competition, 691–693
　　issues, 703–706
　Russia, 486
Retail tariff liberalization, in Australia, 608
Rosenergoatom, 469–470, 472–473
RPM. *see* Reliability pricing model (RPM)
RPM demand curve, 242
RPO. *see* Renewable Purchase Obligation
　　(RPO)
RPS. *see* Renewables Portfolio Standard
　　(RPS)
RTE. *see* Re'seau de Transport de
　　l'Electricite' (RTE)
RusHydro, 472–473
Russia, 461
　age distribution of power plants in, 467, 467*f*
　electricity consumption in (1990–2009), 466*f*
　electricity prices
　　for household consumers, 470*f*
　　for industrial consumers, 471*f*
　as energy superpower, 464–465
　Federal Electricity Law, 463–464, 472, 474
　fuel mix of, 462*f*
　GOELRO plan, 465
　liberalization reform, 476–486
　　balancing market, 480–482
　　retail market, 486
　　transition period, 476–477
　　wholesale capacity market, 482–486
　　wholesale day-ahead market, 477–480,
　　　478*f*
　major events, 463*t*
　overview, 461–465
　privatization and restructuring process
　　of electricity production, 472–474
　　nondiscriminatory treatment by natural
　　　monopolies, 474–476
　　unbundling of network activities, 474, 475*f*
　reform
　　background and drivers, 469–471
　　plans and result, 471–472

transition to low carbon electricity
　production, 486–492
　carbon policies (joint implementation
　　under Kyoto Protocol), 492, 493*t*
　combined heat and power generation,
　　489–491
　demand-side management and end-user
　　energy efficiency, 491–492
　renewable energy, 487–489
Unified Power System of Russian
　Federation, 462*f*
　characteristics, 465–468
　perspectives of development, 469
　structure of, 465
Russian Energy Forecasting Agency, 469
Russian Railways, 473
RWE, 98, 100

S
Salto Grande, binational hydroelectric power
　plants of, 401
San Diego Gas & Electric (SDG&E), 299
SASAC. *see* State-owned Asset Supervision
　and Administration Commission
　(SASAC)
Saskatchewan, 364, 380–382, 381*t*
Sayano–Shushenskaya hydropower plant,
　accidental destruction of, 480
Sberbank, 492
Scarcity rent, 65–66
　nuclear differential rent and, 72–74
　　ARENH mechanism, 74–81
SCED. *see* Security constrained economic
　dispatch (SCED)
SDPC. *see* State Development and Planning
　Commission (SDPC)
SEBs State Electricity Boards (SEBs)
Security constrained economic dispatch
　(SCED), 277
Self-Generation Incentive Program, CPUC,
　312
SEPA. *see* State Environmental Protection
　Agency (SEPA)
SERC. *see* State Electricity Regulatory
　Commission (SERC)
SETC. *see* State Economic and Trade
　Commission (SETC)
Shale gas exploitation issue, in France, 86
Short-term energy markets
　of EU, improving, 41–42
Short-term marginal costs (STMCs),
　133–136, 135*f*, 138–139

SIEPAC project, 406–408
  goal of, 408
  transmission line, 407f
Singapore
  consumption in, 739, 740f
  demand in, 739–740, 741f
  pricing structure under deregulation, 751–752
Singapore Power (SP), 743
Single-Buyer Model, 507
SinoSing Power, 744
Small-Scale Renewable Energy Scheme
    (SRES), 611
Smart grids, 149
  in California, 313–314
  and Korea's electricity industry, 709
Social equity concerns, in China, 557–560
Social welfare maximizing benchmark, and
    disaggregated nodal pricing, 156–162
  financial viability problem, 160–162
  generalized merit order rule and optimal
    consumption rule, 158–160
  of optimal network investment, 156–158
South America, 394–399, 395tsee also
    specific countries
  bilateral initiatives, 401–406, 404f
  electricity exchanges, 397–399, 402t, 403f
  electricity interconnections in, 401–406, 404f
  electrification levels 2009, 395t
  energy mix of countries, 396–397
  gas pipelines in, 404–405, 405f
  gas pipelines network, 426t
  generation capacity by source percent, 400f
  hydroelectric potential in, 398t
  installed power generation capacity, 399f,
    399t
  integration advantages, 415–418
  integration challenges, 414–415
  interconnection lines, 424t
  peak load, 428t
  peak load growth rate in percent, 429t
  power plants, 426t
  technically recoverable shale gas resources
    vs. existing reserves, production, and
    consumption, 423t
  Unasur, 406
South American Defense Council (CDS),
    415–416
Southern California Edison (SCE), 299
South-West Interconnected System (SWIS) in
    Western Australia, 764–765
Spain
  carbon intensity measures for, 34t

SPCC. see State Power Corporation of China
    (SPCC)
Spot markets, Australia, 585–586
SP PowerAssets (SPPA), 745–746, 749
SP PowerGrid (SPPG), 743
SRES. see Small-Scale Renewable Energy
    Scheme (SRES)
State-based reforms, Australia, 582–586
  disaggregation and privatization, 582–584
  early productivity gains, 584–585
  spot markets, 585–586
State Development and Planning Commission
    (SDPC), 537–539, 542, 547
State Economic and Trade Commission
    (SETC), 537–539
State Electricity Boards (SEBs), India, 501
  effects of mismanagement, 502–503
State Electricity Regulatory Commission
    (SERC), 505–506, 539, 542,
    545–546, 560
State Environmental Protection Agency
    (SEPA), 544
State-of-the-art control room, by CAISO, 321
State-owned Asset Supervision and
    Administration Commission (SASAC),
    542–544
State-Owned Enterprises Act 1986 (SOE Act),
    New Zealand, 645
State-owned enterprises (SOEs), 689
State Power Corporation of China (SPCC),
    537–539
State Transmission Utility (STU), 508–509
STMCs. see Short-term marginal costs
    (STMCs)
Strengthened national policies (SNP), 130
Subsidization
  ARENH mechanism and, 75–78
  of renewable electricity, debate about,
    87–89, 88f
SUEK (Russian company), 473
Sweden, 413–414
System marginal cost (SMC), 8

T
Tariffs, China, 535
  calculations of, 536t
  Catalogue system, 535, 536t
  wholesale generation, 545
Tariffs, Singapore, 749
  components, 750
  for domestic consumers, 749–750

TARTAM. *see* Transitory tariff system
   (TARTAM)
Temasek Holdings, 743
TenneT, 103
Ten Year Energy Plan, of EPE, 451, 457
Territorial Generation Companies (TGC),
   472–473
Texas, 265. *see also* Electric Reliability
   Council of Texas (ERCOT)
  challenges, 281–293
    demand response, 281–283, 282*f*
    market design improvements (Nodal 2.0),
      290–291
    resource adequacy, 283–289, 285*t*
    retail market, 292–293
    significant increase in wind capacity,
      289–290
    wholesale market, 291–292
  overview, 265–267
  regional reliability councils, 269*f*
  restructuring and refinements, 268–281
    efficiency and operational improvements,
      275–281
    gradual enhancement of retail operation,
      278–280, 279*t*
    interruptible loads utilization as resource,
      280–281
    legislative and regulatory activities,
      271–275
    transition to nodal market operation,
      276–278
  restructuring model, 335–339, 337*f*, 338*f*
Texas model (no default service), 335–339,
   337*f*, 338*f*
TGC. *see* Territorial Generation Companies
   (TGC)
Tokyo Electric Lighting, 716
Tokyo Electric Power Company (TEPCO),
   733–734
Tradable carbon-intensity targets approach,
   48–53, 49*t*
  benefits of scheme, 51–53
  business friendly, 52
  efficiency and cost-effectiveness, 52
  good for demand side, 52–53
  market neutral, 52
  methods, 50–51, 50*t*
  simplicity, credibility, and predictability for
    producers, 51–52
  technology neutral, 52
  transparency for investors, 52
Tradable Green Certificates, 125

Traditional provinces, Canada, 364
  Nova Scotia, 364, 380–382, 381*t*
  Prince Edward Island, 364, 380–382, 381*t*
  Saskatchewan, 364, 380–382, 381*t*
Transitory tariff system (TARTAM), 75,
   77–78
Transmission grid operators, Germany, 97*f*
Transmission grids failure, in India, 497, 520
Transmission network carrier
  financial viability of, 160–162
  profit maximizing, 163–164. *see also*
    Disaggregated nodal pricing
Transmission Service Organizations (TSOs),
   Germany, 95, 97*f*, 98
Treaty of Rome, 62, 65
TRUenergy, 605–606
TSOs Transmission Service Organizations
   (TSOs)
Tuas Power, 744
Turnaround, Germany
  economic impacts of, 112–118
    industry, trade, and commerce, 112–118,
      113*f*, 114*f*
    private households, 118
  future implications of, 104–112
    competition and markets, 110–111
    distribution, 109–110
    finance, 111–112, 112*f*
    generation, 104–106, 105*f*
    transmission, 106–109, 107*f*
  price effects of, 115–116, 115*f*, 116*f*
20-20-20 *Renewables Directive*, 12
2050 Energy Road Map, 23
TXU Electric, 269–270

U

UK. *see* United Kingdom (UK)
UK's Electricity Supply Act, 1926, 501
"Ultra Mega Power Policy," 517
Unasur, 406
UNFCCC. *see* United Nations Framework
   Convention on Climate Change
   (UNFCCC)
Unforeseen developments
  dual fuel complementarity, 795–796
  generation mix, 791–792
  market design, 793–794
  market integrity, 794–795
  network neutrality, 792–793
  ovrview, 791
Unified Power System of Russian Federation,
   462*f*

characteristics, 465–468
perspectives of development, 469
structure of, 465
Uniform price auction, 186
Uniform Singapore Energy Price (USEP),
  749, 752–753
United Kingdom (UK). *see also* British
  electricity market
carbon intensity measures for, 34*t*
central-buyer approach, 53–55
  advantages, 54–55
  background and objectives, 53–54
climate challenge, 10–15
  GHG emissions and, 11, 11*f*
  pressure for reform, 14–15
  renewable generation, 12–14
CPF, 15–17
EMR, criticisms of, 18–26
  bureaucracy *vs.* market, 24–26
  contract design, 18–22
  credit risk, 22–23
  funding risk, 23–24
  nuclear power, 23
  ROC rebanding and future wind CfDs,
    20–22
evolution of fuel mix in, 4–6, 6*f*
history (first two decades), 4–10
  major events, 7*t*
improved carbon markets, 47–48
liberalization models, 6
market signals for low-carbon investment,
  35–37, 36*t*
overview, 3–4
policy targets, 32–35
restructuring models, 3
short-term energy markets, 41–42
stronger carbon price signals, 37–41
  market-driven approach, 37–41
tradable carbon-intensity targets, 48–53,
  49*t*
  benefits of scheme, 51–53
  business friendly, 52
  efficiency and cost-effectiveness, 52
  good for demand side, 52–53
  market neutral, 52
  methods, 50–53, 50*t*
  simplicity, credibility, and predictability
    for producers, 51–52
  technology neutral, 52
  transparency for investors, 52
United Nations Framework Convention on
  Climate Change (UNFCCC), 492

United States
eligible and competitive retail power sales
  by state, 266*t*
hybrid market, 354–359
market size and competitive activity,
  332–335, 333*f*, 334*f*
PJM capacity markets PJM, capacity
  market
retail choice models, 332*f*
  auction or RFP, 343–354, 343*f*
  California, 357–358, 358*f*
  Connecticut, 344–345, 345*f*, 346*f*
  Illinois, 346–349, 347*f*, 348*f*
  market based on cost pass-through (the
    New York model), 339–343, 340*f*,
    342*f*
  Michigan, 358–359, 359*f*
  no default service (Texas model),
    335–339, 337*f*, 338*f*
  Ohio, 354–356, 356*f*, 357*f*
  Pennsylvania, 349–354, 349*f*, 352*f*
  potential regulatory expansion,
    359–360
wholesale power markets in, 228
Uruguay, 397
installed capacity, 419*t*
main transmission lines length and
  conversion capacity, 421*t*
peak load, 428*t*
peak load growth rate in percent, 429*t*
Salto Grande, binational hydroelectric
  power plants of, 401
technically recoverable shale gas resources
  vs. existing reserves, production, and
  consumption, 423*t*
US Clean Energy Standard, 50–51
USEP. *see* Uniform Singapore Energy Price
  (USEP)

V
Value of lost load (VOLL), 8
Variable generation
  characteristics, 761–764
  fast markets, 764–767
  forecasting, 780–781
  gross pool single platform markets,
    781–783
  large markets, 767–772
  minimizing delay from scheduling to
    dispatch, 766–767
  network congestion, 771–772
  overview, 757–764

Variable generation (*Continued*)
  price signals, 772–774
  regulation reserves
    causer-pays regulation pricing, 779–780
    co-optimization of, 776–777
    dynamic setting, 777–779
Variable resource requirement (VRR), 242
Vattenfall Europe, 98, 100
Venezuela, 397, 409
  gas pipelines, 404–405, 405f
  installed capacity, 419t
  main transmission lines length and
    conversion capacity, 421t
  peak load, 428t
  peak load growth rate in percent, 429t
  technically recoverable shale gas resources
    vs. existing reserves, production, and
    consumption, 423t
Vertical integration, in Australia, 604–606
Virtual power plants (VPPs)
  concept of, 77
VMPs. *see* Voluntary mitigation plans
  (VMPs)
VOLL. *see* Value of lost load (VOLL)
Voluntary capacity market, 144
Voluntary mitigation plans (VMPs), 272
VRR. *see* Variable resource requirement
  (VRR)

W

WACC. *see* Weighted average cost of capital
  (WACC)
Water electrical heating, gains from, 385
Weighted Average Age
  of power plants in Russia, 467, 467f
Weighted average cost of capital (WACC), 15
WEP. *see* Wholesale Electricity Price (WEP)
WGC. *see* Wholesale Generation Companies
  (WGC)
White Paper, 16, 22–23
Wholesale Electricity Price (WEP), 752

Wholesale Generation Companies (WGC),
  472–473
Wholesale market, 227–228
  in Australia, 587–595
    market design and operation, 587–591
    market power concerns, 594–595
    pricing and investment outcomes,
      591–594
  in Brazil, 445–446, 445f
    agents registered on, 448t
  challenges for ERCOT, 254–255
  design issues and market dynamics,
    228–231
  European–Ural price zone, 477
  in Korea, 695–699
    capacity payment, 698
    competition, 691
    features of, 695–696
    issues, 702–703
    market equilibrium and system operation,
      696–698
    market participation, 696
    price controls, 698–699
    purchasing prices, 698
  liquidity in, 41
  participants, nondiscriminatory treatment
    of, 475–476
  reliability regulations, 228
  in Russia
    capacity market, 482–486
    day-ahead market, 477–480, 478f
  Siberian price zon, 477
Wholesale power market, in Japan, 729–730
Wind energy
  generation in Germany, 147–148
  hydro balancing maximization for, 385
  increase in capacity, ERCOT markets and,
    289–290

Y

Young Sam Kim, 689

Printed in the United States
By Bookmasters